Seventh Edition

Introductory Algebra
An Applied Approach

With Math Study Skills Workbook, Second Edition
by Paul D. Nolting

Richard N. Aufmann
Palomar College, California

Vernon C. Barker
Palomar College, California

Joanne S. Lockwood
Plymouth State University, New Hampshire

HOUGHTON MIFFLIN COMPANY   BOSTON   NEW YORK

INTRODUCTORY ALGEBRA: AN APPLIED APPROACH, SEVENTH EDITION
by Richard N. Aufmann, Vernon C. Barker and Joanne S. Lockwood
Copyright © 2006 by Houghton Mifflin Company. All rights reserved.
Editor-in-Chief: Jack Shira
Senior Sponsoring Editor: Lynn Cox
Associate Editor: Melissa Parkin
Editorial Assistant: Noel Kamm
Editorial Assistant: Julia Keller
Manufacturing Manager: Karen Banks
Senior Marketing Manager: Ben Rivera
Marketing Assistant: Lisa Lawler
MATH STUDY SKILLS WORKBOOK, SECOND EDITION
by Paul D. Nolting
Copyright © 2005 by Houghton Mifflin Company. All rights reserved.
Publisher: Jack Shira
Senior Sponsoring Editor: Lynn Cox
Assistant Editor: Melissa Parkin
Associate Project Editor: Kristin Penta
Manufacturing Manager: Florence Cadran
Senior Marketing Manager: Ben Rivera
Marketing Assistant: Lisa Lawler

Custom Publishing Editor: Dan Luciano
Custom Publishing Production Manager: Kathleen McCourt
Custom Publishing Project Coordinator: Christina Battista

Cover photo: © Dave Robertson/Masterfile
Photo Credits from Introductory Algebra: p. 1, Gary Conner/PhotoEdit, Inc.; p. 3, Tony Freeman/PhotoEdit, Inc.; p. 15, Paula Bronstein/Getty Images; p. 40, Bill Ross/CORBIS; p. 43, David Young-Wolff/PhotoEdit Inc.; p. 56, Chris Carroll/CORBIS; p. 64, Royalty Free/CORBIS; p. 65 Richard T. Norwitz/CORBIS; p. 76, Ariel Skelley/CORBIS; p. 77, Bob Daenmmrich/PhotoEdit, Inc.; p. 100, Lester V. Bergman/CORBIS; p. 108, Scott Barbour/Getty Images; p. 108, Guy Motil/CORBIS; p. 110, Shaun Best/Reuters Newmedia Inc./CORBIS; p. 113, A&L Sinibaldi/STONE/Getty Images; p. 119, Terres Du Sud/CORBIS Sygma; p. 130, AP/Wide World Photos; p. 131, Martin Fox/Index Stock Imagery; p. 131, Bill Aron/PhotoEdit, Inc.; p. 132, Lawrence Manning/CORBIS; p. 133, Davis Barber/PhotoEdit Inc.; p. 142, Royalty Free/CORBIS; p. 143, Tony Freeman/PhotoEdit Inc.; p. 159, Steve Prezant/CORBIS; p. 176, Renee Comet/PictureArts Corp./CORBIS; p. 177, Photodisc/Getty Images; p. 186, Pete Seaward/Getty Images; p. 190, Topham/The Image Works, Inc.; p. 191, Jim Richardson/CORBIS; p. 224, Duomo/CORBIS; p. 230, Duomo/CORBIS; p. 234, Royalty Free/CORBIS; p. 235, Pierre Ducharme/Reuters Newmedia Inc./CORBIS; p. 273, Reuters/CORBIS; p. 273, AP/Wide World Photos; p. 275, Bill Aron/PhotoEdit, Inc.; p. 282, Royalty Free/CORBIS; p. 286, Chris Hondros/Newsmakers/Getty Images; p. 287, Jonathan Nourok/PhotoEdit, Inc.; p. 309, Stephen Frink/CORBIS; p. 320, Clayton Sharrard/PhotoEdit, Inc.; p. 324, Royalty Free/CORBIS; p. 324, Alinari Archives/CORBIS; p. 335, Tom Carter/PhotoEdit, Inc.; p. 335, David Young-Wolff/PhotoEdit, Inc.; p. 336, Sheldan Collins/CORBIS; p. 336, Galen Rowell/CORBIS; p. 337, Billy E. Barnes/PhotoEdit, Inc.; p. 338, Lee Cohen/CORBIS; p. 339, Francis G. Mayer/CORBIS; p. 351, AP/Wide World Photos; p. 357, Craig Tuttle/CORBIS; p. 384, Tony Freeman/PhotoEdit, Inc.; p. 403, Jeff Hunter/The Image Bank/Getty Images; p. 421, Michael Newman/PhotoEdit, Inc.; p. 431, Spencer Grant/PhotoEdit, Inc.; p. 433, Eric & David Hosking/CORBIS; p. 434, Jeff Greenberg/PhotoEdit, Inc.; p. 440, Joel W. Rogers/CORBIS; p. 445, Michael Newman/PhotoEdit, Inc.; p. 459, Vic Bider/PhotoEdit, Inc.; p. 460, Spencer Grant/PhotoEdit, Inc.; p. 463, Spencer Grant/PhotoEdit, Inc.; p. 464, Dallas & John Heaton/CORBIS; p. 474, PictureArts/CORBIS; p. 479, AP/Wide World Photos; p. 486, Shaun Egan/Getty Images; p. 499, The Granger Collection; p. 503, PictureArts/CORBIS; p. 504, Robert Brenner/PhotoEdit, Inc.; p. 508, Brian Rea/Images.com/CORBIS; p. 513, AP/Wide World Photos; p. 514, Lori Adamski Peek/STONE/Getty Images; p. 536, Rudi Von/Briel/PhotoEdit, Inc.; p. 539, Bonnie Kamin/PhotoEdit, Inc.; p. 546, David Ponton/Getty Images; p. 548, Grafton Marshall Smith/CORBIS; p. 553, Myrleen Ferguson Cate/PhotoEdit, Inc.; p. 554, Jeff Greenberg/PhotoEdit, Inc.

This book contains select works from existing Houghton Mifflin Company resources and was produced by Houghton Mifflin Custom Publishing for collegiate use. As such, those adopting and/or contributing to this work are responsible for editorial content, accuracy, continuity and completeness.

Compilation copyright © 2006 by Houghton Mifflin Company. All rights reserved.

No part of this work may be reproduced or transmitted in any form or by any means, electronic or mechanical, including photocopying and recording, or by any information storage or retrieval system without the prior written permission of Houghton Mifflin Company unless such copying is expressly permitted by federal copyright law. Address inquiries to College Permissions, Houghton Mifflin Company, 222 Berkeley Street, Boston, MA 02116-3764.

Printed in the United States of America.
ISBN: 0-618-58682-2
N-04494

2 3 4 5 6 7 8 9 – CM – 07 06 05
Houghton Mifflin
Custom Publishing
222 Berkeley Street • Boston, MA 02116
Address all correspondence and order information to the above address.

Contents

Copyright © Houghton Mifflin Company. All rights reserved.

Copyright © Houghton Mifflin Company. All rights reserved.

4   Polynomials   191

5   Factoring   235

Copyright © Houghton Mifflin Company. All rights reserved.

Copyright © Houghton Mifflin Company. All rights reserved.

Copyright © Houghton Mifflin Company. All rights reserved.

10 Radical Expressions 479

11 Quadratic Equations 513

Copyright © Houghton Mifflin Company. All rights reserved.

Copyright © Houghton Mifflin Company. All rights reserved.

Preface

The seventh edition of *Introductory Algebra: An Applied Approach* provides mathematically sound and comprehensive coverage of the topics considered essential in an introductory algebra course. The text has been designed not only to meet the needs of the traditional college student, but also to serve the needs of returning students whose mathematical proficiency may have declined during years away from formal education.

In this new edition of *Introductory Algebra: An Applied Approach*, we have continued to integrate some of the approaches suggested by AMATYC. Each chapter opens with an illustration and a reference to a mathematical application within the chapter. At the end of each section there are Applying the Concepts exercises, which include writing, synthesis, critical thinking, and challenge problems. At the end of each chapter there is a "Focus on Problem Solving," which introduces students to various problem-solving strategies. This is followed by "Projects and Group Activities," which can be used for cooperative-learning activities.

NEW! Changes to This Edition

Chapter 1, now titled "Prealgebra Review," has been rewritten to present a more complete review of topics from prealgebra. Coverage of operations with integers and with rational numbers has been expanded. Included in this chapter is a section on geometry, in which students solve problems involving angle measurement, perimeter, and area. There are now a greater number of applications throughout the sections of this chapter.

The in-text examples are now highlighted by a prominent HOW TO bar. Students looking for a worked-out example can easily locate one of these problems.

As another aid for students, more annotations have been added to the Examples provided in the paired Example/You Try It boxes. This will assist students in understanding what is happening in key steps of the solution to an exercise.

In response to user requests, Section 4 of Chapter 6 now presents two methods of simplifying a complex fraction: (1) multiplying the numerator and denominator of the complex fraction by the least common multiple of the denominators and (2) multiplying the numerator by the reciprocal of the denominator of the complex fraction.

Section 3 in Chapter 7 now introduces the concept of slopes of perpendicular lines.

Throughout the text, data problems have been updated to reflect current data and trends. Also, titles have been added to the application exercises in the exercise sets. These changes emphasize the relevance of mathematics and the variety of problems in real life that require mathematical analysis.

The Chapter Summaries have been remodeled and expanded. Students are provided with definitions, rules, and procedures, along with examples of each. An objective reference and a page reference accompany each entry. We are confident that these will be valuable aids as students review material and study for exams.

In many chapters, the number of exercises in the Chapter Review Exercises has been increased. This will provide students with more practice on the concepts presented in the chapter.

Copyright © Houghton Mifflin Company. All rights reserved.

The calculator appendix has been expanded to include instruction on more functions of the graphing calculator. Notes entitled Integrating Technology appear throughout the book and many refer the student to this appendix. Annotated illustrations of both a scientific calculator and a graphing calculator appear on the inside back cover of this text.

Copyright © Houghton Mifflin Company. All rights reserved.

7 Linear Equations in Two Variables

OBJECTIVES

Section 7.1
A To graph points in a rectangular coordinate system
B To determine ordered-pair solutions of an equation in two variables
C To determine whether a set of ordered pairs is a function
D To evaluate a function written in functional notation

Section 7.2
A To graph an equation of the form $y = mx + b$
B To graph an equation of the form $Ax + By = C$
C To solve application problems

Section 7.3
A To find the x- and y-intercepts of a straight line
B To find the slope of a straight line
C To graph a line using the slope and the y-intercept

Section 7.4
A To find the equation of a line given a point and the slope
B To find the equation of a line given two points
C To solve application problems

This tennis player gets the energy for his workout from carbohydrates. Carbohydrates are the body's primary source of fuel for exercise. They can be released quickly and easily to fulfill the demands that exercise puts on the body. Since carbohydrates also fuel most of our muscular contractions, it is important to eat enough carbohydrates before any rigorous exercise. **Exercise 27 on page 390** presents data on the number of grams of carbohydrates burned as a strenuous tennis workout progresses.

 Need help? For online student resources, such as section quizzes, visit this textbook's website at **math.college.hmco.com/students.**

Copyright © Houghton Mifflin Company. All rights reserved.

Page 351

Chapter Opening Features

NEW! Chapter Opener
New, motivating chapter opener photos and captions have been added, illustrating and referencing a specific application from the chapter.

The [www globe icon] at the bottom of the page lets students know of additional on-line resources at math.college.hmco.com/students.

Objective-Specific Approach
Each chapter begins with a list of learning objectives that form the framework for a complete learning system. The objectives are woven throughout the text (i.e., Exercises, Prep Tests, Chapter Review Exercises, Chapter Tests, Cumulative Review Exercises) as well as through the print and multimedia ancillaries. This results in a seamless learning system delivered in one consistent voice.

Page 78

Prep Test and Go Figure

Prep Tests occur at the beginning of each chapter and test students on previously covered concepts that are required in the coming chapter. Answers are provided in the Answer Section. Objective references are also provided if a student needs to review specific concepts.

The **Go Figure** problem that follows the *Prep Test* is a playful puzzle problem designed to engage students in problem solving.

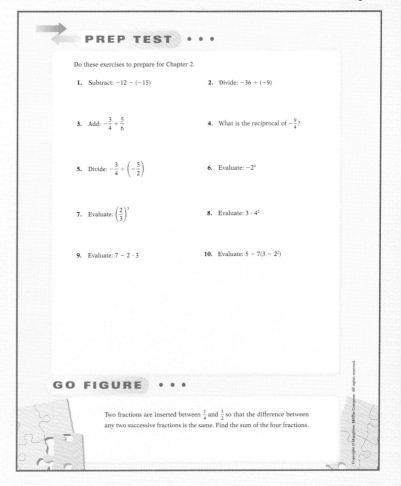

PREP TEST • • •

Do these exercises to prepare for Chapter 2.

1. Subtract: $-12 - (-15)$

2. Divide: $-36 \div (-9)$

3. Add: $-\dfrac{3}{4} + \dfrac{5}{6}$

4. What is the reciprocal of $-\dfrac{9}{4}$?

5. Divide: $-\dfrac{3}{4} \div \left(-\dfrac{5}{2}\right)$

6. Evaluate: -2^4

7. Evaluate: $\left(\dfrac{2}{3}\right)^3$

8. Evaluate: $3 \cdot 4^2$

9. Evaluate: $7 - 2 \cdot 3$

10. Evaluate: $5 - 7(3 - 2^2)$

GO FIGURE • • •

Two fractions are inserted between $\frac{1}{4}$ and $\frac{1}{2}$ so that the difference between any two successive fractions is the same. Find the sum of the four fractions.

Copyright © Houghton Mifflin Company. All rights reserved.

Aufmann Interactive Method (AIM)

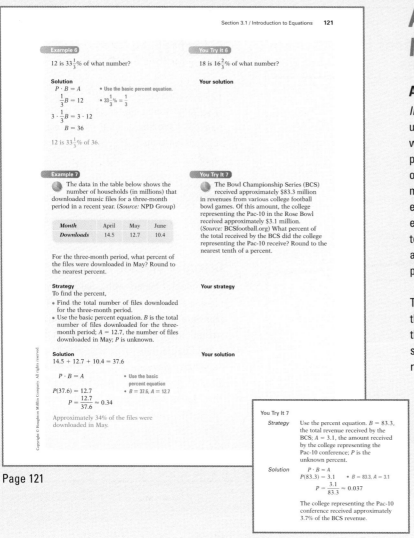

Example 6

12 is $33\frac{1}{3}$% of what number?

Solution

$P \cdot B = A$ • Use the basic percent equation.

$\frac{1}{3}B = 12$ • $33\frac{1}{3}\% = \frac{1}{3}$

$3 \cdot \frac{1}{3}B = 3 \cdot 12$

$B = 36$

12 is $33\frac{1}{3}$% of 36.

You Try It 6

18 is $16\frac{2}{3}$% of what number?

Your solution

Example 7

The data in the table below shows the number of households (in millions) that downloaded music files for a three-month period in a recent year. (*Source:* NPD Group)

Month	April	May	June
Downloads	14.5	12.7	10.4

For the three-month period, what percent of the files were downloaded in May? Round to the nearest percent.

Strategy

To find the percent,
• Find the total number of files downloaded for the three-month period.
• Use the basic percent equation. B is the total number of files downloaded for the three-month period; $A = 12.7$, the number of files downloaded in May; P is unknown.

Solution

$14.5 + 12.7 + 10.4 = 37.6$

$P \cdot B = A$ • Use the basic percent equation

$P(37.6) = 12.7$ • $B = 37.6, A = 12.7$

$P = \frac{12.7}{37.6} \approx 0.34$

Approximately 34% of the files were downloaded in May.

You Try It 7

The Bowl Championship Series (BCS) received approximately $83.3 million in revenues from various college football bowl games. Of this amount, the college representing the Pac-10 in the Rose Bowl received approximately $3.1 million. (*Source:* BCSfootball.org) What percent of the total received by the BCS did the college representing the Pac-10 receive? Round to the nearest tenth of a percent.

Your strategy

Your solution

Copyright © Houghton Mifflin Company. All rights reserved.

Page 121

You Try It 7

Strategy   Use the percent equation. $B = 83.3$, the total revenue received by the BCS; $A = 3.1$, the amount received by the college representing the Pac-10 conference; P is the unknown percent.

Solution $P \cdot B = A$

$P(83.3) = 3.1$ • $B = 83.3, A = 3.1$

$P = \frac{3.1}{83.3} \approx 0.037$

The college representing the Pac-10 conference received approximately 3.7% of the BCS revenue.

Page S5

An Interactive Approach

Introductory Algebra: An Applied Approach uses an interactive style that provides a student with an opportunity to try a skill as it is presented. Each section is divided into objectives, and every objective contains one or more sets of matched-pair examples. The first example in each set is worked out; the second example, called "You Try It," is for the student to work. By solving this problem, the student actively practices concepts as they are presented in the text.

There are complete worked-out solutions to these examples in an appendix. By comparing their solution to the solution in the appendix, students obtain immediate feedback on, and reinforcement of, the concept.

Page xxv

AIM for Success Student Preface

This student 'how to use this book' preface explains what is required of a student to be successful and how this text has been designed to foster student success, including the Aufmann Interactive Method (AIM). *AIM for Success* can be used as a lesson on the first day of class or as a project for students to complete to strengthen their study skills. There are suggestions for teaching this lesson in the *Instructor's Resource Manual*.

AIM for Success

Welcome to *Introductory Algebra: An Applied Approach*. As you begin this course, we know two important facts: (1) We want you to succeed. (2) You want to succeed. To do that requires an effort from each of us. For the next few pages, we are going to show you what is required of you to achieve that success and how you can use the features of this text to be successful.

Motivation   One of the most important keys to success is motivation. We can try to motivate you by offering interesting or important ways mathematics can benefit you. But, in the end, the motivation must come from you. On the first day of class, it is easy to be motivated. Eight weeks into the term, it is harder to keep that motivation.

Copyright © Houghton Mifflin Company. All rights reserved.

Problem Solving

Focus on Problem Solving

At the end of each chapter is a Focus on Problem Solving feature that introduces the student to various successful problem-solving strategies. Strategies such as drawing a diagram, applying solutions to other problems, working backwards, inductive reasoning, and trial and error are some of the techniques that are demonstrated.

Focus on Problem Solving

Negations and If ... then Sentences

The sentence "George Washington was the first president of the United States" is a true sentence. The **negation** of that sentence is "George Washington was **not** the first president of the United States." That sentence is false. In general, the negation of a true sentence is a false statement.

The negation of a false sentence is a true sentence. For instance, the sentence "The moon is made of green cheese" is a false statement. The negation of that sentence, "The moon is **not** made of green cheese," is true.

The words *all*, *no* (or *none*), and *some* are called **quantifiers.** Writing the negation of a sentence that contains these words requires special attention. Consider the sentence "All pets are dogs." This sentence is not true because there are pets that are not dogs; cats, for example, are pets. Because the sentence is false, its negation must be true. You might be tempted to write "All pets are not dogs," but that sentence is not true because some pets are dogs. The correct negation of "All pets are dogs" is "Some pets are not dogs." Note the use of the word *some* in the negation.

Now consider the sentence "Some computers are portable." Because that sentence is true, its negation must be false. Writing "Some computers are not portable" as the negation is not correct, because that sentence is true. The negation of "Some computers are portable" is "No computers are portable."

The sentence "No flowers have red blooms" is false, because there is at least one flower (some roses, for example) that has red blooms. Because the sentence is false, its negation must be true. The negation is "Some flowers have red blooms."

Statement	Negation
All *A* are *B*.	Some *A* are not *B*.
No *A* are *B*.	Some *A* are *B*.
Some *A* are *B*.	No *A* are *B*.
Some *A* are not *B*.	All *A* are *B*.

Write the negation of the sentence.

1. All cats like milk.
2. All computers need people.
3. Some trees are tall.
4. No politicians are honest.
5. No houses have kitchens.
6. All police officers are tall.
7. All lakes are not polluted.
8. Some drivers are unsafe.
9. Some speeches are interesting.
10. All laws are good.
11. All businesses are not profitable.
12. All motorcycles are not large.
13. Some vegetables are good for you to eat.
14. Some banks are not open on Sunday.

Copyright © Houghton Mifflin Company. All rights reserved.

Page 339

Problem-Solving Strategies

The text features a carefully developed approach to problem solving that emphasizes the importance of *strategy* when solving problems. Students are encouraged to develop their own strategies—to draw diagrams, to write out the solution steps in words—as part of their solution to a problem. In each case, model strategies are presented as guides for students to follow as they attempt the "You Try It" problem. Having students provide strategies is a natural way to incorporate writing into the math curriculum.

Copyright © Houghton Mifflin Company. All rights reserved.

Page 171

Example 2

A chemist wishes to make 2 L of an 8% acid solution by mixing a 10% acid solution and a 5% acid solution. How many liters of each solution should the chemist use?

Strategy

• Liters of 10% solution: x
  Liters of 5% solution: $2 - x$

	Amount	Percent	Quantity
10% solution	x	0.10	$0.10x$
5% solution	$2 - x$	0.05	$0.05(2 - x)$
8% solution	2	0.08	$0.08(2)$

• The sum of the quantities before mixing is equal to the quantity after mixing.

Solution

$$0.10x + 0.05(2 - x) = 0.08(2)$$
$$0.10x + 0.10 - 0.05x = 0.16$$
$$0.05x + 0.10 = 0.16$$
$$0.05x = 0.06$$
$$x = 1.2$$

$$2 - x = 2 - 1.2 = 0.8$$

The chemist needs 1.2 L of the 10% solution and 0.8 L of the 5% solution.

You Try It 2

A pharmacist dilutes 5 L of a 12% solution with a 6% solution. How many liters of the 6% solution are added to make an 8% solution?

Your strategy

Your solution

Solution on p. S9

Copyright © Houghton Mifflin Company. All rights reserved.

Real Data and Applications

Applications

One way to motivate an interest in mathematics is through applications. Wherever appropriate, the last objective of a section presents applications that require the student to use problem-solving strategies, along with the skills covered in that section, to solve practical problems. This carefully integrated applied approach generates student awareness of the value of algebra as a real-life tool.

Applications are taken from many disciplines, including agriculture, business, carpentry, chemistry, construction, education, finance, nutrition, real estate, sports, and telecommunications.

Page 16

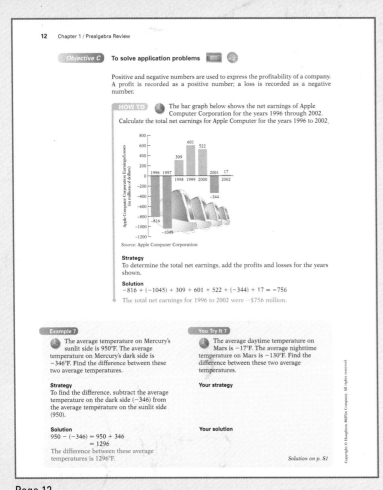

Objective C   To solve application problems

Positive and negative numbers are used to express the profitability of a company. A profit is recorded as a positive number; a loss is recorded as a negative number.

HOW TO   The bar graph below shows the net earnings of Apple Computer Corporation for the years 1996 through 2002. Calculate the total net earnings for Apple Computer for the years 1996 to 2002.

Source: Apple Computer Corporation

Strategy
To determine the total net earnings, add the profits and losses for the years shown.

Solution
$-816 + (-1045) + 309 + 601 + 522 + (-344) + 17 = -756$
The total net earnings for 1996 to 2002 were −$756 million.

Example 7
The average temperature on Mercury's sunlit side is 950°F. The average temperature on Mercury's dark side is −346°F. Find the difference between these two average temperatures.

Strategy
To find the difference, subtract the average temperature on the dark side (−346) from the average temperature on the sunlit side (950).

Solution
$950 - (-346) = 950 + 346$
$= 1296$
The difference between these average temperatures is 1296°F.

You Try It 7
The average daytime temperature on Mars is −17°F. The average nighttime temperature on Mars is −130°F. Find the difference between these two average temperatures.

Your strategy

Your solution

Solution on p. S1

Page 12

Real Data

Real data examples and exercises, identified by , ask students to analyze and solve problems taken from actual situations. Students are often required to work with tables, graphs, and charts drawn from a variety of disciplines.

Geography The graph at the right shows Earth's three deepest ocean trenches and its three tallest mountains. Use this graph for Exercises 87 to 89.

87. Could Mt. Everest fit in the Tonga Trench?

88. What is the difference between the depth of the Philippine Trench and the Mariana Trench?

89. What is the difference between the height of Mt. Everest and the depth of the Mariana Trench?

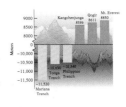

Meteorology A meteorologist may report a wind-chill temperature. This is the equivalent temperature, including the effects of wind and temperature, that a person would feel in calm air conditions. The table below gives the wind-chill temperature for various wind speeds and temperatures. For instance, when the temperature is 5°F and the wind is blowing at 15 mph, the wind-chill temperature is −13°F. Use this table for Exercises 90 and 91.

Wind Speed (in mph)	Thermometer Reading (in degrees Fahrenheit)														
	25	20	15	10	5	0	−5	−10	−15	−20	−25	−30	−35	−40	−45
5	19	13	7	1	−5	−11	−16	−22	−28	−34	−40	−46	−52	−57	−63
10	15	9	3	−4	−10	−16	−22	−28	−35	−41	−47	−53	−59	−66	−72
15	13	6	0	−7	−13	−19	−26	−32	−39	−45	−51	−58	−64	−71	−77
20	11	4	−2	−9	−15	−22	−29	−35	−42	−48	−55	−61	−68	−74	−81
25	9	3	−4	−11	−17	−24	−31	−37	−44	−51	−58	−64	−71	−78	−84
30	8	1	−5	−12	−19	−26	−33	−39	−46	−53	−60	−67	−73	−80	−87
35	7	0	−7	−14	−21	−27	−34	−41	−48	−55	−62	−69	−76	−82	−89
40	6	−1	−8	−15	−22	−29	−36	−43	−50	−57	−64	−71	−78	−84	−91
45	5	−2	−9	−16	−23	−30	−37	−44	−51	−58	−65	−72	−79	−86	−93

90. When the thermometer reading is −5°F, what is the difference between the wind-chill factor when the wind is blowing 10 mph and when the wind is blowing 25 mph?

91. When the thermometer reading is −20°F, what is the difference between the wind-chill factor when the wind is blowing 15 mph and when the wind is blowing 25 mph?

APPLYING THE CONCEPTS

92. If a and b are integers, is the expression $|a + b| = |a| + |b|$ always true, sometimes true, or never true?

93. Is the difference between two integers always smaller than either one of the numbers in the difference? If not, give an example for which the difference between two integers is greater than either integer.

Copyright © Houghton Mifflin Company. All rights reserved.

Student Pedagogy

Icons

The   at each objective head remind students that both a video and a tutorial lesson are available for that objective.

Key Terms and Concepts

Key terms, in bold, emphasize important terms. The key terms are also provided in a **Glossary** at the back of the text.

Key Concepts are presented in orange boxes in order to highlight these important concepts and to provide for easy reference.

Study Tips

These margin notes remind students of study skills presented in the *AIM for Success*; some notes provide page references to the original descriptions. They also provide students with reminders of how to practice good study habits.

HOW TO Examples

HOW TO examples use annotations to explain what is happening in key steps of the complete, worked-out solutions.

Page 531

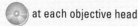

1.2   Addition and Subtraction of Integers

Objective A   To add integers

A number can be represented anywhere along the number line by an arrow. A positive number is represented by an arrow pointing to the right, and a negative number is represented by an arrow pointing to the left. The size of the number is represented by the length of the arrow.

Addition is the process of finding the total of two numbers. The numbers being added are called **addends.** The total is called the **sum.** Addition of integers can be shown on the number line. To add integers, start at zero and draw, above the number line, an arrow representing the first number. At the tip of the first arrow, draw a second arrow representing the second number. The sum is below the tip of the second arrow.

$4 + 2 = 6$

$-4 + (-2) = -6$

$-4 + 2 = -2$

$4 + (-2) = 2$

The pattern for addition shown on the number lines above is summarized in the following rules for adding integers.

> **Addition of Integers**
>
> **To add two numbers with the same sign,** add the absolute values of the numbers. Then attach the sign of the addends.
>
> **To add two numbers with different signs,** find the absolute value of each number. Subtract the smaller of the two numbers from the larger. Then attach the sign of the number with the larger absolute value.

Study Tip
The HOW TO feature indicates an example with explanatory remarks. Using paper and pencil, you should work through the example. See *AIM for Success*, page xxv.

HOW TO   Add: $-12 + (-26)$

$-12 + (-26) = -38$ • The signs are the same. Add the absolute values of the numbers $(12 + 26)$. Attach the sign of the addends.

HOW TO   Add: $-19 + 8$

$|-19| = 19$ $|8| = 8$ • The signs are different. Find the absolute value of each number.

$19 - 8 = 11$ • Subtract the smaller number from the larger.

$-19 + 8 = -11$ • Attach the sign of the number with the larger absolute value.

Page 9

11.4   Graphing Quadratic Equations in Two Variables

Objective A   To graph a quadratic equation of the form $y = ax^2 + bx + c$

TAKE NOTE
For the equation $y = 3x^2 - x + 1$, $a = 3$, $b = -1$, and $c = 1$.

An equation of the form $y = ax^2 + bx + c$, $a \neq 0$, is a **quadratic equation in two variables.** Examples of quadratic equations in two variables are shown at the right.

$y = 3x^2 - x + 1$
$y = -x^2 - 3$
$y = 2x^2 - 5x$

For these equations, y is a function of x, and we can write $f(x) = ax^2 + bx + c$. This equation represents a **quadratic function.**

Point of Interest
Mirrors in some telescopes are ground into the shape of a parabola. The mirror at the Palomar Mountain Observatory is 2 ft thick at the ends and weighs 14.75 tons. The mirror has been ground to a true paraboloid (the three-dimensional version of a parabola) to within 0.0000015 in. A possible equation of the mirror is $y = 2640x^2$.

HOW TO   Evaluate $f(x) = 2x^2 - 3x + 4$ when $x = -2$.

$f(x) = 2x^2 - 3x + 4$
$f(-2) = 2(-2)^2 - 3(-2) + 4$ • Replace x by -2.
$= 2(4) + 6 + 4 = 18$ • Simplify.

The value of the function when $x = -2$ is 18.

TAKE NOTE
One of the equations at the right was written as $y = 2x^2 + 3x - 2$ and the other using functional notation as $f(x) = -x^2 + 3x + 2$. Remember that y and $f(x)$ are different symbols for the same quantity.

The graph of $y = ax^2 + bx + c$ or $f(x) = ax^2 + bx + c$ is a **parabola.** The graph is ∪-shaped and opens up when a is positive, and down when a is negative. The graphs of two parabolas are shown below.

$y = 2x^2 + 3x - 2$
$a = 2$, a positive number
Parabola opens up.

$f(x) = -x^2 + 3x + 2$
$a = -1$, a negative number
Parabola opens down.

HOW TO   Graph $y = x^2 - 2x - 3$.

x	y
-2	5
-1	0
0	-3
1	-4
2	-3
3	0
4	5

• Find several solutions of the equation. Because the graph is not a straight line, several solutions should be found in order to determine the ∪-shape. Record the ordered pairs in a table.

Integrating Technology
One of the Projects and Group Activities at the end of this chapter shows how to graph a quadratic equation by using a graphing calculator. You may want to verify the graphs you draw in this section by drawing them on a graphing calculator.

• Graph the ordered-pair solutions on a rectangular coordinate system. Draw a parabola through the points.

Take Note

These margin notes alert students to a point requiring special attention or are used to amplify the concept under discussion.

Point of Interest

These margin notes contain interesting sidelights about mathematics, its history, or its application.

Integrating Technology

These margin notes provide suggestions for using a calculator or refer the student to an appendix for more complete instructions on using a calculator.

Copyright © Houghton Mifflin Company. All rights reserved.

Copyright © Houghton Mifflin Company. All rights reserved.

Page 374

Objective C   To solve application problems

37. **Emergency Response**   A rescue helicopter is rushing at a constant speed of 150 mph to reach several people stranded in the ocean 11 mi away after their boat sank. The rescuers can determine how far they are from the victims using the equation $D = 11 - 2.5t$, where D is the distance in miles and t is the time elapsed in minutes. Graph this equation for $0 \le t \le 4$. The point $(3, 3.5)$ is on the graph. Write a sentence that describes the meaning of this ordered pair.

38. **Business**   A custom-illustrated sign or banner can be commissioned for a cost of $25 for the material and $10.50 per square foot for the artwork. The equation that represents this cost is given by $y = 10.50x + 25$, where y is the cost and x is the number of square feet in the sign. Graph this equation for $0 \le x \le 20$. The point $(15, 182.5)$ is on the graph. Write a sentence that describes the meaning of this ordered pair.

39. **Veterinary Science**   According to some veterinarians, the age, x, of a dog can be translated to "human years" by using the equation $H = 4x + 16$, where H is the human equivalent age for the dog. Graph this equation for $2 \le x \le 21$. The point whose coordinates are $(6, 40)$ is on this graph. Write a sentence that explains the meaning of this ordered pair.

40. **Business**   Judging on the basis of data from the Consumer Electronics Association, the projected number, N (in millions), of sales of high-definition televisions (HDTVs) can be approximated by $N = 3t + 4$, where $0 \le t \le 4$ and $t = 0$ corresponds to the year 2003. Graph this equation. The point whose coordinates are $(3, 13)$ is on this graph. Write a sentence that explains the meaning of this ordered pair in the context of the problem.

APPLYING THE CONCEPTS

41. Graph $y = 2x - 2$, $y = 2x$, and $y = 2x + 3$. What observation can you make about the graphs?

42. Graph $y = x + 3$, $y = 2x + 3$, and $y = -\frac{1}{2}x + 3$. What observation can you make about the graphs?

43. For the equation $y = 3x + 2$, when the value of x changes from 1 to 2, does the value of y increase or decrease? What is the change in y? Suppose that the value of x chan...

44. For the equation $y = -2x +$ ... does the value of y increase ... pose the value of x changes f...

45. **Telecommunications**   A lon... rate of $.99 for the first 15 ... each additional minute. Th... of the graphs of two linear ... $C = 0.15(t - 15) + 0.99$ whe...
a. What is the cost of a telep...
b. What is the cost of a tele...

Page 374

Page 127

Copyright © Houghton Mifflin Company. All rights reserved.

59. $d + 1.3619 = 2.0148$ 60. $w + 2.932 = 4.801$

61. $-0.813 + x = -1.096$ 62. $-1.926 + t = -1.042$

63. $6.149 = -3.108 + z$ 64. $5.237 = -2.014 + x$

Objective C   To solve an equation of the form $ax = b$

65. Without solving $-\frac{15}{41}x = -\frac{23}{25}$, determine whether x is less than or greater than 0. Explain your answer.

66. Explain why multiplying each side of an equation by the reciprocal of the coefficient of the variable is the same as dividing each side of the equation by the coefficient.

For Exercises 67 to 110, solve and check.

67. $5x = -15$ 68. $4y = -28$ 69. $3b = 0$ 70. $2a = 0$

71. $-3x = 6$ 72. $-5m = 20$ 73. $-3x = -27$ 74. $-\frac{1}{6}n = -30$

Page 127

Exercises and Projects

Exercises

The exercise sets of *Introductory Algebra: An Applied Approach* emphasize skill building, skill maintenance, and applications. Concept-based writing or developmental exercises have been integrated with the exercise sets. Icons identify appropriate writing , data analysis , and calculator exercises.

Included in each exercise set are **Applying the Concepts** that present extensions of topics, require analysis, or offer challenge problems. The writing exercises ask students to explain answers, write about a topic in the section, or research and report on a related topic.

Page 225

Projects and Group Activities

The Projects and Group Activities featured at the end of each chapter can be used as extra credit or for cooperative learning activities. The projects cover various aspects of mathematics, including the use of calculators, collecting data from the Internet, data analysis, and extended applications.

Page 225

Projects and Group Activities

Diagramming the Square of a Binomial

1. Explain why the diagram at the right represents $(a + b)^2 = a^2 + 2ab + b^2$

2. Draw similar diagrams representing each of the following.
$$(x + 2)^2$$
$$(x + 4)^2$$

Pascal's Triangle

Simplifying the power of a binomial is called *expanding the binomial*. The expansions of the first three powers of a binomial are shown below.

$$(a + b)^1 = a + b$$
$$(a + b)^2 = (a + b)(a + b) = a^2 + 2ab + b^2$$
$$(a + b)^3 = (a + b)^2(a + b) = (a^2 + 2ab + b^2)(a + b) = a^3 + 3a^2b + 3ab^2 + b^3$$

Point of Interest

Pascal did not invent the triangle of numbers known as Pascal's Triangle. It was known to mathematicians in China probably as early as 1050 A.D. But Pascal's *Traité du triangle arithmetique* (*Treatise Concerning the Arithmetical Triangle*) brought together all the different aspects of the numbers for the first time.

Find $(a + b)^4$. [*Hint:* $(a + b)^4 = (a + b)^3(a + b)$]

Find $(a + b)^5$. [*Hint:* $(a + b)^5 = (a + b)^4(a + b)$]

If we continue in this way, the results for $(a + b)^6$ are

$$(a + b)^6 = a^6 + 6a^5b + 15a^4b^2 + 20a^3b^3 + 15a^2b^4 + 6ab^5 + b^6$$

Now expand $(a + b)^8$. Before you begin, see whether you can find a pattern that will help you write the expansion of $(a + b)^8$ without having to multiply it out. Here are some hints.

Copyright © Houghton Mifflin Company. All rights reserved.

Chapter 2 Summary

Key Words	Examples
A *variable* is a letter that is used for a quantity that is unknown or that can change. A *variable expression* is an expression that contains one or more variables. [2.1A, p. 79]	$4x + 2y - 6z$ is a variable expression. It contains the variables x, y, and z.
The *terms* of a variable expression are the addends of the expression. Each term is a *variable term* or a *constant term*. [2.1A, p. 79]	The expression $2a^2 - 3b^3 + 7$ has three terms. $2a^2$, $-3b^3$, and 7. $2a^2$ and $-3b^3$ are variable terms. 7 is a constant term.

Page 104

Essential Rules and Procedures	Examples
The Distributive Property [2.2A, p. 83] If a, b, and c are real numbers, then $a(b + c) = ab + ac$.	$5(4 + 7) = 5 \cdot 4 + 5 \cdot 7$ $= 20 + 35 = 55$
The Associative Property of Addition [2.2A, p. 83] If a, b, and c are real numbers, then $(a + b) + c = a + (b + c)$.	$-4 + (2 + 7) = -4 + 9 = 5$ $(-4 + 2) + 7 = -2 + 7 = 5$

Copyright © Houghton Mifflin Company

Page 104

End of Chapter

Chapter Summary

At the end of each chapter there is a Chapter Summary that includes Key Words, Essential Rules and Procedures, and an example of each. Each entry includes an objective reference and a page reference indicating where the concept is introduced. These chapter summaries provide a single point of reference as the student prepares for a test.

Page 106

Chapter 2 Review Exercises

1. Simplify: $3(x^2 - 8x - 7)$

2. Simplify: $7x + 4x$

3. Simplify: $6a - 4b + 2a$

4. Simplify: $(-50n)\left(\dfrac{1}{10}\right)$

Chapter Review Exercises

Chapter Review Exercises are found at the end of each chapter. These exercises are selected to help the student integrate all of the topics presented in the chapter.

Page 109

Chapter 2 Test

1. Simplify: $3x - 5x + 7x$

2. Simplify: $-3(2x^2 - 7y^2)$

3. Simplify: $2x - 3(x - 2)$

4. Simplify: $2x + 3[4 - (3x - 7)]$

Chapter Test

Each Chapter Test is designed to simulate a possible test of the material in the chapter.

Page 111

Cumulative Review Exercises

1. Add: $-4 + 7 + (-10)$

2. Subtract: $-16 - (-25) - 4$

3. Multiply: $(-2)(3)(-4)$

4. Divide: $(-60) \div 12$

Cumulative Review Exercises

Cumulative Review Exercises, which appear at the end of each chapter (beginning with Chapter 2), help students maintain skills learned in previous chapters.

The answers to all Chapter Review Exercises, all Chapter Test exercises, and all Cumulative Review Exercises are given in the Answer Section. Along with the answer, there is a reference to the objective that pertains to the exercise.

Page A4

CUMULATIVE REVIEW EXERCISES

1. -7 [1.2A]   2. 5 [1.2B]   3. 24 [1.3A]   4. -5 [1.3B]   5. $53°$ [1.8A]   6. $\dfrac{11}{48}$ [1.6C]   7. $-\dfrac{1}{6}$ [1.7B]

8. $\dfrac{1}{4}$ [1.7A]   9. 75% [1.7C]   10. -5 [1.4B]   11. $-\dfrac{27}{26}$ [1.7B]   12. 16 [2.1A]   13. $5x^2$ [2.2A]

14. $-7a - 10b$ [2.2A]   15. 153.86 cm^2 [1.8C]   16. 96 ft [1.8B]   17. $24 - 6x$ [2.2C]   18. $6y - 18$ [2.2C]

Copyright © Houghton Mifflin C

Copyright © Houghton Mifflin Company. All rights reserved.

Instructor Resources

Introductory Algebra: An Applied Approach has a complete set of support materials for the instructor.

Instructor's Annotated Edition This edition contains a replica of the student text and additional resources just for the instructor. These include: *Instructor Notes, New Vocabulary/Symbols, etc., Vocabulary/Symbols, etc. to Review, In-Class Examples, Discuss the Concepts, Concept Checks, Optional Student Activities, Suggested Assignments, Quick Quizzes, Answers to Writing Exercises/Focus on Problem Solving/Projects and Group Activities,* and *PowerPoints.* Answers to all exercises are also provided.

Instructor's Solutions Manual The *Instructor's Solutions Manual* contains worked-out solutions for all exercises in the text.

Instructor's Resource Manual with Testing This resource includes eight ready-to-use printed *Chapter Tests* per chapter, suggested *Course Sequences*, and a printout of the *AIM for Success* PowerPoint slide show. All resources are also available on the Instructor website and *Class Prep* CD.

HM ClassPrep with HM Testing CD-ROM *HM ClassPrep* contains a multitude of text-specific resources for instructors to use to enhance the classroom experience. These resources can be easily accessed by chapter or resource type and can also link you to the text's web site. *HM Testing* is our computerized test generator and contains a database of algorithmic test items as well as providing **on-line testing** and **gradebook** functions.

Instructor Text-Specific Website The resources available on the *Class Prep* CD are also available on the instructor website at math.college.hmco.com/instructors. Appropriate items are password protected. Instructors also have access to the student part of the text's website.

WebCT ePacks *WebCT ePacks* provide instructors with a flexible, Internet-based education platform providing multiple ways to present learning materials. The *WebCT ePacks* come with a full array of features to enrich the on-line learning experience.

Blackboard Cartridges The *Houghton Mifflin Blackboard cartridge* allows flexible, efficient, and creative ways to present learning materials and opportunities. In addition to course management benefits, instructors may make use of an electronic grade book, receive papers from students enrolled in the course via the Internet, and track student use of the communication and collaboration functions.

NEW! HM Eduspace® is a powerful course management system powered by Blackboard that makes preparing, presenting, and managing courses easier. You can use this distance-learning platform to customize, create, and deliver course materials and tests online, and easily maintain student portfolios using the grade book, where grades for all assignments are automatically scored, averaged, and saved.

Student Resources

Student Solutions Manual The *Student Solutions Manual* contains complete solutions to all odd-numbered exercises in the text.

Math Study Skills Workbook by Paul D. Nolting This workbook is designed to reinforce skills and minimize frustration for students in any math class, lab, or study skills course. It offers a wealth of study tips and sound advice on note

Copyright © Houghton Mifflin Company. All rights reserved.

taking, time management, and reducing math anxiety. In addition, numerous opportunities for self-assessment enable students to track their own progress.

HM Eduspace® Online Learning Environment *Eduspace* is a text-specific, web-based learning environment which combines an algorithmic tutorial program with homework capabilities. Specific content is available 24 hours a day to help you further understand your textbook.

HM mathSpace® Tutorial CD-ROM This tutorial CD-ROM allows students to practice skills and review concepts as many times as necessary by providing algorithmically-generated exercises and step-by-step solutions for practice.

SMARTHINKING™ Live, Online Tutoring Houghton Mifflin has partnered with SMARTHINKING to provide an easy-to-use and effective on-line tutorial service. **Whiteboard Simulations** and **Practice Area** promote real-time visual interaction.

Three levels of service are offered.

- **Text-specific Tutoring** provides real-time, one-on-one instruction with a specially qualified 'e-structor.'
- **Questions Any Time** allows students to submit questions to the tutor outside the scheduled hours and receive a reply within 24 hours.
- **Independent Study Resources** connect students with around-the-clock access to additional educational services, including interactive websites, diagnostic tests, and Frequently Asked Questions posed to SMARTHINKING e-structors.

Houghton Mifflin Instructional Videos and DVDs Text-specific videos and DVDs, hosted by Dana Mosely, cover all sections of the text and provide a valuable resource for further instruction and review.

Student Text-Specific Website On-line student resources can be found at this text's website at math.college.hmco.com/students.

Acknowledgments

The authors would like to thank the people who have reviewed this manuscript and provided many valuable suggestions.

Dr. Mark L. Campbell, *Slippery Rock University of Pennsylvania, PA*
Lynn M. Irons, *College of Southern Idaho, ID*
Jim Matovina, *Community College of Southern Nevada, NV*
Helen Medley
Christopher P. Reisch, *Jamestown Community College, NY*
Daniel Russow, *Arizona Western College, AZ*
Lauri Semarne
Kenneth Takvorian, *Mount Wachusett Community College, MA*
Joseph Verret, *DeVry University, CA*
Mary L. Wolyniak, *Broome Community College, NY*

With special thanks to Dawn Nuttall for her contributions in the developmental editing of three editions of this textbook series.

Copyright © Houghton Mifflin Company. All rights reserved.

AIM for Success

Welcome to *Introductory Algebra: An Applied Approach*. As you begin this course, we know two important facts: (1) We want you to succeed. (2) You want to succeed. To do that requires an effort from each of us. For the next few pages, we are going to show you what is required of you to achieve that success and how you can use the features of this text to be successful.

Motivation

One of the most important keys to success is motivation. We can try to motivate you by offering interesting or important ways mathematics can benefit you. But, in the end, the motivation must come from you. On the first day of class, it is easy to be motivated. Eight weeks into the term, it is harder to keep that motivation.

To stay motivated, there must be outcomes from this course that are worth your time, money, and energy.

List some reasons you are taking this course.

_____

_____

_____

TAKE NOTE

Motivation alone will not lead to success. For instance, suppose a person who cannot swim is placed in a boat, taken out to the middle of a lake, and then thrown overboard. That person has a lot of motivation but there is a high likelihood the person will drown without some help. Motivation gives us the desire to learn but is not the same as learning.

Although we hope that one of the reasons you listed was an interest in mathematics, we know that many of you are taking this course because it is required to graduate, it is a prerequisite for a course you must take, or because it is required for your major. Although you may not agree that this course is necessary, it is! If you are motivated to graduate or complete the requirements for your major, then use that motivation to succeed in this course. Do not become distracted from your goal to complete your education!

Commitment

To be successful, you must make a commitment to succeed. This means devoting time to math so that you achieve a better understanding of the subject.

List some activities (sports, hobbies, talents such as dance, art, or music) that you enjoy and at which you would like to become better.

ACTIVITY	TIME SPENT	TIME WISHED SPENT

Thinking about these activities, put the number of hours that you spend each week practicing these activities next to the activity. Next to that number, indicate the number of hours per week you would like to spend on these activities.

Whether you listed surfing or sailing, aerobics or restoring cars, or any other activity you enjoy, note how many hours a week you spend doing it. To succeed in math, you must be willing to commit the same amount of time. Success requires some sacrifice.

The "I Can't Do Math" Syndrome

There may be things you cannot do, such as lift a two-ton boulder. You can, however, do math. It is much easier than lifting the two-ton boulder. When you first

Copyright © Houghton Mifflin Company. All rights reserved.

learned the activities you listed above, you probably could not do them well. With practice, you got better. With practice, you will be better at math. Stay focused, motivated, and committed to success.

It is difficult for us to emphasize how important it is to overcome the "I Can't Do Math" Syndrome. If you listen to interviews of very successful athletes after a particularly bad performance, you will note that they focus on the positive aspect of what they did, not the negative. Sports psychologists encourage athletes to always be positive—to have a "Can Do" attitude. Develop this attitude toward math.

Strategies for Success

Textbook Review   Right now, do a 15-minute "textbook review" of this book. Here's how:

First, read the table of contents. Do it in three minutes or less. Next, look through the entire book, page by page. Move quickly. Scan titles, look at pictures, notice diagrams.

A textbook reconnaissance shows you where a course is going. It gives you the big picture. That's useful because brains work best when going from the general to the specific. Getting the big picture before you start makes details easier to recall and understand later on.

Your textbook reconnaissance will work even better if, as you scan, you look for ideas or topics that are interesting to you. List three facts, topics, or problems that you found interesting during your textbook reconnaissance.

_____

_____

_____

The idea behind this technique is simple: It's easier to work at learning material if you know it's going to be useful to you.

Not all the topics in this book will be "interesting" to you. But that is true of any subject. Surfers find that on some days the waves are better than others, musicians find some music more appealing than other music, computer gamers find some computer games more interesting than others, car enthusiasts find some cars more exciting than others. Some car enthusiasts would rather have a completely restored 1957 Chevrolet than a new Ferrari.

Know the Course Requirements   To do your best in this course, you must know exactly what your instructor requires. Course requirements may be stated in a *syllabus*, which is a printed outline of the main topics of the course, or they may be presented orally. When they are listed in a syllabus or on other printed pages, keep them in a safe place. When they are presented orally, make sure to take complete notes. In either case, it is important that you understand them completely and follow them exactly. Be sure you know the answer to each of the following questions.

1. What is your instructor's name?
2. Where is your instructor's office?
3. At what times does your instructor hold office hours?
4. Besides the textbook, what other materials does your instructor require?
5. What is your instructor's attendance policy?
6. If you must be absent from a class meeting, what should you do before returning to class? What should you do when you return to class?

Copyright © Houghton Mifflin Company. All rights reserved.

7. What is the instructor's policy regarding collection or grading of homework assignments?

8. What options are available if you are having difficulty with an assignment? Is there a math tutoring center?

9. If there is a math lab at your school, Where is it located? What hours is it open?

10. What is the instructor's policy if you miss a quiz?

11. What is the instructor's policy if you miss an exam?

12. Where can you get help when studying for an exam?

Remember: Your instructor wants to see you succeed. If you need help, ask! Do not fall behind. If you are running a race and fall behind by 100 yards, you may be able to catch up but it will require more effort than had you not fallen behind.

Time Management   We know that there are demands on your time. Family, work, friends, and entertainment all compete for your time. We do not want to see you receive poor job evaluations because you are studying math. However, it is also true that we do not want to see you receive poor math test scores because you devoted too much time to work. When several competing and important tasks require your time and energy, the only way to manage the stress of being successful at both is to manage your time efficiently.

Instructors often advise students to spend twice the amount of time outside of class studying as they spend in the classroom. Time management is important if you are to accomplish this goal and succeed in school. The following activity is intended to help you structure your time more efficiently.

List the name of each course you are taking this term, the number of class hours each course meets, and the number of hours you should spend studying each subject outside of class. Then fill in a weekly schedule like the one printed below. Begin by writing in the hours spent in your classes, the hours spent at work (if you have a job), and any other commitments that are not flexible with respect to the time that you do them. Then begin to write down commitments that are more flexible, including hours spent studying. Remember to reserve time for activities such as meals and exercise. You should also schedule free time.

TAKE NOTE

Besides time management, there must be realistic ideas of how much time is available. There are very few people who can *successfully* work full-time and go to school full-time. If you work 40 hours a week, take 15 units, spend the recommended study time given at the right, and sleep 8 hours a day, you will use over 80% of the available hours in a week. That leaves less than 20% of the hours in a week for family, friends, eating, recreation, and other activities.

	Monday	Tuesday	Wednesday	Thursday	Friday	Saturday	Sunday
7–8 a.m.							
8–9 a.m.							
9–10 a.m.							
10–11 a.m.							
11–12 p.m.							
12–1 p.m.							
1–2 p.m.							
2–3 p.m.							
3–4 p.m.							
4–5 p.m.							
5–6 p.m.							
6–7 p.m.							
7–8 p.m.							
8–9 p.m.							
9–10 p.m.							
10–11 p.m.							
11–12 a.m.							

Copyright © Houghton Mifflin Company. All rights reserved.

We know that many of you must work. If that is the case, realize that working 10 hours a week at a part-time job is equivalent to taking a three-unit class. If you must work, consider letting your education progress at a slower rate to allow you to be successful at both work and school. There is no rule that says you must finish school in a certain time frame.

Schedule Study Time   As we encouraged you to do by filling out the time management form above, schedule a certain time to study. You should think of this time the way you would the time for work or class—that is, reasons for missing study time should be as compelling as reasons for missing work or class. "I just didn't feel like it" is not a good reason to miss your scheduled study time.

Although this may seem like an obvious exercise, list a few reasons you might want to study.

_____

_____

_____

Of course we have no way of knowing the reasons you listed, but from our experience one reason given quite frequently is "To pass the course." There is nothing wrong with that reason. If that is the most important reason for you to study, then use it to stay focused.

One method of keeping to a study schedule is to form a ***study group***. Look for people who are committed to learning, who pay attention in class, and who are punctual. Ask them to join your group. Choose people with similar educational goals but different methods of learning. You can gain insight from seeing the material from a new perspective. Limit groups to four or five people; larger groups are unwieldy.

There are many ways to conduct a study group. Begin with the following suggestions and see what works best for your group.

1. Test each other by asking questions. Each group member might bring two or three sample test questions to each meeting.
2. Practice teaching each other. Many of us who are teachers learned a lot about our subject when we had to explain it to someone else.
3. Compare class notes. You might ask other students about material in your notes that is difficult for you to understand.
4. Brainstorm test questions.
5. Set an agenda for each meeting. Set approximate time limits for each agenda item and determine a quitting time.

And finally, probably the most important aspect of studying is that it should be done in relatively small chunks. If you can only study three hours a week for this course (probably not enough for most people), do it in blocks of one hour on three separate days, preferably after class. Three hours of studying on a Sunday is not as productive as three hours of paced study.

Text Features That Promote Success   There are 11 chapters in this text. Each chapter is divided into sections, and each section is subdivided into learning objectives. Each learning objective is labeled with a letter from A to E.

Copyright © Houghton Mifflin Company. All rights reserved.

Preparing for a Chapter   Before you begin a new chapter, you should take some time to review previously learned skills. There are two ways to do this. The first is to complete the ***Cumulative Review Exercises***, which occurs after every chapter (except Chapter 1). For instance, turn to page 285. The questions in this review are taken from the previous chapters. The answers for all these exercises can be found on page A11. Turn to that page now and locate the answers for the Chapter 5 Cumulative Review Exercises. After the answer to the first exercise, which is 7, you will see the objective reference [1.2B]. This means that this question was taken from Chapter 1, Section 2, Objective B. If you missed this question, you should return to that objective and restudy the material.

A second way of preparing for a new chapter is to complete the ***Prep Test***. This test focuses on the particular skills that will be required for the new chapter. Turn to page 236 to see a Prep Test. The answers for the Prep Test are the first set of answers in the answer section for a chapter. Turn to page A9 to see the answers for the Chapter 5 Prep Test. Note that an objective reference is given for each question. If you answer a question incorrectly, restudy the objective from which the question was taken.

Before the class meeting in which your professor begins a new section, you should read each objective statement for that section. Next, browse through the objective material, being sure to note each word in bold type. These words indicate important concepts that you must know in order to learn the material. Do not worry about trying to understand all the material. Your professor is there to assist you with that endeavor. The purpose of browsing through the material is so that your brain will be prepared to accept and organize the new information when it is presented to you.

Turn to page 3. Write down the title of the first objective in Section 1.1. Under the title of the objective, write down the words in the objective that are in bold print. It is not necessary for you to understand the meaning of these words. You are in this class to learn their meaning.

_____	_____	_____	_____
_____	_____	_____	_____
_____	_____	_____	_____
_____	_____	_____	_____
_____	_____	_____	_____

Math Is Not a Spectator Sport   To learn mathematics you must be an active participant. Listening and watching your professor do mathematics is not enough. Mathematics requires that you interact with the lesson you are studying. If you filled in the blanks above, you were being interactive. There are other ways this textbook has been designed to help you be an active learner.

Annotated Examples   The HOW TO feature indicates an example with explanatory remarks to the right of the work. Using paper and pencil, you should work along as you go through the example.

Copyright © Houghton Mifflin Company. All rights reserved.

$$\frac{3}{4}x - 2 = -11$$

$$\frac{3}{4}x - 2 + 2 = -11 + 2$$

$$\frac{3}{4}x = -9$$

$$\frac{4}{3} \cdot \frac{3}{4}x = \frac{4}{3}(-9)$$

$$x = -12$$

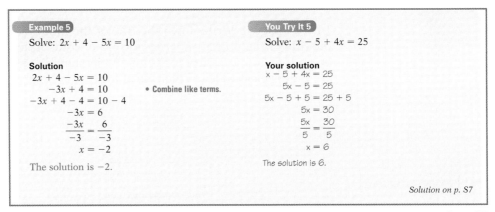

HOW TO Solve: $\frac{3}{4}x - 2 = -11$

The goal is to write the equation in the form *variable = constant*.

$$\frac{3}{4}x - 2 = -11$$

$$\frac{3}{4}x - 2 + 2 = -11 + 2$$ • **Add 2** to each side of the equation.

$$\frac{3}{4}x = -9$$ • Simplify.

$$\frac{4}{3} \cdot \frac{3}{4}x = \frac{4}{3}(-9)$$ • Multiply each side of the equation by $\frac{4}{3}$.

$$x = -12$$ • The equation is in the form *variable = constant*.

The solution is −12.

TAKE NOTE

Check: $\frac{3}{4}x - 2 = -11$

$$\frac{3}{4}(-12) - 2 \mid -11$$

$$-9 - 2 \mid -11$$

$$-11 = -11$$

A true equation

When you complete the example, get a clean sheet of paper. Write down the problem and then try to complete the solution without referring to your notes or the book. When you can do that, move on to the next part of the objective.

Leaf through the book now and write down the page numbers of two other occurrences of a HOW TO example.

You Try Its   One of the key instructional features of this text is the paired examples. Notice that in each example box, the example on the left is completely worked out and the "You Try It" example on the right is not. Study the worked-out example carefully by working through each step. Then work the You Try It. If you get stuck, refer to the page number at the end of the example, which directs you to the place where the You Try It is solved—a complete worked-out solution is provided. Try to use the given solution to get a hint for the step you are stuck on. Then try to complete your solution.

Example 5

Solve: $2x + 4 - 5x = 10$

Solution

$$2x + 4 - 5x = 10$$
$$-3x + 4 = 10$$ • Combine like terms.
$$-3x + 4 - 4 = 10 - 4$$
$$-3x = 6$$
$$\frac{-3x}{-3} = \frac{6}{-3}$$
$$x = -2$$

The solution is −2.

You Try It 5

Solve: $x - 5 + 4x = 25$

Your solution

$$x - 5 + 4x = 25$$
$$5x - 5 = 25$$
$$5x - 5 + 5 = 25 + 5$$
$$5x = 30$$
$$\frac{5x}{5} = \frac{30}{5}$$
$$x = 6$$

The solution is 6.

Solution on p. S7

When you have completed your solution, check your work against the solution we provided. (Turn to page S7 to see the solution of You Try It 5.) Be aware that frequently there is more than one way to solve a problem. Your answer, however, should be the same as the given answer. If you have any question as to whether your method will "always work," check with your instructor or with someone in the math center.

Browse through the textbook and write down the page numbers where two other paired example features occur.

Remember: Be an active participant in your learning process. When you are sitting in class watching and listening to an explanation, you may think that you understand. However, until you actually try to do it, you will have no confirmation of the new knowledge or skill. Most of us have had the experience of sitting in class thinking we knew how to do something only to get home and realize that we didn't.

TAKE NOTE

There is a strong connection between reading and being a successful student in math or any other subject. If you have difficulty reading, consider taking a reading course. Reading is much like other skills. There are certain things you can learn that will make you a better reader.

Copyright © Houghton Mifflin Company. All rights reserved.

Word Problems Word problems are difficult because we must read the problem, determine the quantity we must find, think of a method to do that, and then actually solve the problem. In short, we must formulate a *strategy* to solve the problem and then devise a *solution*.

Note in the paired example below that part of every word problem is a strategy and part is a solution. The strategy is a written description of how we will solve the problem. In the corresponding You Try It, you are asked to formulate a strategy. Do not skip this step, and be sure to write it out.

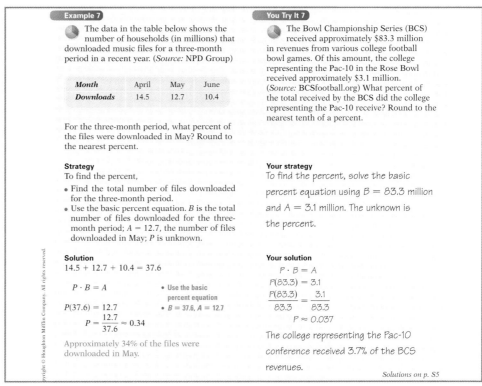

Example 7

The data in the table below shows the number of households (in millions) that downloaded music files for a three-month period in a recent year. (*Source:* NPD Group)

Month	April	May	June
Downloads	14.5	12.7	10.4

For the three-month period, what percent of the files were downloaded in May? Round to the nearest percent.

Strategy
To find the percent,
- Find the total number of files downloaded for the three-month period.
- Use the basic percent equation. B is the total number of files downloaded for the three-month period; $A = 12.7$, the number of files downloaded in May; P is unknown.

Solution
$14.5 + 12.7 + 10.4 = 37.6$

$P \cdot B = A$ • Use the basic percent equation
$P(37.6) = 12.7$ • $B = 37.6$, $A = 12.7$
$P = \dfrac{12.7}{37.6} \approx 0.34$

Approximately 34% of the files were downloaded in May.

You Try It 7

The Bowl Championship Series (BCS) received approximately $83.3 million in revenues from various college football bowl games. Of this amount, the college representing the Pac-10 in the Rose Bowl received approximately $3.1 million. (*Source:* BCSfootball.org) What percent of the total received by the BCS did the college representing the Pac-10 receive? Round to the nearest tenth of a percent.

Your strategy
To find the percent, solve the basic percent equation using $B = 83.3$ million and $A = 3.1$ million. The unknown is the percent.

Your solution
$P \cdot B = A$
$P(83.3) = 3.1$
$\dfrac{P(83.3)}{83.3} = \dfrac{3.1}{83.3}$
$P \approx 0.037$

The college representing the Pac-10 conference received 3.7% of the BCS revenues.

Solutions on p. S5

Copyright © Houghton Mifflin Company. All rights reserved.

TAKE NOTE

If a rule has more than one part, be sure to make a notation to that effect.

Multiply each side of an equation by the same number. Do not use zero.

TAKE NOTE

If you are working at home and need assistance, there is online help available at math.college.hmco.com/students, at this text's website.

Rule Boxes Pay special attention to rules placed in boxes. These rules give you the reasons certain types of problems are solved the way they are. When you see a rule, try to rewrite the rule in your own words.

The equations $2x = 6$, $10x = 30$, and $-8x = -24$ are equivalent equations; each equation has 3 as its solution. These examples suggest that multiplying each side of an equation by the same nonzero number produces an equivalent equation.

Multiplication Property of Equations

Each side of an equation can be multiplied by the same *nonzero* number without changing the solution of the equation. In symbols, if $c \neq 0$, then the equation $a = b$ has the same solutions as the equation $ac = bc$.

Chapter Exercises When you have completed studying an objective, do the exercises in the exercise set that correspond with that objective. The exercises are labeled with the same letter as the objective. Math is a subject that needs to be learned in small sections and practiced continually in order to be mastered. Doing all of the exercises in each exercise set will help you master the problem-solving techniques necessary for success. As you work through the exercises for an objective, check your answers to the odd-numbered exercises with those in the back of the book.

Copyright © Houghton Mifflin Company. All rights reserved.

Preparing for a Test   There are important features of this text that can be used to prepare for a test.

- Chapter Summary
- Chapter Review Exercises
- Chapter Test

After completing a chapter, read the Chapter Summary. (See page 226 for the Chapter 4 Summary.) This summary highlights the important topics covered in the chapter. The page number following each topic refers you to the page in the text on which you can find more information about the concept.

Following the Chapter Summary are Chapter Review Exercises (see page 229) and a Chapter Test (see page 231). Doing the review exercises is an important way of testing your understanding of the chapter. The answer to each review exercise is given at the back of the book, along with its objective reference. After checking your answers, restudy any objective from which a question you missed was taken. It may be helpful to retry some of the exercises for that objective to reinforce your problem-solving techniques.

The Chapter Test should be used to prepare for an exam. We suggest that you try the Chapter Test a few days before your actual exam. Take the test in a quiet place and try to complete the test in the same amount of time you will be allowed for your exam. When taking the Chapter Test, practice the strategies of successful test takers: (1) scan the entire test to get a feel for the questions; (2) read the directions carefully; (3) work the problems that are easiest for you first; and perhaps most importantly, (4) try to stay calm.

When you have completed the Chapter Test, check your answers. If you missed a question, review the material in that objective and rework some of the exercises from that objective. This will strengthen your ability to perform the skills in that objective.

Is it difficult to be successful? YES! Successful music groups, artists, professional athletes, chefs, and _Write your major here_ have to work very hard to achieve their goals. They focus on their goals and ignore distractions. The things we ask you to do to achieve success take time and commitment. We are confident that if you follow our suggestions, you will succeed.

Copyright © Houghton Mifflin Company. All rights reserved.

1 Prealgebra Review

When you take a multiple-choice test, such as a class exam, the ACT, or the SAT, there is usually a point system for scoring your answers. Correct answers receive a positive number of points, and incorrect answers receive a negative number of points. For the ACT, the SAT, and most multiple-choice tests, it is better for your score to leave a question blank if you are unsure of the answer. An unanswered question will cause fewer points to be deducted from your score; sometimes it will not cost you any points at all. **Exercises 67 and 68 on page 23** show how professors can adjust the grading of multiple-choice exams to discourage students from guessing randomly.

Need help? For online student resources, such as section quizzes, visit this textbook's website at **math.college.hmco.com/students**.

Copyright © Houghton Mifflin Company. All rights reserved.

OBJECTIVES

Section 1.1
A To use inequality symbols with integers
B To use opposites and absolute value

Section 1.2
A To add integers
B To subtract integers
C To solve application problems

Section 1.3
A To multiply integers
B To divide integers
C To solve application problems

Section 1.4
A To evaluate exponential expressions
B To use the Order of Operations Agreement to simplify expressions

Section 1.5
A To factor numbers
B To find the prime factorization of a number
C To find the least common multiple and greatest common factor

Section 1.6
A To write a rational number in simplest form and as a decimal
B To add rational numbers
C To subtract rational numbers
D To solve application problems

Section 1.7
A To multiply rational numbers
B To divide rational numbers
C To convert among percents, fractions, and decimals
D To solve application problems

Section 1.8
A To find the measures of angles
B To solve perimeter problems
C To solve area problems

Do these exercises to prepare for Chapter 1.

1. What is 127.1649 rounded to the nearest hundredth?

2. Add: 3416 + 42,561 + 537

3. Subtract: 5004 − 487

4. Multiply: 407 × 28

5. Divide: 11,684 ÷ 23

6. What is the smallest number that both 8 and 12 divide evenly?

7. What is the greatest number that divides both 16 and 20 evenly?

8. Without using 1, write 21 as a product of two whole numbers.

9. Represent the shaded portion of the figure as a fraction.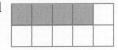

10. Which of the following, if any, is not possible?
a. 6 + 0 **b.** 6 − 0 **c.** 6 × 0 **d.** 6 ÷ 0

GO FIGURE • • •

In a group of 9 coins, one is counterfeit and weighs less than the 8 genuine coins. Using a balance scale, determine the counterfeit coin in two weighings.

Copyright © Houghton Mifflin Company. All rights reserved.

1.1   Introduction to Integers

Objective A   **To use inequality symbols with integers**

It seems to be a human characteristic to group similar items. For instance, a biologist places similar animals in groups called *species*. Nutritionists classify foods according to *food groups*; for example, pasta, crackers, and rice are among the foods in the bread group.

Mathematicians place objects with similar properties in groups called *sets*. A **set** is a collection of objects. The objects in a set are called the **elements of the set.**

The **roster method** of writing sets encloses a list of the elements in braces. Thus the set of sections within an orchestra is written {brass, percussion, string, woodwind}. When the elements of a set are listed, each element is listed only once. For instance, if the list of numbers 1, 2, 3, 2, 3 were placed in a set, the set would be {1, 2, 3}.

The numbers that we use to count objects, such as the students in a classroom or the horses on a ranch, are the *natural numbers*.

$$\textbf{Natural numbers} = \{1, 2, 3, 4, 5, 6, 7, 8, 9, 10, \ldots\}$$

The three dots mean that the list of natural numbers continues on and on and that there is no largest natural number.

The natural numbers alone do not provide all the numbers that are useful in applications. For instance, a meteorologist also needs the number zero and numbers below zero.

$$\textbf{Integers} = \{\ldots, -5, -4, -3, -2, -1, 0, 1, 2, 3, 4, 5, \ldots\}$$

Point of Interest

The Alexandrian astronomer Ptolemy began using *omicron*, O, the first letter of the Greek word that means "nothing," as the symbol for zero in 150 A.D. It was not until the 13th century, however, that Fibonacci introduced 0 to the Western world as a placeholder so that we could distinguish, for example, 45 from 405.

Each integer can be shown on a number line. The integers to the left of zero on the number line are called **negative integers.** The integers to the right of zero are called **positive integers,** or natural numbers. Zero is neither a positive nor a negative integer.

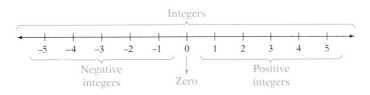

The **graph** of an integer is shown by placing a heavy dot on the number line directly above the number. The graphs of -3 and 4 are shown on the number line below.

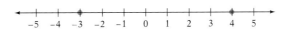

Copyright © Houghton Mifflin Company. All rights reserved.

Consider the following sentences.

> The quarterback threw the football and the receiver caught *it*.
> A student purchased a computer and used *it* to write history papers.

In the first sentence, *it* is used to mean the football; in the second sentence, *it* means the computer. In language, the word *it* can stand for many different objects. Similarly, in mathematics, a letter of the alphabet can be used to stand for a number. Such a letter is called a **variable.** Variables are used in the following definition of inequality symbols.

Point of Interest

The symbols for "is less than" and "is greater than" were introduced by Thomas Harriot around 1630. Before that, ⊏ and ⊐ were used for > and <, respectively.

Inequality Symbols

If *a* and *b* are two numbers and *a* is to the left of *b* on the number line, then *a* **is less than** *b*. This is written $a < b$.

If *a* and *b* are two numbers and *a* is to the right of *b* on the number line, then *a* **is greater than** *b*. This is written $a > b$.

Negative 4 is less than negative 1.

$$-4 < -1$$

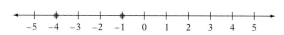

5 is greater than 0.

$$5 > 0$$

There are also inequality symbols for **is less than or equal to (≤)** and **is greater than or equal to (≥).**

$$7 \leq 15$$ 7 is less than or equal to 15.
This is true because $7 < 15$.

$$6 \leq 6$$ 6 is less than or equal to 6.
This is true because $6 = 6$.

Point of Interest

The Greek influence is again felt with the symbol ∈ for "is an element of." Giuseppe Peano used this symbol in 1889 as an abbreviation for the Greek word for "is."

The symbol ∈ means "is an element of." $2 \in B$ is read "2 is an element of set *B*."

Given $C = \{3, 5, 9\}$, then $3 \in C$, $5 \in C$, and $9 \in C$. $7 \notin C$ is read "7 is not an element of set *C*."

Example 1

Use the roster method to write the set of negative integers greater than or equal to -4.

Solution

$A = \{-4, -3, -2, -1\}$ • A set is designated by a capital letter.

You Try It 1

Use the roster method to write the set of positive integers less than 7.

Your solution

Solution on p. S1

Copyright © Houghton Mifflin Company. All rights reserved.

Example 2

Given $A = \{-6, -2, 0\}$, which elements of set A are less than or equal to -2?

Solution

Find the order relation between each element of set A and -2.

$-6 < -2$
$-2 = -2$
$0 > -2$

The elements -6 and -2 are less than or equal to -2.

You Try It 2

Given $B = \{-5, -1, 5\}$, which elements of set B are greater than -1?

Your solution

Solution on p. S1

Objective B   **To use opposites and absolute value**

Two numbers that are the same distance from zero on the number line but are on opposite sides of zero are **opposite numbers,** or **opposites.** The opposite of a number is also called its **additive inverse.**

The opposite of 5 is -5.

The opposite of -5 is 5.

The negative sign can be read "the opposite of."

$$-(2) = -2 \qquad \text{The opposite of 2 is } -2.$$
$$-(-2) = 2 \qquad \text{The opposite of } -2 \text{ is 2.}$$

Copyright © Houghton Mifflin Company. All rights reserved.

Study Tip

Some students think that they can "coast" at the beginning of this course because the topic of Chapter 1 is a review of pre-algebra. However, this chapter lays the foundation for the entire course. Be sure you know and understand all the concepts presented. For example, study the properties of absolute value presented in this lesson.

The **absolute value of a number** is its distance from zero on the number line. Therefore, the absolute value of a number is a positive number or zero. The symbol for absolute value is two vertical bars, $|\;|$.

The distance from 0 to 3 is 3. Therefore, the absolute value of 3 is 3.

$$|3| = 3$$

The distance from 0 to -3 is 3. Therefore, the absolute value of -3 is 3.

$$|-3| = 3$$

Point of Interest

The definition of *absolute value* given in the box is written in what is called rhetorical style. That is, it is written without the use of variables. This is how *all* mathematics was written prior to the Renaissance. During that period from the 14th to the 16th century, the idea of expressing a variable symbolically was developed. In terms of that symbolism, the definition of absolute value is

$$|x| = \begin{cases} x, & x > 0 \\ 0, & x = 0 \\ -x, & x < 0 \end{cases}$$

Absolute Value

The absolute value of a positive number is the number itself. For example, $|9| = 9$.

The absolute value of zero is zero. $|0| = 0$

The absolute value of a negative number is the opposite of the negative number. For example, $|-7| = 7$.

HOW TO   Evaluate: $-|-12|$

$-|-12| = -12$ • The absolute value sign does not affect the negative sign in front of the absolute value sign.

Example 3

Evaluate $|-4|$ and $-|-10|$.

Solution
$|-4| = 4$
$-|-10| = -10$

You Try It 3

Evaluate $|-5|$ and $-|-23|$.

Your solution

Example 4

Given $A = \{-12, 0, 4\}$, find the additive inverse of each element of set A.

Solution
$-(-12) = 12$
$-0 = 0$ • Zero is neither positive nor negative.
$-(4) = -4$

You Try It 4

Given $B = \{-11, 0, 8\}$, find the additive inverse of each element of set B.

Your solution

Example 5

Given $C = \{-17, 0, 14\}$, find the absolute value of each element of set C.

Solution
$|-17| = 17$
$|0| = 0$
$|14| = 14$

You Try It 5

Given $D = \{-37, 0, 29\}$, find the absolute value of each element of set D.

Your solution

Solutions on p. S1

Copyright © Houghton Mifflin Company. All rights reserved.

1.1 Exercises

Objective A   **To use inequality symbols with integers**

1. Explain the difference between the natural numbers and the integers.

2. Name the smallest integer that is larger than any negative integer.

For Exercises 3 to 12, place the correct symbol, $<$ or $>$, between the two numbers.

3. 8 -6

4. -14 16

5. -12 1

6. 35   28

7. 42   19

8. -42 27

9. 0 -31

10. -17 0

11. 53 -46

12. -27 -38

For Exercises 13 to 22, answer True or False.

13. $-13 > 0$

14. $-20 > 3$

15. $12 > -31$

16. $9 > 7$

17. $-5 > -2$

18. $-44 > -21$

19. $-4 > -120$

20. $0 > -8$

21. $-1 \geq -1$

22. $-10 \leq -10$

For Exercises 23 to 28, use the roster method to write the set.

23. The natural numbers less than 9

24. The natural numbers less than or equal to 6

25. The positive integers less than or equal to 8

26. The positive integers less than 4

27. The negative integers greater than -7

28. The negative integers greater than or equal to -5

29. Given $A = \{-7, 0, 2, 5\}$, which elements of set A are greater than 2?

30. Given $B = \{-8, 0, 7, 15\}$, which elements of set B are greater than 7?

31. Given $D = \{-23, -18, -8, 0\}$, which elements of set D are less than -8?

32. Given $C = \{-33, -24, -10, 0\}$, which elements of set C are less than -10?

33. Given $E = \{-35, -13, 21, 37\}$, which elements of set E are greater than -10?

34. Given $F = \{-27, -14, 14, 27\}$, which elements of set F are greater than -15?

35. Given $B = \{-52, -46, 0, 39, 58\}$, which elements of set B are less than or equal to 0?

36. Given $A = \{-12, -9, 0, 12, 34\}$, which elements of set A are greater than or equal to 0?

37. Given $C = \{-23, -17, 0, 4, 29\}$, which elements of set C are greater than or equal to -17?

38. Given $D = \{-31, -12, 0, 11, 45\}$, which elements of set D are less than or equal to -12?

Copyright © Houghton Mifflin Company. All rights reserved.

39. Given that set A is the positive integers less than 10, which elements of set A are greater than or equal to 5?

40. Given that set B is the positive integers less than or equal to 12, which elements of set B are greater than 6?

41. Given that set D is the negative integers greater than or equal to -10, which elements of set D are less than -4?

42. Given that set C is the negative integers greater than -8, which elements of set C are less than or equal to -3?

Objective B   **To use opposites and absolute value**

For Exercises 43 to 47, find the additive inverse.

43. 4   **44.** 8   **45.** -9 **46.** -28 **47.** -36

For Exercises 48 to 57, evaluate.

48. $-(-14)$ **49.** $-(-40)$ **50.** $-(77)$ **51.** $-(39)$ **52.** $-(-13)$

53. $|-74|$ **54.** $|-96|$ **55.** $-|-82|$ **56.** $-|-53|$ **57.** $-|81|$

For Exercises 58 to 65, place the correct symbol, $<$ or $>$, between the values of the two numbers.

58. $|-83|$ $|58|$ **59.** $|22|$ $|-19|$ **60.** $|43|$ $|-52|$ **61.** $|-71|$ $|-92|$

62. $|-68|$ $|-42|$ **63.** $|12|$ $|-31|$ **64.** $|-45|$ $|-61|$ **65.** $|-28|$ $|43|$

66. Use the set $A = \{-8, -5, -2, 1, 3\}$.
  a. Find the opposite of each element of set A.
  b. Find the absolute value of each element of set A.

67. Use the set $B = \{-11, -7, -3, 1, 5\}$.
  a. Find the opposite of each element of set B.
  b. Find the absolute value of each element of set B.

APPLYING THE CONCEPTS

68. If x represents a negative integer, then $-x$ represents a _____ integer.

69. If x is an integer, is the inequality $|x| < -3$ always true, sometimes true, or never true?

70. In your own words, explain the meaning of the absolute value of a number and the additive inverse of a number.

Copyright © Houghton Mifflin Company. All rights reserved.

1.2 Addition and Subtraction of Integers

Objective A   **To add integers**

A number can be represented anywhere along the number line by an arrow. A positive number is represented by an arrow pointing to the right, and a negative number is represented by an arrow pointing to the left. The size of the number is represented by the length of the arrow.

Addition is the process of finding the total of two numbers. The numbers being added are called **addends.** The total is called the **sum.** Addition of integers can be shown on the number line. To add integers, start at zero and draw, above the number line, an arrow representing the first number. At the tip of the first arrow, draw a second arrow representing the second number. The sum is below the tip of the second arrow.

$4 + 2 = 6$ $-4 + (-2) = -6$

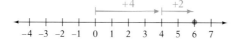   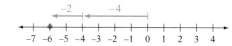

$-4 + 2 = -2$ $4 + (-2) = 2$

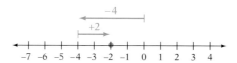   

The pattern for addition shown on the number lines above is summarized in the following rules for adding integers.

> **Addition of Integers**
>
> **To add two numbers with the same sign,** add the absolute values of the numbers. Then attach the sign of the addends.
>
> **To add two numbers with different signs,** find the absolute value of each number. Subtract the smaller of the two numbers from the larger. Then attach the sign of the number with the larger absolute value.

Copyright © Houghton Mifflin Company. All rights reserved.

Study Tip

The HOW TO feature indicates an example with explanatory remarks. Using paper and pencil, you should work through the example. See *AIM for Success,* page xxv.

HOW TO   Add: $-12 + (-26)$

$-12 + (-26) = -38$ • The signs are the same. Add the absolute values of the numbers (12 + 26). Attach the sign of the addends.

HOW TO   Add: $-19 + 8$

$|-19| = 19$ $|8| = 8$ • The signs are different. Find the absolute value of each number.

$19 - 8 = 11$ • Subtract the smaller number from the larger.

$-19 + 8 = -11$ • Attach the sign of the number with the larger absolute value.

HOW TO   Find the sum of -23, 47, -18, and -10.

Recall that a *sum* is the answer to an addition problem.

$-23 + 47 + (-18) + (-10)$

$= 24 + (-18) + (-10)$

$= 6 + (-10)$

$= -4$

• To add more than two numbers, add the first two numbers. Then add the sum to the third number. Continue until all the numbers are added.

Study Tip

One of the key instructional features of this text is the Example/You Try It pairs. Each example is completely worked. You are to solve the You Try It problems. When you are ready, check your solution against the one given in the Solutions section at the back of the book. The solution for You Try It 1 below is on page S1 (see the reference at the bottom right of the You Try It). See *AIM for Success*, page xxv.

The phrase *the sum of* in the example above indicates the operation of addition. All of the phrases below indicate addition.

added to	-6 added to 9	$9 + (-6) = 3$
more than	3 more than -8	$-8 + 3 = -5$
the sum of	the sum of -2 and -8	$-2 + (-8) = -10$
increased by	-7 increased by 5	$-7 + 5 = -2$
the total of	the total of 4 and -9	$4 + (-9) = -5$
plus	6 plus -10	$6 + (-10) = -4$

Example 1   Add: $-52 + (-39)$

Solution
$-52 + (-39) = -91$

You Try It 1   Add: $100 + (-43)$

Your solution

Example 2   Add: $37 + (-52) + (-14)$

Solution
$37 + (-52) + (-14) = -15 + (-14)$
$= -29$

You Try It 2   Add: $-51 + 42 + 17 + (-102)$

Your solution

Example 3   Find 11 more than -23.

Solution $-23 + 11 = -12$

You Try It 3   Find -8 increased by 7.

Your solution

Solutions on p. S1

Objective B   **To subtract integers**

Look at the expressions below. Note that each expression equals the same number.

$8 - 3 = 5$ 8 minus 3 is 5.
$8 + (-3) = 5$ 8 plus the opposite of 3 is 5.

This example suggests the following.

Subtraction of Integers

To subtract one number from another, add the opposite of the second number to the first number.

Copyright © Houghton Mifflin Company. All rights reserved.

HOW TO   Subtract: $-21 - (-40)$

Change this sign to plus.

$$-21 - (-40) = -21 + 40 = 19$$

Change -40 to the opposite of -40.

• Rewrite each subtraction as addition of the opposite. Then add.

HOW TO   Subtract: $15 - 51$

Change this sign to plus.

$$15 - 51 = 15 + (-51) = -36$$

Change 51 to the opposite of 51.

• Rewrite each subtraction as addition of the opposite. Then add.

HOW TO   Subtract: $-12 - (-21) - 15$

$$-12 - (-21) - 15 = -12 + 21 + (-15)$$
$$= 9 + (-15) = -6$$

• Rewrite each subtraction as addition of the opposite. Then add.

HOW TO   Find the difference between -8 and 7.

A *difference* is the answer to a subtraction problem.

$$-8 - 7 = -8 + (-7)$$
$$= -15$$

• Rewrite the subtraction as addition of the opposite.

The phrase *the difference of* in the example above indicates the operation of subtraction. All of the phrases below indicate subtraction.

TAKE NOTE

Note the order in which numbers are subtracted when the phrase *less than* is used. If you have $10 and a friend has $6 less than you do, then your friend has $6 less than $10, or $10 − $6 = $4.

minus	-5 minus 11	$-5 - 11 = -16$
less	-3 less 5	$-3 - 5 = -8$
less than	-8 less than -2	$-2 - (-8) = 6$
the difference between	the difference between -5 and 4	$-5 - 4 = -9$
decreased by	-4 decreased by 9	$-4 - 9 = -13$
subtract . . . from	subtract 8 from -3	$-3 - 8 = -11$

Example 4   Subtract: $-11 - 15$

Solution $-11 - 15 = -11 + (-15) = -26$

You Try It 4   Subtract: $19 - (-32)$

Your solution

Example 5   Subtract: $-14 - 18 - (-21) - 4$

Solution $-14 - 18 - (-21) - 4$
$$= -14 + (-18) + 21 + (-4)$$
$$= -32 + 21 + (-4)$$
$$= -11 + (-4) = -15$$

You Try It 5   Subtract: $-9 - (-12) - 17 - 4$

Your solution

Example 6   Find 9 less than -4.

Solution $-4 - 9 = -4 + (-9) = -13$

You Try It 6   Subtract -12 from -11.

Your solution

Solutions on p. S1

Copyright © Houghton Mifflin Company. All rights reserved.

Objective C To solve application problems

Positive and negative numbers are used to express the profitability of a company. A profit is recorded as a positive number; a loss is recorded as a negative number.

HOW TO The bar graph below shows the net earnings of Apple Computer Corporation for the years 1996 through 2002. Calculate the total net earnings for Apple Computer for the years 1996 to 2002.

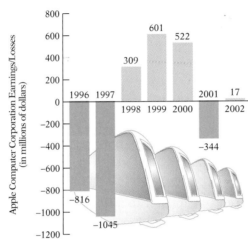

Source: Apple Computer Corporation

Strategy

To determine the total net earnings, add the profits and losses for the years shown.

Solution

$$-816 + (-1045) + 309 + 601 + 522 + (-344) + 17 = -756$$

The total net earnings for 1996 to 2002 were $-\$756$ million.

Example 7

The average temperature on Mercury's sunlit side is 950°F. The average temperature on Mercury's dark side is -346°F. Find the difference between these two average temperatures.

Strategy

To find the difference, subtract the average temperature on the dark side (-346) from the average temperature on the sunlit side (950).

Solution

$$950 - (-346) = 950 + 346$$
$$= 1296$$

The difference between these average temperatures is 1296°F.

You Try It 7

The average daytime temperature on Mars is -17°F. The average nighttime temperature on Mars is -130°F. Find the difference between these two average temperatures.

Your strategy

Your solution

Solution on p. S1

Copyright © Houghton Mifflin Company. All rights reserved.

1.2 Exercises

Objective A   **To add integers**

1. Explain how to add two integers with the same sign.

2. Explain how to add two integers with different signs.

For Exercises 3 to 34, add.

3. $-3 + (-8)$

4. $-6 + (-9)$

5. $-8 + 3$

6. $-9 + 2$

7. $-3 + (-80)$

8. $-12 + (-1)$

9. $-23 + (-23)$

10. $-12 + (-12)$

11. $16 + (-16)$

12. $-17 + 17$

13. $48 + (-53)$

14. $19 + (-41)$

15. $-17 + (-3) + 29$

16. $13 + 62 + (-38)$

17. $-3 + (-8) + 12$

18. $-27 + (-42) + (-18)$

19. $13 + (-22) + 4 + (-5)$

20. $-14 + (-3) + 7 + (-21)$

21. $-22 + 20 + 2 + (-18)$

22. $-6 + (-8) + 14 + (-4)$

23. $-16 + (-17) + (-18) + 10$

24. $-25 + (-31) + 24 + 19$

25. $26 + (-15) + (-11) + (-12)$

26. $-32 + 40 + (-8) + (-19)$

27. $-17 + (-18) + 45 + (-10)$

28. $23 + (-15) + 9 + (-15)$

29. $46 + (-17) + (-13) + (-50)$

30. $-37 + (-17) + (-12) + (-15)$

31. $-14 + (-15) + (-11) + 40$

32. $28 + (-19) + (-8) + (-1)$

33. $-23 + (-22) + (-21) + 5$

34. $-31 + 9 + (-16) + (-15)$

35. Find the sum of -42 and -23.

36. What is 4 more than -8?

37. What is 16 more than -31?

38. Find -17 increased by 12.

39. Find the total of -17, -23, 43, and 19.

40. What is -8 added to -21?

Copyright © Houghton Mifflin Company. All rights reserved.

Objective B   **To subtract integers**

41. ✎ What is the difference between the terms *minus* and *negative*?

42. ✎ Explain how to subtract two integers.

For Exercises 43 to 72, subtract.

43. $16 - 8$

44. $12 - 3$

45. $7 - 14$

46. $6 - 9$

47. $-7 - 2$

48. $-9 - 4$

49. $7 - (-2)$

50. $3 - (-4)$

51. $-6 - (-3)$

52. $-4 - (-2)$

53. $6 - (-12)$

54. $-12 - 16$

55. $-4 - 3 - 2$

56. $4 - 5 - 12$

57. $12 - (-7) - 8$

58. $-12 - (-3) - (-15)$

59. $-19 - (-19) - 18$

60. $-8 - (-8) - 14$

61. $-17 - (-8) - (-9)$

62. $7 - 8 - (-1)$

63. $-30 - (-65) - 29 - 4$

64. $42 - (-82) - 65 - 7$

65. $-16 - 47 - 63 - 12$

66. $42 - (-30) - 65 - (-11)$

67. $-47 - (-67) - 13 - 15$

68. $-18 - 49 - (-84) - 27$

69. $-19 - 17 - (-36) - 12$

70. $48 - 19 - 29 - 51$

71. $21 - (-14) - 43 - 12$

72. $17 - (-17) - 14 - 21$

73. Find the difference between -21 and -36.

74. What is 9 less than -12?

75. What is 12 less than -27?

76. Find -21 decreased by 19.

77. What is -21 minus -37?

78. Subtract 41 from -22.

Copyright © Houghton Mifflin Company. All rights reserved.

Objective C **To solve application problems**

79. **Chemistry** The temperature at which mercury boils is 360°C. Mercury freezes at −39°C. Find the difference between the temperature at which mercury boils and the temperature at which it freezes.

80. **Chemistry** The temperature at which radon boils is −62°C. Radon freezes at −71°C. Find the difference between the temperature at which radon boils and the temperature at which it freezes.

Geography The elevation, or height, of places on Earth is measured in relation to sea level, or the average level of the ocean's surface. The table below shows height above sea level as a positive number and depth below sea level as a negative number. Use the table for Exercises 81 to 84.

Continent	Highest Elevation (in meters)		Lowest Elevation (in meters)	
Africa	Mt. Kilimanjaro	5895	Qattara Depression	−133
Asia	Mt. Everest	8850	Dead Sea	−400
Europe	Mt. Elbrus	5634	Caspian Sea	−28
America	Mt. Aconcagua	6960	Death Valley	−86

Mt. Everest

81. Find the difference in elevation between Mt. Aconcagua and Death Valley.

82. What is the difference in elevation between Mt. Kilimanjaro and the Qattara Depression?

83. For which continent shown is the difference between the highest and lowest elevations greatest?

84. For which continent shown is the difference between the highest and lowest elevations smallest?

Chemistry The table at the right shows the boiling point and the melting point in degrees Celsius of three chemical elements. Use this table for Exercises 85 and 86.

Chemical Element	Boiling Point	Melting Point
Mercury	357	−39
Radon	−62	−71
Xenon	−107	−112

85. Find the difference between the boiling point and the melting point of mercury.

86. Find the difference between the boiling point and the melting point of xenon.

Copyright © Houghton Mifflin Company. All rights reserved.

Geography   The graph at the right shows Earth's three deepest ocean trenches and its three tallest mountains. Use this graph for Exercises 87 to 89.

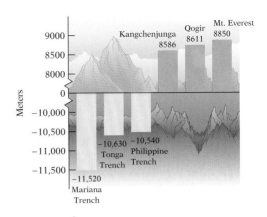

87. Could Mt. Everest fit in the Tonga Trench?

88. What is the difference between the depth of the Philippine Trench and the Mariana Trench?

89. What is the difference between the height of Mt. Everest and the depth of the Mariana Trench?

Meteorology   A meteorologist may report a wind-chill temperature. This is the equivalent temperature, including the effects of wind and temperature, that a person would feel in calm air conditions. The table below gives the wind-chill temperature for various wind speeds and temperatures. For instance, when the temperature is 5°F and the wind is blowing at 15 mph, the wind-chill temperature is −13°F. Use this table for Exercises 90 and 91.

Wind Chill Factors															
Wind Speed (in mph)	Thermometer Reading (in degrees Fahrenheit)														
	25	20	15	10	5	0	−5	−10	−15	−20	−25	−30	−35	−40	−45
5	19	13	7	1	−5	−11	−16	−22	−28	−34	−40	−46	−52	−57	−63
10	15	9	3	−4	−10	−16	−22	−28	−35	−41	−47	−53	−59	−66	−72
15	13	6	0	−7	−13	−19	−26	−32	−39	−45	−51	−58	−64	−71	−77
20	11	4	−2	−9	−15	−22	−29	−35	−42	−48	−55	−61	−68	−74	−81
25	9	3	−4	−11	−17	−24	−31	−37	−44	−51	−58	−64	−71	−78	−84
30	8	1	−5	−12	−19	−26	−33	−39	−46	−53	−60	−67	−73	−80	−87
35	7	0	−7	−14	−21	−27	−34	−41	−48	−55	−62	−69	−76	−82	−89
40	6	−1	−8	−15	−22	−29	−36	−43	−50	−57	−64	−71	−78	−84	−91
45	5	−2	−9	−16	−23	−30	−37	−44	−51	−58	−65	−72	−79	−86	−93

90. When the thermometer reading is −5°F, what is the difference between the wind-chill factor when the wind is blowing 10 mph and when the wind is blowing 25 mph?

91. When the thermometer reading is −20°F, what is the difference between the wind-chill factor when the wind is blowing 15 mph and when the wind is blowing 25 mph?

APPLYING THE CONCEPTS

92. If a and b are integers, is the expression $|a + b| = |a| + |b|$ always true, sometimes true, or never true?

93. Is the difference between two integers always smaller than either one of the numbers in the difference? If not, give an example for which the difference between two integers is greater than either integer.

Copyright © Houghton Mifflin Company. All rights reserved.

1.3 Multiplication and Division of Integers

Objective A   To multiply integers

Point of Interest

The cross × was first used as a symbol for multiplication in 1631 in a book titled *The Key to Mathematics*. Also in that year, another book, *Practice of the Analytical Art*, advocated the use of a dot to indicate multiplication.

Several different symbols are used to indicate multiplication. The numbers being multiplied are called **factors;** for instance, 3 and 2 are factors in each of the examples at the right. The result is called the **product.** Note that when parentheses are used and there is no arithmetic symbol, the operation is multiplication.

$$3 \times 2 = 6$$
$$3 \cdot 2 = 6$$
$$(3)(2) = 6$$
$$3(2) = 6$$
$$(3)2 = 6$$

Multiplication is repeated addition of the same number. The product 3×5 is shown on the number line below.

5   5   5

0 1 2 3 4 5 6 7 8 9 10 11 12 13 14 15

5 is added 3 times.

$$3 \times 5 = 5 + 5 + 5 = 15$$

Now consider the product of a positive and a negative number.

−5 is added 3 times.

$$3(-5) = (-5) + (-5) + (-5) = -15$$

This suggests that the product of a positive number and a negative number is negative. Here are a few more examples.

$$4(-7) = -28 \qquad -6 \cdot 7 = -42 \qquad (-8)7 = -56$$

To find the product of two negative numbers, look at the pattern at the right. As −5 multiplies a sequence of decreasing integers, the products increase by 5.

The pattern can be continued by requiring that the product of two negative numbers be positive.

These numbers decrease by 1.   These numbers increase by 5.

$$-5 \times 3 = -15$$
$$-5 \times 2 = -10$$
$$-5 \times 1 = -5$$
$$-5 \times 0 = 0$$
$$-5 \times (-1) = 5$$
$$-5 \times (-2) = 10$$
$$-5 \times (-3) = 15$$

Multiplication of Integers

To multiply two numbers with the same sign, multiply the absolute values of the numbers. The product is positive.

To multiply two numbers with different signs, multiply the absolute values of the numbers. The product is negative.

HOW TO   Multiply: $5(-12)$.

$$5(-12) = -60$$

• The signs are different. The product is negative.

Copyright © Houghton Mifflin Company. All rights reserved.

> **HOW TO**   Find the product of -8 and -16.
>
> A *product* is the answer to a multiplication problem.
>
> $-8(-16) = 128$ • The signs are the same. The product is positive.

The phrase *the product of* in the example above indicates the operation of multiplication. All of the phrases below indicate multiplication.

times	-7 times -9	$-7(-9) = 63$
the product of	the product of 12 and -8	$12(-8) = -96$
multiplied by	-15 multiplied by 11	$-15(11) = -165$
twice	twice -14	$2(-14) = -28$

> **HOW TO**   Multiply: $-2(5)(-7)(-4)$
>
> $-2(5)(-7)(-4) = -10(-7)(-4)$
>
> $\qquad = 70(-4) = -280$
>
> • To multiply more than two numbers, multiply the first two. Then multiply the product by the third number. Continue until all the numbers are multiplied.

Consider the products shown at the right. Note that when there is an even number of negative factors, the product is positive. When there is an odd number of negative factors, the product is negative.

$$(-3)(-5) = 15$$
$$(-2)(-5)(-6) = -60$$
$$(-4)(-3)(-5)(-7) = 420$$
$$(-3)(-3)(-5)(-4)(-5) = -900$$
$$(-6)(-3)(-4)(-2)(-10)(-5) = 7200$$

This idea can be summarized by the following useful rule: **The product of an even number of negative factors is positive; the product of an odd number of negative factors is negative.**

Example 1   Multiply: $-24(-18)$

Solution $-24(-18) = 432$

You Try It 1   Multiply: $-7(32)$

Your solution

Example 2   Multiply: $(-3)4(-5)$

Solution $(-3)4(-5) = (-12)(-5) = 60$

You Try It 2   Multiply: $8(-9)10$

Your solution

Example 3   Multiply: $12(-4)(-3)(-5)$

Solution $12(-4)(-3)(-5) = (-48)(-3)(-5)$
$\qquad = 144(-5)$
$\qquad = -720$

You Try It 3   Multiply: $(-2)3(-8)7$

Your solution

Example 4   Find the product of -13 and -9.

Solution $-13(-9) = 117$

You Try It 4   What is -9 times 34?

Your solution

Copyright © Houghton Mifflin Company. All rights reserved.

Solutions on p. S1

Copyright © Houghton Mifflin Company. All rights reserved.

Objective B   **To divide integers**

TAKE NOTE

Think of the fraction bar as "divided by." Thus $\frac{8}{2}$ is 8 divided by 2. The number 2 is the **divisor.** The number 8 is the **dividend.** The result of the division, 4, is called the **quotient.**

For every division problem there is a related multiplication problem.

$$\frac{8}{2} = 4 \qquad \text{because} \qquad 4 \cdot 2 = 8.$$

Division   Related multiplication

This fact and the rules for multiplying integers can be used to illustrate the rules for dividing integers.

Note in the following examples that the quotient of two numbers with the same sign is positive.

$$\frac{12}{3} = 4 \text{ because } 4 \cdot 3 = 12. \qquad \frac{-12}{-3} = 4 \text{ because } 4(-3) = -12.$$

The next two examples illustrate that the quotient of two numbers with different signs is negative.

$$\frac{12}{-3} = -4 \text{ because } (-4)(-3) = 12. \qquad \frac{-12}{3} = -4 \text{ because } (-4)3 = -12.$$

Division of Integers

To divide two numbers with the same sign, divide the absolute values of the numbers. The quotient is positive.

To divide two numbers with different signs, divide the absolute values of the numbers. The quotient is negative.

Point of Interest

There was quite a controversy over the date the new millennium started because of the number zero. When our current calendar was created, numbering began with the year 1 because 0 had not yet been invented. Thus at the beginning of year 2, 1 year had elapsed; at the beginning of year 3, 2 years had elapsed; and so on. This means that at the beginning of year 2000, 1999 years had elapsed and it was not until the beginning of year 2001 that 2000 years had elapsed and a new millennium began.

HOW TO   Divide: $-36 \div 9$.

$-36 \div 9 = -4$ • The signs are different. The quotient is negative.

HOW TO   Find the quotient of -63 and -7.

A *quotient* is the answer to a division problem.

$\dfrac{-63}{-7} = 9$ • The signs are the same. The quotient is positive.

TAKE NOTE

We can denote division by $-63 \div (-7)$, $-7\overline{)-63}$, or $\dfrac{-63}{-7}$.

The phrase *the quotient of* in the example above indicates the operation of division. All of the phrases below indicate division.

divided by	15 divided by -3	$15 \div (-3) = -5$
the quotient of	the quotient of -56 and -8	$(-56) \div (-8) = 7$
the ratio of	the ratio of 45 and -5	$45 \div (-5) = -9$
divide ... by ...	divide -100 by -20	$-100 \div (-20) = 5$

HOW TO   Simplify: $-\dfrac{-56}{7}$

$$-\frac{-56}{7} = -\left(\frac{-56}{7}\right) = -(-8) = 8$$

The properties of division are stated below. In these statements, the symbol $\neq$ is read "is not equal to."

Properties of Zero and One in Division

If $a \neq 0$, $\dfrac{0}{a} = 0$.   Zero divided by any number other than zero is zero.

If $a \neq 0$, $\dfrac{a}{a} = 1$.   Any number other than zero divided by itself is one.

$\dfrac{a}{1} = a$.   A number divided by one is the number.

$\dfrac{a}{0}$ is undefined.   Division by zero is not defined.

The fact that $\dfrac{-12}{3} = -4$, $\dfrac{12}{-3} = -4$, and $-\dfrac{12}{3} = -4$ suggests the following rule.

If a and b are integers, and $b \neq 0$, then $\dfrac{-a}{b} = \dfrac{a}{-b} = -\dfrac{a}{b}$.

Example 5   Divide: $(-120) \div (-8)$

Solution $(-120) \div (-8) = 15$

You Try It 5   Divide: $(-135) \div (-9)$

Your solution

Example 6   Divide: $\dfrac{95}{-5}$

Solution $\dfrac{95}{-5} = -19$

You Try It 6   Divide: $\dfrac{-72}{4}$

Your solution

Example 7   Simplify: $-\dfrac{-81}{3}$

Solution $-\dfrac{-81}{3} = -(-27) = 27$

You Try It 7   Simplify: $-\dfrac{36}{-12}$

Your solution

Example 8   Find the quotient of 98 and -14.

Solution $98 \div (-14) = -7$

You Try It 8   What is the ratio of -72 and -8?

Your solution

Solutions on p. S1

Copyright © Houghton Mifflin Company. All rights reserved.

Objective C   **To solve application problems**

In many courses, your course grade depends on the *average* of all your test scores. You compute the average by calculating the sum of all your test scores and then dividing that result by the number of tests. Statisticians call this average an **arithmetic mean.** Besides its application to finding the average of your test scores, the arithmetic mean is used in many other situations.

Stock market analysts calculate the **moving average** of a stock. This is the arithmetic mean of the changes in the value of a stock for a given number of days. To illustrate the procedure, we will calculate the 5-day moving average of a stock. In actual practice, a stock market analyst may use 15 days, 30 days, or some other number.

The table below shows the amount of increase or decrease, in cents, in the closing price of a stock for a 10-day period.

Day 1	Day 2	Day 3	Day 4	Day 5	Day 6	Day 7	Day 8	Day 9	Day 10
+50	−175	+225	0	−275	−75	−50	+50	−475	−50

To calculate the 5-day moving average of this stock, determine the average of the stock for days 1 through 5, days 2 through 6, days 3 through 7, and so on.

Days 1–5	Days 2–6	Days 3–7	Days 4–8	Days 5–9	Days 6–10
+50	−175	+225	0	−275	−75
−175	+225	0	−275	−75	−50
+225	0	−275	−75	−50	+50
0	−275	−75	−50	+50	−475
−275	−75	−50	+50	−475	−50
Sum = −175	Sum = −300	Sum = −175	Sum = −350	Sum = −825	Sum = −600
$Av = \dfrac{-175}{5} = -35$	$Av = \dfrac{-300}{5} = -60$	$Av = \dfrac{-175}{5} = -35$	$Av = \dfrac{-350}{5} = -70$	$Av = \dfrac{-825}{5} = -165$	$Av = \dfrac{-600}{5} = -120$

The 5-day moving average is the list of means: $-35, -60, -35, -70, -165,$ and -120. If the list tends to increase, the price of the stock is showing an upward trend; if it decreases, the price of the stock is showing a downward trend. These trends help an analyst recommend stocks.

Example 9

The daily high temperatures (in degrees Celsius) for six days in Anchorage, Alaska, were −14°, 3°, 0°, −8°, 2°, and −1°. Find the average daily high temperature.

Strategy

To find the average daily high temperature:
* Add the six temperature readings.
* Divide the sum by 6.

Solution

$-14 + 3 + 0 + (-8) + 2 + (-1) = -18$

$-18 \div 6 = -3$

The average daily high temperature was −3°C.

You Try It 9

The daily low temperatures (in degrees Celsius) during one week were recorded as −6°, −7°, 0°, −5°, −8°, −1°, and −1°. Find the average daily low temperature.

Your strategy

Your solution

Solution on p. S1

Copyright © Houghton Mifflin Company. All rights reserved.

1.3 Exercises

Objective A **To multiply integers**

1. Explain how to multiply two integers with the same sign.

2. Explain how to multiply two integers with different signs.

For Exercises 3 to 22, multiply.

3. (14)3

4. 17(6)

5. (−12)(−5)

6. (−13)(−9)

7. −11(23)

8. −8(21)

9. 6(−19)

10. 17(−13)

11. 7(5)(−3)

12. (−3)(−2)8

13. −3(−8)(−9)

14. −7(−6)(−5)

15. (−9)7(5)

16. (−8)7(10)

17. (−3)7(−2)8

18. −9(−4)(−8)(−10)

19. 7(9)(−11)4

20. −12(−4)7(−2)

21. (−14)9(−11)0

22. (−13)(15)(−19)0

23. Find the product of −14 and −25.

24. What is 4 times −8?

25. What is 16 multiplied by −21?

26. Find −22 multiplied by 15.

27. Find the product of 4, −8, and 11.

28. Find the product of −2, −3, −4, and −5.

Objective B **To divide integers**

For Exercises 29 to 56, divide.

29. $12 \div (-6)$

30. $18 \div (-3)$

31. $(-72) \div (-9)$

32. $(-64) \div (-8)$

33. $-42 \div 6$

34. $(-56) \div 8$

35. $(-144) \div 12$

36. $(-93) \div (-3)$

37. $48 \div (-8)$

38. $57 \div (-3)$

39. $\dfrac{-44}{-4}$

40. $\dfrac{-36}{-9}$

41. $\dfrac{98}{-7}$

42. $\dfrac{85}{-5}$

43. $-\dfrac{-120}{8}$

44. $-\dfrac{-72}{4}$

45. $-\dfrac{-80}{-5}$

46. $-\dfrac{-114}{-6}$

47. $0 \div (-9)$

48. $0 \div (-14)$

49. $\dfrac{-261}{9}$

50. $\dfrac{-128}{4}$

51. $9 \div 0$

52. $(-21) \div 0$

53. $\dfrac{132}{-12}$

54. $\dfrac{250}{-25}$

55. $\dfrac{0}{0}$

56. $\dfrac{-58}{0}$

57. Find the quotient of −132 and 11.

58. What is 15 divided by −15?

59. What is −96 divided by −4?

60. What is the ratio of −175 and 25?

61. Divide −196 by −7.

62. Find the quotient of 342 and −9.

Copyright © Houghton Mifflin Company. All rights reserved.

Objective C   **To solve application problems**

63.   **Meteorology**   The high temperatures for a 6-day period in Barrow, Alaska, were $-23°F$, $-29°F$, $-21°F$, $-28°F$, $-28°F$, and $-27°F$. Calculate the average daily high temperature.

64.   **Meteorology**   The low temperatures for a 10-day period in a midwestern city were $-4°F$, $-9°F$, $-5°F$, $-2°F$, $4°F$, $-1°F$, $-1°F$, $-2°F$, $-2°F$, and $2°F$. Calculate the average daily low temperature for this city.

65.   **Investments**   The value of a share of Yahoo's stock on August 21, 2003, was $32.82. The table below shows the approximate increase or decrease, in cents, from the August 21 closing price of the stock for a 10-day period. Calculate the 5-day moving average for this stock.

Day 1	Day 2	Day 3	Day 4	Day 5	Day 6	Day 7	Day 8	Day 9	Day 10
-100	20	0	30	-10	100	80	-60	130	10

66.   **Investments**   The value of a share of Ford Motor Company's stock on August 22, 2003, was $11.13. The table below shows the approximate increase or decrease, in cents, from the August 22 closing price of the stock for a 10-day period. Calculate the 5-day moving average for this stock.

Day 1	Day 2	Day 3	Day 4	Day 5	Day 6	Day 7	Day 8	Day 9	Day 10
-20	30	20	-20	10	90	-40	-10	-10	10

67.   **Testing**   To discourage random guessing on a multiple-choice exam, a professor assigns 5 points for a correct answer, -2 points for an incorrect answer, and 0 points for leaving the question blank. What is the score for a student who had 20 correct answers, had 13 incorrect answers, and left 7 questions blank?

68.   **Testing**   To discourage random guessing on a multiple-choice exam, a professor assigns 7 points for a correct answer, -3 points for an incorrect answer, and -1 point for leaving the question blank. What is the score for a student who had 17 correct answers, had 8 incorrect answers, and left 2 questions blank?

APPLYING THE CONCEPTS

69.   If $x \in \{-6, -2, 7\}$, for which value of x does the expression $-3x$ have the greatest value?

70.   Explain why $0 \div 0$ is not defined.

71.   If $-4x$ equals a positive integer, is x a positive or a negative integer? Explain your answer.

Copyright © Houghton Mifflin Company. All rights reserved.

Copyright © Houghton Mifflin Company. All rights reserved.

1.4 Exponents and the Order of Operations Agreement

Objective A To evaluate exponential expressions

Repeated multiplication of the same factor can be written using an exponent.

$$2 \cdot 2 \cdot 2 \cdot 2 \cdot 2 = 2^5 \leftarrow \text{Exponent} \qquad a \cdot a \cdot a \cdot a = a^4 \leftarrow \text{Exponent}$$
$$\underset{\text{Base}}{\uparrow} \qquad\qquad\qquad\qquad \underset{\text{Base}}{\uparrow}$$

Point of Interest

Rene Descartes (1596–1650) was the first mathematician to use exponential notation extensively as it is used today. However, for some unknown reason, he always used xx for x^2.

The **exponent** indicates how many times the factor, which is called the **base,** occurs in the multiplication. The multiplication $2 \cdot 2 \cdot 2 \cdot 2 \cdot 2$ is in **factored form.** The exponential expression 2^5 is in **exponential form.**

2^1 is read "2 to the first power" or just "2." Usually the exponent 1 is not written.

2^2 is read "2 to the second power" or "2 squared."

2^3 is read "2 to the third power" or "2 cubed."

2^4 is read "2 to the fourth power."

a^4 is read "a to the fourth power."

There is a geometric interpretation of the first three natural-number powers.

 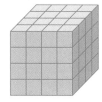

$4^1 = 4$ $4^2 = 16$ $4^3 = 64$
Length: 4 ft   Area: 16 ft^2 Volume: 64 ft^3

To evaluate an exponential expression, write each factor as many times as indicated by the exponent. Then multiply.

HOW TO Evaluate $(-2)^4$.

$(-2)^4 = (-2)(-2)(-2)(-2)$ • Write −2 as a factor 4 times.

$\quad = 16$ • Multiply.

TAKE NOTE

Note the difference between $(-2)^4$ and -2^4.
$(-2)^4 = 16$
$-2^4 = -16$

HOW TO Evaluate -2^4.

$-2^4 = -(2 \cdot 2 \cdot 2 \cdot 2)$ • Write 2 as a factor 4 times.

$\quad = -16$ • Multiply.

Example 1 Evaluate -5^3.

Solution $-5^3 = -(5 \cdot 5 \cdot 5) = -125$

You Try It 1 Evaluate -6^3.

Your solution

Example 2 Evaluate $(-4)^4$.

Solution $(-4)^4 = (-4)(-4)(-4)(-4)$
$= 256$

You Try It 2 Evaluate $(-3)^4$.

Your solution

Example 3 Evaluate $(-3)^2 \cdot 2^3$.

Solution $(-3)^2 \cdot 2^3 = (-3)(-3) \cdot (2)(2)(2)$
$= 9 \cdot 8 = 72$

You Try It 3 Evaluate $(3^3)(-2)^3$.

Your solution

Example 4 Evaluate $(-1)^6$.

Solution The product of an even number of negative factors is positive. Therefore, $(-1)^6 = 1$.

You Try It 4 Evaluate $(-1)^7$.

Your solution

Example 5 Evaluate $-2 \cdot (-3)^2 \cdot (-1)^9$.

Solution $-2 \cdot (-3)^2 \cdot (-1)^9$
$= -2[(-3) \cdot (-3)](-1)$ • $(-1)^9 = -1$
$= -2 \cdot 9 \cdot (-1) = 18$

You Try It 5 Evaluate $-2^2 \cdot (-1)^{12} \cdot (-3)^2$.

Your solution

Solutions on pp. S1–S2

Objective B **To use the Order of Operations Agreement to simplify expressions**

Let's evaluate $2 + 3 \cdot 5$.

There are two arithmetic operations, addition and multiplication, in this expression. The operations could be performed in different orders.

Multiply first.	$2 + \underline{3 \cdot 5}$	Add first.	$\underbrace{2 + 3} \cdot 5$
Then add.	$\underbrace{2 + 15}$	Then multiply.	$\underline{5 \cdot 5}$
	17		25

To prevent there being more than one answer for a numerical expression, an Order of Operations Agreement has been established.

The Order of Operations Agreement

Step 1 Perform operations inside grouping symbols. Grouping symbols include parentheses ( ), brackets [ ], braces { }, and the fraction bar.

Step 2 Simplify exponential expressions.

Step 3 Do multiplication and division as they occur from left to right.

Step 4 Do addition and subtraction as they occur from left to right.

Copyright © Houghton Mifflin Company. All rights reserved.

Integrating Technology

See the Keystroke Guide: *Basic Operations* for instruction on using a calculator to evaluate a numerical expression.

HOW TO Evaluate $12 - 24(8 - 5) \div 2^2$.

$12 - 24(8 - 5) \div 2^2 = 12 - 24(3) \div 2^2$ • Perform operations inside grouping symbols.

$= 12 - 24(3) \div 4$ • Simplify exponential expressions.

$= 12 - 72 \div 4$ • Do multiplication and division as they occur from left to right.

$= 12 - 18$

$= -6$ • Do addition and subtraction as they occur from left to right.

One or more of the steps listed above may not be needed to evaluate an expression. In that case, proceed to the next step in the Order of Operations Agreement.

HOW TO Evaluate $\dfrac{4 + 8}{2 + 1} - (3 - 1) + 2$.

$\dfrac{4 + 8}{2 + 1} - (3 - 1) + 2 = \dfrac{12}{3} - 2 + 2$ • Perform operations above and below the fraction bar and inside parentheses.

$= 4 - 2 + 2$ • Do multiplication and division as they occur from left to right.

$= 2 + 2$ • Do addition and subtraction as they occur from left to right.

$= 4$

Example 6

Evaluate $6 \div [4 - (6 - 8)] - 2^3$.

Solution

$6 \div [4 - (6 - 8)] - 2^3$

$= 6 \div [4 - (-2)] - 2^3$ • Perform operations inside grouping symbols.

$= 6 \div 6 - 2^3$

$= 6 \div 6 - 8$ • Simplify exponential expressions.

$= 1 - 8$ • Do multiplication and division from left to right.

$= -7$ • Do addition and subtraction from left to right.

You Try It 6

Evaluate $7 - 2[2 \cdot 3 - 7 \cdot 2]^2$.

Your solution

Solution on p. S2

Copyright © Houghton Mifflin Company. All rights reserved.

Copyright © Houghton Mifflin Company. All rights reserved.

Example 7

Evaluate $4 - 3[4 - 2(6 - 3)] \div 2$.

Solution

$4 - 3[4 - 2(6 - 3)] \div 2$

$= 4 - 3[4 - 2 \cdot 3] \div 2$ • Perform operations inside grouping symbols.

$= 4 - 3[4 - 6] \div 2$

$= 4 - 3[-2] \div 2$

$= 4 + 6 \div 2$ • Do multiplication and division from left to right.

$= 4 + 3$

$= 7$ • Do addition and subtraction from left to right.

You Try It 7

Evaluate $18 - 5[8 - 2(2 - 5)] \div 10$.

Your solution

Example 8

Evaluate $27 \div (5 - 2)^2 + (-3)^2 \cdot 4$.

Solution

$27 \div (5 - 2)^2 + (-3)^2 \cdot 4$

$= 27 \div 3^2 + (-3)^2 \cdot 4$ • Perform operations inside grouping symbols.

$= 27 \div 9 + 9 \cdot 4$ • Simplify exponential expressions.

$= 3 + 9 \cdot 4$ • Do multiplication and division from left to right.

$= 3 + 36$

$= 39$ • Do addition and subtraction from left to right.

You Try It 8

Evaluate $36 \div (8 - 5)^2 - (-3)^2 \cdot 2$.

Your solution

Solutions on p. S2

1.4 Exercises

Objective A   **To evaluate exponential expressions**

For Exercises 1 to 27, evaluate.

1. 6^2

2. 7^4

3. -7^2

4. -4^3

5. $(-3)^2$

6. $(-2)^3$

7. $(-3)^4$

8. $(-5)^3$

9. -4^4

10. $(-4)^4$

11. $2 \cdot (-3)^2$

12. $-2 \cdot (-4)^2$

13. $(-1)^9 \cdot 3^3$

14. $(-1)^8(-8)^2$

15. $(3)^3 \cdot 2^3$

16. $(5)^2 \cdot 3^3$

17. $(-3) \cdot 2^2$

18. $(-5) \cdot 3^4$

19. $(-2) \cdot (-2)^3$

20. $(-2) \cdot (-2)^2$

21. $2^3 \cdot 3^3 \cdot (-4)$

22. $(-3)^3 \cdot 5^2 \cdot 10$

23. $(-7) \cdot 4^2 \cdot 3^2$

24. $(-2) \cdot 2^3 \cdot (-3)^2$

25. $-3^2 \cdot (-3)^2$

26. $(-2)^3(-3)^2(-1)^7$

27. $8^2 \cdot (-3)^5 \cdot 5$

Objective B   **To use the Order of Operations Agreement to simplify expressions**

28. ✎ Explain why an Order of Operations Agreement is necessary.

29. ✎ Write out the steps of the Order of Operations Agreement.

For Exercises 30 to 62, evaluate by using the Order of Operations Agreement.

30. $4 - 8 \div 2$

31. $2^2 \cdot 3 - 3$

32. $2(3 - 4) - (-3)^2$

33. $16 - 32 \div 2^3$

34. $24 - 18 \div 3 + 2$

35. $8 - (-3)^2 - (-2)$

36. $8 - 2(3)^2$

37. $16 - 16 \cdot 2 \div 4$

38. $12 + 16 \div 4 \cdot 2$

39. $16 - 2 \cdot 4^2$

40. $27 - 18 \div (-3^2)$

41. $4 + 12 \div 3 \cdot 2$

Copyright © Houghton Mifflin Company. All rights reserved.

42. $16 + 15 \div (-5) - 2$

43. $14 - 2^2 - (4 - 7)$

44. $3 - 2[8 - (3 - 2)]$

45. $-2^2 + 4[16 \div (3 - 5)]$

46. $6 + \dfrac{16 - 4}{2^2 + 2} - 2$

47. $24 \div \dfrac{3^2}{8 - 5} - (-5)$

48. $96 \div 2[12 + (6 - 2)] - 3^2$

49. $4[16 - (7 - 1)] \div 10$

50. $18 \div 2 - 4^2 - (-3)^2$

51. $18 \div (9 - 2^3) + (-3)$

52. $16 - 3(8 - 3)^2 \div 5$

53. $4(-8) \div [2(7 - 3)^2]$

54. $\dfrac{(-19) + (-2)}{6^2 - 29} \div (2 - 5)$

55. $16 - 4 \cdot \dfrac{3^3 - 7}{2^3 + 2} - (-2)^2$

56. $7 - 3[1 - (2 - (-3))^2]$

57. $-18 \div (-6) \div (1 - (-2))$

58. $-4 \cdot 2^3 - \dfrac{1 - 13}{2^2 \cdot 3}$

59. $(8 - 3^2)^{10} + (2 \cdot 3 - 7)^{11}$

60. $(2^3 - (-3))(2^3 + (-3))$

61. $-6^2 \cdot 3 - 2^2(1 - 5)^2$

62. $14 - 2^2(3 - 4^2)$

APPLYING THE CONCEPTS

63. ✎ The following was offered as the simplification of $6 + 2(4 - 9)$.

$$6 + 2(4 - 9) = 6 + 2(-5)$$
$$= 8(-5)$$
$$= -40$$

If this is a correct simplification, write yes for the answer. If it is incorrect, write no and explain the incorrect step.

64. ✎ The following was offered as the simplification of $2 \cdot 3^3$.

$$2 \cdot 3^3 = 6^3 = 216$$

If this is a correct simplification, write yes for the answer. If it is incorrect, write no and explain the incorrect step.

Copyright © Houghton Mifflin Company. All rights reserved.

1.5 Factoring Numbers and Prime Factorization

Objective A   To factor numbers

A **factor of a number** is a natural number that divides the number with a remainder of 0.

The factors of 12 are 1, 2, 3, 4, 6, and 12 because each of those numbers divides 12 with a remainder of 0. Note that both the divisor and the quotient are factors of the dividend.

$$12 \div 1 = 12 \qquad 12 \div 4 = 3$$
$$12 \div 2 = 6 \qquad 12 \div 6 = 2$$
$$12 \div 3 = 4 \qquad 12 \div 12 = 1$$

To find the factors of a number, try dividing the number by 1, 2, 3, 4, 5, ... Those numbers that divide the number are its factors. Continue this process until the factors start to repeat.

HOW TO   Find all the factors of 40.

$40 \div 1 = 40$	1 and 40 are factors.
$40 \div 2 = 20$	2 and 20 are factors.
$40 \div 3$	Remainder is not 0.
$40 \div 4 = 10$	4 and 10 are factors.
$40 \div 5 = 8$	5 and 8 are factors.
$40 \div 6$	Remainder is not 0.
$40 \div 7$	Remainder is not 0.
$40 \div 8 = 5$	8 and 5 are factors.

Factors are repeating. All the factors of 40 have been found.

1, 2, 4, 5, 8, 10, 20, and 40 are factors of 40.

The following rules are helpful in finding the factors of a number.

2 is a factor of a number if the last digit of the number is 0, 2, 4, 6, or 8.	528 ends in 8; therefore, 2 is a factor of 528. ($528 \div 2 = 264$)
3 is a factor of a number if the sum of digits of the number is divisible by 3.	The sum of the digits of 378 is $3 + 7 + 8 = 18$. 18 is divisible by 3. Therefore, 3 is a factor of 378. ($378 \div 3 = 126$)
5 is a factor of a number if the last digit of the number is a 0 or a 5.	495 ends in 5; therefore, 5 is a factor of 495. ($495 \div 5 = 99$)

Example 1   Find all the factors of 18.

Solution
$18 \div 1 = 18$
$18 \div 2 = 9$
$18 \div 3 = 6$
$18 \div 4$ Remainder is not 0.
$18 \div 5$ Remainder is not 0.
$18 \div 6 = 3$ The factors are repeating.

1, 2, 3, 6, 9, and 18 are the factors of 18.

You Try It 1   Find the factors of 24.

Your solution

Solution on p. S2

Copyright © Houghton Mifflin Company. All rights reserved.

Copyright © Houghton Mifflin Company. All rights reserved.

Objective B **To find the prime factorization of a number**

A natural number greater than 1 is a **prime number** if its only factors are 1 and the number. For instance, 11 is a prime number because the only factors of 11 are 1 and 11. A natural number greater than 1 that is not a prime number is a **composite number.** An example of a composite number is 6. It has factors of 1, 2, 3, and 6. The number 1 is neither a prime nor a composite number.

Prime numbers less than 50 = 2, 3, 5, 7, 11, 13, 17, 19, 23, 29, 31, 37, 41, 43, 47

The **prime factorization** of a number is the expression of the number as a product of its prime factors. We use a "T-diagram" to find the prime factors of a number. Begin with the smallest prime number as a trial divisor, and continue to use prime numbers as trial divisors until the final quotient is 1.

TAKE NOTE

A prime number that is a factor of a number is called a **prime factor** of the number. For instance, 3 is a prime factor of 18. However, 6 is a factor of 18 but is not a prime factor of 18.

HOW TO Find the prime factorization of 84.

$$\begin{array}{c|c} \multicolumn{2}{c}{84} \\ \hline 2 & 42 \\ 2 & 21 \\ 3 & 7 \\ 7 & 1 \end{array}$$

$84 \div 2 = 42$
$42 \div 2 = 21$
$21 \div 3 = 7$
$7 \div 7 = 1$

The prime factorization of 84 is $2^2 \cdot 3 \cdot 7$.

Finding the prime factorization of larger numbers can be difficult. Try each prime number as a trial divisor until the square of the trial divisor exceeds the number.

Point of Interest

Prime numbers are an important part of cryptology, the study of secret codes. Codes based on prime numbers with hundreds of digits are used to send sensitive information over the Internet.

HOW TO Find the prime factorization of 177.

$$\begin{array}{c|c} \multicolumn{2}{c}{177} \\ \hline 3 & 59 \\ 59 & 1 \end{array}$$

• For 59, only try prime numbers up to 11 because $11^2 = 121 > 59$.

The prime factorization of 177 is $3 \cdot 59$.

Example 2 Find the prime factorization of 132.

Solution

$$\begin{array}{c|c} \multicolumn{2}{c}{132} \\ \hline 2 & 66 \\ 2 & 33 \\ 3 & 11 \\ 11 & 1 \end{array}$$

$132 = 2^2 \cdot 3 \cdot 11$

You Try It 2 Find the prime factorization of 315.

Your solution

Example 3 Find the prime factorization of 141.

Solution

$$\begin{array}{c|c} \multicolumn{2}{c}{141} \\ \hline 3 & 47 \\ 47 & 1 \end{array}$$

• For 47, try prime numbers up to 7 because $7^2 > 47$.

$141 = 3 \cdot 47$

You Try It 3 Find the prime factorization of 326.

Your solution

Solutions on p. S2

Objective C   **To find the least common
multiple and greatest common factor**

The **least common multiple (LCM)** of two or more numbers is the smallest number that is a multiple of all the numbers. For instance, 24 is the LCM of 6 and 8 because it is the smallest number that is divisible by both 6 and 8.

The LCM can be found by first writing each number as a product of prime factors. The LCM must contain all the prime factors of each number.

> **HOW TO**   Find the LCM of 10 and 12.
>
> Determine the prime factorization of each number.
>
> $10 = 2 \cdot 5$ Factors of 10
>
> $12 = 2 \cdot 2 \cdot 3$ LCM = $\underline{2 \cdot 2 \cdot 3 \cdot 5}$ = 60
>
>   Factors of 12
>
> The LCM of 10 and 12 is 60.

> **HOW TO**   Find the LCM of 8, 14, and 18.
>
> Determine the prime factorization of each number.
>
> $8 = 2 \cdot 2 \cdot 2$ $14 = 2 \cdot 7$ $18 = 2 \cdot 3 \cdot 3$
>
> The LCM must contain the prime factors of 8, 14, and 18.
>
> LCM = $2 \cdot 2 \cdot 2 \cdot 3 \cdot 3 \cdot 7 = 504$

The **greatest common factor (GCF)** of two or more numbers is the greatest number that divides evenly into all the numbers. For instance, the GCF of 12 and 18 is 6, the largest number that divides evenly into 12 and 18.

The GCF can be found by first writing each number as a product of prime factors. The GCF contains the prime factors common to each number.

> **HOW TO**   Find the GCF of 36 and 90.
>
> Determine the prime factorization of each number.
>
> $36 = 2 \cdot 2 \cdot 3 \cdot 3$ • **The common factors
> are shown in red.**
> $90 = 2 \cdot 3 \cdot 3 \cdot 5$
>
> The GCF is the product of the prime factors common to each number.
>
> The GCF of 36 and 90 is $2 \cdot 3 \cdot 3 = 18$.

Example 4   Find the LCM of 15, 20, and 30.

Solution
$15 = 3 \cdot 5$ $20 = 2 \cdot 2 \cdot 5$ $30 = 2 \cdot 3 \cdot 5$

LCM = $2 \cdot 2 \cdot 3 \cdot 5 = 60$

You Try It 4   Find the LCM of 20 and 21.

Your solution

Example 5   Find the GCF of 30, 45, and 60.

Solution
$30 = 2 \cdot 3 \cdot 5$ $45 = 3 \cdot 3 \cdot 5$ $60 = 2 \cdot 2 \cdot 3 \cdot 5$

GCF = $3 \cdot 5 = 15$

You Try It 5   Find the GCF of 42 and 63.

Your solution

Solutions on p. S2

Copyright © Houghton Mifflin Company. All rights reserved.

1.5 Exercises

Objective A   To factor numbers

For Exercises 1 to 30, find the factors of the number.

1. 4   **2.** 20   **3.** 12   **4.** 7   **5.** 8   **6.** 9

7. 13   **8.** 30   **9.** 56   **10.** 28   **11.** 45   **12.** 33

13. 29   **14.** 22   **15.** 52   **16.** 37   **17.** 82   **18.** 69

19. 57   **20.** 64   **21.** 48   **22.** 46   **23.** 50   **24.** 54

25. 77   **26.** 66   **27.** 100   **28.** 80   **29.** 85   **30.** 96

Objective B   To find the prime factorization of a number

For Exercises 31 to 60, find the prime factorization of the number.

31. 14   **32.** 6   **33.** 72   **34.** 17   **35.** 24   **36.** 27

37. 36   **38.** 115   **39.** 26   **40.** 18   **41.** 49   **42.** 42

43. 31   **44.** 81   **45.** 62   **46.** 39   **47.** 89   **48.** 101

49. 86   **50.** 66   **51.** 95   **52.** 74   **53.** 78   **54.** 67

55. 144   **56.** 120   **57.** 175   **58.** 160   **59.** 400   **60.** 625

Objective C   To find the least common multiple and greatest common factor

For Exercises 61 to 90, find the LCM.

61. 3, 8   **62.** 5, 11   **63.** 4, 6   **64.** 6, 8   **65.** 9, 12

66. 8, 14   **67.** 14, 20   **68.** 7, 21   **69.** 12, 36   **70.** 6, 10

Copyright © Houghton Mifflin Company. All rights reserved.

71. 48, 60   **72.** 16, 24   **73.** 80, 90   **74.** 35, 42   **75.** 72, 108

76. 5, 12   **77.** 24, 45   **78.** 8, 20   **79.** 32, 80   **80.** 20, 28

81. 3, 8, 12   **82.** 6, 12, 18   **83.** 3, 5, 10   **84.** 6, 12, 24   **85.** 3, 8, 9

86. 4, 10, 14   **87.** 10, 15, 25   **88.** 8, 12, 18   **89.** 18, 27, 36   **90.** 14, 28, 35

For Exercises 91 to 120, find the GCF.

91. 4, 10   **92.** 9, 15   **93.** 5, 11   **94.** 11, 19   **95.** 6, 8

96. 7, 28   **97.** 6, 12   **98.** 14, 42   **99.** 8, 28   **100.** 24, 36

101. 60, 70   **102.** 72, 108   **103.** 40, 56   **104.** 48, 60   **105.** 35, 42

106. 45, 63   **107.** 60, 90   **108.** 45, 55   **109.** 20, 63   **110.** 28, 45

111. 6, 12, 20   **112.** 12, 18, 24   **113.** 6, 12, 18   **114.** 30, 45, 75   **115.** 24, 36, 60

116. 10, 30, 45   **117.** 26, 52, 78   **118.** 100, 150, 200   **119.** 36, 54, 360   **120.** 18, 27, 36

APPLYING THE CONCEPTS

121. Twin primes are two prime numbers that differ by 2. For instance, 17 and 19 are twin primes. Find three sets of twin primes, not including 17 and 19.

122. In 1742, Christian Goldbach conjectured that every even number greater than 2 could be expressed as the sum of two prime numbers. Show this conjecture is true for 12, 24, and 72. (*Note:* Mathematicians have not yet been able to determine whether Goldbach's conjecture is true or false.)

123. Explain why 2 is the only even prime number.

124. Choose some prime numbers and find the square of each number. Now determine the number of factors in the square of the prime number. Make a conjecture as to the number of factors in the square of any prime number.

125. What is the LCM of two consecutive natural numbers?

126. What is the GCF of two consecutive natural numbers?

127. Explain how the LCM of two numbers can be found by using the GCF of the two numbers.

Copyright © Houghton Mifflin Company. All rights reserved.

1.6 Addition and Subtraction of Rational Numbers

Copyright © Houghton Mifflin Company. All rights reserved.

Objective A To write a rational number in simplest form and as a decimal

TAKE NOTE

The numbers $-\frac{4}{9}, \frac{-4}{9}$, and $\frac{4}{-9}$ all represent the same rational number.

A **rational number** is the quotient of two integers. A rational number written in this way is commonly called a fraction. Some examples of rational numbers are shown at the right.

$$\frac{3}{4}, \quad \frac{-4}{9}, \quad \frac{15}{-4}, \quad \frac{8}{1}, \quad -\frac{5}{6}$$

> **Rational Numbers**
>
> A **rational number** is a number that can be written in the form $\frac{a}{b}$, where a and b are integers and $b \neq 0$.

Point of Interest

As early as 630 A.D., the Hindu mathematician Brahmagupta wrote a fraction as one number over another separated by a space. The Arab mathematician al Hassar (around 1050 A.D.) was the first to show a fraction with the horizontal bar separating the numerator and denominator.

Because an integer can be written as the quotient of the integer and 1, every integer is a rational number.

$$6 = \frac{6}{1} \qquad -8 = \frac{-8}{1}$$

A fraction is in **simplest form** when there are no common factors in the numerator and the denominator. The fractions $\frac{4}{6}$ and $\frac{2}{3}$ are equivalent fractions because they represent the same part of a whole. However, the fraction $\frac{2}{3}$ is in simplest form because there are no common factors (other than 1) in the numerator and denominator.

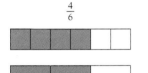

$$\frac{4}{6}$$

$$\frac{2}{3}$$

To write a fraction in simplest form, eliminate the common factors from the numerator and denominator by using the fact that $1 \cdot \frac{a}{b} = \frac{a}{b}$.

$$\frac{4}{6} = \frac{\cancel{2} \cdot 2}{\cancel{2} \cdot 3} = \frac{\cancel{2}}{\cancel{2}} \cdot \frac{2}{3} = 1 \cdot \frac{2}{3} = \frac{2}{3}$$

The process of eliminating common factors is usually written as shown at the right.

$$\frac{4}{6} = \frac{\overset{1}{\cancel{2}} \cdot 2}{\underset{1}{\cancel{2}} \cdot 3} = \frac{2}{3}$$

HOW TO Write $\frac{18}{30}$ in simplest form.

$$\frac{18}{30} = \frac{\overset{1}{\cancel{2}} \cdot \overset{1}{\cancel{3}} \cdot 3}{\underset{1}{\cancel{2}} \cdot \underset{1}{\cancel{3}} \cdot 5} = \frac{3}{5}$$

• To find the common factors, write the numerator and denominator in terms of prime factors.

A rational number can also be written in **decimal notation.**

three tenths $\quad 0.3 = \frac{3}{10}$ $\qquad$ forty-three thousandths $\quad 0.043 = \frac{43}{1000}$

A rational number written as a fraction can be written in decimal notation by dividing the numerator of the fraction by the denominator. Think of the fraction bar as meaning "divided by."

HOW TO Write $\frac{5}{8}$ as a decimal.

$$
\begin{array}{r}
0.625 \\
8\overline{)5.000} \\
-4\,8 \\
\hline
20 \\
-16 \\
\hline
40 \\
-40 \\
\hline
0
\end{array}
$$

• Divide the numerator, 5, by the denominator, 8.

When a resulting remainder is zero, the decimal is called a **terminating decimal.** The decimal 0.625 is a terminating decimal.

$$\frac{5}{8} = 0.625$$

HOW TO Write $\frac{4}{11}$ as a decimal.

$$
\begin{array}{r}
0.3636 \\
11\overline{)4.0000} \\
-3\,3 \\
\hline
70 \\
-66 \\
\hline
40 \\
-33 \\
\hline
70 \\
-66 \\
\hline
4
\end{array}
$$

• Divide the numerator, 4, by the denominator, 11.

No matter how long we continue to divide, the remainder is never zero. The decimal $0.\overline{36}$ is a **repeating decimal.** The bar over the 36 indicates that these digits repeat.

$$\frac{4}{11} = 0.\overline{36}$$

Every rational number can be written as a terminating or a repeating decimal. Some numbers—for example, $\sqrt{7}$ and π—have decimal representations that never terminate or repeat. These numbers are called **irrational numbers.**

$$\sqrt{7} \approx 2.6457513\ldots \qquad \pi \approx 3.1415926\ldots$$

The rational numbers and the irrational numbers taken together are called the **real numbers.**

The diagram below shows the relationship between some of the sets of numbers we have discussed. The arrows indicate that one set is contained completely within the other set.

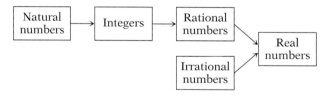

Note that there is no arrow between the rational numbers and the irrational numbers. Any given real number is either a rational number or an irrational number. It cannot be both. However, a natural number such as 7 can also be called an integer, a rational number, and a real number.

> *TAKE NOTE*
>
> Rational numbers are fractions, such as $-\frac{6}{7}$ or $\frac{10}{3}$, in which the numerator and denominator are integers. Rational numbers are also represented by repeating decimals such as 0.25767676... or terminating decimals such as 1.73. An irrational number is neither a terminating decimal nor a repeating decimal. For instance, 2.45454544544445... is an irrational number.

Example 1 Write $\frac{90}{168}$ in simplest form.

Solution
$$\frac{90}{168} = \frac{\overset{1}{\cancel{2}} \cdot \overset{1}{\cancel{3}} \cdot 3 \cdot 5}{\underset{1}{\cancel{2}} \cdot 2 \cdot 2 \cdot \underset{1}{\cancel{3}} \cdot 7} = \frac{15}{28}$$

You Try It 1 Write $\frac{60}{140}$ in simplest form.

Your solution

Solution on p. S2

Copyright © Houghton Mifflin Company. All rights reserved.

Example 2 Write $\frac{3}{20}$ as a decimal.

Solution $\frac{3}{20} = 3 \div 20 = 0.15$

You Try It 2 Write $\frac{4}{9}$ as a decimal. Place a bar over the repeating digits.

Your solution

Solution on p. S2

Objective B   **To add rational numbers**

Point of Interest

One of the earliest written mathematical documents is the Rhind Papyrus. It was discovered in Egypt in 1858 but it is estimated to date from 1650 B.C. The Papyrus shows that the earlier Egyptian method of calculating with fractions was much different from the methods used today. The early Egyptians used *unit fractions*, which are fractions with a numerator of one. With the exception of $\frac{2}{3}$, fractions with numerators other than 1 were written as the sum of two unit fractions. For instance, $\frac{2}{11}$ was written as $\frac{1}{6} + \frac{1}{66}$.

Two of the 7 squares in the rectangle have dark shading. This is $\frac{2}{7}$ of the entire rectangle. Three of the 7 squares in the rectangle have light shading. This is $\frac{3}{7}$ of the entire rectangle. A total of 5 squares are shaded. This is $\frac{5}{7}$ of the entire rectangle.

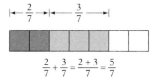

$$\frac{2}{7} + \frac{3}{7} = \frac{2+3}{7} = \frac{5}{7}$$

> **Addition of Fractions**
>
> **To add two fractions with the same denominator,** add the numerators and place the sum over the common denominator.

HOW TO Find the sum of $\frac{3}{8}$ and $\frac{1}{8}$.

$$\frac{3}{8} + \frac{1}{8} = \frac{3+1}{8}$$

- The denominators are the same. **Add the numerators.**

$$= \frac{4}{8} = \frac{1}{2}$$

- Write the answer in simplest form.

To add fractions with different denominators, first rewrite the fractions as equivalent fractions with a common denominator. Then add the fractions. The common denominator is the least common multiple (LCM) of the denominators. The least common multiple of the denominators is frequently called the **least common denominator.**

TAKE NOTE

In this text, we will normally leave improper fractions as the answer and not change them to mixed numbers.

HOW TO Add: $\frac{7}{10} + \frac{11}{12}$

The LCM of 10 and 12 is 60.

$$\frac{7}{10} + \frac{11}{12} = \frac{42}{60} + \frac{55}{60}$$

- Rewrite each fraction as an equivalent fraction with a denominator of 60.

$$= \frac{42+55}{60} = \frac{97}{60}$$

- Add the fractions.

Copyright © Houghton Mifflin Company. All rights reserved.

If one of the addends is a negative rational number, use the same rules as for addition of integers.

> **HOW TO** Add: $-\dfrac{5}{6} + \dfrac{3}{10}$

The LCM of 6 and 10 is 30.

$$-\frac{5}{6} + \frac{3}{10} = -\frac{25}{30} + \frac{9}{30}$$

- Rewrite each fraction as an equivalent fraction with a denominator of 30.

$$= \frac{-25 + 9}{30}$$

- Add the fractions.

$$= \frac{-16}{30} = -\frac{8}{15}$$

> **TAKE NOTE**
> Although we could write the answer as $\dfrac{-8}{15}$, in this text we write $-\dfrac{8}{15}$. That is, we place the negative sign in front of the fraction.

To add decimals, write the numbers so that the decimal points are in a vertical line. Then proceed as in the addition of integers. Write the decimal point in the answer directly below the decimal points in the problem.

> **HOW TO** Add: $-114.03 + 89.254$
>
> $$\begin{array}{r} 114.030 \\ -\ 89.254 \\ \hline 24.776 \end{array}$$
>
> - The signs are different. Find the difference between the absolute values of the numbers $|-114.03| = 114.03;\ |89.254| = 89.254$
>
> $$-114.03 + 89.254 = -24.776$$
>
> - Attach the sign of the number with the larger absolute value. Because $|-114.03| > |89.254|$, use the sign of -114.03.

> **Example 3** Add: $\dfrac{5}{16} + \left(-\dfrac{7}{40}\right)$
>
> **Solution** The LCM of 16 and 40 is 80.
>
> $$\frac{5}{16} + \left(-\frac{7}{40}\right) = \frac{25}{80} + \left(-\frac{14}{80}\right) = \frac{25 + (-14)}{80} = \frac{11}{80}$$

> **You Try It 3** Add: $\dfrac{5}{9} + \left(-\dfrac{11}{12}\right)$
>
> **Your solution**

> **Example 4** Find the total of $\dfrac{3}{4}, \dfrac{1}{6}$, and $\dfrac{5}{8}$.
>
> **Solution** The LCM of 4, 6, and 8 is 24.
>
> $$\frac{3}{4} + \frac{1}{6} + \frac{5}{8} = \frac{18}{24} + \frac{4}{24} + \frac{15}{24}$$
>
> $$= \frac{18 + 4 + 15}{24} = \frac{37}{24}$$

> **You Try It 4** Find $\dfrac{7}{8}$ more than $-\dfrac{5}{6}$.
>
> **Your solution**

> **Example 5** Add: $-4 + 2.37$
>
> **Solution** $-4 + 2.37 = -1.63$

> **You Try It 5** Add: $-6.12 + (-12.881)$
>
> **Your solution**

Solutions on p. S2

Copyright © Houghton Mifflin Company. All rights reserved.

Copyright © Houghton Mifflin Company. All rights reserved.

Objective C   **To subtract rational numbers**

Subtracting fractions is similar to adding fractions in that the denominators must be the same.

> **Subtraction of Fractions**
>
> **To subtract two fractions with the same denominator,** subtract the numerators and place the difference over the common denominator.

HOW TO What is $\frac{3}{10}$ less than $\frac{4}{15}$?

Translate this as the subtraction problem $\frac{4}{15} - \frac{3}{10}$.

$$\frac{4}{15} - \frac{3}{10} = \frac{8}{30} - \frac{9}{30}$$

- The LCM of 15 and 10 is **30**. Rewrite each fraction as an equivalent fraction with a denominator of 30.

$$= \frac{8 - 9}{30} = \frac{8 + (-9)}{30}$$

- Subtract the fractions.

$$= \frac{-1}{30} = -\frac{1}{30}$$

HOW TO Subtract: $-\frac{7}{8} - \left(-\frac{5}{12}\right)$

$$-\frac{7}{8} - \left(-\frac{5}{12}\right) = -\frac{21}{24} - \left(-\frac{10}{24}\right)$$

- The LCM of 8 and 12 is **24**.

$$= \frac{-21 - (-10)}{24} = \frac{-21 + 10}{24}$$

- Subtract the fractions.

$$= \frac{-11}{24} = -\frac{11}{24}$$

HOW TO Subtract: $-2.984 - (-1.45)$

$$-2.984 - (-1.45) = -2.984 + 1.45$$
$$= -1.534$$

Example 6 Subtract: $-\frac{1}{2} - \frac{5}{6} - \left(-\frac{3}{4}\right)$

Solution The LCM of 2, 6, and 4 is 12.

$$-\frac{1}{2} - \frac{5}{6} - \left(-\frac{3}{4}\right) = \frac{6}{12} - \frac{10}{12} - \left(-\frac{9}{12}\right)$$
$$= \frac{-6 - 10 - (-9)}{12} = \frac{-6 - 10 + 9}{12}$$
$$= -\frac{7}{12}$$

You Try It 6 Subtract: $\frac{7}{8} - \left(-\frac{5}{12}\right) - \frac{1}{9}$

Your solution

Example 7 Subtract: $45.2 - 56.89$

Solution $45.2 - 56.89 = -11.69$

You Try It 7 Subtract: $-12.03 - 19.117$

Your solution

Solutions on p. S2

Objective D **To solve application problems**

The graph at the right shows the number of millions of barrels of oil and natural gas that are produced and exported per day for five selected countries. Use this graph for Example 8 and You Try It 8.

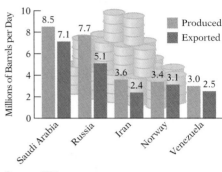

Millions of Barrels per Day — Produced / Exported

8.5 7.1 7.7 5.1 3.6 2.4 3.4 3.1 3.0 2.5

Saudi Arabia Russia Iran Norway Venezuela

Source: IEA

Example 8

Using the graph above, find the total number of barrels of oil and natural gas that are exported by the five countries.

Strategy

To find the total number of barrels:

• Read the numbers from the graph that correspond to exporting (7.1, 5.1, 2.4, 3.1, 2.5).
• Add the numbers.

Solution

Total = 7.1 + 5.1 + 2.4 + 3.1 + 2.5 = 20.2

The five countries export 20.2 million barrels of oil and gas per day.

You Try It 8

For the five countries in the graph above, find the difference between the total amount of oil and gas produced and the total amount of oil and gas exported.

Your strategy

Your solution

Example 9

A cabinet maker is joining two pieces of wood. What is the measure of the cut from the left side of the board so that the pieces fit as shown.

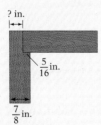

? in. $\frac{5}{16}$ in. $\frac{7}{8}$ in.

Strategy

To find the measure of the cut, subtract $\frac{5}{16}$ in. from $\frac{7}{8}$ in.

Solution

$\frac{7}{8} - \frac{5}{16} = \frac{14}{16} - \frac{5}{16} = \frac{9}{16}$

The cut must be made $\frac{9}{16}$ in. from the left side of the board.

You Try It 9

Barbara Walsh spent $\frac{1}{6}$ of her day studying, $\frac{1}{8}$ of her day in class, and $\frac{1}{4}$ of her day working. What fraction of her day did she spend on these three activities?

Your strategy

Your solution

Solutions on pp. S2–S3

Copyright © Houghton Mifflin Company. All rights reserved.

1.6 Exercises

Objective A   **To write a rational number in simplest form and as a decimal**

For Exercises 1 to 18, write each fraction in simplest form.

1. $\dfrac{7}{21}$ 2. $\dfrac{10}{15}$ 3. $-\dfrac{8}{22}$ 4. $-\dfrac{8}{60}$ 5. $-\dfrac{50}{75}$ 6. $-\dfrac{20}{44}$

7. $\dfrac{12}{8}$ 8. $\dfrac{36}{4}$ 9. $\dfrac{0}{36}$ 10. $\dfrac{12}{18}$ 11. $-\dfrac{60}{100}$ 12. $-\dfrac{14}{45}$

13. $-\dfrac{28}{20}$ 14. $-\dfrac{20}{5}$ 15. $-\dfrac{45}{3}$ 16. $\dfrac{44}{60}$ 17. $\dfrac{23}{46}$ 18. $\dfrac{31}{93}$

For Exercises 19 to 36, write as a decimal. Place a bar over repeating digits.

19. $\dfrac{4}{5}$ 20. $\dfrac{1}{8}$ 21. $\dfrac{1}{6}$ 22. $\dfrac{5}{6}$ 23. $-\dfrac{1}{3}$ 24. $-\dfrac{1}{20}$

25. $-\dfrac{2}{9}$ 26. $-\dfrac{5}{11}$ 27. $-\dfrac{7}{12}$ 28. $\dfrac{7}{8}$ 29. $\dfrac{11}{12}$ 30. $\dfrac{4}{11}$

31. $-\dfrac{7}{18}$ 32. $-\dfrac{17}{18}$ 33. $\dfrac{9}{16}$ 34. $\dfrac{15}{16}$ 35. $-\dfrac{6}{7}$ 36. $\dfrac{5}{13}$

Objective B   **To add rational numbers**

For Exercises 37 to 66, add.

37. $\dfrac{2}{3}+\dfrac{5}{12}$ 38. $\dfrac{1}{2}+\dfrac{3}{8}$ 39. $\dfrac{5}{8}+\dfrac{5}{6}$ 40. $\dfrac{1}{9}+\dfrac{5}{27}$

41. $\dfrac{5}{12}+\left(-\dfrac{3}{8}\right)$ 42. $-\dfrac{5}{6}+\dfrac{5}{9}$ 43. $-\dfrac{6}{13}+\left(-\dfrac{17}{26}\right)$ 44. $\dfrac{3}{5}+\left(-\dfrac{11}{12}\right)$

45. $-\dfrac{3}{4}+\left(-\dfrac{5}{6}\right)$ 46. $-\dfrac{5}{8}+\dfrac{11}{12}$ 47. $\dfrac{1}{3}+\dfrac{5}{6}+\dfrac{2}{9}$ 48. $\dfrac{1}{2}+\dfrac{2}{3}+\dfrac{1}{6}$

Copyright © Houghton Mifflin Company. All rights reserved.

49. $-\dfrac{3}{8} + \dfrac{5}{12} + \left(-\dfrac{3}{16}\right)$

50. $\dfrac{5}{16} + \left(-\dfrac{3}{4}\right) + \left(-\dfrac{7}{8}\right)$

51. $-\dfrac{1}{8} + \left(-\dfrac{11}{12}\right) + \dfrac{1}{3}$

52. $\dfrac{3}{8} + \left(-\dfrac{7}{12}\right) + \left(-\dfrac{5}{9}\right)$

53. $7.56 + 0.462$

54. $1.09 + 6.2$

55. $-32.1 + 6.7$

56. $5.138 + (-8.41)$

57. $-16.92 + 6.956$

58. $48 + (-34.12)$

59. $-19.84 + 17$

60. $-3.739 + (-2.03)$

61. $2.34 + (-3.7) + (-5.601)$

62. $-5.507 + (-4.91) + 15.2$

63. $-7.89 + 12.041 + (-4.151)$

64. $-3.04 + (-2.191) + (-0.06)$

65. $-91.2 + 24.56 + (-42.037)$

66. $81.02 + (-75.603) + (-17.8)$

67. What is $\dfrac{3}{4}$ more than $-\dfrac{5}{6}$?

68. Find the total of $\dfrac{5}{8}$ and $-\dfrac{5}{16}$.

69. Find $-\dfrac{5}{9}$ increased by $\dfrac{1}{6}$.

70. What is $-\dfrac{3}{8}$ added to $-\dfrac{5}{12}$?

71. Find 1.45 more than -7.

72. What is the sum of -4.23 and 3.06?

Objective C **To subtract rational numbers**

For Exercises 73 to 102, subtract.

73. $\dfrac{1}{9} - \dfrac{5}{27}$

74. $\dfrac{5}{8} - \dfrac{5}{6}$

75. $\dfrac{1}{2} - \dfrac{5}{8}$

76. $\dfrac{2}{3} - \dfrac{1}{12}$

77. $-\dfrac{11}{12} - \dfrac{5}{8}$

78. $-\dfrac{7}{13} - \left(-\dfrac{11}{26}\right)$

79. $-\dfrac{5}{6} - \dfrac{4}{9}$

80. $-\dfrac{3}{4} - \left(-\dfrac{5}{6}\right)$

81. $\dfrac{4}{5} - \left(-\dfrac{5}{12}\right)$

82. $\dfrac{3}{8} - \left(-\dfrac{3}{4}\right)$

83. $\dfrac{7}{16} - \left(-\dfrac{3}{4}\right) - \left(-\dfrac{5}{8}\right)$

84. $-\dfrac{1}{8} - \dfrac{5}{12} - \left(-\dfrac{5}{16}\right)$

85. $\dfrac{1}{2} - \dfrac{5}{6} - \dfrac{2}{3}$

86. $-\dfrac{19}{18} - \left(-\dfrac{5}{6}\right) - \left(-\dfrac{2}{9}\right)$

87. $\dfrac{5}{8} - \left(-\dfrac{7}{12}\right) + \dfrac{7}{9}$

Copyright © Houghton Mifflin Company. All rights reserved.

88. $-\dfrac{1}{8} - \left(-\dfrac{11}{12}\right) - \dfrac{1}{3}$ **89.** $6.322 - 9.123$ **90.** $-43.1 - 19.37$

91. $-3.04 - (-5.128)$ **92.** $-25 - (-34.12)$ **93.** $-20.04 - (-41.2)$

94. $0.354 - 16$ **95.** $-1.023 - (-1.023)$ **96.** $-5.0614 - 2.31$

97. $4.32 - (-6.1) - (-4.032)$ **98.** $-1.204 - (-5.027) - 12.3$

99. $-9.2 - 15.02 - (-6.614)$ **100.** $-6.97 - (-3.258) - (-3.712)$

101. $16.48 - 19.283 - (-6.804)$ 4.001 **102.** $12.401 - 17.23 - (-2.1)$

103. What number is $\dfrac{5}{6}$ less than $-\dfrac{3}{8}$?   **104.** Find the difference between $\dfrac{1}{2}$ and $-\dfrac{5}{16}$.

105. What is $\dfrac{2}{3}$ less $\dfrac{3}{4}$?   **106.** What number is $\dfrac{4}{5}$ less than $-\dfrac{2}{15}$?

107. Find the difference between $-\dfrac{5}{9}$ and $-\dfrac{1}{6}$.   **108.** Find $\dfrac{5}{16}$ less $\dfrac{7}{12}$.

 Objective D   **To solve application problems**

 Weights of Coins   For Exercises 109 to 114, use the following table showing the weight, diameter, and thickness of various U.S. coins.

	Penny	Nickel	Dime	Quarter
Weight	2.5 g	5 g	2.268 g	5.67 g
Diameter	19.05 mm	21.21 mm	17.91 mm	24.26 mm
Thickness	1.55 mm	1.95 mm	1.35 mm	1.75 mm

Source: The U.S. Mint

109. List the coins from thinnest to thickest.

Copyright © Houghton Mifflin Company. All rights reserved.

110. List the weight of the coins from heaviest to lightest.

111. What is the sum of the weights of the four coins?

112. What is the largest difference in weight between two coins?

113. What is the largest difference in diameter between two coins?

114. What is the smallest difference in thickness between two coins?

115. Food Science   A recipe calls for $\frac{3}{4}$ c vegetable broth. If a chef has $\frac{2}{3}$ c vegetable broth, how much additional broth is needed for the recipe?

116. Carpentry   A piece of lath $\frac{1}{16}$ in. thick is glued to the edges of a wood strip that is $\frac{3}{4}$ in. width. What is the width of the wood and lath?

Lath

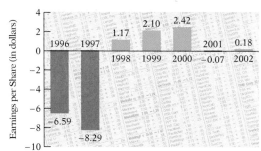

Finance   The graph at the right shows the earnings per share for Apple Computer for the years 1996 through 2002. Use this graph for Exercises 117 to 120.

117. In which years did Apple Computer have negative earnings per share?

118. In which year did Apple Computer have its lowest earnings per share?

Source: Apple Computer Corporation

119. What was the decrease in earnings per share between 1996 and 1997?

120. What was the difference in earnings per share between 2000 and 1997?

Oil Consumption The graph at the right shows the number of millions of barrels of oil per day that are consumed by various countries and the number of millions of barrels of oil per day those countries import. Use this graph for Exercises 121 to 124.

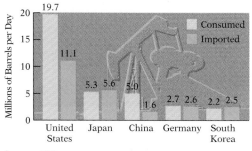

121. How many millions of barrels of oil per day are consumed by these five countries?

Source: IEA

122. How many millions of barrels of oil per day are imported by these five countries?

Copyright © Houghton Mifflin Company. All rights reserved.

123. For these five countries, what is the difference between the number of millions of barrels of oil per day consumed and the number imported?

124. What is the largest difference in the numbers of millions of barrels of oil consumed by these five countries?

APPLYING THE CONCEPTS

For Exercises 125 to 130, simplify.

125. $\dfrac{1}{2} - \dfrac{5}{6} + \dfrac{7}{8}$

126. $-\dfrac{2}{3} + \dfrac{1}{4} - \dfrac{5}{6}$

127. $-\dfrac{3}{4} - \left(-\dfrac{2}{3}\right) + \left(-\dfrac{1}{8}\right)$

128. $-\dfrac{5}{6} + \left(-\dfrac{3}{8}\right) - \dfrac{5}{16}$

129. $\dfrac{4}{5} - \left(\dfrac{1}{4} - \left(-\dfrac{2}{3}\right)\right)$

130. $-\dfrac{5}{6} - \left(\dfrac{2}{3} - \dfrac{5}{12}\right)$

For Exercises 131 to 134, replace the question mark with <, >, or = to make a true statement.

131. $\dfrac{1}{4} - \dfrac{1}{3}$? $0.25 - 0.33$

132. $\dfrac{7}{8} - \dfrac{5}{6}$? $0.875 - 0.83$

133. $0.15 - \dfrac{3}{4}$? $0.75 - \dfrac{27}{20}$

134. $\dfrac{2}{3} - \dfrac{5}{6}$? $0.67 - 0.83$

For Exercises 135 to 138, use a calculator. By experimenting with different fractions, find a fraction as the quotient of integers that is equivalent to each of the following.

135. $0.\overline{1}$

136. $0.08\overline{3}$

137. $0.41\overline{6}$

138. $0.\overline{45}$

139. There are only two possible prime factors of the denominator of a terminating decimal. What are those prime factors?

140. The numerator of a fraction is 1. If the denominator is replaced by -2, -3, -4, -5, . . . , are the resulting fractions getting smaller or larger?

141. The numerator of a fraction is -1. If the denominator is replaced by 2, 3, 4, 5, . . . , are the resulting fractions getting smaller or larger?

Copyright © Houghton Mifflin Company. All rights reserved.

1.7 Multiplication and Division of Rational Numbers

Objective A **To multiply rational numbers**

The product $\frac{2}{3} \times \frac{4}{5}$ can be read "$\frac{2}{3}$ times $\frac{4}{5}$" or "$\frac{2}{3}$ of $\frac{4}{5}$."

Reading the times sign as "of" can help with understanding the procedure for multiplying fractions.

$\frac{4}{5}$ of the bar is shaded.

Shade $\frac{2}{3}$ of the $\frac{4}{5}$ already shaded.

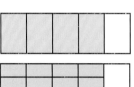

$\frac{8}{15}$ of the bar is then shaded dark yellow.

$$\frac{2}{3} \text{ of } \frac{4}{5} = \frac{2}{3} \times \frac{4}{5} = \frac{2 \times 4}{3 \times 5} = \frac{8}{15}$$

Multiplication of Fractions

The product of two fractions is the product of the numerators over the product of the denominators.

After multiplying two fractions, write the product in simplest form. Use the rules for multiplying integers to determine the sign of the product.

HOW TO Multiply: $-\frac{3}{4} \cdot \frac{10}{21}$

The signs are different. The product is negative.

$$-\frac{3}{4} \cdot \frac{10}{21} = \frac{3 \cdot 10}{4 \cdot 21}$$

- Multiply the numerators. Multiply the denominators.

$$= -\frac{3 \cdot 2 \cdot 5}{2 \cdot 2 \cdot 3 \cdot 7}$$

- Write the prime factorization of each number.

$$= -\frac{\overset{1}{\cancel{3}} \cdot \overset{1}{\cancel{2}} \cdot 5}{2 \cdot 2 \cdot \underset{1}{\cancel{3}} \cdot 7} = -\frac{5}{14}$$

- **Divide by the common factors.** Then multiply the remaining factors in the numerator and in the denominator.

This problem can also be worked by using the greatest common factor (GCF) of the numerator and the denominator.

HOW TO Multiply: $-\frac{3}{4} \cdot \frac{10}{21}$

$$-\frac{3}{4} \cdot \frac{10}{21} = \frac{30}{84}$$

- Multiply the numerators.
- Multiply the denominators.

$$= -\frac{\overset{1}{\cancel{6}} \cdot 5}{\underset{1}{\cancel{6}} \cdot 14} = -\frac{5}{14}$$

- Divide the numerator and denominator by the GCF.

Copyright © Houghton Mifflin Company. All rights reserved.

To multiply decimals, multiply as with integers. Write the decimal point in the product in such a way that the number of decimal places in the product equals the sum of the number of decimal places in the factors.

HOW TO Find the product of 7.43 and −0.00025.

$$
\begin{array}{r}
7.43 \\
\times\ 0.00025 \\
\hline
3715 \\
1486 \\
\hline
0.0018575
\end{array}
$$

2 decimal places
5 decimal places

• Multiply the absolute values.

7 decimal places

$7.43(-0.00025) = -0.0018575$

• The signs are different. The product is negative.

Example 1 Multiply: $-\dfrac{3}{8}\left(-\dfrac{12}{17}\right)$

Solution

$$-\dfrac{3}{8}\left(-\dfrac{12}{17}\right) = \dfrac{3 \cdot 12}{8 \cdot 17}$$

• The signs are the same. The product is positive.

$$= \dfrac{3 \cdot \overset{1}{\cancel{2}} \cdot \overset{1}{\cancel{2}} \cdot 3}{\underset{1}{\cancel{2}} \cdot \underset{1}{\cancel{2}} \cdot 2 \cdot 17}$$

$$= \dfrac{9}{34}$$

• Write the answer in simplest form.

You Try It 1 Multiply: $\dfrac{5}{8}\left(-\dfrac{4}{25}\right)$

Your solution

$-\dfrac{1}{10}$

Example 2

Find the product of $\dfrac{4}{9}$, $\dfrac{3}{10}$ and $-\dfrac{5}{18}$.

Solution

$$\dfrac{4}{9} \cdot \dfrac{3}{10} \cdot \left(-\dfrac{5}{18}\right) = -\dfrac{4 \cdot 3 \cdot 5}{9 \cdot 10 \cdot 18}$$

• The product is negative.

$$= -\dfrac{\overset{1}{\cancel{2}} \cdot \overset{1}{\cancel{2}} \cdot \overset{1}{\cancel{3}} \cdot \overset{1}{\cancel{5}}}{\underset{1}{\cancel{3}} \cdot 3 \cdot \underset{1}{\cancel{2}} \cdot \underset{1}{\cancel{5}} \cdot \underset{1}{\cancel{2}} \cdot 3 \cdot 3}$$

$$= -\dfrac{1}{27}$$

• Write the answer in simplest form.

You Try It 2

Find the product of $-\dfrac{4}{5}$, $-\dfrac{3}{8}$, and $-\dfrac{10}{27}$.

Your solution

Example 3 Multiply: $-4.06(-0.065)$

Solution

The product is positive.

$-4.06(-0.065) = 0.2639$

You Try It 3 Multiply: $0.034(-2.14)$

Your solution

Solutions on p. S3

Copyright © Houghton Mifflin Company. All rights reserved.

Objective B **To divide rational numbers**

The **reciprocal of a fraction** is the fraction with the numerator and denominator interchanged. For instance, the reciprocal of $\frac{3}{4}$ is $\frac{4}{3}$, and the reciprocal of $-\frac{5}{2}$ is $-\frac{2}{5}$.

The product of a number and its reciprocal is 1. This fact is used in the procedure for dividing fractions.

$$\frac{3}{4} \cdot \frac{4}{3} = \frac{12}{12} = 1 \qquad -\frac{5}{2} \cdot \left(-\frac{2}{5}\right) = \frac{10}{10} = 1$$

Study the example below to see how reciprocals are used when dividing fractions.

Divide: $\frac{3}{5} \div \frac{5}{6}$

$$\frac{3}{5} \div \frac{5}{6} = \frac{\dfrac{3}{5}}{\dfrac{5}{6}} = \frac{\dfrac{3}{5} \cdot \dfrac{6}{5}}{\dfrac{5}{6} \cdot \dfrac{6}{5}}$$

- Multiply the numerator and denominator by the **reciprocal of the divisor**.

$$= \frac{\dfrac{3}{5} \cdot \dfrac{6}{5}}{1}$$

- The product of a number and its reciprocal is **1**.

$$= \frac{3}{5} \cdot \frac{6}{5} = \frac{18}{25}$$

- A number divided by 1 is the number.

These steps are summarized by $\frac{3}{5} \div \frac{5}{6} = \frac{3}{5} \cdot \frac{6}{5} = \frac{18}{25}$.

Division of Fractions

To divide two fractions, multiply the dividend by the reciprocal of the divisor.

TAKE NOTE

The method of dividing fractions is sometimes stated, "To divide fractions, invert the divisor and then multiply." Inverting the divisor means writing its reciprocal.

HOW TO Divide: $\frac{3}{10} \div \left(-\frac{18}{25}\right)$

The signs are different. The quotient is negative.

$$\frac{3}{10} \div \left(-\frac{18}{25}\right) = -\left(\frac{3}{10} \div \frac{18}{25}\right) = -\left(\frac{3}{10} \cdot \frac{25}{18}\right)$$

$$= -\frac{3 \cdot 25}{10 \cdot 18} = -\frac{\overset{1}{\cancel{3}} \cdot \overset{1}{\cancel{5}} \cdot 5}{2 \cdot \underset{1}{\cancel{5}} \cdot 2 \cdot \underset{1}{\cancel{3}} \cdot 3} = -\frac{5}{12}$$

To divide decimals, move the decimal point in the divisor to the right so that the divisor becomes a whole number. Move the decimal point in the dividend the same number of places to the right. Place the decimal point in the quotient directly over the decimal point in the dividend. Then divide as with whole numbers.

Copyright © Houghton Mifflin Company. All rights reserved.

Copyright © Houghton Mifflin Company. All rights reserved.

TAKE NOTE

The procedure for dividing decimals as we do at the right can be justified as follows.

$$(-1.4) \div (-0.36)$$
$$= \frac{1.4}{0.36}$$
$$= \frac{1.4}{0.36} \cdot \frac{100}{100}$$
$$= \frac{140}{36}$$
$$\approx 3.9$$

HOW TO   Divide: $(-1.4) \div (-0.36)$. Round to the nearest tenth.

The signs are the same. The quotient is positive.

$$
\begin{array}{r}
3.88 \approx 3.9 \\
0.36.\overline{)1.40.00} \\
\underline{1\ 08} \\
32\ 0 \\
\underline{28\ 8} \\
3\ 20 \\
\underline{2\ 88} \\
32
\end{array}
$$

- **Move the decimal point** 2 places to the right in the divisor and then in the dividend. Place the decimal point in the quotient directly over the decimal point in the dividend.

- Note that the symbol $\approx$ is used to indicate that the quotient is an approximate value that has been rounded off.

$(-1.4) \div (-0.36) \approx 3.9$

Example 4   Divide: $\left(-\dfrac{3}{10}\right) \div \dfrac{9}{16}$

Solution   The quotient is negative.

$$\left(-\frac{3}{10}\right) \div \frac{9}{14} = -\left(\frac{3}{10} \cdot \frac{14}{9}\right)$$

- **Multiply by the reciprocal of the divisor.**

$$= -\frac{3 \cdot 2 \cdot 7}{2 \cdot 5 \cdot 3 \cdot 3}$$

$$= -\frac{7}{15}$$

- Write the answer in simplest form.

You Try It 4   Divide: $\dfrac{5}{8} \div \left(-\dfrac{10}{11}\right)$

Your solution

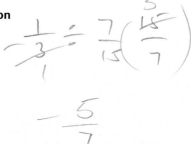

$$-\frac{11}{16}$$

Example 5   Find the quotient of $-\dfrac{5}{8}$ and $-\dfrac{7}{16}$.

Solution   The quotient is positive.

$$-\frac{5}{8} \div \left(-\frac{7}{16}\right) = \frac{5}{8} \cdot \frac{16}{7}$$

- **Multiply by the reciprocal of the divisor.**

$$= \frac{5 \cdot 16}{8 \cdot 7}$$

$$= \frac{10}{7}$$

- Write the answer in simplest form.

You Try It 5   What is $-\dfrac{1}{3}$ divided by $\dfrac{7}{15}$?

Your solution

$$-\frac{1}{3} \div \frac{7}{15} \left(\frac{15}{7}\right)$$

$$-\frac{5}{7}$$

Example 6   Divide: $4.152 \div (-25.2)$. Round to the nearest thousandth.

Solution
Divide the absolute values. The quotient is negative.

$4.152 \div (-25.2) \approx -0.165$

You Try It 6   Divide: $(-34) \div (-9.02)$. Round to the nearest hundredth.

Your solution

Solutions on p. S3

Objective C   **To convert among percents, fractions, and decimals**

"A population growth rate of 3%," "a manufacturer's discount of 25%," and "an 8% increase in pay" are typical examples of the many ways in which *percent* is used in applied problems. **Percent** means "parts of 100." Thus 27% means 27 parts of 100.

In applied problems involving a percent, it may be necessary to rewrite a percent as a fraction or as a decimal, or to rewrite a fraction or a decimal as a percent.

To write a percent as a fraction, remove the percent sign and multiply by $\frac{1}{100}$.

$$27\% = 27\left(\frac{1}{100}\right) = \frac{27}{100}$$

To write a percent as a decimal, remove the percent sign and multiply by 0.01.

$$33\% \qquad = \qquad 33(0.01) \qquad = \qquad 0.33$$

Move the decimal point two places to the left. Then remove the percent sign.

TAKE NOTE

The decimal equivalent of 100% is 1. Therefore, multiplying by 100% is the same as multiplying by 1 and does not change the value of the fraction.

$$\frac{5}{8} = \frac{5}{8}(1) = \frac{5}{8}(100\%)$$

To write a fraction as a percent, multiply by 100%. For example, $\frac{5}{8}$ is changed to a percent as follows:

$$\frac{5}{8} = \frac{5}{8}(100\%) = \frac{500}{8}\% = 62.5\%, \qquad \text{or} \qquad 62\frac{1}{2}\%$$

To write a decimal as a percent, multiply by 100%.

$$0.82 \qquad = \qquad 0.82(100\%) \qquad = \qquad 82\%$$

Move the decimal point two places to the right. Then write the percent sign.

Example 7

Write 130% as a fraction and as a decimal.

Solution

$$130\% = 130\left(\frac{1}{100}\right) = \frac{130}{100} = \frac{13}{10}$$
$$130\% = 130(0.01) = 1.30$$

You Try It 7

Write 125% as a fraction and as a decimal.

Your solution

Example 8   Write $33\frac{1}{3}\%$ as a fraction.

Solution

$$33\frac{1}{3}\% = 33\frac{1}{3}\left(\frac{1}{100}\right) = \frac{100}{3}\left(\frac{1}{100}\right) = \frac{1}{3}$$

You Try It 8   Write $16\frac{2}{3}\%$ as a fraction.

Your solution

Solutions on p. S3

Copyright © Houghton Mifflin Company. All rights reserved.

Example 9   Write $\frac{5}{6}$ as a percent.

Solution

$$\frac{5}{6} = \frac{5}{6}(100\%) = \frac{500}{6}\% = 83\frac{1}{3}\%$$

You Try It 9   Write $\frac{9}{16}$ as a percent.

Your solution

Example 10   Write 0.027 as a percent.

Solution
$$0.027 = 0.027(100\%) = 2.7\%$$

You Try It 10   Write 0.043 as a percent.

Your solution

Solutions on p. S3

Objective D   **To solve application problems**

Example 11

A picture frame is supported by two hooks that are $\frac{1}{3}$ and $\frac{2}{3}$ of the distance from the left-hand side of the frame. If the frame is $31\frac{1}{2}$ in. wide, how far from the left side of the frame are the hooks placed?

Strategy
To find the location of the hooks, multiply the width of the frame, $31\frac{1}{2}$ in., by $\frac{1}{3}$ and $\frac{2}{3}$. Recall that to multiply a mixed number by a fraction, first write the mixed number as an improper fraction: $31\frac{1}{2} = \frac{2 \times 31 + 1}{2} = \frac{63}{2}$

Solution
$$31\frac{1}{2} \cdot \frac{1}{3} = \frac{63}{2} \cdot \frac{1}{3} = \frac{21}{2} = 10\frac{1}{2}$$
$$31\frac{1}{2} \cdot \frac{2}{3} = \frac{63}{2} \cdot \frac{2}{3} = 21$$

The hooks are placed $10\frac{1}{2}$ in. and 21 in. from the left of the frame.

You Try It 11

A piece of fabric 20 ft long is being used to make cushions for outdoor furniture. If each cushion requires $1\frac{1}{2}$ ft of fabric, how many cushions can be cut from the fabric?

Your strategy

Your solution

Solution on p. S3

Copyright © Houghton Mifflin Company. All rights reserved.

1.7 Exercises

Copyright © Houghton Mifflin Company. All rights reserved.

Objective A **To multiply rational numbers**

For Exercises 1 to 32, multiply.

1. $\dfrac{2}{3} \cdot \dfrac{5}{7}$

2. $\dfrac{1}{2}\left(\dfrac{3}{8}\right)$

3. $\dfrac{5}{8} \cdot \left(-\dfrac{3}{10}\right)$

4. $-\dfrac{5}{16} \cdot \dfrac{7}{15}$

5. $\dfrac{5}{12}\left(-\dfrac{3}{10}\right)$

6. $\left(-\dfrac{11}{12}\right)\left(-\dfrac{6}{7}\right)$

7. $\dfrac{6}{13}\left(-\dfrac{26}{27}\right)$

8. $\dfrac{1}{6}\left(-\dfrac{6}{11}\right)$

9. $\left(-\dfrac{3}{5}\right)\left(-\dfrac{3}{10}\right)$

10. $\dfrac{3}{5}\left(-\dfrac{11}{12}\right)$

11. $\left(-\dfrac{3}{4}\right)^2$

12. $\left(-\dfrac{5}{8}\right)^2$

13. $\left(-\dfrac{3}{4}\right)\dfrac{5}{6}\left(-\dfrac{2}{9}\right)$

14. $\left(-\dfrac{1}{2}\right)\left(-\dfrac{2}{3}\right)\left(-\dfrac{6}{7}\right)$

15. $\left(-\dfrac{3}{8}\right)\left(-\dfrac{5}{12}\right)\left(\dfrac{3}{10}\right)$

16. $\dfrac{5}{16}\left(-\dfrac{4}{5}\right)\left(-\dfrac{7}{8}\right)$

17. $\left(-\dfrac{15}{2}\right)\left(-\dfrac{4}{3}\right)\left(-\dfrac{7}{10}\right)$

18. $\left(-\dfrac{5}{8}\right)\left(\dfrac{5}{12}\right)\left(-\dfrac{16}{25}\right)$

19. $\left(\dfrac{8}{9}\right)\left(-\dfrac{11}{12}\right)\left(\dfrac{3}{4}\right)$

20. $\dfrac{3}{8}\left(-\dfrac{7}{10}\right)\left(-\dfrac{5}{9}\right)$

21. $0.46(-3.9)$

22. $-0.78(6.8)$

23. $(-8.23)(-0.09)$

24. $(-0.003)(-0.189)$

25. $-0.48(0.85)$

26. $0.056(-3.425)$

27. $-6.5(0.0341)$

28. $4.237(-0.54)$

29. $(-8.004)(-3.4)$

30. $(-3.739)(-2.03)$

31. $0.089(-1.098)$

32. $(-3.004)(-2.31)$

33. Find the product of $-\dfrac{3}{4}$ and $\dfrac{4}{5}$.

34. Find the product of $-\dfrac{7}{10}$ and $-\dfrac{5}{14}$.

35. Multiply -0.23 by -4.5.

36. Multiply -7.06 by 0.034.

Objective B   **To divide rational numbers**

For Exercises 37 to 48, divide.

37. $\dfrac{3}{8} \div \left(-\dfrac{9}{10}\right)$

38. $\left(-\dfrac{2}{15}\right) \div \dfrac{3}{5}$

39. $\left(-\dfrac{8}{9}\right) \div \left(-\dfrac{4}{5}\right)$

40. $\left(-\dfrac{11}{15}\right) \div \left(-\dfrac{22}{5}\right)$

41. $\left(-\dfrac{11}{12}\right) \div \left(-\dfrac{7}{6}\right)$

42. $\left(-\dfrac{3}{10}\right) \div \dfrac{5}{12}$

43. $\left(-\dfrac{6}{11}\right) \div 6$

44. $\left(-\dfrac{26}{27}\right) \div \dfrac{13}{6}$

45. $\left(-\dfrac{11}{12}\right) \div \dfrac{5}{3}$

46. $\left(-\dfrac{3}{10}\right) \div \left(-\dfrac{5}{3}\right)$

47. $\left(-\dfrac{5}{8}\right) \div \dfrac{15}{16}$

48. $\dfrac{8}{9} \div \left(-\dfrac{4}{3}\right)$

49. Find the quotient of $-\dfrac{7}{9}$ and $-\dfrac{5}{18}$.

50. Find the quotient of $\dfrac{5}{8}$ and $-\dfrac{7}{12}$.

51. What is $\dfrac{1}{2}$ divided by $-\dfrac{1}{4}$?

52. What is $-\dfrac{5}{18}$ divided by $\dfrac{15}{16}$?

For Exercises 53 to 56, divide.

53. $25.61 \div (-5.2)$

54. $(-0.1035) \div (-0.023)$

55. $(-0.2205) \div (-0.21)$

56. $(-0.357) \div 1.02$

For Exercises 57 to 60, divide. Round to the nearest hundredth.

57. $-0.0647 \div 0.75$

58. $-27.981 \div 59.2$

59. $-2.45 \div (-21.44)$

60. $3.2 \div (-45.12)$

For Exercises 61 to 64, divide. Round to the nearest thousandth.

61. $(-75.469) \div (-77.8)$

62. $6.14 \div (-27.1)$

63. $-0.1142 \div (-17.2)$

64. $-0.2246 \div 12.34$

65. Find the quotient of -0.3045 and -0.203.

66. Find the quotient of 3.672 and -3.6.

67. What is -0.00552 divided by 1.2?

68. What is -0.01925 divided by 0.077?

Copyright © Houghton Mifflin Company. All rights reserved.

For Exercises 69 to 86, use the Order of Operations Agreement to simplify the expression.

69. $\dfrac{2}{3} - \dfrac{1}{4}\left(-\dfrac{2}{5}\right)$

70. $-\dfrac{7}{8} - \dfrac{3}{4} \div \dfrac{9}{8}$

71. $\dfrac{3}{4}\left(\dfrac{1}{2}\right)^2 - \dfrac{5}{16}$

72. $\left(-\dfrac{5}{6}\right) \div \dfrac{5}{12} - \dfrac{9}{14}$

73. $\dfrac{7}{12} - \left(\dfrac{2}{3}\right)^2 + \left(-\dfrac{3}{4}\right)$

74. $\left(\dfrac{1}{2} - \dfrac{3}{4}\right)^2 - \left(\dfrac{5}{18} - \dfrac{15}{24}\right)$

75. $\left(\dfrac{2}{3}\right)\left(\dfrac{3}{4}\right) - \left(\dfrac{4}{5}\right)\left(\dfrac{5}{8}\right)$

76. $\dfrac{1}{8} - \left(\dfrac{9}{4}\right)\left(-\dfrac{2}{3}\right)^2$

77. $-\dfrac{3}{4} \div \dfrac{5}{8} - \dfrac{4}{5}$

78. $\dfrac{5}{9}\left(\dfrac{3}{4}\right) - \dfrac{1}{2} \div (-2)$

79. $\left(\dfrac{2}{3}\right)^3 - \left(\dfrac{2}{3}\right)^2$

80. $\left(\dfrac{5}{6} - \dfrac{2}{3}\right)^2 \div \left(\dfrac{7}{18} - \dfrac{7}{12}\right)^2$

81. $1.2 - 2.3^2$

82. $4.01 - 0.2(8.1 - 6.4)$

83. $0.03 \cdot 0.2^2 - 0.5^3$

84. $8.1 - 5.2(3.4 - 5.9)^2$

85. $\dfrac{3.8 - 5.2}{-0.35} - \left(\dfrac{1.2}{0.6}\right)^2$

86. $-0.3^2 + 3.4(-2.01) - (-1.75)$

Objective C   **To convert among percents, fractions, and decimals**

87. Explain how to write a percent as a fraction.

88. Explain how to write a fraction as a percent.

For Exercises 89 to 98, write as a fraction and a decimal.

89. 75%

90. 40%

91. 64%

92. 88%

93. 175%

94. 160%

95. 19%

96. 87%

97. 5%

98. 8%

For Exercises 99 to 108, write as a fraction.

99. $11\dfrac{1}{9}\%$

100. $4\dfrac{2}{7}\%$

101. $12\dfrac{1}{2}\%$

102. $37\dfrac{1}{2}\%$

103. $66\dfrac{2}{3}\%$

104. $\dfrac{1}{4}\%$

105. $\dfrac{1}{2}\%$

106. $6\dfrac{1}{4}\%$

107. $83\dfrac{1}{3}\%$

108. $5\dfrac{3}{4}\%$

Copyright © Houghton Mifflin Company. All rights reserved.

For Exercises 109 to 118, write as a decimal.

109. 7.3%   **110.** 9.1%   **111.** 15.8%   **112.** 16.7%   **113.** 0.3%

114. 0.9%   **115.** 9.9%   **116.** 9.15%   **117.** 121.2%   **118.** 18.23%

For Exercises 119 to 138, write as a percent.

119. 0.15   **120.** 0.37   **121.** 0.05   **122.** 0.02   **123.** 0.175

124. 0.125   **125.** 1.15   **126.** 1.36   **127.** 0.008   **128.** 0.004

129. $\dfrac{27}{50}$ **130.** $\dfrac{83}{100}$ **131.** $\dfrac{1}{3}$ **132.** $\dfrac{3}{8}$ **133.** $\dfrac{5}{11}$

134. $\dfrac{4}{9}$ **135.** $\dfrac{7}{8}$ **136.** $\dfrac{9}{20}$ **137.** $1\dfrac{2}{3}$ **138.** $2\dfrac{1}{2}$

Objective D   **To solve application problems**

139. **Art**   A picture frame measures $18\dfrac{1}{4}$ in. wide by $24\dfrac{1}{2}$ in. high. How far from the left side and how far from the bottom of the frame is its center?

140. **Carpentry**   A carpenter has a board that is 14 ft long. How many pieces $\dfrac{3}{4}$ ft long can the carpenter cut from the board?

141. **Carpentry**   A board $36\dfrac{5}{8}$ in. long is cut into two pieces of equal length. If the saw blade makes a cut $\dfrac{1}{8}$ in. wide, how far from the left side of the board should the cut be made?

Copyright © Houghton Mifflin Company. All rights reserved.

142. **Food Science**   A recipe calls for $\frac{3}{4}$ c of butter. If a chef wants to increase the recipe by one-half, how much butter should the chef use?

143. **Construction**   A stair is made from an 8-inch riser and a $\frac{3}{4}$-inch foot plate. How many inches high is a staircase made from 10 of these stairs?

144. **Interior Design**   An interior designer needs $12\frac{1}{2}$ yd of fabric that costs $5.43 per yard and $5\frac{3}{4}$ yd of fabric that costs $6.94 per yard to recover a large sofa. Find the total cost of the two fabrics.

APPLYING THE CONCEPTS

145. Find a rational number that is one-half the difference between $\frac{5}{11}$ and $\frac{4}{11}$.

146. Find $\left(1 + \frac{1}{1}\right)\left(1 + \frac{1}{2}\right)$, $\left(1 + \frac{1}{1}\right)\left(1 + \frac{1}{2}\right)\left(1 + \frac{1}{3}\right)$, and $\left(1 + \frac{1}{1}\right)\left(1 + \frac{1}{2}\right)\left(1 + \frac{1}{3}\right)\left(1 + \frac{1}{4}\right)$. Based on the pattern of the answers, make a conjecture as to $\left(1 + \frac{1}{1}\right)\left(1 + \frac{1}{2}\right)\left(1 + \frac{1}{3}\right)\left(1 + \frac{1}{4}\right)\left(1 + \frac{1}{5}\right)$.

147. A frog is 2 ft from a wall and jumps halfway to the wall. At that point, the frog jumps one-half of the remaining distance to the wall. If the frog continues to jump one-half the remaining distance to the wall, how far from the wall will the frog be after the fifth jump?

148. $-\frac{2}{3} > -\frac{3}{4}$. Is $-\frac{2+3}{3+4}$ less than $-\frac{3}{4}$, greater than $-\frac{2}{3}$, or between $-\frac{2}{3}$ and $-\frac{3}{4}$?

149. Consider the fraction $-\frac{2}{5}$. Suppose that the same positive number is added to the numerator and denominator. What is the relationship between the new fraction and the original fraction?

150. Given any two different rational numbers, is it always possible to find a rational number between the two given numbers? If so, explain how to find such a number. If not, give two rational numbers for which there is no rational number between them.

Copyright © Houghton Mifflin Company. All rights reserved.

1.8 Concepts from Geometry

Objective A   **To find the measures of angles**

The word geometry comes from the Greek words for "earth" (geo) and "measure." The original purpose of geometry was to measure land. Today, geometry is used in many disciplines such as physics, biology, geology, architecture, art, and astronomy.

Here are some basic geometric concepts.

A **plane** is a flat surface such as a table top that extends indefinitely. Figures that lie entirely in a plane are called **plane figures.**

Space extends in all directions. Objects in space, such as a baseball, house, or a tree, are called **solids.**

A **line** extends indefinitely in two directions in a plane. A line has no width.

A **ray** starts at a point and extends indefinitely in one direction. By placing a point on the ray at the right, we can name the ray *AB*.

A **line segment** is part of a line and has two end-points. The line segment *AB* is designated by its two endpoints.

Lines in a plane can be parallel or intersect. **Parallel lines** never meet. The distance between parallel lines in a plane is always the same. We write $p \parallel q$ to indicate line *p* is parallel to line *q*. **Intersecting lines** cross at a point in the plane.

TAKE NOTE

When using three letters to name an angle, the vertex is always the middle letter. We could also refer to the angle at the right as ∠*CAB*.

An **angle** is formed when two rays start from the same point. Rays *AB* and *AC* start from the same point *A*. The point at which the rays meet is called the **vertex** of the angle. The symbol ∠ is read "angle" and is used to name an angle. We can refer to the angle at the right as ∠*A*, ∠*BAC*, or ∠*x*.

Point of Interest

The Babylonians chose 360° for the measure of one full rotation, probably because they knew there were 365 days in a year and the closest number to 365 with many divisors was 360.

An angle can be measured in **degrees.** The symbol for degree is °. A ray rotated 1 revolution about its beginning point creates an angle of 360°.

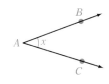

The measure of an angle is symbolized by $m\angle$. For instance, $m\angle C = 40°$. Read this as "the measure of angle *C* is 40°."

Copyright © Houghton Mifflin Company. All rights reserved.

$\frac{1}{4}$ of a revolution is $\frac{1}{4}$ of 360°, or 90°. A 90° angle is called a **right angle.** The symbol ⊾ is used to represent a right angle. **Perpendicular lines** are intersecting lines that form right angles. We write $p \perp q$ to indicate line p is perpendicular to line q.

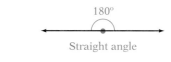

Right angle $p \perp q$

Complementary angles are two angles whose sum is 90°.

$$m\angle A + m\angle B = 35° + 55° = 90°$$

$\angle A$ and $\angle B$ are complementary angles.

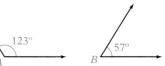

$\frac{1}{2}$ of a revolution is $\frac{1}{2}$ of 360°, or 180°. A 180° angle is called a **straight angle.**

180°

Straight angle

Supplementary angles are two angles whose sum is 180°.

$$m\angle A + m\angle B = 123° + 57° = 180°$$

$\angle A$ and $\angle B$ are supplementary angles.

Example 1   Find the complement of 39°.

Solution   To find the complement of 39°, subtract 39° from 90°.

$90° - 39° = 51°$

51° is the complement of 39°.

You Try It 1   Find the complement of 87°.

Your solution

Example 2   Find the supplement of 122°.

Solution   To find the supplement of 122°, subtract 122° from 180°.

$180° - 122° = 58°$

58° is the complement of 122°.

You Try It 2   Find the supplement of 87°.

Your solution

Example 3

For the figure at the right, find $m\angle AOB$.

Solution
$m\angle AOB$ is the difference between $m\angle AOC$ and $m\angle BOC$.

$m\angle AOB = 95° - 62° = 33°$

$m\angle AOB = 33°$

You Try It 3

For the figure at the right, find $m\angle x$.

Your solution

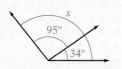

Solutions on p. S3

Copyright © Houghton Mifflin Company. All rights reserved.

Objective B   **To solve perimeter problems**

Perimeter is the distance around a plane figure. Perimeter is used in buying fencing for a yard, wood for the frame of a painting, and rain gutters around a house. The perimeter of a plane figure is the sum of the lengths of the sides of the figure. Formulas for four common geometric figures are given below.

A **triangle** is a three-sided plane figure.

$$\text{Perimeter} = \text{side } 1 + \text{side } 2 + \text{side } 3$$

An **isosceles triangle** has two sides of the same length. An **equilateral triangle** has all three sides the same length.

A **parallelogram** is a four-sided plane figure with opposite sides parallel.

A **rectangle** is a parallelogram that has four right angles.

$$\text{Perimeter} = 2 \cdot \text{length} + 2 \cdot \text{width}$$

A **square** is a rectangle with four equal sides.

$$\text{Perimeter} = 4 \cdot \text{side}$$

A **circle** is a plane figure in which all points are the same distance from point O, the **center of the circle.** The **diameter** of a circle is a line segment across the circle passing through the center. AB is a diameter of the circle at the right. The **radius** of a circle is a line segment from the center of the circle to a point on the circle. OC is a radius of the circle at the right. The perimeter of a circle is called its **circumference.**

$$\text{Diameter} = 2 \cdot \text{radius} \qquad \text{or} \qquad \text{Radius} = \frac{1}{2} \cdot \text{diameter}$$

$$\text{Circumference} = 2 \cdot \pi \cdot \text{radius} \qquad \text{or} \qquad \text{Circumference} = \pi \cdot \text{diameter},$$

where $\pi \approx 3.14$ or $\frac{22}{7}$.

HOW TO   The diameter of the circle is 25 cm. Find the radius of the circle.

$$\text{Radius} = \frac{1}{2} \cdot \text{diameter}$$

$$= \frac{1}{2} \cdot 25 = 12.5$$

The radius is 12.5 cm.

Copyright © Houghton Mifflin Company. All rights reserved.

Example 4 Find the perimeter of a rectangle with a width of 6 ft and a length of 18 feet.

Solution
Perimeter = 2 · length + 2 · width
$$= 2 \cdot 18 \text{ ft} + 2 \cdot 6 \text{ ft}$$
$$= 36 \text{ ft} + 12 \text{ ft} = 48 \text{ ft}$$

You Try It 4 Find the perimeter of a square that has a side of length 4.2 m.

Your solution

Example 5 Find the circumference of a circle with a radius of 23 cm. Use 3.14 for π.

Solution
Circumference = 2 · π · radius
$$\approx 2 \cdot 3.14 \cdot 23 \text{ cm}$$
$$= 144.44 \text{ cm}$$

You Try It 5 Find the circumference of a circle with a diameter of 5 in. Use 3.14 for π.

Your solution

Example 6

A chain-link fence costs $4.37 per foot. How much will it cost to fence a rectangular playground that is 108 ft wide and 195 ft long?

Strategy
To find the cost of the fence:

• Find the perimeter of the playground.
• Multiply the perimeter by the per-foot cost of the fencing.

Solution
Perimeter = 2 · length + 2 · width
$$= 2 \cdot 195 \text{ ft} + 2 \cdot 108 \text{ ft}$$
$$= 390 \text{ ft} + 216 \text{ ft} = 606 \text{ ft}$$
Cost = 606 × $4.37 = $2648.22
The cost is $2648.22.

You Try It 6

A metal strip is being installed around a circular table that has a diameter of 36 in. If the per-foot cost of the metal strip is $3.21, find the cost for the metal strip. Use 3.14 for π. Round to the nearest cent.

Your strategy

Your solution

Solutions on pp. S3–S4

Objective C   **To solve area problems**

Area is a measure of the amount of surface in a region. Area is used to describe the size of a rug, a farm, a house, or a national park.

Area is measured in square units.

A square that is 1 in. on each side has an area of 1 square inch, which is written 1 in^2.

1 in^2 1 in.

1 in.

Copyright © Houghton Mifflin Company. All rights reserved.

1 cm²   1 cm

1 cm

A square that is 1 cm on each side has an area of 1 square centimeter, which is written 1 cm².

Areas of common geometric figures are given by the following formulas.

Rectangle

$$\text{Area} = \text{length} \cdot \text{width}$$
$$= 3 \text{ cm} \cdot 2 \text{ cm}$$
$$= 6 \text{ cm}^2$$

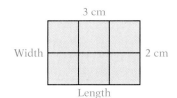

Square

$$\text{Area} = \text{side} \cdot \text{side}$$
$$= 2 \text{ cm} \cdot 2 \text{ cm}$$
$$= 4 \text{ cm}^2$$

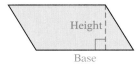

Parallelogram

The **base of a parallelogram** is one of the parallel sides. The **height of a parallelogram** is the distance between the base and the opposite parallel side. It is perpendicular to the base.

$$\text{Area} = \text{base} \cdot \text{height}$$
$$= 5 \text{ ft} \cdot 4 \text{ ft}$$
$$= 20 \text{ ft}^2$$

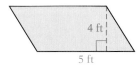

Circle

$$\text{Area} = \pi(\text{radius})^2$$
$$\approx 3.14(4 \text{ in.})^2 = 50.24 \text{ in}^2$$

TAKE NOTE

The height of a triangle is always perpendicular to the base. Sometimes it is necessary to extend the base so that a perpendicular line segment can be drawn. The extension is *not* part of the base.

Triangle

For the triangle at the right, the **base of the triangle** is *AB*; the **height of the triangle** is *CD*. Note that the height is perpendicular to the base.

$$\text{Area} = \frac{1}{2} \cdot \text{base} \cdot \text{height}$$

$$= \frac{1}{2} \cdot 5 \text{ in.} \cdot 4 \text{ in.} = 10 \text{ in}^2$$

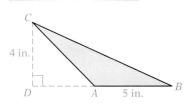

Copyright © Houghton Mifflin Company. All rights reserved.

Example 7 Find the area of a rectangle whose length is 8 in. and whose width is 6 in.

Solution Area = length × width
= 8 in. × 6 in. = 48 in²

You Try It 7 Find the area of a triangle whose base is 5 ft and whose height is 3 ft.

Your solution

Solution on p. S4

Example 8

Find the area of a circle whose diameter is 5 cm. Use 3.14 for π.

Solution

$$\text{Radius} = \frac{1}{2} \cdot \text{diameter}$$

$$= \frac{1}{2} \cdot 5 \text{ cm} = 2.5 \text{ cm}$$

$$\text{Area} = \pi \cdot (\text{radius})^2$$
$$\approx 3.14(2.5 \text{ cm})^2 = 19.625 \text{ cm}^2$$

You Try It 8

Find the area of a circle whose radius is 6 in. Use 3.14 for π.

Your solution

Example 9

Find the area of the parallelogram shown below.

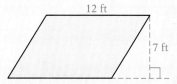

12 ft

7 ft

Solution
$$\text{Area} = \text{base} \cdot \text{height}$$
$$= 12 \text{ ft} \times 7 \text{ ft} = 84 \text{ ft}^2$$

You Try It 9

Find the area of the parallelogram shown below.

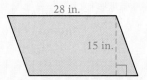

28 in.

15 in.

Your solution

Example 10

To conserve water during a drought, a city's water department is offering homeowners a rebate on their water bill of $1.27 per square foot of lawn that is removed from a yard and replaced with drought-resistant plants. What rebate would a homeowner receive who replaced a rectangular lawn area that is 15 ft wide and 25 ft long?

Strategy   To find the amount of the rebate:
• Find the area of the lawn.
• Multiply the area by the per-square-foot rebate.

Solution
$$\text{Area} = \text{length} \times \text{width}$$
$$= 25 \text{ ft} \times 15 \text{ ft} = 375 \text{ ft}^2$$

$$\text{Rebate} = 375 \times \$1.27 = \$476.25$$

The rebate is $476.25.

You Try It 10

An interior designer is choosing from two hallway rugs. A nylon rug costs $1.25 per square foot and a wool rug costs $1.93 per square foot. If the dimensions of the carpet are 4 ft by 15 ft, how much more expensive is the wool rug than the nylon rug?

Your strategy

Your solution

Solutions on p. S4

Copyright © Houghton Mifflin Company. All rights reserved.

1.8 Exercises

Objective A To find the measures of angles

1. How many degrees are in a right angle?

2. How many degrees are in a straight angle?

3. Find the complement of a 62° angle.

4. Find the complement of a 13° angle.

5. Find the supplement of a 48° angle.

6. Find the supplement of a 106° angle.

7. Find the complement of a 7° angle.

8. Find the complement of a 76° angle.

9. Find the supplement of an 89° angle.

10. Find the supplement of a 21° angle.

11. Angle *AOB* is a straight angle. Find *m*∠*AOC*.

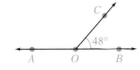

12. Angle *AOB* is a straight angle. Find *m*∠*COB*.

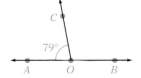

13. Find *m*∠*x*.

14. Find *m*∠*x*.

15. Find *m*∠*AOB*.

16. Find *m*∠*AOB*.

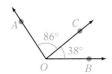

17. Find *m*∠*AOC*.

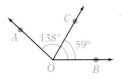

18. Find *m*∠*AOC*.

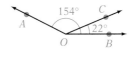

19. Find *m*∠*A*.

20. Find *m*∠*A*.

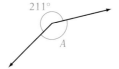

Objective B To solve perimeter problems

21. Find the perimeter of a triangle with sides 2.51 cm, 4.08 cm, and 3.12 cm.

22. Find the perimeter of a triangle with sides 4 ft 5 in., 5 ft 3 in., and 6 ft. 11 in.

Copyright © Houghton Mifflin Company. All rights reserved.

23. Find the perimeter of a rectangle whose length is 4 ft 8 in. and whose width is 2 ft 5 in.

24. Find the perimeter of a rectangle whose dimensions are 5 m by 8 m.

25. Find the perimeter of a square whose side is 13 in.

26. Find the perimeter of a square whose side is 34 cm.

27. Find the circumference of a circle whose radius is 21 cm. Use 3.14 for π.

28. Find the circumference of a circle whose radius is 3.4 m. Use 3.14 for π.

29. Find the circumference of a circle whose diameter is 1.2 m. Use 3.14 for π.

30. Find the circumference of a circle whose diameter is 15 in. Use 3.14 for π.

31. **Art** The wood framing for an art canvas costs $4.81 per foot. How much would the wood framing cost for a rectangular picture that measures 3 ft by 5 ft?

32. **Ceramics** A decorative mosaic tile is being installed on the border of a square wall behind a stove. If one side of the square measures 5 ft and the cost of installing the mosaic tile is $4.86 per foot, find the cost to install the decorative border.

33. **Sewing** To prevent fraying, a binding is attached to the outside of a circular rug whose radius is 3 ft. If the binding costs $1.05 per foot, find the cost of the binding. Use 3.14 for π.

34. **Landscaping** A drip irrigation system is installed around a circular flower garden that is 4 ft in diameter. If the irrigation system costs $2.46 per foot, find the cost to place the irrigation system around the flower garden. Use 3.14 for π.

> **Objective C**   **To solve area problems**

35. Find the area of a rectangle that measures 4 ft by 8 ft.

36. Find the area of a rectangle that measures 3.4 cm by 5.6 cm.

Copyright © Houghton Mifflin Company. All rights reserved.

37. Find the area of a parallelogram whose height is 14 cm and whose base is 27 cm.

38. Find the area of a parallelogram whose height is 7 ft and whose base is 18 ft.

39. Find the area of a circle whose radius is 4 in. Use 3.14 for π.

40. Find the area of a circle whose radius is 8.2 m. Use 3.14 for π.

41. Find the area of a square whose side measures 4.1 m.

42. Find the area of a square whose side measures 5 yd.

43. Find the area of a triangle whose height is 7 cm and whose base is 15 cm.

44. Find the area of a triangle whose height is 8 in. and whose base is 13 in.

45. Find the area of a circle whose diameter is 17 in. Use 3.14 for π.

46. Find the area of a circle whose diameter is 3.6 m. Use 3.14 for π.

47. **Landscaping**   A landscape architect recommends 0.1 gallon of water per day for each square foot of lawn. How many gallons of water should be used per day on a rectangular lawn area that is 33 ft by 42 ft?

48. **Interior Design**   One side of a square room measures 18 ft. How many square yards of carpet are necessary to carpet the room? *Hint*: $1 \text{ yd}^2 = 9 \text{ ft}^2$.

49. **Carpentry**   A circular, inlaid-wood design for a dining table cost $35 per square foot to build. If the radius of the design is 15 in., find the cost to build the design. Use 3.14 for π. Round to the nearest dollar. *Hint*: $144 \text{ in}^2 = 1 \text{ ft}^2$.

50. **Interior Design**   A circular stained glass window cost $48 per square foot to build. If the diameter of the window is 4 ft, find the cost to build the window. Round to the nearest dollar.

51. **Construction**   The cost of plastering the walls of a rectangular room that is 18 ft long, 14 ft wide, and 8 ft high is $2.56 per square foot. If 125 ft^2 are not plastered because of doors and windows, find the cost to plaster the room.

Copyright © Houghton Mifflin Company. All rights reserved.

52. **Interior Design**   A room is 12 ft long, 9 ft wide, and 9 ft high. Two adjacent walls of the rooms are going to be wallpapered using wallpaper that costs $15.25 per square yard. What is the cost to wallpaper the two walls? *Hint*: $1 \text{ yd}^2 = 9 \text{ ft}^2$.

APPLYING THE CONCEPTS

53. Find the perimeter and area of the figure. Use 3.14 for π.

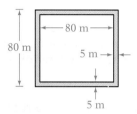

54. Find the perimeter and area of the figure.

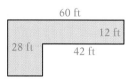

55. Find the outside perimeter and area of the shaded portion of the figure.

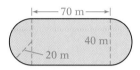

56. Find the area of the shaded portion of the figure.

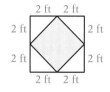

57. A trapezoid is a four-sided plane figure with two parallel sides. The area of a trapezoid is given by Area $= \frac{1}{2} \cdot$ height(base 1 + base 2). See the figure at the right.

    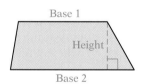

    a. Find the area of a trapezoid for which base 1 is 5 in., base 2 is 8 in., and the height is 6 in.

    b. Find the area of the trapezoid shown at the right.

    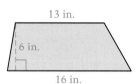

58. ✏   Draw parallelogram *ABCD* or one similar to it and then cut it out. Cut along the dotted line to form the shaded triangle. Slide the triangle so that the slanted side corresponds to the slanted side of the parallelogram as shown. Explain how this demonstrates that the area of a parallelogram is the product of the base and the height.

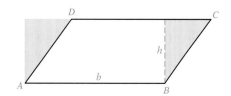

59. ✏   Explain how to draw the height of a triangle.

Copyright © Houghton Mifflin Company. All rights reserved.

Focus on Problem Solving

Inductive Reasoning   Suppose you take 9 credit hours each semester. The total number of credit hours you have taken at the end of each semester can be described by a list of numbers.

$$9, 18, 27, 36, 45, 54, 63, \ldots$$

The list of numbers that indicates the total credit hours is an ordered list of numbers called a **sequence.** Each number in a sequence is called a **term** of the sequence. The list is ordered because the position of a number in the list indicates the semester in which that number of credit hours has been taken. For example, the 7th term of the sequence is 63, and a total of 63 credit hours have been taken after the 7th semester.

Assuming the pattern is continued, find the next three numbers in the pattern

$$-6, -10, -14, -18, \ldots$$

This list of numbers is a sequence. The first step in solving this problem is to observe the pattern in the list of numbers. In this case, each number in the list is 4 less than the previous number. The next three numbers are $-22, -26, -30$.

This process of discovering the pattern in a list of numbers is inductive reasoning. **Inductive reasoning** involves making generalizations from specific examples; in other words, we reach a conclusion by making observations about particular facts or cases.

Try the following exercises. Each exercise requires inductive reasoning.

For Exercises 1 to 4, name the next two terms in the sequence.

1. 1, 3, 5, 7, 1, 3, 5, 7, 1, …

2. 1, 4, 2, 5, 3, 6, 4, …

3. 1, 2, 4, 7, 11, 16, …

4. A, B, C, G, H, I, M, …

For Exercises 5 and 6, draw the next shape in the sequence.

5.

6. |• ‖• ‖•• ‖‖•• ‖‖•••

For Exercises 7 and 8, solve.

7. Convert $\frac{1}{11}, \frac{2}{11}, \frac{3}{11}, \frac{4}{11}$, and $\frac{5}{11}$ to decimals. Then use the pattern you observe to convert $\frac{6}{11}, \frac{7}{11}$, and $\frac{9}{11}$ to decimals.

8. Convert $\frac{1}{33}, \frac{2}{33}, \frac{4}{33}, \frac{5}{33}$, and $\frac{7}{33}$ to decimals. Then use the pattern you observe to convert $\frac{8}{33}, \frac{13}{33}$, and $\frac{19}{33}$ to decimals.

Copyright © Houghton Mifflin Company. All rights reserved.

Projects and Group Activities

The +/− Key on a Calculator

Using your calculator to simplify numerical expressions sometimes requires use of the +/− key or, on some calculators, the negative key, which is frequently shown as (−) . To enter −4:

- For those calculators with +/− , press 4 and then +/− .
- For those calculators with (−) , press (−) and then 4.

Here are the keystrokes for evaluating the expression $3(-4) - (-5)$.

Calculators with +/− key: 3 × 4 +/− − 5 +/− =

Calculators with (−) key: 3 × (−) 4 − (−) 5 =

This example illustrates that calculators make a distinction between negative and minus. To perform the operation $3 - (-3)$, you cannot enter 3 − − 3. This would result in 0, which is not the correct answer. You must enter

3 − 3 +/− =   or   3 − (−) 3 =

For Exercises 1 to 6, use a calculator to evaluate.

1. $-16 \div 2$ **2.** $3(-8)$ **3.** $47 - (-9)$

4. $-50 - (-14)$ **5.** $4 - (-3)^2$ **6.** $-8 + (-6)^2 - 7$

Moving Averages

Objective 1.3C on page 21 describes how to find the moving average of a stock. Use this method to calculate the 5-day moving average for at least three different stocks. Discuss and compare the results for the different stocks.

For this project, you will need to use stock tables, which are printed in the business section of major newspapers. Your college library should have copies of these publications. In a stock table, the column headed "Chg." provides the change in the price of a share of the stock; that is, it gives the difference between the closing price for the day shown and the closing price for the previous day. The symbol + indicates that the change was an increase in price; the symbol − indicates that the change was a decrease in price.

www.fedstats.gov

Information on the history of the federal budget can be found on the web site **www.fedstats.gov.** Find tables entitled "Federal Budget." In the table, find the column that lists each year's surplus or deficit. You will see that a negative sign (−) is used to show a deficit. Note that it states near the top of the screen that the figures in the table are in millions of dollars.

1. During which years shown in the table was there a surplus?

2. During which year was the deficit the greatest?

3. Find the difference between the surplus or deficit this year and the surplus or deficit 5 years ago.

4. What is the difference between the surplus or deficit this year and the surplus or deficit a decade ago?

5. Determine what two numbers in the table are being subtracted in each row in order to arrive at the number in the surplus or deficit column.

6. Describe the trend of the federal deficit over the last 10 years.

Copyright © Houghton Mifflin Company. All rights reserved.

Chapter 1 Summary

Key Words	Examples
The set of *natural numbers* is {1, 2, 3, 4, 5, . . .}. The set of *integers* is {. . . , −3, −2, −1, 0, 1, 2, 3, . . .}. [1.1A, p. 3]	
A number *a is less than* a number *b*, written $a < b$, if *a* is to the left of *b* on a number line. A number *a is greater than* a number *b*, written $a > b$, if *a* is to the right of *b* on a number line. The symbol ≤ means *is less than or equal to*. The symbol ≥ means *is greater than or equal to*. [1.1A, p. 4]	$-5 < -3$ $\quad$ $9 > 0$ $3 \le 3$ $\quad$ $4 \le 7$ $5 \ge 5$ $\quad$ $-6 \ge -9$
Two numbers that are the same distance from zero on the number line but on opposite sides of zero are *opposite numbers* or *opposites*. [1.1B, p. 5]	7 and −7 are opposites. $-\frac{3}{4}$ and $\frac{3}{4}$ are opposites.
The *absolute value* of a number is its distance from 0 on the number line. [1.1B, p. 5]	$\lvert 5 \rvert = 5$ $\quad$ $\lvert -2.3 \rvert = 2.3$ $\quad$ $\lvert 0 \rvert = 0$
An expression of the form a^n is in *exponential form*. The *base* is *a* and the *exponent* is *n*. [1.4A, p. 24]	5^4 is an exponential expression. The base is 5 and the exponent is 4.
A natural number greater than 1 is a *prime number* if its only factors are 1 and the number. [1.5B, p. 31]	3, 17, 23, and 97 are prime numbers.
The *prime factorization* of a number is the expression of the number as a product of its prime factors. [1.5B, p. 31]	$2^3 \cdot 3^2 \cdot 7$ is the prime factorization of 504.
The *least common multiple (LCM)* of two or more numbers is the smallest number that is a multiple of all the numbers. [1.5C, p. 32]	The LCM of 4, 8, and 12 is 24.
The *greatest common factor (GCF)* of two or more numbers is the greatest number that divides evenly into all of the numbers. [1.5C, p. 32]	The GCF of 4, 8, and 12 is 4.
A *rational number* (or fraction) is a number that can be written in the form $\frac{a}{b}$, where *a* and *b* are integers and $b \neq 0$. A fraction is in *simplest form* when there are no common factors in the numerator and denominator. A rational number can be represented as a *terminating* or *repeating decimal*. [1.6A, pp. 35–36]	$\frac{3}{8}, -\frac{9}{2}$ and 4 are rational numbers written in simplest form. $\frac{3}{8}$ is a fraction in simplest form. 1.13 and $0.4\overline{73}$ are also rational numbers.
An *irrational number* is a number that has a decimal representation that never terminates or repeats. [1.6A, p. 36]	π, $\sqrt{2}$, and 1.34334333433334 . . . are irrational numbers.

Copyright © Houghton Mifflin Company. All rights reserved.

The rational numbers and the irrational numbers taken together are the *real numbers*. [1.6A, p. 36]

$\frac{3}{8}, -\frac{9}{2}, 4, 1.13, 0.4\overline{73}, \pi, \sqrt{2}$, and 1.34334333433334 . . . are real numbers.

The *reciprocal of a fraction* is the fraction with the numerator and denominator interchanged. [1.7B, p. 48]

The reciprocal of $\frac{5}{6}$ is $\frac{6}{5}$.

The reciprocal of $-\frac{1}{3}$ is $-\frac{3}{1}$ or -3.

Percent means "parts of 100." [1.7C, p. 50]

72% means 72 of 100 equal parts.

A *plane* is a flat surface that extends indefinitely. A *line* extends indefinitely in two directions in a plane. A *ray* starts at a point and extends indefinitely in one direction. A *line segment* is part of a line and has two endpoints. [1.8A, p. 57]

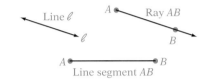

Lines in a plane can be parallel or intersect. *Parallel lines* never meet. The distance between parallel lines in a plane is always the same. *Intersecting lines* cross at a point in the plane. [1.8A, p. 57]

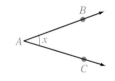

An *angle* is formed when two rays start from the same point. The point at which the rays meet is called the *vertex* of the angle. An angle can be measured in *degrees*. The measure of an angle is symbolized by $m\angle$. [1.8A, p. 57]

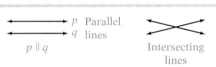

This angle can be named $\angle A$, $\angle BAC$, $\angle CAB$, or $\angle x$.

A *right angle* has a measure of 90°. *Perpendicular lines* are intersecting lines that form right angles. *Complementary angles* are two angles whose sum is 90°. A *straight angle* has a measure of 180°. *Supplementary angles* are two angles whose sum is 180°. [1.8A, p. 58]

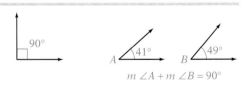

$\angle A$ and $\angle B$ are complementary angles.

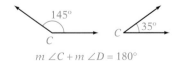

$m\angle C + m\angle D = 180°$

$\angle C$ and $\angle D$ are supplementary angles.

A *circle* is a plane figure in which all points are the same distance from point O, the *center* of the circle. The *diameter* of a circle is a line segment across the circle passing through the center. The *radius* of a circle is a line segment from the center of the circle to a point on the circle. The perimeter of a circle is called its *circumference*. [1.8B, p. 59]

Copyright © Houghton Mifflin Company. All rights reserved.

Essential Rules and Procedures	**Examples**
To add two numbers with the same sign, add the absolute values of the numbers. Then attach the sign of the addends. [1.2A, p. 9]	$7 + 15 = 22$ $-7 + (-15) = -22$
To add two numbers with different signs, find the absolute value of each number. Subtract the smaller of the two numbers from the larger. Then attach the sign of the number with the larger absolute value. [1.2A, p. 9]	$7 + (-15) = -8$ $-7 + 15 = 8$
To subtract one number from another, add the opposite of the second number to the first number. [1.2B, p. 10]	$7 - 19 = 7 + (-19) = -12$ $-6 - (-13) = -6 + 13 = 7$
To multiply two numbers with the same sign, multiply the absolute values of the numbers. The product is positive. [1.3A, p. 17]	$7 \cdot 8 = 56$ $-7(-8) = 56$
To multiply two numbers with different signs, multiply the absolute values of the numbers. The product is negative. [1.3A, p. 17]	$-7 \cdot 8 = -56$ $7(-8) = -56$
To divide two numbers with the same sign, divide the absolute values of the numbers. The quotient is positive. [1.3B, p. 19]	$54 \div 9 = 6$ $(-54) \div (-9) = 6$
To divide two numbers with different signs, divide the absolute values of the numbers. The quotient is negative. [1.3B, p. 19]	$(-54) \div 9 = -6$ $54 \div (-9) = -6$

Properties of Zero and One in Division [1.3B, p. 20]

If $a \neq 0$, $\dfrac{0}{a} = 0$.

$\dfrac{0}{-5} = 0$

If $a \neq 0$, $\dfrac{a}{a} = 1$.

$\dfrac{-12}{-12} = 1$

$\dfrac{a}{1} = a$

$\dfrac{7}{1} = 7$

$\dfrac{a}{0}$ is undefined.

$\dfrac{8}{0}$ is undefined.

Order of Operations Agreement [1.4B, p. 25]

Step 1   Perform operations inside grouping symbols. Grouping symbols include parentheses ( ), brackets [ ], braces { }, and the fraction bar.

Step 2   Simplify exponential expressions.

Step 3   Do multiplication and division as they occur from left to right.

Step 4   Do addition and subtraction as they occur from left to right.

$50 \div (-5)^2 + 2(7 - 16)$
$= 50 \div (-5)^2 + 2(-9)$
$= 50 \div 25 + 2(-9)$
$= 2 + (-18)$
$= -16$

Copyright © Houghton Mifflin Company. All rights reserved.

To add two fractions with the same denominator, add the numerators and place the sum over the common denominator. [1.6B, p. 37]

$$\frac{7}{10} + \frac{1}{10} = \frac{7 + 1}{10} = \frac{8}{10} = \frac{4}{5}$$

To subtract two fractions with the same denominator, subtract the numerators and place the difference over the common denominator. [1.6C, p. 39]

$$\frac{7}{10} - \frac{1}{10} = \frac{7 - 1}{10} = \frac{6}{10} = \frac{3}{5}$$

To multiply two fractions, place the product of the numerators over the product of the denominators. [1.7A, p. 46]

$$-\frac{2}{3} \cdot \frac{5}{6} = -\frac{2 \cdot 5}{3 \cdot 6} = -\frac{10}{18} = -\frac{5}{9}$$

To divide two fractions, multiply the dividend by the reciprocal of the divisor. [1.7B, p. 48]

$$-\frac{4}{5} \div \frac{2}{3} = -\frac{4}{5} \cdot \frac{3}{2} = -\frac{2 \cdot 2 \cdot 3}{5 \cdot 2} = -\frac{6}{5}$$

To write a percent as a fraction, remove the percent sign and multiply by $\frac{1}{100}$. [1.7C, p. 50]

$$60\% = 60\left(\frac{1}{100}\right) = \frac{60}{100} = \frac{3}{5}$$

To write a percent as a decimal, remove the percent sign and multiply by 0.01. [1.7C, p. 50]

$73\% = 73(0.01) = 0.73$
$1.3\% = 1.3(0.01) = 0.013$

To write a decimal or a fraction as a percent, multiply by 100%. [1.7C, p. 50]

$0.3 = 0.3(100\%) = 30\%$
$\frac{5}{8} = \frac{5}{8}(100\%) = \frac{500}{8}\% = 62.5\%$

Diameter $= 2 \cdot$ radius   **Radius** $= \frac{1}{2} \cdot$ diameter

[1.8B, p. 59]

Find the diameter of a circle whose radius is 10 in.

Diameter $= 2 \cdot$ radius
$ = 2(10 \text{ in.}) = 20 \text{ in.}$

Perimeter is the distance around a plane figure. [1.8B, p. 59]

Triangle: Perimeter $=$ side 1 $+$ side 2 $+$ side 3
Rectangle: Perimeter $= 2 \cdot$ length $+ 2 \cdot$ width
Square: Perimeter $= 4 \cdot$ side
Circle: Circumference $= 2 \cdot \pi \cdot$ radius

Find the perimeter of a rectangle whose width is 12 m and whose length is 15 m.

perimeter $= 2 \cdot 15 \text{ m} + 2 \cdot 12 \text{ m} = 54 \text{ m}$

Find the circumference of a circle whose radius is 3 in. Use 3.14 for π.

circumference $= 2 \cdot \pi \cdot 3 \text{ in.} \approx 18.84 \text{ in.}$

Area is a measure of the amount of surface in a region. [1.8C, pp. 60–61]

Triangle: Area $= \frac{1}{2} \cdot$ base $\cdot$ height
Rectangle: Area $=$ length $\cdot$ width
Square: Area $=$ side $\cdot$ side
Parallelogram: Area $=$ base $\cdot$ height
Circle: Area $= \pi(\text{radius})^2$

Find the area of a triangle whose base is 13 m and whose height is 11 m.

area $= \frac{1}{2} \cdot 13 \text{ m} \cdot 11 \text{ m} = 71.5 \text{ m}^2$

Find the area of a circle whose radius is 9 cm.

area $= \pi \cdot (9 \text{ cm})^2 \approx 254.34 \text{ cm}^2$

Copyright © Houghton Mifflin Company. All rights reserved.

Chapter 1 Review Exercises

1. Add: $-13 + 7$

2. Write $\dfrac{7}{25}$ as a decimal.

3. Evaluate -5^2.

4. Evaluate $5 - 2^2 + 9$.

5. Find $m\angle AOB$ for the figure at the right.

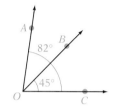

6. Write 6.2% as a decimal.

7. Multiply: $(-6)(7)$

8. Simplify: $\dfrac{1}{3} - \dfrac{1}{6} + \dfrac{5}{12}$

9. Find the complement of a 56° angle.

10. Given $A = \{-4, 0, 11\}$, which elements of set A are less than -1?

11. Find all of the factors of 56.

12. Subtract: $5.17 - 6.238$

13. Write $\dfrac{5}{8}$ as a percent.

14. Write $\dfrac{2}{15}$ as a decimal. Place a bar over the repeating digits of the decimal.

15. Subtract: $9 - 13$

16. What is $\dfrac{2}{5}$ less than $\dfrac{4}{15}$?

17. Find the additive inverse of -4.

18. Find the area of a triangle whose base is 4 cm and whose height is 9 cm.

19. Divide: $-100 \div 5$

20. Write $79\dfrac{1}{2}\%$ as a fraction.

21. Find the prime factorization of 280.

22. Evaluate $-3^2 + 4[18 + (12 - 20)]$.

Copyright © Houghton Mifflin Company. All rights reserved.

23. Add: $-3 + (-12) + 6 + (-4)$

24. Find the sum of $\frac{4}{5}$ and $-\frac{3}{8}$.

25. Write $\frac{19}{35}$ as a percent. Write the remainder in fractional form.

26. Find the area of a circle whose diameter is 6 m. Use 3.14 for π.

27. Multiply: $4.32(-1.07)$

28. Evaluate $-|-5|$.

29. Subtract: $16 - (-3) - 18$

30. Divide: $-\frac{18}{35} \div \frac{27}{28}$

31. Find the supplement of a 28° angle.

32. Find the perimeter of a rectangle whose length is 12 in. and whose width is 10 in.

33. Place the correct symbol, $>$ or $<$, between the two numbers.
$-|6| \quad |-10|$

34. Evaluate $\dfrac{5^2 + 11}{2^2 + 5} \div (2^3 - 2^2)$.

35. **Education**   To discourage random guessing on a multiple-choice exam, a professor assigns 6 points for a correct answer, -4 points for an incorrect answer, and -2 points for leaving a question blank. What is the score for a student who had 21 correct answers, 5 incorrect answers, and left 4 questions blank?

36. 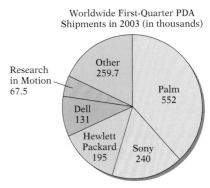 **Business**   The graph at the right shows the number (in thousands) of personal digital assistants (PDAs) shipped worldwide in the first quarter of 2003. What percent of the total number of PDAs shipped did Palm ship? Round to the nearest tenth of a percent.

Worldwide First-Quarter PDA
Shipments in 2003 (in thousands)

Research
in Motion
67.5

Other
259.7

Palm
552

Dell
131

Hewlett
Packard
195

Sony
240

Source: Gartner, Inc.

37. **Chemistry**   The temperature at which mercury boils is 357°C. The temperature at which mercury freezes is $-39°$C. Find the difference between the boiling point and the freezing point of mercury.

38. **Landscaping**   A landscape company is proposing to replace a rectangular flower bed that measures 8 ft by 12 ft with sod that costs $2.51 per square foot. Find the cost to replace the flower bed with the sod.

Copyright © Houghton Mifflin Company. All rights reserved.

Chapter 1 Test

1. Divide: $-561 \div (-33)$

2. Write $\frac{5}{6}$ as a percent. Write the remainder in fractional form.

3. Find the complement of a 28° angle.

4. Multiply: $6.02(-0.89)$

5. Subtract: $16 - 30$

6. Write $37\frac{1}{2}\%$ as a fraction.

7. Subtract: $-\frac{5}{6} - \left(-\frac{7}{8}\right)$

8. Evaluate $\frac{-10 + 2}{2 + (-4)} \div 2 + 6$.

9. Multiply: $-5(-6)(3)$

10. Find the circumference of a circle whose diameter is 27 in. Use 3.14 for π.

11. Evaluate $(-3^3) \cdot 2^2$.

12. Find the area of a parallelogram whose base is 10 cm and whose height is 9 cm.

13. Place the correct symbol, $<$ or $>$, between the two numbers.
$-2 \quad -40$

14. What is $\frac{2}{5}$ more than $-\frac{3}{4}$?

Copyright © Houghton Mifflin Company. All rights reserved.

15. Evaluate $-|-4|$.

16. Write 45% as a fraction and as a decimal.

17. Add: $-22 + 14 + (-8)$

18. Multiply: $-4 \cdot 12$

19. Find the prime factorization of 990.

20. Evaluate $16 \div 2[8 - 3(4 - 2)] + 1$.

21. Subtract: $16 - (-30) - 42$

22. Divide: $\dfrac{5}{12} \div \left(-\dfrac{5}{6}\right)$

23. Find $m\angle x$ for the figure at the right.

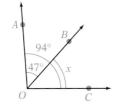

24. Evaluate $3^2 - 4 + 20 \div 5$.

25. Write $\dfrac{7}{9}$ as a decimal. Place a bar over the repeating digit of the decimal.

26. **Finance**   The table below shows the first-quarter profits and losses for 2003 for five companies in the computer chip industry. Profits are shown as positive numbers; losses are shown as negative numbers. One-quarter of a year is 3 months.
  a. If earnings were to continue throughout the year at the same level, what would be the annual profit or loss for LSI Logic Corp.?
  b. For the quarter shown, what was the average monthly profit or loss for National Semiconductor Corp.? Round to the nearest thousand dollars.

Computer Chip Company	First Quarter 2003 Profits (in millions of dollars)
Advanced Micro Devices, Inc.	−125.5
LSI Logic Corp.	−114.4
Micron Technology, Inc.	−183.7
Motorola, Inc.	171.0
National Semiconductor Corp.	−0.2

Source: Yahoo! Finance

27. **Recreation**   The recreation department for a city is enclosing a rectangular playground that measures 150 ft by 200 ft with new fencing that costs $6.52 per foot. Find the cost of the new fencing.

Copyright © Houghton Mifflin Company. All rights reserved.

2 Variable Expressions

Have you ever purchased a product online? How satisfied were you with the experience? A recent survey conducted by Consumer Internet Barometer found that approximately 25% of people who had purchased a product online were extremely satisfied with their experience. See **Exercise 64 on page 100.** In this exercise, you are asked to express the number of people who were extremely satisfied with their online purchase in terms of the number of people who purchased a product on line. To do this, you must use a variable expression. Variable expressions are the topic of this chapter.

Copyright © Houghton Mifflin Company. All rights reserved.

OBJECTIVES

Section 2.1

A To evaluate a variable expression

Section 2.2

A To simplify a variable expression using the Properties of Addition

B To simplify a variable expression using the Properties of Multiplication

C To simplify a variable expression using the Distributive Property

D To simplify general variable expressions

Section 2.3

A To translate a verbal expression into a variable expression, given the variable

B To translate a verbal expression into a variable expression and then simplify

C To translate application problems

Need help? For online student resources, such as section quizzes, visit this textbook's website at **math.college.hmco.com/students.**

Do these exercises to prepare for Chapter 2.

1. Subtract: $-12 - (-15)$

2. Divide: $-36 \div (-9)$

3. Add: $-\dfrac{3}{4} + \dfrac{5}{6}$

4. What is the reciprocal of $-\dfrac{9}{4}$?

5. Divide: $-\dfrac{3}{4} \div \left(-\dfrac{5}{2}\right)$

6. Evaluate: -2^4

7. Evaluate: $\left(\dfrac{2}{3}\right)^3$

8. Evaluate: $3 \cdot 4^2$

9. Evaluate: $7 - 2 \cdot 3$

10. Evaluate: $5 - 7(3 - 2^2)$

GO FIGURE • • •

Two fractions are inserted between $\dfrac{1}{4}$ and $\dfrac{1}{2}$ so that the difference between any two successive fractions is the same. Find the sum of the four fractions.

Copyright © Houghton Mifflin Company. All rights reserved.

2.1 Evaluating Variable Expressions

Objective A   To evaluate a variable expression

Copyright © Houghton Mifflin Company. All rights reserved.

Study Tip

Before you begin a new chapter, you should take some time to review previously learned skills. One way to do this is to complete the Prep Test. See page 78. This test focuses on the particular skills that will be required for the new chapter.

Point of Interest

Historical manuscripts indicate that mathematics is at least 4000 years old. Yet it was only 400 years ago that mathematicians started using variables to stand for numbers. The idea that a letter can stand for some number was a critical turning point in mathematics. Today, x is used by most nations as the standard letter for a single unknown. In fact, x-rays were so named because the scientists who discovered them did not know what they were and thus labeled them the "unknown rays" or x-rays.

Often we discuss a quantity without knowing its exact value—for example, the price of gold next month, the cost of a new automobile next year, or the tuition cost for next semester. Recall that a letter of the alphabet, called a **variable,** is used to stand for a quantity that is unknown or that can change, or *vary*. An expression that contains one or more variables is called a **variable expression.**

A variable expression is shown at the right. The expression can be rewritten by writing subtraction as the addition of the opposite.

$$3x^2 - 5y + 2xy - x - 7$$

$$3x^2 + (-5y) + 2xy + (-x) + (-7)$$

Note that the expression has 5 addends. The **terms** of a variable expression are the addends of the expression. The expression has 5 terms.

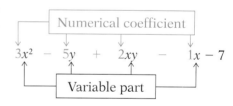

The terms $3x^2$, $-5y$, $2xy$, and $-x$ are **variable terms.**

The term -7 is a **constant term,** or simply a **constant.**

Each variable term is composed of a **numerical coefficient** and a **variable part** (the variable or variables and their exponents).

When the numerical coefficient is 1 or -1, the 1 is usually not written ($x = 1x$ and $-x = -1x$).

Variable expressions occur naturally in science. In a physics lab, a student may discover that a weight of 1 pound will stretch a spring $\frac{1}{2}$ inch. Two pounds will stretch the spring 1 inch. By experimenting, the student can discover that the distance the spring will stretch is found by multiplying the weight by $\frac{1}{2}$. By letting W represent the weight attached to the spring, the student can represent the distance the spring stretches by the variable expression $\frac{1}{2}W$.

With a weight of W pounds, the spring will stretch $\frac{1}{2} \cdot W = \frac{1}{2}W$ inches.

With a weight of 10 pounds, the spring will stretch $\frac{1}{2} \cdot 10 = 5$ inches. The number 10 is called the **value of the variable** W.

With a weight of 3 pounds, the spring will stretch $\frac{1}{2} \cdot 3 = 1\frac{1}{2}$ inches.

Replacing each variable by its value and then simplifying the resulting numerical expression is called **evaluating a variable expression.**

Integrating Technology

See the Keystroke Guide: *Evaluating Variable Expressions* for instructions on using a graphing calculator to evaluate variable expressions.

HOW TO Evaluate $ab - b^2$ when $a = 2$ and $b = -3$.

Replace each variable in the expression by its value. Then use the Order of Operations Agreement to simplify the resulting numerical expression.

$ab - b^2$

$2(-3) - (-3)^2 = -6 - 9$
$= -15$

When $a = 2$ and $b = -3$, the value of $ab - b^2$ is -15.

Example 1 Name the variable terms of the expression $2a^2 - 5a + 7$.

Solution $2a^2$ and $-5a$

You Try It 1 Name the constant term of the expression $6n^2 + 3n - 4$.

Your solution

Example 2 Evaluate $x^2 - 3xy$ when $x = 3$ and $y = -4$.

Solution
$x^2 - 3xy$
$3^2 - 3(3)(-4) = 9 - 3(3)(-4)$
$\qquad = 9 - 9(-4)$
$\qquad = 9 - (-36)$
$\qquad = 9 + 36 = 45$

You Try It 2 Evaluate $2xy + y^2$ when $x = -4$ and $y = 2$.

Your solution

Example 3 Evaluate $\dfrac{a^2 - b^2}{a - b}$ when $a = 3$ and $b = -4$.

Solution
$\dfrac{a^2 - b^2}{a - b}$

$\dfrac{3^2 - (-4)^2}{3 - (-4)} = \dfrac{9 - 16}{3 - (-4)}$

$\qquad = \dfrac{-7}{7} = -1$

You Try It 3 Evaluate $\dfrac{a^2 + b^2}{a + b}$ when $a = 5$ and $b = -3$.

Your solution

Example 4 Evaluate $x^2 - 3(x - y) - z^2$ when $x = 2$, $y = -1$, and $z = 3$.

Solution
$x^2 - 3(x - y) - z^2$
$2^2 - 3[2 - (-1)] - 3^2$
$= 2^2 - 3(3) - 3^2$
$= 4 - 3(3) - 9$
$= 4 - 9 - 9$
$= -5 - 9$
$= -14$

You Try It 4 Evaluate $x^3 - 2(x + y) + z^2$ when $x = 2$, $y = -4$, and $z = -3$.

Your solution

Solutions on p. S4

Copyright © Houghton Mifflin Company. All rights reserved.

2.1 Exercises

Objective A   **To evaluate a variable expression**

For Exercises 1 to 3, name the terms of the variable expression. Then underline the constant term.

1. $2x^2 + 5x - 8$

2. $-3n^2 - 4n + 7$

3. $6 - a^4$

For Exercises 4 to 6, name the variable terms of the expression. Then underline the variable part of each term.

4. $9b^2 - 4ab + a^2$

5. $7x^2y + 6xy^2 + 10$

6. $5 - 8n - 3n^2$

For Exercises 7 to 9, name the coefficients of the variable terms.

7. $x^2 - 9x + 2$

8. $12a^2 - 8ab - b^2$

9. $n^3 - 4n^2 - n + 9$

10. What is the numerical coefficient of a variable term?

11. Explain the meaning of the phrase "evaluate a variable expression."

For Exercises 12 to 32, evaluate the variable expression when $a = 2$, $b = 3$, and $c = -4$.

12. $3a + 2b$

13. $a - 2c$

14. $-a^2$

15. $2c^2$

16. $-3a + 4b$

17. $3b - 3c$

18. $b^2 - 3$

19. $-3c + 4$

20. $16 \div (2c)$

21. $6b \div (-a)$

22. $bc \div (2a)$

23. $b^2 - 4ac$

24. $a^2 - b^2$

25. $b^2 - c^2$

26. $(a + b)^2$

27. $a^2 + b^2$

28. $2a - (c + a)^2$

29. $(b - a)^2 + 4c$

30. $b^2 - \dfrac{ac}{8}$

31. $\dfrac{5ab}{6} - 3cb$

32. $(b - 2a)^2 + bc$

Copyright © Houghton Mifflin Company. All rights reserved.

For Exercises 33 to 50, evaluate the variable expression when $a = -2$, $b = 4$, $c = -1$, and $d = 3$.

33. $\dfrac{b + c}{d}$

34. $\dfrac{d - b}{c}$

35. $\dfrac{2d + b}{-a}$

36. $\dfrac{b + 2d}{b}$

37. $\dfrac{b - d}{c - a}$

38. $\dfrac{2c - d}{-ad}$

39. $(b + d)^2 - 4a$

40. $(d - a)^2 - 3c$

41. $(d - a)^2 \div 5$

42. $3(b - a) - bc$

43. $\dfrac{b - 2a}{bc^2 - d}$

44. $\dfrac{b^2 - a}{ad + 3c}$

45. $\dfrac{1}{3}d^2 - \dfrac{3}{8}b^2$

46. $\dfrac{5}{8}a^4 - c^2$

47. $\dfrac{-4bc}{2a - b}$

48. $-\dfrac{3}{4}b + \dfrac{1}{2}(ac + bd)$

49. $-\dfrac{2}{3}d - \dfrac{1}{5}(bd - ac)$

50. $(b - a)^2 - (d - c)^2$

51. The value of z is the value of $a^2 - 2a$ when $a = -3$. Find the value of z^2.

52. The value of a is the value of $3x^2 - 4x - 5$ when $x = -2$. Find the value of $3a - 4$.

53. The value of c is the value of $a^2 + b^2$ when $a = 2$ and $b = -2$. Find the value of $c^2 - 4$.

APPLYING THE CONCEPTS

For Exercises 54 to 57, evaluate the following expressions for $x = 2$, $y = 3$, and $z = -2$.

54. $3^x - x^3$

55. z^x

56. $x^x - y^y$

57. $y^{(x^2)}$

58. For each of the following, determine the first natural number x, greater than 1, for which the second expression is larger than the first.
   a. $x^3, 3^x$ **b.** $x^4, 4^x$ **c.** $x^5, 5^x$ **d.** $x^6, 6^x$

59. On the basis of your answer to Exercise 58, make a conjecture that appears to be true about the two expressions x^n and n^x, where $n = 3, 4, 5, 6, 7, \ldots$ and x is a natural number greater than 1.

Copyright © Houghton Mifflin Company. All rights reserved.

2.2 Simplifying Variable Expressions

Copyright © Houghton Mifflin Company. All rights reserved.

Objective A   **To simplify a variable expression using the Properties of Addition**

Like terms of a variable expression are terms with the same variable part. (Because $x^2 = x \cdot x$, x^2 and x are not like terms.)

Constant terms are like terms. 4 and 9 are like terms.

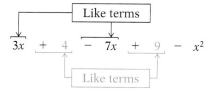

To simplify a variable expression, use the Distributive Property to combine like terms by adding the numerical coefficients. The variable part remains unchanged.

> ***TAKE NOTE***
>
> Here is an example of the Distributive Property with just numbers.
> $2(5 + 9) = 2(5) + 2(9)$
> $= 10 + 18 = 28$
> This is the same result we would obtain using the Order of Operations Agreement.
> $2(5 + 9) = 2(14) = 28$
> The usefulness of the Distributive Property will become more apparent as we explore variable expressions.

Distributive Property

If a, b, and c are real numbers, then $a(b + c) = ab + ac$.

The Distributive Property can also be written $ba + ca = (b + c)a$. This form is used to simplify a variable expression.

To simplify $2x + 3x$, use the Distributive Property to add the numerical coefficients of the like variable terms. This is called **combining like terms.**

$$2x + 3x = (2 + 3)x$$
$$= 5x$$

HOW TO   Simplify: $5y - 11y$

$5y - 11y = (5 - 11)y$ • Use the **Distributive Property**.
$= -6y$

> ***TAKE NOTE***
>
> Simplifying an expression means combining like terms. A constant term (5) and a variable term ($7p$) are not like terms and therefore cannot be combined.

HOW TO   Simplify: $5 + 7p$

The terms 5 and $7p$ are not like terms.

The expression $5 + 7p$ is in simplest form.

The Associative Property of Addition

If a, b, and c are real numbers, then $(a + b) + c = a + (b + c)$.

When three or more terms are added, the terms can be grouped (with parentheses, for example) in any order. The sum is the same. For example,

$$(5 + 7) + 15 = 5 + (7 + 15) \qquad (3x + 5x) + 9x = 3x + (5x + 9x)$$
$$12 + 15 = 5 + 22 \qquad\qquad 8x + 9x = 3x + 14x$$
$$27 = 27 \qquad\qquad\qquad 17x = 17x$$

> **The Commutative Property of Addition**
>
> If a and b are real numbers, then $a + b = b + a$.

When two like terms are added, the terms can be added in either order. The sum is the same. For example,

$$15 + (-28) = (-28) + 15 \qquad\qquad 2x + (-4x) = -4x + 2x$$
$$-13 = -13 \qquad\qquad\qquad\qquad -2x = -2x$$

> **The Addition Property of Zero**
>
> If a is a real number, then $a + 0 = 0 + a = a$.

The sum of a term and zero is the term. For example,

$$-9 + 0 = 0 + (-9) = -9 \qquad\qquad 5x + 0 = 0 + 5x = 5x$$

> **The Inverse Property of Addition**
>
> If a is a real number, then $a + (-a) = (-a) + a = 0$.

The sum of a term and its opposite is zero. Recall that the opposite of a number is called its **additive inverse.**

$$12 + (-12) = (-12) + 12 = 0 \qquad\qquad 7x + (-7x) = -7x + 7x = 0$$

HOW TO Simplify: $8x + 4y - 8x + y$

$8x + 4y - 8x + y$
$= (8x - 8x) + (4y + y)$

$= 0 + 5y = 5y$

- Use the Commutative and Associative Properties of Addition to rearrange and group like terms.
- Combine like terms.

HOW TO Simplify: $4x^2 + 5x - 6x^2 - 2x + 1$

$4x^2 + 5x - 6x^2 - 2x + 1$
$= (4x^2 - 6x^2) + (5x - 2x) + 1$

$= -2x^2 + 3x + 1$

- Use the Commutative and Associative Properties of Addition to rearrange and group like terms.
- Combine like terms.

Example 1 Simplify: $3x + 4y - 10x + 7y$

Solution $3x + 4y - 10x + 7y = -7x + 11y$

You Try It 1 Simplify: $3a - 2b - 5a + 6b$

Your solution

Example 2 Simplify: $x^2 - 7 + 4x^2 - 16$

Solution $x^2 - 7 + 4x^2 - 16 = 5x^2 - 23$

You Try It 2 Simplify: $-3y^2 + 7 + 8y^2 - 14$

Your solution

Solutions on p. S4

Copyright © Houghton Mifflin Company. All rights reserved.

Objective B   **To simplify a variable expression using the Properties of Multiplication**

In simplifying variable expressions, the following Properties of Multiplication are used.

Copyright © Houghton Mifflin Company. All rights reserved.

> **TAKE NOTE**
> The Associative Property of Multiplication allows us to multiply a coefficient by a number. Without this property, the expression $2(3x)$ could not be changed.

> **The Associative Property of Multiplication**
> If a, b, and c are real numbers, then $(ab)c = a(bc)$.

When three or more factors are multiplied, the factors can be grouped in any order. The product is the same.

$$3(5 \cdot 6) = (3 \cdot 5)6 \qquad\qquad 2(3x) = (2 \cdot 3)x$$
$$3(30) = (15)6 \qquad\qquad\qquad = 6x$$
$$90 = 90$$

> **TAKE NOTE**
> The Commutative Property of Multiplication allows us to rearrange factors. This property, along with the Associative Property of Multiplication, allows us to simplify some variable expressions.

> **The Commutative Property of Multiplication**
> If a and b are real numbers, then $ab = ba$.

Two factors can be multiplied in either order. The product is the same.

$$5(-7) = -7(5) \qquad (5x) \cdot 3 = 3 \cdot (5x)$$
$$-35 = -35 \qquad\qquad = (3 \cdot 5)x$$
$$= 15x$$

- Commutative Property of Multiplication
- Associative Property of Multiplication

> **The Multiplication Property of One**
> If a is a real number, then $a \cdot 1 = 1 \cdot a = a$.

The product of a term and one is the term.

$$9 \cdot 1 = 1 \cdot 9 = 9 \qquad\qquad (8x) \cdot 1 = 1 \cdot (8x) = 8x$$

> **The Inverse Property of Multiplication**
> If a is a real number, and a is not equal to zero, then
> $$a \cdot \frac{1}{a} = \frac{1}{a} \cdot a = 1$$

> **TAKE NOTE**
> We must state that $x \neq 0$ because division by zero is undefined.

$\frac{1}{a}$ is called the **reciprocal** of a. $\frac{1}{a}$ is also called the **multiplicative inverse** of a. The product of a number and its reciprocal is one.

$$7 \cdot \frac{1}{7} = \frac{1}{7} \cdot 7 = 1 \qquad\qquad x \cdot \frac{1}{x} = \frac{1}{x} \cdot x = 1, \quad x \neq 0$$

The multiplication properties just discussed are used to simplify variable expressions.

HOW TO   Simplify: $2(-x)$

$$2(-x) = 2(-1 \cdot x)$$
$$= [2(-1)]x$$
$$= -2x$$

- Use the Associative Property of Multiplication to group factors.

HOW TO Simplify: $\dfrac{3}{2}\left(\dfrac{2x}{3}\right)$

$$\dfrac{3}{2}\left(\dfrac{2x}{3}\right) = \dfrac{3}{2}\left(\dfrac{2}{3}x\right)$$

• Note that $\dfrac{2x}{3} = \dfrac{2}{3}x$.

$$= \left(\dfrac{3}{2} \cdot \dfrac{2}{3}\right)x$$

• Use the Associative Property of Multiplication to group factors.

$$= 1 \cdot x$$

$$= x$$

HOW TO Simplify: $(16x)2$

$$(16x)2 = 2(16x)$$
$$= (2 \cdot 16)x$$
$$= 32x$$

• Use the Commutative and Associative Properties of Multiplication to rearrange and group factors.

Example 3 Simplify: $-2(3x^2)$

Solution $-2(3x^2) = -6x^2$

You Try It 3 Simplify: $-5(4y^2)$

Your solution

Example 4 Simplify: $-5(-10x)$

Solution $-5(-10x) = 50x$

You Try It 4 Simplify: $-7(-2a)$

Your solution

Example 5 Simplify: $-\dfrac{3}{4}\left(\dfrac{2}{3}x\right)$

Solution $-\dfrac{3}{4}\left(\dfrac{2}{3}x\right) = -\dfrac{1}{2}x$

You Try It 5 Simplify: $-\dfrac{3}{5}\left(-\dfrac{7}{9}a\right)$

Your solution

Solutions on p. S4

Objective C **To simplify a variable expression using the Distributive Property**

Recall that the Distributive Property states that if a, b, and c are real numbers, then

$$a(b + c) = ab + ac$$

The Distributive Property is used to remove parentheses from a variable expression.

HOW TO Simplify: $3(2x + 7)$

$$3(2x + 7) = 3(2x) + 3(7)$$

$$= 6x + 21$$

• Use the **Distributive Property**. Multiply each term inside the parentheses by **3**.

Copyright © Houghton Mifflin Company. All rights reserved.

HOW TO Simplify: $-5(4x + 6)$

$$-5(4x + 6) = -5(4x) + (-5)(6)$$ • Use the **Distributive Property**.
$$= -20x - 30$$

HOW TO Simplify: $-(2x - 4)$

$$-(2x - 4) = -1(2x - 4)$$ • Use the **Distributive Property**.
$$= -1(2x) - (-1)(4)$$
$$= -2x + 4$$

Note: When a negative sign immediately precedes the parentheses, the sign of each term inside the parentheses is changed.

HOW TO Simplify: $-\dfrac{1}{2}(8x - 12y)$

$$-\frac{1}{2}(8x - 12y) = -\frac{1}{2}(8x) - \left(-\frac{1}{2}\right)(12y)$$ • Use the **Distributive Property**.
$$= -4x + 6y$$

An extension of the Distributive Property is used when an expression contains more than two terms.

HOW TO Simplify: $3(4x - 2y - z)$

$$3(4x - 2y - z) = 3(4x) - 3(2y) - 3(z)$$ • Use the **Distributive Property**.
$$= 12x - 6y - 3z$$

Example 6

Simplify: $7(4 + 2x)$

Solution
$7(4 + 2x) = 28 + 14x$

You Try It 6

Simplify: $5(3 + 7b)$

Your solution

Example 7

Simplify: $(2x - 6)2$

Solution
$(2x - 6)2 = 4x - 12$

You Try It 7

Simplify: $(3a - 1)5$

Your solution

Example 8

Simplify: $-3(-5a + 7b)$

Solution
$-3(-5a + 7b) = 15a - 21b$

You Try It 8

Simplify: $-8(-2a + 7b)$

Your solution

Solutions on p. S4

Copyright © Houghton Mifflin Company. All rights reserved.

Example 9   Simplify: $3(x^2 - x - 5)$

Solution $3(x^2 - x - 5) = 3x^2 - 3x - 15$

You Try It 9   Simplify: $3(12x^2 - x + 8)$

Your solution

Example 10   Simplify: $-2(x^2 + 5x - 4)$

Solution $-2(x^2 + 5x - 4)$
$\qquad = -2x^2 - 10x + 8$

You Try It 10   Simplify: $3(-a^2 - 6a + 7)$

Your solution

Solutions on p. S4

Objective D   **To simplify general variable expressions**

When simplifying variable expressions, use the Distributive Property to remove parentheses and brackets used as grouping symbols.

> **HOW TO**   Simplify: $4(x - y) - 2(-3x + 6y)$
>
> $4(x - y) - 2(-3x + 6y)$
> $\quad = 4x - 4y + 6x - 12y$ • **Use the Distributive Property.**
> $\quad = 10x - 16y$ • **Combine like terms.**

Example 11   Simplify: $2x - 3(2x - 7y)$

Solution $2x - 3(2x - 7y) = 2x - 6x + 21y$
$\qquad\qquad\qquad\qquad\quad = -4x + 21y$

You Try It 11   Simplify: $3y - 2(y - 7x)$

Your solution

Example 12   Simplify:
$7(x - 2y) - (-x - 2y)$

Solution $7(x - 2y) - (-x - 2y)$

$\qquad = 7x - 14y + x + 2y$
$\qquad = 8x - 12y$

You Try It 12   Simplify:
$-2(x - 2y) - (-x + 3y)$

Your solution

Example 13   Simplify: $2x - 3[2x - 3(x + 7)]$

Solution $2x - 3[2x - 3(x + 7)]$

$\qquad = 2x - 3[2x - 3x - 21]$
$\qquad = 2x - 3[-x - 21]$
$\qquad = 2x + 3x + 63$
$\qquad = 5x + 63$

You Try It 13   Simplify: $3y - 2[x - 4(2 - 3y)]$

Your solution

Solutions on pp. S4–S5

Copyright © Houghton Mifflin Company. All rights reserved.

2.3 Translating Verbal Expressions into Variable Expressions

Objective A   **To translate a verbal expression into a variable expression, given the variable**

● **Study Tip** ●

Before the class meeting in which your professor begins a new section, you should read each objective statement for that section. Next, browse through the objective material. The purpose of browsing through the material is so that your brain will be prepared to accept and organize the new information when it is presented to you. See *AIM for Success*, page xxv.

One of the major skills required in applied mathematics is to translate a verbal expression into a variable expression. This requires recognizing the verbal phrases that translate into mathematical operations. A partial list of the verbal phrases used to indicate the different mathematical operations follows.

Addition	added to	6 added to y	$y + 6$
	more than	8 more than x	$x + 8$
	the sum of	the sum of x and z	$x + z$
	increased by	t increased by 9	$t + 9$
	the total of	the total of 5 and y	$5 + y$
Subtraction	minus	x minus 2	$x - 2$
	less than	7 less than t	$t - 7$
	decreased by	m decreased by 3	$m - 3$
	the difference between	the difference between y and 4	$y - 4$
	subtract...from...	subtract 9 from z	$z - 9$
Multiplication	times	10 times t	$10t$
	twice	twice w	$2w$
	of	one-half of x	$\frac{1}{2}x$
	the product of	the product of y and z	yz
	multiplied by	y multiplied by 11	$11y$
Division	divided by	x divided by 12	$\dfrac{x}{12}$
	the quotient of	the quotient of y and z	$\dfrac{y}{z}$
	the ratio of	the ratio of t to 9	$\dfrac{t}{9}$
Power	the square of	the square of x	x^2
	the cube of	the cube of a	a^3

Point of Interest

The way in which expressions are symbolized has changed over time. Here are how some of the expressions shown at the right may have appeared in the early 16th century.

R p. 9 for $x + 9$. The symbol R was used for a variable to the first power. The symbol p. was used for plus.

R m. 3 for $x - 3$. The symbol R is still used for the variable. The symbol m. was used for minus.

The square of a variable was designated by Q and the cube was designated by C. The expression $x^2 + x^3$ was written **Q p. C.**

> **HOW TO**   Translate "14 less than the cube of x" into a variable expression.
>
> 14 <u>less than</u> the <u>cube</u> of x • Identify the words that indicate the mathematical operations.
>
> $x^3 - 14$ • Use the identified operations to write the variable expression.

Copyright © Houghton Mifflin Company. All rights reserved.

Translating a phrase that contains the word *sum, difference, product,* or *quotient* can sometimes cause a problem. In the examples at the right, note where the operation symbol is placed.

the *sum* of x and y $x + y$

the *difference* between x and y $x - y$

the *product* of x and y $x \cdot y$

the *quotient* of x and y $\dfrac{x}{y}$

HOW TO   Translate "the difference between the square of x and the sum of y and z" into a variable expression.

the difference between the square of x and the sum of y and z

$x^2 - (y + z)$

- Identify the words that indicate the mathematical operations.

- Use the identified operations to write the variable expression.

Example 1

Translate "the total of 3 times n and n" into a variable expression.

Solution

the total of 3 times n and n

$3n + n$

You Try It 1

Translate "the difference between twice n and one-third of n" into a variable expression.

Your solution

Example 2

Translate "m decreased by the sum of n and 12" into a variable expression.

Solution

m decreased by the sum of n and 12

$m - (n + 12)$

You Try It 2

Translate "the quotient of 7 less than b and 15" into a variable expression.

Your solution

Solutions on p. S5

Objective B   **To translate a verbal expression into a variable expression and then simplify**

In most applications that involve translating phrases into variable expressions, the variable to be used is not given. To translate these phrases, a variable must be assigned to an unknown quantity before the variable expression can be written.

Copyright © Houghton Mifflin Company. All rights reserved.

HOW TO Translate "a number multiplied by the total of six and the cube of the number" into a variable expression.

the unknown number: n

- Assign a variable to one of the unknown quantities.

the cube of the number: n^3
the total of six and the cube of the number: $6 + n^3$

- Use the assigned variable to write an expression for any other unknown quantity.

$n(6 + n^3)$

- Use the assigned variable to write the variable expression.

Example 3

Translate "a number added to the product of four and the square of the number" into a variable expression.

Solution
the unknown number: n
the square of the number: n^2
the product of four and the square of the number: $4n^2$
$4n^2 + n$

You Try It 3

Translate "negative four multiplied by the total of ten and the cube of a number" into a variable expression.

Your solution

Example 4

Translate "four times the sum of one-half of a number and fourteen" into a variable expression. Then simplify.

Solution
the unknown number: n

one-half of the number: $\dfrac{1}{2}n$

the sum of one-half of the number and fourteen: $\dfrac{1}{2}n + 14$

$4\left(\dfrac{1}{2}n + 14\right)$
$2n + 56$

You Try It 4

Translate "five times the difference between a number and sixty" into a variable expression. Then simplify.

Your solution

Solutions on p. S5

Copyright © Houghton Mifflin Company. All rights reserved.

Objective C   **To translate application problems**

Many of the applications of mathematics require that you identify the unknown quantity, assign a variable to that quantity, and then attempt to express other unknown quantities in terms of the variable.

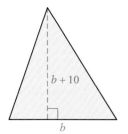

HOW TO   The height of a triangle is 10 ft longer than the base of the triangle. Express the height of the triangle in terms of the base of the triangle.

the base of the triangle: b

- Assign a variable to the base of the triangle.

the height is 10 more than the base: $b + 10$

- Express the height of the triangle in terms of b.

Example 5

The length of a swimming pool is 4 ft less than two times the width. Express the length of the pool in terms of the width.

Solution
the width of the pool: w
the length is 4 ft less than two times the width: $2w - 4$

You Try It 5

The speed of a new jet plane is twice the speed of an older model. Express the speed of the new model in terms of the speed of the older model.

Your solution

Example 6

A banker divided $5000 between two accounts, one paying 10% annual interest and the second paying 8% annual interest. Express the amount invested in the 10% account in terms of the amount invested in the 8% account.

Solution
the amount invested at 8%: x
the amount invested at 10%: $5000 - x$

You Try It 6

A guitar string 6 ft long was cut into two pieces. Express the length of the shorter piece in terms of the length of the longer piece.

Your solution

Solutions on p. S5

Copyright © Houghton Mifflin Company. All rights reserved.

2.3 Exercises

Objective A   **To translate a verbal expression into a variable expression, given the variable**

For Exercises 1 to 26, translate into a variable expression.

1. the sum of 8 and y

2. a less than 16

3. t increased by 10

4. p decreased by 7

5. z added to 14

6. q multiplied by 13

7. 20 less than the square of x

8. 6 times the difference between m and 7

9. the sum of three-fourths of n and 12

10. b decreased by the product of 2 and b

11. 8 increased by the quotient of n and 4

12. the product of -8 and y

13. the product of 3 and the total of y and 7

14. 8 divided by the difference between x and 6

15. the product of t and the sum of t and 16

16. the quotient of 6 less than n and twice n

17. 15 more than one-half of the square of x

18. 19 less than the product of n and -2

19. the total of 5 times the cube of n and the square of n

20. the ratio of 9 more than m to m

21. r decreased by the quotient of r and 3

22. four-fifths of the sum of w and 10

23. the difference between the square of x and the total of x and 17

24. s increased by the quotient of 4 and s

25. the product of 9 and the total of z and 4

26. n increased by the difference between 10 times n and 9

Copyright © Houghton Mifflin Company. All rights reserved.

Objective B   **To translate a verbal expression into a variable expression and then simplify**

For Exercises 27 to 58, translate into a variable expression. Then simplify.

27. twelve minus a number

28. a number divided by eighteen

29. two-thirds of a number

30. twenty more than a number

31. the quotient of twice a number and nine

32. ten times the difference between a number and fifty

33. eight less than the product of eleven and a number

34. the sum of five-eighths of a number and six

35. nine less than the total of a number and two

36. the difference between a number and three more than the number

37. the quotient of seven and the total of five and a number

38. four times the sum of a number and nineteen

39. five increased by one-half of the sum of a number and three

40. the quotient of fifteen and the sum of a number and twelve

41. a number added to the difference between twice the number and four

42. the product of two-thirds and the sum of a number and seven

43. the product of five less than a number and seven

44. the difference between forty and the quotient of a number and twenty

45. the quotient of five more than twice a number and the number

46. the sum of the square of a number and twice the number

47. a number decreased by the difference between three times the number and eight

48. the sum of eight more than a number and one-third of the number

Copyright © Houghton Mifflin Company. All rights reserved.

49. a number added to the product of three and the number

50. a number increased by the total of the number and nine

51. five more than the sum of a number and six

52. a number decreased by the difference between eight and the number

53. a number minus the sum of the number and ten

54. the difference between one-third of a number and five-eighths of the number

55. the sum of one-sixth of a number and four-ninths of the number

56. two more than the total of a number and five

57. the sum of a number divided by three and the number

58. twice the sum of six times a number and seven

Objective C   **To translate application problems**

59. 🔵 **Internet**   According to Brightmail, Inc., one-half of all email filtered by their email program in August 2003 was unsolicited email (spam). Express the amount of spam in terms of the number of emails filtered.

60. 🔵 **Telecommunications**   In 1951, phone companies began using area codes. According to information found at **www.area-code.com,** at the beginning of 2004 there were 207 more area codes than there were in 1951. Express the number of area codes in 2004 in terms of the number of area codes in 1951.

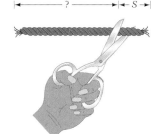

61. **Sports**   A halyard 12 ft long was cut into two pieces of different lengths. Use one variable to express the lengths of the two pieces.

62. **Natural Resources**   Twenty gallons of crude oil were poured into two containers of different sizes. Use one variable to express the amount of oil poured into each container.

Copyright © Houghton Mifflin Company. All rights reserved.

63. Rates of Cars   Two cars start at the same place and travel at different rates in opposite directions. Two hours later the cars are 200 mi apart. Express the distance traveled by the slower car in terms of the distance traveled by the faster car.

64. 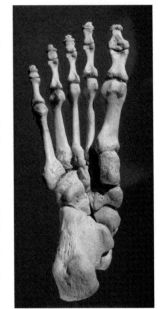 **Online Sales**   A recent survey conducted by Consumer Internet Barometer found that approximately 25% of those people who purchased a product online were extremely satisfied with their experience. Express the number of people that were extremely satisfied with their online purchases in terms of the number of people surveyed.

65. **Medicine**   According to the American Podiatric Medical Association, the bones in your foot account for one-fourth of all the bones in your body. Express the number of bones in your foot in terms of the total number of bones in your body.

66. **Sports**   The diameter of a basketball is approximately 4 times the diameter of a baseball. Express the diameter of a basketball in terms of the diameter of a baseball.

67. **Cost of Living**   A cost-of-living calculator provided by **Realtor.com** shows that a person living in San Francisco, California, would need approximately twice the salary of a person living in Daytona Beach, Florida, to maintain the same standard of living. Express the salary needed in Daytona Beach in terms of the salary needed in San Francisco.

APPLYING THE CONCEPTS

68. Metalwork   A wire whose length is given as x inches is bent into a square. Express the length of a side of the square in terms of x.

69. **Chemistry**   The chemical formula for glucose (sugar) is $C_6H_{12}O_6$. This formula means that there are 12 hydrogen atoms for every 6 carbon atoms and 6 oxygen atoms in each molecule of glucose (see the figure at the right). If x represents the number of atoms of oxygen in a pound of sugar, express the number of hydrogen atoms in the pound of sugar.

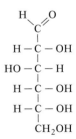

70. Translate the expressions $5x + 8$ and $5(x + 8)$ into phrases.

71. In your own words, explain how variables are used.

72. Explain the similarities and the differences between the expressions "the difference between x and 5" and "5 less than x."

Copyright © Houghton Mifflin Company. All rights reserved.

Focus on Problem Solving

From Concrete to Abstract

In your study of algebra, you will find that the problems are less concrete than those you studied in arithmetic. Problems that are concrete provide information pertaining to a specific instance. Algebra is more abstract. Abstract problems are theoretical; they are stated without reference to a specific instance. Let's look at an example of an abstract problem.

How many minutes are in h hours?

A strategy that can be used to solve this problem is to solve the same problem after substituting a number for the variable.

How many minutes are in 5 hours?

You know that there are 60 minutes in 1 hour. To find the number of minutes in 5 hours, multiply 5 by 60.

$$60 \cdot 5 = 300 \qquad \text{There are 300 minutes in 5 hours.}$$

Use the same procedure to find the number of minutes in h hours: multiply h by 60.

$$60 \cdot h = 60h \qquad \text{There are } 60h \text{ minutes in } h \text{ hours.}$$

This problem might be taken a step further:

If you walk 1 mile in x minutes, how far can you walk in h hours?

Consider the same problem using numbers in place of the variables.

If you walk 1 mile in 20 minutes, how far can you walk in 3 hours?

To solve this problem, you need to calculate the number of minutes in 3 hours (multiply 3 by 60), and divide the result by the number of minutes it takes to walk 1 mile (20 minutes).

$$\frac{60 \cdot 3}{20} = \frac{180}{20} = 9 \qquad \text{If you walk 1 mile in 20 minutes, you can walk 9 miles in 3 hours.}$$

Use the same procedure to solve the related abstract problem. Calculate the number of minutes in h hours (multiply h by 60), and divide the result by the number of minutes it takes to walk 1 mile (x minutes).

$$\frac{60 \cdot h}{x} = \frac{60h}{x} \qquad \text{If you walk 1 mile in } x \text{ minutes, you can walk } \frac{60h}{x} \text{ miles in } h \text{ hours.}$$

At the heart of the study of algebra is the use of variables. It is the variables in the problems above that make them abstract. But it is variables that allow us to generalize situations and state rules about mathematics.

Try each of the following problems.

1. How many hours are in d days?

Copyright © Houghton Mifflin Company. All rights reserved.

2. You earn d dollars an hour. What are your wages for working h hours?
3. If p is the price of one share of stock, how many shares can you purchase with d dollars?
4. A company pays a television station d dollars to air a commercial lasting s seconds. What is the cost per second?
5. After every v videotape rentals, you are entitled to one free rental. You have rented t tapes, where $t < v$. How many more do you need to rent before you are entitled to a free rental?
6. Your car gets g miles per gallon. How many gallons of gasoline does your car consume traveling t miles?
7. If you drink j ounces of juice each day, how many days will q quarts of the juice last?
8. A TV station has m minutes of commercials each hour. How many ads lasting s seconds each can be sold for each hour of programming?
9. A factory worker can assemble p products in m minutes. How many products can the factory worker assemble in h hours?
10. If one candy bar costs n nickels, how many candy bars can be purchased with q quarters?

Projects and Group Activities

Prime and Composite Numbers

Recall that a prime number is a natural number greater than 1 whose only natural-number factors are itself and 1. The number 11 is a prime number because the only natural-number factors of 11 are 11 and 1.

Eratosthenes, a Greek philosopher and astronomer who lived from 270 to 190 B.C., devised a method of identifying prime numbers. It is called the **Sieve of Eratosthenes.** The procedure is illustrated below.

~~1~~	2	3	~~4~~	5	~~6~~	7	~~8~~	~~9~~	~~10~~
11	~~12~~	13	~~14~~	~~15~~	~~16~~	17	~~18~~	19	~~20~~
~~21~~	~~22~~	23	~~24~~	~~25~~	~~26~~	~~27~~	~~28~~	29	~~30~~
31	~~32~~	~~33~~	~~34~~	~~35~~	~~36~~	37	~~38~~	~~39~~	~~40~~
41	~~42~~	43	~~44~~	~~45~~	~~46~~	47	~~48~~	~~49~~	~~50~~
~~51~~	~~52~~	53	~~54~~	~~55~~	~~56~~	~~57~~	~~58~~	59	~~60~~
61	~~62~~	~~63~~	~~64~~	~~65~~	~~66~~	67	~~68~~	~~69~~	~~70~~
71	~~72~~	73	~~74~~	~~75~~	~~76~~	~~77~~	~~78~~	79	~~80~~
~~81~~	~~82~~	83	~~84~~	~~85~~	~~86~~	~~87~~	~~88~~	89	~~90~~
~~91~~	~~92~~	~~93~~	~~94~~	~~95~~	~~96~~	97	~~98~~	~~99~~	~~100~~

Copyright © Houghton Mifflin Company. All rights reserved.

List all the natural numbers from 1 to 100. Cross out the number 1, because it is not a prime number. The number 2 is prime; circle it. Cross out all the other multiples of 2 (4, 6, 8,…), because they are not prime. The number 3 is prime; circle it. Cross out all the other multiples of 3 (6, 9, 12,…) that are not already crossed out. The number 4, the next consecutive number in the list, has already been crossed out. The number 5 is prime; circle it. Cross out all the other multiples of 5 that are not already crossed out. Continue in this manner until all the prime numbers less than 100 are circled.

A composite number is a natural number greater than 1 that has a natural-number factor other than itself and 1. The number 21 is a composite number because it has factors of 3 and 7. All the numbers crossed out in the preceding table, except the number 1, are composite numbers.

1. Use the Sieve of Eratosthenes to find the prime numbers between 100 and 200.

2. How many prime numbers are even numbers?

3. Find the "twin primes" between 100 and 200. Twin primes are two prime numbers whose difference is 2. For instance, 3 and 5 are twin primes; 5 and 7 are also twin primes.

4. **a.** List two prime numbers that are consecutive natural numbers.
   b. Can there be any other pairs of prime numbers that are consecutive natural numbers?

5. Some primes are the sum of a square and 1. For example, $5 = 2^2 + 1$. Find another prime p such that $p = n^2 + 1$, where n is a natural number.

6. Find a prime number p such that $p = n^2 - 1$, where n is a natural number.

7. **a.** 4! (which is read "4 factorial") is equal to $4 \cdot 3 \cdot 2 \cdot 1$. Show that $4! + 2$, $4! + 3$, and $4! + 4$ are all composite numbers.
   b. 5! (which is read "5 factorial") is equal to $5 \cdot 4 \cdot 3 \cdot 2 \cdot 1$. Will $5! + 2$, $5! + 3$, $5! + 4$, and $5! + 5$ generate four consecutive composite numbers?
   c. Use the notation 6! to represent a list of five consecutive composite numbers.

Investigation into Operations with Even and Odd Integers

For Exercises 1 to 10, complete each statement with the word *even* or *odd*.

1. If k is an odd integer, then $k + 1$ is an _____ integer.

2. If k is an odd integer, then $k - 2$ is an _____ integer.

3. If n is an integer, then $2n$ is an _____ integer.

4. If m and n are even integers, then $m - n$ is an _____ integer.

5. If m and n are even integers, then mn is an _____ integer.

6. If m and n are odd integers, then $m + n$ is an _____ integer.

7. If m and n are odd integers, then $m - n$ is an _____ integer.

8. If m and n are odd integers, then mn is an _____ integer.

9. If m is an even integer and n is an odd integer, then $m - n$ is an _____ integer.

10. If m is an even integer and n is an odd integer, then $m + n$ is an _____ integer.

Study Tip

Three important features of this text that can be used to prepare for a test are the:

● Chapter Summary
● Chapter Review Exercises
● Chapter Test

See *AIM for Success*, page xxv.

Copyright © Houghton Mifflin Company. All rights reserved.

Chapter 2 Summary

Key Words	**Examples**
A *variable* is a letter that is used for a quantity that is unknown or that can change. A *variable expression* is an expression that contains one or more variables. [2.1A, p. 79]	$4x + 2y - 6z$ is a variable expression. It contains the variables x, y, and z.
The *terms* of a variable expression are the addends of the expression. Each term is a *variable term* or a *constant term*. [2.1A, p. 79]	The expression $2a^2 - 3b^3 + 7$ has three terms. $2a^2$, $-3b^3$, and 7. $2a^2$ and $-3b^3$ are variable terms. 7 is a constant term.
A variable term is composed of a *numerical coefficient* and a *variable part*. [2.1A, p. 79]	For the expression $-7x^3y^2$, -7 is the coefficient and x^3y^2 is the variable part.
In a variable expression, replacing each variable by its value and then simplifying the resulting numerical expression is called *evaluating the variable expression*. [2.1A, p. 80]	To evaluate $2ab - b^2$ when $a = 3$ and $b = -2$, replace a by 3 and b by -2 then simplify the numerical expression. $2(3)(-2) - (-2)^2 = -16$
Like terms of a variable expression are terms with the same variable part. Constant terms are like terms. [2.2A, p. 83]	For the expressions $3a^2 + 2b - 3$ and $2a^2 - 3a + 4$, $3a^2$ and $2a^2$ are like terms; -3 and 4 are like terms.
To simplify the sum of like variable terms, use the Distributive Property to add the numerical coefficients. This is called *combining like terms*. [2.2A, p. 83]	$5y + 3y = (5 + 3)y$ $= 8y$
The *additive inverse* of a number is the opposite of the number. [2.2A, p. 84]	-4 is the additive inverse of 4. $\frac{2}{3}$ is the additive inverse of $-\frac{2}{3}$. 0 is the additive inverse of 0.
The *multiplicative inverse* of a number is the *reciprocal of the number*. [2.2B, p. 85]	$\frac{3}{4}$ is the multiplicative inverse of $\frac{4}{3}$. $-\frac{1}{4}$ is the multiplicative inverse of -4.

Essential Rules and Procedures	**Examples**
The Distributive Property [2.2A, p. 83] If a, b, and c are real numbers, then $a(b + c) = ab + ac$.	$5(4 + 7) = 5 \cdot 4 + 5 \cdot 7$ $= 20 + 35 = 55$
The Associative Property of Addition [2.2A, p. 83] If a, b, and c are real numbers, then $(a + b) + c = a + (b + c)$.	$-4 + (2 + 7) = -4 + 9 = 5$ $(-4 + 2) + 7 = -2 + 7 = 5$

Copyright © Houghton Mifflin Company. All rights reserved.

The Commutative Property of Addition [2.2A, p. 84]
If a and b are real numbers, then $a + b = b + a$.

$2 + 5 = 7$ and $5 + 2 = 7$

The Addition Property of Zero [2.2A, p. 84]
If a is a real number, then $a + 0 = 0 + a = a$.

$-8 + 0 = -8$ and $0 + (-8) = -8$

The Inverse Property of Addition [2.2A, p. 84]
If a is a real number, then $a + (-a) = (-a) + a = 0$.

$5 + (-5) = 0$ and $(-5) + 5 = 0$

The Associative Property of Multiplication [2.2B, p. 85]
If a, b, and c are real numbers, then $(ab)c = a(bc)$.

$-3 \cdot (5 \cdot 4) = -3(20) = -60$
$(-3 \cdot 5) \cdot 4 = -15 \cdot 4 = -60$

The Commutative Property of Multiplication [2.2B, p. 85]
If a and b are real numbers, then $ab = ba$.

$-3(7) = -21$ and $7(-3) = -21$

The Multiplication Property of One [2.2B, p. 85]
If a is a real number, then $a \cdot 1 = 1 \cdot a = a$.

$-3(1) = -3$ and $1(-3) = -3$

The Inverse Property of Multiplication [2.2B, p. 85]
If a is a real number, and a is not equal to zero, then
$a \cdot \dfrac{1}{a} = \dfrac{1}{a} \cdot a = 1$.

$-3 \cdot -\dfrac{1}{3} = 1$ and $-\dfrac{1}{3} \cdot -3 = 1$

Copyright © Houghton Mifflin Company. All rights reserved.

Chapter 2 Review Exercises

1. Simplify: $3(x^2 - 8x - 7)$

2. Simplify: $7x + 4x$

3. Simplify: $6a - 4b + 2a$

4. Simplify: $(-50n)\left(\dfrac{1}{10}\right)$

5. Evaluate $(5c - 4a)^2 - b$ when $a = -1$, $b = 2$, and $c = 1$.

6. Simplify: $5(2x - 7)$

7. Simplify: $2(6y^2 + 4y - 5)$

8. Simplify: $\dfrac{1}{4}(-24a)$

9. Simplify: $-6(7x^2)$

10. Simplify: $-9(7 + 4x)$

11. Simplify: $12y - 17y$

12. Evaluate $2bc \div (a + 7)$ when $a = 3$, $b = -5$, and $c = 4$.

13. Simplify: $7 - 2(3x + 4)$

14. Simplify: $6 + 2[2 - 5(4a - 3)]$

15. Simplify: $6(8y - 3) - 8(3y - 6)$

16. Simplify: $5c + (-2d) - 3d - (-4c)$

17. Simplify: $5(4x)$

18. Simplify: $-4(2x - 9) + 5(3x + 2)$

19. Evaluate $(b - a)^2 + c$ when $a = -2$, $b = 3$, and $c = 4$.

20. Simplify: $-9r + 2s - 6s + 12s$

Copyright © Houghton Mifflin Company. All rights reserved.

21. Evaluate $(2x - y)^2 + (2x + y)^2$ when $x = -2$ and $y = -3$.

22. Evaluate $b^2 - 4ac$ when $b = -4$, $a = 1$, and $c = -3$.

23. Simplify: $4x - 3x^2 + 2x - x^2$

24. Simplify: $5[2 - 3(6x - 1)]$

25. Simplify: $0.4x + 0.6(250 - x)$

26. Simplify: $\dfrac{2}{3}x - \dfrac{3}{4}x$

27. Simplify: $(7a^2 - 2a + 3)4$

28. Simplify: $18 - (4x - 2)$

29. Evaluate $a^2 - b^2$ when $a = 3$ and $b = 4$.

30. Simplify: $-3(-12y)$

31. Translate "two-thirds of the total of x and 10" into a variable expression.

32. Translate "the product of 4 and x" into a variable expression.

33. Translate "6 less than x" into a variable expression.

34. Translate "a number plus twice the number" into a variable expression. Then simplify.

Copyright © Houghton Mifflin Company. All rights reserved.

35. Translate "the difference between twice a number and one-half of the number" into a variable expression. Then simplify.

36. Translate "three times a number plus the product of five and one less than the number" into a variable expression. Then simplify.

37. Sports A baseball card collection contains five times as many National League players' cards as American League players' cards. Express the number of National League players' cards in the collection in terms of the number of American League players' cards.

38. Finance A club treasurer has some five-dollar bills and some ten-dollar bills. The treasurer has a total of 35 bills. Express the number of five-dollar bills in terms of the number of ten-dollar bills.

39. Nutrition A candy bar contains eight more calories than twice the number of calories in an apple. Express the number of calories in a candy bar in terms of the number of calories in an apple.

40. Architecture The length of the Parthenon is approximately 1.6 times the width. Express the length of the Parthenon in terms of the width.

41. Anatomy Leonardo DaVinci studied various proportions of human anatomy. One of his findings was that the standing height of a person is approximately 1.3 times the kneeling height of the same person. Represent the standing height of a person in terms of kneeling height.

Copyright © Houghton Mifflin Company. All rights reserved.

Chapter 2 Test

1. Simplify: $3x - 5x + 7x$

2. Simplify: $-3(2x^2 - 7y^2)$

3. Simplify: $2x - 3(x - 2)$

4. Simplify: $2x + 3[4 - (3x - 7)]$

5. Simplify: $3x - 7y - 12x$

6. Evaluate $b^2 - 3ab$ when $a = 3$ and $b = -2$.

7. Simplify: $\dfrac{1}{5}(10x)$

8. Simplify: $5(2x + 4) - 3(x - 6)$

9. Simplify: $-5(2x^2 - 3x + 6)$

10. Simplify: $3x + (-12y) - 5x - (-7y)$

11. Evaluate $\dfrac{-2ab}{2b - a}$ when $a = -4$ and $b = 6$.

12. Simplify: $(12x)\left(\dfrac{1}{4}\right)$

13. Simplify: $-7y^2 + 6y^2 - (-2y^2)$

14. Simplify: $-2(2x - 4)$

15. Simplify: $\dfrac{2}{3}(-15a)$

16. Simplify: $-2[x - 2(x - y)] + 5y$

Copyright © Houghton Mifflin Company. All rights reserved.

17. Simplify: $(-3)(-12y)$

18. Simplify: $5(3 - 7b)$

19. Translate "the difference between the squares of a and b" into a variable expression.

20. Translate "ten times the difference between a number and 3" into a variable expression. Then simplify.

21. Translate "the sum of a number and twice the square of the number" into a variable expression.

22. Translate "three less than the quotient of six and a number" into a variable expression.

23. Translate "b decreased by the product of b and seven" into a variable expression.

24. **Sports**   The speed of a pitcher's fastball is twice the speed of the catcher's return throw. Express the speed of the fastball in terms of the speed of the return throw.

25. **Metalwork**   A wire is cut into two lengths. The length of the longer piece is 3 in. less than four times the length of the shorter piece. Express the length of the longer piece in terms of the length of the shorter piece.

Copyright © Houghton Mifflin Company. All rights reserved.

Cumulative Review Exercises

1. Add: $-4 + 7 + (-10)$

2. Subtract: $-16 - (-25) - 4$

3. Multiply: $(-2)(3)(-4)$

4. Divide: $(-60) \div 12$

5. Find the complement of a 37° angle.

6. Simplify: $\dfrac{7}{12} - \dfrac{11}{16} - \left(-\dfrac{1}{3}\right)$

7. Simplify: $-\dfrac{5}{12} \div \dfrac{5}{2}$

8. Simplify: $\left(-\dfrac{9}{16}\right) \cdot \left(\dfrac{8}{27}\right) \cdot \left(-\dfrac{3}{2}\right)$

9. Write $\dfrac{3}{4}$ as a percent.

10. Simplify: $-2^5 \div (3 - 5)^2 - (-3)$

11. Simplify: $\left(-\dfrac{3}{4}\right)^2 \div \left(\dfrac{3}{8} - \dfrac{11}{12}\right)$

12. Evaluate $a^2 - 3b$ when $a = 2$ and $b = -4$.

13. Simplify: $-2x^2 - (-3x^2) + 4x^2$

14. Simplify: $5a - 10b - 12a$

15. Find the area of a circle whose radius is 7 cm. Use 3.14 for π.

16. Find the perimeter of a square whose side measures 24 ft.

17. Simplify: $3(8 - 2x)$

18. Simplify: $-2(-3y + 9)$

Copyright © Houghton Mifflin Company. All rights reserved.

19. Write $37\frac{1}{2}\%$ as a fraction.

20. Write 1.05% as a decimal.

21. Simplify: $-4(2x^2 - 3y^2)$

22. Simplify: $-3(3y^2 - 3y - 7)$

23. Simplify: $-3x - 2(2x - 7)$

24. Simplify: $4(3x - 2) - 7(x + 5)$

25. Simplify: $2x + 3[x - 2(4 - 2x)]$

26. Simplify: $3[2x - 3(x - 2y)] + 3y$

27. Translate "the sum of one-half of b and b" into a variable expression.

28. Translate "10 divided by the difference between y and 2" into a variable expression.

29. Translate "the difference between eight and the quotient of a number and twelve" into a variable expression.

30. Translate "the sum of a number and two more than the number" into a variable expression. Then simplify.

31. **Sports**   A softball diamond is a square with each side measuring 60 ft. Find the area enclosed by the sides of the softball diamond.

32. **Telecommunications**   The speed of a DSL (Digital Subscriber Line) Internet connection is ten times faster than that of a dial-up connection. Express the speed of the DSL connection in terms of the speed of the dial-up connection.

Copyright © Houghton Mifflin Company. All rights reserved.

Copyright © Houghton Mifflin Company. All rights reserved.

chapter

3

Solving Equations

Suppose that this bus is traveling the posted speed limit of 45 mph and is making a 180-mile drive between two cities. How long will the bus ride last? To figure out the answer, you will need to use the formula $d = rt$, where d is the distance traveled, r is the rate of speed, and t is the time spent traveling. In this case, d is 180, and r is 45. When you solve for t, you get the correct answer of 4 hours. The formula $d = rt$ is also used to solve **Exercises 52 and 53 on page 179**, which are more advanced motion problems involving bus travel.

Need help? For online student resources, such as section quizzes, visit this textbook's website at **math.college.hmco.com/students**.

OBJECTIVES

Section 3.1

A To determine whether a given number is a solution of an equation

B To solve an equation of the form $x + a = b$

C To solve an equation of the form $ax = b$

D To solve application problems using the basic percent equation

E To solve uniform motion problems

Section 3.2

A To solve an equation of the form $ax + b = c$

B To solve application problems using formulas

Section 3.3

A To solve an equation of the form $ax + b = cx + d$

B To solve an equation containing parentheses

C To solve application problems using formulas

Section 3.4

A To solve integer problems

B To translate a sentence into an equation and solve

Section 3.5

A To solve problems involving angles

B To solve problems involving the angles of a triangle

Section 3.6

A To solve value mixture problems

B To solve percent mixture problems

C To solve uniform motion problems

Do these exercises to prepare for Chapter 3.

1. Write $\frac{9}{100}$ as a decimal.

2. Write $\frac{3}{4}$ as a percent.

3. Evaluate $3x^2 - 4x - 1$ when $x = -4$.

4. Simplify: $R - 0.35R$

5. Simplify: $\frac{1}{2}x + \frac{2}{3}x$

6. Simplify: $6x - 3(6 - x)$

7. Simplify: $0.22(3x + 6) + x$

8. Translate into a variable expression: "The difference between 5 and twice a number."

9. A new graphics card for computer games is five times faster than a graphics card made two years ago. Express the speed of the new card in terms of the speed of the old card.

10. A board 5 ft long is cut into two pieces. If x represents the length of the longer piece, write an expression for the shorter piece in terms of x.

GO FIGURE • • •

How can a donut be cut into 8 equal pieces with three cuts of a knife?

Copyright © Houghton Mifflin Company. All rights reserved.

3.1 Introduction to Equations

Objective A   **To determine whether a given number is a solution of an equation**

Copyright © Houghton Mifflin Company. All rights reserved.

Point of Interest

One of the most famous equations ever stated is $E = mc^2$. This equation, stated by Albert Einstein, shows that there is a relationship between mass m and energy E. As a side note, the chemical element einsteinium was named in honor of Einstein.

An **equation** expresses the equality of two mathematical expressions. The expressions can be either numerical or variable expressions.

$$\left. \begin{array}{l} 9 + 3 = 12 \\ 3x - 2 = 10 \\ y^2 + 4 = 2y - 1 \\ z = 2 \end{array} \right\} \text{Equations}$$

The equation at the right is true if the variable is replaced by 5.

$x + 8 = 13$
$5 + 8 = 13$ A true equation

The equation is false if the variable is replaced by 7.

$7 + 8 = 13$ A false equation

A **solution of an equation** is a number that, when substituted for the variable, results in a true equation. 5 is a solution of the equation $x + 8 = 13$. 7 is not a solution of the equation $x + 8 = 13$.

HOW TO   Is -2 a solution of $2x + 5 = x^2 - 3$?

TAKE NOTE

The Order of Operations Agreement applies to evaluating $2(-2) + 5$ and $(-2)^2 - 3$.

$$\begin{array}{c|c} \multicolumn{2}{c}{2x + 5 = x^2 - 3} \\ \hline 2(-2) + 5 & (-2)^2 - 3 \\ -4 + 5 & 4 - 3 \\ \multicolumn{2}{c}{1 = 1} \end{array}$$

Yes, -2 is a solution of the equation.

- Replace x by -2.
- Evaluate the numerical expressions.
- If the results are equal, -2 is a solution of the equation. If the results are not equal, -2 is not a solution of the equation.

Example 1   Is -4 a solution of $5x - 2 = 6x + 2$?

Solution
$$\begin{array}{c|c} \multicolumn{2}{c}{5x - 2 = 6x + 2} \\ \hline 5(-4) - 2 & 6(-4) + 2 \\ -20 - 2 & -24 + 2 \\ \multicolumn{2}{c}{-22 = -22} \end{array}$$

Yes, -4 is a solution.

You Try It 1   Is $\frac{1}{4}$ a solution of $5 - 4x = 8x + 2$?

Your solution

Example 2   Is -4 a solution of $4 + 5x = x^2 - 2x$?

Solution
$$\begin{array}{c|c} \multicolumn{2}{c}{4 + 5x = x^2 - 2x} \\ \hline 4 + 5(-4) & (-4)^2 - 2(-4) \\ 4 + (-20) & 16 - (-8) \\ \multicolumn{2}{c}{-16 \neq 24} \end{array}$$

($\neq$ means "is not equal to")

No, -4 is not a solution.

You Try It 2   Is 5 a solution of $10x - x^2 = 3x - 10$?

Your solution

Solutions on p. S5

Copyright © Houghton Mifflin Company. All rights reserved.

Objective B To solve an equation of the form $x + a = b$

Study Tip

To learn mathematics, you must be an active participant. Listening and watching your professor do mathematics is not enough. Take notes in class, mentally think through every question your instructor asks, and try to answer it even if you are not called on to answer it verbally. Ask questions when you have them. See *AIM for Success,* page xxv, for other ways to be an active learner.

To **solve an equation** means to find a solution of the equation. The simplest equation to solve is an equation of the form *variable = constant*, because the constant is the solution.

The solution of the equation $x = 5$ is 5 because $5 = 5$ is a true equation.

The solution of the equation at the right is 7 because $7 + 2 = 9$ is a true equation.

$$x + 2 = 9 \qquad 7 + 2 = 9$$

Note that if 4 is added to each side of the equation $x + 2 = 9$, the solution is still 7.

$$x + 2 = 9$$
$$x + 2 + 4 = 9 + 4$$
$$x + 6 = 13 \qquad 7 + 6 = 13$$

If -5 is added to each side of the equation $x + 2 = 9$, the solution is still 7.

$$x + 2 = 9$$
$$x + 2 + (-5) = 9 + (-5)$$
$$x - 3 = 4 \qquad 7 - 3 = 4$$

Equations that have the same solution are **equivalent equations.** The equations $x + 2 = 9$, $x + 6 = 13$, and $x - 3 = 4$ are equivalent equations; each equation has 7 as its solution. These examples suggest that adding the same number to each side of an equation produces an equivalent equation. This is called the *Addition Property of Equations.*

> **Addition Property of Equations**
>
> The same number can be added to each side of an equation without changing its solution. In symbols, the equation $a = b$ has the same solution as the equation $a + c = b + c$.

In solving an equation, the goal is to rewrite the given equation in the form *variable = constant*. The Addition Property of Equations is used to remove a *term* from one side of the equation by adding the opposite of that term to each side of the equation.

HOW TO Solve: $x - 4 = 2$

$$x - 4 = 2$$
- The goal is to rewrite the equation as *variable = constant*.

$$x - 4 + 4 = 2 + 4$$
- **Add 4** to each side of the equation.

$$x + 0 = 6$$
- Simplify.

$$x = 6$$
- The equation is in the form *variable = constant*.

Check:
$$\frac{x - 4 = 2}{6 - 4 \mid 2}$$
$$2 = 2 \qquad \text{A true equation}$$

The solution is 6.

Because subtraction is defined in terms of addition, the Addition Property of Equations also makes it possible to subtract the same number from each side of an equation without changing the solution of the equation.

Example 6

12 is $33\frac{1}{3}\%$ of what number?

Solution

$P \cdot B = A$ • Use the basic percent equation.

$\frac{1}{3}B = 12$ • $33\frac{1}{3}\% = \frac{1}{3}$

$3 \cdot \frac{1}{3}B = 3 \cdot 12$

$B = 36$

12 is $33\frac{1}{3}\%$ of 36.

You Try It 6

18 is $16\frac{2}{3}\%$ of what number?

Your solution

Example 7

The data in the table below shows the number of households (in millions) that downloaded music files for a three-month period in a recent year. (*Source:* NPD Group)

Month	April	May	June
Downloads	14.5	12.7	10.4

For the three-month period, what percent of the files were downloaded in May? Round to the nearest percent.

Strategy

To find the percent,

• Find the total number of files downloaded for the three-month period.
• Use the basic percent equation. *B* is the total number of files downloaded for the three-month period; $A = 12.7$, the number of files downloaded in May; *P* is unknown.

Solution

$14.5 + 12.7 + 10.4 = 37.6$

$P \cdot B = A$ • Use the basic percent equation

$P(37.6) = 12.7$ • $B = 37.6$, $A = 12.7$

$P = \frac{12.7}{37.6} \approx 0.34$

Approximately 34% of the files were downloaded in May.

You Try It 7

The Bowl Championship Series (BCS) received approximately $83.3 million in revenues from various college football bowl games. Of this amount, the college representing the Pac-10 in the Rose Bowl received approximately $3.1 million. (*Source:* BCSfootball.org) What percent of the total received by the BCS did the college representing the Pac-10 receive? Round to the nearest tenth of a percent.

Your strategy

Your solution

Solutions on p. S5

Copyright © Houghton Mifflin Company. All rights reserved.

Example 8

In April, Marshall Wardell was charged an interest fee of $8.72 on an unpaid credit card balance of $545. Find the annual interest rate on this credit card.

Strategy

The interest is $8.72. Therefore, $I = 8.72$. The unpaid balance is $545. This is the principal on which interest is calculated. Therefore, $P = 545$. The time is 1 month. Because the *annual* interest rate must be found and the time is given as 1 month, we write 1 month as $\frac{1}{12}$ year so $t = \frac{1}{12}$. To find the interest rate, solve $I = Prt$ for r.

Solution

$$I = Prt$$

- Use the simple interest equation.

$$8.72 = 545r\left(\frac{1}{12}\right)$$

- $I = 8.72$, $P = 545$, $t = \frac{1}{12}$.

$$8.72 = \frac{545}{12}r$$

$$\frac{12}{545}(8.72) = \frac{12}{545}\left(\frac{545}{12}r\right)$$

$$0.192 = r$$

The annual interest rate is 19.2%.

You Try It 8

Clarissa Adams purchased a municipal bond for $1000 that earns an annual simple interest rate of 6.4%. How much must she deposit into a bank account that earns 8% annual simple interest so that the interest earned from each account after one year is the same?

Your strategy

Your solution

Example 9

To make a certain color of blue, 4 oz of cyan must be contained in 1 gal of paint. What is the percent concentration of cyan in the paint?

Strategy

The cyan is given in ounces and the amount of paint is given in gallons. We must convert ounces to gallons or gallons to ounces. For this problem, we will convert gallons to ounces: 1 gal = 128 oz. Solve $Q = Ar$ for r with $Q = 4$ and $A = 128$.

Solution

$$Q = Ar$$

- Use the percent mixture equation.

$$4 = 128r$$

- $Q = 4$, $A = 128$.

$$\frac{4}{128} = \frac{128r}{128}$$

$$0.03125 = r$$

The percent concentration of cyan is 3.125%.

You Try It 9

The concentration of sugar in a certain breakfast cereal is 25%. If there are 2 oz of sugar contained in the cereal in a bowl, how many ounces of cereal are in the bowl?

Your strategy

Your solution

Solutions on p. S6

Copyright © Houghton Mifflin Company. All rights reserved.

Objective E   **To solve uniform motion problems**

TAKE NOTE

A car traveling in a *circle* at a constant speed of 45 mph is *not* in uniform motion because the direction of the car is always changing.

Any object that travels at a constant speed in a straight line is said to be in *uniform motion*. **Uniform motion** means that the speed and direction of an object do not change. For instance, a car traveling at a constant speed of 45 mph on a straight road is in uniform motion.

The solution of a uniform motion problem is based on the **uniform motion equation** $d = rt$, where d is the distance traveled, r is the rate of travel, and t is the time spent traveling. For instance, suppose a car travels at 50 mph for 3 h. Because the rate (50 mph) and time (3 h) are known, we can find the distance traveled by solving the equation $d = rt$ for d.

$$d = rt$$
$$d = 50(3) \qquad \bullet \ r = 50, \ t = 3.$$
$$d = 150$$

The car travels a distance of 150 mi.

HOW TO   A jogger runs 3 mi in 45 min. What is the rate of the jogger in miles per hour?

Strategy   • Because the answer must be in miles per *hour* and the given time is in *minutes,* convert 45 min to hours.

• To find the rate of the jogger, solve the equation $d = rt$ for r.

Solution $45 \text{ min} = \dfrac{45}{60} \text{ h} = \dfrac{3}{4} \text{ h}$

$$d = rt$$

$$3 = r\left(\dfrac{3}{4}\right) \qquad \bullet \ d = 3, \ t = \dfrac{3}{4}.$$

$$3 = \dfrac{3}{4}r$$

$$\left(\dfrac{4}{3}\right)3 = \left(\dfrac{4}{3}\right)\dfrac{3}{4}r \qquad \bullet \ \text{Multiply each side of the equation}$$
$$\text{by the } \textbf{reciprocal of } \dfrac{3}{4}.$$

$$4 = r$$

The rate of the jogger is 4 mph.

If two objects are moving in opposite directions, then the rate at which the distance between them is increasing is the sum of the speeds of the two objects. For instance, in the diagram below, two cars start from the same point and travel in opposite directions. The distance between them is changing at 70 mph.

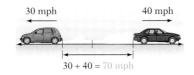

Copyright © Houghton Mifflin Company. All rights reserved.

Similarly, if two objects are moving toward each other, the distance between them is decreasing at a rate that is equal to the sum of the speeds. The rate at which the two planes at the right are approaching one another is 800 mph.

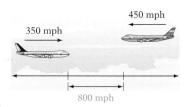

> **HOW TO**   Two cars start from the same point and move in opposite directions. The car moving west is traveling 45 mph, and the car moving east is traveling 60 mph. In how many hours will the cars be 210 mi apart?
>
> **Strategy**   The distance is 210 mi. Therefore, $d = 210$. The cars are moving in opposite directions so the rate at which the distance between them is changing is the sum of the rates of each of the cars. The rate is 45 mph + 60 mph = 105 mph. Therefore, $r = 105$. To find the time, solve the equation $d = rt$ for t.
>
> **Solution**
>
> $$d = rt$$
> $$210 = 105t \qquad \bullet \ d = 210, \ r = 105.$$
> $$\frac{210}{105} = \frac{105t}{105} \qquad \bullet \ \text{Solve for } t.$$
> $$2 = t$$
>
> In 2 h, the cars will be 210 mi apart.

If a motorboat is on a river that is flowing at a rate of 4 mph, then the boat will float down the river at a speed of 4 mph when the motor is not on. Now suppose the motor is turned on and the power adjusted so that the boat would travel 10 mph without the aid of the current. Then, if the boat is moving with the current, its effective speed is the speed of the boat using power plus of the speed of the current: 10 mph + 4 mph = 14 mph. (See the figure below.)

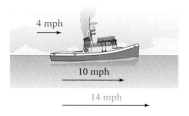

However, if the boat is moving against the current, the current slows the boat down, and the effective speed of the boat is the speed of the boat using power minus the speed of the current: 10 mph − 4 mph = 6 mph. (See the figure below.)

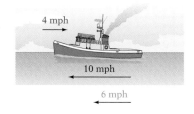

Copyright © Houghton Mifflin Company. All rights reserved.

There are other situations in which the preceding concepts may be applied.

TAKE NOTE

The term ft/s is an abbreviation for "feet per second." Similarly, cm/s is "centimeters per second" and m/s is "meters per second."

HOW TO   An airline passenger is walking between two airline terminals and decides to get on a moving sidewalk that is 150 ft long. If the passenger walks at a rate of 7 ft/s and the moving sidewalk moves at a rate of 9 ft/s, how long, in seconds, will it take the passenger to walk from one end of the moving sidewalk to the other? Round to the nearest thousandth.

Strategy   The distance is 150 ft. Therefore, $d = 150$. The passenger is traveling at 7 ft/s and the moving sidewalk is traveling at 9 ft/s. The rate of the passenger is the sum of the two rates, or 16 ft/s. Therefore, $r = 16$. To find the time, solve the equation $d = rt$ for t.

Solution

$$d = rt$$
$$150 = 16t \qquad \bullet \ d = 150,\ r = 16.$$
$$\frac{150}{16} = \frac{16t}{16} \qquad \bullet \ \text{Solve for } t.$$
$$9.375 = t$$

It will take 9.375 s for the passenger to travel the length of the moving sidewalk.

Example 10

Two cyclists start at the same time at opposite ends of an 80-mile course. One cyclist is traveling 18 mph, and the second cyclist is traveling 14 mph. How long after they begin will they meet?

Strategy
The distance is 80 mi. Therefore, $d = 80$. The cyclists are moving toward each other so the rate at which the distance between them is changing is the sum of the rates of each of the cyclists. The rate is 18 mph + 14 mph = 32 mph. Therefore, $r = 32$. To find the time, solve the equation $d = rt$ for t.

Solution

$$d = rt$$
$$80 = 32t \qquad \bullet \ d = 80,\ r = 32.$$
$$\frac{80}{32} = \frac{32t}{32} \qquad \bullet \ \text{Solve for } t.$$
$$2.5 = t$$

The cyclists will meet in 2.5 h.

You Try It 10

A plane that can normally travel at 250 mph in calm air is flying into a headwind of 25 mph. How far can the plane fly in 3 h?

Your strategy

Your solution

Solution on p. S6

Copyright © Houghton Mifflin Company. All rights reserved.

3.1 Exercises

Objective A **To determine whether a given number is a solution of an equation**

1. What is the difference between an equation and an expression?

2. Explain how to determine whether a given number is a solution of an equation.

3. Is 4 a solution of
$2x = 8$?

4. Is 3 a solution of
$y + 4 = 7$?

5. Is -1 a solution of
$2b - 1 = 3$?

6. Is -2 a solution of
$3a - 4 = 10$?

7. Is 1 a solution of
$4 - 2m = 3$?

8. Is 2 a solution of
$7 - 3n = 2$?

9. Is 5 a solution of
$2x + 5 = 3x$?

10. Is 4 a solution of
$3y - 4 = 2y$?

11. Is -2 a solution of
$3a + 2 = 2 - a$?

12. Is 3 a solution of
$z^2 + 1 = 4 + 3z$?

13. Is 2 a solution of
$2x^2 - 1 = 4x - 1$?

14. Is -1 a solution of
$y^2 - 1 = 4y + 3$?

15. Is 4 a solution of
$x(x + 1) = x^2 + 5$?

16. Is 3 a solution of
$2a(a - 1) = 3a + 3$?

17. Is $-\frac{1}{4}$ a solution of
$8t + 1 = -1$?

18. Is $\frac{1}{2}$ a solution of
$4y + 1 = 3$?

19. Is $\frac{2}{5}$ a solution of
$5m + 1 = 10m - 3$?

20. Is $\frac{3}{4}$ a solution of
$8x - 1 = 12x + 3$?

Objective B **To solve an equation of the form $x + a = b$**

21. Can 0 ever be the solution of an equation? If so, give an example of an equation for which 0 is a solution.

22. Without solving $x + \frac{13}{15} = -\frac{21}{43}$, determine whether x is less than or greater than $-\frac{21}{43}$. Explain your answer.

For Exercises 23 to 64, solve and check.

23. $x + 5 = 7$

24. $y + 3 = 9$

25. $b - 4 = 11$

26. $z - 6 = 10$

27. $2 + a = 8$

28. $5 + x = 12$

29. $n - 5 = -2$

30. $x - 6 = -5$

31. $b + 7 = 7$

32. $y - 5 = -5$

33. $z + 9 = 2$

34. $n + 11 = 1$

35. $10 + m = 3$

36. $8 + x = 5$

37. $9 + x = -3$

38. $10 + y = -4$

Copyright © Houghton Mifflin Company. All rights reserved.

39. $2 = x + 7$ **40.** $-8 = n + 1$ **41.** $4 = m - 11$ **42.** $-6 = y - 5$

43. $12 = 3 + w$ **44.** $-9 = 5 + x$ **45.** $4 = -10 + b$ **46.** $-7 = -2 + x$

47. $m + \dfrac{2}{3} = -\dfrac{1}{3}$ **48.** $c + \dfrac{3}{4} = -\dfrac{1}{4}$ **49.** $x - \dfrac{1}{2} = \dfrac{1}{2}$ **50.** $x - \dfrac{2}{5} = \dfrac{3}{5}$

51. $\dfrac{5}{8} + y = \dfrac{1}{8}$ **52.** $\dfrac{4}{9} + a = -\dfrac{2}{9}$ **53.** $m + \dfrac{1}{2} = -\dfrac{1}{4}$ **54.** $b + \dfrac{1}{6} = -\dfrac{1}{3}$

55. $x + \dfrac{2}{3} = \dfrac{3}{4}$ **56.** $n + \dfrac{2}{5} = \dfrac{2}{3}$ **57.** $-\dfrac{5}{6} = x - \dfrac{1}{4}$ **58.** $-\dfrac{1}{4} = c - \dfrac{2}{3}$

59. $d + 1.3619 = 2.0148$ **60.** $w + 2.932 = 4.801$

61. $-0.813 + x = -1.096$ **62.** $-1.926 + t = -1.042$

63. $6.149 = -3.108 + z$ **64.** $5.237 = -2.014 + x$

Objective C   **To solve an equation of the form $ax = b$**

65.   Without solving $-\dfrac{15}{41}x = -\dfrac{23}{25}$, determine whether x is less than or greater than 0. Explain your answer.

66.   Explain why multiplying each side of an equation by the reciprocal of the coefficient of the variable is the same as dividing each side of the equation by the coefficient.

For Exercises 67 to 110, solve and check.

67. $5x = -15$ **68.** $4y = -28$ **69.** $3b = 0$ **70.** $2a = 0$

71. $-3x = 6$ **72.** $-5m = 20$ **73.** $-3x = -27$ **74.** $-\dfrac{1}{6}n = -30$

Copyright © Houghton Mifflin Company. All rights reserved.

75. $20 = \dfrac{1}{4}c$

76. $18 = 2t$

77. $0 = -5x$

78. $0 = -8a$

79. $49 = -7t$

80. $\dfrac{x}{3} = 2$

81. $\dfrac{x}{4} = 3$

82. $-\dfrac{y}{2} = 5$

83. $-\dfrac{b}{3} = 6$

84. $\dfrac{3}{4}y = 9$

85. $\dfrac{2}{5}x = 6$

86. $-\dfrac{2}{3}d = 8$

87. $-\dfrac{3}{5}m = 12$

88. $\dfrac{2n}{3} = 0$

89. $\dfrac{5x}{6} = 0$

90. $\dfrac{-3z}{8} = 9$

91. $\dfrac{3x}{4} = 2$

92. $\dfrac{3}{4}c = \dfrac{3}{5}$

93. $\dfrac{2}{9} = \dfrac{2}{3}y$

94. $-\dfrac{6}{7} = -\dfrac{3}{4}b$

95. $\dfrac{1}{5}x = -\dfrac{1}{10}$

96. $-\dfrac{2}{3}y = -\dfrac{8}{9}$

97. $-1 = \dfrac{2n}{3}$

98. $-\dfrac{3}{4} = \dfrac{a}{8}$

99. $-\dfrac{2}{5}m = -\dfrac{6}{7}$

100. $5x + 2x = 14$

101. $3n + 2n = 20$

102. $7d - 4d = 9$

103. $10y - 3y = 21$

104. $2x - 5x = 9$

105. $\dfrac{x}{1.46} = 3.25$

106. $\dfrac{z}{2.95} = -7.88$

107. $3.47a = 7.1482$

108. $2.31m = 2.4255$

109. $-3.7x = 7.881$

110. $\dfrac{n}{2.65} = 9.08$

Copyright © Houghton Mifflin Company. All rights reserved.

Objective D **To solve application problems using the basic percent equation**

111. Without solving an equation, indicate whether 40% of 80 is less than, equal to, or greater than 80% of 40.

112. Without solving an equation, indicate whether $\frac{1}{4}$% of 80 is less than, equal to, or greater than 25% of 80.

113. What is 35% of 80?

114. What percent of 8 is 0.5?

115. Find 1.2% of 60.

116. 8 is what percent of 5?

117. 125% of what is 80?

118. What percent of 20 is 30?

119. 12 is what percent of 50?

120. What percent of 125 is 50?

121. Find 18% of 40.

122. What is 25% of 60?

123. 12% of what is 48?

124. 45% of what is 9?

125. What is $33\frac{1}{3}$% of 27?

126. Find $16\frac{2}{3}$% of 30.

127. What percent of 12 is 3?

128. 10 is what percent of 15?

129. 12 is what percent of 6?

130. 20 is what percent of 16?

131. $5\frac{1}{4}$% of what is 21?

132. $37\frac{1}{2}$% of what is 15?

133. Find 15.4% of 50.

134. What is 18.5% of 46?

135. 1 is 0.5% of what?

136. 3 is 1.5% of what?

137. $\frac{3}{4}$% of what is 3?

138. $\frac{1}{2}$% of what is 3?

139. What is 250% of 12?

Copyright © Houghton Mifflin Company. All rights reserved.

140. 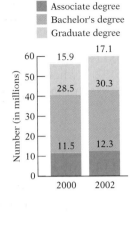 **Education** The stacked-bar graph at the right shows the number of people age 25 or older in the U.S. that have attained some type of degree beyond high school.

   a. In 2002, there were approximately 182.7 million people age 25 or older. What percent of the people age 25 or older had received an associate degree or a bachelor's degree in 2002? Round to the nearest tenth of a percent.

   b. In 2000, there were approximately 177.5 million people age 25 or older. Was the percent of people in 2000 with a graduate degree less than or greater than the percent of people in 2002 with a graduate degree?

■ Associate degree
■ Bachelor's degree
■ Graduate degree

141. **Chemistry** Approximately 21% of air is oxygen. Using this estimate, determine how many liters of oxygen there are in a room containing 21,600 L of air.

142. 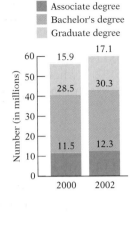 **Record Sales** According to Nielsen SoundScan, there were approximately 680 million record albums sold in the fourth quarter of 2002. This is about 39% of the total number of record albums sold that year. How many record albums were sold in 2002? Round to the nearest million.

143. **Income** According to the U.S. Census Bureau, the median income fell 1.1% between two successive years. If the median income before the decline was $42,900, what was the median income the next year? Round to the nearest dollar.

144. **Government** To override a presidential veto, at least $66\frac{2}{3}$% of the Senate must vote to override the veto. There are 100 senators in the Senate. What is the minimum number of votes needed to override a veto?

145. **Sports** According to **www.superbowl.com,** approximately 138.9 million people watched Super Bowl XXXVIII. What percent of the U.S. population watched Super Bowl XXXVIII? Use a figure of 290 million for the U.S. population. Round to the nearest tenth of a percent.

146. **Advertising** Suppose 9.4 million people watch a 30-second commercial for a new cellular phone during a broadcast of the TV show *CSI*. The cost of that commercial was approximately $470,000. If the cellular phone manufacturer makes a profit of $10 on every phone sold, what percent of the people watching the commercial would have to buy one phone for the company to recover the cost of the commercial? *Source:* Nielsen Media Research/ San Diego Union

147. **School Enrollment** The circle graph at the right shows the percent of the U.S. population over three years old that are enrolled in school. To answer the question "How many people are enrolled in college or graduate school?," what additional piece of information is necessary?

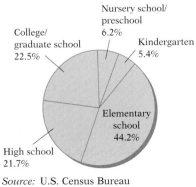

Source: U.S. Census Bureau

Copyright © Houghton Mifflin Company. All rights reserved.

148. **Investment**   If Kachina Caron invested $1200 in a simple interest account and earned $72 in 8 months, what is the annual interest rate?

149. **Investment**   How much money must Andrea invest for two years in an account that earns an annual interest rate of 8% if she wants to earn $300 from the investment?

150. **Investment**   Sal Boxer decided to divide a gift of $3000 into two different accounts. He placed $1000 in one account that earns an annual simple interest rate of 7.5%. The remaining money was placed in an account that earns an annual simple interest rate of 8.25%. How much interest will Sal earn from the two accounts after one year?

151. **Investment**   If Americo invests $2500 at an 8% annual simple interest rate and Octavia invests $3000 at a 7% annual simple interest rate, which of the two will earn the greater amount of interest after one year?

152. **Investment**   Makana invested $900 in a simple interest account that had an interest rate that was 1% more than that of her friend Marlys. If Marlys earned $51 after one year from an investment of $850, how much would Makana earn in one year?

153. **Investment**   A $2000 investment at an annual simple interest rate of 6% earned as much interest after one year as another investment in an account that earns 8% simple interest. How much was invested at 8%?

154. **Investment**   An investor placed $1000 in an account that earns 9% annual simple interest and $1000 in an account that earns 6% annual simple interest. If each investment is left in the account for the same period of time, is the interest rate on the combined investment less than 6%, between 6% and 9%, or greater than 9%?

155. **Metallurgy**   The concentration of platinum in a necklace is 15%. If the necklace weighs 12 g, find the amount of platinum in the necklace.

156. **Dye Mixtures**   A 250-milliliter solution of a fabric dye contains 5 ml of hydrogen peroxide. What is the percent concentration of the hydrogen peroxide?

157. **Fabric Mixtures**   A carpet is made with a blend of wool and other fibers. If the concentration of wool in the carpet is 75% and the carpet weighs 175 lb, how much wool is in the carpet?

158. **Juice Mixtures**   Apple Dan's 32-ounce apple-flavored fruit drink contains 8 oz of apple juice. A 40-ounce generic brand of an apple-flavored fruit drink contains 9 oz of apple juice. Which of the two brands has the greater concentration of apple juice?

Copyright © Houghton Mifflin Company. All rights reserved.

159. **Food Mixtures** Bakers use simple syrup in many of their recipes. Simple syrup is made by combining 500 g of sugar with 500 g of water and mixing it well until the sugar dissolves. What is the percent concentration of sugar in the simple syrup?

160. **Pharmacology** A pharmacist has 50 g of a topical cream that contains 75% glycerine. How many grams of the cream is not glycerine?

161. **Chemistry** A chemist has 100 ml of a solution that is 9% acetic acid. If the chemist adds 50 ml of pure water to this solution, what is the percent concentration of the resulting mixture?

162. **Chemistry** A 500-gram salt and water solution contains 50 g of salt. This mixture is left in the open air and 100 g of water evaporates from the solution. What is the percent concentration of salt in the remaining solution?

Objective E **To solve uniform motion problems**

163. As part of the training program for the Boston Marathon, a runner wants to build endurance by running at a rate of 9 mph for 20 min. How far will the runner travel in that time period?

164. It takes a hospital dietician 40 min to drive from home to the hospital, a distance of 20 mi. What is the dietician's average rate of speed?

165. Marcella leaves home at 9:00 A.M. and drives to school, arriving at 9:45 A.M. If the distance between home and school is 27 mi, what is Marcella's average rate of speed?

166. The Ride for Health Bicycle Club has chosen a 36-mile course for this Saturday's ride. If the riders plan on averaging 12 mph while they are riding, and they have a 1-hour lunch break planned, how long will it take them to complete the trip?

167. Palmer's average running speed is 3 kilometers per hour faster than his walking speed. If Palmer can run around a 30-kilometer course in 2 h, how many hours would it take for Palmer to walk the same course?

168. A shopping mall has a moving sidewalk that takes shoppers from the shopping area to the parking garage, a distance of 250 ft. If your normal walking rate is 5 ft/s and the moving sidewalk is traveling at 3 ft/s, how many seconds would it take for you to walk from one end of the moving sidewalk to the other end?

Copyright © Houghton Mifflin Company. All rights reserved.

169. Two joggers start at the same time from opposite ends of an 8-mile jogging trail and begin running toward each other. One jogger is running at the rate of 5 mph, and the other jogger is running at a rate of 7 mph. How long, in minutes, after they start will the two joggers meet?

170. Two cyclists start from the same point at the same time and move in opposite directions. One cyclist is traveling at 8 mph, and the other cyclist is traveling at 9 mph. After 30 min, how far apart are the two cyclists?

171. Petra and Celine can paddle their canoe at a rate of 10 mph in calm water. How long will it take them to travel 4 mi against the 2 mph current of the river?

172. At 8:00 A.M., a train leaves a station and travels at a rate of 45 mph. At 9:00 A.M., a second train leaves the same station on the same track and travels in the direction of the first train at a speed of 60 mph. At 10:00 A.M., how far apart are the two trains?

APPLYING THE CONCEPTS

173. **Geometry**   Solve for x.

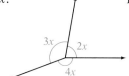

174. **Geometry**   Solve for x.

175. **Geometry**   Solve for x.

176. **Geometry**   Solve for x.

177. **a.** Make up an equation of the form $x + a = b$ that has 2 as a solution.
b. Make up an equation of the form $ax = b$ that has -1 as a solution.

178. Write out the steps for solving the equation $\frac{1}{2}x = -3$. Identify each Property of Real Numbers or Property of Equations as you use it.

179. In your own words, state the Addition Property of Equations and the Multiplication Property of Equations.

180. If a quantity increases by 100%, how many times its original value is the new value?

Copyright © Houghton Mifflin Company. All rights reserved.

3.2 General Equations—Part I

Objective A To solve an equation of the form $ax + b = c$

In solving an equation of the form $ax + b = c$, the goal is to rewrite the equation in the form *variable = constant*. This requires the application of both the Addition and the Multiplication Properties of Equations.

HOW TO Solve: $\dfrac{3}{4}x - 2 = -11$

The goal is to write the equation in the form *variable = constant*.

$$\frac{3}{4}x - 2 = -11$$

$$\frac{3}{4}x - 2 + 2 = -11 + 2 \qquad \bullet \textbf{ Add 2} \text{ to each side of the equation.}$$

$$\frac{3}{4}x = -9 \qquad \bullet \textbf{ Simplify.}$$

$$\frac{4}{3} \cdot \frac{3}{4}x = \frac{4}{3}(-9) \qquad \bullet \textbf{ Multiply} \text{ each side of the equation by } \frac{4}{3}.$$

$$x = -12 \qquad \bullet \text{ The equation is in the form } \textit{variable = constant.}$$

The solution is -12.

TAKE NOTE

Check: $\dfrac{3}{4}x - 2 = -11$

$\dfrac{3}{4}(-12) - 2 \mid -11$

$-9 - 2 \mid -11$

$-11 = -11$

A true equation

Here is an example of solving an equation that contains more than one fraction.

HOW TO Solve: $\dfrac{2}{3}x + \dfrac{1}{2} = \dfrac{3}{4}$

$$\frac{2}{3}x + \frac{1}{2} = \frac{3}{4}$$

$$\frac{2}{3}x + \frac{1}{2} - \frac{1}{2} = \frac{3}{4} - \frac{1}{2} \qquad \bullet \textbf{ Subtract } \frac{1}{2} \text{ from each side of the equation.}$$

$$\frac{2}{3}x = \frac{1}{4} \qquad \bullet \textbf{ Simplify.}$$

$$\frac{3}{2}\left(\frac{2}{3}x\right) = \frac{3}{2}\left(\frac{1}{4}\right) \qquad \bullet \textbf{ Multiply } \text{each side of the equation by } \frac{3}{2},$$
$$\text{the reciprocal of } \frac{2}{3}.$$

$$x = \frac{3}{8}$$

The solution is $\dfrac{3}{8}$.

It may be easier to solve an equation containing two or more fractions by multiplying each side of the equation by the least common multiple (LCM) of the denominators. For the equation above, the LCM of 3, 2, and 4 is 12. The LCM has the property that 3, 2, and 4 will divide evenly into it. Therefore, if both sides of the equation are multiplied by 12, the denominators will divide evenly into 12. The result is an equation that does not contain any fractions. Multiplying each side of an equation that contains fractions by the LCM of the denominators is called **clearing denominators.** It is an alternative method, as we show in the next example, of solving an equation that contains fractions.

Copyright © Houghton Mifflin Company. All rights reserved.

HOW TO   Solve: $\dfrac{2}{3}x + \dfrac{1}{2} = \dfrac{3}{4}$

TAKE NOTE
Observe that after we multiply by the LCM and simplify, the equation no longer contains fractions.

$$\dfrac{2}{3}x + \dfrac{1}{2} = \dfrac{3}{4}$$

$$12\left(\dfrac{2}{3}x + \dfrac{1}{2}\right) = 12\left(\dfrac{3}{4}\right)$$

- Multiply each side of the equation by **12**, the LCM of 3, 2, and 4.

$$12\left(\dfrac{2}{3}x\right) + 12\left(\dfrac{1}{2}\right) = 12\left(\dfrac{3}{4}\right)$$

- Use the Distributive Property.

$$8x + 6 = 9$$

- Simplify.

$$8x + 6 - 6 = 9 - 6$$

- **Subtract 6** from each side of the equation.

$$8x = 3$$

$$\dfrac{8x}{8} = \dfrac{3}{8}$$

- Divide each side of the equation by **8**.

$$x = \dfrac{3}{8}$$

The solution is $\dfrac{3}{8}$.

Note that both methods give exactly the same solution. You may use either method to solve an equation containing fractions.

Example 1   Solve: $3x - 7 = -5$

Solution

$$\begin{aligned}3x - 7 &= -5\\ 3x - 7 + 7 &= -5 + 7 \qquad \text{• \textbf{Add 7} to each side.}\\ 3x &= 2\\ \dfrac{3x}{3} &= \dfrac{2}{3} \qquad\qquad \text{• Divide each side by \textbf{3}.}\\ x &= \dfrac{2}{3}\end{aligned}$$

The solution is $\dfrac{2}{3}$.

You Try It 1   Solve: $5x + 7 = 10$

Your solution

Example 2   Solve: $5 = 9 - 2x$

Solution

$$\begin{aligned}5 &= 9 - 2x\\ 5 - 9 &= 9 - 9 - 2x \qquad \text{• \textbf{Subtract 9} from each side.}\\ -4 &= -2x\\ \dfrac{-4}{-2} &= \dfrac{-2x}{-2} \qquad\qquad \text{• Divide each side by \textbf{−2}.}\\ 2 &= x\end{aligned}$$

The solution is 2.

You Try It 2   Solve: $2 = 11 + 3x$

Your solution

Copyright © Houghton Mifflin Company. All rights reserved.

Solutions on p. S6

Example 3 Solve: $\dfrac{2}{3} - \dfrac{x}{2} = \dfrac{3}{4}$

You Try It 3 Solve: $\dfrac{5}{8} - \dfrac{2x}{3} = \dfrac{5}{4}$

Solution

$$\dfrac{2}{3} - \dfrac{x}{2} = \dfrac{3}{4}$$

$$\dfrac{2}{3} - \dfrac{2}{3} - \dfrac{x}{2} = \dfrac{3}{4} - \dfrac{2}{3}$$ • Subtract $\dfrac{2}{3}$ from each side.

$$-\dfrac{x}{2} = \dfrac{1}{12}$$

$$-2\left(-\dfrac{x}{2}\right) = -2\left(\dfrac{1}{12}\right)$$ • Multiply each side by -2.

$$x = -\dfrac{1}{6}$$

The solution is $-\dfrac{1}{6}$.

Your solution

Example 4 Solve $\dfrac{4}{5}x - \dfrac{1}{2} = \dfrac{3}{4}$ by first clearing denominators.

You Try It 4 Solve $\dfrac{2}{3}x + 3 = \dfrac{7}{2}$ by first clearing denominators.

Solution
The LCM of 5, 2, and 4 is 20.

$$\dfrac{4}{5}x - \dfrac{1}{2} = \dfrac{3}{4}$$

$$20\left(\dfrac{4}{5}x - \dfrac{1}{2}\right) = 20\left(\dfrac{3}{4}\right)$$ • Multiply each side by **20**.

$$20\left(\dfrac{4}{5}x\right) - 20\left(\dfrac{1}{2}\right) = 20\left(\dfrac{3}{4}\right)$$ • Use the Distributive Property.

$$16x - 10 = 15$$

$$16x - 10 + 10 = 15 + 10$$ • Add **10** to each side.

$$16x = 25$$

$$\dfrac{16x}{16} = \dfrac{25}{16}$$ • Divide each side by **16**.

$$x = \dfrac{25}{16}$$

The solution is $\dfrac{25}{16}$.

Your solution

Solutions on p. S6

Copyright © Houghton Mifflin Company. All rights reserved.

Copyright © Houghton Mifflin Company. All rights reserved.

Example 5

Solve: $2x + 4 - 5x = 10$

Solution

$$
\begin{aligned}
2x + 4 - 5x &= 10 \\
-3x + 4 &= 10 \\
-3x + 4 - 4 &= 10 - 4 \\
-3x &= 6 \\
\frac{-3x}{-3} &= \frac{6}{-3} \\
x &= -2
\end{aligned}
$$

$\quad$ • **Combine like terms.**

The solution is -2.

You Try It 5

Solve: $x - 5 + 4x = 25$

Your solution

Solution on p. S7

Objective B   **To solve application problems using formulas**

In this objective we will be solving application problems using formulas. Two of the formulas we will use are related to markup and discount.

Markup

Cost

Selling price

Cost is the price a business pays for a product. **Selling price** is the price for which a business sells a product to a customer. The difference between selling price and cost is called **markup.** Markup is added to the cost to cover the expenses of operating a business. The diagram at the left illustrates these terms. The total length is the selling price. One part of the diagram is the cost, and the other part is the markup.

When the markup is expressed as a percent of the retailer's cost, it is called the **markup rate.**

The **basic markup equations** used by a business are

$$\text{Selling price} = \text{cost} + \text{markup} \qquad \text{Markup} = \text{markup rate} \cdot \text{cost}$$
$$S = C + M \qquad\qquad M = r \cdot C$$

Substituting $r \cdot C$ for M in the first equation results in $S = C + (r \cdot C)$, or $S = C + rC$.

HOW TO The manager of a clothing store buys a jacket for $80 and sells the jacket for $116. Find the markup rate.

$$
\begin{aligned}
S &= C + rC \\
116 &= 80 + 80r \\
36 &= 80r \\
\frac{36}{80} &= \frac{80r}{80} \\
0.45 &= r
\end{aligned}
$$

$\quad$ • Use the equation $S = C + rC$.

$\quad$ • Given: $C = \$80$ and $S = \$116$

$\quad$ • Subtract 80 from each side of the equation.

$\quad$ • Divide both sides of the equation by 80.

The markup rate is 45%.

Discount or markdown

Regular price

Sale price

A retailer may reduce the regular price of a product because the goods are damaged, odd sizes, or discontinued items. The **discount,** or **markdown,** is the amount by which a retailer reduces the regular price of a product. The percent discount is called the **discount rate** and is usually expressed as a percent of the original selling price (the regular price).

The **basic discount equations** used by a business are

$$\frac{\text{Sale}}{\text{price}} = \frac{\text{regular}}{\text{price}} - \text{discount} \qquad \text{Discount} = \frac{\text{discount}}{\text{rate}} \cdot \frac{\text{regular}}{\text{price}}$$

$$S = R - D \qquad\qquad D = r \cdot R$$

Substituting $r \cdot R$ for D in the first equation results in $S = R - (r \cdot R)$, or $S = R - rR$.

> **HOW TO** A portable computer that regularly sells for $1850 is on sale for $1480. Find the discount rate.
>
> $$S = R - rR$$ • Use the equation $S = R - rR$.
>
> $$1480 = 1850 - 1850r$$ • Given: $S = \$1480$ and $R = \$1850$
>
> $$-370 = -1850r$$ • Subtract 1850 from each side of the equation.
>
> $$\frac{-370}{-1850} = \frac{-1850r}{-1850}$$ • Divide each side of the equation by -1850.
>
> $$0.2 = r$$

The discount rate on the portable computer is 20%.

Example 6

A markup rate of 40% was used on a refrigerator that has a selling price of $749. Find the cost of the refrigerator. Use the formula $S = C + rC$.

Strategy

Given: $S = \$749$
$r = 40\% = 0.40$
Unknown: C

Solution

$$S = C + rC$$
$$749 = C + 0.40C \quad \bullet\ C + 0.40C = 1C + 0.40C$$
$$749 = 1.40C \quad\quad \bullet\ \text{Combine like terms.}$$
$$\frac{749}{1.40} = \frac{1.40C}{1.40}$$
$$535 = C$$

The cost of the refrigerator is $535.

You Try It 6

A markup rate of 45% was used on an outboard motor that has a selling price of $986. Find the cost of the outboard motor. Use the formula $S = C + rC$.

Your strategy

Your solution

Solution on p. S7

Copyright © Houghton Mifflin Company. All rights reserved.

Example 7

A necklace that is marked down 35% has a sale price of $292.50. Find the regular price of the necklace. Use the formula $S = R - rR$.

Strategy

Given: $S = 292.50$

$\quad\quad\quad r = 35\% = 0.35$

Unknown: R

Solution

$$S = R - rR$$
$$292.50 = R - 0.35R \quad \bullet\ R - 0.35R = 1R - 0.35R$$
$$292.50 = 0.65R \quad\quad \bullet\ \text{Combine like terms.}$$
$$\frac{292.50}{0.65} = \frac{0.65R}{0.65}$$
$$450 = R$$

The regular price of the necklace is $450.

You Try It 7

A garage door opener, marked down 25%, is on sale for $159. Find the regular price of the garage door opener. Use the formula $S = R - rR$.

Your strategy

Your solution

Example 8

To determine the total cost of production, an economist uses the equation $T = U \cdot N + F$, where T is the total cost, U is the unit cost, N is the number of units made, and F is the fixed cost. Use this equation to find the number of units made during a month when the total cost was $9000, the unit cost was $25, and the fixed cost was $3000.

Strategy

Given: $T = \$9000$

$\quad\quad\quad U = \$25$

$\quad\quad\quad F = \$3000$

Unknown: N

Solution

$$T = U \cdot N + F$$
$$9000 = 25N + 3000$$
$$6000 = 25N$$
$$\frac{6000}{25} = \frac{25N}{25}$$
$$240 = N$$

There were 240 units made.

You Try It 8

The pressure at a certain depth in the ocean can be approximated by the equation $P = 15 + \frac{1}{2}D$, where P is the pressure in pounds per square inch and D is the depth in feet. Use this equation to find the depth when the pressure is 45 pounds per square inch.

Your strategy

Your solution

Solutions on p. S7

Copyright © Houghton Mifflin Company. All rights reserved.

3.2 Exercises

Objective A   To solve an equation of the form $ax + b = c$

For Exercises 1 to 80, solve and check.

1. $3x + 1 = 10$ **2.** $4y + 3 = 11$ **3.** $2a - 5 = 7$ **4.** $5m - 6 = 9$

5. $5 = 4x + 9$ **6.** $2 = 5b + 12$ **7.** $2x - 5 = -11$ **8.** $3n - 7 = -19$

9. $4 - 3w = -2$ **10.** $5 - 6x = -13$ **11.** $8 - 3t = 2$ **12.** $12 - 5x = 7$

13. $4a - 20 = 0$ **14.** $3y - 9 = 0$ **15.** $6 + 2b = 0$ **16.** $10 + 5m = 0$

17. $-2x + 5 = -7$ **18.** $-5d + 3 = -12$ **19.** $-12x + 30 = -6$ **20.** $-13 = -11y + 9$

21. $2 = 7 - 5a$ **22.** $3 = 11 - 4n$ **23.** $-35 = -6b + 1$ **24.** $-8x + 3 = -29$

25. $-3m - 21 = 0$ **26.** $-5x - 30 = 0$ **27.** $-4y + 15 = 15$ **28.** $-3x + 19 = 19$

29. $9 - 4x = 6$ **30.** $3t - 2 = 0$ **31.** $9x - 4 = 0$ **32.** $7 - 8z = 0$

33. $1 - 3x = 0$ **34.** $9d + 10 = 7$ **35.** $12w + 11 = 5$ **36.** $6y - 5 = -7$

37. $8b - 3 = -9$ **38.** $5 - 6m = 2$ **39.** $7 - 9a = 4$ **40.** $9 = -12c + 5$

Copyright © Houghton Mifflin Company. All rights reserved.

41. $10 = -18x + 7$ **42.** $2y + \dfrac{1}{3} = \dfrac{7}{3}$ **43.** $4a + \dfrac{3}{4} = \dfrac{19}{4}$ **44.** $2n - \dfrac{3}{4} = \dfrac{13}{4}$

45. $3x - \dfrac{5}{6} = \dfrac{13}{6}$ **46.** $5y + \dfrac{3}{7} = \dfrac{3}{7}$ **47.** $9x + \dfrac{4}{5} = \dfrac{4}{5}$ **48.** $8 = 7d - 1$

49. $8 = 10x - 5$ **50.** $4 = 7 - 2w$ **51.** $7 = 9 - 5a$ **52.** $8t + 13 = 3$

53. $12x + 19 = 3$ **54.** $-6y + 5 = 13$ **55.** $-4x + 3 = 9$ **56.** $\dfrac{1}{2}a - 3 = 1$

57. $\dfrac{1}{3}m - 1 = 5$ **58.** $\dfrac{2}{5}y + 4 = 6$ **59.** $\dfrac{3}{4}n + 7 = 13$ **60.** $-\dfrac{2}{3}x + 1 = 7$

61. $-\dfrac{3}{8}b + 4 = 10$ **62.** $\dfrac{x}{4} - 6 = 1$ **63.** $\dfrac{y}{5} - 2 = 3$ **64.** $\dfrac{2x}{3} - 1 = 5$

65. $\dfrac{2}{3}x - \dfrac{5}{6} = -\dfrac{1}{3}$ **66.** $\dfrac{5}{4}x + \dfrac{2}{3} = \dfrac{1}{4}$ **67.** $\dfrac{1}{2} - \dfrac{2}{3}x = \dfrac{1}{4}$ **68.** $\dfrac{3}{4} - \dfrac{3}{5}x = \dfrac{19}{20}$

69. $\dfrac{3}{2} = \dfrac{5}{6} + \dfrac{3x}{8}$ **70.** $-\dfrac{1}{4} = \dfrac{5}{12} + \dfrac{5x}{6}$ **71.** $\dfrac{11}{27} = \dfrac{4}{9} - \dfrac{2x}{3}$ **72.** $\dfrac{37}{24} = \dfrac{7}{8} - \dfrac{5x}{6}$

73. $7 = \dfrac{2x}{5} + 4$ **74.** $5 - \dfrac{4c}{7} = 8$ **75.** $7 - \dfrac{5}{9}y = 9$ **76.** $6a + 3 + 2a = 11$

77. $5y + 9 + 2y = 23$ **78.** $7x - 4 - 2x = 6$ **79.** $11z - 3 - 7z = 9$ **80.** $2x - 6x + 1 = 9$

81. Solve $3x + 4y = 13$ when $y = -2$.   **82.** Solve $2x - 3y = 8$ when $y = 0$.

83. Solve $-4x + 3y = 9$ when $x = 0$.   **84.** Solve $5x - 2y = -3$ when $x = -3$.

Copyright © Houghton Mifflin Company. All rights reserved.

85. If $2x - 3 = 7$, evaluate $3x + 4$.

86. If $3x + 5 = -4$, evaluate $2x - 5$.

87. If $4 - 5x = -1$, evaluate $x^2 - 3x + 1$.

88. If $2 - 3x = 11$, evaluate $x^2 + 2x - 3$.

89. If $5x + 3 - 2x = 12$, evaluate $4 - 5x$.

90. If $2x - 4 - 7x = 16$, evaluate $x^2 + 1$.

Objective B   **To solve application problems using formulas**

Business   For Exercises 91 to 98, solve. Use the markup equation $S = C + rC$, where S is selling price, C is cost, and r is the markup rate.

91. A watch costing $98 is sold for $156.80. Find the markup rate on the watch.

92. A set of golf clubs costing $360 is sold for $630. Find the markup rate on the set of golf clubs.

93. A markup rate of 40% was used on a basketball with a selling price of $82.60. Find the cost of the basketball.

94. A portable tape player with a selling price of $57 has a markup rate of 50%. Find the cost of the tape player.

95. A freezer costing $360 is sold for $520. Find the markup rate. Round to the nearest tenth of a percent.

96. A sofa costing $320 is sold for $479. Find the markup rate. Round to the nearest tenth of a percent.

97. A digitally recorded compact disc has a selling price of $11.90. The markup rate is 40%. Find the cost of the CD.

98. A markup rate of 25% is used on a computer that has a selling price of $2187.50. Find the cost of the computer.

Copyright © Houghton Mifflin Company. All rights reserved.

Example 2   Solve: $3x + 4 - 5x = 2 - 4x$

Solution

$3x + 4 - 5x = 2 - 4x$

$-2x + 4 = 2 - 4x$ • Combine like terms.

$-2x + 4x + 4 = 2 - 4x + 4x$ • Add **4x** to each side.

$2x + 4 = 2$

$2x + 4 - 4 = 2 - 4$ • Subtract **4** from each side.

$2x = -2$

$\dfrac{2x}{2} = \dfrac{-2}{2}$ • Divide each side by **2**.

$x = -1$

The solution is -1.

You Try It 2   Solve: $5x - 10 - 3x = 6 - 4x$

Your solution

Solution on p. S7

Objective B   **To solve an equation containing parentheses**

When an equation contains parentheses, one of the steps in solving the equation requires the use of the Distributive Property. The Distributive Property is used to remove parentheses from a variable expression.

HOW TO   Solve: $4 + 5(2x - 3) = 3(4x - 1)$

$4 + 5(2x - 3) = 3(4x - 1)$

$4 + 10x - 15 = 12x - 3$ • Use the Distributive Property. Then simplify.

$10x - 11 = 12x - 3$

$10x - 12x - 11 = 12x - 12x - 3$ • Subtract **12x** from each side of the equation.

$-2x - 11 = -3$ • Simplify.

$-2x - 11 + 11 = -3 + 11$ • Add **11** to each side of the equation.

$-2x = 8$ • Simplify.

$\dfrac{-2x}{-2} = \dfrac{8}{-2}$ • Divide each side of the equation by **−2**.

$x = -4$ • The equation is in the form *variable = constant*.

The solution is -4. You should verify this by checking this solution.

In the next example, we solve an equation with parentheses and decimals.

Copyright © Houghton Mifflin Company. All rights reserved.

HOW TO Solve: $16 + 0.55x = 0.75(x + 20)$

$$16 + 0.55x = 0.75(x + 20)$$
$$16 + 0.55x = 0.75x + 15$$ • Use the Distributive Property.
$$16 + 0.55x - 0.75x = 0.75x - 0.75x + 15$$ • **Subtract 0.75x** from each side of the equation.
$$16 - 0.20x = 15$$ • Simplify.
$$16 - 16 - 0.20x = 15 - 16$$ • **Subtract 16** from each side of the equation.
$$-0.20x = -1$$ • Simplify.
$$\frac{-0.20x}{-0.20} = \frac{-1}{-0.20}$$ • Divide each side of the equation by **-0.20**.
$$x = 5$$ • The equation is in the form *variable = constant*.

The solution is 5.

Example 3

Solve: $3x - 4(2 - x) = 3(x - 2) - 4$

Solution

$$3x - 4(2 - x) = 3(x - 2) - 4$$
$$3x - 8 + 4x = 3x - 6 - 4$$ • Distributive Property
$$7x - 8 = 3x - 10$$
$$7x - 3x - 8 = 3x - 3x - 10$$ • Subtract 3x.
$$4x - 8 = -10$$
$$4x - 8 + 8 = -10 + 8$$ • Add 8.
$$4x = -2$$
$$\frac{4x}{4} = \frac{-2}{4}$$ • Divide by 4.
$$x = -\frac{1}{2}$$

The solution is $-\frac{1}{2}$.

You Try It 3

Solve: $5x - 4(3 - 2x) = 2(3x - 2) + 6$

Your solution

Example 4

Solve: $3[2 - 4(2x - 1)] = 4x - 10$

Solution

$$3[2 - 4(2x - 1)] = 4x - 10$$
$$3[2 - 8x + 4] = 4x - 10$$ • Distributive Property
$$3[6 - 8x] = 4x - 10$$
$$18 - 24x = 4x - 10$$ • Distributive Property
$$18 - 24x - 4x = 4x - 4x - 10$$ • Subtract 4x.
$$18 - 28x = -10$$
$$18 - 18 - 28x = -10 - 18$$ • Subtract 18.
$$-28x = -28$$
$$\frac{-28x}{-28} = \frac{-28}{-28}$$ • Divide by -28.
$$x = 1$$

The solution is 1.

You Try It 4

Solve: $-2[3x - 5(2x - 3)] = 3x - 8$

Your solution

Solutions on p. S7

Copyright © Houghton Mifflin Company. All rights reserved.

Objective C   **To solve application problems using formulas**

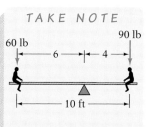

TAKE NOTE

60 lb   90 lb

|← 6 →|← 4 →|

|← 10 ft →|

This system balances because

$F_1x = F_2(d - x)$
$60(6) = 90(10 - 6)$
$60(6) = 90(4)$
$360 = 360$

A lever system is shown at the right. It consists of a lever, or bar; a fulcrum; and two forces, F_1 and F_2. The distance d represents the length of the lever, x represents the distance from F_1 to the fulcrum, and $d - x$ represents the distance from F_2 to the fulcrum.

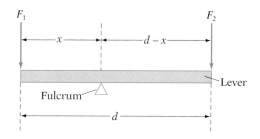

A principle of physics states that when the lever system balances, $F_1x = F_2(d - x)$.

Example 5

A lever is 15 ft long. A force of 50 lb is applied to one end of the lever, and a force of 100 lb is applied to the other end. Where is the fulcrum located when the system balances?

Strategy
Make a drawing.

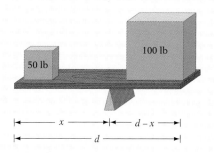

Given: $F_1 = 50$
          $F_2 = 100$
          $d = 15$
Unknown: x

Solution
$$F_1x = F_2(d - x)$$
$$50x = 100(15 - x)$$
$$50x = 1500 - 100x$$
$$50x + 100x = 1500 - 100x + 100x$$
$$150x = 1500$$
$$\frac{150x}{150} = \frac{1500}{150}$$
$$x = 10$$

The fulcrum is 10 ft from the 50-pound force.

You Try It 5

A lever is 25 ft long. A force of 45 lb is applied to one end of the lever, and a force of 80 lb is applied to the other end. Where is the location of the fulcrum when the system balances?

Your strategy

Your solution

Solution on p. S8

Copyright © Houghton Mifflin Company. All rights reserved.

3.3 Exercises

To solve an equation of the form *ax* + *b* = *cx* + *d*

For Exercises 1 to 27, solve and check.

1. $8x + 5 = 4x + 13$

2. $6y + 2 = y + 17$

3. $5x - 4 = 2x + 5$

4. $13b - 1 = 4b - 19$

5. $15x - 2 = 4x - 13$

6. $7a - 5 = 2a - 20$

7. $3x + 1 = 11 - 2x$

8. $n - 2 = 6 - 3n$

9. $2x - 3 = -11 - 2x$

10. $4y - 2 = -16 - 3y$

11. $2b + 3 = 5b + 12$

12. $m + 4 = 3m + 8$

13. $4y - 8 = y - 8$

14. $5a + 7 = 2a + 7$

15. $6 - 5x = 8 - 3x$

16. $10 - 4n = 16 - n$

17. $5 + 7x = 11 + 9x$

18. $3 - 2y = 15 + 4y$

19. $2x - 4 = 6x$

20. $2b - 10 = 7b$

21. $8m = 3m + 20$

22. $9y = 5y + 16$

23. $8b + 5 = 5b + 7$

24. $6y - 1 = 2y + 2$

25. $7x - 8 = x - 3$

26. $2y - 7 = -1 - 2y$

27. $2m - 1 = -6m + 5$

28. If $5x = 3x - 8$, evaluate $4x + 2$.

29. If $7x + 3 = 5x - 7$, evaluate $3x - 2$.

30. If $2 - 6a = 5 - 3a$, evaluate $4a^2 - 2a + 1$.

31. If $1 - 5c = 4 - 4c$, evaluate $3c^2 - 4c + 2$.

32. If $2y + 3 = 5 - 4y$, evaluate $6y - 7$.

33. If $3z + 1 = 1 - 5z$, evaluate $3z^2 - 7z + 8$.

Copyright © Houghton Mifflin Company. All rights reserved.

Objective B   **To solve an equation containing parentheses**

For Exercises 34 to 54, solve and check.

34. $5x + 2(x + 1) = 23$

35. $6y + 2(2y + 3) = 16$

36. $9n - 3(2n - 1) = 15$

37. $12x - 2(4x - 6) = 28$

38. $7a - (3a - 4) = 12$

39. $9m - 4(2m - 3) = 11$

40. $5(3 - 2y) + 4y = 3$

41. $4(1 - 3x) + 7x = 9$

42. $5y - 3 = 7 + 4(y - 2)$

43. $0.22(x + 6) = 0.2x + 1.8$

44. $0.05(4 - x) + 0.1x = 0.32$

45. $0.3x + 0.3(x + 10) = 300$

46. $2a - 5 = 4(3a + 1) - 2$

47. $5 - (9 - 6x) = 2x - 2$

48. $7 - (5 - 8x) = 4x + 3$

49. $3[2 - 4(y - 1)] = 3(2y + 8)$

50. $5[2 - (2x - 4)] = 2(5 - 3x)$

51. $3a + 2[2 + 3(a - 1)] = 2(3a + 4)$

52. $5 + 3[1 + 2(2x - 3)] = 6(x + 5)$

53. $-2[4 - (3b + 2)] = 5 - 2(3b + 6)$

54. $-4[x - 2(2x - 3)] + 1 = 2x - 3$

55. If $4 - 3a = 7 - 2(2a + 5)$, evaluate $a^2 + 7a$.

56. If $9 - 5x = 12 - (6x + 7)$, evaluate $x^2 - 3x - 2$.

57. If $2z - 5 = 3(4z + 5)$, evaluate $\dfrac{z^2}{z - 2}$.

58. If $3n - 7 = 5(2n + 7)$, evaluate $\dfrac{n^2}{2n - 6}$.

Copyright © Houghton Mifflin Company. All rights reserved.

Objective C **To solve application problems using formulas**

Physics For Exercises 59 to 65, solve. Use the lever system equation $F_1x = F_2(d - x)$.

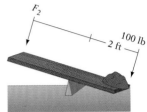

59. A lever 10 ft long is used to move a 100-pound rock. The fulcrum is placed 2 ft from the rock. What force must be applied to the other end of the lever to move the rock?

60. An adult and a child are on a seesaw 14 ft long. The adult weighs 175 lb and the child weighs 70 lb. How many feet from the child must the fulcrum be placed so that the seesaw balances?

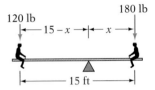

61. Two people are sitting 15 ft apart on a seesaw. One person weighs 180 lb. The second person weighs 120 lb. How far from the 180-pound person should the fulcrum be placed so that the seesaw balances?

62. Two children are sitting on a seesaw that is 12 ft long. One child weighs 60 lb. The other child weighs 90 lb. How far from the 90-pound child should the fulcrum be placed so that the seesaw balances?

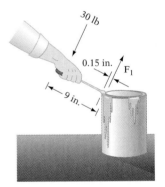

63. In preparation for a stunt, two acrobats are standing on a plank 18 ft long. One acrobat weighs 128 lb and the second acrobat weighs 160 lb. How far from the 128-pound acrobat must the fulcrum be placed so that the acrobats are balanced on the plank?

64. A screwdriver 9 in. long is used as a lever to open a can of paint. The tip of the screwdriver is placed under the lip of the can with the fulcrum 0.15 in. from the lip. A force of 30 lb is applied to the other end of the screwdriver. Find the force on the lip of the can.

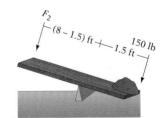

65. A metal bar 8 ft long is used to move a 150-pound rock. The fulcrum is placed 1.5 ft from the rock. What minimum force must be applied to the other end of the bar to move the rock? Round to the nearest tenth.

Business To determine the break-even point, or the number of units that must be sold so that no profit or loss occurs, an economist uses the formula $Px = Cx + F$, where P is the selling price per unit, x is the number of units that must be sold to break even, C is the cost to make each unit, and F is the fixed cost. Use this equation for Exercises 66 to 71.

66. A business analyst has determined that the selling price per unit for a laser printer is $1600. The cost to make one laser printer is $950, and the fixed cost is $211,250. Find the break-even point.

Copyright © Houghton Mifflin Company. All rights reserved.

67. A business analyst has determined that the selling price per unit for a gas barbecue is $325. The cost to make one gas barbecue is $175, and the fixed cost is $39,000. Find the break-even point.

68. A manufacturer of thermostats determines that the cost per unit for a programmable thermostat is $38 and that the fixed cost is $24,400. The selling price for the thermostat is $99. Find the break-even point.

69. A manufacturing engineer determines that the cost per unit for a desk lamp is $12 and that the fixed cost is $19,240. The selling price for the desk lamp is $49. Find the break-even point.

70. A manufacturing engineer determines the cost to make one compact disc to be $3.35 and the fixed cost to be $6180. The selling price for each compact disc is $8.50. Find the number of compact discs that must be sold to break even.

71. To manufacture a softball bat requires two steps. The first step is to cut a rough shape. The second step is to sand the bat to its final form. The cost to rough-shape a bat is $.45, and the cost to sand a bat to final form is $1.05. The total fixed cost for the two steps is $16,500. How many softball bats must be sold at a price of $7.00 to break even?

APPLYING THE CONCEPTS

72. Write an equation of the form $ax + b = cx + d$ that has 4 as the solution.

For Exercises 73 to 76, solve. If the equation has no solution, write "no solution."

73. $3(2x - 1) - (6x - 4) = -9$

74. $7(3x + 6) - 4(3 + 5x) = 13 + x$

75. $\frac{1}{5}(25 - 10a) + 4 = \frac{1}{3}(12a - 15) + 14$

76. $5[m + 2(3 - m)] = 3[2(4 - m) - 5]$

77. The equation $x = x + 1$ has no solution, whereas the solution of the equation $2x + 3 = 3$ is zero. Is there a difference between no solution and a solution of zero? Explain your answer.

Copyright © Houghton Mifflin Company. All rights reserved.

3.4 Translating Sentences into Equations

Objective A To solve integer problems

An equation states that two mathematical expressions are equal. Therefore, to **translate** a sentence into an equation requires recognition of the words or phrases that mean "equals." Some of these phrases are listed below.

$$\left.\begin{array}{l} \text{equals} \\ \text{is} \\ \text{is equal to} \\ \text{amounts to} \\ \text{represents} \end{array}\right\} \text{translate to } =$$

Once the sentence is translated into an equation, the equation can be solved by rewriting the equation in the form *variable* = *constant*.

HOW TO Translate "five less than a number is thirteen" into an equation and solve.

The unknown number: n

• Assign a variable to the unknown number.

| Five less than a number | is | thirteen |

• Find two verbal expressions for the same value.

$$n - 5 \quad = \quad 13$$

• Write a mathematical expression for each verbal expression. Write the equals sign.

$$n - 5 + 5 = 13 + 5$$

• Solve the equation.

$$n = 18$$

The number is 18.

T A K E   N O T E

You can check the solution to a translation problem.

Check:

5 less than 18 is 13
$$18 - 5 \mid 13$$
$$13 = 13$$

Recall that the integers are the numbers $\{\ldots, -4, -3, -2, -1, 0, 1, 2, 3, 4, \ldots\}$. An **even integer** is an integer that is divisible by 2. Examples of even integers are -8, 0, and 22. An **odd integer** is an integer that is not divisible by 2. Examples of odd integers are -17, 1, and 39.

Consecutive integers are integers that follow one another in order. Examples of consecutive integers are shown at the right. (Assume that the variable n represents an integer.)

11, 12, 13
$-8, -7, -6$
$n, n + 1, n + 2$

Examples of **consecutive even integers** are shown at the right. (Assume that the variable n represents an even integer.)

24, 26, 28
$-10, -8, -6$
$n, n + 2, n + 4$

Examples of **consecutive odd integers** are shown at the right. (Assume that the variable n represents an odd integer.)

19, 21, 23
$-1, 1, 3$
$n, n + 2, n + 4$

T A K E   N O T E

Both consecutive even and consecutive odd integers are represented using $n, n + 2, n + 4, \ldots$.

Copyright © Houghton Mifflin Company. All rights reserved.

HOW TO   The sum of three consecutive odd integers is forty-five. Find the integers.

Strategy

- First odd integer: n
  Second odd integer: $n + 2$
  Third odd integer: $n + 4$
- The sum of the three odd integers is 45.

● Represent three consecutive odd integers.

Solution

$$n + (n + 2) + (n + 4) = 45$$ ● Write an equation.

$$3n + 6 = 45$$ ● Solve the equation.

$$3n = 39$$

$$n = 13$$ ● The first odd integer is 13.

$$n + 2 = 13 + 2 = 15$$ ● Find the second odd integer.

$$n + 4 = 13 + 4 = 17$$ ● Find the third odd integer.

The three consecutive odd integers are 13, 15, and 17.

Example 1

The sum of two numbers is sixteen. The difference between four times the smaller number and two is two more than twice the larger number. Find the two numbers.

Solution

The smaller number: n
The larger number: $16 - n$

The difference between four times the smaller and two	is	two more than twice the larger

$$4n - 2 = 2(16 - n) + 2$$
$$4n - 2 = 32 - 2n + 2$$
$$4n - 2 = 34 - 2n$$
$$4n + 2n - 2 = 34 - 2n + 2n$$
$$6n - 2 = 34$$
$$6n - 2 + 2 = 34 + 2$$
$$6n = 36$$
$$\frac{6n}{6} = \frac{36}{6}$$
$$n = 6$$

$$16 - n = 16 - 6 = 10$$

The smaller number is 6.
The larger number is 10.

You Try It 1

The sum of two numbers is twelve. The total of three times the smaller number and six amounts to seven less than the product of four and the larger number. Find the two numbers.

Your solution

Copyright © Houghton Mifflin Company. All rights reserved.

Solution on p. S8

Example 2

Find three consecutive even integers such that three times the second equals four more than the sum of the first and third.

Strategy

• First even integer: n
  Second even integer: $n + 2$
  Third even integer: $n + 4$
• Three times the second equals four more than the sum of the first and third.

Solution

$$3(n + 2) = n + (n + 4) + 4$$
$$3n + 6 = 2n + 8$$
$$3n - 2n + 6 = 2n - 2n + 8$$
$$n + 6 = 8$$
$$n = 2$$
$$n + 2 = 2 + 2 = 4$$
$$n + 4 = 2 + 4 = 6$$

The three integers are 2, 4, and 6.

You Try It 2

Find three consecutive integers whose sum is negative six.

Your strategy

Your solution

Solution on p. S8

Objective B   **To translate a sentence into an equation and solve**

Example 3

A wallpaper hanger charges a fee of $25 plus $12 for each roll of wallpaper used in a room. If the total charge for hanging wallpaper is $97, how many rolls of wallpaper were used?

Strategy

To find the number of rolls of wallpaper used, write and solve an equation using n to represent the number of rolls of wallpaper used.

Solution

$25 plus $12 for each roll of wallpaper	is	$97

$$25 + 12n = 97$$
$$12n = 72$$
$$\frac{12n}{12} = \frac{72}{12}$$
$$n = 6$$

6 rolls of wallpaper were used.

You Try It 3

The fee charged by a ticketing agency for a concert is $3.50 plus $17.50 for each ticket purchased. If your total charge for tickets is $161, how many tickets are you purchasing?

Your strategy

Your solution

Solution on p. S8

Copyright © Houghton Mifflin Company. All rights reserved.

Example 4

A board 20 ft long is cut into two pieces. Five times the length of the shorter piece is 2 ft more than twice the length of the longer piece. Find the length of each piece.

Strategy

Let x represent the length of the shorter piece. Then $20 - x$ represents the length of the longer piece.

Make a drawing.

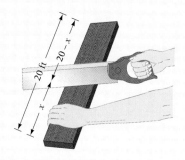

To find the lengths, write and solve an equation using x to represent the length of the shorter piece and $20 - x$ to represent the length of the longer piece.

Solution

Five times the length of the shorter piece	is	2 ft more than twice the length of the longer

$$5x = 2(20 - x) + 2$$
$$5x = 40 - 2x + 2$$
$$5x = 42 - 2x$$
$$5x + 2x = 42 - 2x + 2x$$
$$7x = 42$$
$$\frac{7x}{7} = \frac{42}{7}$$
$$x = 6$$

$20 - x = 20 - 6 = 14$

The length of the shorter piece is 6 ft.
The length of the longer piece is 14 ft.

You Try It 4

A wire 22 in. long is cut into two pieces. The length of the longer piece is 4 in. more than twice the length of the shorter piece. Find the length of each piece.

Your strategy

Your solution

Copyright © Houghton Mifflin Company. All rights reserved.

Solution on p. S8

3.4 Exercises

To solve integer problems

For Exercises 1 to 18, translate into an equation and solve.

1. The difference between a number and fifteen is seven. Find the number.

2. The sum of five and a number is three. Find the number.

3. The product of seven and a number is negative twenty-one. Find the number.

4. The quotient of a number and four is two. Find the number.

5. The difference between nine and a number is seven. Find the number.

6. Three-fifths of a number is negative thirty. Find the number.

7. The difference between five and twice a number is one. Find the number.

8. Four more than three times a number is thirteen. Find the number.

9. The sum of twice a number and five is fifteen. Find the number.

10. The difference between nine times a number and six is twelve. Find the number.

11. Six less than four times a number is twenty-two. Find the number.

12. Four times the sum of twice a number and three is twelve. Find the number.

13. Three times the difference between four times a number and seven is fifteen. Find the number.

14. Twice the difference between a number and twenty-five is three times the number. Find the number.

15. The sum of two numbers is twenty. Three times the smaller is equal to two times the larger. Find the two numbers.

16. The sum of two numbers is fifteen. One less than three times the smaller is equal to the larger. Find the two numbers.

17. The sum of two numbers is fourteen. The difference between two times the smaller and the larger is one. Find the two numbers.

18. The sum of two numbers is eighteen. The total of three times the smaller and twice the larger is forty-four. Find the two numbers.

19. The sum of three consecutive odd integers is fifty-one. Find the integers.

20. Find three consecutive even integers whose sum is negative eighteen.

21. Find three consecutive odd integers such that three times the middle integer is one more than the sum of the first and third.

22. Twice the smallest of three consecutive odd integers is seven more than the largest. Find the integers.

23. Find two consecutive even integers such that three times the first equals twice the second.

24. Find two consecutive even integers such that four times the first is three times the second.

25. Seven times the first of two consecutive odd integers is five times the second. Find the integers.

26. Find three consecutive even integers such that three times the middle integer is four more than the sum of the first and third.

Copyright © Houghton Mifflin Company. All rights reserved.

Objective B   **To translate a sentence into an equation and solve**

27. **Computer Science**   The processor speed of a personal computer is 3.2 gigahertz (GHz). This is three-fourths of the processor speed of a newer model personal computer. Find the processor speed of the newer personal computer.

28. **Computer Science**   The storage capacity of a hard-disk drive is 60 gigabytes. This is one-fourth of the storage capacity of a second hard-disk drive. Find the storage capacity of the second hard-disk drive.

29. **Geometry**   An isosceles triangle has two sides of equal length. The length of the third side is 1 ft less than twice the length of an equal side. Find the length of each side when the perimeter is 23 ft.

30. **Geometry**   An isosceles triangle has two sides of equal length. The length of one of the equal sides is two more than 3 times the length of the third side. If the perimeter is 46 m, find the length of each side.

31. **Union Dues**   A union charges monthly dues of $4.00 plus $.15 for each hour worked during the month. A union member's dues for March were $29.20. How many hours did the union member work during the month of March?

32. **Technical Support**   A technical information hotline charges a customer $15.00 plus $2.00 per minute to answer questions about software. How many minutes did a customer who received a bill for $37 use this service?

33. **Construction**   The total cost to paint the inside of a house was $1346. This cost included $125 for materials and $33 per hour for labor. How many hours of labor were required to paint the inside of the house?

34. **Telecommunications**   The cellular phone service for a business executive is $35 per month plus $.40 per minute of phone use. In a month when the executive's cellular phone bill was $99.80, how many minutes did the executive use the phone?

35. **Computer Science**   A computer screen consists of tiny dots of light called pixels. In a certain graphics mode, there are 1280 horizontal pixels. This is 768 less than twice the number of vertical pixels. Find the number of vertical pixels.

Copyright © Houghton Mifflin Company. All rights reserved.

36. Energy   The cost of electricity in a certain city is \$.08 for each of the first 300 kWh (kilowatt-hours) and \$.13 for each kilowatt-hour over 300 kWh. Find the number of kilowatt-hours used by a family with a \$51.95 electric bill.

37. Geometry   The perimeter of a rectangle is 42 m. The length of the rectangle is 3 m less than twice the width. Find the length and width of the rectangle.

38. Geometry   A rectangular vegetable garden has a perimeter of 64 ft. The length of the garden is 20 ft. Find the width of the garden.

39. Carpentry   A 12-foot board is cut into two pieces. Twice the length of the shorter piece is 3 ft less than the length of the longer piece. Find the length of each piece.

40. Sports   A 14-yard fishing line is cut into two pieces. Three times the length of the longer piece is four times the length of the shorter piece. Find the length of each piece.

41. Education   Seven thousand dollars is divided into two scholarships. Twice the amount of the smaller scholarship is \$1000 less than the larger scholarship. What is the amount of the larger scholarship?

42. Investing   An investment of \$10,000 is divided into two accounts, one for stocks and one for mutual funds. The value of the stock account is \$2000 less than twice the value of the mutual fund account. Find the amount in each account.

APPLYING THE CONCEPTS

43.   Make up two word problems; one that requires solving the equation $6x = 123$ and one that requires solving the equation $8x + 100 = 300$ to find the answer to the problem.

44.   A formula is an equation that relates variables in a known way. Find two examples of formulas that are used in your college major. Explain what each of the variables represents.

45.   It is always important to check the answer to an application problem to be sure the answer makes sense. Consider the following problem. A 4-quart mixture of fruit juices is made from apple juice and cranberry juice. There are 6 more quarts of apple juice than of cranberry juice. Write and solve an equation for the number of quarts of each juice used. Does the answer to this question make sense? Explain.

Copyright © Houghton Mifflin Company. All rights reserved.

3.5 Geometry Problems

Copyright © Houghton Mifflin Company. All rights reserved.

Objective A   **To solve problems involving angles**

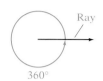

Ray

360°

In Section 1.8, we discussed some basic properties of angles. Recall that a ray that is rotated one complete revolution about its starting point creates an angle of 360°. Recall also that a 90° angle is called a right angle and a 180° angle is called a straight angle.

Point of Interest

The word *degree* first appeared in Chaucer's *Canterbury Tales*, which was written in 1386.

An **acute angle** is an angle whose measure is between 0° and 90°. ∠*A* at the right is an acute angle. An **obtuse angle** is an angle whose measure is between 90° and 180°. ∠*B* at the right is an obtuse angle.

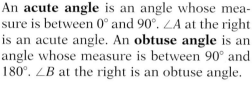
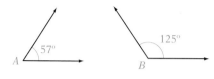

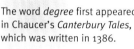

4x   3x

5x

HOW TO   Given the diagram at the left, find *x*.

$$3x + 4x + 5x = 360°$$ • The sum of the measures of three angles is 360°.

$$12x = 360°$$
$$x = 30°$$

The measure of *x* is 30°.

Four angles are formed by the intersection of two lines. If the two lines are perpendicular, then each of the four angles is a right angle. If the two lines are not perpendicular, then two of the angles formed are acute angles and two of the angles are obtuse angles. The two acute angles are always opposite each other, and the two obtuse angles are always opposite each other.

In the figure at the right, ∠*w* and ∠*y* are acute angles, and ∠*x* and ∠*z* are obtuse angles.

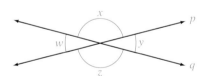

Vertical angles have the same measure.

$$m\angle w = m\angle y$$
$$m\angle x = m\angle z$$

TAKE NOTE

Recall that two angles are supplementary angles if the sum of the measures of the angles is 180°. For instance, angles whose measures are 48° and 132° are supplementary angles because 48° + 132° = 180°.

Two angles that are on opposite sides of the intersection of two lines are called **vertical angles.** In the figure above ∠*w* and ∠*y* are vertical angles. ∠*x* and ∠*z* are vertical angles.

Two angles that share a common side are called **adjacent angles.** In the figure above, ∠*x* and ∠*y* are adjacent angles, as are ∠*y* and ∠*z*, ∠*z* and ∠*w*, and ∠*w* and ∠*x*.

Adjacent angles of intersecting lines are supplementary.

$$m\angle x + m\angle y = 180° \quad m\angle z + m\angle w = 180°$$
$$m\angle y + m\angle z = 180° \quad m\angle w + m\angle x = 180°$$

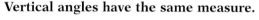

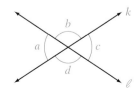

HOW TO   In the diagram at the left, $m\angle b = 115°$. Find $m\angle a$, $m\angle c$, and $m\angle d$.

$$m\angle a + m\angle b = 180°$$
$$m\angle a + 115° = 180°$$
$$m\angle a = 65°$$

• ∠*a* is supplementary to ∠*b* because ∠*a* and ∠*b* are adjacent angles of intersecting lines.

$$m\angle c = 65°$$
$$m\angle d = 115°$$

• $m\angle c = m\angle a$ because ∠*c* and ∠*a* are vertical angles.
• $m\angle d = m\angle b$ because ∠*d* and ∠*b* are vertical angles.

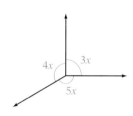

TAKE NOTE
Recall that parallel lines never meet—the distance between them is always the same. Perpendicular lines are intersecting lines that form right angles.

A line that intersects two other lines at different points is called a **transversal.** If the lines cut by a transversal t are parallel lines and the transversal is not perpendicular to the parallel lines, then all four acute angles have the same measure and all four obtuse angles have the same measure.

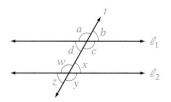

$$m\angle b = m\angle d = m\angle x = m\angle z$$
$$m\angle a = m\angle c = m\angle w = m\angle y$$

Alternate interior angles are two non-adjacent angles that are on opposite sides of the transversal and between the parallel lines. In the figure above, $\angle c$ and $\angle w$ are alternate interior angles, and $\angle d$ and $\angle x$ are alternate interior angles.

Alternate interior angles have the same measure.

$$m\angle c = m\angle w$$
$$m\angle d = m\angle x$$

Alternate exterior angles are two non-adjacent angles that are on opposite sides of the transversal and outside the parallel lines. In the figure above, $\angle a$ and $\angle y$ are alternate exterior angles, and $\angle b$ and $\angle z$ are alternate exterior angles.

Alternate exterior angles have the same measure.

$$m\angle a = m\angle y$$
$$m\angle b = m\angle z$$

Corresponding angles are two angles that are on the same side of the transversal and are both acute angles or are both obtuse angles. In the figure above, there are four pairs of corresponding angles: $\angle a$ and $\angle w$, $\angle d$ and $\angle z$, $\angle b$ and $\angle x$, and $\angle c$ and $\angle y$.

Corresponding angles have the same measure.

$$m\angle a = m\angle w$$
$$m\angle d = m\angle z$$
$$m\angle b = m\angle x$$
$$m\angle c = m\angle y$$

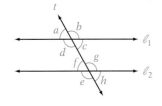

HOW TO In the diagram at the left, $\ell_1 \parallel \ell_2$ and $m\angle f = 58°$. Find $m\angle a$, $m\angle c$, and $m\angle d$.

$m\angle a = m\angle f = 58°$ • $\angle a$ and $\angle f$ are corresponding angles.

$m\angle c = m\angle f = 58°$ • $\angle c$ and $\angle f$ are alternate interior angles.

$m\angle d + m\angle a = 180°$ • $\angle d$ is supplementary to $\angle a$.
$m\angle d + 58° = 180°$
$m\angle d = 122°$

Example 1

Find x.

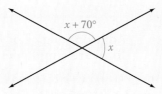

Strategy
The angles labeled are adjacent angles of intersecting lines and are therefore supplementary angles. To find x, write an equation and solve for x.

Solution
$$x + (x + 70°) = 180°$$
$$2x + 70° = 180°$$
$$2x = 110°$$
$$x = 55°$$

You Try It 1

Find x.

Your strategy

Your solution

Solution on p. S9

Copyright © Houghton Mifflin Company. All rights reserved.

Example 2

Given $\ell_1 \parallel \ell_2$, find x.

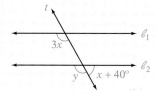

Strategy

$3x = y$ because corresponding angles have the same measure. $y + (x + 40°) = 180°$ because adjacent angles of intersecting lines are supplementary angles. Substitute $3x$ for y and solve for x.

Solution

$$y + (x + 40°) = 180°$$
$$3x + (x + 40°) = 180°$$
$$4x + 40° = 180°$$
$$4x = 140°$$
$$x = 35°$$

You Try It 2

Given $\ell_1 \parallel \ell_2$, find x.

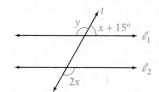

Your strategy

Your solution

Solution on p. S9

Copyright © Houghton Mifflin Company. All rights reserved.

Objective B   **To solve problems involving the angles of a triangle**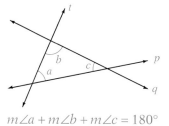

If the lines cut by a transversal are not parallel lines, then the three lines will intersect at three points, forming a triangle. The angles within the region enclosed by the triangle are called **interior angles.** In the figure at the right, angles a, b, and c are interior angles. **The sum of the measures of the interior angles of a triangle is 180°.**

$$m\angle a + m\angle b + m\angle c = 180°$$

An angle adjacent to an interior angle is an **exterior angle.** In the figure at the right, angles m and n are exterior angles for angle a. The sum of the measures of an interior angle of a triangle and an adjacent exterior angle is 180°.

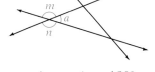

$$m\angle a + m\angle m = 180°$$
$$m\angle a + m\angle n = 180°$$

HOW TO   Given that $m\angle c = 40°$ and $m\angle e = 60°$, find $m\angle d$.

$$m\angle a = m\angle e = 60°$$
     • $\angle a$ and $\angle e$ are vertical angles.

$$m\angle c + m\angle a + m\angle b = 180°$$
$$40° + 60° + m\angle b = 180°$$
$$100° + m\angle b = 180°$$
$$m\angle b = 80°$$
     • The sum of the interior angles is 180°.

$$m\angle b + m\angle d = 180°$$
$$80° + m\angle d = 180°$$
$$m\angle d = 100°$$
     • $\angle b$ and $\angle d$ are supplementary angles.

Example 3

Given that $m\angle a = 45°$ and $m\angle x = 100°$, find the measures of angles b, c, and y.

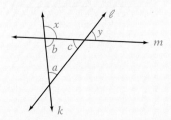

Strategy

- To find the measure of $\angle b$, use the fact that $\angle b$ and $\angle x$ are supplementary angles.
- To find the measure of $\angle c$, use the fact that the sum of the measures of the interior angles of a triangle is 180°.
- To find the measure of $\angle y$, use the fact that $\angle c$ and $\angle y$ are vertical angles.

Solution

$m\angle b + m\angle x = 180°$
$m\angle b + 100° = 180°$
$m\angle b = 80°$

$m\angle a + m\angle b + m\angle c = 180°$
$45° + 80° + m\angle c = 180°$
$125° + m\angle c = 180°$
$m\angle c = 55°$

$m\angle y = m\angle c = 55°$

You Try It 3

Given that $m\angle y = 55°$, find the measures of angles a, b, and d.

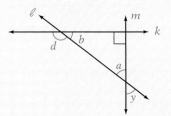

Your strategy

Your solution

Example 4

Two angles of a triangle measure 43° and 86°. Find the measure of the third angle.

Strategy

To find the measure of the third angle, use the fact that the sum of the measures of the interior angles of a triangle is 180°. Write an equation using x to represent the measure of the third angle. Solve the equation for x.

Solution

$x + 43° + 86° = 180°$
$x + 129° = 180°$
$x = 51°$

The measure of the third angle is 51°.

You Try It 4

One angle in a triangle is a right angle, and one angle measures 27°. Find the measure of the third angle.

Your strategy

Your solution

Solutions on p. S9

Copyright © Houghton Mifflin Company. All rights reserved.

3.5 Exercises

Objective A **To solve problems involving angles**

For Exercises 1 and 2, find the measure of ∠*a*.

1.

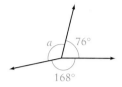

2.

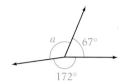

For Exercises 3 to 8, find *x*.

3.

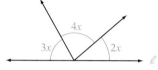

4.

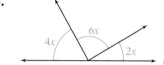

5.

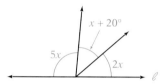

6.

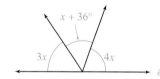

7.

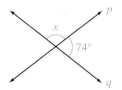

8.

For Exercises 9 to 12, find *x*.

9.

10.

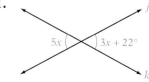

11.

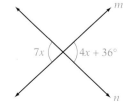

12.

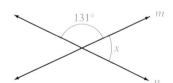

Copyright © Houghton Mifflin Company. All rights reserved.

For Exercises 13 to 16, given that $\ell_1 \parallel \ell_2$, find the measures of angles a and b.

13.

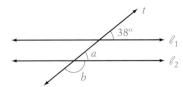

14.

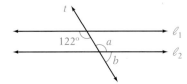

15.

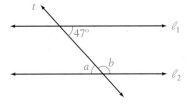

16.

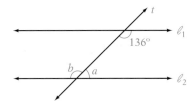

For Exercises 17 to 20, given that $\ell_1 \parallel \ell_2$, find x.

17.

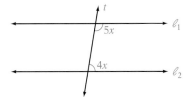

18.

19.

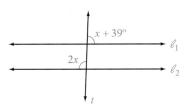

20.

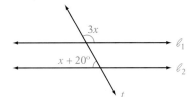

21. Given that $m\angle a = 51°$, find $m\angle b$.

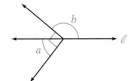

22. Given that $m\angle a = 38°$, find $m\angle b$.

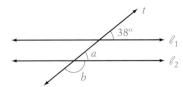 *Objective B*   **To solve problems involving the angles of a triangle**

23. Given that $m\angle a = 95°$ and $m\angle b = 70°$, find $m\angle x$ and $m\angle y$.

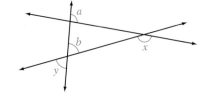

24. Given that $m\angle a = 35°$ and $m\angle b = 55°$, find $m\angle x$ and $m\angle y$.

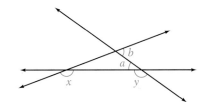

Copyright © Houghton Mifflin Company. All rights reserved.

25. Given that $m\angle y = 45°$, find $m\angle a$ and $m\angle b$.

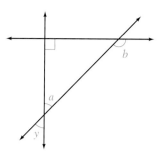

26. Given that $m\angle y = 130°$, find $m\angle a$ and $m\angle b$.

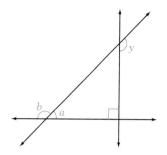

27. Given that $AO \perp OB$, express $m\angle BOC$ in terms of x.

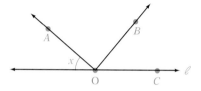

28. Given that $AO \perp OB$, express $m\angle AOC$ in terms of x.

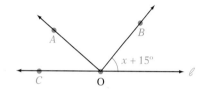

29. A triangle has a 30° angle and a right angle. What is the measure of the third angle?

30. A triangle has a 45° angle and a right angle. Find the measure of the third angle.

31. Two angles of a triangle measure 42° and 103°. Find the measure of the third angle.

32. Two angles of a triangle measure 62° and 45°. Find the measure of the third angle.

33. A triangle has a 13° angle and a 65° angle. What is the measure of the third angle?

34. A triangle has a 105° angle and a 32° angle. What is the measure of the third angle?

APPLYING THE CONCEPTS

35. **Geometry**   A carpenter is carrying a ladder around the corner of a hallway as shown in the figure below. Find $m\angle x$.

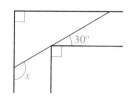

36. **Geometry**   For the figure below, which of the following is true?
(a) $m\angle a < m\angle b$
(b) $m\angle a = m\angle b$
(c) $m\angle a > m\angle b$

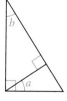

37. **Geometry**   For the figure at the right, find the sum of the measures of angles x, y, and z.

38. ✏ **Geometry**   For the figure at the right, explain why $m\angle a + m\angle b = m\angle x$. Write a rule that describes the relationship between an exterior angle of a triangle and the opposite interior angles. Use the rule to write an equation involving the measures of angles a, c, and z.

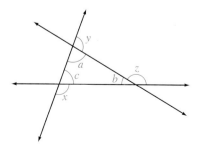

Copyright © Houghton Mifflin Company. All rights reserved.

3.6 Mixture and Uniform Motion Problems

Copyright © Houghton Mifflin Company. All rights reserved.

Objective A   **To solve value mixture problems**

A value mixture problem involves combining two ingredients that have different prices into a single blend. For example, a coffee merchant may blend two types of coffee into a single blend, or a candy manufacturer may combine two types of candy to sell as a variety pack.

The solution of a value mixture problem is based on the **value mixture equation** $AC = V$, where A is the amount of an ingredient, C is the cost per unit of the ingredient, and V is the value of the ingredient.

> **TAKE NOTE**
>
> The equation $AC = V$ is used to find the value of an ingredient. For example, the value of 4 lb of cashews costing $6 per pound is
>
> $$AC = V$$
> $$4 \cdot \$6 = V$$
> $$\$24 = V$$

HOW TO   A coffee merchant wants to make 6 lb of a blend of coffee costing $5 per pound. The blend is made using a $6-per-pound grade and a $3-per-pound grade of coffee. How many pounds of each of these grades should be used?

> **Strategy for Solving a Value Mixture Problem**
>
> **1.** For each ingredient in the mixture, write a numerical or variable expression for the amount of the ingredient used, the unit cost of the ingredient, and the value of the amount used. For the blend, write a numerical or variable expression for the amount, the unit cost of the blend, and the value of the amount. The results can be recorded in a table.

The sum of the amounts is 6 lb.

Amount of $3 coffee: $6 - x$
Amount of $6 coffee: x

> **TAKE NOTE**
>
> Use the information given in the problem to fill in the amount and unit cost columns of the table. Fill in the value column by multiplying the two expressions you wrote in each row. Use the expressions in the last column to write the equation.

	Amount, A	·	Unit Cost, C	=	Value, V
$6 grade	x	·	6	=	$6x$
$3 grade	$6 - x$	·	3	=	$3(6 - x)$
$5 blend	6	·	5	=	$5(6)$

> **2.** Determine how the values of the ingredients are related. Use the fact that the sum of the values of all the ingredients is equal to the value of the blend.

The sum of the values of the $6 grade and the $3 grade is equal to the value of the $5 blend.

$$6x + 3(6 - x) = 5(6)$$
$$6x + 18 - 3x = 30$$
$$3x + 18 = 30$$
$$3x = 12$$
$$x = 4$$

$6 - x = 6 - 4 = 2$ • **Find the amount of the $3 grade coffee.**

The merchant must use 4 lb of the $6 coffee and 2 lb of the $3 coffee.

Copyright © Houghton Mifflin Company. All rights reserved.

Example 1

How many ounces of a silver alloy that costs $4 an ounce must be mixed with 10 oz of an alloy that costs $6 an ounce to make a mixture that costs $4.32 an ounce?

Strategy

- Ounces of $4 alloy: x

	Amount	*Cost*	*Value*
$4 alloy	x	4	$4x$
$6 alloy	10	6	6(10)
$4.32 mixture	$10 + x$	4.32	4.32(10 + x)

- The sum of the values before mixing equals the value after mixing.

Solution

$$4x + 6(10) = 4.32(10 + x)$$
$$4x + 60 = 43.2 + 4.32x$$
$$-0.32x + 60 = 43.2$$
$$-0.32x = -16.8$$
$$x = 52.5$$

52.5 oz of the $4 silver alloy must be used.

You Try It 1

A gardener has 20 lb of a lawn fertilizer that costs $.80 per pound. How many pounds of a fertilizer that costs $.55 per pound should be mixed with this 20 lb of lawn fertilizer to produce a mixture that costs $.75 per pound?

Your strategy

Your solution

Solution on p. S9

Objective B **To solve percent mixture problems**

Recall from Section 3.1 that a percent mixture problem can be solved using the equation $Ar = Q$, where A is the amount of a solution, r is the percent concentration of a substance in the solution, and Q is the quantity of the substance in the solution.

For example, a 500-milliliter bottle is filled with a 4% solution of hydrogen peroxide.

$$Ar = Q$$
$$500(0.04) = Q$$
$$20 = Q$$

The bottle contains 20 ml of hydrogen peroxide.

HOW TO How many gallons of a 20% salt solution must be mixed with 6 gal of a 30% salt solution to make a 22% salt solution?

> **Strategy for Solving a Percent Mixture Problem**
>
> **1.** For each solution, write a numerical or variable expression for the amount of solution, the percent concentration, and the quantity of the substance in the solution. The results can be recorded in a table.

The unknown quantity of 20% solution: x

TAKE NOTE

Use the information given in the problem to fill in the amount and percent columns of the table. Fill in the quantity column by multiplying the two expressions you wrote in each row. Use the expressions in the last column to write the equation.

	Amount of Solution, A	·	*Percent Concentration, r*	=	*Quantity of Substance, Q*
20% solution	x	·	0.20	=	$0.20x$
30% solution	6	·	0.30	=	$0.30(6)$
22% solution	$x + 6$	·	0.22	=	$0.22(x + 6)$

> **2.** Determine how the quantities of the substances in the solutions are related. Use the fact that the sum of the quantities of the substances being mixed is equal to the quantity of the substance after mixing.

The sum of the quantities of the substances in the 20% solution and the 30% solution is equal to the quantity of the substance in the 22% solution.

$$0.20x + 0.30(6) = 0.22(x + 6)$$
$$0.20x + 1.80 = 0.22x + 1.32$$
$$-0.02x + 1.80 = 1.32$$
$$-0.02x = -0.48$$
$$x = 24$$

24 gal of the 20% solution are required.

Copyright © Houghton Mifflin Company. All rights reserved.

Example 2

A chemist wishes to make 2 L of an 8% acid solution by mixing a 10% acid solution and a 5% acid solution. How many liters of each solution should the chemist use?

Strategy

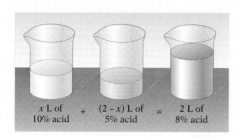

x L of + $(2 - x)$ L of = 2 L of
10% acid   5% acid   8% acid

• Liters of 10% solution: x
  Liters of 5% solution: $2 - x$

	Amount	*Percent*	*Quantity*
10% solution	x	0.10	$0.10x$
5% solution	$2 - x$	0.05	$0.05(2 - x)$
8% solution	2	0.08	$0.08(2)$

• The sum of the quantities before mixing is equal to the quantity after mixing.

Solution

$$0.10x + 0.05(2 - x) = 0.08(2)$$

$$0.10x + 0.10 - 0.05x = 0.16$$

$$0.05x + 0.10 = 0.16$$

$$0.05x = 0.06$$

$$x = 1.2$$

$$2 - x = 2 - 1.2 = 0.8$$

The chemist needs 1.2 L of the 10% solution and 0.8 L of the 5% solution.

You Try It 2

A pharmacist dilutes 5 L of a 12% solution with a 6% solution. How many liters of the 6% solution are added to make an 8% solution?

Your strategy

Your solution

Solution on p. S9

Copyright © Houghton Mifflin Company. All rights reserved.

Copyright © Houghton Mifflin Company. All rights reserved.

Objective C   **To solve uniform motion problems**

Recall from Section 3.1 that an object traveling at a constant speed in a straight line is in *uniform motion*. The solution of a uniform motion problem is based on the equation $rt = d$, where r is the rate of travel, t is the time spent traveling, and d is the distance traveled.

HOW TO   A car leaves a town traveling at 40 mph. Two hours later, a second car leaves the same town, on the same road, traveling at 60 mph. In how many hours will the second car pass the first car?

> **Strategy for Solving a Uniform Motion Problem**
>
> **1.** For each object, write a numerical or variable expression for the rate, time, and distance. The results can be recorded in a table.

The first car traveled 2 h longer than the second car.

Unknown time for the second car: t
Time for the first car: $t + 2$

TAKE NOTE

Use the information given in the problem to fill in the rate and time columns of the table. Find the expression in the distance column by multiplying the two expressions you wrote in each row.

	Rate, r	·	Time, t	=	Distance, d
First car	40	·	$t + 2$	=	$40(t + 2)$
Second car	60	·	t	=	$60t$

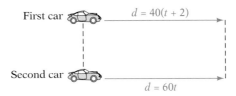

> **2.** Determine how the distances traveled by the two objects are related. For example, the total distance traveled by both objects may be known, or it may be known that the two objects traveled the same distance.

The two cars travel the same distance.

$$40(t + 2) = 60t$$
$$40t + 80 = 60t$$
$$80 = 20t$$
$$4 = t$$

The second car will pass the first car in 4 h.

Example 3

Two cars, one traveling 10 mph faster than the other, start at the same time from the same point and travel in opposite directions. In 3 h they are 300 mi apart. Find the rate of each car.

Strategy

- Rate of 1st car: r
  Rate of 2nd car: $r + 10$

	Rate	Time	Distance
1st car	r	3	$3r$
2nd car	$r + 10$	3	$3(r + 10)$

- The total distance traveled by the two cars is 300 mi.

Solution

$$3r + 3(r + 10) = 300$$
$$3r + 3r + 30 = 300$$
$$6r + 30 = 300$$
$$6r = 270$$
$$r = 45$$

$$r + 10 = 45 + 10 = 55$$

The first car is traveling 45 mph.
The second car is traveling 55 mph.

You Try It 3

Two trains, one traveling at twice the speed of the other, start at the same time on parallel tracks from stations that are 288 mi apart and travel toward each other. In 3 h, the trains pass each other. Find the rate of each train.

Your strategy

Your solution

Example 4

How far can the members of a bicycling club ride out into the country at a speed of 12 mph and return over the same road at 8 mph if they travel a total of 10 h?

Strategy

- Time spent riding out: t
  Time spent riding back: $10 - t$

	Rate	Time	Distance
Out	12	t	$12t$
Back	8	$10 - t$	$8(10 - t)$

- The distance out equals the distance back.

Solution

$$12t = 8(10 - t)$$
$$12t = 80 - 8t$$
$$20t = 80$$
$$t = 4 \quad \text{(The time is 4 h.)}$$

The distance out $= 12t = 12(4) = 48$ mi.

The club can ride 48 mi into the country.

You Try It 4

A pilot flew out to a parcel of land and back in 5 h. The rate out was 150 mph, and the rate returning was 100 mph. How far away was the parcel of land?

Your strategy

Your solution

Copyright © Houghton Mifflin Company. All rights reserved.

Solutions on p. S10

3.6 Exercises

Objective A **To solve value mixture problems**

1. An herbalist has 30 oz of herbs costing $2 per ounce. How many ounces of herbs costing $1 per ounce should be mixed with the 30 oz to produce a mixture costing $1.60 per ounce?

2. The manager of a farmer's market has 500 lb of grain that costs $1.20 per pound. How many pounds of meal costing $.80 per pound should be mixed with the 500 lb of grain to produce a mixture that costs $1.05 per pound?

3. Find the cost per pound of a meatloaf mixture made from 3 lb of ground beef costing $1.99 per pound and 1 lb of ground turkey costing $1.39 per pound.

4. Find the cost per ounce of a sunscreen made from 100 oz of a lotion that costs $2.50 per ounce and 50 oz of a lotion that costs $4.00 per ounce.

5. A snack food is made by mixing 5 lb of popcorn that costs $.80 per pound with caramel that costs $2.40 per pound. How much caramel is needed to make a mixture that costs $1.40 per pound?

6. A wild birdseed mix is made by combining 100 lb of millet seed costing $.60 per pound with sunflower seeds costing $1.10 per pound. How many pounds of sunflower seeds are needed to make a mixture that costs $.70 per pound?

7. Ten cups of a restaurant's house Italian dressing is made by blending olive oil costing $1.50 per cup with vinegar that costs $.25 per cup. How many cups of each are used if the cost of the blend is $.50 per cup?

8. A high-protein diet supplement that costs $6.75 per pound is mixed with a vitamin supplement that costs $3.25 per pound. How many pounds of each should be used to make 5 lb of a mixture that costs $4.65 per pound?

9. Find the cost per ounce of a mixture of 200 oz of a cologne that costs $5.50 per ounce and 500 oz of a cologne that costs $2.00 per ounce.

10. Find the cost per pound of a trail mix made from 40 lb of raisins that cost $4.40 per pound and 100 lb of granola that costs $2.30 per pound.

Copyright © Houghton Mifflin Company. All rights reserved.

11. A 20-ounce alloy of platinum that costs $220 per ounce is mixed with an alloy that costs $400 per ounce. How many ounces of the $400 alloy should be used to make an alloy that costs $300 per ounce?

12. How many liters of a blue dye that costs $1.60 per liter must be mixed with 18 L of anil that costs $2.50 per liter to make a mixture that costs $1.90 per liter?

13. The manager of a specialty food store combined almonds that cost $4.50 per pound with walnuts that cost $2.50 per pound. How many pounds of each were used to make a 100-pound mixture that costs $3.24 per pound?

14. A goldsmith combined an alloy that cost $4.30 per ounce with an alloy that cost $1.80 per ounce. How many ounces of each were used to make a mixture of 200 oz costing $2.50 per ounce?

15. Adult tickets for a play cost $6.00 and children's tickets cost $2.50. For one performance, 370 tickets were sold. Receipts for the performance were $1723. Find the number of adult tickets sold.

16. Tickets for a piano concert sold for $4.50 for each adult. Student tickets sold for $2.00 each. The total receipts for 1720 tickets were $5980. Find the number of adult tickets sold.

17. Find the cost per pound of sugar-coated breakfast cereal made from 40 lb of sugar that costs $1.00 per pound and 120 lb of corn flakes that cost $.60 per pound.

18. Find the cost per pound of a coffee mixture made from 8 lb of coffee that costs $9.20 per pound and 12 lb of coffee that costs $5.50 per pound.

Objective B   **To solve percent mixture problems**

19. Forty ounces of a 30% gold alloy are mixed with 60 oz of a 20% gold alloy. Find the percent concentration of the resulting gold alloy.

20. One hundred ounces of juice that is 50% tomato juice is added to 200 oz of a vegetable juice that is 25% tomato juice. What is the percent concentration of tomato juice in the resulting mixture?

Copyright © Houghton Mifflin Company. All rights reserved.

21. How many gallons of a 15% acid solution must be mixed with 5 gal of a 20% acid solution to make a 16% acid solution?

22. How many pounds of a chicken feed that is 50% corn must be mixed with 400 lb of a feed that is 80% corn to make a chicken feed that is 75% corn?

23. A rug is made by weaving 20 lb of yarn that is 50% wool with a yarn that is 25% wool. How many pounds of the yarn that is 25% wool are used if the finished rug is 35% wool?

24. Five gallons of a light green latex paint that is 20% yellow paint is combined with a darker green latex paint that is 40% yellow paint. How many gallons of the darker green paint must be used to create a green paint that is 25% yellow paint?

25. How many gallons of a plant food that is 9% nitrogen must be combined with another plant food that is 25% nitrogen to make 10 gal of a solution that is 15% nitrogen?

26. A chemist wants to make 50 ml of a 16% acid solution by mixing a 13% acid solution and an 18% acid solution. How many milliliters of each solution should the chemist use?

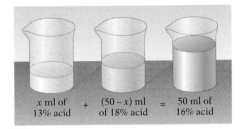

x ml of 13% acid $+$ $(50 - x)$ ml of 18% acid $=$ 50 ml of 16% acid

27. Five grams of sugar are added to a 45-gram serving of a breakfast cereal that is 10% sugar. What is the percent concentration of sugar in the resulting mixture?

28. A goldsmith mixes 8 oz of a 30% gold alloy with 12 oz of a 25% gold alloy. What is the percent concentration of the resulting alloy?

29. How many pounds of coffee that is 40% java beans must be mixed with 80 lb of coffee that is 30% java beans to make a coffee blend that is 32% java beans?

30. The manager of a garden shop mixes grass seed that is 60% rye grass with 70 lb of grass seed that is 80% rye grass to make a mixture that is 74% rye grass. How much of the 60% rye grass is used?

31. A hair dye is made by blending a 7% hydrogen peroxide solution and a 4% hydrogen peroxide solution. How many milliliters of each are used to make a 300-milliliter solution that is 5% hydrogen peroxide?

Copyright © Houghton Mifflin Company. All rights reserved.

11. If a rope 8 m long is cut into two pieces and one of the pieces has length x meters, then the length of the other piece can be represented as $(x - 8)$ meters.

12. An even integer is a multiple of 2.

13. If the first of three consecutive odd integers is n, then the second and third consecutive odd integers are represented by $n + 1$ and $n + 3$.

14. Suppose we are mixing two salt solutions. Then the variable Q in the percent mixture equation $Q = Ar$ represents the amount of salt in a solution.

15. If 100 oz of a silver alloy is 25% silver, then the alloy contains 25 oz of silver.

16. If we combine an alloy that costs $8 an ounce with an alloy that costs $5 an ounce, the cost of the resulting mixture will be greater than $8 an ounce.

17. If we combine a 9% acid solution with a solution that is 4% acid, the resulting solution will be less than 4% acid.

18. If the speed of one train is 20 mph slower than that of a second train, then the speeds of the two trains can be represented as r and $20 - r$.

Projects and Group Activities

Nielsen Ratings

Point of Interest

The 10 top-ranked programs in prime time for the week of September 15 to September 21, 2003, ranked by Nielsen Media Research, were

NFL Monday Night Football
Survivor—Pearl Islands
Emmy Awards
NFL Monday Showcase
Everybody Loves Raymond
ABC News Special
CSI
Friends
Law and Order
CSI: Miami

Nielsen Media Research surveys television viewers to determine the numbers of people watching particular shows. There are an estimated 107.9 million U.S. households with televisions. Each **rating point** represents 1% of that number, or 1,079,000 households. Therefore, for instance, if *CSI* received a rating of 9.2, then 9.2%, or (0.092)(107,900,000) = 9,926,800 households, watched that program.

A rating point does not mean that 1,079,000 people are watching a program. A rating point refers to the number of TV sets tuned to that program; there may be more than one person watching a television set in the household.

Nielsen Media Research also describes a program's *share* of the market. **Share** is the percent of television sets in use that are tuned to a program. Suppose the same week that *CSI* received 9.2 rating points, the show received a share of 25. This would mean that 25% of all households with a television *turned on* were tuned to *CSI*, whereas 9.2% of all households with a television were tuned to the program.

1. If *Law and Order* received a Nielsen rating of 8.8 and a share of 15, how many TV households watched the program that week? How many TV households were watching television during that hour? Round to the nearest hundred thousand.

2. Suppose *Everybody Loves Raymond* received a rating of 9.7 and a share of 15. How many TV households watched the program that week? How many TV households were watching television during that hour? Round to the nearest hundred-thousand.

3. Suppose *NFL Monday Night Football* received a rating of 12.9 during a week in which 34,750,000 people were watching the show. Find the average number of people per TV household who watched the program. Round to the nearest tenth.

Copyright © Houghton Mifflin Company. All rights reserved.

The cost to advertise during a program is related to its Nielsen rating. The sponsor (the company paying for the advertisement) pays a certain number of dollars for each rating point a show receives.

4. Suppose a television network charges $35,000 per rating point for a 30-second commercial on a daytime talk show. Determine the cost for three 30-second commercials if the Nielsen rating of the show is 11.5.

Nielsen Media Research also tracks the exposure of advertisements. For example, it might be reported that commercials for McDonald's had 500,000,000 household exposures during a week when its advertisement was aired 90 times.

5. Information regarding household exposure of advertisements can be found in *USA Today* each Monday. For a recent week, find the information for the top four advertised brands. For each brand, calculate the average household exposure for each time the ad was aired.

Nielsen Media Research has a web site on the Internet. You can locate the site by using a search engine. The site does not list rating points and market share, but these statistics can be found on other web sites by using a search engine.

6. Find the top two prime-time television shows for last week. Calculate the number of TV households that watched each program. Compare these figures with the top two sports programs for last week.

Chapter 3 Summary

Key Words	**Examples**
An *equation* expresses the equality of two mathematical expressions. [3.1A, p. 115]	$3 + 2(4x - 5) = x + 4$ is an equation.
A *solution of an equation* is a number that, when substituted for the variable, results in a true equation. [3.1A, p. 115]	-2 is a solution of $2 - 3x = 8$ because $2 - 3(-2) = 8$ is a true equation.
To *solve an equation* means to find a solution of the equation. The goal is to rewrite the equation in the form *variable = constant*, because the constant is the solution. [3.1B, p. 116]	The equation $x = -3$ is in the form *variable = constant*. The constant, -3, is the solution of the equation.
Cost is the price that a business pays for a product. *Selling price* is the price for which a business sells a product to a customer. *Markup* is the difference between selling price and cost. *Markup rate* is the markup expressed as a percent of the retailer's cost. [3.2B, p. 137]	If a business pays $50 for a product and sells that product for $70, then the cost of the product is $50, the selling price is $70, the markup is $70 - $50 = 20, and the markup rate is $\frac{20}{50} = 40\%$.
Discount is the amount by which a retailer reduces the regular price of a product. *Discount rate* is the discount expressed as a percent of the regular price. [3.2B, p. 138]	The regular price of a product is $25. The product is now on sale for $20. The discount is $25 - $20 = 5. The discount rate is $\frac{5}{25} = 20\%$.

Copyright © Houghton Mifflin Company. All rights reserved.

Consecutive integers follow one another in order. [3.4A, p. 154]

5, 6, 7 are consecutive integers.
−9, −8, −7 are consecutive integers.

An *acute angle* is an angle whose measure is between 0° and 90°.
An *obtuse angle* is an angle whose measure is between 90° and 180°. [3.5A, p. 161]

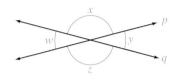

Acute angle   Obtuse angle

Two angles that are on the opposite sides of the intersection of two lines are *vertical angles*. Vertical angles have the same measure. Two angles that share a common side are *adjacent angles*. [3.5A, p. 161]

$$m\angle w = m\angle y$$
$$m\angle x = m\angle z$$

A line that intersects two other lines at two different points is a *transversal*. If the lines cut by a transversal are parallel lines, pairs of equal angles are formed: *alternate exterior angles*, *alternate interior angles*, and *corresponding angles*. [3.5A, p. 162]

$$m\angle b = m\angle d = m\angle x = m\angle z$$
$$m\angle a = m\angle c = m\angle w = m\angle y$$

Essential Rules and Procedures

Examples

Addition Property of Equations [3.1B, p. 116]
The same number can be added to each side of an equation without changing the solution of the equation.

If $a = b$, then $a + c = b + c$.

Multiplication Property of Equations [3.1C, p. 117]
Each side of an equation can be multiplied by the same *nonzero* number without changing the solution of the equation.

If $a = b$ and $c \neq 0$, then $ac = bc$.

Basic Percent Equation [3.1D, p. 119]
Percent · Base = Amount
$$P \cdot B = A$$

30% of what number is 24?

$$PB = A$$
$$0.30B = 24$$
$$\frac{0.30B}{0.30} = \frac{24}{0.30}$$
$$B = 80$$

Simple Interest Equation [3.1D, p. 120]
Interest = Principle · Rate · Time
$$I = Prt$$

A credit card company charges an annual interest rate of 21% on the monthly unpaid balance on a card. Find the amount of interest charged on an unpaid balance of $232 for April.

$$I = Prt$$
$$I = 232(0.21)\left(\frac{1}{12}\right)$$
$$= 4.06$$

Copyright © Houghton Mifflin Company. All rights reserved.

Basic Markup Equation   [3.2B, p. 137]

$S = C + rC$

The manager of a electronics store buys an MP3 player for $200 and sells the player for $250. Find the markup rate.

$250 = 200 + 200r$

Basic Discount Equation   [3.2B, p. 138]

$S = R - rR$

The sale price for a digital phone is $49. This price is 25% off the regular price. Find the regular price.

$49 = R - 0.25R$

Consecutive Integers   [3.4A, p. 154]

$n, n + 1, n + 2, \ldots$

The sum of three consecutive integers is 33.

$n + (n + 1) + (n + 2) = 33$

Consecutive Even or Consecutive Odd Integers   [3.4A, p. 154]

$n, n + 2, n + 4$

The sum of three consecutive odd integers is 33.

$n + (n + 2) + (n + 4) = 33$

Sum of the Angles of a Triangle   [3.5B, p. 163]

The sum of the measures of the angles of a triangle is 180°.

$m\angle a + m\angle b + m\angle c = 180°$

In a right triangle, the measure of one acute angle is twice the measure of the other acute angle.

$90° + x + 2x = 180°$

Value Mixture Equation   [3.6A, p. 168]

Amount · Unit Cost = Value

$AC = V$

An herbalist has 30 oz of herbs costing $4 per ounce. How many ounces of herbs costing $2 per ounce should be mixed with the 30 oz to produce a mixture costing $3.20 per ounce?

$30(4) + 2x = 3.20(30 + x)$

Percent Mixture Equation   [3.1D, 3.6B, p. 170]

Quantity = Amount · Percent Concentration

$Q = Ar$

Forty ounces of a 30% gold alloy are mixed with 60 oz of a 20% gold alloy. Find the percent concentration of the resulting gold alloy.

$0.30(40) + 0.20(60) = x(100)$

Uniform Motion Equation   [3.1E, 3.6C, p. 172]

Distance = Rate · Time

$d = rt$

A boat traveled from a harbor to an island at an average speed of 20 mph. The average speed on the return trip was 15 mph. The total trip took 3.5 h. How long did it take to travel to the island?

$20t = 15(3.5 - t)$

Copyright © Houghton Mifflin Company. All rights reserved.

Chapter 3 Review Exercises

1. Solve: $x + 3 = 24$

2. Solve: $x + 5(3x - 20) = 10(x - 4)$

3. Solve: $5x - 6 = 29$

4. Is 3 a solution of $5x - 2 = 4x + 5$?

5. Solve: $\dfrac{3}{5}a = 12$

6. Solve: $6x + 3(2x - 1) = -27$

7. 30 is what percent of 12?

8. Solve: $5x + 3 = 10x - 17$

9. Solve: $7 - [4 + 2(x - 3)] = 11(x + 2)$

10. Solve: $-6x + 16 = -2x$

11. Business A furniture store uses a markup rate of 60%. The store sells a solid oak curio cabinet for $1074. Find the cost of the curio cabinet. Use the formula $S = C + rC$, where S is the selling price, C is the cost, and r is the markup rate.

12. Geometry Find the measure of x.

$3x + 6$
$2x - 1$

13. Geometry Find the measure of x.

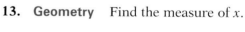
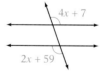

$4x + 7$
$2x + 59$

14. Physics A lever is 12 ft long. At a distance of 2 ft from the fulcrum, a force of 120 lb is applied. How large a force must be applied to the other end so that the system will balance? Use the lever system equation $F_1x = F_2(d - x)$.

15. Travel A bus traveled on a level road for 2 h at an average speed of 20 mph faster than it traveled on a winding road. The time spent on the winding road was 3 h. Find the average speed on the winding road if the total trip was 200 mi.

Copyright © Houghton Mifflin Company. All rights reserved.

16. **Business**   A ceiling fan, which regularly sells for $60, is on sale for $40. Find the discount rate. Use the formula $S = R - rR$, where S is the sale price, R is the regular price, and r is the discount rate.

17. **Geometry**   Given that $m\angle a = 74°$ and $m\angle b = 52°$, find the measures of angles x and y.

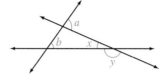

18. **Juice Mixtures**   A health food store combined cranberry juice that cost $1.79 per quart with apple juice that cost $1.19 per quart. How many quarts of each were used to make 10 qt of cranapple juice costing $1.61 per quart?

19. **Number Sense**   Four times the second of three consecutive integers equals the sum of the first and third integers. Find the integers.

20. **Geometry**   One angle of a triangle is $15°$ more than the measure of the second angle. The third angle is $15°$ less than the measure of the second angle. Find the measure of each angle.

21. **Number Sense**   Translate "four less than the product of five and a number is sixteen" into an equation and solve.

22. 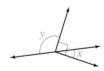 **Building Height**   The Empire State Building is 1472 ft tall. This is 654 ft less than twice the height of the Eiffel Tower. Find the height of the Eiffel Tower.

23. **Geometry**   Given $m\angle y = 115°$, find $m\angle x$.

24. **Geometry**   Given $OA \perp OB$ and $m\angle x = 30°$, find $m\angle y$.

25. **Travel**   A jet plane traveling at 600 mph overtakes a propeller-driven plane that had a 2-hour head start. The propeller-driven plane is traveling at 200 mph. How far from the starting point does the jet overtake the propeller-driven plane?

26. **Number Sense**   The sum of two numbers is twenty-one. Three times the smaller number is two less than twice the larger number. Find the two numbers.

27. **Food Mixtures**   A dairy owner mixed 5 gal of cream containing 30% butterfat with 8 gal of milk containing 4% butterfat. What is the percent of butterfat in the resulting mixture?

Copyright © Houghton Mifflin Company. All rights reserved.

Chapter 3 Test

1. Solve: $3x - 2 = 5x + 8$

2. Solve: $x - 3 = -8$

3. Solve: $3x - 5 = -14$

4. Solve: $4 - 2(3 - 2x) = 2(5 - x)$

5. Is -2 a solution of $x^2 - 3x = 2x - 6$?

6. Solve: $7 - 4x = -13$

7. What is 0.5% of 8?

8. Solve: $5x - 2(4x - 3) = 6x + 9$

9. Solve: $5x + 3 - 7x = 2x - 5$

10. Solve: $\dfrac{3}{4}x = -9$

11. Flour Mixtures   A baker wants to make a 15-pound blend of flour that costs \$.60 per pound. The blend is made using a rye flour that costs \$.70 per pound and a wheat flour that costs \$.40 per pound. How many pounds of each flour should be used?

12. Geometry   Find x.

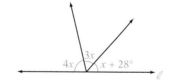

13. Business   A television that regularly sells for \$450 is on sale for \$360. Find the discount rate. Use the formula $S = R - rR$, where S is the sale price, R is the regular price, and r is the discount rate.

14. Finance   A financial manager has determined that the cost per unit for a calculator is \$15 and that the fixed cost per month is \$2000. Find the number of calculators produced during a month in which the total cost was \$5000. Use the equation $T = U \cdot N + F$, where T is the total cost, U is the cost per unit, N is the number of units produced, and F is the fixed cost.

Copyright © Houghton Mifflin Company. All rights reserved.

15. **Geometry** In an isosceles triangle, two angles are equal. The third angle of the triangle is 30° less than one of the equal angles. Find the measure of one of the equal angles.

16. **Consecutive Integers** Find three consecutive even integers whose sum is 36.

17. **Chemistry** How many gallons of water must be mixed with 5 gal of a 20% salt solution to make a 16% salt solution?

18. **Geometry** Given that $\ell_1 \parallel \ell_2$, find the measures of angles a and b.

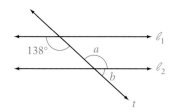

19. **Number Sense** Translate "The difference between three times a number and fifteen is twenty-seven" into an equation and solve.

20. **Sports** A cross-country skier leaves a camp to explore a wilderness area. Two hours later a friend leaves the camp in a snowmobile, traveling 4 mph faster than the skier. This friend meets the skier 1 h later. Find the rate of the snowmobile.

21. **Business** A company makes 140 televisions per day. Three times the number of 15-inch TVs made equals 20 less than the number of 25-inch TVs made. Find the number of 25-inch TVs made each day.

22. **Number Sense** The sum of two numbers is eighteen. The difference between four times the smaller number and seven is equal to the sum of two times the larger number and five. Find the two numbers.

23. **Aviation** As part of flight training, a student pilot was required to fly to an airport and then return. The average speed to the airport was 90 mph, and the average speed returning was 120 mph. Find the distance between the two airports if the total flying time was 7 h.

24. **Geometry** Given that $m\angle a = 50°$ and $m\angle b = 92°$, find the measures of angles x and y.

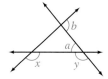

25. **Chemistry** A chemist mixes 100 g of water at 80°C with 50 g of water at 20°C. To find the final temperature of the water after mixing, use the equation $m_1(T_1 - T) = m_2(T - T_2)$, where m_1 is the quantity of water at the hotter temperature, T_1 is the temperature of the hotter water, m_2 is the quantity of water at the cooler temperature, T_2 is the temperature of the cooler water, and T is the final temperature of the water after mixing.

Copyright © Houghton Mifflin Company. All rights reserved.

Cumulative Review Exercises

1. Subtract: $-6 - (-20) - 8$

2. Multiply: $(-2)(-6)(-4)$

3. Subtract: $-\dfrac{5}{6} - \left(-\dfrac{7}{16}\right)$

4. Divide: $-2\dfrac{1}{3} \div 1\dfrac{1}{6}$

5. Simplify: $-4^2 \cdot \left(-\dfrac{3}{2}\right)^3$

6. Simplify: $25 - 3\dfrac{(5-2)^2}{2^3 + 1} - (-2)$

7. Evaluate $3(a - c) - 2ab$ when $a = 2$, $b = 3$, and $c = -4$.

8. Simplify: $3x - 8x + (-12x)$

9. Simplify: $2a - (-3b) - 7a - 5b$

10. Simplify: $(16x)\left(\dfrac{1}{8}\right)$

11. Simplify: $-4(-9y)$

12. Simplify: $-2(-x^2 - 3x + 2)$

13. Simplify: $-2(x - 3) + 2(4 - x)$

14. Simplify: $-3[2x - 4(x - 3)] + 2$

15. Is -3 a solution of $x^2 + 6x + 9 = x + 3$?

16. Is $\dfrac{1}{2}$ a solution of $3 - 8x = 12x - 2$?

17. Find 32% of 60.

18. Solve: $\dfrac{3}{5}x = -15$

19. Solve: $7x - 8 = -29$

20. Solve: $13 - 9x = -14$

Copyright © Houghton Mifflin Company. All rights reserved.

21. Solve: $8x - 3(4x - 5) = -2x - 11$

22. Solve: $6 - 2(5x - 8) = 3x - 4$

23. Solve: $5x - 8 = 12x + 13$

24. Solve: $11 - 4x = 2x + 8$

25. Chemistry   A chemist mixes 300 g of water at 75°C with 100 g of water at 15°C. To find the final temperature of the water after mixing, use the equation $m_1(T_1 - T) = m_2(T - T_2)$, where m_1 is the quantity of water at the hotter temperature, T_1 is the temperature of the hotter water, m_2 is the quantity of water at the cooler temperature, T_2 is the temperature of the cooler water, and T is the final temperature of the water after mixing.

26. Number Sense   Translate "The difference between twelve and the product of five and a number is negative eighteen" into an equation and solve.

27. Construction   The area of a cement foundation of a house is 2000 ft². This is 200 ft² more than three times the area of the garage. Find the area of the garage.

28. Flour Mixtures   How many pounds of an oat flour that costs $.80 per pound must be mixed with 40 lb of a wheat flour that costs $.50 per pound to make a blend that costs $.60 per pound?

29. Metallurgy   How many grams of pure gold must be added to 100 g of a 20% gold alloy to make an alloy that is 36% gold?

30. Geometry   The perimeter of a rectangular office is 44 ft. The length of the office is 2 ft more than the width. Find the dimensions of the office.

31. Geometry   Find the measure of $\angle x$.

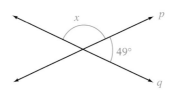

32. Geometry   In an equilateral triangle, all three angles are equal. Find the measure of one of the angles of an equilateral triangle.

33. Sports   A sprinter ran to the end of a track at an average rate of 8 m/s and then jogged back to the starting point at an average rate of 3 m/s. The sprinter took 55 s to run to the end of the track and jog back. Find the length of the track.

Copyright © Houghton Mifflin Company. All rights reserved.

Copyright © Houghton Mifflin Company. All rights reserved.

chapter

4 Polynomials

High-powered telescopes, such as the one shown here, allow scientists to look at objects that are extremely large but very far away. As an astronomer, this man uses scientific notation to describe distances in space. Scientific notation replaces very large or very small numbers with more concise expressions, making these numbers easier to read and write. **Exercises 115 to 122 on pages 217 and 218** describe some situations in which scientific notation would be used.

OBJECTIVES

Section 4.1

A To add polynomials
B To subtract polynomials

Section 4.2

A To multiply monomials
B To simplify powers of monomials

Section 4.3

A To multiply a polynomial by a monomial
B To multiply two polynomials
C To multiply two binomials
D To multiply binomials that have special products
E To solve application problems

Section 4.4

A To divide monomials
B To write a number in scientific notation

Section 4.5

A To divide a polynomial by a monomial
B To divide polynomials

Need help? For online student resources, such as section quizzes, visit this textbook's website at **math.college.hmco.com/students**.

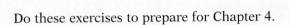
Do these exercises to prepare for Chapter 4.

1. Subtract: $-2 - (-3)$

2. Multiply: $-3(6)$

3. Simplify: $-\dfrac{24}{-36}$

4. Evaluate $3n^4$ when $n = -2$.

5. If $\dfrac{a}{b}$ is a fraction in simplest form, what number is not a possible value of b?

6. Are $2x^2$ and $2x$ like terms?

7. Simplify: $3x^2 - 4x + 1 + 2x^2 - 5x - 7$

8. Simplify: $-4y + 4y$

9. Simplify: $-3(2x - 8)$

10. Simplify: $3xy - 4y - 2(5xy - 7y)$

GO FIGURE • • •

If $x + y$, xy, and $\dfrac{x}{y}$ all equal the same number, find the values of x and y.

Copyright © Houghton Mifflin Company. All rights reserved.

4.1 Addition and Subtraction of Polynomials

Objective A To add polynomials

Copyright © Houghton Mifflin Company. All rights reserved.

TAKE NOTE

The expression $3\sqrt{x}$ is not a monomial because $\sqrt{x}$ cannot be written as a product of variables.

The expression $\dfrac{2x}{y}$ is not a monomial because it is a *quotient* of variables.

A **monomial** is a number, a variable, or a product of numbers and variables. For instance,

7	b	$\dfrac{2}{3}a$	$12xy^2$
A number	A variable	A product of a number and a variable	A product of a number and variables

A **polynomial** is a variable expression in which the terms are monomials.

A polynomial of *one* term is a **monomial.** $\quad -7x^2$ is a monomial.
A polynomial of *two* terms is a **binomial.** $\quad 4x + 2$ is a binomial.
A polynomial of *three* terms is a **trinomial.** $\quad 7x^2 + 5x - 7$ is a trinomial.

The **degree of a polynomial in one variable** is the greatest exponent on a variable. The degree of $4x^3 - 5x^2 + 7x - 8$ is 3; the degree of $2y^4 + y^2 - 1$ is 4. The degree of a nonzero constant is zero. For instance, the degree of 7 is zero. The number zero has no degree.

The terms of a polynomial in one variable are usually arranged so that the exponents of the variable decrease from left to right. This is called **descending order.**

$$5x^3 - 4x^2 + 6x - 1$$
$$7z^4 + 4z^3 + z - 6$$
$$2y^4 + y^3 - 2y^2 + 4y - 1$$

Polynomials can be added, using either a horizontal or a vertical format, by combining like terms.

HOW TO Add $(3x^3 - 7x + 2) + (7x^2 + 2x - 7)$. Use a horizontal format.

$(3x^3 - 7x + 2) + (7x^2 + 2x - 7)$
$\quad = 3x^3 + 7x^2 + (-7x + 2x) + (2 - 7)$ • Use the Commutative and Associative Properties of Addition to rearrange and group like terms.

$\quad = 3x^3 + 7x^2 - 5x - 5$ • Then combine like terms.

HOW TO Add $(-4x^2 + 6x - 9) + (12 - 8x + 2x^3)$. Use a vertical format.

$\quad -4x^2 + 6x - 9$
$\underline{2x^3 - 8x + 12}$ • Arrange the terms of each polynomial in descending order with like terms in the same column.
$2x^3 - 4x^2 - 2x + 3$ • Combine the terms in each column.

Example 1

Use a horizontal format to add
$(8x^2 - 4x - 9) + (2x^2 + 9x - 9)$.

Solution
$(8x^2 - 4x - 9) + (2x^2 + 9x - 9)$
$\quad = (8x^2 + 2x^2) + (-4x + 9x) + (-9 - 9)$
$\quad = 10x^2 + 5x - 18$

You Try It 1

Use a horizontal format to add
$(-4x^3 + 2x^2 - 8) + (4x^3 + 6x^2 - 7x + 5)$.

Your solution

Solution on p. S10

Example 2

Use a vertical format to add
$(-5x^3 + 4x^2 - 7x + 9) + (2x^3 + 5x - 11)$.

Solution $-5x^3 + 4x^2 - 7x + 19$
$2x^3 + 4x^2 + 5x - 11$
$\overline{-3x^3 + 4x^2 - 2x - 2}$

You Try It 2

Use a vertical format to add
$(6x^3 + 2x + 8) + (-9x^3 + 2x^2 - 12x - 8)$.

Your solution

Solution on p. S10

Objective B   **To subtract polynomials**

The **opposite of the polynomial** $(3x^2 - 7x + 8)$ is $-(3x^2 - 7x + 8)$.

To simplify the opposite of a polynomial, $-(3x^2 - 7x + 8) = -3x^2 + 7x - 8$
change the sign of each term to its opposite.

TAKE NOTE

This is the same definition used for subtraction of integers: Subtraction is addition of the opposite.

Polynomials can be subtracted using either a horizontal or a vertical format. To subtract, add the opposite of the second polynomial to the first.

HOW TO   Subtract $(4y^2 - 6y + 7) - (2y^3 - 5y - 4)$. Use a horizontal format.

$(4y^2 - 6y + 7) - (2y^3 - 5y - 4)$
$= (4y^2 - 6y + 7) + (-2y^3 + 5y + 4)$ • **Add the opposite of the second**
$= -2y^3 + 4y^2 + (-6y + 5y) + (7 + 4)$ **polynomial to the first.**
$= -2y^3 + 4y^2 - y + 11$ • **Combine like terms.**

HOW TO   Subtract $(9 + 4y + 3y^3) - (2y^2 + 4y - 21)$. Use a vertical format.

The opposite of $2y^2 + 4y - 21$ is $-2y^2 - 4y + 21$.

$3y^3 + 4y + 9$ • **Arrange the terms of each polynomial in descend-**
$ - 2y^2 - 4y + 21$ **ing order with like terms in the same column.**
$\overline{3y^3 - 2y^2 + 30}$ • **Note that $4y - 4y = 0$, but 0 is not written.**

Example 3

Use a horizontal format to subtract
$(7c^2 - 9c - 12) - (9c^2 + 5c - 8)$.

Solution
$(7c^2 - 9c - 12) - (9c^2 + 5c - 8)$
$= (7c^2 - 9c - 12) + (-9c^2 - 5c + 8)$
$= -2c^2 - 14c - 4$

You Try It 3

Use a horizontal format to subtract
$(-4w^3 + 8w - 8) - (3w^3 - 4w^2 - 2w - 1)$.

Your solution

Example 4

Use a vertical format to subtract
$(3k^2 - 4k + 1) - (k^3 + 3k^2 - 6k - 8)$.

Solution
$ 3k^2 - 4k + 1$ • **Add the opposite of**
$-k^3 - 3k^2 + 6k + 8$ $(k^3 + 3k^2 - 6k - 8)$
$\overline{-k^3 + 2k + 9}$ **to the first polynomial.**

You Try It 4

Use a vertical format to subtract
$(13y^3 - 6y - 7) - (4y^2 - 6y - 9)$.

Your solution

Solutions on p. S10

Copyright © Houghton Mifflin Company. All rights reserved.

4.1 Exercises

Objective A   **To add polynomials**

For Exercises 1 to 8, state whether the expression is a monomial.

1. 17

2. $3x^4$

3. $\dfrac{17}{\sqrt{x}}$

4. xyz

5. $\dfrac{2}{3}y$

6. $\dfrac{xy}{z}$

7. $\sqrt{5}\,x$

8. πx

For Exercises 9 to 16, state whether the expression is a monomial, a binomial, a trinomial, or none of these.

9. $3x + 5$

10. $2y - 3\sqrt{y}$

11. $9x^2 - x - 1$

12. $x^2 + y^2$

13. $\dfrac{2}{x} - 3$

14. $\dfrac{ab}{4}$

15. $6x^2 + 7x$

16. $12a^4 - 3a + 2$

For Exercises 17 to 26, add. Use a vertical format.

17. $(x^2 + 7x) + (-3x^2 - 4x)$

18. $(3y^2 - 2y) + (5y^2 + 6y)$

19. $(y^2 + 4y) + (-4y - 8)$

20. $(3x^2 + 9x) + (6x - 24)$

21. $(2x^2 + 6x + 12) + (3x^2 + x + 8)$

22. $(x^2 + x + 5) + (3x^2 - 10x + 4)$

23. $(-7x + x^3 + 4) + (2x^2 + x - 10)$

24. $(y^2 + 3y^3 + 1) + (-4y^3 - 6y - 3)$

25. $(2a^3 - 7a + 1) + (1 - 4a - 3a^2)$

26. $(5r^3 - 6r^2 + 3r) + (-3 - 2r + r^2)$

For Exercises 27 to 36, add. Use a horizontal format.

27. $(4x^2 + 2x) + (x^2 + 6x)$

28. $(-3y^2 + y) + (4y^2 + 6y)$

29. $(4x^2 - 5xy) + (3x^2 + 6xy - 4y^2)$

30. $(2x^2 - 4y^2) + (6x^2 - 2xy + 4y^2)$

31. $(2a^2 - 7a + 10) + (a^2 + 4a + 7)$

32. $(-6x^2 + 7x + 3) + (3x^2 + x + 3)$

33. $(7x + 5x^3 - 7) + (10x^2 - 8x + 3)$

34. $(4y + 3y^3 + 9) + (2y^2 + 4y - 21)$

35. $(7 - 5r + 2r^2) + (3r^3 - 6r)$

36. $(14 + 4y + 3y^3) + (-4y^2 + 21)$

Copyright © Houghton Mifflin Company. All rights reserved.

| *Objective B* | **To subtract polynomials** |

For Exercises 37 to 46, subtract. Use a vertical format.

37. $(x^2 - 6x) - (x^2 - 10x)$

38. $(y^2 + 4y) - (y^2 + 10y)$

39. $(2y^2 - 4y) - (-y^2 + 2)$

40. $(-3a^2 - 2a) - (4a^2 - 4)$

41. $(x^2 - 2x + 1) - (x^2 + 5x + 8)$

42. $(3x^2 + 2x - 2) - (5x^2 - 5x + 6)$

43. $(4x^3 + 5x + 2) - (1 + 2x - 3x^2)$

44. $(5y^2 - y + 2) - (-3 + 3y - 2y^3)$

45. $(-2y + 6y^2 + 2y^3) - (4 + y^2 + y^3)$

46. $(4 - x - 2x^2) - (-2 + 3x - x^3)$

For Exercises 47 to 56, subtract. Use a horizontal format.

47. $(y^2 - 10xy) - (2y^2 + 3xy)$

48. $(x^2 - 3xy) - (-2x^2 + xy)$

49. $(3x^2 + x - 3) - (4x + x^2 - 2)$

50. $(5y^2 - 2y + 1) - (-y - 2 - 3y^2)$

51. $(-2x^3 + x - 1) - (-x^2 + x - 3)$

52. $(2x^2 + 5x - 3) - (3x^3 + 2x - 5)$

53. $(1 - 2a + 4a^3) - (a^3 - 2a + 3)$

54. $(7 - 8b + b^2) - (4b^3 - 7b - 8)$

55. $(-1 - y + 4y^3) - (3 - 3y - 2y^2)$

56. $(-3 - 2x + 3x^2) - (4 - 2x^2 + 2x^3)$

APPLYING THE CONCEPTS

57. What polynomial must be added to $3x^2 - 6x + 9$ so that the sum is $4x^2 + 3x - 2$?

58. What polynomial must be subtracted from $2x^2 - x - 2$ so that the difference is $5x^2 + 3x + 1$?

59. In your own words, explain the terms *monomial, binomial, trinomial,* and *polynomial.* Give an example of each.

60. Is it possible to subtract two polynomials, each of degree 3, and have the difference be a polynomial of degree 2? If so, give an example. If not, explain why not.

61. Is it possible to add two polynomials, each of degree 3, and have the sum be a polynomial of degree 2? If so, give an example. If not, explain why not.

Copyright © Houghton Mifflin Company. All rights reserved.

4.2 Multiplication of Monomials

Objective A   To multiply monomials

Recall that in an exponential expression such as x^6, x is the base and 6 is the exponent. The exponent indicates the number of times the base occurs as a factor.

The product of exponential expressions with the *same* base can be simplified by writing each expression in factored form and writing the result with an exponent.

$$x^3 \cdot x^2 = \overbrace{(x \cdot x \cdot x)}^{3 \text{ factors}} \cdot \overbrace{(x \cdot x)}^{2 \text{ factors}}$$
$$\underbrace{}_{5 \text{ factors}}$$

$$= x^5$$

Note that adding the exponents results in the same product.

$$x^3 \cdot x^2 = x^{3+2} = x^5$$

Rule for Multiplying Exponential Expressions

If m and n are positive integers, then $x^m \cdot x^n = x^{m+n}$.

HOW TO   Simplify: $y^4 \cdot y \cdot y^3$

$y^4 \cdot y \cdot y^3 = y^{4+1+3}$ • The bases are the same. **Add the**
$\qquad\qquad = y^8$ **exponents.** Recall that $y = y^1$.

HOW TO   Simplify: $(-3a^4b^3)(2ab^4)$

$(-3a^4b^3)(2ab^4) = (-3 \cdot 2)(a^4 \cdot a)(b^3 \cdot b^4)$

• Use the Commutative and Associative Properties of Multiplication to rearrange and group factors.

$$= -6(a^{4+1})(b^{3+4})$$

• To multiply expressions with the same base, **add the exponents**.

$$= -6a^5b^7$$

• Simplify.

TAKE NOTE

The Rule for Multiplying Exponential Expressions requires the bases to be the same. The expression a^5b^7 cannot be simplified.

Example 1   Simplify: $(-5ab^3)(4a^5)$

Solution
$(-5ab^3)(4a^5)$
$= (-5 \cdot 4)(a \cdot a^5)b^3$ • Multiply coefficients. Add
$= -20a^6b^3$ exponents with same base.

You Try It 1   Simplify: $(8m^3n)(-3n^5)$

Your solution

Example 2   Simplify: $(6x^3y^2)(4x^4y^5)$

Solution
$(6x^3y^2)(4x^4y^5)$
$= (6 \cdot 4)(x^3 \cdot x^4)(y^2 \cdot y^5)$ • Multiply coefficients. Add
$= 24x^7y^7$ exponents with same base.

You Try It 2   Simplify: $(12p^4q^3)(-3p^5q^2)$

Your solution

Solutions on p. S10

Copyright © Houghton Mifflin Company. All rights reserved.

Copyright © Houghton Mifflin Company. All rights reserved.

Objective B   **To simplify powers of monomials**

Point of Interest

One of the first symbolic representations of powers was given by Diophantus (c. 250 A.D.) in his book *Arithmetica*. He used Δ^Y for x^2 and κ^Y for x^3. The symbol Δ^Y was the first two letters of the Greek word *dunamis*, which means "power"; κ^Y was from the Greek word *kubos*, which means "cube." He also combined these symbols to denote higher powers. For instance, $\Delta\kappa^Y$ was the symbol for x^5.

The power of a monomial can be simplified by writing the power in factored form and then using the Rule for Multiplying Exponential Expressions.

$$(x^4)^3 = x^4 \cdot x^4 \cdot x^4 \qquad\qquad (a^2b^3)^2 = (a^2b^3)(a^2b^3)$$

- Write in factored form.

$$= x^{4+4+4} \qquad\qquad = a^{2+2}b^{3+3}$$

- Use the Rule for Multiplying Exponential Expressions.

$$= x^{12} \qquad\qquad = a^4b^6$$

Note that multiplying each exponent inside the parentheses by the exponent outside the parentheses results in the same product.

$$(x^4)^3 = x^{4\cdot3} = x^{12} \qquad\qquad (a^2b^3)^2 = a^{2\cdot2}b^{3\cdot2} = a^4b^6$$

- Multiply each exponent inside the parentheses by the exponent outside the parentheses.

Rule for Simplifying the Power of an Exponential Expression

If m and n are positive integers, then $(x^m)^n = x^{mn}$.

Rule for Simplifying the Power of a Product

If m, n, and p are positive integers, then $(x^m y^n)^p = x^{mp} y^{np}$.

HOW TO   Simplify: $(5x^2y^3)^3$

$$(5x^2y^3)^3 = 5^{1\cdot3}x^{2\cdot3}y^{3\cdot3}$$

- Use the **Rule for Simplifying the Power of a Product**. Note that $5 = 5^1$.

$$= 5^3x^6y^9$$

$$= 125x^6y^9$$

- Evaluate 5^3.

Example 3   Simplify: $(-2p^3r)^4$

Solution

$$(-2p^3r)^4 = (-2)^{1\cdot4}p^{3\cdot4}r^{1\cdot4}$$
$$= (-2)^4p^{12}r^4 = 16p^{12}r^4$$

- Use the Rule for Simplifying the Power of a Product.

You Try It 3   Simplify: $(-3a^4bc^2)^3$

Your solution

Example 4   Simplify: $(2a^2b)(2a^3b^2)^3$

Solution

$$(2a^2b)(2a^3b^2)^3$$
$$= (2a^2b)(2^{1\cdot3}a^{3\cdot3}b^{2\cdot3})$$
$$= (2a^2b)(2^3a^9b^6)$$
$$= (2a^2b)(8a^9b^6) = 16a^{11}b^7$$

- Use the Rule for Simplifying the Power of a Product.

You Try It 4   Simplify: $(-xy^4)(-2x^3y^2)^2$

Your solution

Solutions on p. S10

4.2 Exercises

Objective A   **To multiply monomials**

1. ✏️ Explain how to multiply two monomials. Provide an example.

2. ✏️ Explain how to simplify the power of a monomial. Provide an example.

For Exercises 3 to 35, simplify.

3. $(6x^2)(5x)$

4. $(-4y^3)(2y)$

5. $(7c^2)(-6c^4)$

6. $(-8z^5)(5z^8)$

7. $(-3a^3)(-3a^4)$

8. $(-5a^6)(-2a^5)$

9. $(x^2)(xy^4)$

10. $(x^2y^4)(xy^7)$

11. $(-2x^4)(5x^5y)$

12. $(-3a^3)(2a^2b^4)$

13. $(-4x^2y^4)(-3x^5y^4)$

14. $(-6a^2b^4)(-4ab^3)$

15. $(2xy)(-3x^2y^4)$

16. $(-3a^2b)(-2ab^3)$

17. $(x^2yz)(x^2y^4)$

18. $(-ab^2c)(a^2b^5)$

19. $(-a^2b^3)(-ab^2c^4)$

20. $(-x^2y^3z)(-x^3y^4)$

21. $(-5a^2b^2)(6a^3b^6)$

22. $(7xy^4)(-2xy^3)$

23. $(-6a^3)(-a^2b)$

24. $(-2a^2b^3)(-4ab^2)$

25. $(-5y^4z)(-8y^6z^5)$

26. $(3x^2y)(-4xy^2)$

27. $(x^2y)(yz)(xyz)$

28. $(xy^2z)(x^2y)(z^2y^2)$

29. $(3ab^2)(-2abc)(4ac^2)$

30. $(-2x^3y^2)(-3x^2z^2)(-5y^3z^3)$

31. $(4x^4z)(-yz^3)(-2x^3z^2)$

32. $(-a^3b^4)(-3a^4c^2)(4b^3c^4)$

33. $(-2x^2y^3)(3xy)(-5x^3y^4)$

34. $(4a^2b)(-3a^3b^4)(a^5b^2)$

35. $(3a^2b)(-6bc)(2ac^2)$

Objective B   **To simplify powers of monomials**

For Exercises 36 to 66, simplify.

36. $(z^4)^3$

37. $(x^3)^5$

38. $(y^4)^2$

39. $(x^7)^2$

40. $(-y^5)^3$

41. $(-x^2)^4$

42. $(-x^2)^3$

43. $(-y^3)^4$

44. $(-3y)^3$

45. $(-2x^2)^3$

Copyright © Houghton Mifflin Company. All rights reserved.

46. $(a^3b^4)^3$ **47.** $(x^2y^3)^2$ **48.** $(2x^3y^4)^5$ **49.** $(3x^2y)^2$ **50.** $(-2ab^3)^4$

51. $(-3x^3y^2)^5$ **52.** $(3b^2)(2a^3)^4$ **53.** $(-2x)(2x^3)^2$ **54.** $(2y)(-3y^4)^3$

55. $(3x^2y)(2x^2y^2)^3$ **56.** $(a^3b)^2(ab)^3$ **57.** $(ab^2)^2(ab)^2$ **58.** $(-x^2y^3)^2(-2x^3y)^3$

59. $(-2x)^3(-2x^3y)^3$ **60.** $(-3y)(-4x^2y^3)^3$ **61.** $(-2x)(-3xy^2)^2$ **62.** $(-3y)(-2x^2y)^3$

63. $(ab^2)(-2a^2b)^3$ **64.** $(a^2b^2)(-3ab^4)^2$ **65.** $(-2a^3)(3a^2b)^3$ **66.** $(-3b^2)(2ab^2)^3$

APPLYING THE CONCEPTS

For Exercises 67 to 74, simplify.

67. $3x^2 + (3x)^2$ **68.** $4x^2 - (4x)^2$ **69.** $2x^6y^2 + (3x^2y)^2$ **70.** $(x^2y^2)^3 + (x^3y^3)^2$

71. $(2a^3b^2)^3 - 8a^9b^6$ **72.** $4y^2z^4 - (2yz^2)^2$ **73.** $(x^2y^4)^2 + (2xy^2)^4$ **74.** $(3a^3)^2 - 4a^6 + (2a^2)^3$

For Exercises 75 to 78, answer true or false. If the answer is false, correct the right-hand side of the equation.

75. $(-a)^5 = -a^5$ **76.** $(-b)^8 = b^8$ **77.** $(x^2)^5 = x^{2+5} = x^7$ **78.** $x^3 + x^3 = 2x^{3+3} = 2x^6$

79. Evaluate $(2^3)^2$ and $2^{(3^2)}$. Are the results the same? If not, which expression has the larger value?

80. If n is a positive integer and $x^n = y^n$, does $x = y$? Explain your answer.

81. The distance a rock will fall in t seconds is $16t^2$ (neglecting air resistance). Find other examples of quantities that can be expressed in terms of an exponential expression, and explain where the expression is used.

Copyright © Houghton Mifflin Company. All rights reserved.

4.3 Multiplication of Polynomials

Objective A   To multiply a polynomial by a monomial

To multiply a polynomial by a monomial, use the Distributive Property and the Rule for Multiplying Exponential Expressions.

> **HOW TO**   Multiply: $-3a(4a^2 - 5a + 6)$
>
> $-3a(4a^2 - 5a + 6) = -3a(4a^2) - (-3a)(5a) + (-3a)(6)$ • Use the **Distributive Property**.
>
> $= -12a^3 + 15a^2 - 18a$

Example 1

Multiply: $(5x + 4)(-2x)$

Solution

$(5x + 4)(-2x) = -10x^2 - 8x$

You Try It 1

Multiply: $(-2y + 3)(-4y)$

Your solution

Example 2

Multiply: $2a^2b(4a^2 - 2ab + b^2)$

Solution

$2a^2b(4a^2 - 2ab + b^2)$
$= 8a^4b - 4a^3b^2 + 2a^2b^3$

You Try It 2

Multiply: $-a^2(3a^2 + 2a - 7)$

Your solution

Solutions on p. S10

Objective B   To multiply two polynomials

Multiplication of two polynomials requires the repeated application of the Distributive Property.

$$(y - 2)(y^2 + 3y + 1) = (y - 2)(y^2) + (y - 2)(3y) + (y - 2)(1)$$
$$= y^3 - 2y^2 + 3y^2 - 6y + y - 2$$
$$= y^3 + y^2 - 5y - 2$$

A convenient method of multiplying two polynomials is to use a vertical format similar to that used for multiplication of whole numbers.

$$
\begin{array}{r}
y^2 + 3y + 1 \\
y - 2 \\
\hline
-2y^2 - 6y - 2 \\
y^3 + 3y^2 + y \\
\hline
y^3 + y^2 - 5y - 2
\end{array}
$$

$-2y^2 - 6y - 2 = -2(y^2 + 3y + 1)$ • Multiply by **−2**.
$y^3 + 3y^2 + y = y(y + 3y + 1)$ • Multiply by **y**.
• Add the terms in each column.

Copyright © Houghton Mifflin Company. All rights reserved.

HOW TO Multiply: $(2a^3 + a - 3)(a + 5)$

$$
\begin{array}{r}
2a^3 \quad\quad + \ a \ - \ 3 \\
a \ + \ 5 \\
\hline
10a^3 \quad\quad + \ 5a \ - \ 15 \\
2a^4 \quad\quad\quad\quad\ + \ a^2 \ - \ 3a \\
\hline
2a^4 + 10a^3 + a^2 + 2a - 15
\end{array}
$$

- Note that spaces are provided in each product so that like terms are in the same column.

- Add the terms in each column.

Example 3

Multiply: $(2b^3 - b + 1)(2b + 3)$

Solution

$$
\begin{array}{r}
2b^3 \quad\quad - \ b + 1 \\
2b + 3 \\
\hline
6b^3 \quad\quad - \ 3b + 3 = 3(2b^3 - b + 1) \\
4b^4 + \quad\quad - \ 2b^2 + 2b \quad = 2b(2b^3 - b + 1) \\
\hline
4b^4 + 6b^3 - 2b^2 - \ b + 3
\end{array}
$$

You Try It 3

Multiply: $(2y^3 + 2y^2 - 3)(3y - 1)$

Your solution

Solution on p. S10

Objective C To multiply two binomials

It is frequently necessary to find the product of two binomials. The product can be found using a method called **FOIL**, which is based on the Distributive Property. The letters of FOIL stand for **F**irst, **O**uter, **I**nner, and **L**ast. To find the product of two binomials, add the products of the **F**irst terms, the **O**uter terms, the **I**nner terms, and the **L**ast terms.

TAKE NOTE

FOIL is not really a different way of multiplying. It is based on the Distributive Property.

$(2x + 3)(x + 5)$
$= 2x(x + 5) + 3(x + 5)$
$\quad\ \ \text{F} \quad\ \text{O} \quad\ \text{I} \quad\ \text{L}$
$= 2x^2 + 10x + 3x + 15$
$= 2x^2 + 13x + 15$

HOW TO Multiply: $(2x + 3)(x + 5)$

Multiply the First terms.	$(2x + 3)(x + 5)$	$2x \cdot x = 2x^2$
Multiply the Outer terms.	$(2x + 3)(x + 5)$	$2x \cdot 5 = 10x$
Multiply the Inner terms.	$(2x + 3)(x + 5)$	$3 \cdot x = 3x$
Multiply the Last terms.	$(2x + 3)(x + 5)$	$3 \cdot 5 = 15$

$$
\begin{array}{ll}
& \quad\ \text{F} \quad\quad \text{O} \quad\quad \text{I} \quad\ \text{L} \\
\text{Add the products.} \quad (2x + 3)(x + 5) & = 2x^2 + 10x + 3x + 15 \\
\text{Combine like terms.} & = 2x^2 + 13x + 15
\end{array}
$$

HOW TO Multiply: $(4x - 3)(3x - 2)$

$$
\begin{aligned}
(4x - 3)(3x - 2) &= 4x(3x) + 4x(-2) + (-3)(3x) + (-3)(-2) \\
&= 12x^2 - 8x - 9x + 6 \\
&= 12x^2 - 17x + 6
\end{aligned}
$$

HOW TO Multiply: $(3x - 2y)(x + 4y)$

$$
\begin{aligned}
(3x - 2y)(x + 4y) &= 3x(x) + 3x(4y) + (-2y)(x) + (-2y)(4y) \\
&= 3x^2 + 12xy - 2xy - 8y^2 \\
&= 3x^2 + 10xy - 8y^2
\end{aligned}
$$

Copyright © Houghton Mifflin Company. All rights reserved.

Example 4

Multiply: $(2a - 1)(3a - 2)$

Solution

$(2a - 1)(3a - 2) = 6a^2 - 4a - 3a + 2$
$\qquad\qquad\qquad = 6a^2 - 7a + 2$

You Try It 4

Multiply: $(4y - 5)(2y - 3)$

Your solution

Example 5

Multiply: $(3x - 2)(4x + 3)$

Solution

$(3x - 2)(4x + 3) = 12x^2 + 9x - 8x - 6$
$\qquad\qquad\qquad = 12x^2 + x - 6$

You Try It 5

Multiply: $(3b + 2)(3b - 5)$

Your solution

Solutions on p. S11

Objective D   **To multiply binomials that have special products**

Using FOIL, it is possible to find a pattern for the product of the sum and difference of two terms and for the square of a binomial.

Product of the Sum and Difference of the Same Terms

$$(a + b)(a - b) = a^2 - ab + ab - b^2$$
$$= a^2 - b^2$$

Square of the first term ⎯⎯⎯⎯
Square of the second term ⎯⎯⎯⎯

Square of a Binomial

$$(a + b)^2 = (a + b)(a + b) = a^2 + ab + ab + b^2$$
$$= a^2 + 2ab + b^2$$

Square of the first term ⎯⎯⎯
Twice the product of the two terms ⎯⎯
Square of the last term ⎯⎯⎯

HOW TO   Multiply: $(2x + 3)(2x - 3)$

$(2x + 3)(2x - 3) = (2x)^2 - 3^2$ • This is the product of the sum and
$\qquad\qquad\qquad = 4x^2 - 9$ difference of the same terms.

TAKE NOTE

The word *expand* is used frequently to mean "multiply out a power."

HOW TO   Expand: $(3x - 2)^2$

$(3x - 2)^2 = (3x)^2 + 2(3x)(-2) + (-2)^2$ • This is the square of a
$\qquad\qquad = 9x^2 - 12x + 4$ binomial.

Copyright © Houghton Mifflin Company. All rights reserved.

Example 6

Multiply: $(4z - 2w)(4z + 2w)$

Solution

$(4z - 2w)(4z + 2w) = 16z^2 - 4w^2$

You Try It 6

Multiply: $(2a + 5c)(2a - 5c)$

Your solution

Example 7

Expand: $(2r - 3s)^2$

Solution

$(2r - 3s)^2 = 4r^2 - 12rs + 9s^2$

You Try It 7

Expand: $(3x + 2y)^2$

Your solution

Solutions on p. S11

Objective E   **To solve application problems**

Example 8

The length of a rectangle is $(x + 7)$ m. The width is $(x - 4)$ m. Find the area of the rectangle in terms of the variable x.

$x + 7$

$x - 4$

Strategy

To find the area, replace the variables L and W in the equation $A = L \cdot W$ by the given values and solve for A.

Solution

$A = L \cdot W$
$A = (x + 7)(x - 4)$
$A = x^2 - 4x + 7x - 28$
$A = x^2 + 3x - 28$

The area is $(x^2 + 3x - 28)$ m^2.

You Try It 8

The radius of a circle is $(x - 4)$ ft. Use the equation $A = \pi r^2$, where r is the radius, to find the area of the circle in terms of x. Leave the answer in terms of π.

$x - 4$

Your strategy

Your solution

Solution on p. S11

Copyright © Houghton Mifflin Company. All rights reserved.

4.3 Exercises

Objective A   **To multiply a polynomial by a monomial**

For Exercises 1 and 2, replace the question marks to make a true statement.

1. $3(4x - 5) = (?)(4x) - (?)(5) = ?$

2. $2x^2(3x + 7) = (?)(3x) + (?)(7) = ?$

For Exercises 3 to 34, multiply.

3. $x(x - 2)$

4. $y(3 - y)$

5. $-x(x + 7)$

6. $-y(7 - y)$

7. $3a^2(a - 2)$

8. $4b^2(b + 8)$

9. $-5x^2(x^2 - x)$

10. $-6y^2(y + 2y^2)$

11. $-x^3(3x^2 - 7)$

12. $-y^4(2y^2 - y^6)$

13. $2x(6x^2 - 3x)$

14. $3y(4y - y^2)$

15. $(2x - 4)3x$

16. $(3y - 2)y$

17. $(3x + 4)x$

18. $(2x + 1)2x$

19. $-xy(x^2 - y^2)$

20. $-x^2y(2xy - y^2)$

21. $x(2x^3 - 3x + 2)$

22. $y(-3y^2 - 2y + 6)$

23. $-a(-2a^2 - 3a - 2)$

24. $-b(5b^2 + 7b - 35)$

25. $x^2(3x^4 - 3x^2 - 2)$

26. $y^3(-4y^3 - 6y + 7)$

27. $2y^2(-3y^2 - 6y + 7)$

28. $4x^2(3x^2 - 2x + 6)$

29. $(a^2 + 3a - 4)(-2a)$

30. $(b^3 - 2b + 2)(-5b)$

31. $-3y^2(-2y^2 + y - 2)$

32. $-5x^2(3x^2 - 3x - 7)$

33. $xy(x^2 - 3xy + y^2)$

34. $ab(2a^2 - 4ab - 6b^2)$

Objective B   **To multiply two polynomials**

For Exercises 35 to 52, multiply.

35. $(x^2 + 3x + 2)(x + 1)$

36. $(x^2 - 2x + 7)(x - 2)$

37. $(a^2 - 3a + 4)(a - 3)$

Copyright © Houghton Mifflin Company. All rights reserved.

38. $(x^2 - 3x + 5)(2x - 3)$ **39.** $(-2b^2 - 3b + 4)(b - 5)$ **40.** $(-a^2 + 3a - 2)(2a - 1)$

41. $(-2x^2 + 7x - 2)(3x - 5)$ **42.** $(-a^2 - 2a + 3)(2a - 1)$ **43.** $(x^2 + 5)(x - 3)$

44. $(y^2 - 2y)(2y + 5)$ **45.** $(x^3 - 3x + 2)(x - 4)$ **46.** $(y^3 + 4y^2 - 8)(2y - 1)$

47. $(5y^2 + 8y - 2)(3y - 8)$ **48.** $(3y^2 + 3y - 5)(4y - 3)$ **49.** $(5a^3 - 5a + 2)(a - 4)$

50. $(3b^3 - 5b^2 + 7)(6b - 1)$ **51.** $(y^3 + 2y^2 - 3y + 1)(y + 2)$ **52.** $(2a^3 - 3a^2 + 2a - 1)(2a - 3)$

Objective C   **To multiply two binomials**

For Exercises 53 to 84, multiply.

53. $(x + 1)(x + 3)$ **54.** $(y + 2)(y + 5)$ **55.** $(a - 3)(a + 4)$ **56.** $(b - 6)(b + 3)$

57. $(y + 3)(y - 8)$ **58.** $(x + 10)(x - 5)$ **59.** $(y - 7)(y - 3)$ **60.** $(a - 8)(a - 9)$

61. $(2x + 1)(x + 7)$ **62.** $(y + 2)(5y + 1)$ **63.** $(3x - 1)(x + 4)$ **64.** $(7x - 2)(x + 4)$

65. $(4x - 3)(x - 7)$ **66.** $(2x - 3)(4x - 7)$ **67.** $(3y - 8)(y + 2)$ **68.** $(5y - 9)(y + 5)$

69. $(3x + 7)(3x + 11)$ **70.** $(5a + 6)(6a + 5)$ **71.** $(7a - 16)(3a - 5)$ **72.** $(5a - 12)(3a - 7)$

73. $(3a - 2b)(2a - 7b)$ **74.** $(5a - b)(7a - b)$ **75.** $(a - 9b)(2a + 7b)$

Copyright © Houghton Mifflin Company. All rights reserved.

76. $(2a + 5b)(7a - 2b)$　　　　**77.** $(10a - 3b)(10a - 7b)$　　　　**78.** $(12a - 5b)(3a - 4b)$

79. $(5x + 12y)(3x + 4y)$　　　　**80.** $(11x + 2y)(3x + 7y)$　　　　**81.** $(2x - 15y)(7x + 4y)$

82. $(5x + 2y)(2x - 5y)$　　　　**83.** $(8x - 3y)(7x - 5y)$　　　　**84.** $(2x - 9y)(8x - 3y)$

Objective D　　**To multiply binomials that have special products**

For Exercises 85 to 92, multiply.

85. $(y - 5)(y + 5)$　　**86.** $(y + 6)(y - 6)$　　**87.** $(2x + 3)(2x - 3)$　　**88.** $(4x - 7)(4x + 7)$

89. $(3x - 7)(3x + 7)$　　**90.** $(9x - 2)(9x + 2)$　　**91.** $(4 - 3y)(4 + 3y)$　　**92.** $(4x - 9y)(4x + 9y)$

For Exercises 93 to 102, expand.

93. $(x + 1)^2$　　　　**94.** $(y - 3)^2$　　　　**95.** $(3a - 5)^2$　　　　**96.** $(6x - 5)^2$

97. $(2a + b)^2$　　　　　　**98.** $(x + 3y)^2$　　　　　　**99.** $(x - 2y)^2$

100. $(2x - 3y)^2$　　　　**101.** $(5x + 2y)^2$　　　　**102.** $(2a - 9b)^2$

Objective E　　**To solve application problems**

103. Geometry　The length of a rectangle is $(5x)$ ft. The width is $(2x - 7)$ ft. Find the area of the rectangle in terms of the variable x.

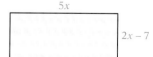

104. Geometry　The width of a rectangle is $(3x + 1)$ in. The length of the rectangle is twice the width. Find the area of the rectangle in terms of the variable x.

Copyright © Houghton Mifflin Company. All rights reserved.

105. Geometry The length of a side of a square is $(2x + 1)$ km. Find the area of the square in terms of the variable x.

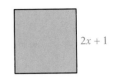

106. Geometry The radius of a circle is $(x + 4)$ cm. Find the area of the circle in terms of the variable x. Leave the answer in terms of π.

107. Geometry The base of a triangle is $(4x)$ m and the height is $(2x + 5)$ m. Find the area of the triangle in terms of the variable x.

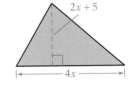

108. Sports A softball diamond has dimensions 45 ft by 45 ft. A base-path border x feet wide lies on both the first-base side and third-base side of the diamond. Express the total area of the softball diamond and the base paths in terms of the variable x.

109. Sports An athletic field has dimensions 30 yd by 100 yd. An end zone that is w yards wide borders each end of the field. Express the total area of the field and the end zones in terms of the variable w.

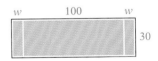

APPLYING THE CONCEPTS

110. Simplify: $(a + b)^2 - (a - b)^2$

111. Expand: $(x^2 + x - 3)^2$

112. Expand: $(a + 3)^3$

113. What polynomial has quotient $3x - 4$ when divided by $4x + 5$?

114. Add $x^2 + 2x - 3$ to the product of $2x - 5$ and $3x + 1$.

115. Subtract $4x^2 - x - 5$ from the product of $x^2 + x + 3$ and $x - 4$.

116. If a polynomial of degree 3 is multiplied by a polynomial of degree 2, what is the degree of the resulting polynomial?

117. Is it possible to multiply a polynomial of degree 2 by a polynomial of degree 2 and have the product be a polynomial of degree 3? If so, give an example. If not, explain why not.

Copyright © Houghton Mifflin Company. All rights reserved.

4.4 Integer Exponents and Scientific Notation

Objective A **To divide monomials**

The quotient of two exponential expressions with the same base can be simplified by writing each expression in factored form, dividing by the common factors, and then writing the result with an exponent.

$$\frac{x^5}{x^2} = \frac{\overset{1}{\cancel{x}} \cdot \overset{1}{\cancel{x}} \cdot x \cdot x \cdot x}{\underset{1}{\cancel{x}} \cdot \underset{1}{\cancel{x}}} = x^3$$

Note that subtracting the exponents gives the same result.

$$\frac{x^5}{x^2} = x^{5-2} = x^3$$

To divide two monomials with the same base, subtract the exponents of the like bases.

HOW TO Simplify: $\dfrac{a^7}{a^3}$

$\dfrac{a^7}{a^3} = a^{7-3}$ • The bases are the same. **Subtract the exponents.**

$\quad = a^4$

HOW TO Simplify: $\dfrac{r^8 t^6}{r^7 t}$

$\dfrac{r^8 t^6}{r^7 t} = r^{8-7} t^{6-1}$ • **Subtract the exponents** of the like bases.

$\quad = r t^5$

HOW TO Simplify: $\dfrac{p^7}{z^4}$

Because the bases are not the same, $\dfrac{p^7}{z^4}$ is already in simplest form.

Consider the expression $\dfrac{x^4}{x^4}$, $x \ne 0$. This expression can be simplified, as shown below, by subtracting exponents or by dividing by common factors.

$$\frac{x^4}{x^4} = x^{4-4} = x^0 \qquad\qquad \frac{x^4}{x^4} = \frac{\overset{1}{\cancel{x}} \cdot \overset{1}{\cancel{x}} \cdot \overset{1}{\cancel{x}} \cdot \overset{1}{\cancel{x}}}{\underset{1}{\cancel{x}} \cdot \underset{1}{\cancel{x}} \cdot \underset{1}{\cancel{x}} \cdot \underset{1}{\cancel{x}}} = 1$$

The equations $\dfrac{x^4}{x^4} = x^0$ and $\dfrac{x^4}{x^4} = 1$ suggest the following definition of x^0.

Definition of Zero as an Exponent

If $x \ne 0$, then $x^0 = 1$. The expression 0^0 is not defined.

Copyright © Houghton Mifflin Company. All rights reserved.

TAKE NOTE

In the example at the right, we indicated that $a \neq 0$. If we try to evaluate $(12a^3)^0$ when $a = 0$, we have

$$[12(0)^3]^0 = [12(0)]^0 = 0^0$$

However, 0^0 is not defined. Therefore, we must assume that $a \neq 0$. To avoid stating this for every example or exercise, we will assume that variables do not have values that result in the expression 0^0.

HOW TO   Simplify: $(12a^3)^0$, $a \neq 0$

$(12a^3)^0 = 1$ • Any nonzero expression to the zero power is **1**.

HOW TO   Simplify: $-(4x^3y^7)^0$

$-(4x^3y^7)^0 = -(1) = -1$

Consider the expression $\dfrac{x^4}{x^6}$, $x \neq 0$. This expression can be simplified, as shown below, by subtracting exponents or by dividing by common factors.

$$\frac{x^4}{x^6} = x^{4-6} = x^{-2} \qquad\qquad \frac{x^4}{x^6} = \frac{\overset{1}{\cancel{x}} \cdot \overset{1}{\cancel{x}} \cdot \overset{1}{\cancel{x}} \cdot \overset{1}{\cancel{x}}}{\underset{1}{\cancel{x}} \cdot \underset{1}{\cancel{x}} \cdot \underset{1}{\cancel{x}} \cdot \underset{1}{\cancel{x}} \cdot x \cdot x} = \frac{1}{x^2}$$

Point of Interest

In the 15th century, the expression $12^{2\overline{m}}$ was used to mean $12x^{-2}$. The use of $\overline{m}$ reflected an Italian influence. In Italy, m was used for minus and p was used for plus. It was understood that $2\overline{m}$ referred to an unnamed variable. Issac Newton, in the 17th century, advocated the negative exponent we currently use.

The equations $\dfrac{x^4}{x^6} = x^{-2}$ and $\dfrac{x^4}{x^6} = \dfrac{1}{x^2}$ suggest that $x^{-2} = \dfrac{1}{x^2}$.

Definition of a Negative Exponent

If $x \neq 0$ and n is a positive integer, then

$$x^{-n} = \frac{1}{x^n} \qquad \text{and} \qquad \frac{1}{x^{-n}} = x^n$$

An exponential expression is in simplest form when it is written with only positive exponents.

TAKE NOTE

Note from the example at the right that 2^{-4} is a *positive* number. A negative exponent does not change the sign of a number.

HOW TO   Evaluate 2^{-4}.

$2^{-4} = \dfrac{1}{2^4}$ • Use the **Definition of a Negative Exponent**.

$\phantom{2^{-4}} = \dfrac{1}{16}$ • Evaluate the expression.

TAKE NOTE

For the expression $3n^{-5}$ the exponent on n is -5 (*negative* 5). The n^{-5} is written in the denominator as n^5. The exponent on 3 is 1 (*positive* 1). The 3 remains in the numerator. Also, we indicated that $n \neq 0$. This is done because division by zero is not defined. In this textbook, we will assume that values of the variables are chosen so that division by zero does not occur.

HOW TO   Simplify: $3n^{-5}$, $n \neq 0$

$3n^{-5} = 3 \cdot \dfrac{1}{n^5} = \dfrac{3}{n^5}$ • Use the **Definition of a Negative Exponent** to rewrite the expression with a positive exponent.

HOW TO   Simplify: $\dfrac{2}{5a^{-4}}$

$\dfrac{2}{5a^{-4}} = \dfrac{2}{5} \cdot \dfrac{1}{a^{-4}} = \dfrac{2}{5} \cdot a^4 = \dfrac{2a^4}{5}$ • Use the **Definition of a Negative Exponent** to rewrite the expression with a positive exponent.

Copyright © Houghton Mifflin Company. All rights reserved.

The expression $\left(\dfrac{x^4}{y^3}\right)^2$, $y \neq 0$, can be simplified by squaring $\dfrac{x^4}{y^3}$ or by multiplying each exponent in the quotient by the exponent outside the parentheses.

$$\left(\frac{x^4}{y^3}\right)^2 = \left(\frac{x^4}{y^3}\right)\left(\frac{x^4}{y^3}\right) = \frac{x^4 \cdot x^4}{y^3 \cdot y^3} = \frac{x^{4+4}}{y^{3+3}} = \frac{x^8}{y^6} \qquad \left(\frac{x^4}{y^3}\right)^2 = \frac{x^{4 \cdot 2}}{y^{3 \cdot 2}} = \frac{x^8}{y^6}$$

Rule for Simplifying the Power of a Quotient

If m, n, and q are integers and $y \neq 0$, then $\left(\dfrac{x^m}{y^n}\right)^p = \dfrac{x^{mp}}{y^{np}}$.

TAKE NOTE

As a reminder, although it is not stated, we are assuming that $a \neq 0$ and $b \neq 0$. This is done to ensure that we do not have division by zero.

HOW TO Simplify: $\left(\dfrac{a^3}{b^2}\right)^{-2}$

$\left(\dfrac{a^3}{b^2}\right)^{-2} = \dfrac{a^{3(-2)}}{b^{2(-2)}}$ • Use the **Rule for Simplifying the Power of a Quotient**.

$= \dfrac{a^{-6}}{b^{-4}} = \dfrac{b^4}{a^6}$ • Use the **Definition of a Negative Exponent** to write the expression with positive exponents.

The example above suggests the following rule.

Rule for Negative Exponents on Fractional Expressions

If $a \neq 0$, $b \neq 0$, and n is a positive integer, then

$$\left(\frac{a}{b}\right)^{-n} = \left(\frac{b}{a}\right)^n$$

Now that zero as an exponent and negative exponents have been defined, a rule for dividing exponential expressions can be stated.

Rule for Dividing Exponential Expressions

If m and n are integers and $x \neq 0$, then $\dfrac{x^m}{x^n} = x^{m-n}$.

HOW TO Evaluate $\dfrac{5^{-2}}{5}$.

$\dfrac{5^{-2}}{5} = 5^{-2-1} = 5^{-3}$ • Use the **Rule for Dividing Exponential Expressions**.

$= \dfrac{1}{5^3} = \dfrac{1}{125}$ • Use the **Definition of a Negative Exponent** to rewrite the expression with a positive exponent. Then evaluate.

Copyright © Houghton Mifflin Company. All rights reserved.

HOW TO   Simplify: $\dfrac{x^4}{x^9}$

$\dfrac{x^4}{x^9} = x^{4-9}$ • Use the **Rule for Dividing Exponential Expressions**.

$\phantom{\dfrac{x^4}{x^9}} = x^{-5}$ • Subtract the exponents.

$\phantom{\dfrac{x^4}{x^9}} = \dfrac{1}{x^5}$ • Use the Definition of a Negative Exponent to rewrite the expression with a positive exponent.

The rules for simplifying exponential expressions and powers of exponential expressions are true for all integers. These rules are restated here, along with the rules for dividing exponential expressions.

Rules of Exponents

If m, n, and p are integers, then

$x^m \cdot x^n = x^{m+n}$ $\qquad (x^m)^n = x^{mn}$ $\qquad (x^m y^n)^p = x^{mp} y^{np}$

$\dfrac{x^m}{x^n} = x^{m-n},\ x \neq 0$ $\qquad \left(\dfrac{x^m}{y^n}\right)^p = \dfrac{x^{mp}}{y^{np}},\ y \neq 0$ $\qquad x^{-n} = \dfrac{1}{x^n},\ x \neq 0$

$x^0 = 1,\ x \neq 0$

HOW TO   Simplify: $(3ab^{-4})(-2a^{-3}b^7)$

$(3ab^{-4})(-2a^{-3}b^7) = [3 \cdot (-2)](a^{1+(-3)}b^{-4+7})$ • When multiplying expressions, add the exponents on like bases.

$\phantom{(3ab^{-4})(-2a^{-3}b^7)} = -6a^{-2}b^3$

$\phantom{(3ab^{-4})(-2a^{-3}b^7)} = -\dfrac{6b^3}{a^2}$

HOW TO   Simplify: $\left[\dfrac{6m^2n^3}{8m^7n^2}\right]^{-3}$

$\left[\dfrac{6m^2n^3}{8m^7n^2}\right]^{-3} = \left[\dfrac{3m^{2-7}n^{3-2}}{4}\right]^{-3}$ • Simplify inside the brackets.

$= \left[\dfrac{3m^{-5}n}{4}\right]^{-3}$ • Subtract the exponents.

$= \dfrac{3^{-3}m^{15}n^{-3}}{4^{-3}}$ • Use the Rule for Simplifying the Power of a Quotient.

$= \dfrac{4^3m^{15}}{3^3n^3} = \dfrac{64m^{15}}{27n^3}$ • Use the Definition of a Negative Exponent to rewrite the expression with positive exponents. Then simplify.

Copyright © Houghton Mifflin Company. All rights reserved.

HOW TO Simplify: $\dfrac{4a^{-2}b^5}{6a^5b^2}$

$\dfrac{4a^{-2}b^5}{6a^5b^2} = \dfrac{2a^{-2}b^5}{3a^5b^2}$ • Divide the coefficients by their common factor.

$= \dfrac{2a^{-2-5}b^{5-2}}{3}$ • Use the Rule for Dividing Exponential Expressions.

$= \dfrac{2a^{-7}b^3}{3} = \dfrac{2b^3}{3a^7}$ • Use the Definition of a Negative Exponent to rewrite the expression with positive exponents.

Example 1 Simplify: $(-2x)(3x^{-2})^{-3}$

Solution
$(-2x)(3x^{-2})^{-3} = (-2x)(3^{-3}x^6)$ • Rule for Simplifying the Power of a Product

$= \dfrac{-2x^{1+6}}{3^3}$

$= -\dfrac{2x^7}{27}$

You Try It 1 Simplify: $(-2x^2)(x^{-3}y^{-4})^{-2}$

Your solution

Example 2 Simplify: $\dfrac{(2r^2t^{-1})^{-3}}{(r^{-3}t^4)^2}$

Solution
$\dfrac{(2r^2t^{-1})^{-3}}{(r^{-3}t^4)^2} = \dfrac{2^{-3}r^{-6}t^3}{r^{-6}t^8}$ • Rule for Simplifying the Power of a Product

$= 2^{-3}r^{-6-(-6)}t^{3-8}$ • Rule for Dividing Exponential Expressions

$= 2^{-3}r^0t^{-5}$

$= \dfrac{1}{2^3t^5}$

$= \dfrac{1}{8t^5}$ • Write answer in simplest form.

You Try It 2 Simplify: $\dfrac{(6a^{-2}b^3)^{-1}}{(4a^3b^{-2})^{-2}}$

Your solution

Example 3 Simplify: $\left[\dfrac{4a^{-2}b^3}{6a^4b^{-2}}\right]^{-3}$

Solution
$\left[\dfrac{4a^{-2}b^3}{6a^4b^{-2}}\right]^{-3} = \left[\dfrac{2a^{-6}b^5}{3}\right]^{-3}$ • Simplify inside brackets.

$= \dfrac{2^{-3}a^{18}b^{-15}}{3^{-3}}$ • Rule for Simplifying the Power of a Quotient

$= \dfrac{27a^{18}}{8b^{15}}$ • Write answer in simplest form.

You Try It 3 Simplify: $\left[\dfrac{6r^3s^{-3}}{9r^3s^{-1}}\right]^{-2}$

Your solution

Solutions on p. S11

Copyright © Houghton Mifflin Company. All rights reserved.

Copyright © Houghton Mifflin Company. All rights reserved.

Objective B   **To write a number in scientific notation**

Integrating Technology

See the Keystroke Guide: *Scientific Notation* for instruction on entering scientific notation on a calculator.

Point of Interest

An electron microscope uses wavelengths that are approximately 4×10^{-12} meter to make images of viruses.

The human eye can detect wavelengths between 4.3×10^{-7} meter and 6.9×10^{-7} meter. Although these are very short, they are approximately 10^5 times longer than the waves used in an electron microscope.

Very large and very small numbers abound in the natural sciences. For example, the mass of an electron is 0.00000000000000000000000000000911 kg. Numbers such as this are difficult to read, so a more convenient system called **scientific notation** is used. In scientific notation, a number is expressed as the product of two factors, one a number between 1 and 10, and the other a power of 10.

To express a number in scientific notation, write it in the form $a \times 10^n$, where a is a number between 1 and 10, and n is an integer.

For numbers greater than or equal to 10, move the decimal point to the right of the first digit. The exponent n is positive and equal to the number of places the decimal point has been moved.

$$240,000 = 2.4 \times 10^5$$

$$93,000,000 = 9.3 \times 10^7$$

For numbers less than 1, move the decimal point to the right of the first nonzero digit. The exponent n is negative. The absolute value of the exponent is equal to the number of places the decimal point has been moved.

$$0.0003 = 3 \times 10^{-4}$$

$$0.0000832 = 8.32 \times 10^{-5}$$

Changing a number written in scientific notation to decimal notation also requires moving the decimal point.

When the exponent is positive, move the decimal point to the right the same number of places as the exponent.

$$3.45 \times 10^6 = 3,450,000$$

$$2.3 \times 10^8 = 230,000,000$$

When the exponent is negative, move the decimal point to the left the same number of places as the absolute value of the exponent.

$$8.1 \times 10^{-3} = 0.0081$$

$$6.34 \times 10^{-7} = 0.000000634$$

Example 4   Write the number 824,300,000 in scientific notation.

Solution $824,300,000 = 8.243 \times 10^8$

You Try It 4   Write the number 0.000000961 in scientific notation.

Your solution

Example 5   Write the number 6.8×10^{-10} in decimal notation.

Solution $6.8 \times 10^{-10} = 0.00000000068$

You Try It 5   Write the number 7.329×10^6 in decimal notation.

Your solution

Solutions on p. S11

4.4 Exercises

Objective A To divide monomials

For Exercises 1 and 2, replace the question marks to make a true statement.

1. $\dfrac{x^7}{x^5} = x^{?-?} = x^?$

2. $\dfrac{a^9}{a^{12}} = a^{?-?} = \dfrac{1}{a^?}$

For Exercises 3 to 10, evaluate.

3. 5^{-2}

4. 3^{-3}

5. $\dfrac{1}{8^{-2}}$

6. $\dfrac{1}{12^{-1}}$

7. $\dfrac{3^{-2}}{3}$

8. $\dfrac{5^{-3}}{5}$

9. $\dfrac{2^{-2}}{2^{-3}}$

10. $\dfrac{3^2}{3^2}$

For Exercises 11 to 94, simplify.

11. x^{-2}

12. y^{-10}

13. $\dfrac{1}{a^{-6}}$

14. $\dfrac{1}{b^{-4}}$

15. $4x^{-7}$

16. $-6y^{-1}$

17. $\dfrac{2}{3}z^{-2}$

18. $\dfrac{4}{5}a^{-4}$

19. $\dfrac{5}{b^{-8}}$

20. $\dfrac{-3}{v^{-3}}$

21. $\dfrac{1}{3x^{-2}}$

22. $\dfrac{2}{5c^{-6}}$

23. $(ab^5)^0$

24. $(32x^3y^4)^0$

25. $-(3p^2q^5)^0$

26. $-\left(\dfrac{2}{3}xy\right)^0$

27. $\dfrac{y^7}{y^3}$

28. $\dfrac{z^9}{z^2}$

29. $\dfrac{a^8}{a^5}$

30. $\dfrac{c^{12}}{c^5}$

31. $\dfrac{p^5}{p}$

32. $\dfrac{w^9}{w}$

33. $\dfrac{4x^8}{2x^5}$

34. $\dfrac{12z^7}{4z^3}$

35. $\dfrac{22k^5}{11k^4}$

36. $\dfrac{14m^{11}}{7m^{10}}$

37. $\dfrac{m^9n^7}{m^4n^5}$

38. $\dfrac{y^5z^6}{yz^3}$

39. $\dfrac{6r^4}{4r^2}$

40. $\dfrac{8x^9}{12x^6}$

41. $\dfrac{-16a^7}{24a^6}$

42. $\dfrac{-18b^5}{27b^4}$

43. $\dfrac{y^3}{y^8}$

44. $\dfrac{z^4}{z^6}$

45. $\dfrac{a^5}{a^{11}}$

46. $\dfrac{m}{m^7}$

Copyright © Houghton Mifflin Company. All rights reserved.

47. $\dfrac{4x^2}{12x^5}$

48. $\dfrac{6y^8}{8y^9}$

49. $\dfrac{-12x}{-18x^6}$

50. $\dfrac{-24c^2}{-36c^{11}}$

51. $\dfrac{x^6y^5}{x^8y}$

52. $\dfrac{a^3b^2}{a^2b^3}$

53. $\dfrac{2m^6n^2}{5m^9n^{10}}$

54. $\dfrac{5r^3t^7}{6r^5t^7}$

55. $\dfrac{pq^3}{p^4q^4}$

56. $\dfrac{a^4b^5}{a^5b^6}$

57. $\dfrac{3x^4y^5}{6x^4y^8}$

58. $\dfrac{14a^3b^6}{21a^5b^6}$

59. $\dfrac{14x^4y^6z^2}{16x^3y^9z}$

60. $\dfrac{24a^2b^7c^9}{36a^7b^5c}$

61. $\dfrac{15mn^9p^3}{30m^4n^9p}$

62. $\dfrac{25x^4y^7z^2}{20x^5y^9z^{11}}$

63. $(-2xy^{-2})^3$

64. $(-3x^{-1}y^2)^2$

65. $(3x^{-1}y^{-2})^2$

66. $(5xy^{-3})^{-2}$

67. $(2x^{-1})(x^{-3})$

68. $(-2x^{-5})x^7$

69. $(-5a^2)(a^{-5})^2$

70. $(2a^{-3})(a^7b^{-1})^3$

71. $(-2ab^{-2})(4a^{-2}b)^{-2}$

72. $(3ab^{-2})(2a^{-1}b)^{-3}$

73. $(-5x^{-2}y)(-2x^{-2}y^2)$

74. $\dfrac{a^{-3}b^{-4}}{a^2b^2}$

75. $\dfrac{3x^{-2}y^2}{6xy^2}$

76. $\dfrac{2x^{-2}y}{8xy}$

77. $\dfrac{3x^{-2}y}{xy}$

78. $\dfrac{2x^{-1}y^4}{x^2y^3}$

79. $\dfrac{2x^{-1}y^{-4}}{4xy^2}$

80. $\dfrac{(x^{-1}y)^2}{xy^2}$

81. $\dfrac{(x^{-2}y)^2}{x^2y^3}$

82. $\dfrac{(x^{-3}y^{-2})^2}{x^6y^8}$

Copyright © Houghton Mifflin Company. All rights reserved.

83. $\dfrac{(a^{-2}y^3)^{-3}}{a^2y}$

84. $\dfrac{12a^2b^3}{-27a^2b^2}$

85. $\dfrac{-16xy^4}{96x^4y^4}$

86. $\dfrac{-8x^2y^4}{44y^2z^5}$

87. $\dfrac{22a^2b^4}{-132b^3c^2}$

88. $\dfrac{-(8a^2b^4)^3}{64a^3b^8}$

89. $\dfrac{-(14ab^4)^2}{28a^4b^2}$

90. $\dfrac{(2a^{-2}b^3)^{-2}}{(4a^2b^{-4})^{-1}}$

91. $\dfrac{(3^{-1}r^4s^{-3})^{-2}}{(6r^2s^{-1}t^{-2})^2}$

92. $\left(\dfrac{6x^{-4}yz^{-1}}{14xy^{-4}z^2}\right)^{-3}$

93. $\left(\dfrac{15m^3n^{-2}p^{-1}}{25m^{-2}n^{-4}}\right)^{-3}$

94. $\left(\dfrac{18a^4b^{-2}c^4}{12ab^{-3}d^2}\right)^{-2}$

Objective B   **To write a number in scientific notation**

95. ✏️ Why might a number be written in scientific notation instead of decimal notation?

96. ✏️ **a.** Explain how to write 0.00000076 in scientific notation.
    b. Explain how to write 4.3×10^8 in decimal notation.

For Exercises 97 to 105, write in scientific notation.

97. 0.00000000324

98. 0.00000012

99. 0.000000000000000003

100. 1,800,000,000

101. 32,000,000,000,000,000

102. 76,700,000,000,000

103. 0.000000000000000000122

104. 0.00137

105. 547,000,000

For Exercises 106 to 114, write in decimal notation.

106. 2.3×10^{-12}

107. 1.67×10^{-4}

108. 2×10^{15}

109. 6.8×10^7

110. 9×10^{-21}

111. 3.05×10^{-5}

112. 9.05×10^{11}

113. 1.02×10^{-9}

114. 7.2×10^{-3}

115. 🥧 **Chemistry**   Avogadro's number is used in chemistry, and its value is approximately 602,300,000,000,000,000,000,000. Express this number in scientific notation.

116. 🥧 **Geology**   5,980,000,000,000,000,000,000,000 kg is the approximate mass of the planet Earth. Write the mass of Earth in scientific notation.

Copyright © Houghton Mifflin Company. All rights reserved.

117. **Physics** The length of an infrared light wave is approximately 0.0000037 m. Write this number in scientific notation.

118. **Physics** Light travels approximately 16,000,000,000 mi in one day. Write this number in scientific notation.

119. **Computer Science** One unit used to measure the speed of a computer is the picosecond. One picosecond is 0.000000001 s. Write this number in scientific notation.

120. **Astronomy** One light-year is the distance traveled by light in 1 year. One light-year is 5,880,000,000,000 mi. Write this number in scientific notation.

121. **Electricity** The electric charge on an electron is 0.00000000000000000016 coulomb. Write this number in scientific notation.

122. **Chemistry** Approximately 35 teragrams (3.5×10^{13} g) of sulfur in the atmosphere is converted to sulfate each year. Write this number in decimal notation.

APPLYING THE CONCEPTS

123. Evaluate 2^x when $x = -2, -1, 0, 1,$ and 2.

124. Evaluate 3^x when $x = -2, -1, 0, 1,$ and 2.

125. Evaluate 2^{-x} when $x = -2, -1, 0, 1,$ and 2.

126. Evaluate 3^{-x} when $x = -2, -1, 0, 1,$ and 2.

For Exercises 127 to 129, determine whether each equation is true or false. If the equation is false, change the right-hand side of the equation to make a true equation.

127. $(2a)^{-3} = \dfrac{2}{a^3}$

128. $\dfrac{x^{-3}}{y^{-3}} = \left(\dfrac{x}{y}\right)^{-3}$

129. $(2 + 3)^{-1} = 2^{-1} + 3^{-1}$

130. Simplify: $\left(\dfrac{6x^4yz^3}{2x^2y^3}\right)\left(\dfrac{2x^2z^3}{4y^2z}\right) \div \left(\dfrac{6x^2y^3}{x^4y^2z}\right)$

131. If x is a nonzero real number, is x^{-2} always positive, always negative, or positive or negative depending on whether x is positive or negative? Explain your answer.

Copyright © Houghton Mifflin Company. All rights reserved.

4.5 Division of Polynomials

Objective A To divide a polynomial by a monomial

To divide a polynomial by a monomial, divide each term in the numerator by the denominator and write the sum of the quotients.

HOW TO Divide: $\dfrac{6x^3 - 3x^2 + 9x}{3x}$

$$\dfrac{6x^3 - 3x^2 + 9x}{3x} = \dfrac{6x^3}{3x} - \dfrac{3x^2}{3x} + \dfrac{9x}{3x}$$

- Divide each term of the polynomial by the monomial.

$$= 2x^2 - x + 3$$

- Simplify each expression.

Example 1 Divide: $\dfrac{12x^2y - 6xy + 4x^2}{2xy}$

Solution

$$\dfrac{12x^2y - 6xy + 4x^2}{2xy} = \dfrac{12x^2y}{2xy} - \dfrac{6xy}{2xy} + \dfrac{4x^2}{2xy} = 6x - 3 + \dfrac{2x}{y}$$

You Try It 1 Divide: $\dfrac{24x^2y^2 - 18xy + 6y}{6xy}$

Your solution

Solution on p. S11

Objective B To divide polynomials

Copyright © Houghton Mifflin Company. All rights reserved.

Study Tip

An important element of success is practice. We cannot do anything well if we do not practice it repeatedly. Practice is crucial to success in mathematics. In this objective you are learning a new skill, how to divide polynomials. You will need to practice this skill over and over again in order to be successful at it.

The procedure for dividing two polynomials is similar to the one for dividing whole numbers. The same equation used to check division of whole numbers is used to check polynomial division.

(Quotient × divisor) + remainder = dividend

HOW TO Divide: $(x^2 - 5x + 8) \div (x - 3)$

Step 1

$$x - 3 \overline{)\begin{array}{r} x \\ x^2 - 5x + 8 \end{array}}$$
$$\underline{x^2 - 3x}$$
$$-2x + 8$$

- Think: $x\overline{)x^2} = \dfrac{x^2}{x} = x$
- Multiply: $x(x - 3) = x^2 - 3x$
- Subtract: $(x^2 - 5x) - (x^2 - 3x) = -2x$
  Bring down the 8.

Step 2

$$x - 3 \overline{)\begin{array}{r} x - 2 \\ x^2 - 5x + 8 \end{array}}$$
$$\underline{x^2 - 3x}$$
$$-2x + 8$$
$$\underline{-2x + 6}$$
$$2$$

- Think: $x\overline{)-2x} = \dfrac{-2x}{x} = -2$
- Multiply: $-2(x - 3) = -2x + 6$
- Subtract: $(-2x + 8) - (-2x + 6) = 2$
- The remainder is 2.

Check: $(x - 2)(x - 3) + 2 = x^2 - 5x + 6 + 2 = x^2 - 5x + 8$

$$(x^2 - 5x + 8) \div (x - 3) = x - 2 + \dfrac{2}{x - 3}$$

If a term is missing from the dividend, a zero can be inserted for that term. This helps keep like terms in the same column.

TAKE NOTE
Recall that a fraction bar means "divided by." Therefore, $6 \div 2$ can be written $\frac{6}{2}$, and $a \div b$ can be written $\frac{a}{b}$.

HOW TO Divide: $\dfrac{6x + 26 + 2x^3}{2 + x}$

$$\dfrac{2x^3 + 6x + 26}{x + 2}$$

• Arrange the terms of each polynomial in descending order.

$$
\begin{array}{r}
2x^2 - 4x + 14 \\
x + 2\overline{\smash{)}2x^3 + 0 + 6x + 26} \\
\underline{2x^3 + 4x^2} \\
-4x^2 + 6x \\
\underline{-4x^2 - 8x} \\
14x + 26 \\
\underline{14x + 28} \\
-2
\end{array}
$$

• There is no x^2 term in $2x^3 + 6x + 26$. Insert a **zero** for the missing term.

Check:
$(2x^2 - 4x + 14)(x + 2) + (-2) = (2x^3 + 6x + 28) + (-2) = 2x^3 + 6x + 26$

$(2x^3 + 6x + 26) \div (x + 2) = 2x^2 - 4x + 14 - \dfrac{2}{x + 2}$

Example 2

Divide: $(8x^2 + 4x^3 + x - 4) \div (2x + 3)$

Solution

$$
\begin{array}{r}
2x^2 + x - 1 \\
2x + 3\overline{\smash{)}4x^3 + 8x^2 + x - 4} \\
\underline{4x^3 + 6x^2} \\
2x^2 + x \\
\underline{2x^2 + 3x} \\
-2x - 4 \\
\underline{-2x - 3} \\
-1
\end{array}
$$

• Write the dividend in descending powers of x.

$(4x^3 + 8x^2 + x - 4) \div (2x + 3)$

$= 2x^2 + x - 1 - \dfrac{1}{2x + 3}$

You Try It 2

Divide: $(2x^3 + x^2 - 8x - 3) \div (2x - 3)$

Your solution

Example 3 Divide: $\dfrac{x^2 - 1}{x + 1}$

Solution

$$
\begin{array}{r}
x - 1 \\
x + 1\overline{\smash{)}x^2 + 0 - 1} \\
\underline{x^2 + x} \\
-x - 1 \\
\underline{-x - 1} \\
0
\end{array}
$$

• Insert a zero for the missing term.

$(x^2 - 1) \div (x + 1) = x - 1$

You Try It 3 Divide: $\dfrac{x^3 - 2x + 1}{x - 1}$

Your solution

Solutions on p. S11

Copyright © Houghton Mifflin Company. All rights reserved.

4.5 Exercises

Objective A   **To divide a polynomial by a monomial**

For Exercises 1 to 24, divide.

1. $\dfrac{10a - 25}{5}$

2. $\dfrac{16b - 40}{8}$

3. $\dfrac{6y^2 + 4y}{y}$

4. $\dfrac{4b^3 - 3b}{b}$

5. $\dfrac{3x^2 - 6x}{3x}$

6. $\dfrac{10y^2 - 6y}{2y}$

7. $\dfrac{5x^2 - 10x}{-5x}$

8. $\dfrac{3y^2 - 27y}{-3y}$

9. $\dfrac{x^3 + 3x^2 - 5x}{x}$

10. $\dfrac{a^3 - 5a^2 + 7a}{a}$

11. $\dfrac{x^6 - 3x^4 - x^2}{x^2}$

12. $\dfrac{a^8 - 5a^5 - 3a^3}{a^2}$

13. $\dfrac{5x^2y^2 + 10xy}{5xy}$

14. $\dfrac{8x^2y^2 - 24xy}{8xy}$

15. $\dfrac{9y^6 - 15y^3}{-3y^3}$

16. $\dfrac{4x^4 - 6x^2}{-2x^2}$

17. $\dfrac{3x^2 - 2x + 1}{x}$

18. $\dfrac{8y^2 + 2y - 3}{y}$

19. $\dfrac{-3x^2 + 7x - 6}{x}$

20. $\dfrac{2y^2 - 6y + 9}{y}$

21. $\dfrac{16a^2b - 20ab + 24ab^2}{4ab}$

22. $\dfrac{22a^2b - 11ab - 33ab^2}{11ab}$

23. $\dfrac{9x^2y + 6xy - 3xy^2}{xy}$

24. $\dfrac{5a^2b - 15ab + 30ab^2}{5ab}$

Objective B   **To divide polynomials**

For Exercises 25 and 26, replace the question marks to make a true statement.

25. If $\dfrac{x^2 - x - 6}{x - 3} = x + 2$, then $x^2 - x - 6 = (?)(?)$.

26. If $\dfrac{x^2 + 2x - 3}{x - 2} = x + 4 + \dfrac{5}{x - 2}$, then $x^2 + 2x - 3 = (?)(?) + ?$.

For Exercises 27 to 53, divide.

27. $(b^2 - 14b + 49) \div (b - 7)$

28. $(x^2 - x - 6) \div (x - 3)$

29. $(y^2 + 2y - 35) \div (y + 7)$

30. $(2x^2 + 5x + 2) \div (x + 2)$

31. $(2y^2 - 13y + 21) \div (y - 3)$

32. $(4x^2 - 16) \div (2x + 4)$

Copyright © Houghton Mifflin Company. All rights reserved.

33. $\dfrac{2y^2 + 7}{y - 3}$

34. $\dfrac{x^2 + 1}{x - 1}$

35. $\dfrac{x^2 + 4}{x + 2}$

36. $\dfrac{6x^2 - 7x}{3x - 2}$

37. $\dfrac{6y^2 + 2y}{2y + 4}$

38. $\dfrac{5x^2 + 7x}{x - 1}$

39. $(6x^2 - 5) \div (x + 2)$

40. $(a^2 + 5a + 10) \div (a + 2)$

41. $(b^2 - 8b - 9) \div (b - 3)$

42. $\dfrac{2y^2 - 9y + 8}{2y + 3}$

43. $\dfrac{3x^2 + 5x - 4}{x - 4}$

44. $(8x + 3 + 4x^2) \div (2x - 1)$

45. $(10 + 21y + 10y^2) \div (2y + 3)$

46. $\dfrac{15a^2 - 8a - 8}{3a + 2}$

47. $\dfrac{12a^2 - 25a - 7}{3a - 7}$

48. $(5 - 23x + 12x^2) \div (4x - 1)$

49. $(24 + 6a^2 + 25a) \div (3a - 1)$

50. $\dfrac{5x + 3x^2 + x^3 + 3}{x + 1}$

51. $\dfrac{7x + x^3 - 6x^2 - 2}{x - 1}$

52. $(x^4 - x^2 - 6) \div (x^2 + 2)$

53. $(x^4 + 3x^2 - 10) \div (x^2 - 2)$

APPLYING THE CONCEPTS

54.   In your own words, explain how to divide exponential expressions.

55. The product of a monomial and $4b$ is $12ab^2$. Find the monomial.

Copyright © Houghton Mifflin Company. All rights reserved.

Focus on Problem Solving

Dimensional Analysis

In solving application problems, it may be useful to include the units in order to organize the problem so that the answer is in the proper units. Using units to organize and check the correctness of an application is called **dimensional analysis.** We use the operations of multiplying units and dividing units in applying dimensional analysis to application problems.

The Rule for Multiplying Exponential Expressions states that we multiply two expressions with the same base by adding the exponents.

$$x^4 \cdot x^6 = x^{4+6} = x^{10}$$

In calculations that involve quantities, the units are operated on algebraically.

HOW TO A rectangle measures 3 m by 5 m. Find the area of the rectangle.

$$A = LW = (3 \text{ m})(5 \text{ m}) = (3 \cdot 5)(\text{m} \cdot \text{m}) = 15 \text{ m}^2$$

The area of the rectangle is 15 m² (square meters).

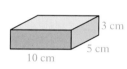

HOW TO A box measures 10 cm by 5 cm by 3 cm. Find the volume of the box.

$$V = LWH = (10 \text{ cm})(5 \text{ cm})(3 \text{ cm}) = (10 \cdot 5 \cdot 3)(\text{cm} \cdot \text{cm} \cdot \text{cm}) = 150 \text{ cm}^3$$

The volume of the box is 150 cm³ (cubic centimeters).

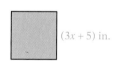

(3x + 5) in.

HOW TO Find the area of a square whose side measures (3x + 5) in.

$$A = s^2 = [(3x + 5) \text{ in.}]^2 = (3x + 5)^2 \text{ in}^2 = (9x^2 + 30x + 25) \text{ in}^2$$

The area of the square is (9x² + 30x + 25) in² (square inches).

Dimensional analysis is used in the conversion of units.

The following example converts the unit miles to feet. The equivalent measures 1 mi = 5280 ft are used to form the following rates, which are called conversion factors: $\dfrac{1 \text{ mi}}{5280 \text{ ft}}$ and $\dfrac{5280 \text{ ft}}{1 \text{ mi}}$. Because 1 mi = 5280 ft, both of the conversion factors $\dfrac{1 \text{ mi}}{5280 \text{ ft}}$ and $\dfrac{5280 \text{ ft}}{1 \text{ mi}}$ are equal to 1.

To convert 3 mi to feet, multiply 3 mi by the conversion factor $\dfrac{5280 \text{ ft}}{1 \text{ mi}}$.

$$3 \text{ mi} = 3 \text{ mi} \cdot 1 = \frac{3 \text{ mi}}{1} \cdot \frac{5280 \text{ ft}}{1 \text{ mi}} = \frac{3 \text{ mi} \cdot 5280 \text{ ft}}{1 \text{ mi}} = 3 \cdot 5280 \text{ ft} = 15,840 \text{ ft}$$

There are two important points in the above illustration. **First,** you can think of dividing the numerator and denominator by the common unit "mile" just as you would divide the numerator and denominator of a fraction by a common factor. **Second,** the conversion factor $\dfrac{5280 \text{ ft}}{1 \text{ mi}}$ is equal to 1, and multiplying an expression by 1 does not change the value of the expression.

Copyright © Houghton Mifflin Company. All rights reserved.

In the application problem that follows, the units are kept in the problem while the problem is worked.

In 2003, a horse named Funny Cide ran a 1.25-mile race in 2.01 min. Find Funny Cide's average speed for that race in miles per hour. Round to the nearest tenth.

Strategy   To find the average speed, use the formula $r = \dfrac{d}{t}$, where r is the speed, d is the distance, and t is the time. Use the conversion factor $\dfrac{60 \text{ min}}{1 \text{ h}}$.

Solution $r = \dfrac{d}{t} = \dfrac{1.25 \text{ mi}}{2.01 \text{ min}} = \dfrac{1.25 \text{ mi}}{2.01 \text{ min}} \cdot \dfrac{60 \text{ min}}{1 \text{ h}}$

$$= \dfrac{75 \text{ mi}}{2.01 \text{ h}} \approx 37.3 \text{ mph}$$

Funny Cide's average speed was 37.3 mph.

Try each of the following problems. Round to the nearest tenth or nearest cent.

1. Convert 88 ft/s to miles per hour.

2. Convert 8 m/s to kilometers per hour (1 km = 1000 m).

3. A carpet is to be placed in a meeting hall that is 36 ft wide and 80 ft long. At $21.50 per square yard, how much will it cost to carpet the meeting hall?

4. A carpet is to be placed in a room that is 20 ft wide and 30 ft long. At $22.25 per square yard, how much will it cost to carpet the area?

5. Find the number of gallons of water in a fish tank that is 36 in. long and 24 in. wide and is filled to a depth of 16 in. (1 gal = 231 in³).

6. Find the number of gallons of water in a fish tank that is 24 in. long and 18 in. wide and is filled to a depth of 12 in. (1 gal = 231 in³).

7. A $\frac{1}{4}$-acre commercial lot is on sale for $2.15 per square foot. Find the sale price of the commercial lot (1 acre = 43,560 ft²).

8. A 0.75-acre industrial parcel was sold for $98,010. Find the parcel's price per square foot (1 acre = 43,560 ft²).

9. A new driveway will require 800 ft³ of concrete. Concrete is ordered by the cubic yard. How much concrete should be ordered?

10. A piston-engined dragster traveled 440 yd in 4.936 s at Ennis, Texas, on October 9, 1988. Find the average speed of the dragster in miles per hour.

11. The Marianas Trench in the Pacific Ocean is the deepest part of the ocean. Its depth is 6.85 mi. The speed of sound under water is 4700 ft/s. Find the time it takes sound to travel from the surface to the bottom of the Marianas Trench and back.

Copyright © Houghton Mifflin Company. All rights reserved.

Projects and Group Activities

Diagramming the Square of a Binomial

1. Explain why the diagram at the right represents
$(a + b)^2 = a^2 + 2ab + b^2$

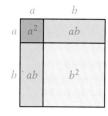

2. Draw similar diagrams representing each of the following.

$$(x + 2)^2$$

$$(x + 4)^2$$

Pascal's Triangle

Simplifying the power of a binomial is called *expanding the binomial*. The expansions of the first three powers of a binomial are shown below.

$$(a + b)^1 = a + b$$

$$(a + b)^2 = (a + b)(a + b) = a^2 + 2ab + b^2$$

$$(a + b)^3 = (a + b)^2(a + b) = (a^2 + 2ab + b^2)(a + b) = a^3 + 3a^2b + 3ab^2 + b^3$$

Point of Interest

Pascal did not invent the triangle of numbers known as Pascal's Triangle. It was known to mathematicians in China probably as early as 1050 A.D. But Pascal's *Traite du triangle arithmetique* (*Treatise Concerning the Arithmetical Triangle*) brought together all the different aspects of the numbers for the first time.

Find $(a + b)^4$. [*Hint:* $(a + b)^4 = (a + b)^3(a + b)$]

Find $(a + b)^5$. [*Hint:* $(a + b)^5 = (a + b)^4(a + b)$]

If we continue in this way, the results for $(a + b)^6$ are

$$(a + b)^6 = a^6 + 6a^5b + 15a^4b^2 + 20a^3b^3 + 15a^2b^4 + 6ab^5 + b^6$$

Now expand $(a + b)^8$. Before you begin, see whether you can find a pattern that will help you write the expansion of $(a + b)^8$ without having to multiply it out. Here are some hints.

1. Write out the variable terms of each binomial expansion from $(a + b)^1$ through $(a + b)^6$. Observe how the exponents on the variables change.

2. Write out the coefficients of all the terms without the variable parts. It will be helpful to make a triangular arrangement as shown at the left. Note that each row begins and ends with a 1. Also note (in the two shaded regions, for example) that any number in a row is the sum of the two closest numbers above it. For instance, $1 + 5 = 6$ and $6 + 4 = 10$.

The triangle of numbers shown at the left is called Pascal's Triangle. To find the expansion of $(a + b)^8$, you need to find the eighth row of Pascal's Triangle. First find row seven. Then find row eight and use the patterns you have observed to write the expansion $(a + b)^8$.

Pascal's Triangle has been the subject of extensive analysis, and many patterns have been found. See whether you can find some of them.

Copyright © Houghton Mifflin Company. All rights reserved.

Chapter 4 Summary

Key Words	**Examples**
A *monomial* is a number, a variable, or a product of numbers and variables. [4.1A, p. 193]	5 is a number, y is a variable. $2a^3b^2$ is a product of numbers and variables. 5, y, and $2a^3b^2$ are monomials.
A *polynomial* is a variable expression in which the terms are monomials. [4.1A, p. 193]	$5x^2y - 3xy^2 + 2$ is a polynomial. Each term of this expression is a monomial.
A polynomial of two terms is a *binomial*. [4.1A, p. 193]	$x + 2$, $y^2 - 3$, and $6a + 5b$ are binomials.
A polynomial of three terms is a *trinomial*. [4.1A, p. 193]	$x^2 - 6x + 7$ is a trinomial.
The *degree of a polynomial in one variable* is the greatest exponent on a variable. [4.1A, p. 193]	The degree of $3x - 4x^3 + 17x^2 + 25$ is 3.
A polynomial in one variable is usually written in *descending order* where the exponents of the variable terms decrease from left to right. [4.1A, p. 193]	The polynomial $2x^4 + 3x^2 - 4x - 7$ is written in descending order.
The *opposite of a polynomial* is the polynomial with the sign of every term changed to its opposite. [4.1B, p. 194]	The opposite of the polynomial $x^2 - 3x + 4$ is $-x^2 + 3x - 4$.

Essential Rules and Procedures	**Examples**
Addition of Polynomials [4.1A, p. 193] To add polynomials, add the coefficients of the like terms.	$(2x^2 + 3x - 4) + (3x^3 - 4x^2 + 2x - 5)$ $= 3x^3 + (2x^2 - 4x^2) + (3x + 2x)$ $\quad + (-4 - 5)$ $= 3x^3 - 2x^2 + 5x - 9$
Subtraction of Polynomials [4.1B, p. 194] To subtract polynomials, add the opposite of the second polynomial to the first.	$(3y^2 - 8y - 9) - (5y^2 - 10y + 3)$ $= (3y^2 - 8y - 9) + (-5y^2 + 10y - 3)$ $= (3y^2 - 5y^2) + (-8y + 10y)$ $\quad + (-9 - 3)$ $= -2y^2 + 2y - 12$
Rule for Multiplying Exponential Expressions [4.2A, p. 197] If m and n are integers, then $x^m \cdot x^n = x^{m+n}$.	$a^3 \cdot a^6 = a^{3+6} = a^9$

Copyright © Houghton Mifflin Company. All rights reserved.

Rule for Simplifying the Power of an Exponential Expression [4.2B, p. 198]
If m and n are integers, then $(x^m)^n = x^{mn}$.

$$(c^3)^4 = c^{3 \cdot 4} = c^{12}$$

Rule for Simplifying the Power of a Product [4.2B, p. 198]
If m, n, and p are integers, then $(x^m y^n)^p = x^{mp} y^{np}$.

$$(a^3 b^2)^4 = a^{3 \cdot 4} b^{2 \cdot 4} = a^{12} b^8$$

To multiply a polynomial by a monomial, use the Distributive Property and the Rule for Multiplying Exponential Expressions. [4.3A, p. 201]

$$\begin{aligned} &(-4y)(5y^2 + 3y - 8) \\ &= (-4y)(5y^2) + (-4y)(3y) - (-4y)(8) \\ &= -20y^3 - 12y^2 + 32y \end{aligned}$$

To multiply two polynomials, multiply each term of one polynomial by each term of the other polynomial. [4.3B, p. 201]

$$\begin{array}{r} x^2 - 5x + 6 \\ x + 4 \\ \hline 4x^2 - 20x + 24 \\ x^3 - 5x^2 + 6x \\ \hline x^3 - x^2 - 14x + 24 \end{array}$$

FOIL Method [4.3C, p. 202]
To find the product of two binomials, add the products of the **F**irst terms, the **O**uter terms, the **I**nner terms, and the **L**ast terms.

$$\begin{aligned} &(2x - 5)(3x + 4) \\ &= (2x)(3x) + (2x)(4) + (-5)(3x) \\ &\quad + (-5)(4) \\ &= 6x^2 + 8x - 15x - 20 \\ &= 6x^2 - 7x - 20 \end{aligned}$$

Product of the Sum and Difference of the Same Terms [4.3D, p. 203]
$(a + b)(a - b) = a^2 - b^2$

$$\begin{aligned} (3x + 4)(3x - 4) &= (3x)^2 - 4^2 \\ &= 9x^2 - 16 \end{aligned}$$

Square of a Binomial [4.3D, p. 203]
$(a + b)^2 = a^2 + 2ab + b^2$
$(a - b)^2 = a^2 - 2ab - b^2$

$$\begin{aligned} (2x + 5)^2 &= (2x^2) + 2(2x)(5) + 5^2 \\ &= 4x^2 + 20x + 25 \\ (3x - 4)^2 &= (3x)^2 - 2(3x)(4) + (-4)^2 \\ &= 9x^2 - 24x + 16 \end{aligned}$$

Definition of Zero as an Exponent [4.4A, p. 209]
If $x \neq 0$ then $x^0 = 1$.

$$17^0 = 1, \ (-6c)^0 = 1, \ c \neq 0$$

Definition of a Negative Exponent [4.4A, p. 210]
If $x \neq 0$ and n is a positive integer, then $x^{-n} = \dfrac{1}{x^n}$ and $\dfrac{1}{x^{-n}} = x^n$.

$$x^{-6} = \frac{1}{x^6} \text{ and } \frac{1}{x^{-6}} = x^6$$

Copyright © Houghton Mifflin Company. All rights reserved.

Rule for Simplifying the Power of a Quotient [4.4A, p. 211]

If m, n, and p are integers and $y \neq 0$, then $\left(\dfrac{x^m}{y^n}\right)^p = \dfrac{x^{mp}}{y^{np}}$.

$\left(\dfrac{c^3}{a^5}\right)^2 = \dfrac{c^{3 \cdot 2}}{a^{5 \cdot 2}} = \dfrac{c^6}{a^{10}}$

Rule for Negative Exponents on Fractional Expressions [4.4A, p. 211]

If $a \neq 0$, $b \neq 0$, and n is a positive integer, then $\left(\dfrac{a}{b}\right)^{-n} = \left(\dfrac{b}{a}\right)^{n}$.

$\left(\dfrac{x}{y}\right)^{-3} = \left(\dfrac{y}{x}\right)^{3}$

Rule for Dividing Exponential Expressions [4.4A, p. 211]

If m and n are integers and $x \neq 0$, then $\dfrac{x^m}{x^n} = x^{m-n}$.

$\dfrac{a^7}{a^2} = a^{7-2} = a^5$

To Express a Number in Scientific Notation [4.4B, p. 214]
To express a number in scientific notation, write it in the form $a \times 10^n$ where $1 \le a < 10$, and n is an integer. If the number is greater than 10, then n is a positive integer. If the number is between 0 and 1, then n is a negative integer.

$367{,}000{,}000 = 3.67 \times 10^8$
$0.0000078 = 7.8 \times 10^{-6}$

To Change a Number in Scientific Notation to Decimal Notation [4.4B, p. 214]
To change a number in scientific notation to decimal notation, move the decimal point to the right if n is positive and to the left if n is negative. Move the decimal point the same number of places as the absolute value of the exponent on 10.

$2.418 \times 10^7 = 24{,}180{,}000$
$9.06 \times 10^{-5} = 0.0000906$

To divide a polynomial by a monomial, divide each term in the numerator by the denominator and write the sum of the quotients. [4.5A, p. 219]

$\dfrac{8xy^3 - 4y^2 + 12y}{4y}$
$= \dfrac{8xy^3}{4y} - \dfrac{4y^2}{4y} + \dfrac{12y}{4y}$
$= 2xy^2 - y + 3$

To check polynomial division, use the same equation used to check division of whole numbers:

(Quotient × divisor) + remainder = dividend

[4.5B, p. 219]

$$\begin{array}{r} x - 4 \\ x + 3 \overline{)\, x^2 - x - 10\,} \\ \underline{x^2 + 3x} \\ -4x - 10 \\ \underline{-4x - 12} \\ 2 \end{array}$$

Check:

$(x - 4)(x + 3) + 2 = x^2 - x - 12 + 2$
$= x^2 - x - 10$

$(x^2 - x - 10) \div (x + 3) = x - 4 + \dfrac{2}{x + 3}$

Copyright © Houghton Mifflin Company. All rights reserved.

Chapter 4 Review Exercises

1. Multiply: $(2b - 3)(4b + 5)$

2. Add: $(12y^2 + 17y - 4) + (9y^2 - 13y + 3)$

3. Simplify: $(xy^5z^3)(x^3y^3z)$

4. Simplify: $\dfrac{8x^{12}}{12x^9}$

5. Multiply: $-2x(4x^2 + 7x - 9)$

6. Simplify: $\dfrac{3ab^4}{-6a^2b^4}$

7. Simplify: $(-2u^3v^4)^4$

8. Evaluate: $(2^3)^2$

9. Subtract: $(5x^2 - 2x - 1) - (3x^2 - 5x + 7)$

10. Simplify: $\dfrac{a^{-1}b^3}{a^3b^{-3}}$

11. Simplify: $(-2x^3)^2(-3x^4)^3$

12. Expand: $(5y - 7)^2$

13. Simplify: $(5a^7b^6)^2(4ab)$

14. Divide: $\dfrac{12b^7 + 36b^5 - 3b^3}{3b^3}$

15. Evaluate: -4^{-2}

16. Subtract: $(13y^3 - 7y - 2) - (12y^2 - 2y - 1)$

17. Divide: $\dfrac{7 - x - x^2}{x + 3}$

18. Multiply: $(2a - b)(x - 2y)$

19. Multiply: $(3y^2 + 4y - 7)(2y + 3)$

20. Divide: $(b^3 - 2b^2 - 33b - 7) \div (b - 7)$

Copyright © Houghton Mifflin Company. All rights reserved.

21. Multiply: $2ab^3(4a^2 - 2ab + 3b^2)$

22. Multiply: $(2a - 5b)(2a + 5b)$

23. Multiply: $(6b^3 - 2b^2 - 5)(2b^2 - 1)$

24. Add: $(2x^3 + 7x^2 + x) + (2x^2 - 4x - 12)$

25. Divide: $\dfrac{16y^2 - 32y}{-4y}$

26. Multiply: $(a + 7)(a - 7)$

27. Write 37,560,000,000 in scientific notation.

28. Write 1.46×10^7 in decimal notation.

29. Simplify: $(2a^{12}b^3)(-9b^2c^6)(3ac)$

30. Divide: $(6y^2 - 35y + 36) \div (3y - 4)$

31. Simplify: $(-3x^{-2}y^{-3})^{-2}$

32. Multiply: $(5a - 7)(2a + 9)$

33. Write 0.000000127 in scientific notation.

34. Write 3.2×10^{-12} in decimal notation.

35. **Geometry** The length of a table-tennis table is 1 ft less than twice the width of the table. Let w represent the width of the table-tennis table. Express the area of the table in terms of the variable w.

36. **Geometry** The side of a checkerboard is $(3x - 2)$ in. Express the area of the checkerboard in terms of the variable x.

Copyright © Houghton Mifflin Company. All rights reserved.

Chapter 4 Test

1. Multiply: $2x(2x^2 - 3x)$

2. Divide: $\dfrac{12x^3 - 3x^2 + 9}{3x^2}$

3. Simplify: $\dfrac{12x^2}{-3x^8}$

4. Simplify: $(-2xy^2)(3x^2y^4)$

5. Divide: $(x^2 + 1) \div (x + 1)$

6. Multiply: $(x - 3)(x^2 - 4x + 5)$

7. Simplify: $(-2a^2b)^3$

8. Simplify: $\dfrac{(3x^{-2}y^3)^3}{3x^4y^{-1}}$

9. Multiply: $(a - 2b)(a + 5b)$

10. Divide: $\dfrac{16x^5 - 8x^3 + 20x}{4x}$

11. Divide: $(x^2 + 6x - 7) \div (x - 1)$

12. Multiply: $-3y^2(-2y^2 + 3y - 6)$

13. Multiply: $(-2x^3 + x^2 - 7)(2x - 3)$

14. Multiply: $(4y - 3)(4y + 3)$

Copyright © Houghton Mifflin Company. All rights reserved.

15. Simplify: $(ab^2)(a^3b^5)$

16. Simplify: $\dfrac{2a^{-1}b}{2^{-2}a^{-2}b^{-3}}$

17. Divide: $\dfrac{20a - 35}{5}$

18. Subtract: $(3a^2 - 2a - 7) - (5a^3 + 2a - 10)$

19. Expand: $(2x - 5)^2$

20. Divide: $(4x^2 - 7) \div (2x - 3)$

21. Simplify: $\dfrac{-(2x^2y)^3}{4x^3y^3}$

22. Multiply: $(2x - 7y)(5x - 4y)$

23. Add: $(3x^3 - 2x^2 - 4) + (8x^2 - 8x + 7)$

24. Write 0.00000000302 in scientific notation.

25. **Geometry** The radius of a circle is $(x - 5)$ m. Use the equation $A = \pi r^2$, where r is the radius, to find the area of the circle in terms of the variable x. Leave the answer in terms of π.

$x - 5$

Copyright © Houghton Mifflin Company. All rights reserved.

Cumulative Review Exercises

1. Simplify: $\dfrac{3}{16} - \left(-\dfrac{5}{8}\right) - \dfrac{7}{9}$

2. Evaluate $-3^2 \cdot \left(\dfrac{2}{3}\right)^3 \cdot \left(-\dfrac{5}{8}\right)$.

3. Simplify: $\left(-\dfrac{1}{2}\right)^3 \div \left(\dfrac{3}{8} - \dfrac{5}{6}\right) + 2$

4. Evaluate $\dfrac{b - (a - b)^2}{b^2}$ when $a = -2$ and $b = 3$.

5. Simplify: $-2x - (-xy) + 7x - 4xy$

6. Simplify: $(12x)\left(-\dfrac{3}{4}\right)$

7. Simplify: $-2[3x - 2(4 - 3x) + 2]$

8. Solve: $12 = -\dfrac{3}{4}x$

9. Solve: $2x - 9 = 3x + 7$

10. Solve: $2 - 3(4 - x) = 2x + 5$

11. 35.2 is what percent of 160?

12. Add: $(4b^3 - 7b^2 - 7) + (3b^2 - 8b + 3)$

13. Subtract: $(3y^3 - 5y + 8) - (-2y^2 + 5y + 8)$

14. Simplify: $(a^3b^5)^3$

15. Simplify: $(4xy^3)(-2x^2y^3)$

16. Multiply: $-2y^2(-3y^2 - 4y + 8)$

Copyright © Houghton Mifflin Company. All rights reserved.

17. Multiply: $(2a - 7)(5a^2 - 2a + 3)$

18. Multiply: $(3b - 2)(5b - 7)$

19. Simplify: $\dfrac{(-2a^2b^3)^2}{8a^4b^8}$

20. Divide: $(a^2 - 4a - 21) \div (a + 3)$

21. Write 6.09×10^{-5} in decimal notation.

22. Translate "the difference between eight times a number and twice the number is eighteen" into an equation and solve.

23. **Juice Mixtures** Fifty ounces of orange juice are added to 200 oz of a fruit punch that is 10% orange juice. What is the percent concentration of orange juice in the resulting mixture?

24. **Transportation** A car traveling at 50 mph overtakes a cyclist who, riding at 10 mph, has had a 2-hour head start. How far from the starting point does the car overtake the cyclist?

25. **Geometry** The width of a rectangle is 40% of the length. The perimeter of the rectangle is 42 m. Find the length and width of the rectangle.

Copyright © Houghton Mifflin Company. All rights reserved.

Copyright © Houghton Mifflin Company. All rights reserved.

chapter

5

Factoring

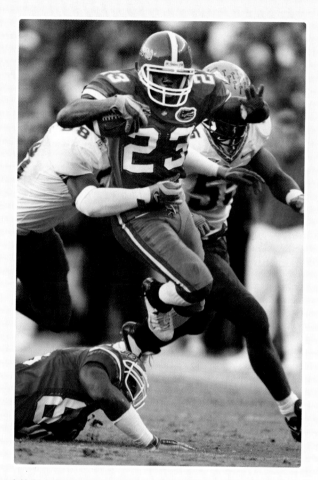

A University of Florida receiver is being tackled during the play in the photo above. The University of Florida is part of the Southeastern conference of the NCAA. Imagine you have just gotten a job as the manager for this conference. One of your first tasks would be to organize the game schedule for the upcoming season. There are 12 teams in the Southeastern conference. How would you go about creating this schedule? **Exercises 71 and 72 on page 273** show you how to use a formula to determine the number of league games that must be scheduled if each team is to play each other team once.

OBJECTIVES

Section 5.1

A To factor a monomial from a polynomial
B To factor by grouping

Section 5.2

A To factor a trinomial of the form $x^2 + bx + c$
B To factor completely

Section 5.3

A To factor a trinomial of the form $ax^2 + bx + c$ by using trial factors
B To factor a trinomial of the form $ax^2 + bx + c$ by grouping

Section 5.4

A To factor the difference of two squares and perfect-square trinomials
B To factor completely

Section 5.5

A To solve equations by factoring
B To solve application problems

Need help? For online student resources, such as section quizzes, visit this textbook's website at **math.college.hmco.com/students.**

Do these exercises to prepare for Chapter 5.

1. Write 30 as a product of prime numbers.

2. Simplify: $-3(4y - 5)$

3. Simplify: $-(a - b)$

4. Simplify: $2(a - b) - 5(a - b)$

5. Solve: $4x = 0$

6. Solve: $2x + 1 = 0$

7. Multiply: $(x + 4)(x - 6)$

8. Multiply: $(2x - 5)(3x + 2)$

9. Simplify: $\dfrac{x^5}{x^2}$

10. Simplify: $\dfrac{6x^4y^3}{2xy^2}$

GO FIGURE • • •

Without using a calculator (it probably won't work anyway), how many digits are in the product $4^{54} \cdot 5^{100}$?

Copyright © Houghton Mifflin Company. All rights reserved.

5.1 Common Factors

Copyright © Houghton Mifflin Company. All rights reserved.

Objective A   To factor a monomial from a polynomial

In Section 1.5C we discussed how to find the greatest common factor (GCF) of two or more integers. The **greatest common factor (GCF) of two or more monomials** is the product of the GCF of the coefficients and the common variable factors.

$$6x^3y = 2 \cdot 3 \cdot x \cdot x \cdot x \cdot y$$
$$8x^2y^2 = 2 \cdot 2 \cdot 2 \cdot x \cdot x \cdot y \cdot y$$
$$GCF = 2 \cdot x \cdot x \cdot y = 2x^2y$$

Note that the exponent of each variable in the GCF is the same as the *smallest* exponent of that variable in either of the monomials.

The GCF of $6x^3y$ and $8x^2y^2$ is $2x^2y$.

HOW TO   Find the GCF of $12a^4b$ and $18a^2b^2c$.

The common variable factors are a^2 and b; c is not a common variable factor.

$$12a^4b = 2 \cdot 2 \cdot 3 \cdot a^4 \cdot b$$
$$18a^2b^2c = 2 \cdot 3 \cdot 3 \cdot a^2 \cdot b^2 \cdot c$$
$$GCF = 2 \cdot 3 \cdot a^2 \cdot b = 6a^2b$$

To **factor a polynomial** means to write the polynomial as a product of other polynomials. In the example at the right, $2x$ is the GCF of the terms $2x^2$ and $10x$.

HOW TO   Factor: $5x^3 - 35x^2 + 10x$

Find the GCF of the terms of the polynomial.

$$5x^3 = 5 \cdot x^3$$
$$35x^2 = 5 \cdot 7 \cdot x^2$$
$$10x = 2 \cdot 5 \cdot x$$

The GCF is $5x$.

Rewrite the polynomial, expressing each term as a product with the GCF as one of the factors.

$$5x^3 - 35x^2 + 10x = 5x(x^2) + 5x(-7x) + 5x(2)$$
$$= 5x(x^2 - 7x + 2)$$

• Use the Distributive Property to write the polynomial as a product of factors.

TAKE NOTE

At the right, the factors in parentheses are determined by dividing each term of the trinomial by the GCF, $5x$.

$$\frac{5x^3}{5x} = x^2,$$

$$\frac{-35x^2}{5x} = -7x, \text{ and}$$

$$\frac{10x}{5x} = 2$$

HOW TO Factor: $21x^2y^3 - 6xy^5 + 15x^4y^2$

Find the GCF of the terms of the polynomial.

$21x^2y^3 = 3 \cdot 7 \cdot x^2 \cdot y^3$

$6xy^5 = 2 \cdot 3 \cdot x \cdot y^5$

$15x^4y^2 = 3 \cdot 5 \cdot x^4 \cdot y^2$

The GCF is $3xy^2$.

Rewrite the polynomial, expressing each term as a product with the GCF as one of the factors.

$21x^2y^3 - 6xy^5 + 15x^4y^2 = 3xy^2(7xy) + 3xy^2(-2y^3) + 3xy^2(5x^3)$

$\qquad\qquad\qquad\qquad\quad = 3xy^2(7xy - 2y^3 + 5x^3)$

• Use the Distributive Property to write the polynomial as a product of factors.

Example 1

Factor: $8x^2 + 2xy$

Solution
The GCF is $2x$.

$8x^2 + 2xy = 2x(4x) + 2x(y)$
$\qquad\qquad = 2x(4x + y)$

You Try It 1

Factor: $14a^2 - 21a^4b$

Your solution

Example 2

Factor: $n^3 - 5n^2 + 2n$

Solution
The GCF is n.

$n^3 - 5n^2 + 2n = n(n^2) + n(-5n) + n(2)$
$\qquad\qquad\qquad = n(n^2 - 5n + 2)$

You Try It 2

Factor: $27b^2 + 18b + 9$

Your solution

Example 3

Factor: $16x^2y + 8x^4y^2 - 12x^4y^5$

Solution
The GCF is $4x^2y$.

$16x^2y + 8x^4y^2 - 12x^4y^5$
$\quad = 4x^2y(4) + 4x^2y(2x^2y) + 4x^2y(-3x^2y^4)$
$\quad = 4x^2y(4 + 2x^2y - 3x^2y^4)$

You Try It 3

Factor: $6x^4y^2 - 9x^3y^2 + 12x^2y^4$

Your solution

Solutions on pp. S11–S12

Copyright © Houghton Mifflin Company. All rights reserved.

Objective B   **To factor by grouping**

A factor that has two terms is called a **binomial factor.** In the examples at the right, the binomials in parentheses are binomial factors.

$$2a(a + b)^2$$
$$3xy(x - y)$$

The Distributive Property is used to factor a common binomial factor from an expression.

The common binomial factor of the expression $6x(x - 3) + y(x - 3)$ is $(x - 3)$. To factor that expression, use the Distributive Property to write the expression as a product of factors.

$$6x(x - 3) + y(x - 3) = (x - 3)(6x + y)$$

Consider the following simplification of $-(a - b)$.

$$-(a - b) = -1(a - b) = -a + b = b - a$$

Thus

$$b - a = -(a - b)$$

This equation is sometimes used to factor a common binomial from an expression.

HOW TO   Factor: $2x(x - y) + 5(y - x)$

$$2x(x - y) + 5(y - x) = 2x(x - y) - 5(x - y)$$
$$= (x - y)(2x - 5)$$

• $5(y - x) = 5[(-1)(x - y)]$
  $\quad = -5(x - y)$

A polynomial can be **factored by grouping** if its terms can be grouped and factored in such a way that a common binomial factor is found.

HOW TO   Factor: $ax + bx - ay - by$

$$ax + bx - ay - by = (ax + bx) - (ay + by)$$

$$= x(a + b) - y(a + b)$$
$$= (a + b)(x - y)$$

• Group the first two terms and the last two terms. Note that $-ay - by = -(ay + by)$.
• Factor each group.
• Factor the GCF, $(a + b)$, from each group.

HOW TO   Factor: $6x^2 - 9x - 4xy + 6y$

$$6x^2 - 9x - 4xy + 6y = (6x^2 - 9x) - (4xy - 6y)$$

$$= 3x(2x - 3) - 2y(2x - 3)$$
$$= (2x - 3)(3x - 2y)$$

• Group the first two terms and the last two terms. Note that $-4xy + 6y = -(4xy - 6y)$.
• Factor each group.
• Factor the GCF, $(2x - 3)$, from each group.

Copyright © Houghton Mifflin Company. All rights reserved.

Example 4

Factor: $4x(3x - 2) - 7(3x - 2)$

Solution

$4x(3x - 2) - 7(3x - 2)$ • **$3x - 2$ is the common binomial factor.**

$= (3x - 2)(4x - 7)$

You Try It 4

Factor: $2y(5x - 2) - 3(2 - 5x)$

Your solution

Example 5

Factor: $9x^2 - 15x - 6xy + 10y$

Solution

$9x^2 - 15x - 6xy + 10y$

$= (9x^2 - 15x) - (6xy - 10y)$ • **$-6xy + 10y = -(6xy - 10y)$**

$= 3x(3x - 5) - 2y(3x - 5)$ • **$3x - 5$ is the common factor.**

$= (3x - 5)(3x - 2y)$

You Try It 5

Factor: $a^2 - 3a + 2ab - 6b$

Your solution

Example 6

Factor: $3x^2y - 4x - 15xy + 20$

Solution

$3x^2y - 4x - 15xy + 20$

$= (3x^2y - 4x) - (15xy - 20)$ • **$-15xy + 20 = -(15xy - 20)$**

$= x(3xy - 4) - 5(3xy - 4)$ • **$3xy - 4$ is the common factor.**

$= (3xy - 4)(x - 5)$

You Try It 6

Factor: $2mn^2 - n + 8mn - 4$

Your solution

Example 7

Factor: $4ab - 6 + 3b - 2ab^2$

Solution

$4ab - 6 + 3b - 2ab^2$

$= (4ab - 6) + (3b - 2ab^2)$

$= 2(2ab - 3) + b(3 - 2ab)$

$= 2(2ab - 3) - b(2ab - 3)$ • **$3 - 2ab = -(2ab - 3)$**

$= (2ab - 3)(2 - b)$ • **$2ab - 3$ is the common factor.**

You Try It 7

Factor: $3xy - 9y - 12 + 4x$

Your solution

Solutions on p. S12

Copyright © Houghton Mifflin Company. All rights reserved.

5.1 Exercises

1. Explain the meaning of "a common monomial factor of a polynomial."

2. Explain the meaning of "a factor" and the meaning of "to factor."

For Exercises 3 to 41, factor.

3. $5a + 5$

4. $7b - 7$

5. $16 - 8a^2$

6. $12 + 12y^2$

7. $8x + 12$

8. $16a - 24$

9. $30a - 6$

10. $20b + 5$

11. $7x^2 - 3x$

12. $12y^2 - 5y$

13. $3a^2 + 5a^5$

14. $9x - 5x^2$

15. $14y^2 + 11y$

16. $6b^3 - 5b^2$

17. $2x^4 - 4x$

18. $3y^4 - 9y$

19. $10x^4 - 12x^2$

20. $12a^5 - 32a^2$

21. $8a^8 - 4a^5$

22. $16y^4 - 8y^7$

23. $x^2y^2 - xy$

24. $a^2b^2 + ab$

25. $3x^2y^4 - 6xy$

26. $12a^2b^5 - 9ab$

27. $x^2y - xy^3$

28. $3x^3 + 6x^2 + 9x$

29. $5y^3 - 20y^2 + 5y$

30. $2x^4 - 4x^3 + 6x^2$

31. $3y^4 - 9y^3 - 6y^2$

32. $2x^3 + 6x^2 - 14x$

33. $3y^3 - 9y^2 + 24y$

34. $2y^5 - 3y^4 + 7y^3$

35. $6a^5 - 3a^3 - 2a^2$

36. $x^3y - 3x^2y^2 + 7xy^3$

37. $2a^2b - 5a^2b^2 + 7ab^2$

38. $5y^3 + 10y^2 - 25y$

39. $4b^5 + 6b^3 - 12b$

40. $3a^2b^2 - 9ab^2 + 15b^2$

41. $8x^2y^2 - 4x^2y + x^2$

For Exercises 42 to 68, factor.

42. $x(b + 4) + 3(b + 4)$

43. $y(a + z) + 7(a + z)$

44. $a(y - x) - b(y - x)$

45. $3r(a - b) + s(a - b)$

46. $x(x - 2) + y(2 - x)$

47. $t(m - 7) + 7(7 - m)$

Copyright © Houghton Mifflin Company. All rights reserved.

48. $2x(7 + b) - y(b + 7)$

49. $2y(4a - b) - (b - 4a)$

50. $8c(2m - 3n) + (3n - 2m)$

51. $x^2 + 2x + 2xy + 4y$

52. $x^2 - 3x + 4ax - 12a$

53. $p^2 - 2p - 3rp + 6r$

54. $t^2 + 4t - st - 4s$

55. $ab + 6b - 4a - 24$

56. $xy - 5y - 2x + 10$

57. $2z^2 - z + 2yz - y$

58. $2y^2 - 10y + 7xy - 35x$

59. $8v^2 - 12vy + 14v - 21y$

60. $21x^2 + 6xy - 49x - 14y$

61. $2x^2 - 5x - 6xy + 15y$

62. $4a^2 + 5ab - 10b - 8a$

63. $3y^2 - 6y - ay + 2a$

64. $2ra + a^2 - 2r - a$

65. $3xy - y^2 - y + 3x$

66. $2ab - 3b^2 - 3b + 2a$

67. $3st + t^2 - 2t - 6s$

68. $4x^2 + 3xy - 12y - 16x$

APPLYING THE CONCEPTS

69. **Number Sense**   A natural number is a *perfect number* if it is the sum of all its factors less than itself. For example, 6 is a perfect number because all the factors of 6 that are less than 6 are 1, 2, and 3, and $1 + 2 + 3 = 6$.
a. Find the one perfect number between 20 and 30.
b. Find the one perfect number between 490 and 500.

70. **Geometry**   In the equation $P = 2L + 2W$, what is the effect on P when the quantity $L + W$ doubles?

71. **Geometry**   Write an expression in factored form for each of the shaded portions in the following diagrams. Use the equation for the area of a rectangle ($A = LW$) and the equation for the area of a circle ($A = \pi r^2$).

a.
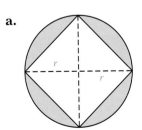

b.
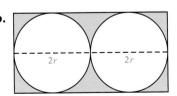

c.
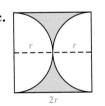

Copyright © Houghton Mifflin Company. All rights reserved.

5.2 Factoring Polynomials of the Form $x^2 + bx + c$

Objective A **To factor a trinomial of the form $x^2 + bx + c$**

Trinomials of the form $x^2 + bx + c$, where b and c are integers, are shown at the right.

$$x^2 + 8x + 12;\ b = 8,\ c = 12$$
$$x^2 - 7x + 12;\ b = -7,\ c = 12$$
$$x^2 - 2x - 15;\ b = -2,\ c = -15$$

To factor a trinomial of this form means to express the trinomial as the product of two binomials.

Trinomials expressed as the product of binomials are shown at the right.

$$x^2 + 8x + 12 = (x + 6)(x + 2)$$
$$x^2 - 7x + 12 = (x - 3)(x - 4)$$
$$x^2 - 2x - 15 = (x + 3)(x - 5)$$

The method by which factors of a trinomial are found is based on FOIL. Consider the following binomial products, noting the relationship between the constant terms of the binomials and the terms of the trinomials.

The signs in the binomials are the same.

$$(x + 6)(x + 2) = x^2 + 2x + 6x + (6)(2) = x^2 + 8x + 12$$
Sum of 6 and 2
Product of 6 and 2

$$(x - 3)(x - 4) = x^2 - 4x - 3x + (-3)(-4) = x^2 - 7x + 12$$
Sum of -3 and -4
Product of -3 and -4

The signs in the binomials are opposites.

$$(x + 3)(x - 5) = x^2 - 5x + 3x + (3)(-5) = x^2 - 2x - 15$$
Sum of 3 and -5
Product of 3 and -5

$$(x - 4)(x + 6) = x^2 + 6x - 4x + (-4)(6) = x^2 + 2x - 24$$
Sum of -4 and 6
Product of -4 and 6

Factoring $x^2 + bx + c$: IMPORTANT RELATIONSHIPS

1. When the constant term of the trinomial is positive, the constant terms of the binomials have the same sign. They are both positive when the coefficient of the x term in the trinomial is positive. They are both negative when the coefficient of the x term in the trinomial is negative.

2. When the constant term of the trinomial is negative, the constant terms of the binomials have opposite signs.

3. In the trinomial, the coefficient of x is the sum of the constant terms of the binomials.

4. In the trinomial, the constant term is the product of the constant terms of the binomials.

Copyright © Houghton Mifflin Company. All rights reserved.

> **HOW TO** Factor: $x^2 - 7x + 10$

Because the constant term is positive and the coefficient of x is negative, the binomial constants will be negative. Find two negative factors of 10 whose sum is -7. The results can be recorded in a table.

Negative Factors of 10	Sum
$-1, -10$	-11
$-2, -5$	-7

• These are the correct factors.

$x^2 - 7x + 10 = (x - 2)(x - 5)$ • Write the trinomial as a product of its factors.

TAKE NOTE
Always check your proposed factorization to ensure accuracy.

You can check the proposed factorization by multiplying the two binomials.

$$Check: (x - 2)(x - 5) = x^2 - 5x - 2x + 10$$
$$= x^2 - 7x + 10$$

> **HOW TO** Factor: $x^2 - 9x - 36$

The constant term is negative. The binomial constants will have opposite signs. Find two factors of -36 whose sum is -9.

Factors of -36	Sum
$+1, -36$	-35
$-1, +36$	35
$+2, -18$	-16
$-2, +18$	16
$+3, -12$	-9

• Once the correct factors are found, it is not necessary to try the remaining factors.

$x^2 - 9x - 36 = (x + 3)(x - 12)$ • Write the trinomial as a product of its factors.

For some trinomials it is not possible to find integer factors of the constant term whose sum is the coefficient of the middle term. A polynomial that does not factor using only integers is **nonfactorable over the integers.**

> **HOW TO** Factor: $x^2 + 7x + 8$

The constant term is positive and the coefficient of x is positive. The binomial constants will be positive. Find two positive factors of 8 whose sum is 7.

TAKE NOTE
Just as 17 is a prime number, $x^2 + 7x + 8$ is a **prime polynomial.** Binomials of the form $x - a$ and $x + a$ are also prime polynomials.

Positive Factors of 8	Sum
$1, 8$	9
$2, 4$	6

• There are no positive integer factors of 8 whose sum is 7.

$x^2 + 7x + 8$ is nonfactorable over the integers.

Example 1   Factor: $x^2 - 8x + 15$

Solution
Find two negative factors of 15 whose sum is -8.

Factors	Sum
$-1, -15$	-16
$-3, -5$	-8

$x^2 - 8x + 15 = (x - 3)(x - 5)$

You Try It 1   Factor: $x^2 + 9x + 20$

Your solution

Solution on p. S12

Copyright © Houghton Mifflin Company. All rights reserved.

Example 2   Factor: $x^2 + 6x - 27$

Solution
Find two factors of
-27 whose sum is 6.

Factors	Sum
$+1, -27$	-26
$-1, +27$	26
$+3, -9$	-6
$-3, +9$	6

$x^2 + 6x - 27 = (x - 3)(x + 9)$

You Try It 2   Factor: $x^2 + 7x - 18$

Your solution

Solution on p. S12

Objective B   **To factor completely**

A polynomial is **factored completely** when it is written as a product of factors that are nonfactorable over the integers.

TAKE NOTE

The first step in *any* factoring problem is to determine whether the terms of the polynomial have a *common factor*. If they do, factor it out first.

HOW TO   Factor: $4y^3 - 4y^2 - 24y$

$4y^3 - 4y^2 - 24y = 4y(y^2) - 4y(y) - 4y(6)$

• The GCF is **4y**.

$= 4y(y^2 - y - 6)$

• Use the Distributive Property to factor out the GCF.

$= 4y(y + 2)(y - 3)$

• Factor $y^2 - y - 6$. The two factors of -6 whose sum is -1 are 2 and -3.

It is always possible to check the proposed factorization by multiplying the polynomials. Here is the check for the last example.

Check: $4y(y + 2)(y - 3) = 4y(y^2 - 3y + 2y - 6)$
$= 4y(y^2 - y - 6)$
$= 4y^3 - 4y^2 - 24y$

• This is the original polynomial.

HOW TO   Factor: $5x^2 + 60xy + 100y^2$

$5x^2 + 60xy + 100y^2 = 5(x^2) + 5(12xy) + 5(20y^2)$

• The GCF is **5**.

$= 5(x^2 + 12xy + 20y^2)$

• Use the Distributive Property to factor out the GCF.

$= 5(x + 2y)(x + 10y)$

• Factor $x^2 + 12xy + 20y^2$. The two factors of 20 whose sum is 12 are 2 and 10.

TAKE NOTE

$2y$ and $10y$ are placed in the binomials. This is necessary so that the middle term contains xy and the last term contains y^2.

Note that $2y$ and $10y$ were placed in the binomials. The following check shows that this was necessary.

Check: $5(x + 2y)(x + 10y) = 5(x^2 + 10xy + 2xy + 20y^2)$
$= 5(x^2 + 12xy + 20y^2)$
$= 5x^2 + 60xy + 100y^2$

• This is the original polynomial.

Copyright © Houghton Mifflin Company. All rights reserved.

HOW TO   Factor: $15 - 2x - x^2$

Because the coefficient of x^2 is -1, factor -1 from the trinomial and then write the resulting trinomial in descending order.

$15 - 2x - x^2 = -(x^2 + 2x - 15)$

$= -(x + 5)(x - 3)$

• $15 - 2x - x^2 = -1(-15 + 2x + x^2)$
  $= -(x^2 + 2x - 15)$
• Factor $x^2 + 2x - 15$. The two factors of -15 whose sum is 2 are 5 and -3.

TAKE NOTE
When the coefficient of the highest power in a polynomial is negative, consider factoring out a negative GCF. Example 3 is another example of this technique.

Check: $-(x + 5)(x - 3) = -(x^2 + 2x - 15)$
$= -x^2 - 2x + 15$
$= 15 - 2x - x^2$ • This is the original polynomial.

Example 3

Factor: $-3x^3 + 9x^2 + 12x$

Solution
The GCF is $-3x$.
$-3x^3 + 9x^2 + 12x = -3x(x^2 - 3x - 4)$
Factor the trinomial $x^2 - 3x - 4$. Find two factors of -4 whose sum is -3.

Factors	Sum
$+1, -4$	-3

$-3x^3 + 9x^2 + 12x = -3x(x + 1)(x - 4)$

You Try It 3

Factor: $-2x^3 + 14x^2 - 12x$

Your solution

Example 4

Factor: $4x^2 - 40xy + 84y^2$

Solution
The GCF is 4.
$4x^2 - 40xy + 84y^2 = 4(x^2 - 10xy + 21y^2)$
Factor the trinomial $x^2 - 10xy + 21y^2$. Find two negative factors of 21 whose sum is -10.

Factors	Sum
$-1, -21$	-22
$-3, -7$	-10

$4x^2 - 40xy + 84y^2 = 4(x - 3y)(x - 7y)$

You Try It 4

Factor: $3x^2 - 9xy - 12y^2$

Your solution

Solutions on p. S12

Copyright © Houghton Mifflin Company. All rights reserved.

5.2 Exercises

Objective A   **To factor a trinomial of the form $x^2 + bx + c$**

1. Fill in the blank. In factoring a trinomial, if the constant term is positive, then the signs in both binomial factors will be _____ .

2. Fill in the blanks. To factor $x^2 + 8x - 48$, we must find two numbers whose product is _____ and whose sum is _____ .

For Exercises 3 to 75, factor.

3. $x^2 + 3x + 2$ **4.** $x^2 + 5x + 6$ **5.** $x^2 - x - 2$ **6.** $x^2 + x - 6$

7. $a^2 + a - 12$ **8.** $a^2 - 2a - 35$ **9.** $a^2 - 3a + 2$ **10.** $a^2 - 5a + 4$

11. $a^2 + a - 2$ **12.** $a^2 - 2a - 3$ **13.** $b^2 - 6b + 9$ **14.** $b^2 + 8b + 16$

15. $b^2 + 7b - 8$ **16.** $y^2 - y - 6$ **17.** $y^2 + 6y - 55$ **18.** $z^2 - 4z - 45$

19. $y^2 - 5y + 6$ **20.** $y^2 - 8y + 15$ **21.** $z^2 - 14z + 45$ **22.** $z^2 - 14z + 49$

23. $z^2 - 12z - 160$ **24.** $p^2 + 2p - 35$ **25.** $p^2 + 12p + 27$ **26.** $p^2 - 6p + 8$

27. $x^2 + 20x + 100$ **28.** $x^2 + 18x + 81$ **29.** $b^2 + 9b + 20$ **30.** $b^2 + 13b + 40$

31. $x^2 - 11x - 42$ **32.** $x^2 + 9x - 70$ **33.** $b^2 - b - 20$ **34.** $b^2 + 3b - 40$

35. $y^2 - 14y - 51$ **36.** $y^2 - y - 72$ **37.** $p^2 - 4p - 21$ **38.** $p^2 + 16p + 39$

39. $y^2 - 8y + 32$ **40.** $y^2 - 9y + 81$ **41.** $x^2 - 20x + 75$ **42.** $x^2 - 12x + 11$

Copyright © Houghton Mifflin Company. All rights reserved.

43. $p^2 + 24p + 63$

44. $x^2 - 15x + 56$

45. $x^2 + 21x + 38$

46. $x^2 + x - 56$

47. $x^2 + 5x - 36$

48. $a^2 - 21a - 72$

49. $a^2 - 7a - 44$

50. $a^2 - 15a + 36$

51. $a^2 - 21a + 54$

52. $z^2 - 9z - 136$

53. $z^2 + 14z - 147$

54. $c^2 - c - 90$

55. $c^2 - 3c - 180$

56. $z^2 + 15z + 44$

57. $p^2 + 24p + 135$

58. $c^2 + 19c + 34$

59. $c^2 + 11c + 18$

60. $x^2 - 4x - 96$

61. $x^2 + 10x - 75$

62. $x^2 - 22x + 112$

63. $x^2 + 21x - 100$

64. $b^2 + 8b - 105$

65. $b^2 - 22b + 72$

66. $a^2 - 9a - 36$

67. $a^2 + 42a - 135$

68. $b^2 - 23b + 102$

69. $b^2 - 25b + 126$

70. $a^2 + 27a + 72$

71. $z^2 + 24z + 144$

72. $x^2 + 25x + 156$

73. $x^2 - 29x + 100$

74. $x^2 - 10x - 96$

75. $x^2 + 9x - 112$

Copyright © Houghton Mifflin Company. All rights reserved.

Objective B   **To factor completely**

For Exercises 76 to 135, factor.

76. $2x^2 + 6x + 4$

77. $3x^2 + 15x + 18$

78. $18 + 7x - x^2$

79. $12 - 4x - x^2$

80. $ab^2 + 2ab - 15a$

81. $ab^2 + 7ab - 8a$

82. $xy^2 - 5xy + 6x$

83. $xy^2 + 8xy + 15x$

84. $z^3 - 7z^2 + 12z$

85. $-2a^3 - 6a^2 - 4a$

86. $-3y^3 + 15y^2 - 18y$

87. $4y^3 + 12y^2 - 72y$

88. $3x^2 + 3x - 36$

89. $2x^3 - 2x^2 + 4x$

90. $5z^2 - 15z - 140$

91. $6z^2 + 12z - 90$

92. $2a^3 + 8a^2 - 64a$

93. $3a^3 - 9a^2 - 54a$

94. $x^2 - 5xy + 6y^2$

95. $x^2 + 4xy - 21y^2$

96. $a^2 - 9ab + 20b^2$

97. $a^2 - 15ab + 50b^2$

98. $x^2 - 3xy - 28y^2$

99. $s^2 + 2st - 48t^2$

100. $y^2 - 15yz - 41z^2$

101. $x^2 + 85xy + 36y^2$

102. $z^4 - 12z^3 + 35z^2$

103. $z^4 + 2z^3 - 80z^2$

104. $b^4 - 22b^3 + 120b^2$

105. $b^4 - 3b^3 - 10b^2$

106. $2y^4 - 26y^3 - 96y^2$

107. $3y^4 + 54y^3 + 135y^2$

108. $-x^4 - 7x^3 + 8x^2$

109. $-x^4 + 11x^3 + 12x^2$

110. $4x^2y + 20xy - 56y$

111. $3x^2y - 6xy - 45y$

Copyright © Houghton Mifflin Company. All rights reserved.

112. $c^3 + 18c^2 - 40c$

113. $-3x^3 + 36x^2 - 81x$

114. $-4x^3 - 4x^2 + 24x$

115. $x^2 - 8xy + 15y^2$

116. $y^2 - 7xy - 8x^2$

117. $a^2 - 13ab + 42b^2$

118. $y^2 + 4yz - 21z^2$

119. $y^2 + 8yz + 7z^2$

120. $y^2 - 16yz + 15z^2$

121. $3x^2y + 60xy - 63y$

122. $4x^2y - 68xy - 72y$

123. $3x^3 + 3x^2 - 36x$

124. $4x^3 + 12x^2 - 160x$

125. $4z^3 + 32z^2 - 132z$

126. $5z^3 - 50z^2 - 120z$

127. $4x^3 + 8x^2 - 12x$

128. $5x^3 + 30x^2 + 40x$

129. $5p^2 + 25p - 420$

130. $4p^2 - 28p - 480$

131. $p^4 + 9p^3 - 36p^2$

132. $p^4 + p^3 - 56p^2$

133. $t^2 - 12ts + 35s^2$

134. $a^2 - 10ab + 25b^2$

135. $a^2 - 8ab - 33b^2$

APPLYING THE CONCEPTS

For Exercises 136 to 138, factor.

136. $2 + c^2 + 9c$

137. $x^2y - 54y - 3xy$

138. $45a^2 + a^2b^2 - 14a^2b$

For Exercises 139 to 141, find all integers k such that the trinomial can be factored over the integers.

139. $x^2 + kx + 35$

140. $x^2 + kx + 18$

141. $x^2 + kx + 21$

For Exercises 142 to 147, determine the positive integer values of k for which the following polynomials are factorable over the integers.

142. $y^2 + 4y + k$

143. $z^2 + 7z + k$

144. $a^2 - 6a + k$

145. $c^2 - 7c + k$

146. $x^2 - 3x + k$

147. $y^2 + 5y + k$

148.   In Exercises 142 to 147, there was the stated requirement that $k > 0$. If k is allowed to be any integer, how many different values of k are possible for each polynomial?

Copyright © Houghton Mifflin Company. All rights reserved.

5.3 Factoring Polynomials of the Form $ax^2 + bx + c$

Objective A **To factor a trinomial of the form $ax^2 + bx + c$ by using trial factors**

Trinomials of the form $ax^2 + bx + c$, where a, b, and c are integers, are shown at the right.

$3x^2 - x + 4; a = 3, b = -1, c = 4$
$6x^2 + 2x - 3; a = 6, b = 2, c = -3$

These trinomials differ from those in the preceding section in that the coefficient of x^2 is not 1. There are various methods of factoring these trinomials. The method described in this objective is factoring polynomials using trial factors.

To reduce the number of trial factors that must be considered, remember the following:

1. Use the signs of the constant term and the coefficient of x in the trinomial to determine the signs of the binomial factors. If the constant term is positive, the signs of the binomial factors will be the same as the sign of the coefficient of x in the trinomial. If the sign of the constant term is negative, the constant terms in the binomials have opposite signs.

2. If the terms of the trinomial do not have a common factor, then the terms of neither of the binomial factors will have a common factor.

HOW TO Factor: $2x^2 - 7x + 3$

The terms have no common factor. The constant term is positive. The coefficient of x is negative. The binomial constants will be negative.

Positive Factors of 2 (coefficient of x^2)	Negative Factors of 3 (constant term)
1, 2	−1, −3

Write trial factors. Use the **O**uter and **I**nner products of FOIL to determine the middle term, $-7x$, of the trinomial.

Trial Factors	Middle Term
$(x - 1)(2x - 3)$	$-3x - 2x = -5x$
$(x - 3)(2x - 1)$	$-x - 6x = -7x$

Write the factors of the trinomial. $2x^2 - 7x + 3 = (x - 3)(2x - 1)$

HOW TO Factor: $3x^2 + 14x + 15$

The terms have no common factor. The constant term is positive. The coefficient of x is positive. The binomial constants will be positive.

Positive Factors of 3 (coefficient of x^2)	Positive Factors of 15 (constant term)
1, 3	1, 15
	3, 5

Write trial factors. Use the **O**uter and **I**nner products of FOIL to determine the middle term, $14x$, of the trinomial.

Trial Factors	Middle Term
$(x + 1)(3x + 15)$	Common factor
$(x + 15)(3x + 1)$	$x + 45x = 46x$
$(x + 3)(3x + 5)$	$5x + 9x = 14x$
$(x + 5)(3x + 3)$	Common factor

Write the factors of the trinomial. $3x^2 + 14x + 15 = (x + 3)(3x + 5)$

Copyright © Houghton Mifflin Company. All rights reserved.

HOW TO Factor: $6x^3 + 14x^2 - 12x$

Factor the GCF, $2x$, from the terms.

$$6x^3 + 14x^2 - 12x = 2x(3x^2 + 7x - 6)$$

Factor the trinomial. The constant term is negative. The binomial constants will have opposite signs.

Positive Factors of 3	*Factors of −6*
1, 3	1, −6
	−1, 6
	2, −3
	−2, 3

Write trial factors. Use the **O**uter and **I**nner products of FOIL to determine the middle term, $7x$, of the trinomial.

It is not necessary to test trial factors that have a common factor.

Trial Factors	*Middle Term*
$(x + 1)(3x - 6)$	Common factor
$(x - 6)(3x + 1)$	$x - 18x = -17x$
$(x - 1)(3x + 6)$	Common factor
$(x + 6)(3x - 1)$	$-x + 18x = 17x$
$(x + 2)(3x - 3)$	Common factor
$(x - 3)(3x + 2)$	$2x - 9x = -7x$
$(x - 2)(3x + 3)$	Common factor
$(x + 3)(3x - 2)$	$-2x + 9x = 7x$

Write the factors of the trinomial.

$$6x^3 + 14x^2 - 12x = 2x(x + 3)(3x - 2)$$

For this example, all the trial factors were listed. Once the correct factors have been found, however, the remaining trial factors can be omitted. For the examples and solutions in this text, all trial factors except those that have a common factor will be listed.

Example 1 Factor: $3x^2 + x - 2$

Solution

Positive factors of 3: 1, 3

Factors of −2: 1, −2
　　　　　　−1, 2

Trial Factors	*Middle Term*
$(x + 1)(3x - 2)$	$-2x + 3x = x$
$(x - 2)(3x + 1)$	$x - 6x = -5x$
$(x - 1)(3x + 2)$	$2x - 3x = -x$
$(x + 2)(3x - 1)$	$-x + 6x = 5x$

$3x^2 + x - 2 = (x + 1)(3x - 2)$

You Try It 1 Factor: $2x^2 - x - 3$

Your solution

Example 2 Factor: $-12x^3 - 32x^2 + 12x$

Solution

The GCF is $-4x$.

$-12x^3 - 32x^2 + 12x = -4x(3x^2 + 8x - 3)$

Factor the trinomial.

Positive factors of 3: 1, 3

Factors of −3: 1, −3
　　　　　　−1, 3

Trial Factors	*Middle Term*
$(x - 3)(3x + 1)$	$x - 9x = -8x$
$(x + 3)(3x - 1)$	$-x + 9x = 8x$

$-12x^3 - 32x^2 + 12x = -4x(x + 3)(3x - 1)$

You Try It 2 Factor: $-45y^3 + 12y^2 + 12y$

Your solution

Solutions on pp. S12–S13

Copyright © Houghton Mifflin Company. All rights reserved.

Copyright © Houghton Mifflin Company. All rights reserved.

Objective B   **To factor a trinomial of the form $ax^2 + bx + c$ by grouping**

In the preceding objective, trinomials of the form $ax^2 + bx + c$ were factored by using trial factors. In this objective, these trinomials will be factored by grouping.

To factor $ax^2 + bx + c$, first find two factors of $a \cdot c$ whose sum is b. Then use factoring by grouping to write the factorization of the trinomial.

HOW TO   Factor: $2x^2 + 13x + 15$

Find two positive factors of 30 $(2 \cdot 15)$ whose sum is 13.

Positive Factors of 30	Sum
1, 30	31
2, 15	17
3, 10	13

• Once the required sum has been found, the remaining factors need not be checked.

$2x^2 + 13x + 15 = 2x^2 + 3x + 10x + 15$

$\qquad = (2x^2 + 3x) + (10x + 15)$

$\qquad = x(2x + 3) + 5(2x + 3)$

$\qquad = (2x + 3)(x + 5)$

• Use the factors of 30 whose sum is 13 to write $13x$ as $3x + 10x$.
• Factor by grouping.

Check: $(2x + 3)(x + 5) = 2x^2 + 10x + 3x + 15$

$\qquad\qquad\qquad\quad = 2x^2 + 13x + 15$

HOW TO   Factor: $6x^2 - 11x - 10$

Find two factors of -60 $[6(-10)]$ whose sum is -11.

Factors of −60	Sum
1, −60	−59
−1, 60	59
2, −30	−28
−2, 30	28
3, −20	−17
−3, 20	17
4, −15	−11

$6x^2 - 11x - 10 = 6x^2 + 4x - 15x - 10$

$\qquad = (6x^2 + 4x) - (15x + 10)$

$\qquad = 2x(3x + 2) - 5(3x + 2)$

$\qquad = (3x + 2)(2x - 5)$

• Use the factors of -60 whose sum is -11 to write $-11x$ as $4x - 15x$.
• Factor by grouping. Recall that $-15x - 10 = -(15x + 10)$.

Check: $(3x + 2)(2x - 5) = 6x^2 - 15x + 4x - 10$

$\qquad\qquad\qquad\quad = 6x^2 - 11x - 10$

HOW TO Factor: $3x^2 - 2x - 4$

Find two factors of -12 [$3(-4)$] whose sum is -2.

Factors of -12	Sum
1, -12	-11
-1, 12	11
2, -6	-4
-2, 6	4
3, -4	-1
-3, 4	1

TAKE NOTE
$3x^2 - 2x - 4$ is a prime polynomial.

Because no integer factors of -12 have a sum of -2, $3x^2 - 2x - 4$ is nonfactorable over the integers.

Example 3

Factor: $2x^2 + 19x - 10$

Solution

Factors of -20 [$2(-10)$]	Sum
$-1, 20$	19

$$2x^2 + 19x - 10 = 2x^2 - x + 20x - 10$$
$$= (2x^2 - x) + (20x - 10)$$
$$= x(2x - 1) + 10(2x - 1)$$
$$= (2x - 1)(x + 10)$$

Example 4

Factor: $24x^2y - 76xy + 40y$

Solution
The GCF is $4y$.
$$24x^2y - 76xy + 40y = 4y(6x^2 - 19x + 10)$$

Negative Factors of 60 [$6(10)$]	Sum
$-1, -60$	-61
$-2, -30$	-32
$-3, -20$	-23
$-4, -15$	-19

$$6x^2 - 19x + 10 = 6x^2 - 4x - 15x + 10$$
$$= (6x^2 - 4x) - (15x - 10)$$
$$= 2x(3x - 2) - 5(3x - 2)$$
$$= (3x - 2)(2x - 5)$$

$$24x^2y - 76xy + 40y = 4y(6x^2 - 19x + 10)$$
$$= 4y(3x - 2)(2x - 5)$$

You Try It 3

Factor: $2a^2 + 13a - 7$

Your solution

You Try It 4

Factor: $15x^3 + 40x^2 - 80x$

Your solution

Solutions on p. S13

Copyright © Houghton Mifflin Company. All rights reserved.

5.3 Exercises

Objective A   **To factor a trinomial of the form $ax^2 + bx + c$ by using trial factors**

For Exercises 1 to 70, factor by using trial factors.

1. $2x^2 + 3x + 1$

2. $5x^2 + 6x + 1$

3. $2y^2 + 7y + 3$

4. $3y^2 + 7y + 2$

5. $2a^2 - 3a + 1$

6. $3a^2 - 4a + 1$

7. $2b^2 - 11b + 5$

8. $3b^2 - 13b + 4$

9. $2x^2 + x - 1$

10. $4x^2 - 3x - 1$

11. $2x^2 - 5x - 3$

12. $3x^2 + 5x - 2$

13. $2t^2 - t - 10$

14. $2t^2 + 5t - 12$

15. $3p^2 - 16p + 5$

16. $6p^2 + 5p + 1$

17. $12y^2 - 7y + 1$

18. $6y^2 - 5y + 1$

19. $6z^2 - 7z + 3$

20. $9z^2 + 3z + 2$

21. $6t^2 - 11t + 4$

22. $10t^2 + 11t + 3$

23. $8x^2 + 33x + 4$

24. $7x^2 + 50x + 7$

25. $5x^2 - 62x - 7$

26. $9x^2 - 13x - 4$

27. $12y^2 + 19y + 5$

28. $5y^2 - 22y + 8$

29. $7a^2 + 47a - 14$

30. $11a^2 - 54a - 5$

31. $3b^2 - 16b + 16$

32. $6b^2 - 19b + 15$

33. $2z^2 - 27z - 14$

34. $4z^2 + 5z - 6$

35. $3p^2 + 22p - 16$

36. $7p^2 + 19p + 10$

Copyright © Houghton Mifflin Company. All rights reserved.

37. $4x^2 + 6x + 2$ **38.** $12x^2 + 33x - 9$ **39.** $15y^2 - 50y + 35$ **40.** $30y^2 + 10y - 20$

41. $2x^3 - 11x^2 + 5x$ **42.** $2x^3 - 3x^2 - 5x$ **43.** $3a^2b - 16ab + 16b$ **44.** $2a^2b - ab - 21b$

45. $3z^2 + 95z + 10$ **46.** $8z^2 - 36z + 1$ **47.** $36x - 3x^2 - 3x^3$ **48.** $-2x^3 + 2x^2 + 4x$

49. $80y^2 - 36y + 4$ **50.** $24y^2 - 24y - 18$ **51.** $8z^3 + 14z^2 + 3z$ **52.** $6z^3 - 23z^2 + 20z$

53. $6x^2y - 11xy - 10y$ **54.** $8x^2y - 27xy + 9y$ **55.** $10t^2 - 5t - 50$

56. $16t^2 + 40t - 96$ **57.** $3p^3 - 16p^2 + 5p$ **58.** $6p^3 + 5p^2 + p$

59. $26z^2 + 98z - 24$ **60.** $30z^2 - 87z + 30$ **61.** $10y^3 - 44y^2 + 16y$

62. $14y^3 + 94y^2 - 28y$ **63.** $4yz^3 + 5yz^2 - 6yz$ **64.** $12a^3 + 14a^2 - 48a$

65. $42a^3 + 45a^2 - 27a$ **66.** $36p^2 - 9p^3 - p^4$ **67.** $9x^2y - 30xy^2 + 25y^3$

68. $8x^2y - 38xy^2 + 35y^3$ **69.** $9x^3y - 24x^2y^2 + 16xy^3$ **70.** $9x^3y + 12x^2y + 4xy$

Copyright © Houghton Mifflin Company. All rights reserved.

Objective B   **To factor a trinomial of the form $ax^2 + bx + c$ by grouping**

For Exercises 71 to 130, factor by grouping.

71. $6x^2 - 17x + 12$ **72.** $15x^2 - 19x + 6$ **73.** $5b^2 + 33b - 14$ **74.** $8x^2 - 30x + 25$

75. $6a^2 + 7a - 24$ **76.** $14a^2 + 15a - 9$ **77.** $4z^2 + 11z + 6$ **78.** $6z^2 - 25z + 14$

79. $22p^2 + 51p - 10$ **80.** $14p^2 - 41p + 15$ **81.** $8y^2 + 17y + 9$ **82.** $12y^2 - 145y + 12$

83. $18t^2 - 9t - 5$ **84.** $12t^2 + 28t - 5$ **85.** $6b^2 + 71b - 12$ **86.** $8b^2 + 65b + 8$

87. $9x^2 + 12x + 4$ **88.** $25x^2 - 30x + 9$ **89.** $6b^2 - 13b + 6$ **90.** $20b^2 + 37b + 15$

91. $33b^2 + 34b - 35$ **92.** $15b^2 - 43b + 22$ **93.** $18y^2 - 39y + 20$ **94.** $24y^2 + 41y + 12$

95. $15a^2 + 26a - 21$ **96.** $6a^2 + 23a + 21$ **97.** $8y^2 - 26y + 15$ **98.** $18y^2 - 27y + 4$

99. $8z^2 + 2z - 15$ **100.** $10z^2 + 3z - 4$ **101.** $15x^2 - 82x + 24$ **102.** $13z^2 + 49z - 8$

103. $10z^2 - 29z + 10$ **104.** $15z^2 - 44z + 32$ **105.** $36z^2 + 72z + 35$ **106.** $16z^2 + 8z - 35$

107. $3x^2 + xy - 2y^2$ **108.** $6x^2 + 10xy + 4y^2$ **109.** $3a^2 + 5ab - 2b^2$ **110.** $2a^2 - 9ab + 9b^2$

Copyright © Houghton Mifflin Company. All rights reserved.

111. $4y^2 - 11yz + 6z^2$ **112.** $2y^2 + 7yz + 5z^2$ **113.** $28 + 3z - z^2$ **114.** $15 - 2z - z^2$

115. $8 - 7x - x^2$ **116.** $12 + 11x - x^2$ **117.** $9x^2 + 33x - 60$ **118.** $16x^2 - 16x - 12$

119. $24x^2 - 52x + 24$ **120.** $60x^2 + 95x + 20$ **121.** $35a^4 + 9a^3 - 2a^2$

122. $15a^4 + 26a^3 + 7a^2$ **123.** $15b^2 - 115b + 70$ **124.** $25b^2 + 35b - 30$

125. $3x^2 - 26xy + 35y^2$ **126.** $4x^2 + 16xy + 15y^2$ **127.** $216y^2 - 3y - 3$

128. $360y^2 + 4y - 4$ **129.** $21 - 20x - x^2$ **130.** $18 + 17x - x^2$

APPLYING THE CONCEPTS

131. In your own words, explain how the signs of the last terms of the two binomial factors of a trinomial are determined.

For Exercises 132 to 137, factor.

132. $(x + 1)^2 - (x + 1) - 6$ **133.** $(x - 2)^2 + 3(x - 2) + 2$ **134.** $(y + 3)^2 - 5(y + 3) + 6$

135. $2(y + 2)^2 - (y + 2) - 3$ **136.** $3(a + 2)^2 - (a + 2) - 4$ **137.** $4(y - 1)^2 - 7(y - 1) - 2$

For Exercises 138 to 143, find all integers k such that the trinomial can be factored over the integers.

138. $2x^2 + kx + 3$ **139.** $2x^2 + kx - 3$ **140.** $3x^2 + kx + 2$

141. $3x^2 + kx - 2$ **142.** $2x^2 + kx + 5$ **143.** $2x^2 + kx - 5$

Copyright © Houghton Mifflin Company. All rights reserved.

5.4 Special Factoring

Objective A **To factor the difference of two squares and perfect-square trinomials**

A polynomial in the form $a^2 - b^2$ is called a **difference of two squares.** Recall the following relationship from Objective 4.3D.

Sum and difference of the same terms		Difference of two squares
$(a + b)(a - b)$	$=$	$a^2 - b^2$

TAKE NOTE

Note that the polynomial $x^2 + y^2$ is the *sum* of two squares. The sum of two squares is nonfactorable over the integers.

Factoring the Difference of Two Squares

The difference of two squares factors as the sum and difference of the same terms.
$$a^2 - b^2 = (a + b)(a - b)$$

HOW TO   Factor: $x^2 - 16$

$x^2 - 16 = (x)^2 - (4)^2$ • $x^2 - 16$ is the difference of two squares.

$\qquad = (x + 4)(x - 4)$ • Factor the difference of squares.

Check: $(x + 4)(x - 4) = x^2 - 4x + 4x - 16$

$\qquad\qquad\qquad\qquad = x^2 - 16$

HOW TO   Factor: $8x^3 - 18x$

$8x^3 - 18x = 2x(4x^2 - 9)$ • The GCF is $2x$.

$\qquad = 2x[(2x)^2 - 3^2]$ • $4x^2 - 9$ is the difference of two squares.

$\qquad = 2x(2x + 3)(2x - 3)$ • Factor the difference of squares.

You should check the factorization.

HOW TO   Factor: $x^2 - 10$

Because 10 cannot be written as the square of an integer, $x^2 - 10$ is nonfactorable over the integers.

A trinomial that can be written as the square of a binomial is called a **perfect-square trinomial.** Recall the pattern for finding the square of a binomial.

$$(a + b)^2 = a^2 + 2ab + b^2$$

Square of the first term ——┘   └—— Square of the last term

Twice the product of the two terms

Copyright © Houghton Mifflin Company. All rights reserved.

> **Factoring a Perfect-Square Trinomial**
>
> A perfect-square trinomial factors as the square of a binomial.
>
> $$a^2 + 2ab + b^2 = (a + b)^2$$
> $$a^2 - 2ab + b^2 = (a - b)^2$$

HOW TO Factor: $4x^2 - 20x + 25$

Because the first and last terms are squares $[(2x)^2 = 4x^2; 5^2 = 25]$, try to factor this as the square of a binomial. Check the factorization.

$$4x^2 - 20x + 25 = (2x - 5)^2$$

Check: $(2x - 5)^2 = (2x)^2 + 2(2x)(-5) + 5^2$
$$= 4x^2 - 20x + 25$$

• **The factorization is correct.**

$$4x^2 - 20x + 25 = (2x - 5)^2$$

HOW TO Factor: $4x^2 + 37x + 9$

Because the first and last terms are squares $[(2x)^2 = 4x^2; 3^2 = 9]$, try to factor this as the square of a binomial. Check the proposed factorization.

$$4x^2 + 37x + 9 = (2x + 3)^2$$

Check: $(2x + 3)^2 = (2x)^2 + 2(2x)(3) + 3^2$
$$= 4x^2 + 12x + 9$$

Because $4x^2 + 12x + 9 \neq 4x^2 + 37x + 9$, the proposed factorization is not correct. In this case, the polynomial is not a perfect-square trinomial. It may, however, still factor. In fact, $4x^2 + 37x + 9 = (4x + 1)(x + 9)$.

Example 1

Factor: $16x^2 - y^2$

Solution
$16x^2 - y^2 = (4x)^2 - y^2$ • **The difference of two squares**

$\qquad = (4x + y)(4x - y)$ • **Factor.**

You Try It 1

Factor: $25a^2 - b^2$

Your solution

Example 2

Factor: $z^4 - 16$

Solution
$z^4 - 16 = (z^2)^2 - 4^2$ • **The difference of two squares**

$\qquad = (z^2 + 4)(z^2 - 4)$ • $z^2 - 4$ **is the difference**
$\qquad = (z^2 + 4)(z^2 - 2^2)$ **of two squares.**
$\qquad = (z^2 + 4)(z + 2)(z - 2)$ • **Factor.**

You Try It 2

Factor: $n^4 - 81$

Your solution

Solutions on p. S13

Copyright © Houghton Mifflin Company. All rights reserved.

Example 3

Factor: $9x^2 - 30x + 25$

Solution
$9x^2 = (3x)^2, \; 25 = (5)^2$
$9x^2 - 30x + 25 = (3x - 5)^2$

Check: $(3x - 5)^2 = (3x)^2 + 2(3x)(-5) + 5^2$
$= 9x^2 - 30x + 25$

You Try It 3

Factor: $16y^2 + 8y + 1$

Your solution

Example 4

Factor: $9x^2 + 40x + 16$

Solution
Because $9x^2 = (3x)^2$, $16 = 4^2$, and
$40x \neq 2(3x)(4)$, the trinomial is not
a perfect-square trinomial.

Try to factor by another method.

$9x^2 + 40x + 16 = (9x + 4)(x + 4)$

You Try It 4

Factor: $x^2 + 15x + 36$

Your solution

Solutions on p. S13

Objective B   **To factor completely**

Study Tip

You have now learned to factor many different types of polynomials. You will need to be able to recognize each of the situations described in the box at the right. To test yourself, you might do the exercises in the Chapter Review.

General Factoring Strategy

1. Is there a common factor? If so, factor out the common factor.
2. Is the polynomial the difference of two perfect squares? If so, factor.
3. Is the polynomial a perfect-square trinomial? If so, factor.
4. Is the polynomial a trinomial that is the product of two binomials? If so, factor.
5. Does the polynomial contain four terms? If so, try factoring by grouping.
6. Is each binomial factor nonfactorable over the integers? If not, factor the binomial.

HOW TO   Factor: $z^3 + 4z^2 - 9z - 36$

$z^3 + 4z^2 - 9z - 36 = (z^3 + 4z^2) - (9z + 36)$

$= z^2(z + 4) - 9(z + 4)$

$= (z + 4)(z^2 - 9)$

$= (z + 4)(z + 3)(z - 3)$

- Factor by grouping. Recall that $-9z - 36 = -(9z + 36)$.
- $z^3 + 4z^2 = z^2(z + 4)$
  $9z + 36 = 9(z + 4)$
- Factor out the common binomial factor $(z + 4)$.
- Factor the difference of squares.

Copyright © Houghton Mifflin Company. All rights reserved.

Example 5

Factor: $3x^2 - 48$

Solution

The GCF is 3.

$3x^2 - 48 = 3(x^2 - 16)$
$\qquad\qquad = 3(x + 4)(x - 4)$ • Factor the difference of two squares.

You Try It 5

Factor: $12x^3 - 75x$

Your solution

Example 6

Factor: $x^3 - 3x^2 - 4x + 12$

Solution

Factor by grouping.

$x^3 - 3x^2 - 4x + 12$
$\quad = (x^3 - 3x^2) - (4x - 12)$ • Factor by grouping.
$\quad = x^2(x - 3) - 4(x - 3)$ • **x − 3** is the common factor.
$\quad = (x - 3)(x^2 - 4)$ • $x^2 - 4$ is the difference of two squares.
$\quad = (x - 3)(x + 2)(x - 2)$ • Factor.

You Try It 6

Factor: $a^2b - 7a^2 - b + 7$

Your solution

Example 7

Factor: $4x^2y^2 + 12xy^2 + 9y^2$

Solution

The GCF is y^2.

$4x^2y^2 + 12xy^2 + 9y^2$
$\quad = y^2(4x^2 + 12x + 9)$ • Factor the GCF, y^2.
$\quad = y^2(2x + 3)^2$ • Factor the perfect-square trinomial.

You Try It 7

Factor: $4x^3 + 28x^2 - 120x$

Your solution

Solutions on p. S13

Copyright © Houghton Mifflin Company. All rights reserved.

5.4 Exercises

Objective A   **To factor the difference of two squares and perfect-square trinomials**

1. **a.** Provide an example of a binomial that is the difference of two squares.

    b. Provide an example of a perfect-square trinomial.

2. Explain why a binomial that is the sum of two squares is nonfactorable over the integers.

For Exercises 3 to 48, factor.

3. $x^2 - 4$

4. $x^2 - 9$

5. $a^2 - 81$

6. $a^2 - 49$

7. $y^2 + 2y + 1$

8. $y^2 + 14y + 49$

9. $a^2 - 2a + 1$

10. $x^2 - 12x + 36$

11. $4x^2 - 1$

12. $9x^2 - 16$

13. $x^6 - 9$

14. $y^{12} - 4$

15. $x^2 + 8x - 16$

16. $z^2 - 18z - 81$

17. $x^2 + 2xy + y^2$

18. $x^2 + 6xy + 9y^2$

19. $4a^2 + 4a + 1$

20. $25x^2 + 10x + 1$

21. $9x^2 - 1$

22. $1 - 49x^2$

23. $1 - 64x^2$

24. $t^2 + 36$

25. $x^2 + 64$

26. $64a^2 - 16a + 1$

27. $9a^2 + 6a + 1$

28. $x^4 - y^2$

29. $b^4 - 16a^2$

30. $16b^2 + 8b + 1$

31. $4a^2 - 20a + 25$

32. $4b^2 + 28b + 49$

33. $9a^2 - 42a + 49$

34. $9x^2 - 16y^2$

35. $25z^2 - y^2$

36. $x^2y^2 - 4$

37. $a^2b^2 - 25$

38. $16 - x^2y^2$

Copyright © Houghton Mifflin Company. All rights reserved.

39. $25x^2 - 1$

40. $25a^2 + 30ab + 9b^2$

41. $4a^2 - 12ab + 9b^2$

42. $49x^2 + 28xy + 4y^2$

43. $4y^2 - 36yz + 81z^2$

44. $64y^2 - 48yz + 9z^2$

45. $\dfrac{1}{x^2} - 4$

46. $\dfrac{9}{a^2} - 16$

47. $9a^2b^2 - 6ab + 1$

48. $16x^2y^2 - 24xy + 9$

> **Objective B**   **To factor completely**

For Exercises 49 to 126, factor.

49. $8y^2 - 2$

50. $12n^2 - 48$

51. $3a^3 + 6a^2 + 3a$

52. $4rs^2 - 4rs + r$

53. $m^4 - 256$

54. $81 - t^4$

55. $9x^2 + 13x + 4$

56. $x^2 + 10x + 16$

57. $16y^4 + 48y^3 + 36y^2$

58. $36c^4 - 48c^3 + 16c^2$

59. $y^8 - 81$

60. $32s^4 - 2$

61. $25 - 20p + 4p^2$

62. $9 + 24a + 16a^2$

63. $(4x - 3)^2 - y^2$

64. $(2x + 5)^2 - 25$

65. $(x^2 - 4x + 4) - y^2$

66. $(4x^2 + 12x + 9) - 4y^2$

Copyright © Houghton Mifflin Company. All rights reserved.

67. $5x^2 - 5$

68. $2x^2 - 18$

69. $x^3 + 4x^2 + 4x$

70. $y^3 - 10y^2 + 25y$

71. $x^4 + 2x^3 - 35x^2$

72. $a^4 - 11a^3 + 24a^2$

73. $5b^2 + 75b + 180$

74. $6y^2 - 48y + 72$

75. $3a^2 + 36a + 10$

76. $5a^2 - 30a + 4$

77. $2x^2y + 16xy - 66y$

78. $3a^2b + 21ab - 54b$

79. $x^3 - 6x^2 - 5x$

80. $b^3 - 8b^2 - 7b$

81. $3y^2 - 36$

82. $3y^2 - 147$

83. $20a^2 + 12a + 1$

84. $12a^2 - 36a + 27$

85. $x^2y^2 - 7xy^2 - 8y^2$

86. $a^2b^2 + 3a^2b - 88a^2$

87. $10a^2 - 5ab - 15b^2$

88. $16x^2 - 32xy + 12y^2$

89. $50 - 2x^2$

90. $72 - 2x^2$

91. $a^2b^2 - 10ab^2 + 25b^2$

92. $a^2b^2 + 6ab^2 + 9b^2$

93. $12a^3b - a^2b^2 - ab^3$

94. $2x^3y - 7x^2y^2 + 6xy^3$

95. $12a^3 - 12a^2 + 3a$

96. $18a^3 + 24a^2 + 8a$

97. $243 + 3a^2$

98. $75 + 27y^2$

99. $12a^3 - 46a^2 + 40a$

100. $24x^3 - 66x^2 + 15x$

101. $4a^3 + 20a^2 + 25a$

102. $2a^3 - 8a^2b + 8ab^2$

Copyright © Houghton Mifflin Company. All rights reserved.

103. $27a^2b - 18ab + 3b$ **104.** $a^2b^2 - 6ab^2 + 9b^2$ **105.** $48 - 12x - 6x^2$

106. $21x^2 - 11x^3 - 2x^4$ **107.** $x^4 - x^2y^2$ **108.** $b^4 - a^2b^2$

109. $18a^3 + 24a^2 + 8a$ **110.** $32xy^2 - 48xy + 18x$ **111.** $2b + ab - 6a^2b$

112. $15y^2 - 2xy^2 - x^2y^2$ **113.** $4x^4 - 38x^3 + 48x^2$ **114.** $3x^2 - 27y^2$

115. $x^4 - 25x^2$ **116.** $y^3 - 9y$ **117.** $a^4 - 16$

118. $15x^4y^2 - 13x^3y^3 - 20x^2y^4$ **119.** $45y^2 - 42y^3 - 24y^4$ **120.** $a(2x - 2) + b(2x - 2)$

121. $4a(x - 3) - 2b(x - 3)$ **122.** $x^2(x - 2) - (x - 2)$ **123.** $y^2(a - b) - (a - b)$

124. $a(x^2 - 4) + b(x^2 - 4)$ **125.** $x(a^2 - b^2) - y(a^2 - b^2)$ **126.** $4(x - 5) - x^2(x - 5)$

APPLYING THE CONCEPTS

For Exercises 127 to 132, find all integers k such that the trinomial is a perfect-square trinomial.

127. $4x^2 - kx + 9$ **128.** $x^2 + 6x + k$ **129.** $64x^2 + kxy + y^2$

130. $x^2 - 2x + k$ **131.** $25x^2 - kx + 1$ **132.** $x^2 + 10x + k$

133.   **Number Sense** Select any odd integer greater than 1, square it, and then subtract 1. Is the result evenly divisible by 8? Prove that this procedure always produces a number divisible by 8. (*Suggestion:* Any odd integer greater than 1 can be expressed as $2n + 1$, where n is a natural number.)

Copyright © Houghton Mifflin Company. All rights reserved.

5.5 Solving Equations

Objective A **To solve equations by factoring**

The Multiplication Property of Zero states that the product of a number and zero is zero. This property is stated below.

$$\text{If } a \text{ is a real number, then } a \cdot 0 = 0 \cdot a = 0.$$

Now consider $x \cdot y = 0$. For this to be a true equation, then either $x = 0$ or $y = 0$.

Principle of Zero Products

If the product of two factors is zero, then at least one of the factors must be zero.

If $a \cdot b = 0$, then $a = 0$ or $b = 0$.

The Principle of Zero Products is used to solve some equations.

HOW TO   Solve: $(x - 2)(x - 3) = 0$

$(x - 2)(x - 3) = 0$

$x - 2 = 0 \quad x - 3 = 0$ • Let each factor equal zero (the Principle of Zero Products).

$\qquad x = 2 \qquad\quad x = 3$ • Solve each equation for x.

Check:

$(x - 2)(x - 3) = 0$	
$(2 - 2)(2 - 3)$	0
$0(-1)$	0
$0 = 0$	• A true equation

$(x - 2)(x - 3) = 0$	
$(3 - 2)(3 - 3)$	0
$(1)(0)$	0
$0 = 0$	• A true equation

The solutions are 2 and 3.

An equation that can be written in the form $ax^2 + bx + c = 0, a \neq 0$, is a **quadratic equation.** A quadratic equation is in **standard form** when the polynomial is in descending order and equal to zero. The quadratic equations at the right are in standard form.

$3x^2 + 2x + 1 = 0$
$a = 3, b = 2, c = 1$

$4x^2 - 3x + 2 = 0$
$a = 4, b = -3, c = 2$

Copyright © Houghton Mifflin Company. All rights reserved.

HOW TO   Solve: $2x^2 + x = 6$

$$2x^2 + x = 6$$

$$2x^2 + x - 6 = 0$$ • **Write the equation in standard form.**

$$(2x - 3)(x + 2) = 0$$ • **Factor.**

$$2x - 3 = 0 \qquad x + 2 = 0$$ • **Use the Principle of Zero Products.**

$$2x = 3 \qquad\qquad x = -2$$ • **Solve each equation for x.**

$$x = \frac{3}{2}$$

Check: $\frac{3}{2}$ and -2 check as solutions.

The solutions are $\frac{3}{2}$ and -2.

Example 1

Solve: $x(x - 3) = 0$

Solution
$x(x - 3) = 0$

$x = 0 \qquad\qquad x - 3 = 0$ • **Use the Principle**
$\qquad\qquad\qquad\quad x = 3$ **of Zero Products.**

The solutions are 0 and 3.

You Try It 1

Solve: $2x(x + 7) = 0$

Your solution

Example 2

Solve: $2x^2 - 50 = 0$

Solution
$\qquad 2x^2 - 50 = 0$
$\qquad 2(x^2 - 25) = 0$ • **Factor the GCF, 2.**
$2(x + 5)(x - 5) = 0$ • **Factor the difference of two squares.**

$x + 5 = 0 \qquad x - 5 = 0$ • **Use the Principle of**
$\quad x = -5 \qquad\quad x = 5$ **Zero Products.**

The solutions are -5 and 5.

You Try It 2

Solve: $4x^2 - 9 = 0$

Your solution

Example 3

Solve: $(x - 3)(x - 10) = -10$

Solution
$(x - 3)(x - 10) = -10$
$\quad x^2 - 13x + 30 = -10$ • **Multiply $(x - 3)(x - 10)$.**
$\quad x^2 - 13x + 40 = 0$ • **Add 10 to each side of**
$\quad (x - 8)(x - 5) = 0$ **the equation. The equation**
is now in standard form.

$x - 8 = 0 \qquad x - 5 = 0$
$\quad x = 8 \qquad\quad x = 5$

The solutions are 8 and 5.

You Try It 3

Solve: $(x + 2)(x - 7) = 52$

Your solution

Solutions on pp. S13–S14

Copyright © Houghton Mifflin Company. All rights reserved.

Objective B   To solve application problems

Example 4	You Try It 4
The sum of the squares of two consecutive positive even integers is equal to 100. Find the two integers.	The sum of the squares of two consecutive positive integers is 61. Find the two integers.

Strategy

First positive even integer: n
Second positive even integer: $n + 2$

The sum of the square of the first positive even integer and the square of the second positive even integer is 100.

Your strategy

Solution

$$n^2 + (n + 2)^2 = 100$$
$$n^2 + n^2 + 4n + 4 = 100$$
$$2n^2 + 4n + 4 = 100$$
$$2n^2 + 4n - 96 = 0$$
$$2(n^2 + 2n - 48) = 0$$
$$2(n - 6)(n + 8) = 0$$

$$n - 6 = 0 \qquad n + 8 = 0 \qquad \bullet \text{ Principle of}$$
$$n = 6 \qquad\qquad n = -8 \qquad \text{Zero Products.}$$

Because -8 is not a positive even integer, it is not a solution.

$$n = 6$$
$$n + 2 = 6 + 2 = 8$$

The two integers are 6 and 8.

Your solution

Copyright © Houghton Mifflin Company. All rights reserved.

Solution on p. S14

Example 5

A stone is thrown into a well with an initial speed of 4 ft/s. The well is 420 ft deep. How many seconds later will the stone hit the bottom of the well? Use the equation $d = vt + 16t^2$, where d is the distance in feet that the stone travels in t seconds when its initial speed is v feet per second.

Strategy

To find the time for the stone to drop to the bottom of the well, replace the variables d and v by their given values and solve for t.

Solution

$$d = vt + 16t^2$$
$$420 = 4t + 16t^2$$
$$0 = -420 + 4t + 16t^2$$
$$0 = 16t^2 + 4t - 420$$
$$0 = 4(4t^2 + t - 105)$$
$$0 = 4(4t + 21)(t - 5)$$

$$4t + 21 = 0 \qquad t - 5 = 0 \quad \bullet \text{ Principle of}$$
$$4t = -21 \qquad t = 5 \qquad \text{Zero Products.}$$
$$t = -\frac{21}{4}$$

Because the time cannot be a negative number, $-\frac{21}{4}$ is not a solution.

The stone will hit the bottom of the well 5 s later.

You Try It 5

The length of a rectangle is 4 in. longer than twice the width. The area of the rectangle is 96 in². Find the length and width of the rectangle.

Your strategy

Your solution

Solution on p. S14

Copyright © Houghton Mifflin Company. All rights reserved.

5.5 Exercises

Objective A **To solve equations by factoring**

1. ✏️ In your own words, explain the Principle of Zero Products.

2. Fill in the blanks. If $(x + 5)(2x - 7) = 0$, then _____ $= 0$ or
   _____ $= 0$.

For Exercises 3 to 60, solve.

3. $(y + 3)(y + 2) = 0$ 4. $(y - 3)(y - 5) = 0$ 5. $(z - 7)(z - 3) = 0$ 6. $(z + 8)(z - 9) = 0$

7. $x(x - 5) = 0$ 8. $x(x + 2) = 0$ 9. $a(a - 9) = 0$ 10. $a(a + 12) = 0$

11. $y(2y + 3) = 0$ 12. $t(4t - 7) = 0$ 13. $2a(3a - 2) = 0$ 14. $4b(2b + 5) = 0$

15. $(b + 2)(b - 5) = 0$ 16. $(b - 8)(b + 3) = 0$ 17. $x^2 - 81 = 0$ 18. $x^2 - 121 = 0$

19. $4x^2 - 49 = 0$ 20. $16x^2 - 1 = 0$ 21. $9x^2 - 1 = 0$ 22. $16x^2 - 49 = 0$

23. $x^2 + 6x + 8 = 0$ 24. $x^2 - 8x + 15 = 0$ 25. $z^2 + 5z - 14 = 0$ 26. $z^2 + z - 72 = 0$

27. $2a^2 - 9a - 5 = 0$ 28. $3a^2 + 14a + 8 = 0$ 29. $6z^2 + 5z + 1 = 0$ 30. $6y^2 - 19y + 15 = 0$

31. $x^2 - 3x = 0$ 32. $a^2 - 5a = 0$ 33. $x^2 - 7x = 0$ 34. $2a^2 - 8a = 0$

35. $a^2 + 5a = -4$ 36. $a^2 - 5a = 24$ 37. $y^2 - 5y = -6$ 38. $y^2 - 7y = 8$

39. $2t^2 + 7t = 4$ 40. $3t^2 + t = 10$ 41. $3t^2 - 13t = -4$ 42. $5t^2 - 16t = -12$

43. $x(x - 12) = -27$ 44. $x(x - 11) = 12$ 45. $y(y - 7) = 18$ 46. $y(y + 8) = -15$

Copyright © Houghton Mifflin Company. All rights reserved.

47. $p(p + 3) = -2$ **48.** $p(p - 1) = 20$ **49.** $y(y + 4) = 45$ **50.** $y(y - 8) = -15$

51. $x(x + 3) = 28$ **52.** $p(p - 14) = 15$ **53.** $(x + 8)(x - 3) = -30$ **54.** $(x + 4)(x - 1) = 14$

55. $(z - 5)(z + 4) = 52$ **56.** $(z - 8)(z + 4) = -35$ **57.** $(z - 6)(z + 1) = -10$

58. $(a + 3)(a + 4) = 72$ **59.** $(a - 4)(a + 7) = -18$ **60.** $(2x + 5)(x + 1) = -1$

Objective B   **To solve application problems**

61. Number Sense   The square of a positive number is six more than five times the positive number. Find the number.

62. Number Sense   The square of a negative number is fifteen more than twice the negative number. Find the number.

63. Number Sense   The sum of two numbers is six. The sum of the squares of the two numbers is twenty. Find the two numbers.

64. Number Sense   The sum of two numbers is eight. The sum of the squares of the two numbers is thirty-four. Find the two numbers.

65. Number Sense   The sum of the squares of two consecutive positive integers is forty-one. Find the two integers.

66. Number Sense   The sum of the squares of two consecutive positive even integers is one hundred. Find the two integers.

67. Number Sense   The sum of two numbers is ten. The product of the two numbers is twenty-one. Find the two numbers.

68. Number Sense   The sum of two numbers is thirteen. The product of the two numbers is forty. Find the two numbers.

Copyright © Houghton Mifflin Company. All rights reserved.

Sum of Natural Numbers   The formula $S = \frac{n^2 + n}{2}$ gives the sum, S, of the first n natural numbers. Use this formula for Exercises 69 and 70.

69. How many consecutive natural numbers beginning with 1 will give a sum of 78?

70. How many consecutive natural numbers beginning with 1 will give a sum of 171?

Sports   The formula $N = \frac{t^2 - t}{2}$ gives the number, N, of football games that must be scheduled in a league with t teams if each team is to play every other team once. Use this formula for Exercises 71 and 72.

71. How many teams are in a league that schedules 15 games in such a way that each team plays every other team once?

72. How many teams are in a league that schedules 45 games in such a way that each team plays every other team once?

Physics   The distance, s, in feet, that an object will fall (neglecting air resistance) in t seconds is given by $s = vt + 16t^2$, where v is the initial velocity of the object in feet per second. Use this formula for Exercises 73 and 74.

73. An object is released from the top of a building 192 ft high. The initial velocity is 16 ft/s, and air resistance is neglected. How many seconds later will the object hit the ground?

74. In October 2003, the world's tallest building, Taipei 101, was completed. The top of the spire is 1667 ft above ground. If an object is released from this building at a point 640 ft above the ground at an initial velocity of 48 ft/s, assuming no air resistance, how many seconds later will the object reach the ground?

Sports   The height, h, in feet, an object will attain (neglecting air resistance) in t seconds is given by $h = vt - 16t^2$, where v is the initial velocity of the object in feet per second. Use this formula for Exercises 75 and 76.

75. A golf ball is thrown onto a cement surface and rebounds straight up. The initial velocity of the rebound is 60 ft/s. How many seconds later will the golf ball return to the ground?

76. A foul ball leaves a bat, hits home plate, and travels straight up with an initial velocity of 64 ft/s. How many seconds later will the ball be 64 ft above the ground?

77. Geometry   The length of a rectangle is 5 in. more than twice its width. Its area is 75 in². Find the length and width of the rectangle.

Copyright © Houghton Mifflin Company. All rights reserved.

78. Geometry The width of a rectangle is 5 ft less than the length. The area of the rectangle is 176 ft². Find the length and width of the rectangle.

79. Geometry The height of a triangle is 4 m more than twice the length of the base. The area of the triangle is 35 m². Find the height of the triangle.

80. Geometry The length of each side of a square is extended 5 in. The area of the resulting square is 64 in². Find the length of a side of the original square.

81. Publishing The page of a book measures 6 in. by 9 in. A uniform border around the page leaves 28 in² for type. What are the dimensions of the type area?

82. Gardening A small garden measures 8 ft by 10 ft. A uniform border around the garden increases the total area to 143 ft². What is the width of the border?

83. Landscaping A landscape designer decides to increase the radius of a circular lawn by 3 ft. This increases the area of the lawn by 100 ft². Find the radius of the original circular lawn. Round to the nearest hundredth.

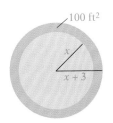

84. Geometry A circle has a radius of 10 in. Find the increase in area that occurs when the radius is increased by 2 in. Round to the nearest hundredth.

APPLYING THE CONCEPTS

85. Find $3n^2$ if $n(n + 5) = -4$.

86. Find $2n^2$ if $n(n + 3) = 4$.

For Exercises 87 to 90, solve.

87. $2y(y + 4) = -5(y + 3)$

88. $(b + 5)^2 = 16$

89. $p^3 = 9p^2$

90. $(x + 3)(2x - 1) = (3 - x)(5 - 3x)$

91. Explain the error made in solving the equation at the right. Solve the equation correctly.

$$(x + 2)(x - 3) = 6$$
$$x + 2 = 6 \quad x - 3 = 6$$
$$x = 4 \quad\quad x = 9$$

92. Explain the error made in solving the equation at the right. Solve the equation correctly.

$$x^2 = x$$
$$\frac{x^2}{x} = \frac{x}{x}$$
$$x = 1$$

Copyright © Houghton Mifflin Company. All rights reserved.

Focus on Problem Solving

Making a Table

There are six students using a gym. The wall on the gym has six lockers that are numbered 1, 2, 3, 4, 5, and 6. After a practice, the first student goes by and opens all the lockers. The second student shuts every second locker, the third student changes every third locker (opens a locker if it is shut, shuts a locker if it is open), the fourth student changes every fourth locker, the fifth student changes every fifth locker, and the sixth student changes every sixth locker. After the sixth student makes changes, which lockers are open?

One method of solving this problem would be to create a table as shown below.

Locker \ Student	1	2	3	4	5	6
1	O	O	O	O	O	O
2	O	C	C	C	C	C
3	O	O	C	C	C	C
4	O	C	C	O	O	O
5	O	O	O	O	C	C
6	O	C	O	O	O	C

From this table, lockers 1 and 4 are open after the sixth student passes through.

Now extend this to more lockers and students. In each case, the nth student changes multiples of the nth locker. For instance, the 8th student would change the 8th, 16th, 24th, . . .

1. Suppose there were 10 lockers and 10 students. Which lockers would remain open?

2. Suppose there were 16 lockers and 16 students. Which lockers would remain open?

3. Suppose there are 25 lockers and 25 students. Which lockers would remain open?

4. Suppose there are 40 lockers and 40 students. Which lockers would remain open?

5. Suppose there are 50 lockers and 50 students. Which lockers would remain open?

6. Make a conjecture as to which lockers would be open if there were 100 lockers and 100 students.

7. Give a reason why your conjecture should be true. [*Hint:* Consider how many factors there are for the door numbers that remain open and those that remain closed. For instance, with 40 lockers and 40 students, locker 36 (which remains open) has factors 1, 2, 3, 4, 6, 9, 12, 18, 36—an odd number of factors. Locker 28, a closed locker, has factors 1, 2, 4, 7, 14, 28—an even number of factors.]

Copyright © Houghton Mifflin Company. All rights reserved.

Projects and Group Activities

Evaluating Polynomials Using a Graphing Calculator

A graphing calculator can be used to evaluate a polynomial. To illustrate the method, consider the polynomial $2x^3 - 3x^2 + 4x - 7$. The keystrokes below are for a TI-83 Plus calculator, but the keystrokes for other calculators will closely follow these keystrokes.

Press the **Y=** key. You will see a screen similar to the one below. Press **CLEAR** to erase any expression next to Y1.

Enter the polynomial as follows. The **^** key is used to enter an exponent.

2 **X,T,θ,n** **^** 3 – 3 **X,T,θ,n** **^** 2 + 4 **X,T,θ,n** – 7

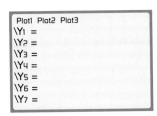

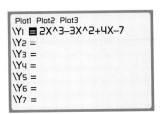

TAKE NOTE
Once the polynomial has been entered in Y1, there are several methods to evaluate it. We will show just one option.

To evaluate the polynomial when $x = 3$, first return to what is called the HOME screen by pressing **2nd** QUIT.

Enter the following keystrokes. Sample screens are shown at the right.

(1) 3 **STO→** **X,T,θ,n** **ENTER**

(2) **VARS** ▶ **ENTER**

(3) **ENTER**

(4) **ENTER**

The value of the polynomial when $x = 3$ is 32.

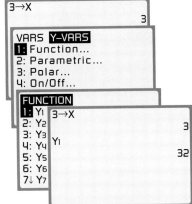

To evaluate the polynomial at a different value of x, repeat Steps 1 through 4. For instance, to evaluate the polynomial when $x = -4$, we would have

(1) **(-)** 4 **STO→** **X,T,θ,n** **ENTER**

(2) **VARS** ▶ **ENTER**

(3) **ENTER**

(4) **ENTER**

The value of the polynomial when $x = -4$ is -199.

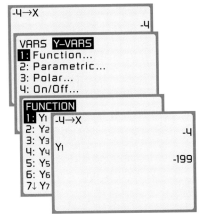

Here are some practice exercises.
Evaluate for the given value.

1. $2x^2 - 3x + 7; x = 4$

2. $3x^2 + 7x - 12; x = -3$

3. $3x^3 - 2x^2 + 6x - 8; x = 3$

4. $2x^3 + 4x^2 - x - 2; x = 2$

5. $x^4 - 3x^3 + 6x^2 + 5x - 1; x = 2$

6. $x^5 - x^3 + 2x - 7; x = -4$

Copyright © Houghton Mifflin Company. All rights reserved.

Exploring Integers   *Number theory* is a branch of mathematics that focuses on integers and the relationships that exist among the integers. Some of the results from this field of study have important, practical applications for sending sensitive information such as credit card numbers over the Internet. In this project, you will be asked to discover some of those relationships.

1. If n is an integer, explain why the product $n(n + 1)$ is always an even number.

2. If n is an integer, explain why $2n$ is always an even integer.

3. If n is an integer, explain why $2n + 1$ is always an odd integer.

4. Select any odd integer greater than 1, square it, and then subtract 1. Try this for various odd integers greater than 1. Is the result always evenly divisible by 8?

5. Prove the assertion in Exercise 4. [*Suggestion:* From Exercise 2, an odd integer can be represented as $2n + 1$. Therefore, the assertion in Exercise 4 can be stated "$(2n + 1)^2 - 1$ is evenly divisible by 8." Expand this expression and explain why the result must be divisible by 8. You will need to use the result from Exercise 1.

6. The integers 2 and 3 are consecutive prime numbers. Are there any other consecutive prime numbers? Why?

7. If n is a positive integer, for what values of n is $n^2 - 1$ a prime number?

8. A Mersenne prime number is a prime that can be written in the form $2^n - 1$, where n is also a prime number. For instance, $2^5 - 1 = 32 - 1 = 31$. Because 5 and 31 are prime numbers, 31 is a Mersenne prime number. On the other hand, $2^{11} - 1 = 2048 - 1 = 2047$. In this case, although 11 is a prime number, $2047 = 23 \cdot 89$, which is not a prime number. Find two Mersenne prime numbers other than 31.

Using the World Wide Web

At the address **http://www.utm.edu/research/primes/mersenne/,** you can find more information on Mersenne prime numbers. By searching other web sites, you can also find information about various topics in math.

The web site **http://mathforum.org/dr.math/** is an especially rich source of information. You can even submit math questions to this site and get an answer from Dr. Math. One student posed the question, "What is the purpose of the number zero?" Here is the reply from Dr. Math.

The invention of zero was one of the most important breakthroughs in the history of civilization. More important, in my opinion, than the invention of the wheel. I think that it's a fairly deep concept.

One crucial purpose that zero holds is as a placeholder in our system of notation. When we write the number 408, we're really using a shorthand notation. What we really mean by 408 is "4 times 100, plus 0 times 10, plus 8 times 1." Without the number zero, we wouldn't be able to tell the numbers 408, 48, 480, 408,000, and 4800 apart. So yes, zero is important.

Another crucial role that zero plays in mathematics is that of an "additive identity element." What this means is that when you add zero to any number, you get the number that you started with. For instance, $5 + 0 = 5$. That may seem obvious and trivial, but

Copyright © Houghton Mifflin Company. All rights reserved.

it's actually quite important to have such a number. For instance, when you're manipulating some numerical quantity and you want to change its form but not its value, you
might add some fancy version of zero to it, like this:

$$
\begin{aligned}
x^2 + y^2 &= x^2 + y^2 + 2xy - 2xy \\
&= x^2 + 2xy + y^2 - 2xy \\
&= (x + y)^2 - 2xy
\end{aligned}
$$

Now if we wanted to, we could use this as a proof that $(x + y)^2$ is always greater than
$2xy$; the expression we started with was positive, so the one we ended up with must be
positive, too. Therefore, subtracting $2xy$ from $(x + y)^2$ must leave us with a positive
number. Neat stuff.*

*Ask Dr. Math. Copyright © 1994–1997 *The Math Forum.* Used by permission.

Chapter 5 Summary

Key Words	Examples
The *greatest common factor (GCF) of two or more monomials* is the product of the GCF of the coefficients and the common variable factors. [5.1A, p. 237]	The GCF of $8x^2y$ and $12xyz$ is $4xy$.
To *factor a polynomial* means to write the polynomial as a product of other polynomials. [5.1A, p. 237]	To factor $x^2 + 3x + 2$ means to write it as the product $(x + 1)(x + 2)$.
A factor that has two terms is called a *binomial factor.* [5.1B, p. 239]	$(x + 1)$ is a binomial factor of $3x(x + 1)$.
A polynomial that does not factor using only integers is *nonfactorable over the integers.* [5.2A, p. 244]	The trinomial $x^2 + x + 4$ is nonfactorable over the integers. There are no integers whose product is 4 and whose sum is 1.
A polynomial is *factored completely* if it is written as a product of factors that are nonfactorable over the integers. [5.2B, p. 245]	The polynomial $3y^3 + 9y^2 - 12y$ is factored completely as $3y(y + 4)(y - 1)$.

Copyright © Houghton Mifflin Company. All rights reserved.

An equation that can be written in the form $ax^2 + bx + c = 0$, $a \neq 0$, is a *quadratic equation*. A quadratic equation is in *standard form* when the polynomial is written in descending order and equal to zero. [5.5A, p. 267]

The equation $2x^2 - 3x + 7 = 0$ is a quadratic equation in standard form.

Essential Rules and Procedures

Examples

Factoring by Grouping [5.1B, p. 239]
A polynomial can be factored by grouping if its terms can be grouped and factored in such a way that a common binomial factor is found.

$3a^2 - a - 15ab + 5b$
$= (3a^2 - a) - (15ab - 5b)$
$= a(3a - 1) - 5b(3a - 1)$
$= (3a - 1)(a - 5b)$

Factoring $x^2 + bx + c$: IMPORTANT RELATIONSHIPS
[5.2A, p. 243]

1. When the constant term of the trinomial is positive, the constant terms of the binomials have the same sign. They are both positive when the coefficient of the x term in the trinomial is positive. They are both negative when the coefficient of the x term in the trinomial is negative.

$x^2 + 6x + 8 = (x + 4)(x + 2)$

$x^2 - 6x + 5 = (x - 5)(x - 1)$

2. When the constant term of the trinomial is negative, the constant terms of the binomials have opposite signs.

$x^2 - 4x - 21 = (x + 3)(x - 7)$

3. In the trinomial, the coefficient of x is the sum of the constant terms of the binomials.

In the three examples above, note that $6 = 4 + 2$, $-6 = -5 + (-1)$, and $-4 = 3 + (-7)$.

4. In the trinomial, the constant term is the product of the constant terms of the binomials.

In the three examples above, note that $8 = 4 \cdot 2$, $5 = -5(-1)$, and $-21 = 3(-7)$.

To factor $ax^2 + bx + c$ by grouping [5.3B, p. 253]

First find two factors of $a \cdot c$ whose sum is b. Then use factoring by grouping to write the factorization of the trinomial.

$3x^2 - 11x - 20$

$a \cdot c = 3(-20) = -60$

The product of 4 and -15 is -60.

The sum of 4 and -15 is -11.

$3x^2 + 4x - 15x - 20$
$= (3x^2 + 4x) - (15x + 20)$
$= x(3x + 4) - 5(3x + 4)$
$= (3x + 4)(x - 5)$

Copyright © Houghton Mifflin Company. All rights reserved.

Factoring the Difference of Two Squares   [5.4A, p. 259]

The difference of two squares factors as the sum and difference of the same terms.

$a^2 - b^2 = (a + b)(a - b)$

$$x^2 - 64 = (x + 8)(x - 8)$$
$$4x^2 - 81 = (2x)^2 - 9^2$$
$$= (2x + 9)(2x - 9)$$

Factoring a Perfect-Square Trinomial   [5.4A, p. 260]

A perfect-square trinomial is the square of a binomial.

$a^2 + 2ab + b^2 = (a + b)^2$
$a^2 - 2ab + b^2 = (a - b)^2$

$$x^2 + 14x + 49 = (x + 7)^2$$
$$x^2 - 10x + 25 = (x - 5)^2$$

General Factoring Strategy [5.4B, p. 261]

1. Is there a common factor? Is so, factor out the common factor.

   $6x^2 - 8x = 2x(3x - 4)$

2. Is the polynomial the difference of two perfect squares? If so, factor.

   $9x^2 - 25 = (3x + 5)(3x - 5)$

3. Is the polynomial a perfect-square trinomial? If so, factor.

   $9x^2 + 6x + 1 = (3x + 1)^2$

4. Is the polynomial a trinomial that is the product of two binomials? If so, factor.

   $6x^2 + 5x - 6 = (3x - 2)(2x + 3)$

5. Does the polynomial contain four terms? If so, try factoring by grouping.

   $$x^3 - 3x^2 + 2x - 6$$
   $$= (x^3 - 3x^2) + (2x - 6)$$
   $$= x^2(x - 3) + 2(x - 3)$$
   $$= (x - 3)(x^2 + 2)$$

6. Is each binomial factor nonfactorable over the integers? If not, factor the binomial.

   $$x^4 - 16 = (x^2 + 4)(x^2 - 4)$$
   $$= (x^2 + 4)(x + 2)(x - 2)$$

Principle of Zero Products   [5.5A, p. 267]

If the product of two factors is zero, then at least one of the factors must be zero.

If $a \cdot b = 0$, then $a = 0$ or $b = 0$.

The Principle of Zero Products is used to solve a quadratic equation by factoring.

$$x^2 + x = 12$$
$$x^2 + x - 12 = 0$$
$$(x - 3)(x + 4) = 0$$

$$x - 3 = 0 \qquad x + 4 = 0$$
$$x = 3 \qquad\qquad x = -4$$

Copyright © Houghton Mifflin Company. All rights reserved.

Chapter 5 Review Exercises

1. Factor: $b^2 - 13b + 30$

2. Factor: $4x(x - 3) - 5(3 - x)$

3. Factor $2x^2 - 5x + 6$ by using trial factors.

4. Factor: $5x^3 + 10x^2 + 35x$

5. Factor: $14y^9 - 49y^6 + 7y^3$

6. Factor: $y^2 + 5y - 36$

7. Factor $6x^2 - 29x + 28$ by using trial factors.

8. Factor: $12a^2b + 3ab^2$

9. Factor: $a^6 - 100$

10. Factor: $n^4 - 2n^3 - 3n^2$

11. Factor $12y^2 + 16y - 3$ by using trial factors.

12. Factor: $12b^3 - 58b^2 + 56b$

13. Factor: $9y^4 - 25z^2$

14. Factor: $c^2 + 8c + 12$

15. Factor $18a^2 - 3a - 10$ by grouping.

16. Solve: $4x^2 + 27x = 7$

17. Factor: $4x^3 - 20x^2 - 24x$

18. Factor: $3a^2 - 15a - 42$

Copyright © Houghton Mifflin Company. All rights reserved.

19. Factor $2a^2 - 19a - 60$ by grouping.

20. Solve: $(x + 1)(x - 5) = 16$

21. Factor: $21ax - 35bx - 10by + 6ay$

22. Factor: $a^2b^2 - 1$

23. Factor: $10x^2 + 25x + 4xy + 10y$

24. Factor: $5x^2 - 5x - 30$

25. Factor: $3x^2 + 36x + 108$

26. Factor $3x^2 - 17x + 10$ by grouping.

27. Sports   The length of the field in field hockey is 20 yd less than twice the width of the field. The area of the field in field hockey is 6000 yd². Find the length and width of the field.

28. Image Projection   The size, S, of an image from a slide projector depends on the distance, d, of the screen from the projector and is given by $S = d^2$. Find the distance between the projector and the screen when the size of the picture is 400 ft².

29. Photography   A rectangular photograph has dimensions 15 in. by 12 in. A picture frame around the photograph increases the total area to 270 in². What is the width of the frame?

30. Gardening   The length of each side of a square garden plot is extended 4 ft. The area of the resulting square is 576 ft². Find the length of a side of the original garden plot.

Copyright © Houghton Mifflin Company. All rights reserved.

Chapter 5 Test

1. Factor: $ab + 6a - 3b - 18$

2. Factor: $2y^4 - 14y^3 - 16y^2$

3. Factor $8x^2 + 20x - 48$ by grouping.

4. Factor $6x^2 + 19x + 8$ by using trial factors.

5. Factor: $a^2 - 19a + 48$

6. Factor: $6x^3 - 8x^2 + 10x$

7. Factor: $x^2 + 2x - 15$

8. Solve: $4x^2 - 1 = 0$

9. Factor: $5x^2 - 45x - 15$

10. Factor: $p^2 + 12p + 36$

11. Solve: $x(x - 8) = -15$

12. Factor: $3x^2 + 12xy + 12y^2$

13. Factor: $b^2 - 16$

14. Factor $6x^2y^2 + 9xy^2 + 3y^2$ by grouping.

Copyright © Houghton Mifflin Company. All rights reserved.

15. Factor: $p^2 + 5p + 6$

16. Factor: $a(x - 2) + b(x - 2)$

17. Factor: $x(p + 1) - (p + 1)$

18. Factor: $3a^2 - 75$

19. Factor $2x^2 + 4x - 5$ by using trial factors.

20. Factor: $x^2 - 9x - 36$

21. Factor: $4a^2 - 12ab + 9b^2$

22. Factor: $4x^2 - 49y^2$

23. Solve: $(2a - 3)(a + 7) = 0$

24. **Number Sense**   The sum of two numbers is ten. The sum of the squares of the two numbers is fifty-eight. Find the two numbers.

25. **Geometry**   The length of a rectangle is 3 cm longer than twice its width. The area of the rectangle is 90 cm². Find the length and width of the rectangle.

Copyright © Houghton Mifflin Company. All rights reserved.

Cumulative Review Exercises

1. Subtract: $-2 - (-3) - 5 - (-11)$

2. Simplify: $(3 - 7)^2 \div (-2) - 3 \cdot (-4)$

3. Evaluate $-2a^2 \div (2b) - c$ when $a = -4$, $b = 2$, and $c = -1$.

4. Simplify: $-\dfrac{3}{4}(-20x^2)$

5. Simplify: $-2[4x - 2(3 - 2x) - 8x]$

6. Solve: $-\dfrac{5}{7}x = -\dfrac{10}{21}$

7. Solve: $3x - 2 = 12 - 5x$

8. Solve: $-2 + 4[3x - 2(4 - x) - 3] = 4x + 2$

9. 120% of what number is 54?

10. Simplify: $(-3a^3b^2)^2$

11. Multiply: $(x + 2)(x^2 - 5x + 4)$

12. Divide: $(8x^2 + 4x - 3) \div (2x - 3)$

13. Simplify: $(x^{-4}y^3)^2$

14. Factor: $3a - 3b - ax + bx$

15. Factor: $15xy^2 - 20xy^4$

16. Factor: $x^2 - 5xy - 14y^2$

17. Factor: $p^2 - 9p - 10$

18. Factor: $18a^3 + 57a^2 + 30a$

Copyright © Houghton Mifflin Company. All rights reserved.

19. Factor: $36a^2 - 49b^2$

20. Factor: $4x^2 + 28xy + 49y^2$

21. Factor: $9x^2 + 15x - 14$

22. Factor: $18x^2 - 48xy + 32y^2$

23. Factor: $3y(x - 3) - 2(x - 3)$

24. Solve: $3x^2 + 19x - 14 = 0$

25. Carpentry   A board 10 ft long is cut into two pieces. Four times the length of the shorter piece is 2 ft less than three times the length of the longer piece. Find the length of each piece.

26. Business   A portable MP3 player that regularly sells for $165 is on sale for $99. Find the discount rate. Use the formula $S = R - rR$.

27. Geometry Given that lines ℓ_1 and ℓ_2 are parallel, find the measures of angles a and b.

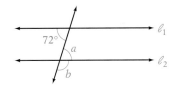

28. Travel   A family drove to a resort at an average speed of 42 mph and later returned over the same road at an average speed of 56 mph. Find the distance to the resort if the total driving time was 7 h.

29. Consecutive Integers   Find three consecutive even integers such that five times the middle integer is twelve more than twice the sum of the first and third integers.

30. Geometry   The length of the base of a triangle is three times the height. The area of the triangle is 24 in². Find the length of the base of the triangle.

Copyright © Houghton Mifflin Company. All rights reserved.

Copyright © Houghton Mifflin Company. All rights reserved.

6

Rational Expressions

Deep-sea diving is not only a recreational sport enjoyed by tourists and adventure-seekers. Commercial deep-sea divers work on a variety of projects, including pipeline systems, offshore oilfield sites, and search-and-recovery missions. For any deep-sea diver, whether diving for recreation or for business, underwater survival depends on an oxygen tank. On land, the average adult at rest inhales and exhales about one cubic foot of air every four minutes. This air contains approximately 21% oxygen. The percent of oxygen needed by a person changes when he or she is underwater. **Exercise 82 on page 309** uses a variable expression that gives the recommended percent of oxygen for a diver as a function of the diver's depth.

Need help? For online student resources, such as section quizzes, visit this textbook's website at **math.college.hmco.com/students.**

OBJECTIVES

Section 6.1

A To simplify a rational expression
B To multiply rational expressions
C To divide rational expressions

Section 6.2

A To find the least common multiple (LCM) of two or more polynomials
B To express two fractions in terms of the LCM of their denominators

Section 6.3

A To add or subtract rational expressions with the same denominators
B To add or subtract rational expressions with different denominators

Section 6.4

A To simplify a complex fraction

Section 6.5

A To solve an equation containing fractions

Section 6.6

A To solve a proportion
B To solve application problems
C To solve problems involving similar triangles

Section 6.7

A To solve a literal equation for one of the variables

Section 6.8

A To solve work problems
B To use rational expressions to solve uniform motion problems

Do these exercises to prepare for Chapter 6.

1. Find the least common multiple (LCM) of 12 and 18.

2. Simplify: $\dfrac{9x^3y^4}{3x^2y^7}$

3. Subtract: $\dfrac{3}{4} - \dfrac{8}{9}$

4. Divide: $\left(-\dfrac{8}{11}\right) \div \dfrac{4}{5}$

5. If a is a nonzero number, are the following two quantities equal: $\dfrac{0}{a}$ and $\dfrac{a}{0}$?

6. Solve: $\dfrac{2}{3}x - \dfrac{3}{4} = \dfrac{5}{6}$

7. Line l_1 is parallel to line l_2. Find the measure of angle a.

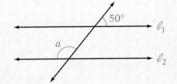

8. Factor: $x^2 - 4x - 12$

9. Factor: $2x^2 - x - 3$

10. At 9:00 AM, Anthony begins walking on a park trail at a rate of 9 m/min. Ten minutes later his sister Jean begins walking the same trail in pursuit of her brother at a rate of 12 m/min. At what time will Jean catch up to Anthony?

GO FIGURE • • •

A mouse begins at corner A of a 12 ft by 12 ft square maze traveling clockwise at a constant speed of 2 feet per second. Six seconds later a second mouse starts from the same corner A and travels clockwise at a constant speed of 3 feet per second. How far apart are the two mice 18 seconds after the second mouse begins?

Copyright © Houghton Mifflin Company. All rights reserved.

6.1 Multiplication and Division of Rational Expressions

Objective A   **To simplify a rational expression**

A fraction in which the numerator and denominator is a polynomial is called a **rational expression.** Examples of rational expressions are shown at the right.

$$\frac{5}{z}, \quad \frac{x^2 + 1}{2x - 1}, \quad \frac{y^2 + y - 1}{4y^2 + 1}$$

Care must be exercised with a rational expression to ensure that when the variables are replaced with numbers, the resulting denominator is not zero. Consider the rational expression at the right. The value of x cannot be 3 because the denominator would then be zero.

$$\frac{4x^2 - 9}{2x - 6}$$

$$\frac{4(3)^2 - 9}{2(3) - 6} = \frac{27}{0} \quad \text{Not a real number}$$

In the **simplest form of a rational expression,** the numerator and denominator have no common factors. The Multiplication Property of One is used to write a rational expression in simplest form.

HOW TO   Simplify: $\dfrac{x^2 - 4}{x^2 - 2x - 8}$

$$\frac{x^2 - 4}{x^2 - 2x - 8} = \frac{(x - 2)(x + 2)}{(x - 4)(x + 2)}$$

• Factor the numerator and denominator.

$$= \frac{x - 2}{x - 4} \cdot \boxed{\frac{x + 2}{x + 2}} = \frac{x - 2}{x - 4} \cdot 1$$

$$= \frac{x - 2}{x - 4}, \, x \neq -2, 4$$

• The restrictions, $x \neq -2$ or 4, are necessary to prevent division by zero.

This simplification is usually shown with slashes through the common factors:

$$\frac{x^2 - 4}{x^2 - 2x - 8} = \frac{(x - 2)\overset{1}{\cancel{(x + 2)}}}{(x - 4)\underset{1}{\cancel{(x + 2)}}}$$

• Factor the numerator and denominator.

$$= \frac{x - 2}{x - 4}, \, x \neq -2, 4$$

• Divide by the common factors. The restrictions, $x \neq -2$ or 4, are necessary to prevent division by zero.

In summary, to simplify a rational expression, factor the numerator and denominator. Then divide the numerator and denominator by the common factors.

HOW TO   Simplify: $\dfrac{10 + 3x - x^2}{x^2 - 4x - 5}$

$$\frac{10 + 3x - x^2}{x^2 - 4x - 5} = \frac{-(x^2 - 3x - 10)}{x^2 - 4x - 5}$$

• Because the coefficient of x^2 in the numerator is -1, factor -1 from the numerator.

$$= \frac{-\overset{1}{\cancel{(x - 5)}}(x + 2)}{\underset{1}{\cancel{(x - 5)}}(x + 1)}$$

• Factor the numerator and denominator. Divide by the common factors.

$$= -\frac{x + 2}{x + 1}, \, x \neq -1, 5$$

Copyright © Houghton Mifflin Company. All rights reserved.

For the remaining examples, we will omit the restrictions on the variables that prevent division by zero and assume the values of the variables are such that division by zero is not possible.

Copyright © Houghton Mifflin Company. All rights reserved.

Example 1

Simplify: $\dfrac{4x^3y^4}{6x^4y}$

Solution

$\dfrac{4x^3y^4}{6x^4y} = \dfrac{2y^3}{3x}$ • **Use rules of exponents.**

You Try It 1

Simplify: $\dfrac{6x^5y}{12x^2y^3}$

Your solution

Example 2

Simplify: $\dfrac{9 - x^2}{x^2 + x - 12}$

Solution

$\dfrac{9 - x^2}{x^2 + x - 12} = \dfrac{\overset{-1}{\cancel{(3 - x)}}(3 + x)}{\underset{1}{\cancel{(x - 3)}}(x + 4)}$ • $(3 - x) = -1(x - 3)$

$\qquad = -\dfrac{x + 3}{x + 4}$

You Try It 2

Simplify: $\dfrac{x^2 + 2x - 24}{16 - x^2}$

Your solution

Example 3

Simplify: $\dfrac{x^2 + 2x - 15}{x^2 - 7x + 12}$

Solution

$\dfrac{x^2 + 2x - 15}{x^2 - 7x + 12} = \dfrac{(x + 5)\overset{1}{\cancel{(x - 3)}}}{\underset{1}{\cancel{(x - 3)}}(x - 4)} = \dfrac{x + 5}{x - 4}$

You Try It 3

Simplify: $\dfrac{x^2 + 4x - 12}{x^2 - 3x + 2}$

Your solution

Solutions on p. S14

Objective B   **To multiply rational expressions**

The product of two fractions is a fraction whose numerator is the product of the numerators of the two fractions and whose denominator is the product of the denominators of the two fractions.

> **Multiplying Rational Expressions**
>
> Multiply the numerators. $\dfrac{a}{b} \cdot \dfrac{c}{d} = \dfrac{ac}{bd}$
> Multiply the denominators.

$$\dfrac{2}{3} \cdot \dfrac{4}{5} = \dfrac{8}{15} \qquad \dfrac{3x}{y} \cdot \dfrac{2}{z} = \dfrac{6x}{yz} \qquad \dfrac{x + 2}{x} \cdot \dfrac{3}{x - 2} = \dfrac{3x + 6}{x^2 - 2x}$$

HOW TO   Multiply: $\dfrac{x^2 + 3x}{x^2 - 3x - 4} \cdot \dfrac{x^2 - 5x + 4}{x^2 + 2x - 3}$

$\dfrac{x^2 + 3x}{x^2 - 3x - 4} \cdot \dfrac{x^2 - 5x + 4}{x^2 + 2x - 3}$

$= \dfrac{x(x + 3)}{(x - 4)(x + 1)} \cdot \dfrac{(x - 4)(x - 1)}{(x + 3)(x - 1)}$ • Factor the numerator and denominator of each fraction.

$= \dfrac{x\cancel{(x + 3)}\cancel{(x - 4)}\cancel{(x - 1)}}{\cancel{(x - 4)}(x + 1)\cancel{(x + 3)}\cancel{(x - 1)}}$ • Multiply. Then divide by the common factors.

$= \dfrac{x}{x + 1}$ • Write the answer in simplest form.

Example 4

Multiply: $\dfrac{10x^2 - 15x}{12x - 8} \cdot \dfrac{3x - 2}{20x - 25}$

Solution

$\dfrac{10x^2 - 15x}{12x - 8} \cdot \dfrac{3x - 2}{20x - 25}$

$= \dfrac{5x(2x - 3)}{4(3x - 2)} \cdot \dfrac{(3x - 2)}{5(4x - 5)}$ • Factor.

$= \dfrac{\cancel{5}x(2x - 3)\cancel{(3x - 2)}}{4\cancel{(3x - 2)}\cancel{5}(4x - 5)}$ • Divide by common factors.

$= \dfrac{x(2x - 3)}{4(4x - 5)}$

You Try It 4

Multiply: $\dfrac{12x^2 + 3x}{10x - 15} \cdot \dfrac{8x - 12}{9x + 18}$

Your solution

Example 5

Multiply: $\dfrac{x^2 + x - 6}{x^2 + 7x + 12} \cdot \dfrac{x^2 + 3x - 4}{4 - x^2}$

Solution

$\dfrac{x^2 + x - 6}{x^2 + 7x + 12} \cdot \dfrac{x^2 + 3x - 4}{4 - x^2}$

$= \dfrac{(x + 3)(x - 2)}{(x + 3)(x + 4)} \cdot \dfrac{(x + 4)(x - 1)}{(2 - x)(2 + x)}$ • Factor.

$= \dfrac{\cancel{(x + 3)}\cancel{(x - 2)}\cancel{(x + 4)}(x - 1)}{\cancel{(x + 3)}\cancel{(x + 4)}\cancel{(2 - x)}(2 + x)}$ • Divide by common factors.

$= -\dfrac{x - 1}{x + 2}$

You Try It 5

Multiply: $\dfrac{x^2 + 2x - 15}{9 - x^2} \cdot \dfrac{x^2 - 3x - 18}{x^2 - 7x + 6}$

Your solution

Solutions on pp. S14–S15

Copyright © Houghton Mifflin Company. All rights reserved.

Objective C **To divide rational expressions**

The **reciprocal of a rational expression** is the rational expression with the numerator and denominator interchanged.

$$
\text{Fraction} \left\{ \begin{array}{ccc} \dfrac{a}{b} & & \dfrac{b}{a} \\[2mm] x^2 = \dfrac{x^2}{1} & & \dfrac{1}{x^2} \\[2mm] \dfrac{x+2}{x} & & \dfrac{x}{x+2} \end{array} \right\} \text{Reciprocal}
$$

Dividing Rational Expressions

Multiply the dividend by the reciprocal of the divisor.

$$\frac{a}{b} \div \frac{c}{d} = \frac{a}{b} \cdot \frac{d}{c} = \frac{ad}{bc}$$

$$\frac{4}{x} \div \frac{y}{5} = \frac{4}{x} \cdot \frac{5}{y} = \frac{20}{xy}$$

$$\frac{x+4}{x} \div \frac{x-2}{4} = \frac{x+4}{x} \cdot \frac{4}{x-2} = \frac{4(x+4)}{x(x-2)}$$

The basis for the division rule is shown at the right.

$$\frac{a}{b} \div \frac{c}{d} = \frac{\dfrac{a}{b}}{\dfrac{c}{d}} = \frac{\dfrac{a}{b} \cdot \dfrac{d}{c}}{\dfrac{c}{d} \cdot \dfrac{d}{c}} = \frac{\dfrac{a}{b} \cdot \dfrac{d}{c}}{1} = \frac{a}{b} \cdot \frac{d}{c}$$

Example 6

Divide: $\dfrac{xy^2 - 3x^2y}{z^2} \div \dfrac{6x^2 - 2xy}{z^3}$

Solution

$$\frac{xy^2 - 3x^2y}{z^2} \div \frac{6x^2 - 2xy}{z^3}$$

$$= \frac{xy^2 - 3x^2y}{z^2} \cdot \frac{z^3}{6x^2 - 2xy}$$ • Multiply by the reciprocal.

$$= \frac{xy(y - 3x) \cdot z^3}{z^2 \cdot 2x(3x - y)} = -\frac{yz}{2}$$

You Try It 6

Divide: $\dfrac{a^2}{4bc^2 - 2b^2c} \div \dfrac{a}{6bc - 3b^2}$

Your solution

Example 7

Divide: $\dfrac{2x^2 + 5x + 2}{2x^2 + 3x - 2} \div \dfrac{3x^2 + 13x + 4}{2x^2 + 7x - 4}$

Solution

$$\frac{2x^2 + 5x + 2}{2x^2 + 3x - 2} \div \frac{3x^2 + 13x + 4}{2x^2 + 7x - 4}$$

$$= \frac{2x^2 + 5x + 2}{2x^2 + 3x - 2} \cdot \frac{2x^2 + 7x - 4}{3x^2 + 13x + 4}$$ • Multiply by the reciprocal.

$$= \frac{(2x + 1)(x + 2) \cdot (2x - 1)(x + 4)}{(2x - 1)(x + 2) \cdot (3x + 1)(x + 4)} = \frac{2x + 1}{3x + 1}$$

You Try It 7

Divide: $\dfrac{3x^2 + 26x + 16}{3x^2 - 7x - 6} \div \dfrac{2x^2 + 9x - 5}{x^2 + 2x - 15}$

Your solution

Solutions on p. S15

Copyright © Houghton Mifflin Company. All rights reserved.

6.1 Exercises

Objective A   **To simplify a rational expression**

1. ✏️ Explain the procedure for writing a rational expression in simplest form.

2. ✏️ Explain why the following simplification is incorrect.

$$\frac{x+3}{x} = \frac{\overset{1}{\cancel{x}}+3}{\underset{1}{\cancel{x}}} = 4$$

For Exercises 3 to 29, simplify.

3. $\dfrac{9x^3}{12x^4}$

4. $\dfrac{16x^2y}{24xy^3}$

5. $\dfrac{(x+3)^2}{(x+3)^3}$

6. $\dfrac{(2x-1)^5}{(2x-1)^4}$

7. $\dfrac{3n-4}{4-3n}$

8. $\dfrac{5-2x}{2x-5}$

9. $\dfrac{6y(y+2)}{9y^2(y+2)}$

10. $\dfrac{12x^2(3-x)}{18x(3-x)}$

11. $\dfrac{6x(x-5)}{8x^2(5-x)}$

12. $\dfrac{14x^3(7-3x)}{21x(3x-7)}$

13. $\dfrac{a^2+4a}{ab+4b}$

14. $\dfrac{x^2-3x}{2x-6}$

15. $\dfrac{4-6x}{3x^2-2x}$

16. $\dfrac{5xy-3y}{9-15x}$

17. $\dfrac{y^2-3y+2}{y^2-4y+3}$

18. $\dfrac{x^2+5x+6}{x^2+8x+15}$

19. $\dfrac{x^2+3x-10}{x^2+2x-8}$

20. $\dfrac{a^2+7a-8}{a^2+6a-7}$

21. $\dfrac{x^2+x-12}{x^2-6x+9}$

22. $\dfrac{x^2+8x+16}{x^2-2x-24}$

23. $\dfrac{x^2-3x-10}{25-x^2}$

24. $\dfrac{4-y^2}{y^2-3y-10}$

25. $\dfrac{2x^3+2x^2-4x}{x^3+2x^2-3x}$

26. $\dfrac{3x^3-12x}{6x^3-24x^2+24x}$

27. $\dfrac{6x^2-7x+2}{6x^2+5x-6}$

28. $\dfrac{2n^2-9n+4}{2n^2-5n-12}$

29. $\dfrac{x^2+3x-28}{24-2x-x^2}$

Copyright © Houghton Mifflin Company. All rights reserved.

Copyright © Houghton Mifflin Company. All rights reserved.

Objective B   **To multiply rational expressions**

For Exercises 30 to 55, multiply.

30. $\dfrac{8x^2}{9y^3} \cdot \dfrac{3y^2}{4x^3}$

31. $\dfrac{14a^2b^3}{15x^5y^2} \cdot \dfrac{25x^3y}{16ab}$

32. $\dfrac{12x^3y^4}{7a^2b^3} \cdot \dfrac{14a^3b^4}{9x^2y^2}$

33. $\dfrac{18a^4b^2}{25x^2y^3} \cdot \dfrac{50x^5y^6}{27a^6b^2}$

34. $\dfrac{3x - 6}{5x - 20} \cdot \dfrac{10x - 40}{27x - 54}$

35. $\dfrac{8x - 12}{14x + 7} \cdot \dfrac{42x + 21}{32x - 48}$

36. $\dfrac{3x^2 + 2x}{2xy - 3y} \cdot \dfrac{2xy^3 - 3y^3}{3x^3 + 2x^2}$

37. $\dfrac{4a^2x - 3a^2}{2by + 5b} \cdot \dfrac{2b^3y + 5b^3}{4ax - 3a}$

38. $\dfrac{x^2 + 5x + 4}{x^3y^2} \cdot \dfrac{x^2y^3}{x^2 + 2x + 1}$

39. $\dfrac{x^2 + x - 2}{xy^2} \cdot \dfrac{x^3y}{x^2 + 5x + 6}$

40. $\dfrac{x^4y^2}{x^2 + 3x - 28} \cdot \dfrac{x^2 - 49}{xy^4}$

41. $\dfrac{x^5y^3}{x^2 + 13x + 30} \cdot \dfrac{x^2 + 2x - 3}{x^7y^2}$

42. $\dfrac{2x^2 - 5x}{2xy + y} \cdot \dfrac{2xy^2 + y^2}{5x^2 - 2x^3}$

43. $\dfrac{3a^3 + 4a^2}{5ab - 3b} \cdot \dfrac{3b^3 - 5ab^3}{3a^2 + 4a}$

44. $\dfrac{x^2 - 2x - 24}{x^2 - 5x - 6} \cdot \dfrac{x^2 + 5x + 6}{x^2 + 6x + 8}$

45. $\dfrac{x^2 - 8x + 7}{x^2 + 3x - 4} \cdot \dfrac{x^2 + 3x - 10}{x^2 - 9x + 14}$

46. $\dfrac{x^2 + 2x - 35}{x^2 + 4x - 21} \cdot \dfrac{x^2 + 3x - 18}{x^2 + 9x + 18}$

47. $\dfrac{y^2 + y - 20}{y^2 + 2y - 15} \cdot \dfrac{y^2 + 4y - 21}{y^2 + 3y - 28}$

48. $\dfrac{x^2 - 3x - 4}{x^2 + 6x + 5} \cdot \dfrac{x^2 + 5x + 6}{8 + 2x - x^2}$

49. $\dfrac{25 - n^2}{n^2 - 2n - 35} \cdot \dfrac{n^2 - 8n - 20}{n^2 - 3n - 10}$

50. $\dfrac{12x^2 - 6x}{x^2 + 6x + 5} \cdot \dfrac{2x^4 + 10x^3}{4x^2 - 1}$

51. $\dfrac{8x^3 + 4x^2}{x^2 - 3x + 2} \cdot \dfrac{x^2 - 4}{16x^2 + 8x}$

52. $\dfrac{16 + 6x - x^2}{x^2 - 10x - 24} \cdot \dfrac{x^2 - 6x - 27}{x^2 - 17x + 72}$

53. $\dfrac{x^2 - 11x + 28}{x^2 - 13x + 42} \cdot \dfrac{x^2 + 7x + 10}{20 - x - x^2}$

54. $\dfrac{2x^2 + 5x + 2}{2x^2 + 7x + 3} \cdot \dfrac{x^2 - 7x - 30}{x^2 - 6x - 40}$

55. $\dfrac{x^2 - 4x - 32}{x^2 - 8x - 48} \cdot \dfrac{3x^2 + 17x + 10}{3x^2 - 22x - 16}$

Objective C   To divide rational expressions

56. What is the reciprocal of a rational expression?

57. Explain how to divide rational expressions.

For Exercises 58 to 77, divide.

58. $\dfrac{4x^2y^3}{15a^2b^3} \div \dfrac{6xy}{5a^3b^5}$

59. $\dfrac{9x^3y^4}{16a^4b^2} \div \dfrac{45x^4y^2}{14a^7b}$

60. $\dfrac{6x - 12}{8x + 32} \div \dfrac{18x - 36}{10x + 40}$

61. $\dfrac{28x + 14}{45x - 30} \div \dfrac{14x + 7}{30x - 20}$

62. $\dfrac{6x^3 + 7x^2}{12x - 3} \div \dfrac{6x^2 + 7x}{36x - 9}$

63. $\dfrac{5a^2y + 3a^2}{2x^3 + 5x^2} \div \dfrac{10ay + 6a}{6x^3 + 15x^2}$

64. $\dfrac{x^2 + 4x + 3}{x^2y} \div \dfrac{x^2 + 2x + 1}{xy^2}$

65. $\dfrac{x^3y^2}{x^2 - 3x - 10} \div \dfrac{xy^4}{x^2 - x - 20}$

66. $\dfrac{x^2 - 49}{x^4y^3} \div \dfrac{x^2 - 14x + 49}{x^4y^3}$

67. $\dfrac{x^2y^5}{x^2 - 11x + 30} \div \dfrac{xy^6}{x^2 - 7x + 10}$

Copyright © Houghton Mifflin Company. All rights reserved.

68. $\dfrac{4ax - 8a}{c^2} \div \dfrac{2y - xy}{c^3}$

69. $\dfrac{3x^2y - 9xy}{a^2b} \div \dfrac{3x^2 - x^3}{ab^2}$

70. $\dfrac{x^2 - 5x + 6}{x^2 - 9x + 18} \div \dfrac{x^2 - 6x + 8}{x^2 - 9x + 20}$

71. $\dfrac{x^2 + 3x - 40}{x^2 + 2x - 35} \div \dfrac{x^2 + 2x - 48}{x^2 + 3x - 18}$

72. $\dfrac{x^2 + 2x - 15}{x^2 - 4x - 45} \div \dfrac{x^2 + x - 12}{x^2 - 5x - 36}$

73. $\dfrac{y^2 - y - 56}{y^2 + 8y + 7} \div \dfrac{y^2 - 13y + 40}{y^2 - 4y - 5}$

74. $\dfrac{8 + 2x - x^2}{x^2 + 7x + 10} \div \dfrac{x^2 - 11x + 28}{x^2 - x - 42}$

75. $\dfrac{x^2 - x - 2}{x^2 - 7x + 10} \div \dfrac{x^2 - 3x - 4}{40 - 3x - x^2}$

76. $\dfrac{2x^2 - 3x - 20}{2x^2 - 7x - 30} \div \dfrac{2x^2 - 5x - 12}{4x^2 + 12x + 9}$

77. $\dfrac{6n^2 + 13n + 6}{4n^2 - 9} \div \dfrac{6n^2 + n - 2}{4n^2 - 1}$

APPLYING THE CONCEPTS

78. Given the expression $\dfrac{9}{x^2 + 1}$, choose some values of x and evaluate the expression for those values. Is it possible to choose a value of x for which the value of the expression is greater than 10? If so, what is that value of x? If not, explain why it is not possible.

79. Given the expression $\dfrac{1}{y - 3}$, choose some values of y and evaluate the expression for those values. Is it possible to choose a value of y for which the value of the expression is greater than 10,000,000? If so, what is that value of y? If not, explain why it is not possible.

Geometry For Exercises 80 and 81, write in simplest form the ratio of the shaded area of the figure to the total area of the figure.

80.

81.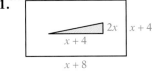

For Exercises 82 to 84, complete the simplification.

82. $\dfrac{8x}{9y} \div \underline{\ ?\ } = \dfrac{10y}{3}$

83. $\dfrac{n}{n + 3} \div \underline{\ ?\ } = \dfrac{n}{n - 2}$

84. $\underline{\ ?\ } \div \dfrac{n - 1}{4n^3} = 2n^2(n + 1)$

Copyright © Houghton Mifflin Company. All rights reserved.

Copyright © Houghton Mifflin Company. All rights reserved.

6.2 Expressing Fractions in Terms of the Least Common Multiple (LCM)

Objective A   **To find the least common multiple (LCM) of two or more polynomials**

Recall that the least common multiple (LCM) of two or more numbers is the smallest number that contains the prime factorization of each number.

The LCM of 12 and 18 is 36 because 36 contains the prime factors of 12 and the prime factors of 18.

$$12 = 2 \cdot 2 \cdot 3$$
$$18 = 2 \cdot 3 \cdot 3$$

$$\text{LCM} = 36 = \overbrace{2 \cdot \underbrace{2 \cdot 3} \cdot 3}^{\text{Factors of 12}}_{\text{Factors of 18}}$$

The **least common multiple (LCM) of two or more polynomials** is the polynomial of least degree that contains all the factors of each polynomial.

To find the LCM of two or more polynomials, first factor each polynomial completely. The LCM is the product of each factor the greatest number of times it occurs in any one factorization.

TAKE NOTE

The LCM must contain the factors of each polynomial. As shown with the braces at the right, the LCM contains the factors of $4x^2 + 4x$ and the factors of $x^2 + 2x + 1$.

HOW TO   Find the LCM of $4x^2 + 4x$ and $x^2 + 2x + 1$.

The LCM of the polynomials is the product of the LCM of the numerical coefficients and each variable factor the greatest number of times it occurs in any one factorization.

$$4x^2 + 4x = 4x(x + 1) = 2 \cdot 2 \cdot x(x + 1)$$
$$x^2 + 2x + 1 = (x + 1)(x + 1)$$

$$\text{LCM} = \overbrace{2 \cdot 2 \cdot x\underbrace{(x + 1)(x + 1)}} = 4x(x + 1)(x + 1)$$
$$\text{Factors of } 4x^2 + 4x$$
$$\text{Factors of } x^2 + 2x + 1$$

Example 1

Find the LCM of $4x^2y$ and $6xy^2$.

Solution
$4x^2y = 2 \cdot 2 \cdot x \cdot x \cdot y$
$6xy^2 = 2 \cdot 3 \cdot x \cdot y \cdot y$
$\text{LCM} = 2 \cdot 2 \cdot 3 \cdot x \cdot x \cdot y \cdot y = 12x^2y^2$

You Try It 1

Find the LCM of $8uv^2$ and $12uw$.

Your solution

Example 2

Find the LCM of $x^2 - x - 6$ and $9 - x^2$.

Solution
$x^2 - x - 6 = (x - 3)(x + 2)$
$9 - x^2 = -(x^2 - 9) = -(x + 3)(x - 3)$
$\text{LCM} = (x - 3)(x + 2)(x + 3)$

You Try It 2

Find the LCM of $m^2 - 6m + 9$ and $m^2 - 2m - 3$.

Your solution

Solutions on p. S15

> **Objective B**   **To express two fractions in terms of the LCM of their denominators**

When adding and subtracting fractions, it is frequently necessary to express two or more fractions in terms of a common denominator. This common denominator is the LCM of the denominators of the fractions.

HOW TO Write the fractions $\dfrac{x+1}{4x^2}$ and $\dfrac{x-3}{6x^2-12x}$ in terms of the LCM of the denominators.

Find the LCM of the denominators.

The LCM is $12x^2(x-2)$.

For each fraction, multiply the numerator and the denominator by the factors whose product with the denominator is the LCM.

$$\dfrac{x+1}{4x^2} = \dfrac{x+1}{4x^2} \cdot \dfrac{3(x-2)}{3(x-2)} = \dfrac{3x^2-3x-6}{12x^2(x-2)} \leftarrow$$

$$\dfrac{x-3}{6x^2-12x} = \dfrac{x-3}{6x(x-2)} \cdot \dfrac{2x}{2x} = \dfrac{2x^2-6x}{12x^2(x-2)} \leftarrow$$

LCM

Example 3

Write the fractions $\dfrac{x+2}{3x^2}$ and $\dfrac{x-1}{8xy}$ in terms of the LCM of the denominators.

Solution
The LCM is $24x^2y$.

$$\dfrac{x+2}{3x^2} = \dfrac{x+2}{3x^2} \cdot \dfrac{8y}{8y} = \dfrac{8xy+16y}{24x^2y}$$

$$\dfrac{x-1}{8xy} = \dfrac{x-1}{8xy} \cdot \dfrac{3x}{3x} = \dfrac{3x^2-3x}{24x^2y}$$

You Try It 3

Write the fractions $\dfrac{x-3}{4xy^2}$ and $\dfrac{2x+1}{9y^2z}$ in terms of the LCM of the denominators.

Your solution

Example 4

Write the fractions $\dfrac{2x-1}{2x-x^2}$ and $\dfrac{x}{x^2+x-6}$ in terms of the LCM of the denominators.

Solution
$$\dfrac{2x-1}{2x-x^2} = \dfrac{2x-1}{-(x^2-2x)} = -\dfrac{2x-1}{x^2-2x}$$

The LCM is $x(x-2)(x+3)$.

$$\dfrac{2x-1}{2x-x^2} = -\dfrac{2x-1}{x(x-2)} \cdot \dfrac{x+3}{x+3} = -\dfrac{2x^2+5x-3}{x(x-2)(x+3)}$$

$$\dfrac{x}{x^2+x-6} = \dfrac{x}{(x-2)(x+3)} \cdot \dfrac{x}{x} = \dfrac{x^2}{x(x-2)(x+3)}$$

You Try It 4

Write the fractions $\dfrac{x+4}{x^2-3x-10}$ and $\dfrac{2x}{25-x^2}$ in terms of the LCM of the denominators.

Your solution

Solutions on p. S15

Copyright © Houghton Mifflin Company. All rights reserved.

6.2 Exercises

Copyright © Houghton Mifflin Company. All rights reserved.

Objective A   **To find the least common multiple (LCM) of two or more polynomials**

For Exercises 1 to 33, find the LCM of the polynomials.

1. $8x^3y$
$12xy^2$

2. $6ab^2$
$18ab^3$

3. $10x^4y^2$
$15x^3y$

4. $12a^2b$
$18ab^3$

5. $8x^2$
$4x^2 + 8x$

6. $6y^2$
$4y + 12$

7. $2x^2y$
$3x^2 + 12x$

8. $4xy^2$
$6xy^2 + 12y^2$

9. $9x(x + 2)$
$12(x + 2)^2$

10. $8x^2(x - 1)^2$
$10x^3(x - 1)$

11. $3x + 3$
$2x^2 + 4x + 2$

12. $4x - 12$
$2x^2 - 12x + 18$

13. $(x - 1)(x + 2)$
$(x - 1)(x + 3)$

14. $(2x - 1)(x + 4)$
$(2x + 1)(x + 4)$

15. $(2x + 3)^2$
$(2x + 3)(x - 5)$

16. $(x - 7)(x + 2)$
$(x - 7)^2$

17. $x - 1$
$x - 2$
$(x - 1)(x - 2)$

18. $(x + 4)(x - 3)$
$x + 4$
$x - 3$

19. $x^2 - x - 6$
$x^2 + x - 12$

20. $x^2 + 3x - 10$
$x^2 + 5x - 14$

21. $x^2 + 5x + 4$
$x^2 - 3x - 28$

22. $x^2 - 10x + 21$
$x^2 - 8x + 15$

23. $x^2 - 2x - 24$
$x^2 - 36$

24. $x^2 + 7x + 10$
$x^2 - 25$

25. $x^2 - 7x - 30$
$x^2 - 5x - 24$

26. $2x^2 - 7x + 3$
$2x^2 + x - 1$

27. $3x^2 - 11x + 6$
$3x^2 + 4x - 4$

28. $2x^2 - 9x + 10$
$2x^2 + x - 15$

29. $6 + x - x^2$
$x + 2$
$x - 3$

30. $15 + 2x - x^2$
$x - 5$
$x + 3$

31. $5 + 4x - x^2$
$x - 5$
$x + 1$

32. $x^2 + 3x - 18$
$3 - x$
$x + 6$

33. $x^2 - 5x + 6$
$1 - x$
$x - 6$

Objective B	To express two fractions in terms of the LCM of their denominators

For Exercises 34 to 53, write the fraction in terms of the LCM of the denominators.

34. $\dfrac{4}{x}, \dfrac{3}{x^2}$

35. $\dfrac{5}{ab^2}, \dfrac{6}{ab}$

36. $\dfrac{x}{3y^2}, \dfrac{z}{4y}$

37. $\dfrac{5y}{6x^2}, \dfrac{7}{9xy}$

38. $\dfrac{y}{x(x-3)}, \dfrac{6}{x^2}$

39. $\dfrac{a}{y^2}, \dfrac{6}{y(y+5)}$

40. $\dfrac{9}{(x-1)^2}, \dfrac{6}{x(x-1)}$

41. $\dfrac{a^2}{y(y+7)}, \dfrac{a}{(y+7)^2}$

42. $\dfrac{3}{x-3}, \dfrac{5}{x(3-x)}$

43. $\dfrac{b}{y(y-4)}, \dfrac{b^2}{4-y}$

44. $\dfrac{3}{(x-5)^2}, \dfrac{2}{5-x}$

45. $\dfrac{3}{7-y}, \dfrac{2}{(y-7)^2}$

46. $\dfrac{3}{x^2+2x}, \dfrac{4}{x^2}$

47. $\dfrac{2}{y-3}, \dfrac{3}{y^3-3y^2}$

48. $\dfrac{x-2}{x+3}, \dfrac{x}{x-4}$

49. $\dfrac{x^2}{2x-1}, \dfrac{x+1}{x+4}$

50. $\dfrac{3}{x^2+x-2}, \dfrac{x}{x+2}$

51. $\dfrac{3x}{x-5}, \dfrac{4}{x^2-25}$

52. $\dfrac{x}{x^2+x-6}, \dfrac{2x}{x^2-9}$

53. $\dfrac{x-1}{x^2+2x-15}, \dfrac{x}{x^2+6x+5}$

APPLYING THE CONCEPTS

54. ✏ When is the LCM of two polynomials equal to their product?

For Exercises 55 to 60, write each fraction in terms of the LCM of the denominators.

55. $\dfrac{8}{10^3}, \dfrac{9}{10^5}$

56. $3, \dfrac{2}{n}$

57. $x, \dfrac{x}{x^2-1}$

58. $\dfrac{x^2+1}{(x-1)^3}, \dfrac{x+1}{(x-1)^2}, \dfrac{1}{x-1}$

59. $\dfrac{c}{6c^2+7cd+d^2}, \dfrac{d}{3c^2-3d^2}$

60. $\dfrac{1}{ab+3a-3b-b^2}, \dfrac{1}{ab+3a+3b+b^2}$

Copyright © Houghton Mifflin Company. All rights reserved.

6.3 Addition and Subtraction of Rational Expressions

Objective A To add or subtract rational expressions with the same denominators

When adding rational expressions in which the denominators are the same, add the numerators. The denominator of the sum is the common denominator.

$$\frac{5x}{18} + \frac{7x}{18} = \frac{5x + 7x}{18} = \frac{12x}{18} = \frac{2x}{3}$$

$$\frac{x}{x^2 - 1} + \frac{1}{x^2 - 1} = \frac{x + 1}{x^2 - 1} = \frac{\overset{1}{\cancel{(x + 1)}}}{(x - 1)\cancel{(x + 1)}} = \frac{1}{x - 1}$$

Note that the sum is written in simplest form.

When subtracting rational expressions with like denominators, subtract the numerators. The denominator of the difference is the common denominator. Write the answer in simplest form.

$$\frac{2x}{x - 2} - \frac{4}{x - 2} = \frac{2x - 4}{x - 2} = \frac{2\overset{1}{\cancel{(x - 2)}}}{\cancel{x - 2}} = 2$$

$$\frac{3x - 1}{x^2 - 5x + 4} - \frac{2x + 3}{x^2 - 5x + 4} = \frac{(3x - 1) - (2x + 3)}{x^2 - 5x + 4} = \frac{3x - 1 - 2x - 3}{x^2 - 5x + 4}$$

$$= \frac{x - 4}{x^2 - 5x + 4} = \frac{\overset{1}{\cancel{(x - 4)}}}{\cancel{(x - 4)}(x - 1)} = \frac{1}{x - 1}$$

> **Adding and Subtracting Rational Expressions with the Same Denominator**
>
> Add or subtract the numerators. Place the result over the common denominator.
> $$\frac{a}{b} + \frac{c}{b} = \frac{a + c}{b} \qquad \frac{a}{b} - \frac{c}{b} = \frac{a - c}{b}$$

Example 1

Subtract: $\dfrac{3x^2}{x^2 - 1} - \dfrac{x + 4}{x^2 - 1}$

Solution

$$\frac{3x^2}{x^2 - 1} - \frac{x + 4}{x^2 - 1} = \frac{3x^2 - (x + 4)}{x^2 - 1}$$

$$= \frac{3x^2 - x - 4}{x^2 - 1}$$

$$= \frac{(3x - 4)\overset{1}{\cancel{(x + 1)}}}{(x - 1)\cancel{(x + 1)}} = \frac{3x - 4}{x - 1}$$

You Try It 1

Subtract: $\dfrac{2x^2}{x^2 - x - 12} - \dfrac{7x + 4}{x^2 - x - 12}$

Your solution

Solution on p. S15

Copyright © Houghton Mifflin Company. All rights reserved.

Example 2

Simplify:

$$\dfrac{2x^2 + 5}{x^2 + 2x - 3} - \dfrac{x^2 - 3x}{x^2 + 2x - 3} + \dfrac{x - 2}{x^2 + 2x - 3}$$

Solution

$$\dfrac{2x^2 + 5}{x^2 + 2x - 3} - \dfrac{x^2 - 3x}{x^2 + 2x - 3} + \dfrac{x - 2}{x^2 + 2x - 3}$$

$$= \dfrac{(2x^2 + 5) - (x^2 - 3x) + (x - 2)}{x^2 + 2x - 3}$$

$$= \dfrac{2x^2 + 5 - x^2 + 3x + x - 2}{x^2 + 2x - 3}$$

$$= \dfrac{x^2 + 4x + 3}{x^2 + 2x - 3}$$

$$= \dfrac{\overset{1}{\cancel{(x + 3)}}(x + 1)}{\underset{1}{\cancel{(x + 3)}}(x - 1)} = \dfrac{x + 1}{x - 1}$$

You Try It 2

Simplify:

$$\dfrac{x^2 - 1}{x^2 - 8x + 12} - \dfrac{2x + 1}{x^2 - 8x + 12} + \dfrac{x}{x^2 - 8x + 12}$$

Your solution

Solution on p. S15

Objective B   To add or subtract rational expressions with different denominators

Before two fractions with unlike denominators can be added or subtracted, each fraction must be expressed in terms of a common denominator. This common denominator is the LCM of the denominators of the fractions.

HOW TO   Add: $\dfrac{x - 3}{x^2 - 2x} + \dfrac{6}{x^2 - 4}$

The LCM is $x(x - 2)(x + 2)$.   • Find the LCM of the denominators.

$$\dfrac{x - 3}{x^2 - 2x} + \dfrac{6}{x^2 - 4}$$

$$= \dfrac{x - 3}{x(x - 2)} \cdot \dfrac{x + 2}{x + 2} + \dfrac{6}{(x - 2)(x + 2)} \cdot \dfrac{x}{x}$$ • Write each fraction in terms of the LCM.

$$= \dfrac{x^2 - x - 6}{x(x - 2)(x + 2)} + \dfrac{6x}{x(x - 2)(x + 2)}$$ • Multiply the factors in the numerators.

$$= \dfrac{(x^2 - x - 6) + 6x}{x(x - 2)(x + 2)}$$ • Add the fractions.

$$= \dfrac{x^2 + 5x - 6}{x(x - 2)(x + 2)}$$ • Simplify.

$$= \dfrac{(x + 6)(x - 1)}{x(x - 2)(x + 2)}$$ • Factor.

Copyright © Houghton Mifflin Company. All rights reserved.

After combining the numerators over the common denominator, the last step is to factor the numerator to determine whether there are common factors in the numerator and denominator. For the previous example, there are no common factors, so the answer is in simplest form.

The process of adding and subtracting rational expressions is summarized below.

Adding and Subtracting Rational Expressions

1. Find the LCM of the denominators.
2. Write each fraction as an equivalent fraction using the LCM as the denominator.
3. Add or subtract the numerators and place the result over the common denominator.
4. Write the answer in simplest form.

Example 3

Simplify: $\dfrac{y}{x} - \dfrac{4y}{3x} + \dfrac{3y}{4x}$

Solution
The LCM of the denominators is $12x$.

$\dfrac{y}{x} - \dfrac{4y}{3x} + \dfrac{3y}{4x}$

$= \dfrac{y}{x} \cdot \dfrac{12}{12} - \dfrac{4y}{3x} \cdot \dfrac{4}{4} + \dfrac{3y}{4x} \cdot \dfrac{3}{3}$ • Write each fraction using the LCM.

$= \dfrac{12y}{12x} - \dfrac{16y}{12x} + \dfrac{9y}{12x}$

$= \dfrac{12y - 16y + 9y}{12x} = \dfrac{5y}{12x}$ • Combine the numerators.

You Try It 3

Simplify: $\dfrac{z}{8y} - \dfrac{4z}{3y} + \dfrac{5z}{4y}$

Your solution

Solution on p. S15

Copyright © Houghton Mifflin Company. All rights reserved.

Example 4

Subtract: $\dfrac{2x}{x - 3} - \dfrac{5}{3 - x}$

Solution

Remember $3 - x = -(x - 3)$.

Therefore, $\dfrac{5}{3 - x} = \dfrac{5}{-(x - 3)} = \dfrac{-5}{x - 3}$.

$\dfrac{2x}{x - 3} - \dfrac{5}{3 - x}$

$= \dfrac{2x}{x - 3} - \dfrac{-5}{x - 3}$ • The LCM is $x - 3$.

$= \dfrac{2x - (-5)}{x - 3} = \dfrac{2x + 5}{x - 3}$ • Combine the numerators.

You Try It 4

Add: $\dfrac{5x}{x - 2} + \dfrac{3}{2 - x}$

Your solution

Example 5

Subtract: $\dfrac{2x}{2x - 3} - \dfrac{1}{x + 1}$

Solution

The LCM is $(2x - 3)(x + 1)$.

$\dfrac{2x}{2x - 3} - \dfrac{1}{x + 1}$

$= \dfrac{2x}{2x - 3} \cdot \dfrac{x + 1}{x + 1} - \dfrac{1}{x + 1} \cdot \dfrac{2x - 3}{2x - 3}$

$= \dfrac{2x^2 + 2x}{(2x - 3)(x + 1)} - \dfrac{2x - 3}{(2x - 3)(x + 1)}$

$= \dfrac{(2x^2 + 2x) - (2x - 3)}{(2x - 3)(x + 1)}$

$= \dfrac{2x^2 + 2x - 2x + 3}{(2x - 3)(x + 1)} = \dfrac{2x^2 + 3}{(2x - 3)(x + 1)}$

You Try It 5

Add: $\dfrac{4x}{3x - 1} + \dfrac{9}{x + 4}$

Your solution

Example 6

Add: $1 + \dfrac{3}{x^2}$

Solution

The LCM is x^2.

$1 + \dfrac{3}{x^2} = 1 \cdot \dfrac{x^2}{x^2} + \dfrac{3}{x^2} = \dfrac{x^2}{x^2} + \dfrac{3}{x^2} = \dfrac{x^2 + 3}{x^2}$

You Try It 6

Subtract: $2 - \dfrac{1}{x - 3}$

Your solution

Solutions on pp. S15–S16

Copyright © Houghton Mifflin Company. All rights reserved.

Example 7

Add: $\dfrac{x + 3}{x^2 - 2x - 8} + \dfrac{3}{4 - x}$

Solution

Recall: $\dfrac{3}{4 - x} = \dfrac{-3}{x - 4}$

The LCM is $(x - 4)(x + 2)$.

$\dfrac{x + 3}{x^2 - 2x - 8} + \dfrac{3}{4 - x}$

$= \dfrac{x + 3}{(x - 4)(x + 2)} + \dfrac{(-3)}{x - 4}$

$= \dfrac{x + 3}{(x - 4)(x + 2)} + \dfrac{(-3)}{x - 4} \cdot \dfrac{x + 2}{x + 2}$

$= \dfrac{x + 3}{(x - 4)(x + 2)} + \dfrac{(-3)(x + 2)}{(x - 4)(x + 2)}$

$= \dfrac{(x + 3) + (-3)(x + 2)}{(x - 4)(x + 2)}$

$= \dfrac{x + 3 - 3x - 6}{(x - 4)(x + 2)}$

$= \dfrac{-2x - 3}{(x - 4)(x + 2)}$

You Try It 7

Add: $\dfrac{2x - 1}{x^2 - 25} + \dfrac{2}{5 - x}$

Your solution

Example 8

Simplify: $\dfrac{3x + 2}{2x^2 - x - 1} - \dfrac{3}{2x + 1} + \dfrac{4}{x - 1}$

Solution

The LCM is $(2x + 1)(x - 1)$.

$\dfrac{3x + 2}{2x^2 - x - 1} - \dfrac{3}{2x + 1} + \dfrac{4}{x - 1}$

$= \dfrac{3x + 2}{(2x + 1)(x - 1)} - \dfrac{3}{2x + 1} \cdot \dfrac{x - 1}{x - 1} + \dfrac{4}{x - 1} \cdot \dfrac{2x + 1}{2x + 1}$

$= \dfrac{3x + 2}{(2x + 1)(x - 1)} - \dfrac{3x - 3}{(2x + 1)(x - 1)} + \dfrac{8x + 4}{(2x + 1)(x - 1)}$

$= \dfrac{(3x + 2) - (3x - 3) + (8x + 4)}{(2x + 1)(x - 1)}$

$= \dfrac{3x + 2 - 3x + 3 + 8x + 4}{(2x + 1)(x - 1)}$

$= \dfrac{8x + 9}{(2x + 1)(x - 1)}$

You Try It 8

Simplify: $\dfrac{2x - 3}{3x^2 - x - 2} + \dfrac{5}{3x + 2} - \dfrac{1}{x - 1}$

Your solution

Solutions on p. S16

Copyright © Houghton Mifflin Company. All rights reserved.

6.3 Exercises

Objective A **To add or subtract rational expressions with the same denominators**

For Exercises 1 to 20, simplify.

1. $\dfrac{3}{y^2} + \dfrac{8}{y^2}$

2. $\dfrac{6}{ab} - \dfrac{2}{ab}$

3. $\dfrac{3}{x+4} - \dfrac{10}{x+4}$

4. $\dfrac{x}{x+6} - \dfrac{2}{x+6}$

5. $\dfrac{3x}{2x+3} + \dfrac{5x}{2x+3}$

6. $\dfrac{6y}{4y+1} - \dfrac{11y}{4y+1}$

7. $\dfrac{2x+1}{x-3} + \dfrac{3x+6}{x-3}$

8. $\dfrac{4x+3}{2x-7} + \dfrac{3x-8}{2x-7}$

9. $\dfrac{5x-1}{x+9} - \dfrac{3x+4}{x+9}$

10. $\dfrac{6x-5}{x-10} - \dfrac{3x-4}{x-10}$

11. $\dfrac{x-7}{2x+7} - \dfrac{4x-3}{2x+7}$

12. $\dfrac{2n}{3n+4} - \dfrac{5n-3}{3n+4}$

13. $\dfrac{x}{x^2+2x-15} - \dfrac{3}{x^2+2x-15}$

14. $\dfrac{3x}{x^2+3x-10} - \dfrac{6}{x^2+3x-10}$

15. $\dfrac{2x+3}{x^2-x-30} - \dfrac{x-2}{x^2-x-30}$

16. $\dfrac{3x-1}{x^2+5x-6} - \dfrac{2x-7}{x^2+5x-6}$

17. $\dfrac{4y+7}{2y^2+7y-4} - \dfrac{y-5}{2y^2+7y-4}$

18. $\dfrac{x+1}{2x^2-5x-12} + \dfrac{x+2}{2x^2-5x-12}$

19. $\dfrac{2x^2+3x}{x^2-9x+20} + \dfrac{2x^2-3}{x^2-9x+20} - \dfrac{4x^2+2x+1}{x^2-9x+20}$

20. $\dfrac{2x^2+3x}{x^2-2x-63} - \dfrac{x^2-3x+21}{x^2-2x-63} - \dfrac{x-7}{x^2-2x-63}$

Copyright © Houghton Mifflin Company. All rights reserved.

Objective B   **To add or subtract rational expressions with different denominators**

21. ✎   Explain the process of writing equivalent rational expressions with the LCM of the denominators of the rational expressions as the new denominator.

22. ✎   Explain the process of adding rational expressions with different denominators.

For Exercises 23 to 80, simplify.

23. $\dfrac{4}{x} + \dfrac{5}{y}$

24. $\dfrac{7}{a} + \dfrac{5}{b}$

25. $\dfrac{12}{x} - \dfrac{5}{2x}$

26. $\dfrac{5}{3a} - \dfrac{3}{4a}$

27. $\dfrac{1}{2x} - \dfrac{5}{4x} + \dfrac{7}{6x}$

28. $\dfrac{7}{4y} + \dfrac{11}{6y} - \dfrac{8}{3y}$

29. $\dfrac{5}{3x} - \dfrac{2}{x^2} + \dfrac{3}{2x}$

30. $\dfrac{6}{y^2} + \dfrac{3}{4y} - \dfrac{2}{5y}$

31. $\dfrac{2}{x} - \dfrac{3}{2y} + \dfrac{3}{5x} - \dfrac{1}{4y}$

32. $\dfrac{5}{2a} + \dfrac{7}{3b} - \dfrac{2}{b} - \dfrac{3}{4a}$

33. $\dfrac{2x + 1}{3x} + \dfrac{x - 1}{5x}$

34. $\dfrac{4x - 3}{6x} + \dfrac{2x + 3}{4x}$

35. $\dfrac{x - 3}{6x} + \dfrac{x + 4}{8x}$

36. $\dfrac{2x - 3}{2x} + \dfrac{x + 3}{3x}$

37. $\dfrac{2x + 9}{9x} - \dfrac{x - 5}{5x}$

38. $\dfrac{3y - 2}{12y} - \dfrac{y - 3}{18y}$

39. $\dfrac{x + 4}{2x} - \dfrac{x - 1}{x^2}$

40. $\dfrac{x - 2}{3x^2} - \dfrac{x + 4}{x}$

41. $\dfrac{x - 10}{4x^2} + \dfrac{x + 1}{2x}$

42. $\dfrac{x + 5}{3x^2} + \dfrac{2x + 1}{2x}$

43. $\dfrac{4}{x + 4} - x$

44. $2x + \dfrac{1}{x}$

45. $5 - \dfrac{x - 2}{x + 1}$

46. $3 + \dfrac{x - 1}{x + 1}$

Copyright © Houghton Mifflin Company. All rights reserved.

47. $\dfrac{x+3}{6x} - \dfrac{x-3}{8x^2}$

48. $\dfrac{x+2}{xy} - \dfrac{3x-2}{x^2y}$

49. $\dfrac{3x-1}{xy^2} - \dfrac{2x+3}{xy}$

50. $\dfrac{4x-3}{3x^2y} + \dfrac{2x+1}{4xy^2}$

51. $\dfrac{5x+7}{6xy^2} - \dfrac{4x-3}{8x^2y}$

52. $\dfrac{x-2}{8x^2} - \dfrac{x+7}{12xy}$

53. $\dfrac{3x-1}{6y^2} - \dfrac{x+5}{9xy}$

54. $\dfrac{4}{x-2} + \dfrac{5}{x+3}$

55. $\dfrac{2}{x-3} + \dfrac{5}{x-4}$

56. $\dfrac{6}{x-7} - \dfrac{4}{x+3}$

57. $\dfrac{3}{y+6} - \dfrac{4}{y-3}$

58. $\dfrac{2x}{x+1} + \dfrac{1}{x-3}$

59. $\dfrac{3x}{x-4} + \dfrac{2}{x+6}$

60. $\dfrac{4x}{2x-1} - \dfrac{5}{x-6}$

61. $\dfrac{6x}{x+5} - \dfrac{3}{2x+3}$

62. $\dfrac{2a}{a-7} + \dfrac{5}{7-a}$

63. $\dfrac{4x}{6-x} + \dfrac{5}{x-6}$

64. $\dfrac{x}{x^2-9} + \dfrac{3}{x-3}$

65. $\dfrac{y}{y^2-16} + \dfrac{1}{y-4}$

66. $\dfrac{2x}{x^2-x-6} - \dfrac{3}{x+2}$

67. $\dfrac{(x-1)^2}{(x+1)^2} - 1$

68. $1 - \dfrac{(y-2)^2}{(y+2)^2}$

69. $\dfrac{x}{1-x^2} - 1 + \dfrac{x}{1+x}$

70. $\dfrac{y}{x-y} + 2 - \dfrac{x}{y-x}$

Copyright © Houghton Mifflin Company. All rights reserved.

71. $\dfrac{3x - 1}{x^2 - 10x + 25} - \dfrac{3}{x - 5}$

72. $\dfrac{2a + 3}{a^2 - 7a + 12} - \dfrac{2}{a - 3}$

73. $\dfrac{x + 4}{x^2 - x - 42} + \dfrac{3}{7 - x}$

74. $\dfrac{x + 3}{x^2 - 3x - 10} + \dfrac{2}{5 - x}$

75. $\dfrac{1}{x + 1} + \dfrac{x}{x - 6} - \dfrac{5x - 2}{x^2 - 5x - 6}$

76. $\dfrac{x}{x - 4} + \dfrac{5}{x + 5} - \dfrac{11x - 8}{x^2 + x - 20}$

77. $\dfrac{3x + 1}{x - 1} - \dfrac{x - 1}{x - 3} + \dfrac{x + 1}{x^2 - 4x + 3}$

78. $\dfrac{4x + 1}{x - 8} - \dfrac{3x + 2}{x + 4} - \dfrac{49x + 4}{x^2 - 4x - 32}$

79. $\dfrac{2x + 9}{3 - x} + \dfrac{x + 5}{x + 7} - \dfrac{2x^2 + 3x - 3}{x^2 + 4x - 21}$

80. $\dfrac{3x + 5}{x + 5} - \dfrac{x + 1}{2 - x} - \dfrac{4x^2 - 3x - 1}{x^2 + 3x - 10}$

APPLYING THE CONCEPTS

81. Transportation   Suppose that you drive about 12,000 mi per year and that the cost of gasoline averages $1.70 per gallon.

   a. Let x represent the number of miles per gallon your car gets. Write a variable expression for the amount you spend on gasoline in 1 year.

   b. Write and simplify a variable expression for the amount of money you will save each year if you can increase your gas mileage by 5 miles per gallon.

   c. If you currently get 25 miles per gallon and you increase your gas mileage by 5 miles per gallon, how much will you save in 1 year?

82. Deep-Sea Diving   A recommended percent of oxygen (by volume) in the air that a deep-sea diver breathes is given by $\dfrac{660}{d + 33}$, where d is the depth, in feet, at which the diver is working.

   a. What is the recommended percent (to the nearest percent) of oxygen for a diver working at a depth of 50 ft?

   b. As the depth of the diver increases, does the recommended amount of oxygen increase or decrease?

   c. At sea level, the oxygen content of air is approximately 21%. Is this less than or more than the recommended amount of oxygen for a diver working at the water's surface?

Copyright © Houghton Mifflin Company. All rights reserved.

6.4 Complex Fractions

Objective A To simplify a complex fraction

Point of Interest

There are many instances of complex fractions in application problems. The fraction $\dfrac{1}{\dfrac{1}{r_1} + \dfrac{1}{r_2}}$ is used to determine the total resistance in certain electric circuits.

A **complex fraction** is a fraction whose numerator or denominator contains one or more fractions. Examples of complex fractions are shown at the right.

$$\dfrac{3}{2 - \dfrac{1}{2}}, \quad \dfrac{4 + \dfrac{1}{x}}{3 + \dfrac{2}{x}}, \quad \dfrac{\dfrac{1}{x-1} + x + 3}{x - 3 + \dfrac{1}{x+4}}$$

To simplify a complex fraction, use one of the following methods.

TAKE NOTE

You may use either method to simplify a complex fraction. The result will be the same.

Simplifying Complex Fractions

Method 1: Multiply by 1 in the form $\dfrac{\text{LCM}}{\text{LCM}}$.

1. Determine the LCM of the denominators of the fractions in the numerator and denominator of the complex fraction.
2. Multiply the numerator and denominator of the complex fraction by the LCM.
3. Simplify.

Method 2: Multiply the numerator by the reciprocal of the denominator.

1. Simplify the numerator to a single fraction and simplify the denominator to a single fraction.
2. Using the definition for dividing fractions, multiply the numerator by the reciprocal of the denominator.
3. Simplify.

Here is an example of using Method 1.

HOW TO Simplify: $\dfrac{9 - \dfrac{4}{x^2}}{3 + \dfrac{2}{x}}$

The LCM of x and x^2 is x^2.

- Find the **LCM** of the denominators of the fractions in the numerator and the denominator.

$$\dfrac{9 - \dfrac{4}{x^2}}{3 + \dfrac{2}{x}} = \dfrac{9 - \dfrac{4}{x^2}}{3 + \dfrac{2}{x}} \cdot \dfrac{x^2}{x^2}$$

- Multiply the numerator and denominator by the **LCM**.

$$= \dfrac{9 \cdot x^2 - \dfrac{4}{x^2} \cdot x^2}{3 \cdot x^2 + \dfrac{2}{x} \cdot x^2} = \dfrac{9x^2 - 4}{3x^2 + 2x}$$

- Use the **Distributive Property**.

$$= \dfrac{(3x - 2)(3x + 2)}{x(3x + 2)} = \dfrac{3x - 2}{x}$$

- Simplify.

Copyright © Houghton Mifflin Company. All rights reserved.

Here is the same example using the second method.

HOW TO   Simplify: $\dfrac{9 - \dfrac{4}{x^2}}{3 + \dfrac{2}{x}}$

$\dfrac{9 - \dfrac{4}{x^2}}{3 + \dfrac{2}{x}} = \dfrac{\dfrac{9x^2}{x^2} - \dfrac{4}{x^2}}{\dfrac{3x}{x} + \dfrac{2}{x}} = \dfrac{\dfrac{9x^2 - 4}{x^2}}{\dfrac{3x + 2}{x}}$

- Simplify the numerator to a single fraction and simplify the denominator to a single fraction.

$= \dfrac{9x^2 - 4}{x^2} \cdot \dfrac{x}{3x + 2}$

- Multiply the numerator by the reciprocal of the denominator.

$= \dfrac{x(3x - 2)\overset{1}{\cancel{(3x + 2)}}}{x^2 \underset{1}{\cancel{(3x + 2)}}}$

- Simplify.

$= \dfrac{3x - 2}{x}$

For the examples below, we will use the first method.

Example 1

Simplify: $\dfrac{\dfrac{1}{x} + \dfrac{1}{2}}{\dfrac{1}{x^2} - \dfrac{1}{4}}$

Solution
The LCM of x, 2, x^2, and 4 is $4x^2$.

$\dfrac{\dfrac{1}{x} + \dfrac{1}{2}}{\dfrac{1}{x^2} - \dfrac{1}{4}} = \dfrac{\dfrac{1}{x} + \dfrac{1}{2}}{\dfrac{1}{x^2} - \dfrac{1}{4}} \cdot \dfrac{4x^2}{4x^2}$

- Multiply by the LCM.

$= \dfrac{\dfrac{1}{x} \cdot 4x^2 + \dfrac{1}{2} \cdot 4x^2}{\dfrac{1}{x^2} \cdot 4x^2 - \dfrac{1}{4} \cdot 4x^2}$

- Distributive Property

$= \dfrac{4x + 2x^2}{4 - x^2}$

- Simplify.

$= \dfrac{2x\overset{1}{\cancel{(2 + x)}}}{(2 - x)\underset{1}{\cancel{(2 + x)}}}$

$= \dfrac{2x}{2 - x}$

You Try It 1

Simplify: $\dfrac{\dfrac{1}{3} - \dfrac{1}{x}}{\dfrac{1}{9} - \dfrac{1}{x^2}}$

Your solution

Solution on p. S16

Copyright © Houghton Mifflin Company. All rights reserved.

Example 2

Simplify: $\dfrac{1 - \dfrac{2}{x} - \dfrac{15}{x^2}}{1 - \dfrac{11}{x} + \dfrac{30}{x^2}}$

You Try It 2

Simplify: $\dfrac{1 + \dfrac{4}{x} + \dfrac{3}{x^2}}{1 + \dfrac{10}{x} + \dfrac{21}{x^2}}$

Solution

The LCM of x and x^2 is x^2.

$$\dfrac{1 - \dfrac{2}{x} - \dfrac{15}{x^2}}{1 - \dfrac{11}{x} + \dfrac{30}{x^2}} = \dfrac{1 - \dfrac{2}{x} - \dfrac{15}{x^2}}{1 - \dfrac{11}{x} + \dfrac{30}{x^2}} \cdot \dfrac{x^2}{x^2}$$

• Multiply by the LCM.

$$= \dfrac{1 \cdot x^2 - \dfrac{2}{x} \cdot x^2 - \dfrac{15}{x^2} \cdot x^2}{1 \cdot x^2 - \dfrac{11}{x} \cdot x^2 + \dfrac{30}{x^2} \cdot x^2}$$

• Distributive Property

$$= \dfrac{x^2 - 2x - 15}{x^2 - 11x + 30}$$

$$= \dfrac{\overset{1}{\cancel{(x - 5)}}(x + 3)}{\underset{1}{\cancel{(x - 5)}}(x - 6)} = \dfrac{x + 3}{x - 6}$$

• Simplify.

Your solution

Example 3

Simplify: $\dfrac{x - 8 + \dfrac{20}{x + 4}}{x - 10 + \dfrac{24}{x + 4}}$

You Try It 3

Simplify: $\dfrac{x + 3 - \dfrac{20}{x - 5}}{x + 8 + \dfrac{30}{x - 5}}$

Solution

The LCM is $x + 4$.

$$\dfrac{x - 8 + \dfrac{20}{x + 4}}{x - 10 + \dfrac{24}{x + 4}}$$

$$= \dfrac{x - 8 + \dfrac{20}{x + 4}}{x - 10 + \dfrac{24}{x + 4}} \cdot \dfrac{x + 4}{x + 4}$$

• Multiply by the LCM.

$$= \dfrac{(x - 8)(x + 4) + \dfrac{20}{x + 4} \cdot (x + 4)}{(x - 10)(x + 4) + \dfrac{24}{x + 4} \cdot (x + 4)}$$

• Distributive Property

$$= \dfrac{x^2 - 4x - 32 + 20}{x^2 - 6x - 40 + 24} = \dfrac{x^2 - 4x - 12}{x^2 - 6x - 16}$$

• Simplify.

$$= \dfrac{(x - 6)\overset{1}{\cancel{(x + 2)}}}{(x - 8)\underset{1}{\cancel{(x + 2)}}} = \dfrac{x - 6}{x - 8}$$

Your solution

Solutions on p. S16

Copyright © Houghton Mifflin Company. All rights reserved.

6.4 Exercises

Objective A   To simplify a complex fraction

For Exercises 1 to 30, simplify.

1. $\dfrac{1 + \dfrac{3}{x}}{1 - \dfrac{9}{x^2}}$

2. $\dfrac{1 + \dfrac{4}{x}}{1 - \dfrac{16}{x^2}}$

3. $\dfrac{2 - \dfrac{8}{x + 4}}{3 - \dfrac{12}{x + 4}}$

4. $\dfrac{5 - \dfrac{25}{x + 5}}{1 - \dfrac{3}{x + 5}}$

5. $\dfrac{1 + \dfrac{5}{y - 2}}{1 - \dfrac{2}{y - 2}}$

6. $\dfrac{2 - \dfrac{11}{2x - 1}}{3 - \dfrac{17}{2x - 1}}$

7. $\dfrac{4 - \dfrac{2}{x + 7}}{5 + \dfrac{1}{x + 7}}$

8. $\dfrac{5 + \dfrac{3}{x - 8}}{2 - \dfrac{1}{x - 8}}$

9. $\dfrac{1 - \dfrac{1}{x} - \dfrac{6}{x^2}}{1 - \dfrac{9}{x^2}}$

10. $\dfrac{1 + \dfrac{4}{x} + \dfrac{4}{x^2}}{1 - \dfrac{2}{x} - \dfrac{8}{x^2}}$

11. $\dfrac{1 - \dfrac{5}{x} - \dfrac{6}{x^2}}{1 + \dfrac{6}{x} + \dfrac{5}{x^2}}$

12. $\dfrac{1 - \dfrac{7}{a} + \dfrac{12}{a^2}}{1 + \dfrac{1}{a} - \dfrac{20}{a^2}}$

13. $\dfrac{1 - \dfrac{6}{x} + \dfrac{8}{x^2}}{\dfrac{4}{x^2} + \dfrac{3}{x} - 1}$

14. $\dfrac{1 + \dfrac{3}{x} - \dfrac{18}{x^2}}{\dfrac{21}{x^2} - \dfrac{4}{x} - 1}$

15. $\dfrac{x - \dfrac{4}{x + 3}}{1 + \dfrac{1}{x + 3}}$

16. $\dfrac{y + \dfrac{1}{y - 2}}{1 + \dfrac{1}{y - 2}}$

17. $\dfrac{1 - \dfrac{x}{2x + 1}}{x - \dfrac{1}{2x + 1}}$

18. $\dfrac{1 - \dfrac{2x - 2}{3x - 1}}{x - \dfrac{4}{3x - 1}}$

Copyright © Houghton Mifflin Company. All rights reserved.

19. $\dfrac{x - 5 + \dfrac{14}{x + 4}}{x + 3 - \dfrac{2}{x + 4}}$

20. $\dfrac{a + 4 + \dfrac{5}{a - 2}}{a + 6 + \dfrac{15}{a - 2}}$

21. $\dfrac{x + 3 - \dfrac{10}{x - 6}}{x + 2 - \dfrac{20}{x - 6}}$

22. $\dfrac{x - 7 + \dfrac{5}{x - 1}}{x - 3 + \dfrac{1}{x - 1}}$

23. $\dfrac{y - 6 + \dfrac{22}{2y + 3}}{y - 5 + \dfrac{11}{2y + 3}}$

24. $\dfrac{x + 2 - \dfrac{12}{2x - 1}}{x + 1 - \dfrac{9}{2x - 1}}$

25. $\dfrac{x - \dfrac{2}{2x - 3}}{2x - 1 - \dfrac{8}{2x - 3}}$

26. $\dfrac{x + 3 - \dfrac{18}{2x + 1}}{x - \dfrac{6}{2x + 1}}$

27. $\dfrac{\dfrac{1}{x} - \dfrac{2}{x - 1}}{\dfrac{3}{x} + \dfrac{1}{x - 1}}$

28. $\dfrac{\dfrac{3}{n + 1} + \dfrac{1}{n}}{\dfrac{2}{n + 1} + \dfrac{3}{n}}$

29. $\dfrac{\dfrac{3}{2x - 1} - \dfrac{1}{x}}{\dfrac{4}{x} + \dfrac{2}{2x - 1}}$

30. $\dfrac{\dfrac{4}{3x + 1} + \dfrac{3}{x}}{\dfrac{6}{x} - \dfrac{2}{3x + 1}}$

APPLYING THE CONCEPTS

For Exercises 31 to 39, simplify.

31. $1 + \dfrac{1}{1 + \dfrac{1}{2}}$

32. $1 + \dfrac{1}{1 + \dfrac{1}{1 + \dfrac{1}{2}}}$

33. $1 - \dfrac{1}{1 - \dfrac{1}{x}}$

34. $\dfrac{a^{-1} - b^{-1}}{a^{-2} - b^{-2}}$

35. $\left(\dfrac{y}{4} - \dfrac{4}{y}\right) \div \left(\dfrac{4}{y} - 3 + \dfrac{y}{2}\right)$

36. $\left(\dfrac{b}{8} - \dfrac{8}{b}\right) \div \left(\dfrac{8}{b} - 5 + \dfrac{b}{2}\right)$

37. $\dfrac{1 + x^{-1}}{1 - x^{-1}}$

38. $\dfrac{x + x^{-1}}{x - x^{-1}}$

39. $\dfrac{x^{-1}}{y^{-1}} + \dfrac{x}{y}$

Copyright © Houghton Mifflin Company. All rights reserved.

6.5 Solving Equations Containing Fractions

Objective A   **To solve an equation containing fractions**

Recall that to solve an equation containing fractions, clear denominators by multiplying each side of the equation by the LCM of the denominators. Then solve for the variable.

HOW TO   Solve: $\dfrac{3x - 1}{4} + \dfrac{2}{3} = \dfrac{7}{6}$

$$\dfrac{3x - 1}{4} + \dfrac{2}{3} = \dfrac{7}{6}$$

$$12\left(\dfrac{3x - 1}{4} + \dfrac{2}{3}\right) = 12 \cdot \dfrac{7}{6}$$

- The LCM is **12**. To clear denominators, multiply each side of the equation by the LCM.

$$12\left(\dfrac{3x - 1}{4}\right) + 12 \cdot \dfrac{2}{3} = 12 \cdot \dfrac{7}{6}$$

- Simplify using the **Distributive Property** and the Properties of Fractions.

$$\dfrac{\overset{3}{\cancel{12}}}{1}\left(\dfrac{3x - 1}{\cancel{4}}\right) + \dfrac{\overset{4}{\cancel{12}}}{1} \cdot \dfrac{2}{\cancel{3}} = \dfrac{\overset{2}{\cancel{12}}}{1} \cdot \dfrac{7}{\cancel{6}}$$

$$9x - 3 + 8 = 14$$

- Solve for x.

$$9x + 5 = 14$$

$$9x = 9$$

$$x = 1$$

1 checks as a solution. The solution is 1.

Occasionally, a value of the variable that appears to be a solution of an equation will make one of the denominators zero. In this case, that value is not a solution of the equation.

HOW TO   Solve: $\dfrac{2x}{x - 2} = 1 + \dfrac{4}{x - 2}$

$$\dfrac{2x}{x - 2} = 1 + \dfrac{4}{x - 2}$$

$$(x - 2)\dfrac{2x}{x - 2} = (x - 2)\left(1 + \dfrac{4}{x - 2}\right)$$

- The LCM is $x - 2$. Multiply each side of the equation by the LCM.

$$(x - 2)\dfrac{2x}{x - 2} = (x - 2) \cdot 1 + (x - 2)\dfrac{4}{x - 2}$$

- Simplify using the **Distributive Property** and the Properties of Fractions.

$$\dfrac{\overset{1}{\cancel{(x - 2)}}}{1} \cdot \dfrac{2x}{\cancel{x - 2}} = (x - 2) \cdot 1 + \dfrac{\overset{1}{\cancel{(x - 2)}}}{1} \cdot \dfrac{4}{\cancel{x - 2}}$$

$$2x = x - 2 + 4$$

- Solve for x.

$$2x = x + 2$$

$$x = 2$$

When x is replaced by 2, the denominators of $\dfrac{2x}{x - 2}$ and $\dfrac{4}{x - 2}$ are zero.

Therefore, the equation has no solution.

Copyright © Houghton Mifflin Company. All rights reserved.

Example 1

Solve: $\dfrac{x}{x+4} = \dfrac{2}{x}$

Solution
The LCM is $x(x+4)$.

$$\dfrac{x}{x+4} = \dfrac{2}{x}$$

$$x(x+4)\left(\dfrac{x}{x+4}\right) = x(x+4)\left(\dfrac{2}{x}\right)$$ • Multiply by the LCM.

$$\dfrac{x\overset{1}{\cancel{(x+4)}}}{1} \cdot \dfrac{x}{\underset{1}{\cancel{x+4}}} = \dfrac{\overset{1}{\cancel{x}}(x+4)}{1} \cdot \dfrac{2}{\underset{1}{\cancel{x}}}$$ • Divide by common factors.

$$x^2 = (x+4)2$$ • Simplify.

$$x^2 = 2x + 8$$

Solve the quadratic equation by factoring.

$$x^2 - 2x - 8 = 0$$ • Write in standard form.

$$(x-4)(x+2) = 0$$ • Factor.

$$x - 4 = 0 \qquad x + 2 = 0$$ • Principle of Zero Products

$$x = 4 \qquad\quad x = -2$$

Both 4 and -2 check as solutions.
The solutions are 4 and -2.

You Try It 1

Solve: $\dfrac{x}{x+6} = \dfrac{3}{x}$

Your solution

Example 2

Solve: $\dfrac{3x}{x-4} = 5 + \dfrac{12}{x-4}$

Solution
The LCM is $x - 4$.

$$\dfrac{3x}{x-4} = 5 + \dfrac{12}{x-4}$$

$$(x-4)\left(\dfrac{3x}{x-4}\right) = (x-4)\left(5 + \dfrac{12}{x-4}\right)$$ • Clear denominators.

$$\dfrac{\overset{1}{\cancel{(x-4)}}}{1} \cdot \dfrac{3x}{\underset{1}{\cancel{x-4}}} = (x-4)5 + \dfrac{\overset{1}{\cancel{(x-4)}}}{1} \cdot \dfrac{12}{\underset{1}{\cancel{x-4}}}$$

$$3x = (x-4)5 + 12$$ • Solve for x.

$$3x = 5x - 20 + 12$$

$$3x = 5x - 8$$

$$-2x = -8$$

$$x = 4$$

4 does not check as a solution.
The equation has no solution.

You Try It 2

Solve: $\dfrac{5x}{x+2} = 3 - \dfrac{10}{x+2}$

Your solution

Solutions on p. S17

Copyright © Houghton Mifflin Company. All rights reserved.

6.5 Exercises

Objective A   To solve an equation containing fractions

1.   Can 2 be a solution of the equation $\dfrac{6x}{x+1} - \dfrac{x}{x-2} = 4$? Explain your answer.

2.   After multiplying each side of an equation by a variable expression, why must we check the solution?

For Exercises 3 to 35, solve.

3. $\dfrac{2x}{3} - \dfrac{5}{2} = -\dfrac{1}{2}$

4. $\dfrac{x}{3} - \dfrac{1}{4} = \dfrac{1}{12}$

5. $\dfrac{x}{3} - \dfrac{1}{4} = \dfrac{x}{4} - \dfrac{1}{6}$

6. $\dfrac{2y}{9} - \dfrac{1}{6} = \dfrac{y}{9} + \dfrac{1}{6}$

7. $\dfrac{2x-5}{8} + \dfrac{1}{4} = \dfrac{x}{8} + \dfrac{3}{4}$

8. $\dfrac{3x+4}{12} - \dfrac{1}{3} = \dfrac{5x+2}{12} - \dfrac{1}{2}$

9. $\dfrac{6}{2a+1} = 2$

10. $\dfrac{12}{3x-2} = 3$

11. $\dfrac{9}{2x-5} = -2$

12. $\dfrac{6}{4-3x} = 3$

13. $2 + \dfrac{5}{x} = 7$

14. $3 + \dfrac{8}{n} = 5$

15. $1 - \dfrac{9}{x} = 4$

16. $3 - \dfrac{12}{x} = 7$

17. $\dfrac{2}{y} + 5 = 9$

18. $\dfrac{6}{x} + 3 = 11$

19. $\dfrac{3}{x-2} = \dfrac{4}{x}$

20. $\dfrac{5}{x+3} = \dfrac{3}{x-1}$

21. $\dfrac{2}{3x-1} = \dfrac{3}{4x+1}$

22. $\dfrac{5}{3x-4} = \dfrac{-3}{1-2x}$

23. $\dfrac{-3}{2x+5} = \dfrac{2}{x-1}$

Copyright © Houghton Mifflin Company. All rights reserved.

24. $\dfrac{4}{5y - 1} = \dfrac{2}{2y - 1}$

25. $\dfrac{4x}{x - 4} + 5 = \dfrac{5x}{x - 4}$

26. $\dfrac{2x}{x + 2} - 5 = \dfrac{7x}{x + 2}$

27. $2 + \dfrac{3}{a - 3} = \dfrac{a}{a - 3}$

28. $\dfrac{x}{x + 4} = 3 - \dfrac{4}{x + 4}$

29. $\dfrac{x}{x - 1} = \dfrac{8}{x + 2}$

30. $\dfrac{x}{x + 12} = \dfrac{1}{x + 5}$

31. $\dfrac{2x}{x + 4} = \dfrac{3}{x - 1}$

32. $\dfrac{5}{3n - 8} = \dfrac{n}{n + 2}$

33. $x + \dfrac{6}{x - 2} = \dfrac{3x}{x - 2}$

34. $x - \dfrac{6}{x - 3} = \dfrac{2x}{x - 3}$

35. $\dfrac{8}{y} = \dfrac{2}{y - 2} + 1$

APPLYING THE CONCEPTS

36. Explain the procedure for solving an equation containing fractions. Include in your discussion how the LCM of the denominators is used to eliminate fractions in the equation.

For Exercises 37 to 42, solve.

37. $\dfrac{3}{5}y - \dfrac{1}{3}(1 - y) = \dfrac{2y - 5}{15}$

38. $\dfrac{3}{4}a = \dfrac{1}{2}(3 - a) + \dfrac{a - 2}{4}$

39. $\dfrac{b + 2}{5} = \dfrac{1}{4}b - \dfrac{3}{10}(b - 1)$

40. $\dfrac{x}{2x^2 - x - 1} = \dfrac{3}{x^2 - 1} + \dfrac{3}{2x + 1}$

41. $\dfrac{x + 1}{x^2 + x - 2} = \dfrac{x + 2}{x^2 - 1} + \dfrac{3}{x + 2}$

42. $\dfrac{y + 2}{y^2 - y - 2} + \dfrac{y + 1}{y^2 - 4} = \dfrac{1}{y + 1}$

Copyright © Houghton Mifflin Company. All rights reserved.

6.6 Ratio and Proportion

Objective A   To solve a proportion

Quantities such as 4 meters, 15 seconds, and 8 gallons are number quantities written with units. In these examples the units are meters, seconds, and gallons.

A **ratio** is the quotient of two quantities that have the same unit.

The length of a living room is 16 ft and the width is 12 ft. The ratio of the length to the width is written

$$\frac{16 \text{ ft}}{12 \text{ ft}} = \frac{16}{12} = \frac{4}{3}$$ A ratio is in simplest form when the two numbers do not have a common factor. Note that the units are not written.

A **rate** is the quotient of two quantities that have different units.

There are 2 lb of salt in 8 gal of water. The salt-to-water rate is

$$\frac{2 \text{ lb}}{8 \text{ gal}} = \frac{1 \text{ lb}}{4 \text{ gal}}$$ A rate is in simplest form when the two numbers do not have a common factor. The units are written as part of the rate.

A **proportion** is an equation that states the equality of two ratios or rates. Examples of proportions are shown at the right.

$$\frac{30 \text{ mi}}{4 \text{ h}} = \frac{15 \text{ mi}}{2 \text{ h}} \qquad \frac{4}{6} = \frac{8}{12} \qquad \frac{3}{4} = \frac{x}{8}$$

HOW TO   Solve the proportion $\dfrac{4}{x} = \dfrac{2}{3}$.

$$\frac{4}{x} = \frac{2}{3}$$

$$3x\left(\frac{4}{x}\right) = 3x\left(\frac{2}{3}\right)$$ • The LCM of the denominators is **3x**. To clear denominators, multiply each side of the proportion by the LCM.

$$12 = 2x$$ • Solve the equation.

$$6 = x$$

The solution is 6.

Example 1   Solve: $\dfrac{8}{x + 3} = \dfrac{4}{x}$

Solution

$$\frac{8}{x + 3} = \frac{4}{x}$$

$$x(x + 3)\frac{8}{x + 3} = x(x + 3)\frac{4}{x}$$ • Clear denominators.

$$8x = 4(x + 3)$$ • Solve for **x**.

$$8x = 4x + 12$$

$$4x = 12$$

$$x = 3$$

The solution is 3.

You Try It 1   Solve: $\dfrac{2}{x + 3} = \dfrac{6}{5x + 5}$

Your solution

Solution on p. S17

Copyright © Houghton Mifflin Company. All rights reserved.

Copyright © Houghton Mifflin Company. All rights reserved.

Objective B   **To solve application problems**

Example 2

The monthly loan payment for a car is $28.35 for each $1000 borrowed. At this rate, find the monthly payment for a $6000 car loan.

Strategy
To find the monthly payment, write and solve a proportion, using P to represent the monthly car payment.

Solution

$$\frac{28.35}{1000} = \frac{P}{6000}$$ • Write a proportion.

$$6000\left(\frac{28.35}{1000}\right) = 6000\left(\frac{P}{6000}\right)$$ • Clear denominators.

$$170.10 = P$$

The monthly payment is $170.10.

You Try It 2

Sixteen ceramic tiles are needed to tile a 9-square-foot area. At this rate, how many square feet can be tiled using 256 ceramic tiles?

Your strategy

Your solution

Solution on p. S17

Objective C   **To solve problems involving similar triangles**

Similar objects have the same shape but not necessarily the same size. A tennis ball is similar to a basketball. A model ship is similar to an actual ship.

Similar objects have corresponding parts; for example, the rudder on the model ship corresponds to the rudder on the actual ship. The relationship between the sizes of each of the corresponding parts can be written as a ratio, and each ratio will be the same. If the rudder on the model ship is $\frac{1}{100}$ the size of the rudder on the actual ship, then the model wheelhouse is $\frac{1}{100}$ the size of the actual wheelhouse, the width of the model is $\frac{1}{100}$ the width of the actual ship, and so on.

The two triangles *ABC* and *DEF* shown at the right are similar. Side *AB* corresponds to *DE*, side *BC* corresponds to *EF*, and side *AC* corresponds to *DF*. The height *CH* corresponds to the height *FK*. The ratios of corresponding parts of similar triangles are equal.

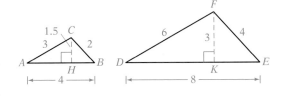

$$\frac{AB}{DE} = \frac{4}{8} = \frac{1}{2}, \qquad \frac{AC}{DF} = \frac{3}{6} = \frac{1}{2}, \qquad \frac{BC}{EF} = \frac{2}{4} = \frac{1}{2}, \qquad \text{and} \qquad \frac{CH}{FK} = \frac{1.5}{3} = \frac{1}{2}$$

Because the ratios of corresponding parts are equal, three proportions can be formed using the sides of the triangles.

$$\frac{AB}{DE} = \frac{AC}{DF}, \qquad \frac{AB}{DE} = \frac{BC}{EF}, \qquad \text{and} \qquad \frac{AC}{DF} = \frac{BC}{EF}$$

Three proportions can also be formed by using the sides and height of the triangles.

$$\frac{AB}{DE} = \frac{CH}{FK}, \qquad \frac{AC}{DF} = \frac{CH}{FK}, \qquad \text{and} \qquad \frac{BC}{EF} = \frac{CH}{FK}$$

The measures of the corresponding angles in similar triangles are equal. Therefore,

$$m\angle A = m\angle D, \qquad m\angle B = m\angle E, \qquad \text{and} \qquad m\angle C = m\angle F$$

It is also true that if the measures of the three angles of one triangle are equal, respectively, to the measures of the three angles of another triangle, then the two triangles are similar.

T A K E N O T E

Vertical angles of intersecting lines, corresponding angles of parallel lines, and angles of a triangle are discussed in Section 3.5.

A line DE is drawn parallel to the base AB in the triangle at the right. $m\angle x = m\angle m$ and $m\angle y = m\angle n$ because corresponding angles are equal. $m\angle C = m\angle C$; thus the measures of the three angles of triangle DEC are equal, respectively, to the measures of the three angles of triangle ABC. Triangle DEC is similar to triangle ABC.

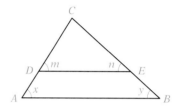

The sum of the measures of the three angles of a triangle is 180°. If two angles of one triangle are equal in measure to two angles of another triangle, then the third angles must be equal in measure. Thus we can say that if two angles of one triangle are equal in measure to two angles of another triangle, then the two triangles are similar.

HOW TO   The line segments AB and CD intersect at point O in the figure at the right. Angles C and D are right angles. Find the length of DO.

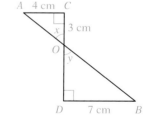

First we must determine whether triangle AOC is similar to triangle BOD.

$m\angle C = m\angle D$ because they are right angles.

$m\angle x = m\angle y$ because they are vertical angles.

Triangle AOC is similar to triangle BOD because two angles of one triangle are equal in measure to two angles of the other triangle.

$$\frac{AC}{DB} = \frac{CO}{DO}$$
• Use a proportion to find the length of the unknown side.

$$\frac{4}{7} = \frac{3}{DO}$$
• $AC = 4$, $CO = 3$, and $DB = 7$.

$$7(DO)\frac{4}{7} = 7(DO)\frac{3}{DO}$$
• To clear denominators, multiply each side of the proportion by $7(DO)$.

$$4(DO) = 7(3)$$
• Solve for DO.

$$4(DO) = 21$$

$$DO = 5.25$$

Copyright © Houghton Mifflin Company. All rights reserved.

HOW TO Triangles ABC and DEF at the right are similar. Find the area of triangle ABC.

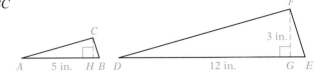

$\dfrac{AB}{DE} = \dfrac{CH}{FG}$ • Solve a proportion to find the height of triangle **ABC**.

$\dfrac{5}{12} = \dfrac{CH}{3}$ • **AB** = 5, **DE** = 12, and **FG** = 3.

$12 \cdot \dfrac{5}{12} = 12 \cdot \dfrac{CH}{3}$ • To clear denominators, multiply each side of the proportion by **12**.

$5 = 4(CH)$ • Solve for **CH**.

$1.25 = CH$ • The height is 1.25 in. The base is 5 in.

$A = \dfrac{1}{2}bh = \dfrac{1}{2}(5)(1.25) = 3.125$ • Use the formula for the area of a triangle.

The area of triangle ABC is 3.125 in^2.

Example 3

In the figure below, AB is parallel to DC, and angles B and D are right angles. $AB = 12$ m, $DC = 4$ m, and $AC = 18$ m. Find the length of CO.

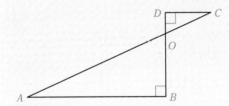

Strategy

Triangle AOB is similar to triangle COD. Solve a proportion to find the length of CO. Let x represent the length of CO and $18 - x$ represent the length of AO.

Solution

$\dfrac{DC}{AB} = \dfrac{CO}{AO}$ • Write a proportion.

$\dfrac{4}{12} = \dfrac{x}{18 - x}$ • Substitute.

$12(18 - x) \cdot \dfrac{4}{12} = 12(18 - x) \cdot \dfrac{x}{18 - x}$ • Clear denominators.

$4(18 - x) = 12x$ • Solve for x.

$72 - 4x = 12x$

$72 = 16x$

$4.5 = x$ • x is the length of **CO**.

The length of CO is 4.5 m.

You Try It 3

In the figure below, AB is parallel to DC, and angles A and D are right angles. $AB = 10$ cm, $CD = 4$ cm, and $DO = 3$ cm. Find the area of triangle AOB.

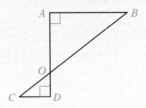

Your strategy

Your solution

Solution on p. S17

Copyright © Houghton Mifflin Company. All rights reserved.

6.6 Exercises

Objective A   **To solve a proportion**

1. What is a proportion?

2. Explain a method for solving a proportion.

For Exercises 3 to 17, solve.

3. $\dfrac{x}{12} = \dfrac{3}{4}$

4. $\dfrac{6}{x} = \dfrac{2}{3}$

5. $\dfrac{4}{9} = \dfrac{x}{27}$

6. $\dfrac{16}{9} = \dfrac{64}{x}$

7. $\dfrac{x+3}{12} = \dfrac{5}{6}$

8. $\dfrac{3}{5} = \dfrac{x-4}{10}$

9. $\dfrac{18}{x+4} = \dfrac{9}{5}$

10. $\dfrac{2}{11} = \dfrac{20}{x-3}$

11. $\dfrac{2}{x} = \dfrac{4}{x+1}$

12. $\dfrac{16}{x-2} = \dfrac{8}{x}$

13. $\dfrac{x+3}{4} = \dfrac{x}{8}$

14. $\dfrac{x-6}{3} = \dfrac{x}{5}$

15. $\dfrac{2}{x-1} = \dfrac{6}{2x+1}$

16. $\dfrac{9}{x+2} = \dfrac{3}{x-2}$

17. $\dfrac{2x}{7} = \dfrac{x-2}{14}$

Objective B   **To solve application problems**

18. **Cooking**   Simple syrup used in making some desserts requires 2 c of sugar for every $\frac{2}{3}$ c of boiling water. At this rate, how many cups of sugar are required for 2 c boiling water?

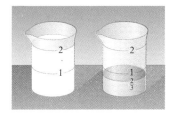

19. **Surveys**   An exit poll survey showed that 4 out of every 7 voters cast a ballot in favor of an amendment to a city charter. At this rate, how many voters voted in favor of the amendment if 35,000 people voted?

20. **Surveys**   In a city of 25,000 homes, a survey was taken to determine the number with cable television. Of the 300 homes surveyed, 210 had cable television. Estimate the number of homes in the city that have cable television.

21. **Cartography**   On a map, two cities are $2\frac{5}{8}$ in. apart. If $\frac{3}{8}$ in. on the map represents 25 mi, find the number of miles between the two cities.

Copyright © Houghton Mifflin Company. All rights reserved.

22. Business A company decides to accept a large shipment of 10,000 computer chips if there are 2 or fewer defects in a sample of 100 randomly chosen chips. Assuming that there are 300 defective chips in the shipment and that the rate of defective chips in the sample is the same as the rate in the shipment, will the shipment be accepted?

23. Taxes The sales tax on a car that sold for $12,000 is $780. At this rate, how much higher is the sales tax on a car that sells for $13,500?

24. Art Leonardo da Vinci measured various distances on the human body in order to make accurate drawings. He determined that generally the ratio of the kneeling height of a person to the standing height of that person was $\frac{3}{4}$. Using this ratio, determine how tall a person is who has a kneeling height of 48 in.

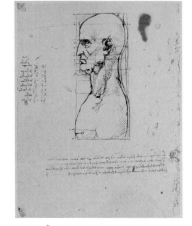

25. Art In one of Leonardo da Vinci's notebooks, he wrote that "…from the top to the bottom of the chin is the sixth part of a face, and it is the fifty-fourth part of the man." Suppose the distance from the top to the bottom of the chin of a person is 1.25 in. Using da Vinci's measurements, find the height of this person.

26. Conservation As part of a conservation effort for a lake, 40 fish are caught, tagged, and then released. Later 80 fish are caught. Four of the 80 fish are found to have tags. Estimate the number of fish in the lake.

27. Conservation In a wildlife preserve, 10 elk are captured, tagged, and then released. Later 15 elk are captured and 2 are found to have tags. Estimate the number of elk in the preserve.

28. Rocketry The engine of a small rocket burns 170,000 lb of fuel in 1 min. At this rate, how many pounds of fuel does the rocket burn in 45 s?

Objective C **To solve problems involving similar triangles**

Triangles *ABC* and *DEF* in Exercises 29 to 36 are similar. Round answers to the nearest tenth.

29. Find side *AC*.

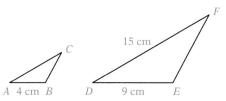

30. Find side *DE*.

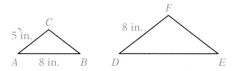

Copyright © Houghton Mifflin Company. All rights reserved.

31. Find the height of triangle *ABC*.

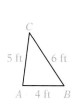

32. Find the height of triangle *DEF*.

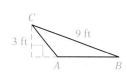
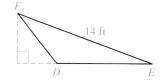

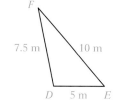

33. Find the perimeter of triangle *DEF*.

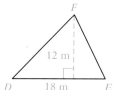

34. Find the perimeter of triangle *ABC*.

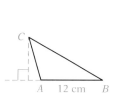
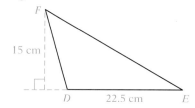

35. Find the area of triangle *ABC*.

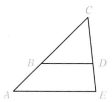

36. Find the area of triangle *ABC*.

37. Given *BD* ∥ *AE*, *BD* measures 5 cm, *AE* measures 8 cm, and *AC* measures 10 cm, find the length of *BC*.

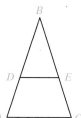

38. Given *AC* ∥ *DE*, *BD* measures 8 m, *AD* measures 12 m, and *BE* measures 6 m, find the length of *BC*.

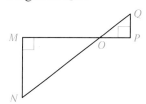

39. Given *DE* ∥ *AC*, *DE* measures 6 in., *AC* measures 10 in., and *AB* measures 15 in., find the length of *DA*.

40. Given *MP* and *NQ* intersect at *O*, *NO* measures 25 ft, *MO* measures 20 ft, and *PO* measures 8 ft, find the length of *QO*.

Copyright © Houghton Mifflin Company. All rights reserved.

41. Given *MP* and *NQ* intersect at *O*, *NO* measures 24 cm, *MN* measures 10 cm, *MP* measures 39 cm, and *QO* measures 12 cm, find the length of *OP*.

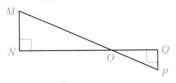

42. Given *MQ* and *NP* intersect at *O*, *NO* measures 12 m, *MN* measures 9 m, *PQ* measures 3 m, and *MQ* measures 20 m, find the perimeter of triangle *OPQ*.

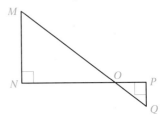

43. Indirect Measurement Similar triangles can be used as an indirect way to measure inaccessible distances. The diagram at the right represents a river of width *DC*. The triangles *AOB* and *DOC* are similar. The distances *AB*, *BO*, and *OC* can be measured. Find the width of the river.

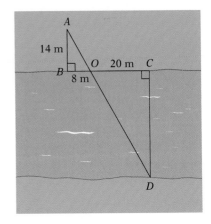

44. Indirect Measurement The sun's rays cast a shadow as shown in the diagram at the right. Find the height of the flagpole. Write the answer in terms of feet.

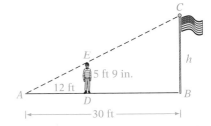

APPLYING THE CONCEPTS

45. Lottery Tickets Three people put their money together to buy lottery tickets. The first person put in $25, the second person put in $30, and the third person put in $35. One of their tickets was a winning ticket. If they won $4.5 million, what was the first person's share of the winnings?

46. Clubs No one belongs to both the Math Club and the Photography Club, but the two clubs join to hold a car wash. Ten members of the Math Club and 6 members of the Photography Club participate. The profits from the car wash are $120. If each club's profits are proportional to the number of members participating, what share of the profits does the Math Club receive?

47. Sports A basketball player has made 5 out of every 6 foul shots attempted in one year of play. If 42 foul shots were missed that year, how many foul shots did the basketball player make?

Copyright © Houghton Mifflin Company. All rights reserved.

6.7 Literal Equations

Objective A   **To solve a literal equation for one of the variables**

A **literal equation** is an equation that contains more than one variable. Examples of literal equations are shown at the right.

$$2x + 3y = 6$$
$$4w - 2x + z = 0$$

Formulas are used to express a relationship among physical quantities. A **formula** is a literal equation that states a rule about measurements. Examples of formulas are shown at the right.

$$\frac{1}{R_1} + \frac{1}{R_2} = \frac{1}{R} \qquad \text{(Physics)}$$
$$s = a + (n - 1)d \qquad \text{(Mathematics)}$$
$$A = P + Prt \qquad \text{(Business)}$$

The Addition and Multiplication Properties can be used to solve a literal equation for one of the variables. The goal is to rewrite the equation so that the variable being solved for is alone on one side of the equation and all the other numbers and variables are on the other side.

HOW TO   Solve $A = P(1 + i)$ for i.

The goal is to rewrite the equation so that i is on one side of the equation and all other variables are on the other side.

$$A = P(1 + i)$$
$$A = P + Pi$$ • Use the **Distributive Property** to remove parentheses.
$$A - P = P - P + Pi$$ • Subtract P from each side of the equation.
$$A - P = Pi$$
$$\frac{A - P}{P} = \frac{Pi}{P}$$ • Divide each side of the equation by P.
$$\frac{A - P}{P} = i$$

Example 1

Solve $3x - 4y = 12$ for y.

Solution

$$3x - 4y = 12$$
$$3x - 3x - 4y = -3x + 12 \qquad \text{• Subtract } 3x.$$
$$-4y = -3x + 12$$
$$\frac{-4y}{-4} = \frac{-3x + 12}{-4} \qquad \text{• Divide by } -4.$$
$$y = \frac{3}{4}x - 3$$

You Try It 1

Solve $5x - 2y = 10$ for y.

Your solution

Solution on p. S17

Copyright © Houghton Mifflin Company. All rights reserved.

Example 2

Solve $I = \dfrac{E}{R + r}$ for R.

Solution

$$I = \dfrac{E}{R + r}$$

$$(R + r)I = (R + r)\dfrac{E}{R + r} \qquad \bullet \text{ Multiply by } (R + r).$$

$$RI + rI = E$$

$$RI + rI - rI = E - rI \qquad \bullet \text{ Subtract } rI.$$

$$RI = E - rI$$

$$\dfrac{RI}{I} = \dfrac{E - rI}{I} \qquad \bullet \text{ Divide by } I.$$

$$R = \dfrac{E - rI}{I}$$

You Try It 2

Solve $s = \dfrac{A + L}{2}$ for L.

Your solution

Example 3

Solve $L = a(1 + ct)$ for c.

Solution

$$L = a(1 + ct)$$

$$L = a + act \qquad \bullet \text{ Distributive Property}$$

$$L - a = a - a + act \qquad \bullet \text{ Subtract } a.$$

$$L - a = act$$

$$\dfrac{L - a}{at} = \dfrac{act}{at} \qquad \bullet \text{ Divide by } at.$$

$$\dfrac{L - a}{at} = c$$

You Try It 3

Solve $S = a + (n - 1)d$ for n.

Your solution

Example 4

Solve $S = C - rC$ for C.

Solution

$$S = C - rC$$

$$S = (1 - r)C \qquad \bullet \text{ Factor.}$$

$$\dfrac{S}{1 - r} = \dfrac{(1 - r)C}{1 - r} \qquad \bullet \text{ Divide by } (1 - r).$$

$$\dfrac{S}{1 - r} = C$$

You Try It 4

Solve $S = rS + C$ for S.

Your solution

Solutions on p. S18

Copyright © Houghton Mifflin Company. All rights reserved.

6.7 Exercises

Objective A   To solve a literal equation for one of the variables

For Exercises 1 to 15, solve for y.

1. $3x + y = 10$

2. $2x + y = 5$

3. $4x - y = 3$

4. $5x - y = 7$

5. $3x + 2y = 6$

6. $2x + 3y = 9$

7. $2x - 5y = 10$

8. $5x - 2y = 4$

9. $2x + 7y = 14$

10. $6x - 5y = 10$

11. $x + 3y = 6$

12. $x + 2y = 8$

13. $y - 2 = 3(x + 2)$

14. $y + 4 = -2(x - 3)$

15. $y - 1 = -\dfrac{2}{3}(x + 6)$

For Exercises 16 to 24, solve for x.

16. $x + 3y = 6$

17. $x + 6y = 10$

18. $3x - y = 3$

19. $2x - y = 6$

20. $2x + 5y = 10$

21. $4x + 3y = 12$

22. $x - 2y + 1 = 0$

23. $x - 4y - 3 = 0$

24. $5x + 4y + 20 = 0$

For Exercises 25 to 40, solve the formula for the given variable.

25. $d = rt$; t (Physics)

26. $E = IR$; R (Physics)

27. $PV = nRT$; T (Chemistry)

28. $A = bh$; h (Geometry)

Copyright © Houghton Mifflin Company. All rights reserved.

29. $P = 2l + 2w; l$ (Geometry)

30. $F = \dfrac{9}{5}C + 32; C$ (Temperature conversion)

31. $A = \dfrac{1}{2}h(b_1 + b_2); b_1$ (Geometry)

32. $C = \dfrac{5}{9}(F - 32); F$ (Temperature conversion)

33. $V = \dfrac{1}{3}Ah; h$ (Geometry)

34. $P = R - C; C$ (Business)

35. $R = \dfrac{C - S}{t}; S$ (Business)

36. $P = \dfrac{R - C}{n}; R$ (Business)

37. $A = P + Prt; P$ (Business)

38. $T = fm - gm; m$ (Engineering)

39. $A = Sw + w; w$ (Physics)

40. $a = S - Sr; S$ (Mathematics)

APPLYING THE CONCEPTS

Business   Break-even analysis is a method used to determine the sales volume required for a company to break even, or experience neither a profit nor a loss on the sale of a product. The break-even point represents the number of units that must be made and sold for income from sales to equal the cost of the product. The break-even point can be calculated using the formula $B = \dfrac{F}{S - V}$, where F is the fixed costs, S is the selling price per unit, and V is the variable costs per unit. Use this information for Exercise 41.

41. a. Solve the formula $B = \dfrac{F}{S - V}$ for S.

  b. Use your answer to part **a** to find the selling price per unit required for a company to break even. The fixed costs are $20,000, the variable costs per unit are $80, and the company plans to make and sell 200 desks.

  c. Use your answer to part **a** to find the selling price per unit required for a company to break even. The fixed costs are $15,000, the variable costs per unit are $50, and the company plans to make and sell 600 cameras.

Copyright © Houghton Mifflin Company. All rights reserved.

6.8 Application Problems

Objective A   To solve work problems

If a painter can paint a room in 4 h, then in 1 h the painter can paint $\frac{1}{4}$ of the room. The painter's rate of work is $\frac{1}{4}$ of the room each hour. The **rate of work** is the part of a task that is completed in 1 unit of time.

A pipe can fill a tank in 30 min. This pipe can fill $\frac{1}{30}$ of the tank in 1 min. The rate of work is $\frac{1}{30}$ of the tank each minute. If a second pipe can fill the tank in x min, the rate of work for the second pipe is $\frac{1}{x}$ of the tank each minute.

In solving a work problem, the goal is to determine the time it takes to complete a task. The basic equation that is used to solve work problems is

Rate of work × time worked = part of task completed

For example, if a faucet can fill a sink in 6 min, then in 5 min the faucet will fill $\frac{1}{6} \times 5 = \frac{5}{6}$ of the sink. In 5 min the faucet completes $\frac{5}{6}$ of the task.

• **Study Tip** •

Note in the examples in this section that solving a word problem includes stating a strategy and using the strategy to find a solution. If you have difficulty with a word problem, write down the known information. Be very specific. Write out a phrase or sentence that states what you are trying to find. See *AIM for Success,* page xxv.

TAKE NOTE

Use the information given in the problem to fill in the "Rate" and "Time" columns of the table. Fill in the "Part Completed" column by multiplying the two expressions you wrote in each row.

Copyright © Houghton Mifflin Company. All rights reserved.

HOW TO   A painter can paint a wall in 20 min. The painter's apprentice can paint the same wall in 30 min. How long will it take them to paint the wall when they work together?

Strategy for Solving a Work Problem

1. For each person or machine, write a numerical or variable expression for the rate of work, the time worked, and the part of the task completed. The results can be recorded in a table.

Unknown time to paint the wall working together: t

	Rate of Work	·	*Time Worked*	=	*Part of Task Completed*
Painter	$\frac{1}{20}$	·	t	=	$\frac{t}{20}$
Apprentice	$\frac{1}{30}$	·	t	=	$\frac{t}{30}$

2. Determine how the parts of the task completed are related. Use the fact that the sum of the parts of the task completed must equal 1, the complete task.

$$\frac{t}{20} + \frac{t}{30} = 1$$

• The sum of the part of the task completed by the painter and the part of the task completed by the apprentice is **1**.

$$60\left(\frac{t}{20} + \frac{t}{30}\right) = 60 \cdot 1$$

• Multiply by the **LCM of 20 and 30**.

$$3t + 2t = 60$$

• Distributive Property

$$5t = 60$$

$$t = 12$$

Working together, they will paint the wall in 12 min.

Example 1

A small water pipe takes three times longer to fill a tank than does a large water pipe. With both pipes open it takes 4 h to fill the tank. Find the time it would take the small pipe, working alone, to fill the tank.

You Try It 1

Two computer printers that work at the same rate are working together to print the payroll checks for a large corporation. After they work together for 2 h, one of the printers quits. The second requires 3 h more to complete the payroll checks. Find the time it would take one printer, working alone, to print the payroll.

Strategy

• Time for large pipe to fill the tank: t
  Time for small pipe to fill the tank: $3t$

Your strategy

Fills tank in $3t$ hours Fills tank in t hours

Fills $\frac{4}{3t}$ of the tank in 4 hours Fills $\frac{4}{t}$ of the tank in 4 hours

	Rate	*Time*	*Part*
Small pipe	$\frac{1}{3t}$	4	$\frac{4}{3t}$
Large pipe	$\frac{1}{t}$	4	$\frac{4}{t}$

• The sum of the parts of the task completed by each pipe must equal 1.

Solution

$$\frac{4}{3t} + \frac{4}{t} = 1$$

$$3t\left(\frac{4}{3t} + \frac{4}{t}\right) = 3t \cdot 1$$

   • **Multiply by the LCM of 3t and t.**

$$4 + 12 = 3t$$

   • **Distributive Property**

$$16 = 3t$$

$$\frac{16}{3} = t$$

$$3t = 3\left(\frac{16}{3}\right) = 16$$

The small pipe working alone takes 16 h to fill the tank.

Your solution

Solution on p. S18

Copyright © Houghton Mifflin Company. All rights reserved.

Objective B   **To use rational expressions to solve uniform motion problems**

A car that travels constantly in a straight line at 30 mph is in uniform motion. **Uniform motion** means that the speed or direction of an object does not change.

The basic equation used to solve uniform motion problems is

$$\textbf{Distance = rate} \times \textbf{time}$$

An alternative form of this equation can be written by solving the equation for time.

$$\frac{\textbf{Distance}}{\textbf{Rate}} = \textbf{time}$$

This form of the equation is useful when the total time of travel for two objects or the time of travel between two points is known.

HOW TO   The speed of a boat in still water is 20 mph. The boat traveled 75 mi down a river in the same amount of time it took to travel 45 mi up the river. Find the rate of the river's current.

> **Strategy for Solving a Uniform Motion Problem**
>
> **1.** For each object, write a numerical or variable expression for the distance, rate, and time. The results can be recorded in a table.

The unknown rate of the river's current: r

TAKE NOTE

Use the information given in the problem to fill in the "Distance" and "Rate" columns of the table. Fill in the "Time" column by dividing the two expressions you wrote in each row.

	Distance	÷	*Rate*	=	*Time*
Down river	75	÷	$20 + r$	=	$\dfrac{75}{20 + r}$
Up river	45	÷	$20 - r$	=	$\dfrac{45}{20 - r}$

> **2.** Determine how the times traveled by each object are related. For example, it may be known that the times are equal, or the total time may be known.

$$\frac{75}{20 + r} = \frac{45}{20 - r}$$

• The time down the river is equal to the time up the river.

$$(20 + r)(20 - r)\frac{75}{20 + r} = (20 + r)(20 - r)\frac{45}{20 - r}$$

• Multiply by the **LCM**.

$$(20 - r)75 = (20 + r)45$$
$$1500 - 75r = 900 + 45r$$

• Distributive Property

$$-120r = -600$$
$$r = 5$$

The rate of the river's current is 5 mph.

Copyright © Houghton Mifflin Company. All rights reserved.

Copyright © Houghton Mifflin Company. All rights reserved.

Example 2

A cyclist rode the first 20 mi of a trip at a constant rate. For the next 16 mi, the cyclist reduced the speed by 2 mph. The total time for the 36 mi was 4 h. Find the rate of the cyclist for each leg of the trip.

Strategy

• Rate for the first 20 mi: r
  Rate for the next 16 mi: $r - 2$

	Distance	Rate	Time
First 20 mi	20	r	$\dfrac{20}{r}$
Next 16 mi	16	$r - 2$	$\dfrac{16}{r - 2}$

• The total time for the trip was 4 h.

Solution

$$\frac{20}{r} + \frac{16}{r - 2} = 4$$

$$r(r - 2)\left[\frac{20}{r} + \frac{16}{r - 2}\right] = r(r - 2) \cdot 4$$

$$(r - 2)20 + 16r = 4r^2 - 8r$$

$$20r - 40 + 16r = 4r^2 - 8r$$

$$36r - 40 = 4r^2 - 8r$$

Solve the quadratic equation by factoring.

$$0 = 4r^2 - 44r + 40$$

$$0 = 4(r^2 - 11r + 10)$$

$$0 = 4(r - 10)(r - 1)$$

$$r - 10 = 0 \qquad r - 1 = 0$$

$$r = 10 \qquad r = 1$$

The solution $r = 1$ mph is not possible, because the rate on the last 16 mi would then be -1 mph.

10 mph was the rate for the first 20 mi.
8 mph was the rate for the next 16 mi.

You Try It 2

The total time it took for a sailboat to sail back and forth across a lake 6 km wide was 2 h. The rate sailing back was three times the rate sailing across. Find the rate sailing out across the lake.

Your strategy

Your solution

• The total time was 4 h.

• Multiply by the LCM.

• Distributive Property

• Standard form

• Factor.

• Principle of Zero Products

Solution on p. S18

6.8 Exercises

Copyright © Houghton Mifflin Company. All rights reserved.

Objective A   **To solve work problems**

1.   Explain the meaning of the phrase "rate of work."

2.   If $\frac{2}{5}$ of a room can be painted in 1 h, what is the rate of work? At the same rate, how long will it take to paint the entire room?

3.   A park has two sprinklers that are used to fill a fountain. One sprinkler can fill the fountain in 3 h, whereas the second sprinkler can fill the fountain in 6 h. How long will it take to fill the fountain with both sprinklers operating?

4.   One grocery clerk can stock a shelf in 20 min, whereas a second clerk requires 30 min to stock the same shelf. How long would it take to stock the shelf if the two clerks worked together?

5.   One person with a skiploader requires 12 h to remove a large quantity of earth. A second, larger skiploader can remove the same amount of earth in 4 h. How long would it take to remove the earth with both skiploaders working together?

6.   An experienced painter can paint a fence twice as fast as an inexperienced painter. Working together, the painters require 4 h to paint the fence. How long would it take the experienced painter, working alone, to paint the fence?

7.   One computer can solve a complex prime factorization problem in 75 h. A second computer can solve the same problem in 50 h. How long would it take both computers, working together, to solve the problem?

8.   A new machine can make 10,000 aluminum cans three times faster than an older machine. With both machines working, 10,000 cans can be made in 9 h. How long would it take the new machine, working alone, to make the 10,000 cans?

9.   A small air conditioner can cool a room 5° in 75 min. A larger air conditioner can cool the room 5° in 50 min. How long would it take to cool the room 5° with both air conditioners working?

10.   One printing press can print the first edition of a book in 55 min, whereas a second printing press requires 66 min to print the same number of copies. How long would it take to print the first edition with both presses operating?

11.   Two oil pipelines can fill a small tank in 30 min. Using one of the pipelines would require 45 min to fill the tank. How long would it take the second pipeline, working alone, to fill the tank?

12. Working together, two dock workers can load a crate in 6 min. One dock worker, working alone, can load the crate in 15 min. How long would it take the second dock worker, working alone, to load the crate?

13. A mason can construct a retaining wall in 10 h. With the mason's apprentice assisting, the task takes 6 h. How long would it take the apprentice, working alone, to construct the wall?

14. A mechanic requires 2 h to repair a transmission, whereas an apprentice requires 6 h to make the same repairs. The mechanic worked alone for 1 h and then stopped. How long will it take the apprentice, working alone, to complete the repairs?

15. One computer technician can wire a modem in 4 h, whereas it takes 6 h for a second technician to do the same job. After working alone for 2 h, the first technician quit. How long will it take the second technician to complete the wiring?

16. A wallpaper hanger requires 2 h to hang the wallpaper on one wall of a room. A second wallpaper hanger requires 4 h to hang the same amount of paper. The first wallpaper hanger worked alone for 1 h and then quit. How long will it take the second wallpaper hanger, working alone, to complete the wall?

17. Two welders who work at the same rate are welding the girders of a building. After they work together for 10 h, one of the welders quits. The second welder requires 20 more hours to complete the welds. Find the time it would have taken one of the welders, working alone, to complete the welds.

18. A large and a small heating unit are being used to heat the water of a pool. The larger unit, working alone, requires 8 h to heat the pool. After both units have been operating for 2 h, the larger unit is turned off. The small unit requires 9 h more to heat the pool. How long would it take the small unit, working alone, to heat the pool?

19. Two machines that fill cereal boxes work at the same rate. After they work together for 7 h, one machine breaks down. The second machine requires 14 h more to finish filling the boxes. How long would it have taken one of the machines, working alone, to fill the boxes?

20. A large and a small drain are opened to drain a pool. The large drain can empty the pool in 6 h. After both drains have been open for 1 h, the large drain becomes clogged and is closed. The smaller drain remains open and requires 9 h more to empty the pool. How long would it have taken the small drain, working alone, to empty the pool?

Copyright © Houghton Mifflin Company. All rights reserved.

Objective B   **To use rational expressions to solve uniform motion problems**

21. Running at a constant speed, a jogger ran 24 mi in 3 h. How far did the jogger run in 2 h?

22. For uniform motion, distance = rate · time. How is time related to distance and rate? How is rate related to distance and time?

23. Commuting from work to home, a lab technician traveled 10 mi at a constant rate through congested traffic. On reaching the expressway, the technician increased the speed by 20 mph. An additional 20 mi was traveled at the increased speed. The total time for the trip was 1 h. Find the rate of travel through the congested traffic.

24. The president of a company traveled 1800 mi by jet and 300 mi on a prop plane. The rate of the jet was four times the rate of the prop plane. The entire trip took a total of 5 h. Find the rate of the jet plane.

25. As part of a conditioning program, a jogger ran 8 mi in the same amount of time a cyclist rode 20 mi. The rate of the cyclist was 12 mph faster than the rate of the jogger. Find the rate of the jogger and that of the cyclist.

26. An express train travels 600 mi in the same amount of time it takes a freight train to travel 360 mi. The rate of the express train is 20 mph faster than that of the freight train. Find the rate of each train.

27. To assess the damage done by a fire, a forest ranger traveled 1080 mi by jet and then an additional 180 mi by helicopter. The rate of the jet was four times the rate of the helicopter. The entire trip took a total of 5 h. Find the rate of the jet.

28. A twin-engine plane can fly 800 mi in the same time that it takes a single-engine plane to fly 600 mi. The rate of the twin-engine plane is 50 mph faster than that of the single-engine plane. Find the rate of the twin-engine plane.

29. As part of an exercise plan, Camille Ellison walked for 40 min and then ran for 20 min. If Camille runs 3 mph faster than she walks and covered 5 mi during the 1-hour exercise period, what is her walking speed?

30. A car and a bus leave a town at 1 P.M. and head for a town 300 mi away. The rate of the car is twice the rate of the bus. The car arrives 5 h ahead of the bus. Find the rate of the car.

Copyright © Houghton Mifflin Company. All rights reserved.

31. A car is traveling at a rate that is 36 mph faster than the rate of a cyclist. The car travels 384 mi in the same time it takes the cyclist to travel 96 mi. Find the rate of the car.

32. A backpacker hiking into a wilderness area walked 9 mi at a constant rate and then reduced this rate by 1 mph. Another 4 mi was hiked at this reduced rate. The time required to hike the 4 mi was 1 h less than the time required to walk the 9 mi. Find the rate at which the hiker walked the first 9 mi.

33. A plane can fly 180 mph in calm air. Flying with the wind, the plane can fly 600 mi in the same amount of time it takes to fly 480 mi against the wind. Find the rate of the wind.

34. A commercial jet can fly 550 mph in calm air. Traveling with the jet stream, the plane flew 2400 mi in the same amount of time it takes to fly 2000 mi against the jet stream. Find the rate of the jet stream.

35. A cruise ship can sail at 28 mph in calm water. Sailing with the gulf current, the ship can sail 170 mi in the same amount of time that it can sail 110 mi against the gulf current. Find the rate of the gulf current.

36. Rowing with the current of a river, a rowing team can row 25 mi in the same amount of time it takes to row 15 mi against the current. The rate of the rowing team in calm water is 20 mph. Find the rate of the current.

37. On a recent trip, a trucker traveled 330 mi at a constant rate. Because of road construction, the trucker then had to reduce the speed by 25 mph. An additional 30 mi was traveled at the reduced rate. The total time for the entire trip was 7 h. Find the rate of the trucker for the first 330 mi.

APPLYING THE CONCEPTS

38. Work   One pipe can fill a tank in 2 h, a second pipe can fill the tank in 4 h, and a third pipe can fill the tank in 5 h. How long will it take to fill the tank with all three pipes working?

39. Transportation   Because of bad weather, a bus driver reduced the usual speed along a 150-mile bus route by 10 mph. The bus arrived only 30 min later than its usual arrival time. How fast does the bus usually travel?

Copyright © Houghton Mifflin Company. All rights reserved.

Focus on Problem Solving

Negations and If ... then Sentences

The sentence "George Washington was the first president of the United States" is a true sentence. The **negation** of that sentence is "George Washington was **not** the first president of the United States." That sentence is false. In general, the negation of a true sentence is a false statement.

The negation of a false sentence is a true sentence. For instance, the sentence "The moon is made of green cheese" is a false statement. The negation of that sentence, "The moon is **not** made of green cheese," is true.

The words *all, no* (or *none*), and *some* are called **quantifiers.** Writing the negation of a sentence that contains these words requires special attention. Consider the sentence "All pets are dogs." This sentence is not true because there are pets that are not dogs; cats, for example, are pets. Because the sentence is false, its negation must be true. You might be tempted to write "All pets are not dogs," but that sentence is not true because some pets are dogs. The correct negation of "All pets are dogs" is "Some pets are not dogs." Note the use of the word *some* in the negation.

Now consider the sentence "Some computers are portable." Because that sentence is true, its negation must be false. Writing "Some computers are not portable" as the negation is not correct, because that sentence is true. The negation of "Some computers are portable" is "No computers are portable."

The sentence "No flowers have red blooms" is false, because there is at least one flower (some roses, for example) that has red blooms. Because the sentence is false, its negation must be true. The negation is "Some flowers have red blooms."

Statement	Negation
All *A* are *B*.	Some *A* are not *B*.
No *A* are *B*.	Some *A* are *B*.
Some *A* are *B*.	No *A* are *B*.
Some *A* are not *B*.	All *A* are *B*.

Write the negation of the sentence.

1. All cats like milk.

2. All computers need people.

3. Some trees are tall.

4. No politicians are honest.

5. No houses have kitchens.

6. All police officers are tall.

7. All lakes are not polluted.

8. Some drivers are unsafe.

9. Some speeches are interesting.

10. All laws are good.

11. All businesses are not profitable.

12. All motorcycles are not large.

13. Some vegetables are good for you to eat.

14. Some banks are not open on Sunday.

Copyright © Houghton Mifflin Company. All rights reserved.

A **premise** is a known or assumed fact. A premise can be stated using one of the quantifiers (*all*, *no*, *none*, or *some*) or using an *If … then* sentence. For instance, the sentence "All triangles have three sides" can be written "*If* a figure is a triangle, *then* it has three sides."

We can write the sentence "No whole numbers are negative numbers" as an *If … then* sentence: If a number is a whole number, then it is not a negative number.

Write the sentence as an *If … then* sentence.

15. All students at Barlock College must take a life science course.

16. All baseballs are round.

17. All computers need people.

18. All cats like milk.

19. No odd number is evenly divisible by 2.

20. No prime number greater than 2 is an even number.

21. No rectangles have five sides.

22. All roads lead to Rome.

23. All dogs have fleas.

24. No triangle has four angles.

Projects and Group Activities

Intensity of Illumination

You are already aware that the standard unit of length in the metric system is the meter (m) and that the standard unit of mass in the metric system is the gram (g). You may not know that the standard unit of light intensity is the **candela (cd).**

The rate at which light falls on a 1-square-unit area of surface is called the **intensity of illumination.** Intensity of illumination is measured in **lumens (lm).** A lumen is defined in the following illustration.

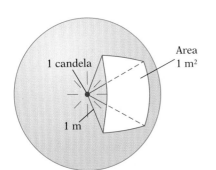

Picture a source of light equal to 1 cd positioned at the center of a hollow sphere that has a radius of 1 m. The rate at which light falls on 1 m² of the inner surface of the sphere is equal to 1 lm. If a light source equal to 4 cd is positioned at the center of the sphere, each square meter of the inner surface receives four times as much illumination, or 4 lm.

Light rays diverge as they leave a light source. The light that falls on an area of 1 m² at a distance of 1 m from the source of light spreads out over an area of 4 m² when it is 2 m from the source. The same light spreads out over an area of 9 m² when it is 3 m from the light source and over an area of 16 m² when it is 4 m from the light source. Therefore, as a surface moves farther away from the source of light, the intensity of illumination on the surface decreases from its value at 1 m to $\left(\frac{1}{2}\right)^2$, or $\frac{1}{4}$, that value at 2 m; to $\left(\frac{1}{3}\right)^2$, or $\frac{1}{9}$, that value at 3 m; and to $\left(\frac{1}{4}\right)^2$, or $\frac{1}{16}$, that value at 4 m.

Copyright © Houghton Mifflin Company. All rights reserved.

The formula for the intensity of illumination is

$$I = \frac{s}{r^2}$$

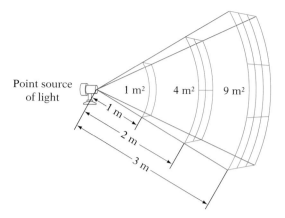

Point source of light

1 m² 4 m² 9 m²

1 m

2 m

3 m

where I is the intensity of illumination in lumens, s is the strength of the light source in candelas, and r is the distance in meters between the light source and the illuminated surface.

A 30-candela lamp is 0.5 m above a desk. Find the illumination on the desk.

$$I = \frac{s}{r^2}$$

$$I = \frac{30}{(0.5)^2} = 120$$

The illumination on the desk is 120 lm.

1. A 100-candela light is hanging 5 m above a floor. What is the intensity of the illumination on the floor beneath it?

2. A 25-candela source of light is 2 m above a desk. Find the intensity of illumination on the desk.

3. How strong a light source is needed to cast 20 lm of light on a surface 4 m from the source?

4. How strong a light source is needed to cast 80 lm of light on a surface 5 m from the source?

5. How far from the desk surface must a 40-candela light source be positioned if the desired intensity of illumination is 10 lm?

6. Find the distance between a 36-candela light source and a surface if the intensity of illumination on the surface is 0.01 lm.

7. Two lights cast the same intensity of illumination on a wall. One light is 6 m from the wall and has a rating of 36 cd. The second light is 8 m from the wall. Find the candela rating of the second light.

8. A 40-candela light source and a 10-candela light source both throw the same intensity of illumination on a wall. The 10-candela light is 6 m from the wall. Find the distance from the 40-candela light to the wall.

Copyright © Houghton Mifflin Company. All rights reserved.

Chapter 6 Summary

Key Words	Examples

A *rational expression* is a fraction in which the numerator and denominator are polynomials. A rational expression is in *simplest form* when the numerator and denominator have no common factors. [6.1A, p. 289]

$\frac{2x+1}{x^2+4}$ is a rational expression in simplest form.

The *reciprocal of a rational expression* is the rational expression with the numerator and denominator interchanged. [6.1C, p. 292]

The reciprocal of $\frac{3x-y}{x+4}$ is $\frac{x+4}{3x-y}$.

The *least common multiple (LCM) of two or more polynomials* is the polynomial of least degree that contains all the factors of each polynomial. [6.2A, p. 297]

The LCM of $3x^2 - 6x$ and $x^2 - 4$ is $3x(x-2)(x+2)$, because it contains the factors of $3x^2 - 6x = 3x(x-2)$ and the factors of $x^2 - 4 = (x-2)(x+2)$.

A *complex fraction* is a fraction whose numerator or denominator contains one or more fractions. [6.4A, p. 310]

$\dfrac{x - \frac{2}{x+1}}{1 - \frac{4}{x}}$ is a complex fraction.

A *ratio* is the quotient of two quantities that have the same unit. A *rate* is the ratio of two quantities that have different units. [6.6A, p. 319]

$\frac{9}{4}$ is a ratio. $\frac{60 \text{ m}}{12 \text{ s}}$ is a rate.

A *proportion* is an equation that states the equality of two ratios or rates. [6.6A, p. 319]

$\frac{3}{8} = \frac{12}{32}$ and $\frac{x \text{ ft}}{12 \text{ s}} = \frac{15 \text{ ft}}{160 \text{ s}}$ are proportions.

A *literal equation* is an equation that contains more than one variable. A *formula* is a literal equation that states a rule about measurements. [6.7A, p. 327]

$3x - 4y = 12$ is a literal equation. $A = LW$ is a literal equation that is also the formula for the area of a rectangle.

Essential Rules and Procedures	Examples

Simplifying Rational Expressions [6.1A, p. 289]
Factor the numerator and denominator. Divide the numerator and denominator by the common factors.

$$\frac{x^2 - 3x - 10}{x^2 - 5} = \frac{(x+2)(x-5)}{(x+5)(x-5)}$$
$$= \frac{x+2}{x+5}$$

Multiplying Rational Expressions [6.1B, p. 290]
Multiply the numerators. Multiply the denominators. Write the answer in simplest form.

$$\frac{a}{b} \cdot \frac{c}{d} = \frac{ac}{bd}$$

$$\frac{x^2 - 3x}{x^2 + x} \cdot \frac{x^2 + 5x + 4}{x^2 - 4x + 3}$$
$$= \frac{x(x-3)}{x(x+1)} \cdot \frac{(x+1)(x+4)}{(x-3)(x-1)}$$
$$= \frac{x(x-3)(x+1)(x+4)}{x(x+1)(x-3)(x-1)}$$
$$= \frac{x+4}{x-1}$$

Copyright © Houghton Mifflin Company. All rights reserved.

Dividing Rational Expressions [6.1C, p. 292]
Multiply the dividend by the reciprocal of the divisor. Write the answer in simplest form.

$$\frac{a}{b} \div \frac{c}{d} = \frac{a}{b} \cdot \frac{d}{c} = \frac{ad}{bc}$$

$$\frac{4x + 16}{3x - 6} \div \frac{x^2 + 6x + 8}{x^2 - 4}$$

$$= \frac{4x + 16}{3x - 6} \cdot \frac{x^2 - 4}{x^2 + 6x + 8}$$

$$= \frac{4(x + 4)}{3(x - 2)} \cdot \frac{(x - 2)(x + 2)}{(x + 4)(x + 2)}$$

$$= \frac{4}{3}$$

Adding and Subtracting Rational Expressions [6.3B, p. 303]

1. Find the LCM of the denominators.

2. Write each fraction as an equivalent fraction using the LCM as the denominator.

3. Add or subtract the numerators and place the result over the common denominator.

4. Write the answer in simplest form.

$$\frac{a}{b} + \frac{c}{b} = \frac{a + c}{b} \qquad \frac{a}{b} - \frac{c}{b} = \frac{a - c}{b}$$

$$\frac{x}{x + 1} - \frac{x + 3}{x - 2}$$

$$= \frac{x}{x + 1} \cdot \frac{x - 2}{x - 2} - \frac{x + 3}{x - 2} \cdot \frac{x + 1}{x + 1}$$

$$= \frac{x(x - 2)}{(x + 1)(x - 2)} - \frac{(x + 3)(x + 1)}{(x + 1)(x - 2)}$$

$$= \frac{x(x - 2) - (x + 3)(x + 1)}{(x + 1)(x - 2)}$$

$$= \frac{(x^2 - 2x) - (x^2 + 4x + 3)}{(x + 1)(x - 2)}$$

$$= \frac{-6x - 3}{(x + 1)(x - 2)}$$

Simplifying Complex Fractions [6.4A, p. 310]

Method 1: Multiply by 1 in the form $\dfrac{\text{LCM}}{\text{LCM}}$.

1. Determine the LCM of the denominators of the fractions in the numerator and denominator of the complex fraction.

2. Multiply the numerator and denominator of the complex fraction by the LCM.

3. Simplify.

Method 1: $\dfrac{\dfrac{1}{x} + \dfrac{1}{y}}{\dfrac{1}{x} - \dfrac{1}{y}} = \dfrac{\dfrac{1}{x} + \dfrac{1}{y}}{\dfrac{1}{x} - \dfrac{1}{y}} \cdot \dfrac{xy}{xy}$

$$= \frac{\dfrac{1}{x} \cdot xy + \dfrac{1}{y} \cdot xy}{\dfrac{1}{x} \cdot xy - \dfrac{1}{y} \cdot xy}$$

$$= \frac{y + x}{y - x}$$

Method 2: Multiply the numerator by the reciprocal of the denominator.

1. Simplify the numerator to a single fraction and simplify the denominator to a single fraction.

2. Using the definition for dividing fractions, multiply the numerator by the reciprocal of the denominator.

3. Simplify.

Method 2: $\dfrac{\dfrac{1}{x} + \dfrac{1}{y}}{\dfrac{1}{x} - \dfrac{1}{y}} = \dfrac{\dfrac{y + x}{xy}}{\dfrac{y - x}{xy}}$

$$= \frac{y + x}{xy} \cdot \frac{xy}{y - x}$$

$$= \frac{y + x}{y - x}$$

Copyright © Houghton Mifflin Company. All rights reserved.

Solving Equations Containing Fractions [6.5A, p. 315]

Clear denominators by multiplying each side of the equation by the LCM of the denominators. Then solve for the variable.

$$\frac{1}{2a} = \frac{2}{a} - \frac{3}{8}$$

$$8a\left(\frac{1}{2a}\right) = 8a\left(\frac{2}{a}\right) - 8a\left(\frac{3}{8}\right)$$

$$4 = 16 - 3a$$

$$-12 = -3a$$

$$4 = a$$

Similar Triangles [6.6C, pp. 320–321]

Similar triangles have the same shape but not necessarily the same size. The ratios of corresponding parts of similar triangles are equal. The measures of the corresponding angles of similar triangles are equal.

Triangles ABC and DFE are similar triangles. The ratios of corresponding parts are equal to $\frac{2}{3}$.

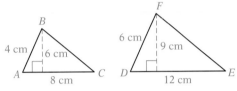

If two angles of one triangle are equal in measure to two angles of another triangle, then the two triangles are similar.

Triangles AOB and COD are similar because $m\angle AOB = m\angle COD$ and $m\angle B = m\angle D$.

Solving Literal Equations [6.7A, p. 327]

Rewrite the equation so that the letter being solved for is alone on one side of the equation and all numbers and other variables are on the other side.

Solve $2x + ax = 5$ for x.

$$2x + ax = 5$$

$$x(2 + a) = 5$$

$$\frac{x(2 + a)}{2 + a} = \frac{5}{2 + a}$$

$$x = \frac{5}{2 + a}$$

Work Problems [6.8A, p. 331]

Rate of work × time worked = part of task completed

Pat can do a certain job in 3 h. Chris can do the same job in 5 h. How long would it take them, working together, to get the job done?

$$\frac{t}{3} + \frac{t}{5} = 1$$

Uniform Motion Problems with Rational Expressions
[6.8B, p. 333]

$$\frac{\text{Distance}}{\text{Rate}} = \text{time}$$

Train A's speed is 15 mph faster than train B's speed. Train A travels 150 mi in the same amount of time it takes train B to travel 120 mi. Find the rate of train B.

$$\frac{120}{r} = \frac{150}{r + 15}$$

Copyright © Houghton Mifflin Company. All rights reserved.

Chapter 6 Review Exercises

1. Divide: $\dfrac{6a^2b^7}{25x^3y} \div \dfrac{12a^3b^4}{5x^2y^2}$

2. Add: $\dfrac{x+7}{15x} + \dfrac{x-2}{20x}$

3. Multiply: $\dfrac{3x^3+9x^2}{6xy^2-18y^2} \cdot \dfrac{4xy^3-12y^3}{5x^2+15x}$

4. Divide: $\dfrac{2x(x-y)}{x^2y(x+y)} \div \dfrac{3(x-y)}{x^2y^2}$

5. Simplify: $\dfrac{x-\dfrac{16}{5x-2}}{3x-4-\dfrac{88}{5x-2}}$

6. Simplify: $\dfrac{x^2+x-30}{15+2x-x^2}$

7. Simplify: $\dfrac{16x^5y^3}{24xy^{10}}$

8. Solve: $\dfrac{20}{x+2} = \dfrac{5}{16}$

9. Divide: $\dfrac{10-23y+12y^2}{6y^2-y-5} \div \dfrac{4y^2-13y+10}{18y^2+3y-10}$

10. Solve $3ax - x = 5$ for x.

11. Solve: $\dfrac{2}{x} + \dfrac{3}{4} = 1$

12. Add: $\dfrac{x}{y} + \dfrac{3}{x}$

13. Solve $5x + 4y = 20$ for y.

14. Multiply: $\dfrac{8ab^2}{15x^3y} \cdot \dfrac{5xy^4}{16a^2b}$

15. Simplify: $\dfrac{1-\dfrac{1}{x}}{1-\dfrac{8x-7}{x^2}}$

16. Write each fraction in terms of the LCM of the denominators.

$\dfrac{x}{12x^2+16x-3}, \dfrac{4x^2}{6x^2+7x-3}$

17. Solve $T = 2(ab + bc + ca)$ for a.

18. Solve: $\dfrac{5}{7} + \dfrac{x}{2} = 2 - \dfrac{x}{7}$

19. Simplify: $\dfrac{2+\dfrac{1}{x}}{3-\dfrac{2}{x}}$

20. Subtract: $\dfrac{2x}{x-5} - \dfrac{x+1}{x-2}$

21. Solve $i = \dfrac{100m}{c}$ for c.

22. Solve: $\dfrac{x+8}{x+4} = 1 + \dfrac{5}{x+4}$

23. Divide: $\dfrac{20x^2-45x}{6x^3+4x^2} \div \dfrac{40x^3-90x^2}{12x^2+8x}$

24. Add: $\dfrac{2y}{5y-7} + \dfrac{3}{7-5y}$

Copyright © Houghton Mifflin Company. All rights reserved.

25. Subtract: $\dfrac{5x + 3}{2x^2 + 5x - 3} - \dfrac{3x + 4}{2x^2 + 5x - 3}$

26. Find the LCM of $10x^2 - 11x + 3$ and $20x^2 - 17x + 3$.

27. Solve $4x + 9y = 18$ for y.

28. Multiply: $\dfrac{2x^2 - 5x - 3}{3x^2 - 7x - 6} \cdot \dfrac{3x^2 + 8x + 4}{x^2 + 4x + 4}$

29. Solve: $\dfrac{20}{2x + 3} = \dfrac{17x}{2x + 3} - 5$

30. Add: $\dfrac{x - 1}{x + 2} + \dfrac{3x - 2}{5 - x} + \dfrac{5x^2 + 15x - 11}{x^2 - 3x - 10}$

31. Solve: $\dfrac{6}{x - 7} = \dfrac{8}{x - 6}$

32. Solve: $\dfrac{3}{20} = \dfrac{x}{80}$

33. **Geometry**   Given that MP and NQ intersect at O, NQ measures 25 cm, MO measures 6 cm, and PO measures 9 cm, find the length of QO.

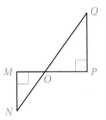

34. **Geometry**   Triangles ABC and DEF are similar triangles. Find the area of triangle DEF.

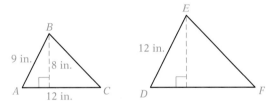

35. **Work**   One hose can fill a pool in 15 h. The second hose can fill the pool in 10 h. How long would it take to fill the pool using both hoses?

36. **Travel**   A car travels 315 mi in the same amount of time in which a bus travels 245 mi. The rate of the car is 10 mph faster than that of the bus. Find the rate of the car.

37. **Travel**   The rate of a jet is 400 mph in calm air. Traveling with the wind, the jet can fly 2100 mi in the same amount of time it takes to fly 1900 mi against the wind. Find the rate of the wind.

38. **Sports**   A pitcher's earned run average (ERA) is the average number of runs allowed in 9 innings of pitching. If a pitcher allows 15 runs in 100 innings, find the pitcher's ERA.

Copyright © Houghton Mifflin Company. All rights reserved.

Chapter 6 Test

1. Subtract: $\dfrac{x}{x + 3} - \dfrac{2x - 5}{x^2 + x - 6}$

2. Solve: $\dfrac{3}{x + 4} = \dfrac{5}{x + 6}$

3. Multiply: $\dfrac{x^2 + 2x - 3}{x^2 + 6x + 9} \cdot \dfrac{2x^2 - 11x + 5}{2x^2 + 3x - 5}$

4. Simplify: $\dfrac{16x^5 y}{24x^2 y^4}$

5. Solve $d = s + rt$ for t.

6. Solve: $\dfrac{6}{x} - 2 = 1$

7. Simplify: $\dfrac{x^2 + 4x - 5}{1 - x^2}$

8. Find the LCM of $6x - 3$ and $2x^2 + x - 1$.

9. Subtract: $\dfrac{2}{2x - 1} - \dfrac{3}{3x + 1}$

10. Divide: $\dfrac{x^2 + 3x + 2}{x^2 + 5x + 4} \div \dfrac{x^2 - x - 6}{x^2 + 2x - 15}$

11. Simplify: $\dfrac{1 + \dfrac{1}{x} - \dfrac{12}{x^2}}{1 + \dfrac{2}{x} - \dfrac{8}{x^2}}$

12. Write each fraction in terms of the LCM of the denominators.

$$\dfrac{3}{x^2 - 2x}, \dfrac{x}{x^2 - 4}$$

Copyright © Houghton Mifflin Company. All rights reserved.

13.   Subtract: $\dfrac{2x}{x^2 + 3x - 10} - \dfrac{4}{x^2 + 3x - 10}$

14.   Solve $3x - 8y = 16$ for y.

15.   Solve: $\dfrac{2x}{x + 1} - 3 = \dfrac{-2}{x + 1}$

16.   Multiply: $\dfrac{x^3 y^4}{x^2 - 4x + 4} \cdot \dfrac{x^2 - x - 2}{x^6 y^4}$

17.   Geometry   Given $AE \parallel BD$, AB measures 5 ft, ED measures 8 ft, and BC measures 3 ft, find the length of CE.

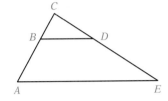

18.   Chemistry   A saltwater solution is formed by mixing 4 lb of salt with 10 gal of water. At this rate, how many additional pounds of salt are required for 15 gal of water?

19.   Work   A pool can be filled with one pipe in 6 h, whereas a second pipe requires 12 h to fill the pool. How long would it take to fill the pool with both pipes turned on?

20.   Travel   A small plane can fly at 110 mph in calm air. Flying with the wind, the plane can fly 260 mi in the same amount of time it takes to fly 180 mi against the wind. Find the rate of the wind.

21.   Landscaping   A landscape architect uses three sprinklers for each 200 ft² of lawn. At this rate, how many sprinklers are needed for a 3600-square-foot lawn?

Copyright © Houghton Mifflin Company. All rights reserved.

Cumulative Review Exercises

1. Evaluate: $\left(\dfrac{2}{3}\right)^2 \div \left(\dfrac{3}{2} - \dfrac{2}{3}\right) + \dfrac{1}{2}$

2. Evaluate $-a^2 + (a - b)^2$ when $a = -2$ and $b = 3$.

3. Simplify: $-2x - (-3y) + 7x - 5y$

4. Simplify: $2[3x - 7(x - 3) - 8]$

5. Solve: $4 - \dfrac{2}{3}x = 7$

6. Solve: $3[x - 2(x - 3)] = 2(3 - 2x)$

7. Find $16\dfrac{2}{3}\%$ of 60.

8. Simplify: $(a^2b^5)(ab^2)$

9. Multiply: $(a - 3b)(a + 4b)$

10. Divide: $\dfrac{15b^4 - 5b^2 + 10b}{5b}$

11. Divide: $(x^3 - 8) \div (x - 2)$

12. Factor: $12x^2 - x - 1$

13. Factor: $y^2 - 7y + 6$

14. Factor: $2a^3 + 7a^2 - 15a$

15. Factor: $4b^2 - 100$

16. Solve: $(x + 3)(2x - 5) = 0$

17. Simplify: $\dfrac{12x^4y^2}{18xy^7}$

18. Simplify: $\dfrac{x^2 - 7x + 10}{25 - x^2}$

Copyright © Houghton Mifflin Company. All rights reserved.

19. Divide: $\dfrac{x^2 - x - 56}{x^2 + 8x + 7} \div \dfrac{x^2 - 13x + 40}{x^2 - 4x - 5}$

20. Subtract: $\dfrac{2}{2x - 1} - \dfrac{1}{x + 1}$

21. Simplify: $\dfrac{1 - \dfrac{2}{x} - \dfrac{15}{x^2}}{1 - \dfrac{25}{x^2}}$

22. Solve: $\dfrac{3x}{x - 3} - 2 = \dfrac{10}{x - 3}$

23. Solve: $\dfrac{2}{x - 2} = \dfrac{12}{x + 3}$

24. Solve $f = v + at$ for t.

25. **Number Sense**   Translate "the difference between five times a number and thirteen is the opposite of eight" into an equation and solve.

26. **Metallurgy**   A silversmith mixes 60 g of an alloy that is 40% silver with 120 g of another silver alloy. The resulting alloy is 60% silver. Find the percent of silver in the 120-gram alloy.

27. **Geometry**   The length of the base of a triangle is 2 in. less than twice the height. The area of the triangle is 30 in². Find the base and height of the triangle.

28. **Insurance**   A life insurance policy costs $16 for every $1000 of coverage. At this rate, how much money would a policy of $5000 cost?

29. **Work**   One water pipe can fill a tank in 9 min, whereas a second pipe requires 18 min to fill the tank. How long would it take both pipes, working together, to fill the tank?

30. **Travel**   The rower of a boat can row at a rate of 5 mph in calm water. Rowing with the current, the boat travels 14 mi in the same amount of time it takes to travel 6 mi against the current. Find the rate of the current.

Copyright © Houghton Mifflin Company. All rights reserved.

7 Linear Equations in Two Variables

This tennis player gets the energy for his workout from carbohydrates. Carbohydrates are the body's primary source of fuel for exercise. They can be released quickly and easily to fulfill the demands that exercise puts on the body. Since carbohydrates also fuel most of our muscular contractions, it is important to eat enough carbohydrates before any rigorous exercise. **Exercise 27 on page 390** presents data on the number of grams of carbohydrates burned as a strenuous tennis workout progresses.

OBJECTIVES

Section 7.1

A To graph points in a rectangular coordinate system

B To determine ordered-pair solutions of an equation in two variables

C To determine whether a set of ordered pairs is a function

D To evaluate a function written in functional notation

Section 7.2

A To graph an equation of the form $y = mx + b$

B To graph an equation of the form $Ax + By = C$

C To solve application problems

Section 7.3

A To find the x- and y-intercepts of a straight line

B To find the slope of a straight line

C To graph a line using the slope and the y-intercept

Section 7.4

A To find the equation of a line given a point and the slope

B To find the equation of a line given two points

C To solve application problems

Need help? For online student resources, such as section quizzes, visit this textbook's website at **math.college.hmco.com/students.**

Copyright © Houghton Mifflin Company. All rights reserved.

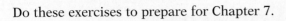

Do these exercises to prepare for Chapter 7.

1. Simplify: $-\dfrac{5 - (-7)}{4 - 8}$

2. Evaluate $\dfrac{a - b}{c - d}$ when $a = 3$, $b = -2$, $c = -3$, and $d = 2$.

3. Simplify: $-3(x - 4)$

4. Solve: $3x + 6 = 0$

5. Solve $4x + 5y = 20$ when $y = 0$.

6. Solve $3x - 7y = 11$ when $x = -1$.

7. Divide: $\dfrac{12x - 15}{-3}$

8. Solve: $\dfrac{2x + 1}{3} = \dfrac{3x}{4}$

9. Solve $3x - 5y = 15$ for y.

10. Solve $y + 3 = -\dfrac{1}{2}(x + 4)$ for y.

GO FIGURE • • •

Points A, B, C, and D lie on the same line and in that order. The ratio of AB to AC is $\dfrac{1}{4}$ and the ratio of BC to CD is $\dfrac{1}{2}$. Find the ratio of AB to CD.

Copyright © Houghton Mifflin Company. All rights reserved.

7.1 The Rectangular Coordinate System

Objective A   ### To graph points in a rectangular coordinate system

Before the 15th century, geometry and algebra were considered separate branches of mathematics. That all changed when René Descartes, a French mathematician who lived from 1596 to 1650, founded **analytic geometry.** In this geometry, a *coordinate system* is used to study relationships between variables.

A **rectangular coordinate system** is formed by two number lines, one horizontal and one vertical, that intersect at the zero point of each line. The point of intersection is called the **origin.** The two lines are called **coordinate axes,** or simply **axes.** The axes determine a **plane,** which can be thought of as a large, flat sheet of paper. The two axes divide the plane into four regions called **quadrants,** which are numbered counterclockwise from I to IV.

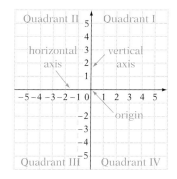

Each point in the plane can be identified by a pair of numbers called an **ordered pair.** The first number of the pair measures a horizontal distance and is called the **abscissa.** The second number of the pair measures a vertical distance and is called the **ordinate.** The **coordinates of a point** are the numbers in the ordered pair associated with the point. The abscissa is also called the **first coordinate** of the ordered pair, and the ordinate is also called the **second coordinate** of the ordered pair.

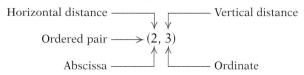

When drawing a rectangular coordinate system, we often label the horizontal axis x and the vertical axis y. In this case, the coordinate system is called an **xy-coordinate system.** The coordinates of the points are given by ordered pairs (x, y), where the abscissa is called the **x-coordinate** and the ordinate is called the **y-coordinate.**

To **graph or plot a point in the plane,** place a dot at the location given by the ordered pair. The **graph of an ordered pair** (x, y) is the dot drawn at the coordinates of the point in the plane. The points whose coordinates are $(3, 4)$ and $(-2.5, -3)$ are graphed in the figures below.

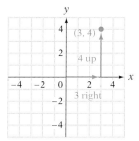 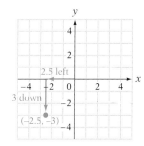

Copyright © Houghton Mifflin Company. All rights reserved.

TAKE NOTE

This is very important. An **ordered pair** is a *pair* of coordinates, and the *order* in which the coordinates appear is crucial.

The points whose coordinates are (3, −1) and (−1, 3) are graphed at the right. Note that the graphed points are in different locations. *The order of the coordinates of an ordered pair is important.*

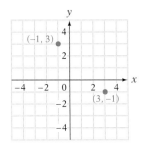

Each point in the plane is associated with an ordered pair, and each ordered pair is associated with a point in the plane. Although only the labels for integers are given on a coordinate grid, the graph of any ordered pair can be approximated. For example, the points whose coordinates are (−2.3, 4.1) and (π, 1) are shown on the graph at the right.

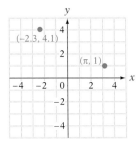

Example 1 Graph the ordered pairs (−2, −3), (3, −2), (0, −2), and (3, 0).

Solution

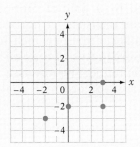

You Try It 1 Graph the ordered pairs (−4, 1), (3, −3), (0, 4), and (−3, 0).

Your solution

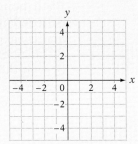

Example 2 Give the coordinates of the points labeled *A* and *B*. Give the abscissa of point *C* and the ordinate of point *D*.

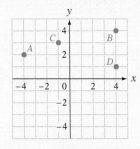

Solution The coordinates of *A* are (−4, 2).
The coordinates of *B* are (4, 4).
The abscissa of *C* is −1.
The ordinate of *D* is 1.

You Try It 2 Give the coordinates of the points labeled *A* and *B*. Give the abscissa of point *D* and the ordinate of point *C*.

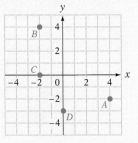

Your solution

Solutions on p. S18

Copyright © Houghton Mifflin Company. All rights reserved.

Objective B   **To determine ordered-pair**
solutions of an equation in two variables

An *xy*-coordinate system is used to study the relationship between two variables. Frequently this relationship is given by an equation. Examples of equations in two variables include

$$y = 2x - 3 \qquad\qquad 3x + 2y = 6 \qquad\qquad x^2 - y = 0$$

A **solution of an equation in two variables** is an ordered pair (x, y) whose coordinates make the equation a true statement.

> **HOW TO**   Is $(-3, 7)$ a solution of $y = -2x + 1$?
>
> $$\begin{array}{c|l} y = -2x + 1 & \\ \hline 7 & -2(-3) + 1 \\ & 6 + 1 \\ 7 = 7 & \end{array}$$
>
> • Replace *x* by **−3**; replace *y* by **7**.
>
> • The results are equal.
>
> $(-3, 7)$ is a solution of the equation $y = -2x + 1$.

Besides $(-3, 7)$, there are many other ordered-pair solutions of $y = -2x + 1$. For example, $(0, 1)$, $\left(-\dfrac{3}{2}, 4\right)$, and $(4, -7)$ are also solutions. In general, an equation in two variables has an infinite number of solutions. By choosing any value of x and substituting that value into the equation, we can calculate a corresponding value of y.

> **HOW TO**   Find the ordered-pair solution of $y = \dfrac{2}{3}x - 3$ that corresponds to $x = 6$.
>
> $$y = \frac{2}{3}x - 3$$
>
> $$= \frac{2}{3}(6) - 3 \qquad \text{• Replace } x \text{ by } \mathbf{6}.$$
>
> $$= 4 - 3 = 1 \qquad \text{• Simplify.}$$
>
> The ordered-pair solution is $(6, 1)$.

The solutions of an equation in two variables can be graphed in an *xy*-coordinate system.

> **HOW TO**   Graph the ordered-pair solutions of $y = -2x + 1$ when $x = -2, -1, 0, 1,$ and 2.

Use the values of x to determine ordered-pair solutions of the equation. It is convenient to record these in a table.

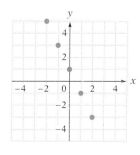

x	$y = -2x + 1$	y	(x, y)
-2	$-2(-2) + 1$	5	$(-2, 5)$
-1	$-2(-1) + 1$	3	$(-1, 3)$
0	$-2(0) + 1$	1	$(0, 1)$
1	$-2(1) + 1$	-1	$(1, -1)$
2	$-2(2) + 1$	-3	$(2, -3)$

Copyright © Houghton Mifflin Company. All rights reserved.

Example 3

Is $(3, -2)$ a solution of $3x - 4y = 15$?

Solution

$$3x - 4y = 15$$

$$\begin{array}{c|c} 3(3) - 4(-2) & 15 \\ 9 + 8 & \\ & 17 \neq 15 \end{array}$$

• Replace x by 3 and y by -2.

No. $(3, -2)$ is not a solution of $3x - 4y = 15$.

You Try It 3

Is $(-2, 4)$ a solution of $x - 3y = -14$?

Your solution

Example 4

Graph the ordered-pair solutions of $2x - 3y = 6$ when $x = -3, 0, 3,$ and 6.

Solution

$$2x - 3y = 6$$
$$-3y = -2x + 6$$
$$y = \frac{2}{3}x - 2$$

• Solve $2x - 3y = 6$ for y.

Replace x in $y = \frac{2}{3}x - 2$ by $-3, 0, 3,$ and 6. For each value of x, determine the value of y.

x	$y = \frac{2}{3}x = 2$	y	(x, y)
-3	$\frac{2}{3}(-3) - 2$	-4	$(-3, -4)$
0	$\frac{2}{3}(0) - 2$	-2	$(0, -2)$
3	$\frac{2}{3}(3) - 2$	0	$(3, 0)$
6	$\frac{2}{3}(6) - 2$	2	$(6, 2)$

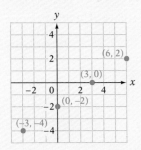

You Try It 4

Graph the ordered-pair solutions of $x + 2y = 4$ when $x = -4, -2, 0,$ and 2.

Your solution

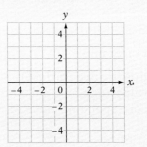

Solutions on pp. S18–S19

Copyright © Houghton Mifflin Company. All rights reserved.

Copyright © Houghton Mifflin Company. All rights reserved.

Objective C   **To determine whether a set of ordered pairs is a function**

Discovering a relationship between two variables is an important task in the application of mathematics. Here are some examples.

- Botanists study the relationship between the number of bushels of wheat yielded per acre and the amount of watering per acre.
- Environmental scientists study the relationship between the incidents of skin cancer and the amount of ozone in the atmosphere.
- Business analysts study the relationship between the price of a product and the number of products that are sold at that price.

Each of these relationships can be described by a set of ordered pairs.

> **Definition of a Relation**
>
> A **relation** is any set of ordered pairs.

The following table shows the number of hours that each of 9 students spent studying for a midterm exam and the grade that each of these 9 students received.

Hours	3	3.5	2.75	2	4	4.5	3	2.5	5
Grade	78	75	70	65	85	85	80	75	90

This information can be written as the relation

{(3, 78), (3.5, 75), (2.75, 70), (2, 65), (4, 85), (4.5, 85), (3, 80), (2.5, 75), (5, 90)}

where the first coordinate of the ordered pair is the hours spent studying and the second coordinate is the score on the midterm.

The **domain** of a relation is the set of first coordinates of the ordered pairs; the **range** is the set of second coordinates. For the relation above,

Domain = {2, 2.5, 2.75, 3, 3.5, 4, 4.5, 5}   Range = {65, 70, 75, 78, 80, 85, 90}

The **graph of a relation** is the graph of the ordered pairs that belong to the relation. The graph of the relation given above is shown at the right. The horizontal axis represents the hours spent studying (the domain); the vertical axis represents the test score (the range). The axes could be labeled H for hours studied and S for test score.

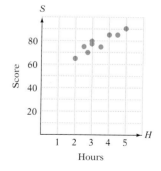

A *function* is a special type of relation in which no two ordered pairs have the same first coordinate.

> **Definition of a Function**
>
> A **function** is a relation in which no two ordered pairs have the same first coordinate.

The table at the right is the grading scale for a 100-point test. This table defines a relationship between the *score* on the test and a *letter grade*. Some of the ordered pairs of this function are (78, C), (97, A), (84, B), and (82, B).

Score	Grade
90–100	A
80–89	B
70–79	C
60–69	D
0–59	F

The grading-scale table defines a function, because no two ordered pairs can have the *same* first coordinate and *different* second coordinates. For instance, it is not possible to have the ordered pairs (72, C), and (72, B)—same first coordinate (test score) but different second coordinates (test grade). The domain of this function is {0, 1, 2,..., 99, 100}. The range is {A, B, C, D, F}.

The example of hours spent studying and test score given earlier is *not* a function, because (3, 78) and (3, 80) are ordered pairs of the relation that have the *same* first coordinate but *different* second coordinates.

Consider, again, the grading-scale example. Note that (84, B) and (82, B) are ordered pairs of the function. Ordered pairs of a function may have the same *second* coordinates but not the same first coordinates.

Although relations and functions can be given by tables, they are frequently given by an equation in two variables.

The equation $y = 2x$ expresses the relationship between a number, x, and twice the number, y. For instance, if $x = 3$, then $y = 6$, which is twice 3. To indicate exactly which ordered pairs are determined by the equation, the domain (values of x) is specified. If $x \in \{-2, -1, 0, 1, 2\}$, then the ordered pairs determined by the equation are $\{(-2, -4), (-1, -2), (0, 0), (1, 2), (2, 4)\}$. This relation is a function because no two ordered pairs have the same first coordinate.

The graph of the function $y = 2x$ with domain $\{-2, -1, 0, 1, 2\}$ is shown at the right. The horizontal axis (domain) is labeled x; the vertical axis (range) is labeled y.

The domain $\{-2, -1, 0, 1, 2\}$ was chosen arbitrarily. Other domains could have been selected. The type of application usually influences the choice of the domain.

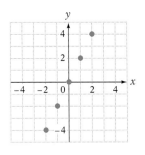

For the equation $y = 2x$, we say that "y is a function of x" because the set of ordered pairs is a function.

Not all equations, however, define a function. For instance, the equation $|y| = x + 2$ does not define y as a function of x. The ordered pairs (2, 4) and (2, −4) both satisfy the equation. Thus there are two ordered pairs with the same first coordinate but different second coordinates.

Copyright © Houghton Mifflin Company. All rights reserved.

Example 5

The number of tournaments and the total earnings of the top five Ladies Professional Golf Association (LPGA) players is given in the following table.

Player	Tournaments	Winnings
A. Sorenstam	16	$1,914,506
Se Ri Pak	25	$1,561,928
Grace Park	25	$1,374,702
Hee-Won Han	26	$1,101,060
Juli Inkster	20	$1,012,455

Write a relation where the first coordinate is the number of tournaments played and the second coordinate is the winnings per tournament rounded to the nearest dollar. Is the relation a function?

Solution

Find the winnings per tournament for each player by dividing the player's winnings by the number of tournaments played.

Sorenstam: $1,914,506 \div 16 \approx 119,657$
Pak: $1,561,928 \div 25 \approx 62,477$
Park: $1,374,702 \div 25 \approx 54,988$
Han: $1,101,060 \div 26 \approx 42,348$
Inkster: $1,012,455 \div 20 \approx 50,623$

The relation is {(16, 119,657), (25, 62,477), (25, 54,988), (26, 42,348), (20, 50,623)}. The relation is not a function. Two ordered pairs have the same first coordinate but different second coordinates.

You Try It 5

Six students decided to go on a diet and fitness program over the summer. Their weights (in pounds) at the beginning and end of the program are given in the table below.

Beginning	End
145	140
140	125
150	130
165	150
140	130
165	160

Write a relation wherein the first coordinate is the weight at the beginning of the summer and the second coordinate is the weight at the end of the summer. Is the relation a function?

Your solution

Example 6

Does $y = x^2 + 3$, where $x \in \{-2, -1, 1, 3\}$, define y as a function of x?

Solution

Determine the ordered pairs defined by the equation. Replace x in $y = x^2 + 3$ by the given values and solve for y.

{(−2, 7), (−1, 4), (1, 4), (3, 12)}

No two ordered pairs have the same first coordinate. Therefore, the relation is a function and the equation $y = x^2 + 3$ defines y as a function of x.

Note that (−1, 4) and (1, 4) are ordered pairs that belong to this function. Ordered pairs of a function may have the same *second* coordinate but not the same *first* coordinate.

You Try It 6

Does $y = \frac{1}{2}x + 1$, where $x \in \{-4, 0, 2\}$, define y as a function of x?

Your solution

Solutions on p. S19

Copyright © Houghton Mifflin Company. All rights reserved.

Objective D **To evaluate a function written in functional notation**

When an equation defines y as a function of x, **functional notation** is frequently used to emphasize that the relation is a function. In this case, it is common to replace y in the function's equation with the symbol $f(x)$, where

$f(x)$ is read "f of x" or "the value of f at x."

For instance, the equation $y = x^2 + 3$ from Example 6 defined y as a function of x. The equation can also be written in functional notation as

$$f(x) = x^2 + 3$$

where y has been replaced by $f(x)$.

The symbol $f(x)$ is called the **value of a function at x** because it is the result of evaluating a variable expression. For instance, $f(4)$ means to replace x by 4 and then simplify the resulting numerical expression.

$$f(x) = x^2 + 3$$
$$f(4) = 4^2 + 3 \qquad \bullet \text{ Replace } x \text{ by } 4.$$
$$= 16 + 3 = 19$$

This process is called **evaluating a function.**

> **HOW TO** Given $f(x) = x^2 + x - 3$, find $f(-2)$.
>
> $$f(x) = x^2 + x - 3$$
> $$f(-2) = (-2)^2 + (-2) - 3 \qquad \bullet \text{ Replace } x \text{ by } -2.$$
> $$= 4 - 2 - 3 = -1$$
> $$f(-2) = -1$$

In this example, $f(-2)$ is the second coordinate of an ordered pair of the function; the first coordinate is -2. Therefore, an ordered pair of this function is $(-2, f(-2))$, or, because $f(-2) = -1$, $(-2, -1)$.

For the function given by $y = f(x) = x^2 + x - 3$, y is called the **dependent variable** because its value depends on the value of x. The **independent variable** is x.

Functions can be written using other letters or even combinations of letters. For instance, some calculators use $ABS(x)$ for the absolute-value function. Thus the equation $y = |x|$ would be written $ABS(x) = |x|$, where $ABS(x)$ replaces y.

Example 7

Given $G(t) = \dfrac{3t}{t + 4}$, find $G(1)$.

Solution

$$G(t) = \dfrac{3t}{t + 4}$$

$$G(1) = \dfrac{3(1)}{1 + 4} \qquad \bullet \text{ Replace } t \text{ by } 1. \text{ Then simplify.}$$

$$G(1) = \dfrac{3}{5}$$

You Try It 7

Given $H(x) = \dfrac{x}{x - 4}$, find $H(8)$.

Your solution

Solution on p. S19

Copyright © Houghton Mifflin Company. All rights reserved.

7.1 Exercises

Objective A To graph points in a rectangular coordinate system

1. Graph $(-2, 1)$, $(3, -5)$, $(-2, 4)$, and $(0, 3)$.

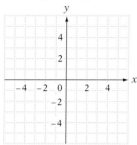

2. Graph $(5, -1)$, $(-3, -3)$, $(-1, 0)$, and $(1, -1)$.

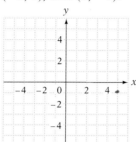

3. Graph $(0, 0)$, $(0, -5)$, $(-3, 0)$, and $(0, 2)$.

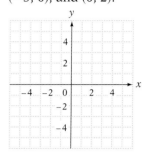

4. Graph $(-4, 5)$, $(-3, 1)$, $(3, -4)$, and $(5, 0)$.

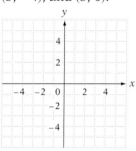

5. Graph $(-1, 4)$, $(-2, -3)$, $(0, 2)$, and $(4, 0)$.

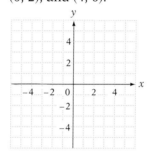

6. Graph $(5, 2)$, $(-4, -1)$, $(0, 0)$, and $(0, 3)$.

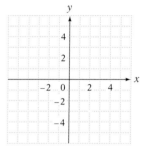

7. Find the coordinates of each of the points.

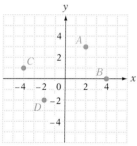

8. Find the coordinates of each of the points.

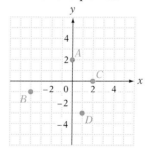

9. Find the coordinates of each of the points.

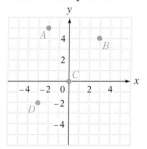

10. Find the coordinates of each of the points.

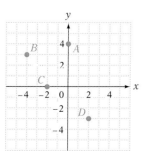

11. **a.** Name the abscissas of points A and C.
b. Name the ordinates of points B and D.

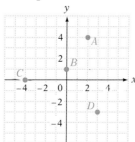

12. **a.** Name the abscissas of points A and C.
b. Name the ordinates of points B and D.

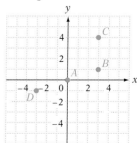

Copyright © Houghton Mifflin Company. All rights reserved.

13. Suppose you are helping a student who is having trouble graphing ordered pairs. The work of the student is at the right. What can you say to this student to correct the error that is being made?

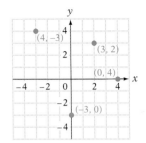

14. **a.** What are the signs of the coordinates of a point in the third quadrant?

   b. What are the signs of the coordinates of a point in the fourth quadrant?

   c. On an *xy*-coordinate system, what is the name of the axis for which all the *x*-coordinates are zero?

   d. On an *xy*-coordinate system, what is the name of the axis for which all the *y*-coordinates are zero?

Objective B **To determine ordered-pair solutions of an equation in two variables**

15. Is $(3, 4)$ a solution of $y = -x + 7$?

16. Is $(2, -3)$ a solution of $y = x + 5$?

17. Is $(-1, 2)$ a solution of $y = \frac{1}{2}x - 1$?

18. Is $(1, -3)$ a solution of $y = -2x - 1$?

19. Is $(4, 1)$ a solution of $2x - 5y = 4$?

20. Is $(-5, 3)$ a solution of $3x - 2y = 9$?

21. Is $(0, 4)$ a solution of $3x - 4y = -4$?

22. Is $(-2, 0)$ a solution of $x + 2y = -1$?

For Exercises 23 to 28, graph the ordered-pair solutions of the equation for the given values of *x*.

23. $y = 2x; x = -2, -1, 0, 2$

24. $y = -2x; x = -2, -1, 0, 2$

25. $y = \frac{2}{3}x + 1; x = -3, 0, 3$

26. $y = -\frac{1}{3}x - 2; x = -3, 0, 3$

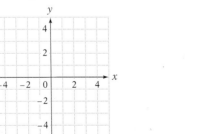

27. $2x + 3y = 6; x = -3, 0, 3$

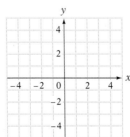

28. $x - 2y = 4; x = -2, 0, 2$

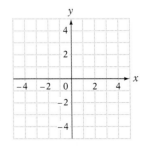

Copyright © Houghton Mifflin Company. All rights reserved.

Objective C   To determine whether a set of ordered pairs is a function

29.   **Biology**   The table below shows the length, in centimeters, of the humerus (the long bone of the forelimb, from shoulder to elbow) and the total wingspan, in centimeters, of several pterosaurs, which are extinct flying reptiles of the order Pterosauria. Write a relation where the first coordinate is the length of the humerus and the second is the wingspan. Is the relation a function?

Humerus (in centimeters)	24	32	22	15	4.4	17	15	4.4
Wingspan (in centimeters)	600	750	430	300	68	370	310	55

30.   **Environmental Science**   The table below, based in part on data from the National Oceanic and Atmospheric Administration, shows the average annual concentration of atmospheric carbon dioxide (in parts per million) and the average sea surface temperature (in degrees Celsius) for eight consecutive years. Write a relation wherein the first coordinate is the carbon dioxide concentration and the second coordinate is the average sea surface temperature. Is the relation a function?

Carbon dioxide concentration (in parts per million)	352	353	354	355	356	358	360	361
Surface sea temperature (in degrees Celsius)	15.4	15.4	15.1	15.1	15.2	15.4	15.3	15.5

31.   **Sports**   The table at the right shows the number of home runs and the number of at-bats for the top five home runs leaders in the National League for the 2003 season. Write a relation where the first coordinate is the number of at-bats and the second coordinate is the number of home runs per at-bats rounded to the nearest thousand. Is the relation a function?

Player	At-bats	Home runs
Barry Bonds	390	45
Albert Pujois	591	43
Sammy Sosa	517	40
Gary Sheffield	576	39
Jeff Bagwell	605	39

32.   **Nielsen Ratings**   The ratings (each rating point is 1,055,000 households) and share (the percentage of television sets in use tuned to a specific program) for selected television shows for a week in November 2003 are shown in the table at the right. Write a relation where the first coordinate is the ratings and the second coordinate is the share. Is the relation a function?

Television Show	Rating	Share
CSI	18.1	27.0
E.R.	13.6	22.0
Friends	13.4	21.0
CSI: Miami	13.2	21.0
60 Minutes	11.3	18.0

33. Does $y = -2x - 3$, where $x \in \{-2, -1, 0, 3\}$, define y as a function of x?

34. Does $y = 2x + 3$, where $x \in \{-2, -1, 1, 4\}$, define y as a function of x?

35. Does $|y| = x - 1$, where $x \in \{1, 2, 3, 4\}$, define y as a function of x?

36. Does $|y| = x + 2$, where $x \in \{-2, -1, 0, 3\}$, define y as a function of x?

37. Does $y = x^2$, where $x \in \{-2, -1, 0, 1, 2\}$, define y as a function of x?

38. Does $y = x^2 - 1$, where $x \in \{-2, -1, 0, 1, 2\}$, define y as a function of x?

Copyright © Houghton Mifflin Company. All rights reserved.

> **Objective D**   To evaluate a function written in functional notation

39. Given $f(x) = 3x - 4$, find $f(4)$.

40. Given $f(x) = 5x + 1$, find $f(2)$.

41. Given $f(x) = x^2$, find $f(3)$.

42. Given $f(x) = x^2 - 1$, find $f(1)$.

43. Given $G(x) = x^2 + x$, find $G(-2)$.

44. Given $H(x) = x^2 - x$, find $H(-2)$.

45. Given $s(t) = \dfrac{3}{t - 1}$, find $s(-2)$.

46. Given $P(x) = \dfrac{4}{2x + 1}$, find $P(-2)$.

47. Given $h(x) = 3x^2 - 2x + 1$, find $h(3)$.

48. Given $Q(r) = 4r^2 - r - 3$, find $Q(2)$.

49. Given $f(x) = \dfrac{x}{x + 5}$, find $f(-3)$.

50. Given $v(t) = \dfrac{2t}{2t + 1}$, find $v(3)$.

51. Given $g(x) = x^3 - x^2 + 2x - 7$, find $g(0)$.

52. Given $F(z) = \dfrac{z}{z^2 + 1}$, find $F(0)$.

APPLYING THE CONCEPTS

53. Write a few sentences that describe the similarities and differences between relations and functions.

54. The graph of $y^2 = x$, where $x \in \{0, 1, 4, 9\}$, is shown at the right. Is this the graph of a function? Explain your answer.

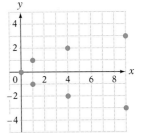

55. Is it possible to evaluate $f(x) = \dfrac{5}{x - 1}$ when $x = 1$? If so, what is $f(1)$? If not, explain why not.

Copyright © Houghton Mifflin Company. All rights reserved.

7.2 Linear Equations in Two Variables

Copyright © Houghton Mifflin Company. All rights reserved.

Objective A   **To graph an equation of the form $y = mx + b$**

The **graph of an equation in two variables** is a graph of the ordered-pair solutions of the equation.

Consider $y = 2x + 1$. Choosing $x = -2$, -1, 0, 1, and 2 and determining the corresponding values of y produces some of the ordered pairs of the equation. These are recorded in the table at the right. See the graph of the ordered pairs in Figure 1.

x	$y = 2x + 1$	y	(x, y)
-2	$2(-2) + 1$	-3	$(-2, -3)$
-1	$2(-1) + 1$	-1	$(-1, -1)$
0	$2(0) + 1$	1	$(0, 1)$
1	$2(1) + 1$	3	$(1, 3)$
2	$2(2) + 1$	5	$(2, 5)$

Choosing values of x that are not integers produces more ordered pairs to graph, such as $\left(-\frac{5}{2}, -4\right)$ and $\left(\frac{3}{2}, 4\right)$, as shown in Figure 2. Choosing still other values of x would result in more and more ordered pairs being graphed. The result would be so many dots that the graph would appear as the straight line shown in Figure 3, which is the graph of $y = 2x + 1$.

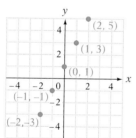

Figure 1

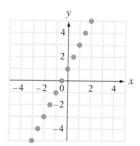

Figure 2

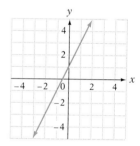

Figure 3

Equations in two variables have characteristic graphs. The equation $y = 2x + 1$ is an example of a *linear equation*, or *linear function*, because its graph is a straight line. It is also called a *first-degree equation* in two variables because the exponent on each variable is the first power.

> **Linear Equation in Two Variables**
>
> Any equation of the form $y = mx + b$, where m is the coefficient of x and b is a constant, is a **linear equation in two variables**, or a **first-degree equation in two variables** or a **linear function**. The graph of a linear equation in two variables is a straight line.

Examples of linear equations are shown at the right. These equations represent linear functions because there is only one possible y for each x. Note that for $y = 3 - 2x$, m is the coefficient of x and b is the constant.

$$y = 2x + 1 \qquad (m = 2, b = 1)$$
$$y = x - 4 \qquad (m = 1, b = -4)$$
$$y = -\frac{3}{4}x \qquad \left(m = -\frac{3}{4}, b = 0\right)$$
$$y = 3 - 2x \qquad (m = -2, b = 3)$$

The equation $y = x^2 + 4x + 3$ is not a linear equation in two variables because there is a term with a variable squared. The equation $y = \frac{3}{x - 4}$ is not a linear equation because a variable occurs in the denominator of a fraction.

Copyright © Houghton Mifflin Company. All rights reserved.

Integrating Technology

The Projects and Group Activities at the end of this chapter contain information on using calculators to graph an equation.

To graph a linear equation, choose some values of x and then find the corresponding values of y. Because a straight line is determined by two points, it is sufficient to find only two ordered-pair solutions. However, it is recommended that at least three ordered-pair solutions be used to ensure accuracy.

HOW TO Graph $y = -\dfrac{3}{2}x + 2$.

This is a linear equation with $m = -\dfrac{3}{2}$ and $b = 2$. Find at least three solutions. Because m is a fraction, choose values of x that will simplify the calculations. We have chosen $-2, 0,$ and 4 for x. (Any values of x could have been selected.)

x	$y = -\dfrac{3}{2}x = 2$	y	(x, y)
-2	$-\dfrac{3}{2}(-2) + 2$	5	$(-2, 5)$
0	$-\dfrac{3}{2}(0) + 2$	2	$(0, 2)$
4	$-\dfrac{3}{2}(4) + 2$	-4	$(4, -4)$

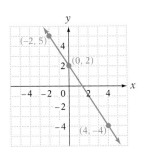

The graph of $y = -\dfrac{3}{2}x + 2$ is shown at the right.

Remember that a graph is a drawing of the ordered-pair solutions of the equation. Therefore, every point on the graph is a solution of the equation, and every solution of the equation is a point on the graph.

The graph at the right is the graph of $y = x + 2$. Note that $(-4, -2)$ and $(1, 3)$ are points on the graph and that these points are solutions of $y = x + 2$. The point whose coordinates are $(4, 1)$ is not a point on the graph and is not a solution of the equation.

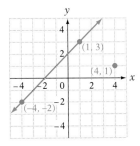

Example 1 Graph $y = 3x - 2$.

Solution

x	y
0	-2
-1	-5
2	4

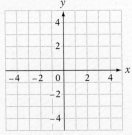

You Try It 1 Graph $y = 3x + 1$.

Your solution

Solution on p. S19

Example 2 Graph $y = 2x$.

Solution

x	y
0	0
2	4
-2	-4

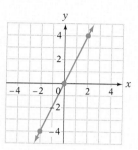

You Try It 2 Graph $y = -2x$.

Your solution

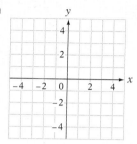

Example 3 Graph $y = \dfrac{1}{2}x - 1$.

Solution

x	y
0	-1
2	0
-2	-2

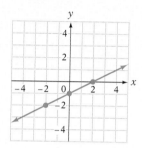

You Try It 3 Graph $y = \dfrac{1}{3}x - 3$.

Your solution

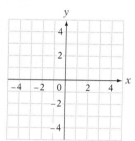

Solutions on p. S19

Objective B

To graph an equation of the form $Ax + By = C$

The equation $Ax + By = C$, where A and B are coefficients and C is a constant, is called the **standard form of a linear equation in two variables.** Examples are shown at the right.

$2x + 3y = 6$	$(A = 2, B = 3, C = 6)$
$x - 2y = -4$	$(A = 1, B = -2, C = -4)$
$2x + y = 0$	$(A = 2, B = 1, C = 0)$
$4x - 5y = 2$	$(A = 4, B = -5, C = 2)$

To graph an equation of the form $Ax + By = C$, first solve the equation for y. Then follow the same procedure used for graphing $y = mx + b$.

Study Tip

Remember that a How To example indicates a worked-out example. Using paper and pencil, work through the example. See *AIM for Success,* page xxv.

HOW TO Graph $3x + 4y = 12$.

$$3x + 4y = 12$$
$$4y = -3x + 12$$

$$y = -\frac{3}{4}x + 3$$

x	y
0	3
4	0
-4	6

- Solve for y.
- Subtract $3x$ from each side of the equation.

- Divide each side of the equation by 4.

- Find three ordered-pair solutions of the equation.

- Graph the ordered pairs and then draw a line through the points.

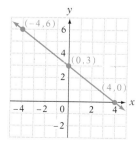

Copyright © Houghton Mifflin Company. All rights reserved.

The graph of a linear equation with one of the variables missing is either a horizontal or a vertical line.

The equation $y = 2$ could be written $0 \cdot x + y = 2$. Because $0 \cdot x = 0$ for any value of x, the value of y is always 2 no matter what value of x is chosen. For instance, replace x by -4, by -1, by 0, and by 3. In each case, $y = 2$.

$$0x + y = 2$$
$$0(-4) + y = 2 \qquad (-4, 2) \text{ is a solution.}$$
$$0(-1) + y = 2 \qquad (-1, 2) \text{ is a solution.}$$
$$0(0) + y = 2 \qquad (0, 2) \text{ is a solution.}$$
$$0(3) + y = 2 \qquad (3, 2) \text{ is a solution.}$$

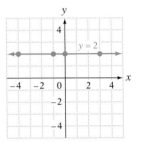

The solutions are plotted in the graph at the right, and a line is drawn through the plotted points. Note that the line is horizontal.

> **Graph of a Horizontal Line**
>
> The graph of $y = b$ is a horizontal line passing through $(0, b)$.

The equation $x = -2$ could be written $x + 0 \cdot y = -2$. Because $0 \cdot y = 0$ for any value of y, the value of x is always -2 no matter what value of y is chosen. For instance, replace y by -2, by 0, by 2, and by 3. In each case, $x = -2$.

$$x + 0y = -2$$
$$x + 0(-2) = -2 \qquad (-2, -2) \text{ is a solution.}$$
$$x + 0(0) = -2 \qquad (-2, 0) \text{ is a solution.}$$
$$x + 0(2) = -2 \qquad (-2, 2) \text{ is a solution.}$$
$$x + 0(3) = -2 \qquad (-2, 3) \text{ is a solution.}$$

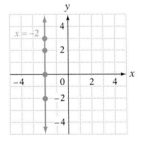

The solutions are plotted in the graph at the right, and a line is drawn through the plotted points. Note that the line is vertical.

> **Graph of a Vertical Line**
>
> The graph of $x = a$ is a vertical line passing through $(a, 0)$.

HOW TO Graph $x = -3$ and $y = 1$ on the same coordinate grid.

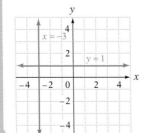

- The graph of $x = -3$ is a vertical line passing through $(-3, 0)$.

- The graph of $y = 1$ is a horizontal line passing through $(0, 1)$.

Copyright © Houghton Mifflin Company. All rights reserved.

Example 4   Graph $2x - 5y = 10$.

Solution   Solve $2x - 5y = 10$ for y.
$$2x - 5y = 10$$
$$-5y = -2x + 10$$
$$y = \frac{2}{5}x - 2$$

x	y
0	-2
5	0
-5	-4

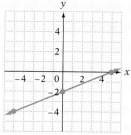

You Try It 4   Graph $5x - 2y = 10$.

Your solution

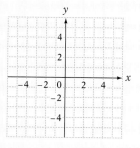

Example 5   Graph $x + 2y = 6$.

Solution   Solve $x + 2y = 6$ for y.
$$x + 2y = 6$$
$$2y = -x + 6$$
$$y = -\frac{1}{2}x + 3$$

x	y
0	3
-2	4
4	1

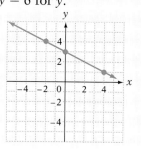

You Try It 5   Graph $x - 3y = 9$.

Your solution

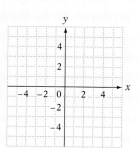

Example 6   Graph $y = -2$.

Solution
The graph of an equation of the form $y = b$ is a horizontal line passing through the point $(0, b)$.

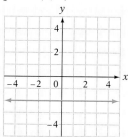

You Try It 6   Graph $y = 3$.

Your solution

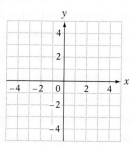

Example 7   Graph $x = 3$.

Solution
The graph of an equation of the form $x = a$ is a vertical line passing through the point $(a, 0)$.

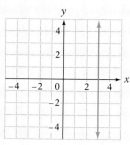

You Try It 7   Graph $x = -4$.

Your solution

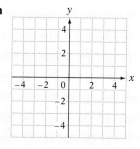

Copyright © Houghton Mifflin Company. All rights reserved.

Solutions on p. S19

Objective C **To solve application problems**

There are a variety of applications of linear functions.

HOW TO Solve: The temperature of a cup of water that has been placed in a microwave oven to be heated can be approximated by the equation $T = 0.7s + 65$, where T is the temperature of the water s seconds after the microwave oven is turned on.

a. Graph this equation for $0 \le s \le 220$. (Note: In many applications, the domain of the variable is given so that the equation makes sense. For instance, it would not be sensible to have values of s that are less than 0. This would correspond to negative time. The choice of 220 is somewhat arbitrary and was chosen so that the water would not boil over.)

b. The point whose coordinates are (120, 149) is on the graph of this equation. Write a sentence that describes the meaning of this ordered pair.

Solution

a.
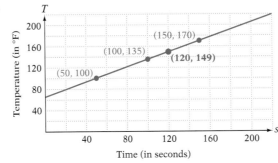

- Choosing $s = 50, 100$, and 150, you find the corresponding ordered pairs (50, 100), (100, 135), and (150, 170). Plot these points and draw a line through the points.

b. The point whose coordinates are (120, 149) means that 120 s (2 min) after the oven is turned on, the water temperature is 149°F.

Example 8

The number of kilobytes, K, of an MP3 file that remain to be downloaded t seconds after starting the download is given by $K = 935 - 5.5t$. Graph this equation for $0 \le t \le 170$. The point whose coordinates are (50, 660) are on this graph. Write a sentence that describes the meaning of this ordered pair.

Solution
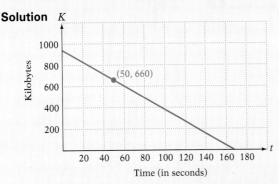

The ordered pair (50, 660) means that after 50 s, there are 660 K remaining to be downloaded.

You Try It 8

A car is traveling at a uniform speed of 40 mph. The distance, d, the car travels in t hours is given by $d = 40t$. Graph this equation for $0 \le t \le 5$. The point whose coordinates are (3, 120) is on the graph. Write a sentence that describes the meaning of this ordered pair.

Your solution
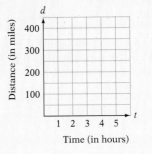

Solution on p. S19

Copyright © Houghton Mifflin Company. All rights reserved.

7.2 Exercises

Copyright © Houghton Mifflin Company. All rights reserved.

Objective A   **To graph an equation of the form $y = mx + b$**

For Exercises 1 to 18, graph.

1. $y = 2x - 3$

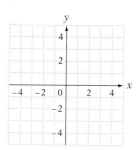

2. $y = -2x + 2$

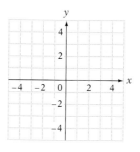

3. $y = \dfrac{1}{3}x$

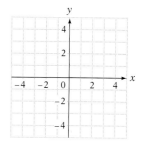

4. $y = -3x$

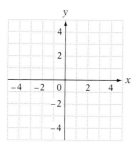

5. $y = \dfrac{2}{3}x - 1$

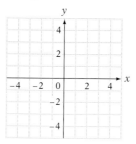

6. $y = \dfrac{3}{4}x + 2$

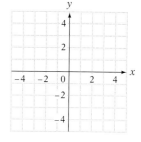

7. $y = -\dfrac{1}{4}x + 2$

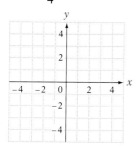

8. $y = -\dfrac{1}{3}x + 1$

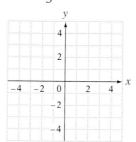

9. $y = -\dfrac{2}{5}x + 1$

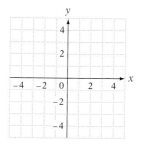

10. $y = -\dfrac{1}{2}x + 3$

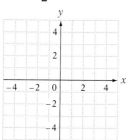

11. $y = 2x - 4$

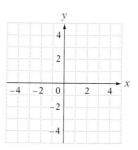

12. $y = 3x - 4$

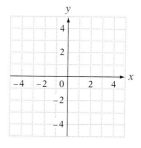

13. $y = x - 3$

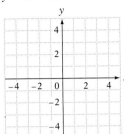

14. $y = x + 2$

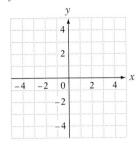

15. $y = -x + 2$

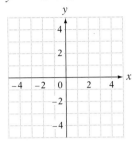

16. $y = -x - 1$

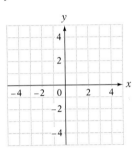

17. $y = -\dfrac{2}{3}x + 1$

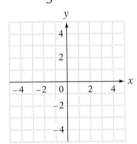

18. $y = 5x - 4$

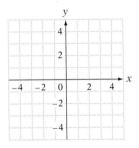

Objective B   **To graph an equation of the form $Ax = By + C$**

For Exercises 19 to 36, graph.

19. $3x + y = 3$

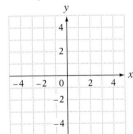

20. $2x + y = 4$

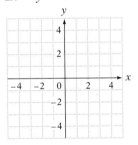

21. $2x + 3y = 6$

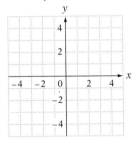

22. $3x + 2y = 4$

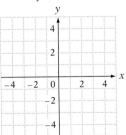

23. $x - 2y = 4$

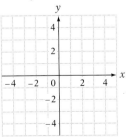

24. $x - 3y = 6$

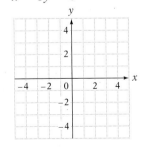

Copyright © Houghton Mifflin Company. All rights reserved.

25. $2x - 3y = 6$

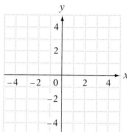

26. $3x - 2y = 8$

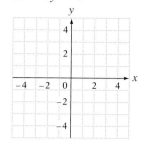

27. $2x + 5y = 10$

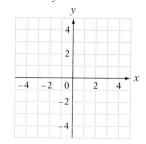

28. $3x + 4y = 12$

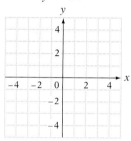

29. $x = 3$

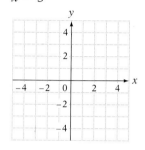

30. $y = -4$

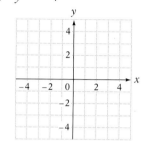

31. $x + 4y = 4$

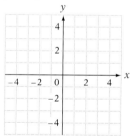

32. $4x - 3y = 12$

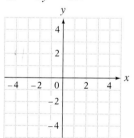

33. $y = 4$

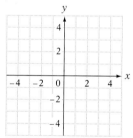

34. $x = -2$

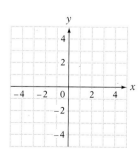

35. $\dfrac{x}{5} + \dfrac{y}{4} = 1$

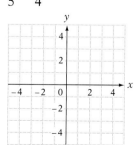

36. $\dfrac{x}{4} - \dfrac{y}{3} = 1$

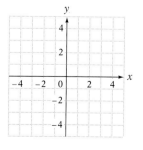

Copyright © Houghton Mifflin Company. All rights reserved.

Copyright © Houghton Mifflin Company. All rights reserved.

Objective C   **To solve application problems**

37. **Emergency Response**   A rescue helicopter is rushing at a constant speed of 150 mph to reach several people stranded in the ocean 11 mi away after their boat sank. The rescuers can determine how far they are from the victims using the equation $D = 11 - 2.5t$, where D is the distance in miles and t is the time elapsed in minutes. Graph this equation for $0 \le t \le 4$. The point $(3, 3.5)$ is on the graph. Write a sentence that describes the meaning of this ordered pair.

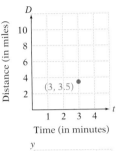

38. **Business**   A custom-illustrated sign or banner can be commissioned for a cost of $25 for the material and $10.50 per square foot for the artwork. The equation that represents this cost is given by $y = 10.50x + 25$, where y is the cost and x is the number of square feet in the sign. Graph this equation for $0 \le x \le 20$. The point $(15, 182.5)$ is on the graph. Write a sentence that describes the meaning of this ordered pair.

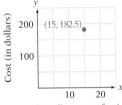

39. **Veterinary Science**   According to some veterinarians, the age, x, of a dog can be translated to "human years" by using the equation $H = 4x + 16$, where H is the human equivalent age for the dog. Graph this equation for $2 \le x \le 21$. The point whose coordinates are $(6, 40)$ is on this graph. Write a sentence that explains the meaning of this ordered pair.

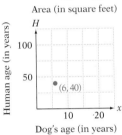

40. **Business**   Judging on the basis of data from the Consumer Electronics Association, the projected number, N (in millions), of sales of high-definition televisions (HDTVs) can be approximated by $N = 3t + 4$, where $0 \le t \le 4$ and $t = 0$ corresponds to the year 2003. Graph this equation. The point whose coordinates are $(3, 13)$ is on this graph. Write a sentence that explains the meaning of this ordered pair in the context of the problem.

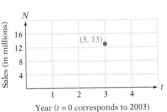

APPLYING THE CONCEPTS

41. Graph $y = 2x - 2$, $y = 2x$, and $y = 2x + 3$. What observation can you make about the graphs?

42. Graph $y = x + 3$, $y = 2x + 3$, and $y = -\frac{1}{2}x + 3$. What observation can you make about the graphs?

43. For the equation $y = 3x + 2$, when the value of x changes from 1 to 2, does the value of y increase or decrease? What is the change in y? Suppose that the value of x changes from 13 to 14. What is the change in y?

44. For the equation $y = -2x + 1$, when the value of x changes from 1 to 2, does the value of y increase or decrease? What is the change in y? Suppose the value of x changes from 13 to 14. What is the change in y?

45. **Telecommunications**   A long-distance telephone company offers a flat rate of $.99 for the first 15 minutes of a phone call and then $.15 for each additional minute. The graph of this situation is a combination of the graphs of two linear equations: $C = 0.99$ when $0 < t \le 15$ and $C = 0.15(t - 15) + 0.99$ when $t > 15$. The graph is shown at the right.
a. What is the cost of a telephone call that lasts 5 minutes?
b. What is the cost of a telephone call that lasts 20 minutes?

7.3 Intercepts and Slopes of Straight Lines

Objective A To find the *x*- and *y*-intercepts of a straight line

The graph of the equation $2x + 3y = 6$ is shown at the right. The graph crosses the *x*-axis at the point (3, 0) and crosses the *y*-axis at the point (0, 2). The point at which a graph crosses the *x*-axis is called the **x-intercept.** At the *x*-intercept, the *y*-coordinate is 0. The point at which a graph crosses the *y*-axis is called the **y-intercept.** At the *y*-intercept, the *x*-coordinate is 0.

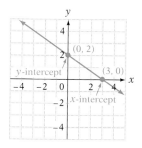

HOW TO Find the *x*- and *y*-intercepts of the graph of the equation $2x - 3y = 12$.

To find the *x*-intercept, let $y = 0$. (Any point on the *x*-axis has *y*-coordinate 0.)

$$2x - 3y = 12$$
$$2x - 3(0) = 12$$
$$2x = 12$$
$$x = 6$$

The *x*-intercept is (6, 0).

To find the *y*-intercept, let $x = 0$. (Any point on the *y*-axis has *x*-coordinate 0.)

$$2x - 3y = 12$$
$$2(0) - 3y = 12$$
$$-3y = 12$$
$$y = -4$$

The *y*-intercept is (0, −4).

HOW TO Find the *y*-intercept of $y = 3x + 4$.

$$y = 3x + 4 = 3(0) + 4 = 4 \qquad \bullet \text{ Let } x = 0.$$

The *y*-intercept is (0, 4).

For any equation of the form $y = mx + b$, the *y*-intercept is (0, b).

Some linear equations can be graphed by finding the *x*- and *y*-intercepts and then drawing a line through these two points.

> **TAKE NOTE**
>
> To find the *x*-intercept, let $y = 0$ and solve for *x*.
> To find the *y*-intercept, let $x = 0$ and solve for *y*.

Example 1 Find the *x*- and *y*-intercepts for $x - 2y = 4$. Graph the line.

Solution
To find the *x*-intercept, let $y = 0$ and solve for *x*.
$$x - 2y = 4$$
$$x - 2(0) = 4$$
$$x = 4 \qquad (4, 0)$$

To find the *y*-intercept, let $x = 0$ and solve for *y*.
$$x - 2y = 4$$
$$0 - 2y = 4$$
$$-2y = 4$$
$$y = -2 \qquad (0, -2)$$

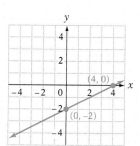

You Try It 1 Find the *x*- and *y*-intercepts for $y = 2x - 4$. Graph the line.

Your solution

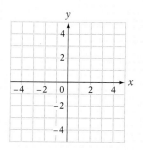

Solution on p. S20

Copyright © Houghton Mifflin Company. All rights reserved.

Objective B **To find the slope of a straight line**

The graphs of $y = \frac{2}{3}x + 1$ and $y = 2x + 1$ are shown in Figure 1. Each graph crosses the y-axis at the point $(0, 1)$, but the graphs have different slants. The **slope** of a line is a measure of the slant of the line. The symbol for slope is m.

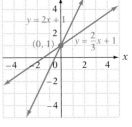

Figure 1

TAKE NOTE

The change in the y values can be thought of as the *rise* of the line, and the change in the x values can be thought of as the *run*. Then

$$\text{Slope} = m = \frac{\text{rise}}{\text{run}}$$

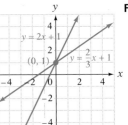

$$m = \frac{\text{rise}}{\text{run}}$$

The slope of a line containing two points is the ratio of the change in the y values of the two points to the change in the x values. The line containing the points $(-2, -3)$ and $(6, 1)$ is graphed in Figure 2. The change in the y values is the difference between the two ordinates.

$$\text{Change in } y = 1 - (-3) = 4$$

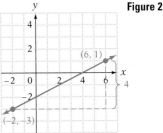

Figure 2

The change in the x values is the difference between the two abscissas (Figure 3).

$$\text{Change in } x = 6 - (-2) = 8$$

$$\text{Slope} = m = \frac{\text{change in } y}{\text{change in } x} = \frac{4}{8} = \frac{1}{2}$$

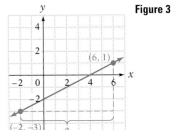

Figure 3

Slope Formula

If $P_1(x_1, y_1)$ and $P_2(x_2, y_2)$ are two points on a line and $x_1 \neq x_2$, then $m = \dfrac{y_2 - y_1}{x_2 - x_1}$ (Figure 4). If $x_1 = x_2$, the slope is undefined.

Figure 4

HOW TO Find the slope of the line containing the points $(-1, 1)$ and $(2, 3)$.

Let P_1 be $(-1, 1)$ and P_2 be $(2, 3)$. Then $x_1 = -1, y_1 = 1, x_2 = 2,$ and $y_2 = 3$.

$$m = \frac{y_2 - y_1}{x_2 - x_1} = \frac{3 - 1}{2 - (-1)} = \frac{2}{3}$$

The slope is $\frac{2}{3}$.

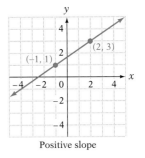

Positive slope

TAKE NOTE

Positive slope means that the value of y increases as the value of x increases.

A line that slants upward to the right always has a **positive slope.**

Note that you obtain the same results if the points are named oppositely. Let P_1 be $(2, 3)$ and P_2 be $(-1, 1)$. Then $x_1 = 2, y_1 = 3, x_2 = -1,$ and $y_2 = 1$.

$$m = \frac{y_2 - y_1}{x_2 - x_1} = \frac{1 - 3}{-1 - 2} = \frac{-2}{-3} = \frac{2}{3}$$

The slope is $\frac{2}{3}$. Therefore, it does not matter which point is named P_1 and which is named P_2; the slope remains the same.

Copyright © Houghton Mifflin Company. All rights reserved.

Copyright © Houghton Mifflin Company. All rights reserved.

TAKE NOTE

Negative slope means that the value of y decreases as x increases. Compare this to positive slope.

HOW TO Find the slope of the line containing the points $(-3, 4)$ and $(2, -2)$.

Let P_1 be $(-3, 4)$ and P_2 be $(2, -2)$.

$$m = \frac{y_2 - y_1}{x_2 - x_1} = \frac{-2 - 4}{2 - (-3)} = \frac{-6}{5} = -\frac{6}{5}$$

The slope is $-\frac{6}{5}$.

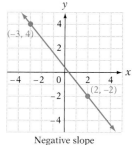
Negative slope

A line that slants downward to the right always has a **negative slope.**

HOW TO Find the slope of the line containing the points $(-1, 3)$ and $(4, 3)$.

Let P_1 be $(-1, 3)$ and P_2 be $(4, 3)$.

$$m = \frac{y_2 - y_1}{x_2 - x_1} = \frac{3 - 3}{4 - (-1)} = \frac{0}{5} = 0$$

The slope is 0.

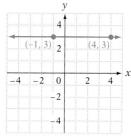
Zero slope

A horizontal line has **zero slope.**

HOW TO Find the slope of the line containing the points $(2, -2)$ and $(2, 4)$.

Let P_1 be $(2, -2)$ and P_2 be $(2, 4)$.

$$m = \frac{y_2 - y_1}{x_2 - x_1} = \frac{4 - (-2)}{2 - 2} = \frac{6}{0} \qquad \text{Division by zero is not defined.}$$

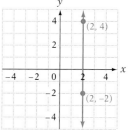
Undefined slope

A vertical line has undefined slope.

Two lines in the plane that never intersect are called parallel lines. The lines l_1 and l_2 in the figure at the right are parallel. Calculating the slope of each line, we have

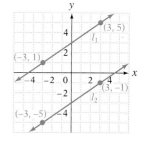

Slope of l_1: $m_1 = \dfrac{y_2 - y_1}{x_2 - x_1} = \dfrac{5 - 1}{3 - (-3)} = \dfrac{4}{6} = \dfrac{2}{3}$

Slope of l_2: $m_2 = \dfrac{y_2 - y_1}{x_2 - x_1} = \dfrac{-1 - (-5)}{3 - (-3)} = \dfrac{4}{6} = \dfrac{2}{3}$

Note that these parallel lines have the same slope. This is always true.

TAKE NOTE

We must separate the description of parallel lines at the right because vertical lines in the plane are parallel but their slopes are undefined.

Parallel Lines

Two nonvertical lines in the plane are parallel if and only if they have the same slope. Vertical lines in the plane are parallel.

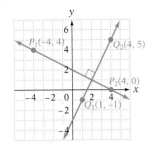

Two lines that intersect at a 90° angle (right angle) are perpendicular lines. The lines at the left are perpendicular.

> **Perpendicular Lines**
>
> Two nonvertical lines in the plane are perpendicular if and only if the product of their slopes is -1. A vertical and horizontal line are perpendicular.

The slope of the line between P_1 and P_2 is $m_1 = \dfrac{0 - 4}{4 - (-4)} = -\dfrac{4}{8} = -\dfrac{1}{2}$. The slope of the line between Q_1 and Q_2 is $m_2 = \dfrac{5 - (-1)}{4 - 1} = \dfrac{6}{3} = 2$. The product of the slopes is $\left(-\dfrac{1}{2}\right)2 = -1$. Because the product of the slopes is -1, the graphs are perpendicular.

There are many applications of the concept of slope. Here is a possibility.

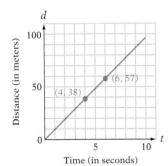

When Florence Griffith-Joyner set the world record for the 100-meter dash, her average rate of speed was approximately 9.5 meters per second. The graph at the right shows the distance she ran during her record-setting run. From the graph, note that after 4 s she had traveled 38 m and that after 6 s she had traveled 57 m. The slope of the line between these two points is

$$m = \frac{57 - 38}{6 - 4} = \frac{19}{2} = 9.5$$

Note that the slope of the line is the same as the rate she was running, 9.5 meters per second. The average speed of an object is related to slope.

Example 2

Find the slope of the line containing the points with coordinates $(-2, -3)$ and $(3, 4)$.

Solution

Let $P_1 = (-2, -3)$ and $P_2 = (3, 4)$.

$m = \dfrac{y_2 - y_1}{x_2 - x_1} = \dfrac{4 - (-3)}{3 - (-2)}$ • $y_2 = 4,\ y_1 = -3,$
                                  $x_2 = 3,\ x_1 = -2$

$= \dfrac{7}{5}$

The slope is $\dfrac{7}{5}$.

You Try It 2

Find the slope of the line containing the points with coordinates $(1, 4)$ and $(-3, 8)$.

Your solution

Example 3

Find the slope of the line containing the points with coordinates $(-1, 4)$ and $(-1, 0)$.

Solution

Let $P_1 = (-1, 4)$ and $P_2 = (-1, 0)$.

$m = \dfrac{y_2 - y_1}{x_2 - x_1} = \dfrac{0 - 4}{-1 - (-1)}$ • $y_2 = 0,\ y_1 = 4,$
                                    $x_2 = -1,\ x_1 = -1$

$= \dfrac{-4}{0}$

The slope is undefined.

You Try It 3

Find the slope of the line containing the points with coordinates $(-1, 2)$ and $(4, 2)$.

Your solution

Solutions on p. S20

Copyright © Houghton Mifflin Company. All rights reserved.

Example 4

The graph below shows the height of a plane above an airport during its 30-minute descent from cruising altitude to landing. Find the slope of the line. Write a sentence that explains the meaning of the slope.

Solution

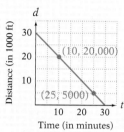

$$m = \frac{5000 - 20{,}000}{25 - 10} = \frac{-15{,}000}{15}$$

$$= -1000$$

A slope of -1000 means that the height of the plane is *decreasing* at the rate of 1000 ft/min.

You Try It 4

The graph below shows the approximate decline in the value of a used car over a 5-year period. Find the slope of the line. Write a sentence that states the meaning of the slope.

Your solution

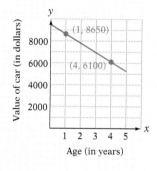

Solution on p. S20

Objective C — **To graph a line using the slope and the *y*-intercept**

The graph of the equation $y = \frac{2}{3}x + 1$ is shown at the right. The points $(-3, -1)$ and $(3, 3)$ are on the graph. The slope of the line between the two points is

$$m = \frac{3 - (-1)}{3 - (-3)} = \frac{4}{6} = \frac{2}{3}$$

Observe that the slope of the line is the coefficient of x in the equation $y = \frac{2}{3}x + 1$. Also recall that the y-intercept is $(0, 1)$, where 1 is the constant term of the equation.

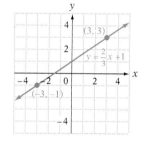

> **TAKE NOTE**
>
> Here are some equations in slope-intercept form.
>
> $y = 2x - 3$: Slope is 2; y-intercept is $(0, -3)$.
>
> $y = -x + 2$: Slope is -1 (recall that $-x = -1x$); y-intercept is $(0, 2)$.
>
> $y = \frac{x}{2}$: Because $\frac{x}{2} = \frac{1}{2}x$, slope is $\frac{1}{2}$; y-intercept is $(0, 0)$.

Slope-Intercept Form of a Linear Equation

An equation of the form $y = mx + b$ is called the **slope-intercept form** of a straight line. The slope of the line is m, the coefficient of x. The y-intercept is $(0, b)$, where b is the constant term of the equation.

If a linear equation is not in slope-intercept form, solve the equation for y.

HOW TO Find the slope and y-intercept for the graph of $3x + 2y = 12$.

$3x + 2y = 12$

$\quad\quad 2y = -3x + 12$ • Write the equation in slope-intercept

$\quad\quad\ \ y = -\frac{3}{2}x + 6$ form by solving for y.

The slope is $-\frac{3}{2}$; the y-intercept is $(0, 6)$.

When an equation of a line is in slope-intercept form, the graph can be drawn using the slope and the y-intercept. First locate the y-intercept. Use the slope to find a second point on the line. Then draw a line through the two points.

Copyright © Houghton Mifflin Company. All rights reserved.

HOW TO Graph $y = 2x - 3$.

y-intercept $= (0, b) = (0, -3)$

$$m = 2 = \frac{2}{1} = \frac{\text{change in } y}{\text{change in } x}$$

Beginning at the y-intercept, move right 1 unit (change in x) and then up 2 units (change in y).

$(1, -1)$ is a second point on the graph.

Draw a line through the two points $(0, -3)$ and $(1, -1)$.

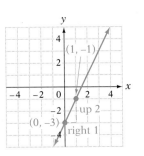

Example 5   Graph $y = -\frac{2}{3}x + 1$ by using the slope and y-intercept.

Solution y-intercept $= (0, b) = (0, 1)$

$$m = -\frac{2}{3} = \frac{-2}{3} = \frac{\text{change in } y}{\text{change in } x}$$

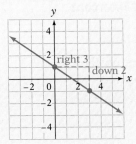

You Try It 5   Graph $y = -\frac{1}{4}x - 1$ by using the slope and y-intercept.

Your solution

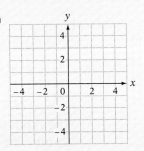

Example 6   Graph $2x - 3y = 6$ by using the slope and y-intercept.

Solution   Solve the equation for y.

$$2x - 3y = 6$$
$$-3y = -2x + 6$$
$$y = \frac{2}{3}x - 2$$

y-intercept $= (0, -2)$; $m = \frac{2}{3}$

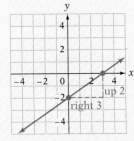

You Try It 6   Graph $x - 2y = 4$ by using the slope and y-intercept.

Your solution

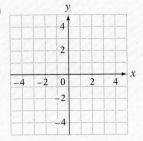

Solutions on p. S20

Copyright © Houghton Mifflin Company. All rights reserved.

HOW TO Graph $y = 2x - 3$.

y-intercept $= (0, b) = (0, -3)$

$m = 2 = \dfrac{2}{1} = \dfrac{\text{change in } y}{\text{change in } x}$

Beginning at the y-intercept, move right 1 unit (change in x) and then up 2 units (change in y).

$(1, -1)$ is a second point on the graph.

Draw a line through the two points $(0, -3)$ and $(1, -1)$.

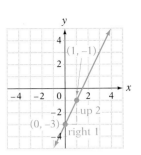

Example 5 Graph $y = -\dfrac{2}{3}x + 1$ by using the slope and y-intercept.

Solution y-intercept $= (0, b) = (0, 1)$

$m = -\dfrac{2}{3} = \dfrac{-2}{3} = \dfrac{\text{change in } y}{\text{change in } x}$

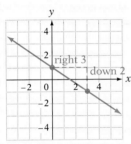

You Try It 5 Graph $y = -\dfrac{1}{4}x - 1$ by using the slope and y-intercept.

Your solution

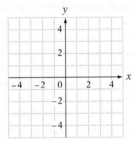

Example 6 Graph $2x - 3y = 6$ by using the slope and y-intercept.

Solution Solve the equation for y.

$2x - 3y = 6$

$-3y = -2x + 6$

$y = \dfrac{2}{3}x - 2$

y-intercept $= (0, -2)$; $m = \dfrac{2}{3}$

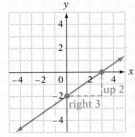

You Try It 6 Graph $x - 2y = 4$ by using the slope and y-intercept.

Your solution

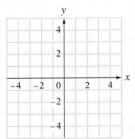

Solutions on p. S20

Copyright © Houghton Mifflin Company. All rights reserved.

Example 4

The graph below shows the height of a plane above an airport during its 30-minute descent from cruising altitude to landing. Find the slope of the line. Write a sentence that explains the meaning of the slope.

Solution

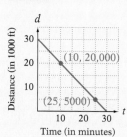

$$m = \frac{5000 - 20,000}{25 - 10} = \frac{-15,000}{15}$$

$$= -1000$$

A slope of -1000 means that the height of the plane is *decreasing* at the rate of 1000 ft/min.

You Try It 4

The graph below shows the approximate decline in the value of a used car over a 5-year period. Find the slope of the line. Write a sentence that states the meaning of the slope.

Your solution

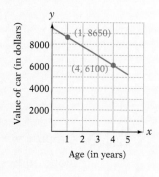

Solution on p. S20

Objective C   **To graph a line using the slope and the y-intercept**

The graph of the equation $y = \frac{2}{3}x + 1$ is shown at the right. The points $(-3, -1)$ and $(3, 3)$ are on the graph. The slope of the line between the two points is

$$m = \frac{3 - (-1)}{3 - (-3)} = \frac{4}{6} = \frac{2}{3}$$

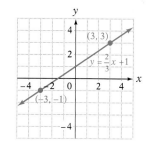

Observe that the slope of the line is the coefficient of x in the equation $y = \frac{2}{3}x + 1$. Also recall that the y-intercept is $(0, 1)$, where 1 is the constant term of the equation.

> **TAKE NOTE**
>
> Here are some equations in slope-intercept form.
>
> $y = 2x - 3$: Slope is 2; y-intercept is $(0, -3)$.
>
> $y = -x + 2$: Slope is -1 (recall that $-x = -1x$); y-intercept is $(0, 2)$.
>
> $y = \frac{x}{2}$: Because $\frac{x}{2} = \frac{1}{2}x$, slope is $\frac{1}{2}$; y-intercept is $(0, 0)$.

Slope-Intercept Form of a Linear Equation

An equation of the form $y = mx + b$ is called the **slope-intercept form** of a straight line. The slope of the line is m, the coefficient of x. The y-intercept is $(0, b)$, where b is the constant term of the equation.

If a linear equation is not in slope-intercept form, solve the equation for y.

HOW TO   Find the slope and y-intercept for the graph of $3x + 2y = 12$.

$$3x + 2y = 12$$
$$2y = -3x + 12$$
$$y = -\frac{3}{2}x + 6$$

• Write the equation in slope-intercept form by solving for y.

The slope is $-\frac{3}{2}$; the y-intercept is $(0, 6)$.

When an equation of a line is in slope-intercept form, the graph can be drawn using the slope and the y-intercept. First locate the y-intercept. Use the slope to find a second point on the line. Then draw a line through the two points.

Copyright © Houghton Mifflin Company. All rights reserved.

7.3 Exercises

To find the *x*- and *y*-intercepts of a straight line

For Exercises 1 to 12, find the *x*- and *y*-intercepts.

1. $x - y = 3$ **2.** $3x + 4y = 12$ **3.** $y = 3x - 6$ **4.** $y = 2x + 10$

5. $x - 5y = 10$ **6.** $3x + 2y = 12$ **7.** $y = 3x + 12$ **8.** $y = 5x + 10$

9. $2x - 3y = 0$ **10.** $3x + 4y = 0$ **11.** $y = -\dfrac{1}{2}x + 3$ **12.** $y = \dfrac{2}{3}x - 4$

For Exercises 13 to 18, find the *x*- and *y*-intercepts and then graph.

13. $5x + 2y = 10$

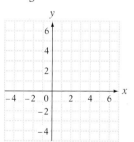

14. $x - 3y = 6$

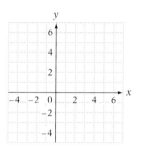

15. $y = \dfrac{3}{4}x - 3$

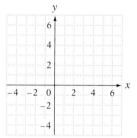

16. $y = \dfrac{2}{5}x - 2$

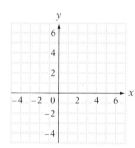

17. $5y - 3x = 15$

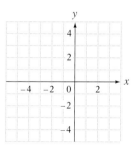

18. $9y - 4x = 18$

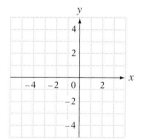

To find the slope of a straight line

19. Explain how to find the slope of a line given two points on the line.

20. What is the difference between a line that has zero slope and one that has undefined slope?

Copyright © Houghton Mifflin Company. All rights reserved.

For Exercises 21 to 32, find the slope of the line containing the given points.

21. $P_1(4, 2), P_2(3, 4)$ **22.** $P_1(2, 1), P_2(3, 4)$ **23.** $P_1(-1, 3), P_2(2, 4)$ **24.** $P_1(-2, 1), P_2(2, 2)$

25. $P_1(2, 4), P_2(4, -1)$ **26.** $P_1(1, 3), P_2(5, -3)$ **27.** $P_1(3, -4), P_2(3, 5)$ **28.** $P_1(-1, 2), P_2(-1, 3)$

29. $P_1(4, -2), P_2(3, -2)$ **30.** $P_1(5, 1), P_2(-2, 1)$ **31.** $P_1(0, -1), P_2(3, -2)$ **32.** $P_1(3, 0), P_2(2, -1)$

For Exercises 33 to 40, determine whether the line through P_1 and P_2 is parallel, perpendicular, or neither parallel nor perpendicular to the line through Q_1 and Q_2.

33. $P_1(-3, 4), P_2(2, -5); Q_1(3, 6), Q_2(-2, -3)$ **34.** $P_1(4, -5), P_2(6, -9); Q_1(5, -4), Q_2(1, 4)$

35. $P_1(0, 1), P_2(2, 4); Q_1(-4, -7), Q_2(2, 5)$ **36.** $P_1(5, 1), P_2(3, -2); Q_1(0, -2), Q_2(3, -4)$

37. $P_1(-2, 4), P_2(2, 4); Q_1(-3, 6), Q_2(4, 6)$ **38.** $P_1(1, -1), P_2(3, -2); Q_1(-4, 1), Q_2(2, -5)$

39. $P_1(7, -1), P_2(-4, 6); Q_1(3, 0), Q_2(-5, 3)$ **40.** $P_1(5, -2), P_2(-1, 3); Q_1(3, 4), Q_2(-2, -2)$

41. **Business** The graph at the right is based on data from *InfoSync World*. It shows the projected camera-phone sales worldwide through 2008. Find the slope of the line. Write a sentence that states the meaning of the slope in the context of this problem.

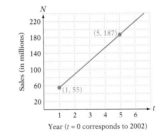

42. **Deep-Sea Diving** The pressure, in pounds per square inch, on a diver is shown in the graph at the right. Find the slope of the line. Write a sentence that explains the meaning of the slope.

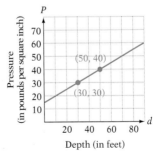

43. **Depreciation** The graph at the right, based on data from *Kelley Blue Book*, shows the decline in value of a 2002 Porsche Boxster as the number of miles the car is driven increases. Find the slope of the line. Write a sentence that states the meaning of the slope in the context of this problem.

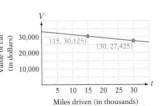

Copyright © Houghton Mifflin Company. All rights reserved.

44. 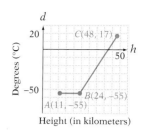 **Environmental Science** The stratosphere extends from approximately 11 km to 50 km above Earth. The graph at the right shows how the temperature, in degrees Celsius, changes in the stratosphere. Explain the meaning of the horizontal line segment from A to B. Find the slope of the line from B to C and explain its meaning.

Objective C   **To graph a line using the slope and the *y*-intercept**

For Exercises 45 to 50, find the slope and *y*-intercept for the graph of the equation.

45. $2x - 3y = 6$

46. $4x + 3y = 12$

47. $2x + 5y = 10$

48. $2x + y = 0$

49. $x - 4y = 0$

50. $2x + 3y = 8$

For Exercises 51 to 65, graph by using the slope and *y*-intercept.

51. $y = 3x + 1$

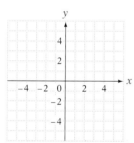

52. $y = -2x - 1$

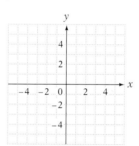

53. $y = \dfrac{2}{5}x - 2$

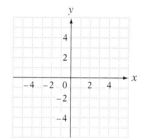

54. $y = \dfrac{3}{4}x + 1$

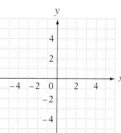

55. $2x + y = 3$

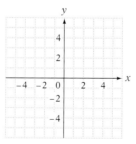

56. $3x - y = 1$

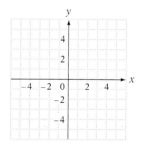

57. $x - 2y = 4$

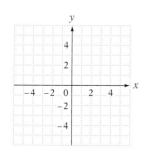

58. $x + 3y = 6$

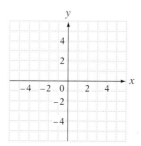

59. $y = \dfrac{2}{3}x$

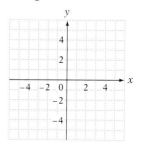

Copyright © Houghton Mifflin Company. All rights reserved.

60. $y = \dfrac{1}{2}x$

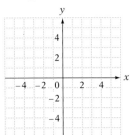

61. $y = -x + 1$

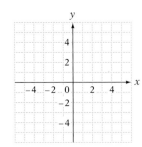

62. $y = -x - 3$

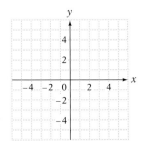

63. $3x - 4y = 12$

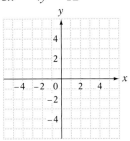

64. $5x - 2y = 10$

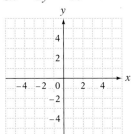

65. $y = -4x + 2$

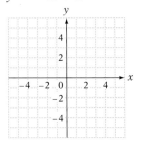

APPLYING THE CONCEPTS

66. Do all straight lines have a y-intercept? If not, give an example of one that does not.

67. If two lines have the same slope and the same y-intercept, must the graphs of the lines be the same? If not, give an example.

68. **a.** Graph: $\dfrac{x}{3} + \dfrac{y}{4} = 1$ **b.** Graph: $\dfrac{x}{2} - \dfrac{y}{3} = 1$ **c.** Graph: $-\dfrac{x}{4} + \dfrac{y}{2} = 1$

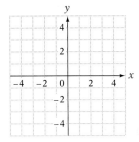

d. What observation can you make about the x- and y-intercepts of these graphs and the coefficients of x and y? Use this observation to draw the graph of $\dfrac{x}{4} - \dfrac{y}{3} = 1$.

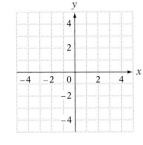

69. **Safety**   What does the highway sign at the right have to do with slope?

Copyright © Houghton Mifflin Company. All rights reserved.

7.4 Equations of Straight Lines

Objective A   **To find the equation of a line given a point and the slope**

In earlier sections, the equation of a line was given and you were asked to determine some properties of the line, such as its intercepts and slope. Here, the process is reversed. Given properties of a line, you will determine its equation.

If the slope and y-intercept of a line are known, the equation of the line can be determined by using the slope-intercept form of a straight line.

> **HOW TO**   Find the equation of the line with slope $-\dfrac{1}{2}$ and y-intercept $(0, 3)$.
>
> $y = mx + b$ • Use the slope-intercept formula.
>
> $y = -\dfrac{1}{2}x + 3$ • $m = -\dfrac{1}{2}$; $(0, b) = (0, 3)$, so $b = 3$.
>
> The equation of the line is $y = -\dfrac{1}{2}x + 3$.

When the slope and the coordinates of a point other than the y-intercept are known, the equation of the line can be found by using the formula for slope.

Suppose a line passes through the point $(3, 1)$ and has a slope of $\dfrac{2}{3}$. The equation of the line with these properties is determined by letting (x, y) be the coordinates of an unknown point on the line. Because the slope of the line is known, use the slope formula to write an equation. Then solve for y.

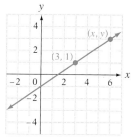

$$\frac{y - 1}{x - 3} = \frac{2}{3}$$ • $\dfrac{y_2 - y_1}{x_2 - x_1} = m; \ m = \dfrac{2}{3}; (x_2, y_2) = (x, y); (x_1, y_1) = (3, 1)$

$$\frac{y - 1}{x - 3}(x - 3) = \frac{2}{3}(x - 3)$$ • Multiply each side by $(x - 3)$.

$$y - 1 = \frac{2}{3}x - 2$$ • Simplify.

$$y = \frac{2}{3}x - 1$$ • Solve for y. Add 1 to each side.

The equation of the line is $y = \dfrac{2}{3}x - 1$.

The same procedure that was used above is used to derive the *point-slope formula*. We use this formula to determine the equation of a line when we are given the coordinates of a point on the line and the slope of the line.

Let (x_1, y_1) be the given coordinates of a point on a line, m the given slope of the line, and (x, y) the coordinates of an unknown point on the line. Then

$$\frac{y - y_1}{x - x_1} = m$$ • Formula for slope.

$$\frac{y - y_1}{x - x_1}(x - x_1) = m(x - x_1)$$ • Multiply each side by $x - x_1$.

$$y - y_1 = m(x - x_1)$$ • Simplify.

Copyright © Houghton Mifflin Company. All rights reserved.

> **Point-Slope Formula**
>
> If (x_1, y_1) is a point on a line with slope m, then $y - y_1 = m(x - x_1)$.

HOW TO Find the equation of the line that passes through the point $(2, 3)$ and has slope -2.

$$y - y_1 = m(x - x_1)$$ • Use the point-slope formula.

$$y - 3 = -2(x - 2)$$ • $m = -2$; $(x_1, y_1) = (2, 3)$

$$y - 3 = -2x + 4$$ • Solve for y.

$$y = -2x + 7$$

The equation of the line is $y = -2x + 7$.

Example 1

Find the equation of the line whose slope is $-\frac{2}{3}$ and whose y-intercept is $(0, -1)$.

Solution

Because the slope and y-intercept are known, use the slope-intercept formula, $y = mx + b$.

$$y = -\frac{2}{3}x - 1 \qquad • \ m = -\frac{2}{3}; b = -1$$

You Try It 1

Find the equation of the line whose slope is $\frac{5}{3}$ and whose y-intercept is $(0, 2)$.

Your solution

Example 2

Use the point-slope formula to find the equation of the line that passes through the point $(-2, -1)$ and has slope $\frac{3}{2}$.

Solution

$$y - y_1 = m(x - x_1)$$

$$y - (-1) = \frac{3}{2}[x - (-2)] \qquad • \ m = \frac{3}{2};$$
$$(x_1, y_1) = (-2, -1)$$

$$y + 1 = \frac{3}{2}(x + 2)$$

$$y + 1 = \frac{3}{2}x + 3$$

$$y = \frac{3}{2}x + 2$$

You Try It 2

Use the point-slope formula to find the equation of the line that passes through the point $(4, -2)$ and has slope $\frac{3}{4}$.

Your solution

Solutions on p. S20

Objective B   **To find the equation of a line given two points**

The point-slope formula is used to find the equation of a line when a point on the line and the slope of the line are known. But this formula can also be used to find the equation of a line given two points on the line. In this case,

1. Use the slope formula to determine the slope of the line between the points.

2. Use the point-slope formula, the slope you just calculated, and one of the given points to find the equation of the line.

Copyright © Houghton Mifflin Company. All rights reserved.

HOW TO Find the equation of the line that passes through the points whose coordinates are $(-3, -1)$ and $(3, 3)$.

Use the slope formula to determine the slope of the line between the points.

$$m = \frac{y_2 - y_1}{x_2 - x_1} = \frac{3 - (-1)}{3 - (-3)} = \frac{4}{6} = \frac{2}{3}$$ • $(x_1, y_1) = (-3, -1); (x_2, y_2) = (3, 3)$

Use the point-slope formula, the slope you just calculated, and one of the given points to find the equation of the line.

$$y - y_1 = m(x - x_1)$$ • Point-slope formula

$$y - (-1) = \frac{2}{3}[x - (-3)]$$ • $m = \frac{2}{3}; (x_1, y_1) = (-3, -1)$

$$y + 1 = \frac{2}{3}(x + 3)$$

$$y + 1 = \frac{2}{3}x + 2$$

$$y = \frac{2}{3}x + 1$$

Check:

$y = \frac{2}{3}x + 1$			$y = \frac{2}{3}x + 1$	
-1	$\frac{2}{3}(-3) + 1$	• $(x, y) = (-3, -1)$	3	$\frac{2}{3}(3) + 1$	• $(x, y) = (3, 3)$
-1	$-2 + 1$		3	$2 + 1$	
$-1 = -1$			$3 = 3$		

TAKE NOTE
You can verify that the equation $y = \frac{2}{3}x + 1$ passes through the points $(-3, -1)$ and $(3, 3)$ by substituting the coordinates of these points into the equation.

The equation of the line that passes through the two points is $y = \frac{2}{3}x + 1$.

Example 3

Find the equation of the line that passes through the points $(-4, 0)$ and $(2, -3)$.

Solution

Find the slope of the line between the two points.

$$m = \frac{y_2 - y_1}{x_2 - x_1} = \frac{-3 - 0}{2 - (-4)} = \frac{-3}{6} = -\frac{1}{2}$$

Use the point-slope formula.

$$y - y_1 = m(x - x_1)$$ • Point-slope formula

$$y - 0 = -\frac{1}{2}[x - (-4)]$$ • $m = -\frac{1}{2}; (x_1, y_1) = (-4, 0)$

$$y = -\frac{1}{2}(x + 4)$$

$$y = -\frac{1}{2}x - 2$$

The equation of the line is $y = -\frac{1}{2}x - 2$.

You Try It 3

Find the equation of the line that passes through the points $(-6, -2)$ and $(3, 1)$.

Your solution

Solution on p. S20

Copyright © Houghton Mifflin Company. All rights reserved.

Objective C **To solve application problems**

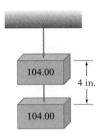

A **linear model** is a first-degree equation that is used to describe a relationship between quantities. In many cases, a linear model is used to approximate collected data. The data are graphed as points in a coordinate system, and then a line is drawn that approximates the data. The graph of the points is called a **scatter diagram;** the line is called the **line of best fit.**

Consider an experiment to determine the weight required to stretch a spring a certain distance. Data from such an experiment are shown in the table below.

Distance (in inches)	2.5	4	2	3.5	1	4.5
Weight (in pounds)	63	104	47	85	27	115

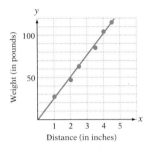

The accompanying graph shows the scatter diagram, which is the plotted points, and the line of best fit, which is the line that approximately goes through the plotted points. The equation of the line of best fit is $y = 25.6x - 1.3$, where x is the number of inches the spring is stretched and y is the weight in pounds.

The table below shows the values that the model would predict to the nearest tenth. Good linear models should predict values that are close to the actual values. A more thorough analysis of lines of best fit is undertaken in statistics courses.

Distance, x	2.5	4	2	3.5	1	4.5
Weight predicted using $y = 25.6x - 1.3$	62.7	101.1	49.9	88.3	24.3	113.9

Example 4

The data in the table below show the percent of people in the U.S. that purchase their music from the Internet. (*Source*: The Recording Industry Association of America) The line of best fit is $y = 0.62x + 0.667$, where x is the year (with 1997 corresponding to $x = 0$) and y the percent of people in the U.S. purchasing music from the Internet.

Year	0	1	2	3	4	5
Percent	0.3	1.1	2.4	3.2	2.9	3.4

Graph the data and the line of best fit in the coordinate system below. Write a sentence that describes the meaning of the slope of the line.

Solution

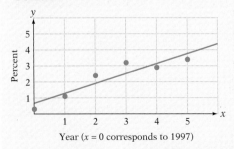

Year (x = 0 corresponds to 1997)

The slope of the line means that the percent of people purchasing music from the Internet is increasing 0.62% per year.

You Try It 4

The data in the table below show a reading test grade and the final exam grade in a history class. The line of best fit is $y = 8.3x - 7.8$, where x is the reading test score and y is the history test score.

Reading	8.5	9.4	10.0	11.4	12.0
History	64	68	76	87	92

Graph the data and the line of best fit in the coordinate system below. Write a sentence that describes the meaning of the slope of the line of best fit.

Your solution

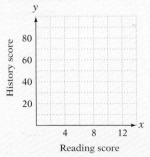

Reading score

Solution on p. S20

Copyright © Houghton Mifflin Company. All rights reserved.

7.4 Exercises

Objective A   **To find the equation of a line given a point and the slope**

1. ✎ What is the point-slope formula and how is it used?

2. ✎ Can the point-slope formula be used to find the equation of any line? If not, equations for which types of lines cannot be found using this formula?

3. Find the equation of the line that contains the point $(0, 2)$ and has slope 2.

4. Find the equation of the line that contains the point $(0, -1)$ and has slope -2.

5. Find the equation of the line that contains the point $(-1, 2)$ and has slope -3.

6. Find the equation of the line that contains the point $(2, -3)$ and has slope 3.

7. Find the equation of the line that contains the point $(3, 1)$ and has slope $\frac{1}{3}$.

8. Find the equation of the line that contains the point $(-2, 3)$ and has slope $\frac{1}{2}$.

9. Find the equation of the line that contains the point $(4, -2)$ and has slope $\frac{3}{4}$.

10. Find the equation of the line that contains the point $(2, 3)$ and has slope $-\frac{1}{2}$.

11. Find the equation of the line that contains the point $(5, -3)$ and has slope $-\frac{3}{5}$.

12. Find the equation of the line that contains the point $(5, -1)$ and has slope $\frac{1}{5}$.

13. Find the equation of the line that contains the point $(2, 3)$ and has slope $\frac{1}{4}$.

14. Find the equation of the line that contains the point $(-1, 2)$ and has slope $-\frac{1}{2}$.

Objective B   **To find the equation of a line given two points**

15. Find the equation of the line that passes through the points $(1, -1)$ and $(-2, -7)$.

16. Find the equation of the line that passes through the points $(2, 3)$ and $(3, 2)$.

17. Find the equation of the line that passes through the points $(-2, 1)$ and $(1, -5)$.

18. Find the equation of the line that passes through the points $(-1, -3)$ and $(2, -12)$.

19. Find the equation of the line that passes through the points $(0, 0)$ and $(-3, -2)$.

20. Find the equation of the line that passes through the points $(0, 0)$ and $(-5, 1)$.

Copyright © Houghton Mifflin Company. All rights reserved.

21. Find the equation of the line that passes through the points (2, 3) and (−4, 0).

22. Find the equation of the line that passes through the points (3, −1) and (0, −3).

23. Find the equation of the line that passes through the points (−4, 1) and (4, −5).

24. Find the equation of the line that passes through the points (−5, 0) and (10, −3).

25. Find the equation of the line that passes through the points (−2, 1) and (2, 4).

26. Find the equation of the line that passes through the points (3, −2) and (−3, −3).

Objective C   **To solve application problems**

27. **Sports**   The data in the table below show the number of carbohydrates used for various amounts of time during a strenuous tennis workout. The line of best fit is $y = 1.55x + 1.45$, where x is the time of the workout in minutes and y is the number of carbohydrates used in grams.

Time of workout, x (in minutes)	5	10	20	30	60
Carbohydrates used, y (in grams)	10	15	33	49	94

Graph the data and the line of best fit in the coordinate system at the right. Write a sentence that describes the meaning of the slope of the line of best fit in the context of this problem.

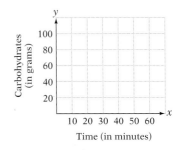

28. **Sports**   The data in the table below show the amount of water a professional tennis player loses for various times during a tennis match. The line of best fit is $y = 34.6x + 207$, where x is the time of the workout in minutes and y is the milliliters of water lost during the match.

Time of workout, x (in minutes)	10	20	30	40	50	60
Water lost, y (in milliliters)	600	900	1200	1500	2000	2300

Graph the data and the line of best fit in the coordinate system at the right. Write a sentence that describes the meaning of the slope of the line of best fit in the context of this problem.

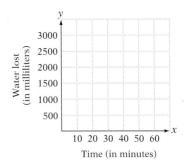

29. **Entertainment**   The data in the table below show the decline in the percent of music purchased in stores in the U.S. (*Source:* RIAA) The line of best fit is $y = -3x + 55$, where x is the year (with $x = 0$ corresponding to 1997) and y is the percent of music purchased in stores in the U.S.

Year, x	1	2	3	4	5	6
Percent, y	52	51	45	42	42	37

Graph the data and the line of best fit in the coordinate system at the right. Write a sentence that describes the meaning of the slope of the line of best fit in the context of this problem.

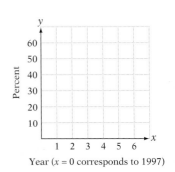

Copyright © Houghton Mifflin Company. All rights reserved.

30. 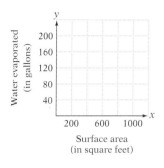 **Evaporation** The data in the table below show the amount of water that evaporates from swimming pools of various surface areas. The line of best fit is $y = 0.17x - 1$, where x is the surface area of the swimming pool and y is the number of gallons of water that evaporates in one day.

Surface area, x (in square feet)	100	200	300	400	600	1000
Water evaporated, y (in gallons)	25	30	45	60	100	170

Graph the data and the line of best fit in the coordinate system at the right. Write a sentence that describes the meaning of the slope of the line of best fit in the context of this problem.

APPLYING THE CONCEPTS

In Exercises 31 to 34, the first two given points are on a line. Determine whether the third point is on the line.

31. $(-3, 2)$, $(4, 1)$; $(-1, 0)$

32. $(2, -2)$, $(3, 4)$; $(-1, 5)$

33. $(-3, -5)$, $(1, 3)$; $(4, 9)$

34. $(-3, 7)$, $(0, -2)$; $(1, -5)$

35. If $(-2, 4)$ are the coordinates of a point on the line whose equation is $y = mx + 1$, what is the slope of the line?

36. If $(3, 1)$ are the coordinates of a point on the line whose equation is $y = mx - 3$, what is the slope of the line?

37. If $(0, -3)$, $(6, -7)$, and $(3, n)$ are coordinates of points on the same line, determine n.

38. If $(-4, 11)$, $(2, -4)$, and $(6, n)$ are coordinates of points on the same line, determine n.

The formula $y - y_1 = \frac{y_2 - y_1}{x_2 - x_1}(x - x_1)$, where $x_1 \neq x_2$, is called the **two-point formula** for a straight line. This formula can be used to find the equation of a line given two points. Use this formula for Exercises 39 and 40.

39. Find the equation of the line passing through $(-2, 3)$ and $(4, -1)$.

40. Find the equation of the line passing through $(3, -1)$ and $(4, -3)$.

41. Explain why the condition $x_1 \neq x_2$ is placed on the two-point formula given above.

42. Explain how the two-point formula given above can be derived from the point-slope formula.

Copyright © Houghton Mifflin Company. All rights reserved.

Focus on Problem Solving

Counterexamples

Some of the exercises in this text ask you to determine whether a statement is true or false. For instance, the statement "Every real number has a reciprocal" is false because 0 is a real number and 0 does not have a reciprocal.

Finding an example, such as "0 has no reciprocal," to show that a statement is not always true is called finding a counterexample. A **counterexample** is an example that shows that a statement is not always true.

Here are some counterexamples to the statement "The square of a number is always larger than the number."

$$\left(\frac{1}{2}\right)^2 = \frac{1}{4} \quad \text{but} \quad \frac{1}{4} < \frac{1}{2} \qquad 1^2 = 1 \quad \text{but} \quad 1 = 1$$

For Exercises 1 to 9, answer true if the statement is always true. If there is an instance when the statement is false, give a counterexample.

1. The product of two integers is always a positive number.

2. The sum of two prime numbers is never a prime number.

3. For all real numbers, $|x + y| = |x| + |y|$.

4. If x and y are nonzero real numbers and $x > y$, then $x^2 > y^2$.

5. The quotient of any two nonzero real numbers is less than either one of the numbers.

6. The reciprocal of a positive number is always smaller than the number.

7. If $x < 0$, then $|x| = -x$.

8. For any two real numbers x and y, $x + y > x - y$.

9. The list of numbers 1, 11, 111, 1111, 11111, ... contains infinitely many composite numbers. (*Hint:* A number is divisible by 3 if the sum of the digits of the number is divisible by 3.)

Projects and Group Activities

Graphing Linear Equations with a Graphing Utility

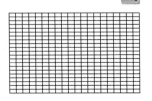

A computer or graphing calculator screen is divided into *pixels*. There are approximately 6000 to 790,000 pixels available on the screen (depending on the computer or calculator). The greater the number of pixels, the smoother a graph will appear. A portion of a screen is shown at the left. Each little rectangle represents one pixel.

The graphing utilities that are used by computers or calculators to graph an equation do basically what we have shown in the text: They choose values of x and, for each, calculate the corresponding value of y. The pixel corresponding to the ordered pair is then turned on. The graph is jagged because pixels are much larger than the dots we draw on paper.

Copyright © Houghton Mifflin Company. All rights reserved.

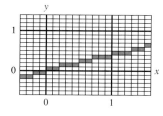

The graph of $y = 0.45x$ is shown at the left as the calculator drew it (jagged). The x- and y-axes have been chosen so that each pixel represents $\frac{1}{10}$ of a unit. Consider the region of the graph where $x = 1$, 1.1, and 1.2.

The corresponding values of y are 0.45, 0.495, and 0.54. Because the y-axis is in tenths, the numbers 0.45, 0.495, and 0.54 are rounded to the nearest tenth before plotting. Rounding 0.45, 0.495, and 0.54 to the nearest tenth results in 0.5 for each number. Thus the ordered pairs (1, 0.45), (1.1, 0.495), and (1.2, 0.54) are graphed as (1, 0.5), (1.1, 0.5), and (1.2, 0.5). These points appear as three illuminated horizontal pixels. However, if you use the TRACE feature of the calculator (see the Appendix), the actual y-coordinate for each value of x is displayed.

TAKE NOTE

Xmin and Xmax are the smallest and largest values of x that will be shown on the screen. Ymin and Ymax are the smallest and largest values of y that will be shown on the screen.

Here are the keystrokes for a TI-83 calculator to graph $y = \frac{2}{3}x + 1$. First the equation is entered. Then the domain (Xmin to Xmax) and the range (Ymin to Ymax) are entered. This is called the **viewing window.**

By changing the keystrokes 2 [X,T,θ,n] [÷] 3 [+] 1, you can graph different equations.

Integrating Technology

See the Keystroke Guide: [Y=] and [WINDOW] for assistance.

For Exercises 1 to 4, graph on a graphing calculator.

1. $y = 2x + 1$ **2.** $y = -\frac{1}{2}x - 2$ **3.** $3x + 2y = 6$ **4.** $4x + 3y = 75$

Graphs of Motion

A graph can be useful in analyzing the motion of a body. For example, consider an airplane in uniform motion traveling at 100 m/s. The table at the right shows the distance, in meters, traveled by the plane at the end of each of five one-second intervals.

Time (in seconds)	Distance (in meters)
0	0
1	100
2	200
3	300
4	400
5	500

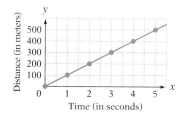

These data can be graphed on a rectangular coordinate system and a straight line drawn through the points plotted. The travel time is shown along the horizontal axis, and the distance traveled by the plane is shown along the vertical axis. (Note that the units along the two axes are not the same length.)

To write the equation for the line just graphed, use the coordinates of any two points on the line to find the slope. The y-intercept is (0, 0).

Let $(x_1, y_1) = (1, 100)$ and $(x_2, y_2) = (2, 200)$.

$$m = \frac{y_2 - y_1}{x_2 - x_1} = \frac{200 - 100}{2 - 1} = 100$$

$$y = mx + b$$
$$y = 100x + 0$$
$$y = 100x$$

Note that the slope of the line, 100, is equal to the speed, 100 m/s. *The slope of a distance-time graph represents the speed of the object.*

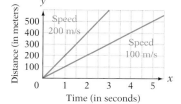

The distance-time graphs for two planes are shown at the left. One plane is traveling at 100 m/s, and the other is traveling at 200 m/s. The slope of the line representing the faster plane is greater than the slope of the line representing the slower plane.

Copyright © Houghton Mifflin Company. All rights reserved.

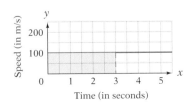

Speed (in m/s) / Time (in seconds)

In the speed-time graph at the left, the time a plane has been flying at 100 m/s is shown along the horizontal axis, and its speed is shown along the vertical axis. Because the speed is constant, the graph is a horizontal line.

The area between the horizontal line graphed and the horizontal axis is equal to the distance traveled by the plane up to that time. For example, the area of the shaded region on the graph is

$$\text{Length} \cdot \text{width} = (3 \text{ s})(100 \text{ m/s}) = 300 \text{ m}$$

The distance traveled by the plane in 3 s is equal to 300 m.

1. A car in uniform motion is traveling at 20 m/s.
   a. Prepare a distance-time graph for the car for 0 s to 5 s.
   b. Find the slope of the line.
   c. Find the equation of the line.
   d. Prepare a speed-time graph for the car for 0 s to 5 s.
   e. Find the distance traveled by the car after 3 s.

2. One car in uniform motion is traveling at 10 m/s. A second car in uniform motion is traveling at 15 m/s.
   a. Prepare one distance-time graph for both cars for 0 s to 5 s.
   b. Find the slope of each line.
   c. Find the equation of each line graphed.
   d. Assuming that the cars started at the same point at 0 s, find the distance between the cars at the end of 5 s.

Chapter 7 Summary

Key Words

A *rectangular coordinate system* is formed by two number lines, one horizontal and one vertical, that intersect at the zero point of each line. The number lines that make up a rectangular coordinate system are called the *coordinate axes*, or simply *axes*. The *origin* is the point of intersection of the two coordinate axes. Generally, the horizontal axis is labeled the *x*-axis and the vertical axis is labeled the *y*-axis. The coordinate system divides the plane into four regions called *quadrants*. The *coordinates of a point* in the plane are given by an *ordered pair* (x, y). The first number in the ordered pair is called the *abscissa* or *x*-coordinate. The second number in the ordered pair is the *ordinate* or *y*-coordinate. The *graph of an ordered pair* (x, y) is the dot drawn at the coordinates of the point in the plane.
[7.1A, p. 353]

Examples

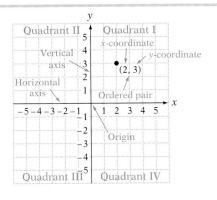

A *solution of an equation in two variables* is an ordered pair (x, y) that makes the equation a true statement.
[7.1B, p. 355]

The ordered pair $(-1, 1)$ is a solution of the equation $y = 2x + 3$ because when -1 is substituted for x and 1 is substituted for y, the result is a true equation.

A *relation* is any set of ordered pairs. The *domain* of a relation is the set of first coordinates of the ordered pairs. The *range* is the set of second coordinates of the ordered pairs.
[7.1C, p. 357]

For the relation $\{(-1, 2), (2, 4), (3, 5), (3, 7)\}$, the domain is $\{-1, 2, 3\}$; the range is $\{2, 4, 5, 7\}$.

Copyright © Houghton Mifflin Company. All rights reserved.

A *function* is a relation in which no two ordered pairs have the same first coordinate. [7.1C, p. 357]

The relation $\{(-2, -3), (0, 4), (1, 5)\}$ is a function. No two ordered pairs have the same first coordinate.

The *graph of an equation in two variables* is a graph of the ordered-pair solutions of the equation. An equation of the form $y = mx + b$ is a *linear equation in two variables*. [7.2A, p. 365]

$y = 2x + 3$ is a linear equation in two variables. Its graph is shown at the right.

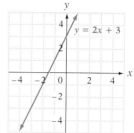

An equation written in the form $Ax + By = C$ is the *standard form of a linear equation in two variables*. [7.2B, p. 367]

$2x + 7y = 10$ is an example of a linear equation in two variables written in standard form.

The point at which a graph crosses the *x*-axis is called the *x-intercept*. At the *x*-intercept, the *y*-coordinate is 0. The point at which a graph crosses the *y*-axis is called the *y-intercept*. At the *y*-intercept, the *x*-coordinate is 0. [7.3A, p. 375]

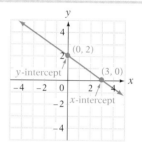

The *slope* of a line is a measure of the slant of the line. The symbol for slope is m. A line with *positive slope* slants upward to the right. A line with *negative slope* slants downward to the right. A horizontal line has *zero slope*. A vertical line has an *undefined slope*. [7.3A, pp. 376–377]

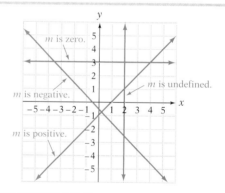

When data are graphed as points in a coordinate system, the graph is called a *scatter diagram*. A line drawn to approximate the data is called the *line of best fit*. [7.4C, p. 388]

The graph shown at the right is the scatter diagram and line of best fit for the spring data on page 388.

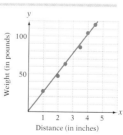

Essential Rules and Procedures

Examples

Functional Notation [7.1D, p. 360]
The equation of a function is written in functional notation when y is replaced by the symbol $f(x)$, where $f(x)$ is read "f of x" or "the value of f at x." To evaluate a function at a given value of x, replace x by the given value and then simplify the resulting numerical expression to find the value of $f(x)$.

$y = x^2 + 2x - 1$ is written in functional notation as
$f(x) = x^2 + 2x - 1$. To evaluate
$f(x) = x^2 + 2x - 1$ at $x = -3$, find
$f(-3)$.

$$f(-3) = (-3)^2 + 2(-3) - 1$$
$$= 9 - 6 - 1 = 2$$

Copyright © Houghton Mifflin Company. All rights reserved.

Horizontal and Vertical Lines [7.2B, p. 368]
The graph of $y = b$ is a horizontal line passing through $(0, b)$.
The graph of $x = a$ is a vertical line passing through $(a, 0)$.

The graph of $y = -2$ is a horizontal line passing through $(0, -2)$. The graph of $x = 3$ is a vertical line passing through $(3, 0)$.

To find the x-intercept, let $y = 0$ and solve for x.
To find the y-intercept, let $x = 0$ and solve for y.
[7.3A, p. 375]

To find the x-intercept of $4x - 5y = 20$, let $y = 0$ and solve for x. To find the y-intercept, let $x = 0$ and solve for y.

$$4x - 5y = 20 \qquad 4x - 5y = 20$$
$$4x - 5(0) = 20 \qquad 4(0) - 5y = 20$$
$$4x = 20 \qquad -5y = 20$$
$$x = 5 \qquad y = -4$$

The x-intercept is $(5, 0)$.   The y-intercept is $(0, -4)$.

Slope Formula [7.3B, p. 376]
If $P_1(x_1, y_1)$ and $P_2(x_2, y_2)$ are two points on a line and $x_1 \neq x_2$, then

$$m = \frac{y_2 - y_1}{x_2 - x_1}$$

To find the slope of the line between the points $(1, -2)$ and $(-3, -1)$, let $P_1 = (1, -2)$ and $P_2 = (-3, -1)$. Then

$$m = \frac{y_2 - y_1}{x_2 - x_1} = \frac{-1 - (-2)}{-3 - 1} = \frac{1}{-4} = -\frac{1}{4}.$$

Parallel Lines [7.3B, p. 377]
Two nonvertical lines in the plane are parallel if and only if they have the same slope. Vertical lines in the plane are parallel.

The slope of the line through $P_1(3, -6)$ and $P_2(5, -10)$ is $m_1 = \frac{-10 - (-6)}{5 - 3} = -2$.
The slope of the line through $Q_1(4, -5)$ and $Q_2(0, 3)$ is $m_2 = \frac{3 - (-5)}{0 - 4} = -2$.
Because $m_1 = m_2$, the lines are parallel.

Perpendicular Lines [7.3B, p. 378]
Two nonvertical lines in the plane are perpendicular if and only if the product of their slopes is -1. A vertical and horizontal line are perpendicular.

The slope of the line through $P_1(5, -3)$ and $P_2(2, -1)$ is $m_1 = \frac{-1 - (-3)}{2 - 5} = -\frac{2}{3}$.
The slope of the line through $Q_1(1, -4)$ and $Q_2(3, -1)$ is $m_2 = \frac{-1 - (-4)}{3 - 1} = \frac{3}{2}$.
Because $m_1 m_2 = \left(-\frac{2}{3}\right)\left(\frac{3}{2}\right) = -1$, the lines are perpendicular.

Slope-Intercept Form of a Linear Equation [7.3C, p. 379]
An equation of the form $y = mx + b$ is called the slope-intercept form of a straight line. The slope of the line is m, the coefficient of x. The y-intercept is $(0, b)$, where b is the constant term of the equation.

For the line with equation $y = -3x + 2$, the slope is -3 and the y-intercept is $(0, 2)$.

Point-Slope Formula [7.4A, p. 386]
If (x_1, y_1) is a point on a line with slope m, then

$$y - y_1 = m(x - x_1)$$

The equation of the line that passes through the point $(5, -3)$ and has slope -2 is:

$$y - y_1 = m(x - x_1)$$
$$y - (-3) = -2(x - 5)$$
$$y + 3 = -2x + 10$$
$$y = -2x + 7$$

Copyright © Houghton Mifflin Company. All rights reserved.

Chapter 7 Review Exercises

1. **a.** Graph the ordered pairs $(-2, 4)$ and $(3, -2)$.
    b. Name the abscissa of point A.
    c. Name the ordinate of point B.

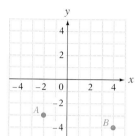

2. Graph the ordered-pair solutions of $y = -\frac{1}{2}x - 2$ when $x \in \{-4, -2, 0, 2\}$.

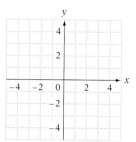

3. Determine the equation of the line that passes through the points $(-1, 3)$ and $(2, -5)$.

4. Determine the equation of the line that passes through the point $(6, 1)$ and has slope $-\frac{5}{2}$.

5. Graph $y = \frac{1}{4}x + 3$.

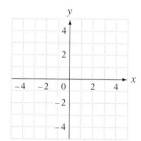

6. Graph $5x + 3y = 15$.

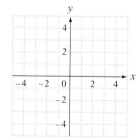

7. Is the line that passes through $(7, -5)$ and $(6, -1)$ parallel, perpendicular, or neither parallel nor perpendicular to the line that passes through $(4, 5)$ and $(2, -3)$?

8. Given $f(x) = x^2 - 2$, find $f(-1)$.

9. Determine the equation of the line that passes through the points $(-2, 5)$ and $(4, 1)$.

10. Does $y = -x + 3$, where $x \in \{-2, 0, 3, 5\}$, define y as a function of x?

11. Find the slope of the line containing the points $(9, 8)$ and $(-2, 1)$.

12. Find the x- and y-intercepts of $3x - 2y = 24$.

13. Find the slope of the line containing the points $(-2, -3)$ and $(4, -3)$.

Copyright © Houghton Mifflin Company. All rights reserved.

14. Graph the line that has slope $\frac{1}{2}$ and y-intercept $(0, -1)$.

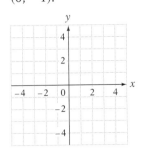

15. Graph $x = -3$.

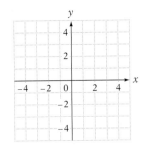

16. Graph the line that has slope $-\frac{2}{3}$ and y-intercept $(0, 2)$.

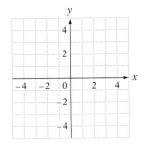

17. Graph $y = -2x - 1$.

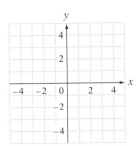

18. Graph the line that has slope 2 and y-intercept $(0, -4)$.

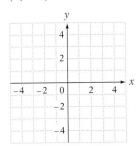

19. Graph $3x - 2y = -6$.

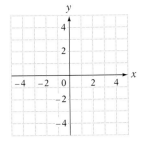

20. **Health** The height and weight of 8 seventh-grade students are shown in the following table. Write a relation where the first coordinate is height, in inches, and the second coordinate is weight, in pounds. Is the relation a function?

Height (in inches)	55	57	53	57	60	61	58	54
Weight (in pounds)	95	101	94	98	100	105	97	95

21. **Business** An online research service charges a monthly access fee of $75 plus $.45 per minute to use the service. An equation that represents the monthly cost to use this service is $C = 0.45x + 75$, where C is the monthly cost and x is the number of minutes of access used. Graph this equation for $0 \leq x \leq 100$. The point $(50, 97.5)$ is on the graph. Write a sentence that describes the meaning of this ordered pair.

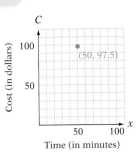

22. **Telecommunications** The data in the table below are estimates of the projected annual increase in average telephone bills for a family. The line of best fit is $y = 34x + 657$, where x is the year (with $x = 0$ corresponding to 1999) and y is the annual cost, in dollars, of telephone bills.

Year, x	0	1	2	3	4	5	6
Telephone bills, y (in dollars)	658	690	708	772	809	830	849

Graph the data and the line of best fit in the coordinate system at the right. Write a sentence that describes the meaning of the slope of the line of best fit.

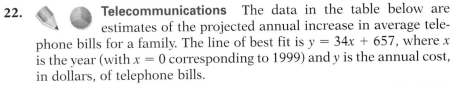

Copyright © Houghton Mifflin Company. All rights reserved.

Chapter 7 Test

1. Find the ordered-pair solution of $2x - 3y = 15$ corresponding to $x = 3$.

2. Graph the ordered-pair solutions of $y = -\frac{3}{2}x + 1$ for $x \in \{-2, 0, 4\}$.

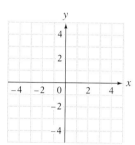

3. Does $y = \frac{1}{2}x - 3$ define y as a function of x for $x \in \{-2, 0, 4\}$?

4. Given $f(t) = t^2 + t$, find $f(2)$.

5. Given $f(x) = x^2 - 2x$, find $f(-1)$.

6. **Emergency Response** The distance a house is from a fire station and the amount of damage that the house sustained in a fire are given in the following table. Write a relation wherein the first coordinate of the ordered pair is the distance, in miles, from the fire station and the second coordinate is the amount of damage in thousands of dollars. Is the relation a function?

Distance (in miles)	3.5	4.0	5.2	5.0	4.0	6.3	5.4
Damage (in thousands of dollars)	25	30	45	38	42	12	34

7. Graph $y = 3x + 1$.

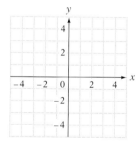

8. Graph $y = -\frac{3}{4}x + 3$.

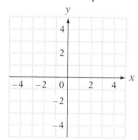

9. Graph $3x - 2y = 6$.

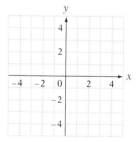

10. Graph $x + 3 = 0$.

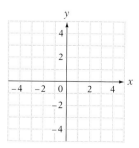

11. Graph the line that has slope $-\frac{2}{3}$ and y-intercept $(0, 4)$.

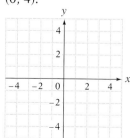

12. Graph the line that has slope 2 and y-intercept -2.

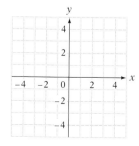

Copyright © Houghton Mifflin Company. All rights reserved.

13. 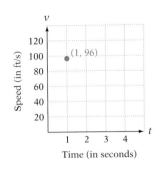 **Sports**   The equation for the speed of a ball that is thrown straight up with an initial speed of 128 ft/s is $v = 128 - 32t$, where v is the speed of the ball after t seconds. Graph this equation for $0 \leq t \leq 4$. The point whose coordinates are (1, 96) is on the graph. Write a sentence that describes this ordered pair.

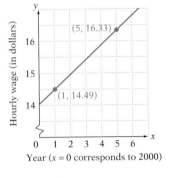

14. 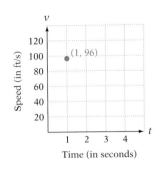 **Wages**   The graph at the right shows the projected increase in the average hourly wage of a U.S. worker for the years 2000 through 2006 (with $x = 0$ corresponding to 2000). Find the slope of the line. Write a sentence that states the meaning of the slope.

15. 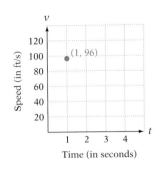 **Tuition**   The data in the table below are the average annual tuition costs for 4-year private colleges in the United States. The line of best fit is $y = 809x + 12,195$, where x is the year (with $x = 0$ corresponding to 1995) and y is the annual tuition cost in dollars rounded to the nearest 100.

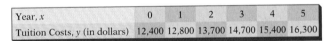

Year, x	0	1	2	3	4	5
Tuition Costs, y (in dollars)	12,400	12,800	13,700	14,700	15,400	16,300

Graph the data and the line of best fit in the coordinate system at the right. Write a sentence that describes the meaning of the slope of the line of best fit.

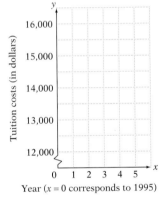

16. Find the x- and y-intercepts for $6x - 4y = 12$.

17. Find the x- and y-intercepts for $y = \frac{1}{2}x + 1$.

18. Find the slope of the line containing the points $(2, -3)$ and $(4, 1)$.

19. Is the line that passes through $(2, 5)$ and $(-1, 1)$ parallel, perpendicular, or neither parallel nor perpendicular to the line that passes through $(-2, 3)$ and $(4, 11)$?

20. Find the slope of the line containing the points $(-5, 2)$ and $(-5, 7)$.

21. Find the slope of the line whose equation is $2x + 3y = 6$.

22. Find the equation of the line that contains the point $(0, -1)$ and has slope 3.

23. Find the equation of the line that contains the point $(-3, 1)$ and has slope $\frac{2}{3}$.

24. Find the equation of the line that passes through the points $(5, -4)$ and $(-3, 1)$.

25. Find the equation of the line that passes through the points $(-2, 0)$ and $(5, -2)$.

Copyright © Houghton Mifflin Company. All rights reserved.

Cumulative Review Exercises

1. Simplify: $12 - 18 \div 3 \cdot (-2)^2$

2. Evaluate $\dfrac{a - b}{a^2 - c}$ when $a = -2$, $b = 3$, and $c = -4$.

3. Given $f(x) = \dfrac{2}{x - 1}$, find $f(-2)$.

4. Solve: $2x - \dfrac{2}{3} = \dfrac{7}{3}$

5. Solve: $3x - 2[x - 3(2 - 3x)] = x - 7$

6. Write $6\dfrac{2}{3}\%$ as a fraction.

7. Simplify: $(-2x^2y)^3(2xy^2)^2$

8. Simplify: $\dfrac{-15x^7}{5x^5}$

9. Divide: $(x^2 - 4x - 21) \div (x - 7)$

10. Factor: $5x^2 + 15x + 10$

11. Factor: $x(a + 2) + y(a + 2)$

12. Solve: $x(x - 2) = 8$

13. Multiply: $\dfrac{x^5y^3}{x^2 - x - 6} \cdot \dfrac{x^2 - 9}{x^2y^4}$

14. Subtract: $\dfrac{3x}{x^2 + 5x - 24} - \dfrac{9}{x^2 + 5x - 24}$

15. Solve: $3 - \dfrac{1}{x} = \dfrac{5}{x}$

16. Solve $4x - 5y = 15$ for y.

Copyright © Houghton Mifflin Company. All rights reserved.

17. Find the ordered-pair solution of $y = 2x - 1$ corresponding to $x = -2$.

18. Find the slope of the line that contains the points $(2, 3)$ and $(-2, 3)$.

19. Find the equation of the line that contains the point $(2, -1)$ and has slope $\frac{1}{2}$.

20. Find the equation of the line that contains the point $(0, 2)$ and has slope -3.

21. Find the equation of the line that contains the point $(-1, 0)$ and has slope 2.

22. Find the equation of the line that contains the point $(6, 1)$ and has slope $\frac{2}{3}$.

23. **Business** A suit that regularly sells for $89 is on sale for 30% off the regular price. Find the sale price.

24. **Geometry** The measure of the first angle of a triangle is 3° more than the measure of the second angle. The measure of the third angle is 5° more than twice the measure of the second angle. Find the measure of each angle.

25. **Taxes** The real estate tax for a home that costs $50,000 is $625. At this rate, what is the value of a home for which the real estate tax is $1375?

26. **Business** An electrician requires 6 h to wire a garage. An apprentice can do the same job in 10 h. How long would it take to wire the garage if both the electrician and the apprentice were working?

27. Graph $y = \frac{1}{2}x - 1$.

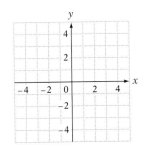

28. Graph the line that has slope $-\frac{2}{3}$ and y-intercept 2.

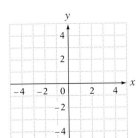

Copyright © Houghton Mifflin Company. All rights reserved.

8

Systems of Linear Equations

Recall from Section 3.1 that if a boat is traveling with the current, the effective speed of the boat is increased. If the boat is traveling against the current, the current slows the boat down. The same is true for airplanes and the wind. If the plane in the photo above is flying with a tailwind, its effective speed is increased. If the plane is flying against a headwind, the wind will slow the plane down. In **Exercises 2, 5, 6, and 8 on page 433**, you will use systems of equations to determine the rates of planes in calm air and rates of the wind.

OBJECTIVES

Section 8.1

A To solve a system of linear equations by graphing

Section 8.2

A To solve a system of linear equations by the substitution method
B To solve investment problems

Section 8.3

A To solve a system of linear equations by the addition method

Section 8.4

A To solve rate-of-wind or rate-of-current problems
B To solve application problems using two variables

Need help? For online student resources, such as section quizzes, visit this textbook's website at **math.college.hmco.com/students**.

Copyright © Houghton Mifflin Company. All rights reserved.

Do these exercises to prepare for Chapter 8.

1. Solve $3x - 4y = 24$ for y.

2. Solve: $50 + 0.07x = 0.05(x + 1400)$

3. Simplify: $-3(2x - 7y) + 3(2x + 4y)$

4. Simplify: $4x + 2(3x - 5)$

5. Is $(-4, 2)$ a solution of $3x - 5y = -22$?

6. Find the x- and y-intercepts for $3x - 4y = 12$.

7. Are the graphs of $3x + y = 6$ and $y = -3x - 4$ parallel?

8. Graph: $y = \dfrac{5}{4}x - 2$

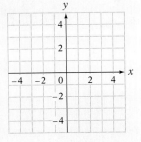

9. Pharmacology   A pharmacist has 20 ml of an 80% acetic acid solution. How many milliliters of a 55% acetic acid solution should be mixed with the 20-milliliter solution to produce a solution that is 75% acetic acid?

10. Hiking   One hiker starts along a trail walking at 3 mph. One-half hour later, another hiker starts on the same walking trail at a speed of 4 mph. How long after the second hiker started will the two hikers be side-by-side?

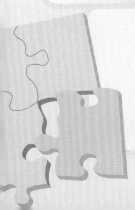

GO FIGURE • • •

Two children are running from school to home. Carla runs half the time and walks half the time. James runs half the distance and walks half the distance. Assuming the children run and walk at the same rate, which child arrives home first?

Copyright © Houghton Mifflin Company. All rights reserved.

8.1 Solving Systems of Linear Equations by Graphing

Objective A   To solve a system of linear equations by graphing

Two or more equations considered together are called a **system of equations.** Three examples of *linear* systems of equations in *two* variables are shown below, along with the graphs of the equations of each system.

System I	**System II**	**System III**
$x - 2y = -8$	$4x + 2y = 6$	$4x + 6y = 12$
$2x + 5y = 11$	$y = -2x + 3$	$6x + 9y = -9$

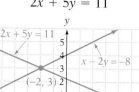

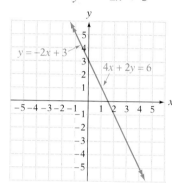

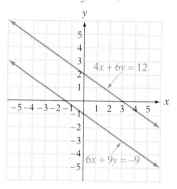

Copyright © Houghton Mifflin Company. All rights reserved.

TAKE NOTE

The systems of equations above are *linear systems of equations* because each of the equations in the system has a graph that is a line. Also, each equation has two variables. In future math courses, you will study equations that contain more than two variables.

For system I, the two lines intersect at a single point, $(-2, 3)$. Because this point lies on both lines, it is a solution of each equation of the system of equations. We can check this by replacing x by -2 and y by 3. The check is shown below.

$$\begin{array}{c|c}
x - 2y = -8 \\
\hline
-2 - 2(3) & -8 \\
-2 - 6 & -8 \\
-8 = -8 \ \checkmark
\end{array} \qquad \begin{array}{c|c}
2x + 5y = 11 \\
\hline
2(-2) + 5(3) & 11 \\
-4 + 15 & 11 \\
11 = 11 \ \checkmark
\end{array}$$

• Replace x by -2 and replace y by **3**.

A **solution of a system of equations in two variables** is an ordered pair that is a solution of each equation of the system. The ordered pair $(-2, 3)$ is a solution of system I.

HOW TO   Is $(-1, 4)$ a solution of the system of equations? $7x + 3y = 5$
                                                                    $3x - 2y = 12$

$$\begin{array}{c|c}
7x + 3y = 5 \\
\hline
7(-1) + 3(4) & 5 \\
-7 + 12 & 5 \\
5 = 5 \ \checkmark
\end{array} \qquad \begin{array}{c|c}
3x - 2y = 12 \\
\hline
3(-1) - 2(4) & 12 \\
-3 - 8 & 12 \\
-11 \ne 12
\end{array}$$

• Replace x by -1 and replace y by **4**.

• Does not check

Because $(-1, 4)$ is not a solution of both equations, $(-1, 4)$ is not a solution of the system of equations.

Using the system of equations above and the graph at the right, note that the graph of the ordered pair $(-1, 4)$ lies on the graph of $7x + 3y = 5$ but not on *both* lines. The ordered pair $(-1, 4)$ is *not* a solution of the system of equations. The graph of the ordered pair $(2, -3)$ does lie on both lines and therefore the ordered pair $(2, -3)$ is a solution of the system of equations.

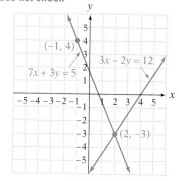

TAKE NOTE

The fact that there is an infinite number of ordered pairs that are solutions of the system at the right does not mean *every* ordered pair is a solution. For instance, $(0, 3)$, $(-2, 7)$, and $(2, -1)$ are solutions. However, $(3, 1)$, $(-1, 4)$, and $(1, 6)$ are not solutions. You should verify these statements.

System II from the preceding page and the graph of the equations of that system are shown again at the right. Note that the graph of $y = -2x + 3$ lies directly on top of the graph of $4x + 2y = 6$. Thus the two lines intersect at an infinite number of points. The graphs intersect at an infinite number of points, so there are an infinite number of solutions of this system of equations. Because each equation represents the same set of points, the solutions of the system of equations can be stated by using the ordered pairs of either one of the equations. Therefore, we can say, "The solutions are the ordered pairs that satisfy $4x + 2y = 6$," or we can say "The solutions are the ordered pairs that satisfy $y = -2x + 3$."

$$4x + 2y = 6$$
$$y = -2x + 3$$

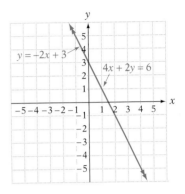

System III from the preceding page and the graph of the equations of that system are shown again at the right. Note that in this case, the graphs of the lines are parallel and do not intersect. Because the graphs do not intersect, there is no point that is on both lines. Therefore, the system of equations has no solution.

$$4x + 6y = 12$$
$$6x + 9y = -9$$

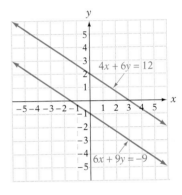

The preceding examples illustrate three types of systems of linear equations. An **independent system** has exactly one solution—the graphs intersect at one point. A **dependent system** has an infinite number of solutions—the graphs are the same line. An **inconsistent system** has no solution—the graphs are parallel lines.

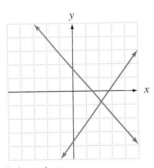

Independent:
one solution

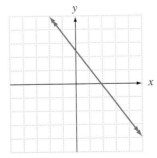

Dependent:
infinitely many solutions

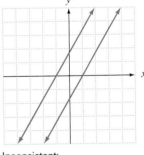

Inconsistent:
no solutions

HOW TO The graphs of the equations for the system of equations below are shown at the right. What is the solution of the system of equations?

$$2x + 3y = 6$$
$$2x + y = -2$$

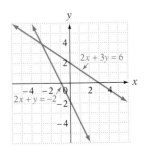

The graphs intersect at $(-3, 4)$. This is an *independent* system of equations. The solution of the system of equations is $(-3, 4)$.

Copyright © Houghton Mifflin Company. All rights reserved.

Copyright © Houghton Mifflin Company. All rights reserved.

TAKE NOTE

Because both equations represent the same ordered pairs, we can also say that the solutions of the system of equations are the ordered pairs that satisfy $x = \frac{1}{2}y + 1$. Either answer is correct.

HOW TO The graphs of the equations for the system of equations at the right are shown below. What is the solution of the system of equations?

$$y = 2x - 2$$
$$x = \frac{1}{2}y + 1$$

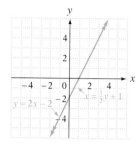

The two graphs lie directly on top of one another. Thus the two lines intersect at an infinite number of points, and the system of equations has an infinite number of solutions. This is a *dependent* system of equations. The solutions of the system of equations are the ordered pairs that satisfy $y = 2x - 2$.

Integrating Technology

The Projects and Group Activities at the end of this chapter discusses using a calculator to approximate the solution of an independent system of equations. Also see the Keystroke Guide: *Intersect*.

Solving a system of equations means finding the ordered-pair solutions of the system. One way to do this is to draw the graphs of the equations in the system of equations and determine where the graphs intersect.

To solve a system of linear equations in two variables by graphing, graph each equation on the same coordinate system, and then determine the points of intersection.

HOW TO Solve by graphing: $2x - y = -1$
$$x + 2y = 7$$

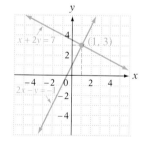

Graph each line.

The point of intersection of the two graphs lies on both lines and is therefore the solution of the system of equations.

The system of equations is independent. (1, 3) is a solution of each equation.

The solution is (1, 3).

HOW TO Solve by graphing: $y = 2x + 2$
$$4x - 2y = 4$$

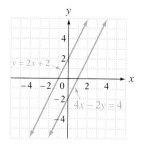

Graph each line.

The graphs do not intersect. The system of equations is inconsistent.

The system of equations has no solution.

Example 1

Is $(1, -3)$ a solution of the following system?
$3x + 2y = -3$
$x - 3y = 6$

Solution
Replace x by 1 and y by -3.

$3x + 2y = -3$		$x - 3y = 6$	
$3 \cdot 1 + 2(-3)$	-3	$1 - 3(-3)$	6
$3 + (-6)$	-3	$1 - (-9)$	6
$-3 = -3$		$10 \neq 6$	

No, $(1, -3)$ is not a solution of the system of equations.

Example 2

Solve by graphing:
$x - 2y = 2$
$x + y = 5$

Solution

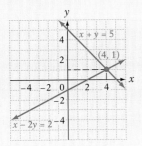

The solution is $(4, 1)$.

Example 3

Solve by graphing:
$4x - 2y = 6$
$\quad\quad y = 2x - 3$

Solution

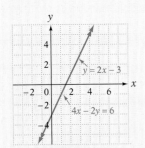

The solutions are the ordered pairs that satisfy the equation $y = 2x - 3$.

You Try It 1

Is $(-1, -2)$ a solution of the following system?
$2x - 5y = 8$
$-x + 3y = -5$

Your solution

You Try It 2

Solve by graphing:
$x + 3y = 3$
$-x + y = 5$

Your solution

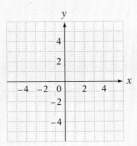

You Try It 3

Solve by graphing:
$\quad\quad y = 3x - 1$
$6x - 2y = -6$

Your solution

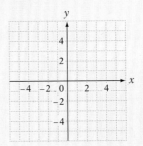

Solutions on p. S21

Copyright © Houghton Mifflin Company. All rights reserved.

Copyright © Houghton Mifflin Company. All rights reserved.

TAKE NOTE

Because both equations represent the same ordered pairs, we can also say that the solutions of the system of equations are the ordered pairs that satisfy

$x = \frac{1}{2}y + 1$. Either

answer is correct.

HOW TO The graphs of the equations for the system of equations at the right are shown below. What is the solution of the system of equations?

$$y = 2x - 2$$
$$x = \frac{1}{2}y + 1$$

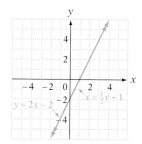

The two graphs lie directly on top of one another. Thus the two lines intersect at an infinite number of points, and the system of equations has an infinite number of solutions. This is a *dependent* system of equations. The solutions of the system of equations are the ordered pairs that satisfy $y = 2x - 2$.

Integrating Technology

The Projects and Group Activities at the end of this chapter discusses using a calculator to approximate the solution of an independent system of equations. Also see the Keystroke Guide: *Intersect.*

Solving a system of equations means finding the ordered-pair solutions of the system. One way to do this is to draw the graphs of the equations in the system of equations and determine where the graphs intersect.

To solve a system of linear equations in two variables by graphing, graph each equation on the same coordinate system, and then determine the points of intersection.

HOW TO Solve by graphing: $2x - y = -1$
$\qquad\qquad\qquad\qquad\qquad x + 2y = 7$

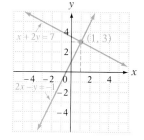

Graph each line.

The point of intersection of the two graphs lies on both lines and is therefore the solution of the system of equations.

The system of equations is independent. (1, 3) is a solution of each equation.

The solution is (1, 3).

HOW TO Solve by graphing: $\quad y = 2x + 2$
$\qquad\qquad\qquad\qquad\qquad 4x - 2y = 4$

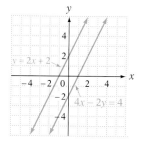

Graph each line.

The graphs do not intersect. The system of equations is inconsistent.

The system of equations has no solution.

Example 1

Is $(1, -3)$ a solution of the following system?
$3x + 2y = -3$
$x - 3y = 6$

Solution
Replace x by 1 and y by -3.

$$
\begin{array}{c|c}
3x + 2y = -3 \\
\hline
3 \cdot 1 + 2(-3) & -3 \\
3 + (-6) & -3 \\
-3 = -3 &
\end{array}
\qquad
\begin{array}{c|c}
x - 3y = 6 \\
\hline
1 - 3(-3) & 6 \\
1 - (-9) & 6 \\
10 \neq 6 &
\end{array}
$$

No, $(1, -3)$ is not a solution of the system of equations.

You Try It 1

Is $(-1, -2)$ a solution of the following system?
$2x - 5y = 8$
$-x + 3y = -5$

Your solution

Example 2

Solve by graphing:
$x - 2y = 2$
$x + y = 5$

Solution

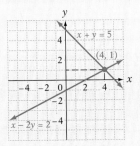

The solution is $(4, 1)$.

You Try It 2

Solve by graphing:
$x + 3y = 3$
$-x + y = 5$

Your solution

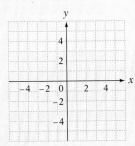

Example 3

Solve by graphing:
$4x - 2y = 6$
$y = 2x - 3$

Solution

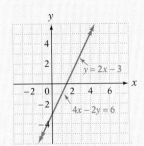

The solutions are the ordered pairs that satisfy the equation $y = 2x - 3$.

You Try It 3

Solve by graphing:
$y = 3x - 1$
$6x - 2y = -6$

Your solution

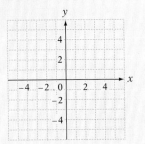

Solutions on p. S21

Copyright © Houghton Mifflin Company. All rights reserved.

8.1 Exercises

Objective A   **To solve a system of linear equations by graphing**

For Exercises 1 and 2, match each system of equations (systems I, II, and III) with (a) independent, (b) dependent, or (c) inconsistent.

1. I 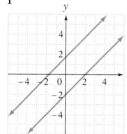   II 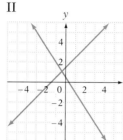   III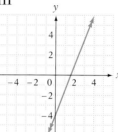

2. I 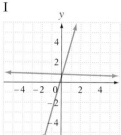   II 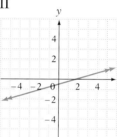   III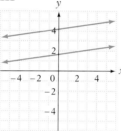

For Exercises 3 to 10, use the graphs of the equations of the system of equations to find the solution of the system of equations.

3. 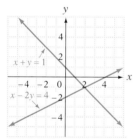   **4.** 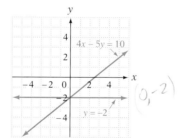   **5.**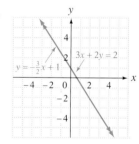

6. 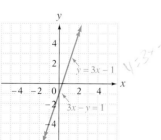   **7.** 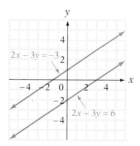   **8.**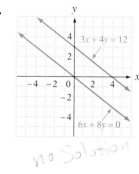

9. 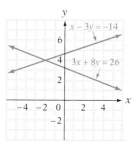   **10.**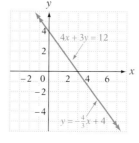

Copyright © Houghton Mifflin Company. All rights reserved.

11. Is $(2, 3)$ a solution of $\begin{aligned} 3x + 4y &= 18 \\ 2x - y &= 1 \end{aligned}$?

12. Is $(2, -1)$ a solution of $\begin{aligned} x - 2y &= 4 \\ 2x + y &= 3 \end{aligned}$?

13. Is $(4, 3)$ a solution of $\begin{aligned} 5x - 2y &= 14 \\ x + y &= 8 \end{aligned}$?

14. Is $(2, 5)$ a solution of $\begin{aligned} 3x + 2y &= 16 \\ 2x - 3y &= 4 \end{aligned}$?

15. Is $(2, -3)$ a solution of $\begin{aligned} y &= 2x - 7 \\ 3x - y &= 9 \end{aligned}$?

16. Is $(-1, -2)$ a solution of $\begin{aligned} 3x - 4y &= 5 \\ y &= x - 1 \end{aligned}$?

17. Is $(0, 0)$ a solution of $\begin{aligned} 3x + 4y &= 0 \\ y &= x \end{aligned}$?

18. Is $(3, -4)$ a solution of $\begin{aligned} 5x - 2y &= 23 \\ 2x - 5y &= 25 \end{aligned}$?

For Exercises 19 to 38, solve by graphing.

19. $x - y = 3$
    $x + y = 5$

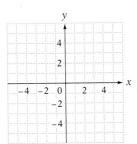

20. $2x - y = 4$
    $x + y = 5$

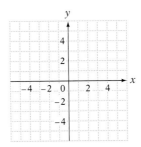

21. $x + 2y = 6$
    $x - y = 3$

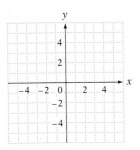

22. $3x - y = 3$
    $2x + y = 2$

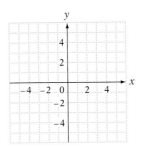

23. $3x - 2y = 6$
    $y = 3$

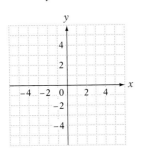

24. $x = 2$
    $3x + 2y = 4$

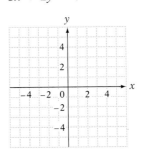

Copyright © Houghton Mifflin Company. All rights reserved.

25. $x = 3$
$y = -2$

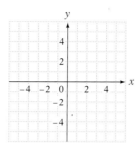

26. $x + 1 = 0$
$y - 3 = 0$

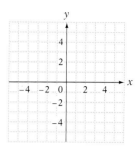

27. $y = 2x - 6$
$x + y = 0$

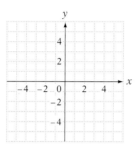

28. $5x - 2y = 11$
$y = 2x - 5$

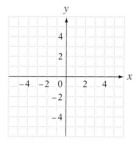

29. $2x + y = -2$
$6x + 3y = 6$

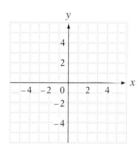

30. $x + y = 5$
$3x + 3y = 6$

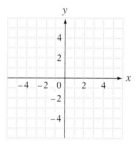

31. $y = 2x - 2$
$4x - 2y = 4$

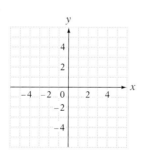

32. $y = -\dfrac{1}{3}x + 1$
$2x + 6y = 6$

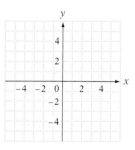

33. $x - y = 5$
$2x - y = 6$

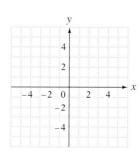

34. $5x - 2y = 10$
$3x + 2y = 6$

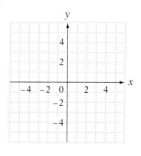

Copyright © Houghton Mifflin Company. All rights reserved.

35. $3x + 4y = 0$
    $2x - 5y = 0$

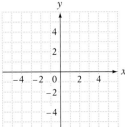

36. $2x - 3y = 0$
    $y = -\dfrac{1}{3}x$

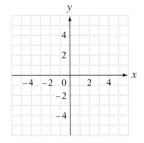

37. $x - 3y = 3$
    $2x - 6y = 12$

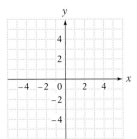

38. $4x + 6y = 12$
    $6x + 9y = 18$

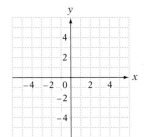

APPLYING THE CONCEPTS

39. Determine whether the statement is always true, sometimes true, or never true.
   a. A solution of a system of two equations in two variables is a point in the plane.
   b. Two parallel lines have the same slope.
   c. Two different lines with the same y-intercept are parallel.
   d. Two different lines with the same slope are parallel.

40. Explain how you can determine from the graph of a system of two linear equations in two variables whether it is an independent system of equations.

41. Explain how you can determine from the graph of a system of two linear equations in two variables whether it is an inconsistent system of equations.

42. Write a system of equations that has $(-2, 4)$ as its only solution.

43. Write a system of equations for which there is no solution.

44. Write a system of equations that is a dependent system of equations.

Copyright © Houghton Mifflin Company. All rights reserved.

8.2 Solving Systems of Linear Equations by the Substitution Method

Copyright © Houghton Mifflin Company. All rights reserved.

Objective A   **To solve a system of linear equations by the substitution method**

A graphical solution of a system of equations is based on approximating the coordinates of a point of intersection. Algebraic methods can be used to find an exact solution of a system of equations. To solve a system of equations by the **substitution method,** one variable must be written in terms of the other variable.

> **HOW TO**   Solve by the substitution method: (1) $2x + 5y = -11$
> (2) $y = 3x - 9$
>
> Equation (2) states that $y = 3x - 9$. Substitute $3x - 9$ for y in Equation (1). Then solve for x.
>
> $2x + 5y = -11$ • This is Equation (1).
>
> $2x + 5(3x - 9) = -11$ • From Equation (2), substitute **3x − 9** for y.
>
> $2x + 15x - 45 = -11$ • Solve for x.
>
> $17x - 45 = -11$
>
> $17x = 34$
>
> $x = 2$
>
> Now substitute the value of x into Equation (2) and solve for y.
>
> $y = 3x - 9$ • This is Equation (2).
>
> $y = 3(2) - 9$ • Substitute **2** for x.
>
> $y = 6 - 9 = -3$
>
> The solution is the ordered pair $(2, -3)$.

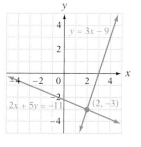

The graph of the equations in this system of equations is shown at the right. Note that the lines intersect at the point whose coordinates are $(2, -3)$, which is the algebraic solution we determined by the substitution method.

To solve a system of equations by the substitution method, we may need to solve one of the equations in the system of equations for one of its variables. For instance, the first step in solving the system of equations

$$(1)   x + 2y = -3$$
$$(2)   2x - 3y = 5$$

is to solve an equation of the system for one of its variables. Either equation can be used.

Solving Equation (1) for x:

$x + 2y = -3$

$x = -2y - 3$

Solving Equation (2) for x:

$2x - 3y = 5$

$2x = 3y + 5$

$x = \dfrac{3y + 5}{2} = \dfrac{3}{2}y + \dfrac{5}{2}$

Because solving Equation (1) for x does not result in fractions, it is the easier of the two equations to use.

Here is the solution of the system of equations given on the preceding page.

HOW TO   Solve by the substitution method: (1) $x + 2y = -3$
                                                     (2) $2x - 3y = 5$

To use the substitution method, we must solve an equation for one of its variables. Equation (1) is used here because solving it for x does not result in fractions.

$$x + 2y = -3$$
(3) $$x = -2y - 3$$ • Solve for x. This is Equation (3).

Now substitute $-2y - 3$ for x in Equation (2) and solve for y.

$$2x - 3y = 5$$ • This is Equation (2).
$$2(-2y - 3) - 3y = 5$$ • From Equation (3), substitute $-2y - 3$ for x.
$$-4y - 6 - 3y = 5$$ • Solve for y.
$$-7y - 6 = 5$$
$$-7y = 11$$
$$y = -\frac{11}{7}$$

Substitute the value of y into Equation (3) and solve for x.

$$x = -2y - 3$$ • This is Equation (3).
$$= -2\left(-\frac{11}{7}\right) - 3$$ • Substitute $-\frac{11}{7}$ for y.
$$= \frac{22}{7} - 3 = \frac{22}{7} - \frac{21}{7} = \frac{1}{7}$$

The solution is $\left(\frac{1}{7}, -\frac{11}{7}\right)$.

The graph of the system of equations given above is shown at the right. It would be difficult to determine the exact solution of this system of equations from the graphs of the equations.

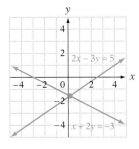

HOW TO   Solve by the substitution method: (1) $y = 3x - 1$
                                                     (2) $y = -2x - 6$

$$y = -2x - 6$$
$$3x - 1 = -2x - 6$$ • Substitute $3x - 1$ for y in Equation (2).
$$5x = -5$$ • Solve for x.
$$x = -1$$

Substitute this value of x into Equation (1) or Equation (2) and solve for y. Equation (1) is used here.

$$y = 3x - 1$$
$$y = 3(-1) - 1 = -4$$ • Substitute -1 for x.

The solution is $(-1, -4)$.

Copyright © Houghton Mifflin Company. All rights reserved.

The substitution method can be used to analyze inconsistent and dependent systems of equations. If, when solving a system of equations algebraically, the variable is eliminated and the result is a false equation, such as $0 = 4$, the system of equations is inconsistent. If the variable is eliminated and the result is a true equation, such as $12 = 12$, the system of equations is dependent.

HOW TO   Solve by the substitution method: (1) $2x + 3y = 3$

(2) $y = -\dfrac{2}{3}x + 3$

$2x + 3y = 3$ • This is Equation (1).

$2x + 3\left(-\dfrac{2}{3}x + 3\right) = 3$ • From Equation (2), replace y with $-\dfrac{2}{3}x + 3$.

$2x - 2x + 9 = 3$ • Solve for x.

$9 = 3$ • This is not a true equation.

Because $9 = 3$ is not a true equation, the system of equations has no solution.

Solving Equation (1) above for y, we have $y = -\dfrac{2}{3}x + 1$.

Comparing this with Equation (2) reveals that the slopes are equal and the y-intercepts are different. The graphs of the equations that make up this system of equations are parallel and thus never intersect. Because the graphs do not intersect, there are no solutions of the system of equations. The system of equations is inconsistent.

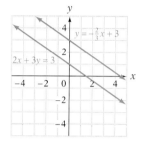

HOW TO   Solve by the substitution method: (1) $x = 2y + 3$
(2) $4x - 8y = 12$

$4x - 8y = 12$ • This is Equation (2).

$4(2y + 3) - 8y = 12$ • From Equation (1), replace x by $2y + 3$.

$8y + 12 - 8y = 12$ • Solve for y.

$12 = 12$ • This is a true equation.

Copyright © Houghton Mifflin Company. All rights reserved.

TAKE NOTE

As we mentioned in the previous section, when the system of equations is dependent, either equation can be used to write the ordered-pair solutions. Thus we could have said, "The solutions are the ordered pairs (x, y) that are solutions of $4x - 8y = 12$." Also note that, as we show at the right, if we solve each equation for y, the equations have the same slope-intercept form. This means we could also say, "The solutions are the ordered pairs (x, y) that are solutions of $y = \dfrac{1}{2}x - \dfrac{3}{2}$." When a system of equations is dependent, there are many ways in which the solutions can be stated.

The true equation $12 = 12$ indicates that any ordered pair (x, y) that satisfies one equation of the system satisfies the other equation. Therefore, the system of equations has an infinite number of solutions. The solutions are the ordered pairs (x, y) that are solutions of $x = 2y + 3$.

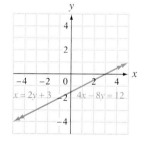

If we write Equation (1) and Equation (2) in slope-intercept form, we have

$$x = 2y + 3 \qquad\qquad 4x - 8y = 12$$
$$-2y = -x + 3 \qquad\qquad -8y = -4x + 12$$
$$y = \dfrac{1}{2}x - \dfrac{3}{2} \qquad\qquad y = \dfrac{1}{2}x - \dfrac{3}{2}$$

The slope-intercept forms of the equations are the same, and therefore the graphs are the same. If we graph these two equations, we essentially graph one over the other. Accordingly, the graphs intersect at an infinite number of points.

Example 1 Solve by substitution:

(1) $3x + 4y = -2$
(2) $-x + 2y = 4$

Solution

$-x + 2y = 4$ • Solve Equation (2) for x.
   $-x = -2y + 4$
   $x = 2y - 4$

Substitute in Equation (1).
(1) $3x + 4y = -2$
    $3(2y - 4) + 4y = -2$ • $x = 2y - 4$
    $6y - 12 + 4y = -2$ • Solve for y.
      $10y - 12 = -2$
         $10y = 10$
           $y = 1$

Substitute in $x = 2y - 4$.
   $x = 2y - 4$
   $x = 2(1) - 4$ • $y = 1$
   $x = 2 - 4$
   $x = -2$

The solution is $(-2, 1)$.

You Try It 1 Solve by substitution:

(1) $7x - y = 4$
(2) $3x + 2y = 9$

Your solution

Example 2 Solve by substitution:

$4x + 2y = 5$
$y = -2x + 1$

Solution

      $4x + 2y = 5$
$4x + 2(-2x + 1) = 5$ • $y = -2x + 1$
   $4x - 4x + 2 = 5$ • Solve for x.
         $2 = 5$ • Not a true equation

The system of equations is inconsistent and therefore does not have a solution.

You Try It 2 Solve by substitution:

$3x - y = 4$
$y = 3x + 2$

Your solution

Example 3 Solve by substitution:

$y = 3x - 2$
$6x - 2y = 4$

Solution

      $6x - 2y = 4$
$6x - 2(3x - 2) = 4$ • $y = 3x - 2$
   $6x - 6x + 4 = 4$ • Solve for x.
         $4 = 4$ • A true equation

The system of equations is dependent. The solutions are the ordered pairs that satisfy the equation $y = 3x - 2$.

You Try It 3 Solve by substitution:

$y = -2x + 1$
$6x + 3y = 3$

Your solution

Solutions on p. S21

Copyright © Houghton Mifflin Company. All rights reserved.

Copyright © Houghton Mifflin Company. All rights reserved.

Objective B   ## To solve investment problems

The annual simple interest that an investment earns is given by the equation $Pr = I$, where P is the principal, or the amount invested, r is the simple interest rate, and I is the simple interest.

For instance, if you invest \$750 at a simple interest rate of 6%, then the interest earned after 1 year is calculated as follows:

$$Pr = I$$
$$750(0.06) = I \qquad \bullet \text{ Replace } P \text{ by } \mathbf{750} \text{ and } r \text{ by } \mathbf{0.06}\ (6\%).$$
$$45 = I \qquad \qquad \bullet \text{ Simplify.}$$

The amount of interest earned is \$45.

Study Tip

Word problems are difficult because we must read the problem, determine the quantity we must find, think of a method to find it, actually solve the problem, and then check the answer. In short, we must devise a *strategy* and then use that strategy to find the *solution*. See *AIM for Success*, page xxv.

HOW TO   A medical lab technician decided to open an Individual Retirement Account (IRA) by placing \$2000 in two simple interest accounts. On one account, a corporate bond fund, the annual simple interest rate is 7.5%. On the second account, a real estate investment trust, the annual simple interest rate is 9%. If the technician wants to have annual earnings of \$168 from these two investments, how much must be invested in each account?

Strategy for Solving Simple-Interest Investment Problems

1. For each amount invested, use the equation $Pr = I$. Write a numerical or variable expression for the principal, the interest rate, and the interest earned.

Amount invested at 7.5%: x
Amount invested at 9%: y

	Principal, P	·	Interest rate, r	=	Interest earned, I
Amount at 7.5%	x	·	0.075	=	$0.075x$
Amount at 9%	y	·	0.09	=	$0.09y$

2. Write a system of equations. One equation will express the relationship among the amounts invested. The second equation will express the relationship among the amounts of interest earned by the investments.

The total amount invested is \$2000: $x + y = 2000$
The total annual interest earned is \$168: $0.075x + 0.09y = 168$

Solve the system of equations.
(1) $\qquad\qquad x + y = 2000$
(2) $\quad 0.075x + 0.09y = 168$

Solve Equation (1) for y and substitute into Equation (2).
(3) $\quad y = -x + 2000$

$$0.075x + 0.09(-x + 2000) = 168 \qquad \bullet \text{ Substitute } -x + 2000 \text{ for } y.$$
$$0.075x - 0.09x + 180 = 168$$
$$-0.015x = -12$$
$$x = 800$$

Substitute the value of x into Equation (3) and solve for y.
$$y = -x + 2000$$
$$y = -800 + 2000 = 1200 \qquad \bullet \text{ Substitute } -800 \text{ for } x.$$

The amount invested at 7.5% is \$800. The amount invested at 9% is \$1200.

Example 4

A hair stylist invested some money at an annual simple interest rate of 5.2%. A second investment, $1000 more than the first, was invested at an annual simple interest rate of 7.2% so that the total interest earned was $320. How much was invested in each account?

Strategy

• Amount invested at 5.2%: x
  Amount invested at 7.2%: y

	Principal	Rate	Interest
Amount at 5.2%	x	0.052	$0.052x$
Amount at 7.2%	y	0.072	$0.072y$

• The second investment is $1000 more than the first investment:

$y = x + 1000$

The sum of the interest earned at 5.2% and the interest earned at 7.2% equals $320.

$0.052x + 0.072y = 320$

Solution

(1) $\qquad\qquad y = x + 1000$
(2) $0.052x + 0.072y = 320$

Replace y in Equation (2) by $x + 1000$ from Equation (1). Then solve for x.

$$0.052x + 0.072y = 320$$
$$0.052x + 0.072(x + 1000) = 320 \quad • \; y = x + 1000$$
$$0.052x + 0.072x + 72 = 320 \quad • \text{ Solve for } x.$$
$$0.124x + 72 = 320$$
$$0.124x = 248$$
$$x = 2000$$

$y = x + 1000$
$\quad = 2000 + 1000 \qquad • \; x = 2000$
$\quad = 3000$

$2000 was invested at an annual simple interest rate of 5.2%; $3000 was invested at 7.2%.

You Try It 4

The manager of a city's investment income wished to place $330,000 in two simple interest accounts. The first account earns 6.5% annual interest and the second account earns 4.5%. How much should be invested in each account so that both accounts earn the same annual interest?

Your strategy

Your solution

Solution on pp. S21–S22

Copyright © Houghton Mifflin Company. All rights reserved.

8.2 Exercises

Objective A | To solve a system of linear equations by the substitution method

1. Describe in your own words the process of solving a system of equations by the substitution method.

2. When you solve a system of equations by the substitution method, how do you determine whether the system of equations is dependent?

For Exercises 3 to 32, solve by substitution.

3. $2x + 3y = 7$
$x = 2$

4. $y = 3$
$3x - 2y = 6$

5. $y = x - 3$
$x + y = 5$

6. $y = x + 2$
$x + y = 6$

7. $x - y - 2$
$x + 3y = 2$

8. $x = y + 1$
$x + 2y = 7$

9. $y = 4 - 3x$
$3x + y = 5$

10. $y = 2 - 3x$
$6x + 2y = 7$

11. $x = 3y + 3$
$2x - 6y = 12$

12. $x = 2 - y$
$3x + 3y = 6$

13. $3x + 5y = -6$
$x = 5y + 3$

14. $y = 2x + 3$
$4x - 3y = 1$

15. $3x + y = 4$
$4x - 3y = 1$

16. $x - 4y = 9$
$2x - 3y = 11$

17. $3x - y = 6$
$x + 3y = 2$

18. $4x - y = -5$
$2x + 5y = 13$

19. $3x - y = 5$
$2x + 5y = -8$

20. $3x + 4y = 18$
$2x - y = 1$

21. $4x + 3y = 0$
$2x - y = 0$

22. $5x + 2y = 0$
$x - 3y = 0$

23. $2x - y = 2$
$6x - 3y = 6$

Copyright © Houghton Mifflin Company. All rights reserved.

24. $3x + y = 4$
      $9x + 3y = 12$

25. $x = 3y + 2$
      $y = 2x + 6$

26. $x = 4 - 2y$
      $y = 2x - 13$

27. $y = 2x + 11$
      $y = 5x - 19$

28. $y = 2x - 8$
      $y = 3x - 13$

29. $y = -4x + 2$
      $y = -3x - 1$

30. $x = 3y + 7$
      $x = 2y - 1$

31. $x = 4y - 2$
      $x = 6y + 8$

32. $x = 3 - 2y$
      $x = 5y - 10$

Objective B   **To solve investment problems**

33. An investment of $3500 is divided between two simple interest accounts. On one account, the annual simple interest rate is 5%, and on the second account, the annual simple interest rate is 7.5%. How much should be invested in each account so that the total interest from the two accounts is $215?

34. A mortgage broker purchased two trust deeds for a total of $250,000. One trust deed earns 7% simple annual interest, and the second one earns 8% simple annual interest. If the total annual interest from the two trust deeds is $18,500, what was the purchase price of each trust deed?

35. When Sara Whitehorse changed jobs, she rolled over the $6000 in her retirement account into two simple interest accounts. On one account, the annual simple interest rate is 9%; on the second account, the annual simple interest rate is 6%. How much must be invested in each account if the accounts earn the same amount of annual interest?

36. An animal trainer decided to take the $15,000 won on a game show and deposit it in two simple interest accounts. Part of the winnings were placed in an account paying 7% annual simple interest, and the remainder was used to purchase a government bond that earns 6.5% annual simple interest. The amount of interest earned for 1 year was $1020. How much was invested in each account?

37. A police officer has chosen a high-yield stock fund that earns 8% annual simple interest for part of a $6000 investment. The remaining portion is used to purchase a preferred stock that earns 11% annual simple interest. How much should be invested in each account so that the amount earned on the 8% account is twice the amount earned on the 11% account?

Copyright © Houghton Mifflin Company. All rights reserved.

38. To plan for the purchase of a new car, a deposit was made into an account that earns 7% annual simple interest. Another deposit, $1500 less than the first deposit, was placed in a certificate of deposit earning 9% annual simple interest. The total interest earned on both accounts for 1 year was $505. How much money was deposited in the certificate of deposit?

39. The Pacific Investment Group invested some money in a certificate of deposit (CD) that earns 6.5% annual simple interest. Twice the amount invested at 6.5% was invested in a second CD that earns 8.5% annual simple interest. If the total annual interest earned from the two investments was $4935, how much was invested at 6.5%?

40. A corporation gave a university $300,000 to support product safety research. The university deposited some of the money in a 10% simple interest account and the remainder in an 8.5% simple interest account. How much should be deposited in each account so that the annual interest earned is $28,500?

41. Ten co-workers formed an investment club, and each deposited $2000 in the club's account. They decided to take the total amount and invest some of it in preferred stock that pays 8% annual simple interest and the remainder in a municipal bond that pays 7% annual simple interest. The amount of interest earned each year from the investments was $1520. How much was invested in each?

42. A financial consultant advises a client to invest part of $30,000 in municipal bonds that earn 6.5% annual simple interest and the remainder of the money in 8.5% corporate bonds. How much should be invested in each so that the total interest earned each year is $2190?

43. Alisa Rhodes placed some money in a real estate investment trust that earns 7.5% annual simple interest. A second investment, which was one-half the amount placed in the real estate investment trust, was used to purchase a trust deed that earns 9% annual simple interest. If the total annual interest earned from the two investments was $900, how much was invested in the trust deed?

APPLYING THE CONCEPTS

44. Business   Suppose a breadmaker costs $180 and that the ingredients and electricity needed to make one loaf of bread cost $.95. If a comparable loaf of bread at a grocery store costs $1.55, how many loaves of bread must you make before the breadmaker pays for itself?

45. Business   Suppose a natural gas clothes dryer costs $240 and uses $.45 worth of gas to dry a load of clothes for one hour. If a laundromat charges $1.75 to use a dryer for one hour, how many loads of clothes must you dry before purchasing a gas dryer is more economical?

For Exercises 46 to 48, find the value of k for which the system of equations has no solution.

46. $2x - 3y = 7$
$kx - 3y = 4$

47. $8x - 4y = 1$
$2x - ky = 3$

48. $x = 4y + 4$
$kx - 8y = 4$

Copyright © Houghton Mifflin Company. All rights reserved.

49. The following was offered as a solution of the system of equations

(1) $\qquad y = \dfrac{1}{2}x + 2$

(2) $\quad 2x + 5y = 10$

$$2x + 5y = 10 \qquad \bullet \text{ Equation (2)}$$

$$2x + 5\left(\dfrac{1}{2}x + 2\right) = 10 \qquad \bullet \text{ Substitute } \dfrac{1}{2}x + 2 \text{ for } y.$$

$$2x + \dfrac{5}{2}x + 10 = 10 \qquad \bullet \text{ Solve for } x.$$

$$\dfrac{9}{2}x = 0$$

$$x = 0$$

At this point the student stated that because $x = 0$, the system of equations has no solution. If this assertion is correct, is the system of equations independent, dependent, or inconsistent? If the assertion is not correct, what is the correct solution?

50. When you solve a system of equations by the substitution method, how do you determine whether the system of equations is inconsistent?

51. **Investments**   A sales representative invests in a stock paying a 9% dividend. A research consultant invests $5000 more than the sales representative in bonds paying 8% annual simple interest. The research consultant's income from the investment is equal to the sales representative's. Find the amount of the research consultant's investment.

52. **Investments**   A plant manager invested $3000 more in stocks than in bonds. The stocks paid 8% annual simple interest, and the bonds paid 9.5% annual simple interest. Both investments yielded the same income. Find the total annual interest received on both investments.

53. **Compound Interest**   The exercises in this objective were based on annual *simple* interest, r, which means that the amount of interest earned after 1 year is given by $I = Pr$. For **compound interest,** the interest earned for a certain period of time (usually daily or monthly) is added to the principal before the interest for the next period is calculated. The compound interest earned in 1 year is given by the formula

$I = P\left[\left(1 + \dfrac{r}{n}\right)^{n} - 1\right]$, where n is the number of times per year that

interest is compounded. For instance, if interest is compounded daily, then $n = 365$; if interest is compounded monthly, then $n = 12$. Suppose an investment of $5000 is made into three different accounts. The first account earns 8% annual simple interest, the second earns 8% compounded monthly ($n = 12$), and the third earns 8% compounded daily ($n = 365$). Find the amount of interest earned from each account.

Copyright © Houghton Mifflin Company. All rights reserved.

8.3 Solving Systems of Equations by the Addition Method

Objective A   **To solve a system of linear equations by the addition method**

Another method of solving a system of equations is called the **addition method.** This method is based on the Addition Property of Equations.

Note, for the system of equations at the right, the effect of adding Equation (2) to Equation (1). Because $2y$ and $-2y$ are opposites, adding the equations results in an equation with only one variable.

(1) $5x + 2y = 11$
(2) $3x - 2y = 13$
$8x + 0y = 24$
$8x = 24$

Solving $8x = 24$ for x gives the first coordinate of the ordered-pair solution of the system of equations.

$$\frac{8x}{8} = \frac{24}{8}$$
$$x = 3$$

The second coordinate is found by substituting the value of x into Equation (1) or Equation (2) and then solving for y. Equation (1) is used here.

(1) $5x + 2y = 11$
$5(3) + 2y = 11$
$15 + 2y = 11$
$2y = -4$
$y = -2$

The solution is $(3, -2)$.

Sometimes adding the two equations does not eliminate one of the variables. In this case, use the Multiplication Property of Equations to rewrite one or both of the equations so that the coefficients of one variable are opposites. Then add the equations and solve for the variables.

HOW TO   Solve by the addition method: (1) $4x + y = 5$
(2) $2x - 5y = 19$

Multiply Equation (2) by -2. The coefficients of x will then be opposites.

$-2(2x - 5y) = -2 \cdot 19$ • Multiply Equation (2) by **−2**.
(3) $-4x + 10y = -38$ • Simplify. This is Equation (3).

Add Equation (1) to Equation (3). Then solve for y.

(1) $4x + y = 5$
(3) $-4x + 10y = -38$ • Note that the coefficients of x are opposites.
$11y = -33$ • Add the two equations.
$y = -3$ • Solve for y.

Substitute the value of y into Equation (1) or Equation (2) and solve for x. Equation (1) is used here.

(1) $4x + y = 5$
$4x + (-3) = 5$ • Substitute **−3** for y.
$4x - 3 = 5$ • Solve for x.
$4x = 8$
$x = 2$

The solution is $(2, -3)$.

Copyright © Houghton Mifflin Company. All rights reserved.

Sometimes each equation of a system of equations must be multiplied by a constant so that the coefficients of one of the variable terms are opposites.

HOW TO Solve by the addition method: (1) $3x + 7y = 2$
(2) $5x - 3y = -26$

To eliminate x, multiply Equation (1) by 5 and Equation (2) by -3. Note at the right how the constants are chosen.

$5(3x + 7y) = 5 \cdot 2$

$-3(5x - 3y) = -3(-26)$

• The negative is used so that the coefficients will be opposites.

$\begin{array}{ll} 15x + 35y = 10 & \text{• 5 times Equation (1)} \\ -15x + 9y = 78 & \text{• } -3 \text{ times Equation (2)} \\ \hline \qquad\; 44y = 88 & \text{• Add the equations.} \\ \qquad\quad\; y = 2 & \text{• Solve for } y. \end{array}$

Substitute the value of y into Equation (1) or Equation (2) and solve for x. Equation (1) is used here.

$\begin{array}{ll} (1) \qquad 3x + 7y = 2 \\ \qquad\;\; 3x + 7(2) = 2 & \text{• Substitute 2 for } y. \\ \qquad\;\; 3x + 14 = 2 & \text{• Solve for } x. \\ \qquad\qquad\; 3x = -12 \\ \qquad\qquad\;\; x = -4 \end{array}$

The solution is $(-4, 2)$.

For the above system of equations, the value of x was determined by substitution. This value can also be determined by eliminating y from the system.

$\begin{array}{ll} 9x + 21y = 6 & \text{• 3 times Equation (1)} \\ 35x - 21y = -182 & \text{• 7 times Equation (2)} \\ \hline 44x \qquad\quad = -176 & \text{• Add the equations.} \\ \qquad\; x = -4 & \text{• Solve for } x. \end{array}$

Note that this is the same value of x as was determined by using substitution.

TAKE NOTE

When you use the addition method to solve a system of equations and the result is an equation that is always true (like the one at the right), the system of equations is dependent. Compare this with the following example.

HOW TO Solve by the addition method: (1) $2x - 5y = 4$
(2) $4x - 10y = 8$

Eliminate x. Multiply Equation (1) by -2.

$\begin{array}{ll} -2(2x - 5y) = -2(4) & \text{• } -2 \text{ times Equation (1)} \\ (3) \qquad -4x + 10y = -8 & \text{• This is Equation (3).} \end{array}$

Add Equation (3) to Equation (2) and solve for y.

$\begin{array}{ll} (2) \qquad\; 4x - 10y = 8 \\ (3) \qquad -4x + 10y = -8 \\ \hline \qquad\quad\; 0x + 0y = 0 \\ \qquad\qquad\qquad\; 0 = 0 \end{array}$

The equation $0 = 0$ means that the system of equations is dependent. Therefore, the solutions of the system of equations are the ordered pairs that satisfy $2x - 5y = 4$.

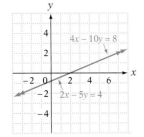

The graphs of the two equations in the system of equations above are shown at the left. Note that one line is on top of the other line and therefore they intersect infinitely often. The system of equations is dependent and the solutions are the ordered pairs that belong to either one of the equations.

Copyright © Houghton Mifflin Company. All rights reserved.

HOW TO Solve by the addition method: (1) $2x + y = 2$
(2) $4x + 2y = -5$

Eliminate y. Multiply Equation (1) by -2.

(1) $-2(2x + y) = -2 \cdot 2$ • -2 times Equation (1)
(3) $-4x - 2y = -4$ • This is Equation (3).

Add Equation (2) to Equation (3) and solve for x.

(3) $-4x - 2y = -4$
(2) $\underline{4x + 2y = -5}$
          $0x + 0y = -9$ • Add Equation (2) to Equation (3).
              $0 = -9$ • This is not a true equation.

The system of equations is inconsistent and therefore does not have a solution.

The graphs of the two equations in the system of equations above are shown at the left. Note that the graphs are parallel and therefore do not intersect. Thus the system of equations has no solution.

Example 1

Solve by the addition method:
(1) $2x + 4y = 7$
(2) $5x - 3y = -2$

Solution
Eliminate x.
  $5(2x + 4y) = 5 \cdot 7$ • 5 times Equation (1)
  $-2(5x - 3y) = -2(-2)$ • -2 times Equation (2)

  $10x + 20y = 35$
  $\underline{-10x + 6y = 4}$
          $26y = 39$ • Add the equations.

          $y = \dfrac{39}{26} = \dfrac{3}{2}$ • Solve for y.

Substitute $\dfrac{3}{2}$ for y in Equation (1).

(1) $2x + 4y = 7$

          $2x + 4\left(\dfrac{3}{2}\right) = 7$ • Replace y by $\dfrac{3}{2}$.

          $2x + 6 = 7$ • Solve for x.
          $2x = 1$

          $x = \dfrac{1}{2}$

The solution is $\left(\dfrac{1}{2}, \dfrac{3}{2}\right)$.

You Try It 1

Solve by the addition method:
(1) $x - 2y = 1$
(2) $2x + 4y = 0$

Your solution

Solution on p. S22

Copyright © Houghton Mifflin Company. All rights reserved.

Example 2

Solve by the addition method:
(1) $6x + 9y = 15$
(2) $4x + 6y = 10$

Solution

Eliminate x.

$\begin{array}{ll} 4(6x + 9y) = 4 \cdot 15 & \bullet \textbf{ 4 times Equation (1)} \\ -6(4x + 6y) = -6 \cdot 10 & \bullet \textbf{ −6 times Equation (2)} \end{array}$

$\begin{array}{l} 24x + 36y = 60 \\ \underline{-24x - 36y = -60} \\ 0x + 0y = 0 \qquad \bullet \textbf{ Add the equations.} \\ 0 = 0 \end{array}$

The system of equations is dependent. The solutions are the ordered pairs that satisfy the equation $6x + 9y = 15$.

You Try It 2

Solve by the addition method:
$2x - 3y = 4$
$-4x + 6y = -8$

Your solution

Example 3

Solve by the addition method:
(1) $2x = y + 8$
(2) $3x + 2y = 5$

Solution

Write Equation (1) in the form
$Ax + By = C$.

$\begin{array}{ll} \quad 2x = y + 8 \\ (3) \quad 2x - y = 8 & \bullet \textbf{ This is Equation (3).} \end{array}$

Eliminate y.

$\begin{array}{ll} 2(2x - y) = 2 \cdot 8 & \bullet \textbf{ 2 times Equation (3)} \\ 3x + 2y = 5 & \bullet \textbf{ This is Equation (2).} \end{array}$

$\begin{array}{l} 4x - 2y = 16 \\ \underline{3x + 2y = 5} \\ 7x = 21 \qquad \bullet \textbf{ Add the equations.} \\ x = 3 \end{array}$

Replace x in Equation (1).
$\begin{array}{ll} (1) \quad 2x = y + 8 \\ \quad 2 \cdot 3 = y + 8 & \bullet \textbf{ Replace } x \textbf{ by 3.} \\ 6 = y + 8 \\ -2 = y \end{array}$

The solution is $(3, -2)$.

You Try It 3

Solve by the addition method:
$4x + 5y = 11$
$3y = x + 10$

Your solution

Solutions on p. S22

Copyright © Houghton Mifflin Company. All rights reserved.

8.3 Exercises

Objective A **To solve a system of linear equations by the addition method**

For Exercises 1 to 36, solve by the addition method.

1. $x + y = 4$
 $x - y = 6$

2. $2x + y = 3$
 $x - y = 3$

3. $x + y = 4$
 $2x + y = 5$

4. $x - 3y = 2$
 $x + 2y = -3$

5. $2x - y = 1$
 $x + 3y = 4$

6. $x - 2y = 4$
 $3x + 4y = 2$

7. $4x - 5y = 22$
 $x + 2y = -1$

8. $3x - y = 11$
 $2x + 5y = 13$

9. $2x - y = 1$
 $4x - 2y = 2$

10. $x + 3y = 2$
 $3x + 9y = 6$

11. $4x + 3y = 15$
 $2x - 5y = 1$

12. $3x - 7y = 13$
 $6x + 5y = 7$

13. $2x - 3y = 1$
 $4x - 6y = 2$

14. $2x + 4y = 6$
 $3x + 6y = 9$

15. $3x - 6y = -1$
 $6x - 4y = 2$

16. $5x + 2y = 3$
 $3x - 10y = -1$

17. $5x + 7y = 10$
 $3x - 14y = 6$

18. $7x + 10y = 13$
 $4x + 5y = 6$

19. $3x - 2y = 0$
 $6x + 5y = 0$

20. $5x + 2y = 0$
 $3x + 5y = 0$

21. $2x - 3y = 16$
 $3x + 4y = 7$

22. $3x + 4y = 10$
 $4x + 3y = 11$

23. $5x + 3y = 7$
 $2x + 5y = 1$

24. $-2x + 7y = 9$
 $3x + 2y = -1$

Copyright © Houghton Mifflin Company. All rights reserved.

25. $3x + 4y = 4$
$5x + 12y = 5$

26. $2x + 5y = 2$
$3x + 3y = 1$

27. $8x - 3y = 11$
$6x - 5y = 11$

28. $4x - 8y = 36$
$3x - 6y = 15$

29. $5x + 15y = 20$
$2x + 6y = 12$

30. $y = 2x - 3$
$3x + 4y = -1$

31. $3x = 2y + 7$
$5x - 2y = 13$

32. $2y = 4 - 9x$
$9x - y = 25$

33. $2x + 9y = 16$
$5x = 1 - 3y$

34. $3x - 4 = y + 18$
$4x + 5y = -21$

35. $2x + 3y = 7 - 2x$
$7x + 2y = 9$

36. $5x - 3y = 3y + 4$
$4x + 3y = 11$

APPLYING THE CONCEPTS

37. Describe in your own words the process of solving a system of equations by the addition method.

38. The point of intersection of the graphs of the equations $Ax + 2y = 2$ and $2x + By = 10$ is $(2, -2)$. Find A and B.

39. The point of intersection of the graphs of the equations $Ax - 4y = 9$ and $4x + By = -1$ is $(-1, -3)$. Find A and B.

40. For what value of k is the system of equations dependent?

a. $2x + 3y = 7$
$4x + 6y = k$

b. $y = \dfrac{2}{3}x - 3$
$y = kx - 3$

c. $x = ky - 1$
$y = 2x + 2$

41. For what value of k is the system of equations inconsistent?

a. $x + y = 7$
$kx + y = 3$

b. $x + 2y = 4$
$kx + 3y = 2$

c. $2x + ky = 1$
$x + 2y = 2$

Copyright © Houghton Mifflin Company. All rights reserved.

8.4 Application Problems in Two Variables

Objective A   To solve rate-of-wind or rate-of-current problems

We normally need two variables to solve motion problems that involve an object moving with or against a wind or current.

HOW TO   Flying with the wind, a small plane can fly 600 mi in 3 h. Against the wind, the plane can fly the same distance in 4 h. Find the rate of the plane in calm air and the rate of the wind.

> **Strategy for Solving Rate-of-Wind or Rate-of-Current Problems**
>
> Choose one variable to represent the rate of the object in calm conditions and a second variable to represent the rate of the wind or current. Using these variables, express the rate of the object with and against the wind or current. Use the equation $rt = d$ to write expressions for the distance traveled by the object. The results can be recorded in a table.

Rate of plane in calm air: p
Rate of wind: w

	Rate ·	**Time** =	**Distance**
With the wind	$p + w$ ·	3 =	$3(p + w)$
Against the wind	$p - w$ ·	4 =	$4(p - w)$

Determine how the expressions for distance are related.

The distance traveled with the wind is 600 mi. $3(p + w) = 600$
The distance traveled against the wind is 600 mi. $4(p - w) = 600$

Solve the system of equations.

$$3(p + w) = 600 \quad \rightarrow \quad \frac{1}{3} \cdot 3(p + w) = \frac{1}{3} \cdot 600 \quad \rightarrow \quad p + w = 200$$

$$4(p - w) = 600 \qquad\qquad \frac{1}{4} \cdot 4(p - w) = \frac{1}{4} \cdot 600 \qquad\qquad p - w = 150$$

$$2p = 350$$
$$p = 175$$

$$p + w = 200$$
$$175 + w = 200 \qquad \bullet \; p = \mathbf{175}$$
$$w = 25$$

The rate of the plane in calm air is 175 mph.
The rate of the wind is 25 mph.

Copyright © Houghton Mifflin Company. All rights reserved.

Example 1

A 450-mile trip from one city to another takes 3 h when a plane is flying with the wind. The return trip, against the wind, takes 5 h. Find the rate of the plane in still air and the rate of the wind.

Strategy

• Rate of the plane in still air: p
  Rate of the wind: w

	Rate	Time	Distance
With wind	$p + w$	3	$3(p + w)$
Against wind	$p - w$	5	$5(p - w)$

• The distance traveled with the wind is 450 mi. The distance traveled against the wind is 450 mi.

Solution

$$3(p + w) = 450 \qquad \frac{1}{3} \cdot 3(p + w) = \frac{1}{3} \cdot 450$$

$$5(p - w) = 450 \qquad \frac{1}{5} \cdot 5(p - w) = \frac{1}{5} \cdot 450$$

$$p + w = 150$$
$$p - w = 90$$

$$2p = 240$$
$$p = 120$$

$$p + w = 150$$
$$120 + w = 150 \qquad • \; p = \mathbf{120}$$
$$w = 30$$

The rate of the plane in still air is 120 mph.
The rate of the wind is 30 mph.

You Try It 1

A canoeist paddling with the current can travel 15 mi in 3 h. Against the current, it takes the canoeist 5 h to travel the same distance. Find the rate of the current and the rate of the canoeist in calm water.

Your strategy

Your solution

Solution on pp. S22–S23

Objective B   **To solve application problems using two variables**

The application problems in this section are varieties of those problems solved earlier in the text. Each of the strategies for the problems in this section will result in a system of equations.

Copyright © Houghton Mifflin Company. All rights reserved.

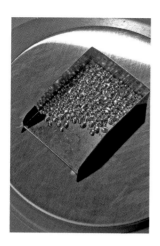

HOW TO   A jeweler purchased 5 oz of a gold alloy and 20 oz of a silver alloy for a total cost of $540. The next day, at the same prices per ounce, the jeweler purchased 4 oz of the gold alloy and 25 oz of the silver alloy for a total cost of $450. Find the cost per ounce of the gold and silver alloys.

> **Strategy for Solving an Application Problem in Two Variables**
>
> Choose one variable to represent one of the unknown quantities and a second variable to represent the other unknown quantity. Write numerical or variable expressions for all of the remaining quantities. These results can be recorded in two tables, one for each of the conditions.

Point of Interest

The Babylonians had a method for solving a system of equations. Here is an adaptation of a problem from an ancient (around 1500 B.C.) Babylonian text. "There are two silver blocks. The sum of $\frac{1}{7}$ of the first block and $\frac{1}{11}$ of the second block is one sheqel (a weight). The first block diminished by $\frac{1}{7}$ of its weight equals the second diminished by $\frac{1}{11}$ of its weight. What are the weights of the two blocks?"

Cost per ounce of gold: g
Cost per ounce of silver: s

First day:

	Amount	·	*Unit Cost*	=	*Value*
Gold	5	·	g	=	$5g$
Silver	20	·	s	=	$20s$

Second day:

	Amount	·	*Unit Cost*	=	*Value*
Gold	4	·	g	=	$4g$
Silver	25	·	s	=	$25s$

> Determine a system of equations. Each table will give one equation of the system.

The total value of the purchase on the first day was $540. $5g + 20s = 540$

The total value of the purchase on the second day was $450. $4g + 25s = 450$

Solve the system of equations.

$5g + 20s = 540$ $4(5g + 20s) = 4 \cdot 540$ $20g + 80s = 2160$
$4g + 25s = 450$ $-5(4g + 25s) = -5 \cdot 450$ $\underline{-20g - 125s = -2250}$
                                                        $-45s = -90$
                                                        $s = 2$

$5g + 20s = 540$
$5g + 20(2) = 540$ • $s = 2$
$5g + 40 = 540$
$5g = 500$
$g = 100$

The cost per ounce of the gold alloy was $100.
The cost per ounce of the silver alloy was $2.

Copyright © Houghton Mifflin Company. All rights reserved.

Copyright © Houghton Mifflin Company. All rights reserved.

Example 2

A store owner purchased 20 incandescent light bulbs and 30 fluorescent bulbs for a total cost of $40. A second purchase, at the same prices, included 30 incandescent bulbs and 10 fluorescent bulbs for a total cost of $25. Find the cost of an incandescent bulb and of a fluorescent bulb.

Strategy

Cost of an incandescent bulb: b
Cost of a fluorescent bulb: f

First purchase:

	Amount	Unit Cost	Value
Incandescent	20	b	$20b$
Fluorescent	30	f	$30f$

Second purchase:

	Amount	Unit Cost	Value
Incandescent	30	b	$30b$
Fluorescent	10	f	$10f$

The total cost of the first purchase was $40.
The total cost of the second purchase was $25.

Solution

$$20b + 30f = 40 \qquad 3(20b + 30f) = 3 \cdot 40$$
$$30b + 10f = 25 \qquad -2(30b + 10f) = -2 \cdot 25$$

$$60b + 90f = 120$$
$$\underline{-60b - 20f = -50}$$
$$70f = 70$$
$$f = 1$$

$$20b + 30f = 40$$
$$20b + 30(1) = 40 \qquad \bullet\ f = 1$$
$$20b = 10$$
$$b = \frac{1}{2}$$

The cost of an incandescent bulb was $.50.
The cost of a fluorescent bulb was $1.00.

You Try It 2

A citrus grower purchased 25 orange trees and 20 grapefruit trees for $290. The next week, at the same prices, the grower bought 20 orange trees and 30 grapefruit trees for $330. Find the cost of an orange tree and the cost of a grapefruit tree.

Your strategy

Your solution

Solution on p. S23

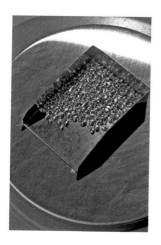

HOW TO A jeweler purchased 5 oz of a gold alloy and 20 oz of a silver alloy for a total cost of $540. The next day, at the same prices per ounce, the jeweler purchased 4 oz of the gold alloy and 25 oz of the silver alloy for a total cost of $450. Find the cost per ounce of the gold and silver alloys.

> **Strategy for Solving an Application Problem in Two Variables**
>
> Choose one variable to represent one of the unknown quantities and a second variable to represent the other unknown quantity. Write numerical or variable expressions for all of the remaining quantities. These results can be recorded in two tables, one for each of the conditions.

Cost per ounce of gold: g
Cost per ounce of silver: s

First day:

	Amount	·	Unit Cost	=	Value
Gold	5	·	g	=	$5g$
Silver	20	·	s	=	$20s$

Second day:

	Amount	·	Unit Cost	=	Value
Gold	4	·	g	=	$4g$
Silver	25	·	s	=	$25s$

> Determine a system of equations. Each table will give one equation of the system.

The total value of the purchase on the first day was $540. $5g + 20s = 540$

The total value of the purchase on the second day was $450. $4g + 25s = 450$

Solve the system of equations.

$$5g + 20s = 540 \qquad 4(5g + 20s) = 4 \cdot 540 \qquad 20g + 80s = 2160$$
$$4g + 25s = 450 \qquad -5(4g + 25s) = -5 \cdot 450 \qquad \underline{-20g - 125s = -2250}$$
$$-45s = -90$$
$$s = 2$$

$$5g + 20s = 540$$
$$5g + 20(2) = 540 \qquad \bullet\ s = 2$$
$$5g + 40 = 540$$
$$5g = 500$$
$$g = 100$$

The cost per ounce of the gold alloy was $100.
The cost per ounce of the silver alloy was $2.

Point of Interest

The Babylonians had a method for solving a system of equations. Here is an adaptation of a problem from an ancient (around 1500 B.C.) Babylonian text. "There are two silver blocks. The sum of $\frac{1}{7}$ of the first block and $\frac{1}{11}$ of the second block is one sheqel (a weight). The first block diminished by $\frac{1}{7}$ of its weight equals the second diminished by $\frac{1}{11}$ of its weight. What are the weights of the two blocks?"

Copyright © Houghton Mifflin Company. All rights reserved.

Copyright © Houghton Mifflin Company. All rights reserved.

Example 2

A store owner purchased 20 incandescent light bulbs and 30 fluorescent bulbs for a total cost of $40. A second purchase, at the same prices, included 30 incandescent bulbs and 10 fluorescent bulbs for a total cost of $25. Find the cost of an incandescent bulb and of a fluorescent bulb.

Strategy

Cost of an incandescent bulb: b
Cost of a fluorescent bulb: f

First purchase:

	Amount	Unit Cost	Value
Incandescent	20	b	$20b$
Fluorescent	30	f	$30f$

Second purchase:

	Amount	Unit Cost	Value
Incandescent	30	b	$30b$
Fluorescent	10	f	$10f$

The total cost of the first purchase was $40.
The total cost of the second purchase was $25.

Solution

$$20b + 30f = 40 \qquad 3(20b + 30f) = 3 \cdot 40$$
$$30b + 10f = 25 \qquad -2(30b + 10f) = -2 \cdot 25$$

$$60b + 90f = 120$$
$$-60b - 20f = -50$$
$$\overline{\hphantom{-60b - 20f = -50}}$$
$$70f = 70$$
$$f = 1$$

$$20b + 30f = 40$$
$$20b + 30(1) = 40 \qquad \bullet\ f = 1$$
$$20b = 10$$
$$b = \frac{1}{2}$$

The cost of an incandescent bulb was $.50.
The cost of a fluorescent bulb was $1.00.

You Try It 2

A citrus grower purchased 25 orange trees and 20 grapefruit trees for $290. The next week, at the same prices, the grower bought 20 orange trees and 30 grapefruit trees for $330. Find the cost of an orange tree and the cost of a grapefruit tree.

Your strategy

Your solution

Solution on p. S23

8.4 Exercises

Objective A   **To solve rate-of-wind or rate-of-current problems**

1. A whale swimming against an ocean current traveled 60 mi in 2 h. Swimming in the opposite direction, with the current, the whale was able to travel the same distance in 1.5 h. Find the speed of the whale in calm water and the rate of the ocean current.

2. A plane flying with the jet stream flew from Los Angeles to Chicago, a distance of 2250 mi, in 5 h. Flying against the jet stream, the plane could fly only 1750 mi in the same amount of time. Find the rate of the plane in calm air and the rate of the wind.

3. A rowing team rowing with the current traveled 40 km in 2 h. Rowing against the current, the team could travel only 16 km in 2 h. Find the rowing rate in calm water and the rate of the current.

4. The bird capable of the fastest flying speed is the swift. A swift flying with the wind to a favorite feeding spot traveled 26 mi in 0.2 h. On returning, now against the wind, the swift was able to travel only 16 mi in the same amount of time. What is the rate of the swift in calm air and what was the rate of the wind?

5. A private Learjet 31A was flying with a tailwind and traveled 1120 mi in 2 h. Flying against the wind on the return trip, the jet was able to travel only 980 mi in 2 h. Find the speed of the jet in calm air and the rate of the wind.

6. A plane flying with a tailwind flew 300 mi in 2 h. Against the wind, it took 3 h to travel the same distance. Find the rate of the plane in calm air and the rate of the wind.

7. A Boeing Apache Longbow military helicopter traveling directly into a strong headwind was able to travel 450 mi in 2.5 h. The return trip, now with a tailwind, took 1 h 40 min. Find the speed of the helicopter in calm air and the rate of the wind.

8. A seaplane pilot flying with the wind flew from an ocean port to a lake, a distance of 240 mi, in 2 h. Flying against the wind, the pilot flew from the lake to the ocean port in 2 h 40 min. Find the rate of the plane in calm air and the rate of the wind.

9. Rowing with the current, a canoeist paddled 14 mi in 2 h. Against the current, the canoeist could paddle only 10 mi in the same amount of time. Find the rate of the canoeist in calm water and the rate of the current.

Objective B   **To solve application problems using two variables**

10. **Internet Services**   A computer online service charges one hourly price for regular use but a higher hourly rate for designated "premium" areas. One customer was charged $28 after spending 2 h in premium areas and 9 regular hours; another spent 3 h in the premium areas and 6 regular hours and was charged $27. What does the online service charge per hour for regular and premium services?

11. **Flour Mixtures**   A baker purchased 12 lb of wheat flour and 15 lb of rye flour for a total cost of $18.30. A second purchase, at the same prices, included 15 lb of wheat flour and 10 lb of rye flour. The cost of the second purchase was $16.75. Find the cost per pound of the wheat flour and of the rye flour.

Copyright © Houghton Mifflin Company. All rights reserved.

12. **Investments**   An investor owned 300 shares of an oil company and 200 shares of a movie company. The quarterly dividend from the two stocks was $165. After the investor sold 100 shares of the oil company and bought an additional 100 shares of the movie company, the quarterly dividend became $185. Find the dividend per share for each stock.

13. **Chemistry**   A lab technician mixed 40 ml of reagent I with 60 ml of reagent II to produce a solution that was 31% hydrochloric acid. A second solution, which was 31.5% hydrochloric acid, was created by mixing 35 ml of reagent I with 65 ml of reagent II. What is the percent hydrochloric acid in the two reagents?

14. **Food Mixtures**   A pastry chef created a 50-ounce sugar solution that was 34% sugar from a 20% sugar solution and a 40% sugar solution. How much of the 20% sugar solution and how much of the 40% sugar solution were used?

15. **Fuel Mixtures**   The octane number of 87 on gasoline means that it will fight engine "knock" as effectively as a reference fuel that is 87% isooctane, a type of gas. Suppose you want to fill an empty 18-gallon tank with some 87-octane gasoline and some 93-octane fuel to produce a mixture that is 89-octane. How much of each type of gasoline must you use?

APPLYING THE CONCEPTS

16. **Geometry**   Two angles are supplementary. The measure of the larger angle is 15° more than twice the measure of the smaller angle. Find the measures of the two angles. (Supplementary angles are two angles whose sum is 180°.)

17. **Coin Problem**   The value of the nickels and dimes in a coin bank is $.25. If the number of nickels and the number of dimes were doubled, the value of the coins would be $.50. How many nickels and how many dimes are in the bank?

18. **Investments**   An investor has $5000 to invest in two accounts. The first account earns 8% annual simple interest, and the second account earns 10% annual simple interest. How much money should be invested in each account so that the annual simple interest earned is $600?

19. **Ancient Problem**   Solve the following problem, which dates from a Chinese manuscript called the Jinzhang that is approximately 2100 years old. "The price of 1 acre of good land is 300 pieces of gold; the price of 7 acres of bad land is 500 pieces of gold. One has purchased altogether 100 acres. The price was 10,000 pieces of gold. How much good land and how much bad land was bought?" Adapted from Victor J. Katz, *A History of Mathematics, An Introduction* (New York: Harper-Collins, 1993), p. 15.

20. **Coin Problem**   A coin bank contains only nickels or dimes, but there are no more than 27 coins. The value of the coins is $2.10. How many different combinations of nickels and dimes could be in the coin bank?

Copyright © Houghton Mifflin Company. All rights reserved.

Focus on Problem Solving

Using a Table and Searching for a Pattern

Consider the numbers 10, 12, and 28 and the sum of the proper factors (the natural number factors less than the number) of those numbers.

$$10: 1 + 2 + 5 = 8 \qquad 12: 1 + 2 + 3 + 4 + 6 = 16 \qquad 28: 1 + 2 + 4 + 7 + 14 = 28$$

10 is called a **deficient number** because the sum of its proper factors is less than the number ($8 < 10$). 12 is called an **abundant number** because the sum of its proper factors is greater than the number ($16 > 12$), and 28 is called a **perfect number** because the sum of its proper divisors equals the number ($28 = 28$).

Our goal for this Focus on Problem Solving is to try to find a method that will determine whether a number is deficient, abundant, or perfect without having to first find all the factors and then add them up. We will use a table and search for a pattern.

Before we begin, recall that a prime number is a number greater than 1 whose only factors are itself and 1, and that each natural number greater than 1 has a unique prime factorization. For instance, the prime factorization of 36 is given by $36 = 2^2 \cdot 3^2$. Note that the proper factors of 36 (1, 2, 3, 4, 6, 9, 12, 18) can be represented in terms of the same prime numbers.

$$1 = 2^0, \quad 2 = 2^1, \quad 3 = 3^1, \quad 4 = 2^2, \quad 6 = 2 \cdot 3, \quad 9 = 3^2, \quad 12 = 2^2 \cdot 3, \quad 18 = 2 \cdot 3^2$$

Now let us consider a trial problem of determining whether 432 is deficient, abundant, or perfect.

TAKE NOTE

This table contains the factor $432 = 2^4 \cdot 3^3$, which is not a proper factor.

We write the prime factorization of 432 as $2^4 \cdot 3^3$ and place the factors of 432 in a table as shown at the right. This table contains *all* the factors of 432 represented in terms of the prime number factors. The sum of each column is shown at the bottom.

	1	$1 \cdot 3$	$1 \cdot 3^2$	$1 \cdot 3^3$
	2	$2 \cdot 3$	$2 \cdot 3^2$	$2 \cdot 3^3$
	2^2	$2^2 \cdot 3$	$2^2 \cdot 3^2$	$2^2 \cdot 3^3$
	2^3	$2^3 \cdot 3$	$2^3 \cdot 3^2$	$2^3 \cdot 3^3$
	2^4	$2^4 \cdot 3$	$2^4 \cdot 3^2$	$2^4 \cdot 3^3$
Sum	31	$31 \cdot 3$	$31 \cdot 3^2$	$31 \cdot 3^3$

Here is the calculation of the sum for the column headed by $1 \cdot 3$.

$$1 \cdot 3 + 2 \cdot 3 + 2^2 \cdot 3 + 2^3 \cdot 3 + 2^4 \cdot 3 = (1 + 2 + 2^2 + 2^3 + 2^4)3 = 31(3)$$

For the column headed by $1 \cdot 3^2$ there is a similar situation.

$$1 \cdot 3^2 + 2 \cdot 3^2 + 2^2 \cdot 3^2 + 2^3 \cdot 3^2 + 2^4 \cdot 3^2 = (1 + 2 + 2^2 + 2^3 + 2^4)3^2 = 31(3^2)$$

The sum of *all* the factors (including 432) is the sum of the last row.

$$\text{Sum of all factors} = 31 + 31 \cdot 3 + 31 \cdot 3^2 + 31 \cdot 3^3 = 31(1 + 3 + 3^2 + 3^3) = 31(40) = 1240$$

To find the sum of the *proper* factors, we must subtract 432 from 1240; we get 808 (see Take Note). Thus 432 is abundant.

We now look for some pattern for the sum of all the factors. Note that

$$\underset{\text{column}}{\text{Sum of left}} \quad 1 + 2 + 2^2 + 2^3 + 2^4 = \overbrace{}^{} \quad \overbrace{}^{} = 1 + 3 + 3^2 + 3^3 \quad \underset{\text{top row}}{\text{Sum of}}$$

$$31(40) = 1240$$

This suggests that the sum of the proper factors can be found by finding the sum of all the prime power factors for each prime, multiplying those numbers, and then subtracting the original number. Although we have not proved this for all cases, it is a true statement.

Copyright © Houghton Mifflin Company. All rights reserved.

For instance, to find the sum of the proper factors of 3240, first find the prime factorization.

$$3240 = 2^3 \cdot 3^4 \cdot 5$$

Now find the following sums:

$$1 + 2 + 2^2 + 2^3 = 15 \qquad 1 + 3 + 3^2 + 3^3 + 3^4 = 121 \qquad 1 + 5 = 6$$

The sum of the proper factors $= (15)(121)6 - 3240 = 7650$. Thus 3240 is abundant.

Determine whether the number is deficient, abundant, or perfect.

1. 200 **2.** 3125 **3.** 8128 **4.** 10,000

5. Is a prime number deficient, abundant, or perfect?

Projects and Group Activities

Finding a Pattern   The Focus on Problem Solving involved finding the sum of $1 + 3 + 3^2$ and $1 + 2 + 2^2 + 2^3$. For sums that contain a larger number of terms, it may be difficult or time consuming to try to evaluate the sum. Perhaps there is a pattern for these sums that we can use to calculate them without having to evaluate each exponential expression and then add the results.

Look at the following from the calculation of the sum of the factors of 3240.

$$1 + 2 + 2^2 + 2^3 = 1 + 2 + 4 + 8 = 15$$

$$\frac{2^{3+1} - 1}{2 - 1} = \frac{2^4 - 1}{1} = 16 - 1 = 15$$

$$1 + 3 + 3^2 + 3^3 + 3^4 = 1 + 3 + 9 + 27 + 81 = 121$$

$$\frac{3^{4+1} - 1}{3 - 1} = \frac{3^5 - 1}{2} = \frac{243 - 1}{2} = \frac{242}{2} = 121$$

Consider another sum of this type.

$$1 + 7 + 7^2 + 7^3 + 7^4 + 7^5 = 1 + 7 + 49 + 343 + 2401 + 16,807 = 19,608$$

$$\frac{7^{5+1} - 1}{7 - 1} = \frac{7^6 - 1}{6} = \frac{117,649 - 1}{6} = \frac{117,648}{6} = 19,608$$

On the basis of the examples shown above, make a conjecture about the value of $1 + n + n^2 + n^3 + n^4 + \cdots + n^k$, where n and k are natural numbers greater than 1.

Solving a System of Equations with a Graphing Calulator   A graphing calculator can be used to approximate the solution of a system of equations in two variables. Graph each equation of the system of equations, and then approximate the coordinates of the point of intersection. The process by which you approximate the solution depends on what model of calculator you have. In all cases, however, you must first solve each equation in the system of equations for y.

Copyright © Houghton Mifflin Company. All rights reserved.

Solve: $2x - 5y = 9$
$4x + 3y = 2$

$$2x - 5y = 9 \qquad\qquad 4x + 3y = 2$$
$$-5y = -2x + 9 \qquad\quad 3y = -4x + 2$$
$$y = \frac{2}{5}x - \frac{9}{5} \qquad\quad y = -\frac{4}{3}x + \frac{2}{3}$$

• Solve each equation for *y*.

TAKE NOTE

The graphing calculator screens shown here are taken from a TI-83 Plus. Similar screens would appear if we used a different model graphing calculator.

For the TI-83 and TI-83 Plus, press ⟨ Y= ⟩. Enter one equation as Y1 and the other as Y2. The result should be similar to the screen at the left below. Press ⟨ GRAPH ⟩. The graphs of the two equations should appear on the screen, as shown at the right below. If the point of intersection is not on the screen, adjust the viewing window by pressing the ⟨ WINDOW ⟩ key.

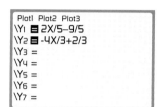

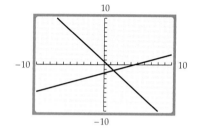

Integrating Technology

See the Keystroke Guide: *Intersect* for instructions on using a graphing calculator to solve systems of equations.

Press ⟨ 2nd ⟩ CALC 5 ⟨ ENTER ⟩ ⟨ ENTER ⟩ ⟨ ENTER ⟩. After a few seconds, the point of intersection will show on the bottom of the screen as X = 1.4230769, Y = −1.230769.

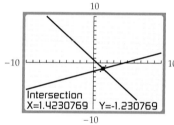

For Exercises 1 to 4, solve by using a graphing calculator.

1. $4x - 5y = 8$
$ 5x + 7y = 7$

2. $3x + 2y = 11$
$ 7x - 6y = 13$

3. $x = 3y + 2$
$ y = 4x - 2$

4. $x = 2y - 5$
$ x = 3y + 2$

Chapter 8 Summary

Key Words

Two or more equations considered together are called a *system of equations.* [8.1A, p. 405]

Examples

An example of a system of equations is
$$2x - 3y = 9$$
$$3x + 4y = 5$$

A *solution of a system of equations in two variables* is an ordered pair that is a solution of each equation of the system. [8.1A, p. 405]

The solution of the system of equations shown above is the ordered pair $(3, -1)$ because it is a solution of each equation of the system of equations.

An *independent system* of linear equations has exactly one solution. The graphs of the equations in an independent system of linear equations intersect at one point. [8.1A, p. 406]

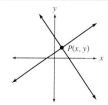

Copyright © Houghton Mifflin Company. All rights reserved.

A *dependent system* of linear equations has an infinite number of solutions. The graphs of the equations in a dependent system of linear equations are the same line. [8.1A, p. 406]

If, when solving a system of equations algebraically, the variable is eliminated and the result is a true equation, such as $5 = 5$, the system of equations is dependent. [8.2A, p. 415]

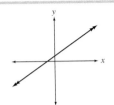

An *inconsistent system* of linear equations has no solution. The graphs of the equations of an inconsistent system of linear equations are parallel lines. [8.1A, p. 406]

If, when solving a system of equations algebraically, the variable is eliminated and the result is a false equation, such as $0 = 4$, the system of equations is inconsistent. [8.2A, p. 415]

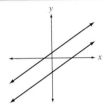

Essential Rules and Procedures

Examples

To solve a system of linear equations in two variables by graphing, graph each equation on the same coordinate system, and then determine the points of intersection. [8.1A, p. 407]

Solve by graphing: $x + 2y = 4$
$2x + y = -1$

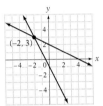

The solution is $(-2, 3)$.

To solve a system of equations by the substitution method, one variable must be written in terms of the other variable. [8.2A, p. 413]

Solve by substitution: $2x + y = 5$ (1)
$3x - 2y = 11$ (2)

$2x + y = 5$
    $y = -2x + 5$ • Solve Equation (1) for *y*.

$3x - 2y = 11$
$3x - 2(-2x + 5) = 11$ • Substitute for *y*
$3x + 4x - 10 = 11$ in Equation (2).
$7x - 10 = 11$
$7x = 21$
$x = 3$

$y = -2x + 5$
$y = -2(3) + 5$
$y = -1$
The solution is $(3, -1)$.

To solve a system of linear equations by the addition method, use the Multiplication Property of Equations to rewrite one or both of the equations so that the coefficients of one variable are opposites. Then add the equations and solve for the variables. [8.3A, p. 423]

Solve by the addition method:
$2x + 5y = 8$ (1)
$3x - 4y = -11$ (2)

$6x + 15y = 24$ • 3 times Equation (1)
$\underline{-6x + 8y = 22}$ • −2 times Equation (2)
$23y = 46$ • Add the equations.
$y = 2$ • Solve for *y*.

$2x + 5y = 8$
$2x + 5(2) = 8$ • Replace *y* by 2 in Equation (1).
$2x + 10 = 8$ • Solve for *x*.
$2x = -2$
$x = -1$
The solution is $(-1, 2)$.

Copyright © Houghton Mifflin Company. All rights reserved.

Chapter 8 Review Exercises

1. Is $(-1, -3)$ a solution of this system of equations? $5x + 4y = -17$
$2x - y = 1$

2. Is $(-2, 0)$ a solution of this system of equations? $-x + 9y = 2$
$6x - 4y = 12$

3. Solve by graphing:
$3x - y = 6$
$y = -3$

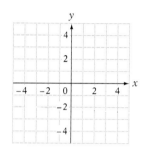

4. Solve by graphing:
$4x - 2y = 8$
$y = 2x - 4$

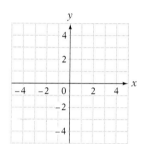

5. Solve by graphing:
$x + 2y = 3$
$y = -\dfrac{1}{2}x + 1$

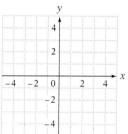

6. Solve by substitution:
$4x + 7y = 3$
$x = y - 2$

7. Solve by substitution:
$6x - y = 0$
$7x - y = 1$

8. Solve by the addition method:
$3x + 8y = -1$
$x - 2y = -5$

9. Solve by the addition method:
$6x + 4y = -3$
$12x - 10y = -15$

10. Solve by substitution:
$12x - 9y = 18$
$y = \dfrac{4}{3}x - 3$

11. Solve by substitution:
$8x - y = 2$
$y = 5x + 1$

12. Solve by the addition method:
$4x - y = 9$
$2x + 3y = -13$

13. Solve by the addition method:
$5x + 7y = 21$
$20x + 28y = 63$

14. Solve by substitution:
$4x + 3y = 12$
$y = -\dfrac{4}{3}x + 4$

Copyright © Houghton Mifflin Company. All rights reserved.

15. Solve by substitution:

$7x + 3y = -16$
$x - 2y = 5$

16. Solve by the addition method:

$3x + y = -2$
$-9x - 3y = 6$

17. Solve by the addition method:

$6x - 18y = 7$
$9x + 24y = 2$

18. Sculling   A sculling team rowing with the current went 24 mi in 2 h. Rowing against the current, the sculling team went 18 mi in 3 h. Find the rate of the sculling team in calm water and the rate of the current.

19. Investments   An investor bought 1500 shares of stock, some at $6 per share and the rest at $25 per share. If $12,800 worth of stock was purchased, how many shares of each kind did the investor buy?

20. Travel   A flight crew flew 420 km in 3 h with a tailwind. Flying against the wind, the flight crew flew 440 km in 4 h. Find the rate of the flight crew in calm air and the rate of the wind.

21. Travel   A small plane flying with the wind flew 360 mi in 3 h. Against a headwind, the plane took 4 h to fly the same distance. Find the rate of the plane in calm air and the rate of the wind.

22. Postage   A small wood-carving company mailed 190 advertisements, some requiring $.25 postage and others requiring $.45. The total cost for mailing was $59.50. Find the number of advertisements mailed at each rate.

23. Investments   Terra Cotta Art Center receives an annual income of $915 from two simple interest investments. One investment, in a corporate bond fund, earns 8.5% annual simple interest. The second investment, in a real estate investment trust, earns 7% annual simple interest. If the total amount invested in the two accounts is $12,000, how much is invested in each account?

24. Grain Mixtures   A silo contains a mixture of lentils and corn. If 50 bushels of lentils were added, there would be twice as many bushels of lentils as of corn. If 150 bushels of corn were added instead, there would be the same amount of corn as of lentils. How many bushels of each were originally in the silo?

25. Investments   Mosher Children's Hospital received a $300,000 donation that it invested in two simple interest accounts, one earning 5.4% and the other earning 6.6%. If each account earned the same amount of annual interest, how much was invested in each account?

Copyright © Houghton Mifflin Company. All rights reserved.

Chapter 8 Test

1. Is $(-2, 3)$ a solution of this system?
$$2x + 5y = 11$$
$$x + 3y = 7$$

2. Is $(1, -3)$ a solution of this system?
$$3x - 2y = 9$$
$$4x + y = 1$$

3. Solve by graphing: $3x + 2y = 6$
$$5x + 2y = 2$$

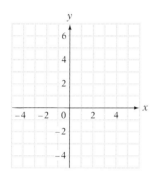

4. Solve by substitution:
$$4x - y = 11$$
$$y = 2x - 5$$

5. Solve by substitution:
$$x = 2y + 3$$
$$3x - 2y = 5$$

6. Solve by substitution:
$$3x + 5y = 1$$
$$2x - y = 5$$

7. Solve by substitution:
$$3x - 5y = 13$$
$$x + 3y = 1$$

8. Solve by substitution:
$$2x - 4y = 1$$
$$y = \frac{1}{2}x + 3$$

9. Solve by the addition method:
$$4x + 3y = 11$$
$$5x - 3y = 7$$

10. Solve by the addition method:
$$2x - 5y = 6$$
$$4x + 3y = -1$$

Copyright © Houghton Mifflin Company. All rights reserved.

11. Solve by the addition method:
$$x + 2y = 8$$
$$3x + 6y = 24$$

12. Solve by the addition method:
$$7x + 3y = 11$$
$$2x - 5y = 9$$

13. Solve by the addition method:
$$5x + 6y = -7$$
$$3x + 4y = -5$$

14. **Travel**   With the wind, a plane flies 240 mi in 2 h. Against the wind, the plane requires 3 h to fly the same distance. Find the rate of the plane in calm air and the rate of the wind.

15. **Entertainment**   For the first performance of a play in a community theater, 50 reserved-seat tickets and 80 general-admission tickets were sold. The total receipts were $980. For the second performance, 60 reserved-seat tickets and 90 general-admission tickets were sold. The total receipts were $1140. Find the price of a reserved-seat ticket and the price of a general-admission ticket.

16. **Investments**   Bernardo Community Library received a $28,000 donation that it invested in two accounts, one earning 7.6% simple interest and the other earning 6.4% simple interest. If both accounts earned the same amount of annual interest, how much was invested in each account?

Copyright © Houghton Mifflin Company. All rights reserved.

Cumulative Review Exercises

1. Evaluate $\dfrac{a^2 - b^2}{2a}$ when $a = 4$ and $b = -2$.

2. Solve: $-\dfrac{3}{4}x = \dfrac{9}{8}$

3. Given $f(x) = x^2 + 2x - 1$, find $f(2)$.

4. Multiply: $(2a^2 - 3a + 1)(2 - 3a)$

5. Simplify: $\dfrac{(-2x^2y)^4}{-8x^3y^2}$

6. Divide: $(4b^2 - 8b + 4) \div (2b - 3)$

7. Simplify: $\dfrac{8x^{-2}y^5}{-2xy^4}$

8. Factor: $4x^2y^4 - 64y^2$

9. Solve: $(x - 5)(x + 2) = -6$

10. Divide: $\dfrac{x^2 - 6x + 8}{2x^3 + 6x^2} \div \dfrac{2x - 8}{4x^3 + 12x^2}$

11. Add: $\dfrac{x - 1}{x + 2} + \dfrac{2x + 1}{x^2 + x - 2}$

12. Simplify: $\dfrac{x + 4 - \dfrac{7}{x - 2}}{x + 8 + \dfrac{21}{x - 2}}$

13. Solve: $\dfrac{x}{2x - 3} + 2 = \dfrac{-7}{2x - 3}$

14. Solve $A = P + Prt$ for r.

15. Find the x- and y-intercepts for $2x - 3y = 12$.

16. Find the slope of the line that passes through the points $(2, -3)$ and $(-3, 4)$.

17. Find the equation of the line that passes through the point $(-2, 3)$ and has slope $-\dfrac{3}{2}$.

18. Is $(2, 0)$ a solution of this system?
$$5x - 3y = 10$$
$$4x + 7y = 8$$

Copyright © Houghton Mifflin Company. All rights reserved.

19. Solve by substitution:
$$3x - 5y = -23$$
$$x + 2y = -4$$

20. Solve by the addition method:
$$5x - 3y = 29$$
$$4x + 7y = -5$$

21. **Investments**   A total of $8750 is invested in two accounts. On one account, the annual simple interest rate is 9.6%; on the second account, the annual simple interest rate is 7.2%. How much should be invested in each account so that both accounts earn the same interest?

22. **Travel**   A passenger train leaves a train depot $\frac{1}{2}$ h after a freight train leaves the same depot. The freight train is traveling 8 mph slower than the passenger train. Find the rate of each train if the passenger train overtakes the freight train in 3 h.

23. **Geometry**   The length of each side of a square is extended 4 in. The area of the resulting square is 144 in². Find the length of a side of the original square.

24. **Travel**   A plane can travel 160 mph in calm air. Flying with the wind, the plane can fly 570 mi in the same amount of time as it takes to fly 390 mi against the wind. Find the rate of the wind.

25. Graph $2x - 3y = 6$.

26. Solve by graphing: $3x + 2y = 6$
$3x - 2y = 6$

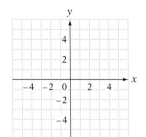

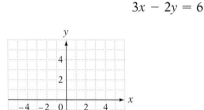

27. **Travel**   With the current, a motorboat can travel 48 mi in 3 h. Against the current, the boat requires 4 h to travel the same distance. Find the rate of the boat in calm water.

28. **Food Mixtures**   A child adds 8 g of sugar to a 50-gram serving of a breakfast cereal that is 25% sugar. What is now the percent concentration of sugar in the mixture? Round to the nearest tenth of a percent.

Copyright © Houghton Mifflin Company. All rights reserved.

9

Inequalities

You know how to determine the average of four exam scores when each exam has the same weight: add the four scores and divide the sum by 4. But do you know how to determine what score you must receive on a fifth exam to earn a specific average on the five exams? In this chapter you will learn how to use an inequality to determine that fifth and final score. **Exercises 86 and 87 on page 460** give you different scores for the first four tests, and ask you to determine the score needed for the fifth test in order to achieve a specific average.

Copyright © Houghton Mifflin Company. All rights reserved.

OBJECTIVES

Section 9.1

A To write a set using the roster method
B To write a set using set-builder notation
C To graph an inequality on the number line

Section 9.2

A To solve an inequality using the Addition Property of Inequalities
B To solve an inequality using the Multiplication Property of Inequalities
C To solve application problems

Section 9.3

A To solve general inequalities
B To solve application problems

Section 9.4

A To graph an inequality in two variables

Need help? For online student resources, such as section quizzes, visit this textbook's website at **math.college.hmco.com/students.**

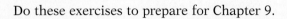

Do these exercises to prepare for Chapter 9.

1. Place the correct symbol, $<$ or $>$, between the two numbers.

$-45 \quad -27$

2. Simplify: $3x - 5(2x - 3)$

3. State the Addition Property of Equations.

4. State the Multiplication Property of Equations.

5. **Nutrition** A certain grade of hamburger contains 15% fat. How many pounds of fat are in 3 lb of this hamburger?

6. Solve: $4x - 5 = -7$

7. Solve: $4 = 2 - \dfrac{3}{4}x$

8. Solve: $7 - 2(2x - 3) = 3x - 1$

9. Graph: $y = \dfrac{2}{3}x - 3$

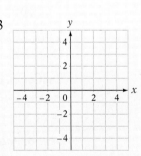

10. Graph: $3x + 4y = 12$

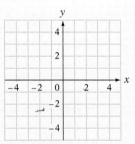

GO FIGURE • • •

Without using a calculator, which is the largest number: 2^{150}, 3^{100}, or 5^{50}?

Copyright © Houghton Mifflin Company. All rights reserved.

9.1 Sets

Copyright © Houghton Mifflin Company. All rights reserved.

Objective A   **To write a set using the roster method**

Recall that a *set* is a collection of objects, which are called the *elements* of the set. The roster method of writing a set encloses a list of the elements in braces.

The set of the last three letters of the alphabet is written {x, y, z}.

The set of the positive integers less than 5 is written {1, 2, 3, 4}.

HOW TO   Use the roster method to write the set of integers between 0 and 10.

$A = \{1, 2, 3, 4, 5, 6, 7, 8, 9\}$ • A set can be designated by a capital letter. Note that 0 and 10 are not elements of the set.

HOW TO   Use the roster method to write the set of natural numbers.

$A = \{1, 2, 3, 4, \ldots\}$ • The three dots mean that the pattern of numbers continues without end.

The **empty set,** or **null set,** is the set that contains no elements. The symbol ∅ or { } is used to represent the empty set.

The set of people who have run a 2-minute mile is an empty set.

Union of Two Sets

The **union** of two sets, written $A \cup B$, is the set that contains the elements of A and the elements of B.

HOW TO   Find $A \cup B$, given $A = \{1, 2, 3, 4\}$ and $B = \{3, 4, 5, 6\}$.

$A \cup B = \{1, 2, 3, 4, 5, 6\}$ • The union of A and B contains all the elements of A and all the elements of B. Elements in both sets are listed only once.

Intersection of Two Sets

The **intersection** of two sets, written $A \cap B$, is the set that contains the elements that are common to both A and B.

HOW TO   Find $A \cap B$, given $A = \{1, 2, 3, 4\}$ and $B = \{3, 4, 5, 6\}$.

$A \cap B = \{3, 4\}$ • The intersection of A and B contains the elements common to A and B.

Example 1

Use the roster method to write the set of the odd positive integers less than 12.

Solution
$A = \{1, 3, 5, 7, 9, 11\}$

You Try It 1

Use the roster method to write the set of the odd negative integers greater than −10.

Your solution

Solution on p. S23

Example 2

Use the roster method to write the set of the even positive integers.

Solution
$A = \{2, 4, 6, \ldots\}$

You Try It 2

Use the roster method to write the set of the odd positive integers.

Your solution

Example 3

Find $D \cup E$, given
$D = \{6, 8, 10, 12\}$ and
$E = \{-8, -6, 10, 12\}$.

Solution
$D \cup E = \{-8, -6, 6, 8, 10, 12\}$

You Try It 3

Find $A \cup B$, given
$A = \{-2, -1, 0, 1, 2\}$ and
$B = \{0, 1, 2, 3, 4\}$.

Your solution

Example 4

Find $A \cap B$, given
$A = \{5, 6, 9, 11\}$ and
$B = \{5, 9, 13, 15\}$.

Solution
$A \cap B = \{5, 9\}$

You Try It 4

Find $C \cap D$, given
$C = \{10, 12, 14, 16\}$ and
$D = \{10, 16, 20, 26\}$.

Your solution

Example 5

Find $A \cap B$, given
$A = \{1, 2, 3, 4\}$ and
$B = \{8, 9, 10, 11\}$.

Solution
$A \cap B = \varnothing$

You Try It 5

Find $A \cap B$, given
$A = \{-5, -4, -3, -2\}$ and
$B = \{2, 3, 4, 5\}$.

Your solution

Solutions on p. S23

Objective B   **To write a set using set-builder notation**

Point of Interest

The symbol $\in$ was first used in the book *Arithmeticae Principia*, published in 1889. It is the first letter of the Greek word $\varepsilon\sigma\tau\iota$ which means "is." The symbols for union and intersection were also introduced at that time.

Another method of representing sets is called **set-builder notation.** This method of writing a set uses a rule to describe the elements of the set. Using set-builder notation, we represent the set of all positive integers less than 10 as

$\{x \mid x < 10, x \in \text{positive integers}\}$, which is read "the set of all x such that x is less than 10 and x is an element of the positive integers."

HOW TO   Use set-builder notation to write the set of real numbers greater than 4.

$\{x \mid x > 4, x \in \text{real numbers}\}$ • "$x \in$ real numbers" is read "x is an element of the real numbers."

Copyright © Houghton Mifflin Company. All rights reserved.

Example 6

Use set-builder notation to write the set of negative integers greater than -100.

Solution

$\{x \mid x > -100, x \in \text{negative integers}\}$

Example 7

Use set-builder notation to write the set of real numbers less than 60.

Solution

$\{x \mid x < 60, x \in \text{real numbers}\}$

You Try It 6

Use set-builder notation to write the set of positive even integers less than 59.

Your solution

You Try It 7

Use set-builder notation to write the set of real numbers greater than -3.

Your solution

Solutions on p. S23

Objective C   **To graph an inequality on the number line**

An expression that contains the symbol $>$, $<$, $\geq$ (is greater than or equal to), or $\leq$ (is less than or equal to) is called an **inequality.** An inequality expresses the relative order of two mathematical expressions. The expressions can be either numerical or variable.

$$\left.\begin{array}{l} 4 > 2 \\ 3x \leq 7 \\ x^2 - 2x > y + 4 \end{array}\right\} \text{Inequalities}$$

The **solution set of an inequality** is the set of numbers each element of which, when substituted for the variable, results in a true inequality. The solution set of an inequality can be graphed on the number line.

Copyright © Houghton Mifflin Company. All rights reserved.

TAKE NOTE

For the remainder of this section, all variables will represent real numbers. Given this convention, the expression $\{x \mid x > 1, x \in \text{real numbers}\}$ will be written as $\{x \mid x > 1\}$, as shown in the example at the right.

HOW TO   Graph: $\{x \mid x > 1\}$

The graph is the real numbers greater than 1. The parenthesis at 1 indicates that 1 is not included in the graph.

HOW TO   Graph: $\{x \mid x \geq 1\}$

The bracket at 1 indicates that 1 is included in the solution set.

HOW TO   Graph: $\{x \mid -1 > x\}$

$-1 > x$ is equivalent to $x < -1$. The numbers less than -1 are to the left of -1 on the number line.

The union of two sets is the set that contains all the elements of each set.

HOW TO   Graph: $\{x \mid x > 4\} \cup \{x \mid x < 1\}$

The graph is the numbers greater than 4 and the numbers less than 1.

The intersection of two sets is the set that contains the elements common to both sets.

HOW TO Graph: $\{x \mid x > -1\} \cap \{x \mid x < 2\}$

The graphs of $\{x \mid x > -1\}$ and $\{x \mid x < 2\}$ are shown at the right.

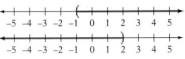

The graph of $\{x \mid x > -1\} \cap \{x \mid x < 2\}$ is the numbers between -1 and 2.

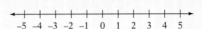

Example 8

Graph: $\{x \mid x < 5\}$

Solution

The solution set is the numbers less than 5.

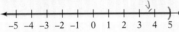

You Try It 8

Graph: $\{x \mid -2 < x\}$

Your solution

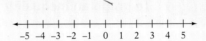

Example 9

Graph: $\{x \mid x > -2\} \cap \{x \mid x < 1\}$

Solution

The solution set is the numbers between -2 and 1.

You Try It 9

Graph: $\{x \mid x > -1\} \cup \{x \mid x < -3\}$

Your solution

Example 10

Graph: $\{x \mid x > 3\} \cup \{x \mid x < 1\}$

Solution

The solution set is the numbers greater than 3 and the numbers less than 1.

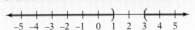

You Try It 10

Graph: $\{x \mid x \leq 4\} \cap \{x \mid x \geq -4\}$

Your solution

Example 11

Graph: $\{x \mid x \leq 5\} \cup \{x \mid x \geq -3\}$

Solution

The solution set is the real numbers.

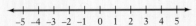

You Try It 11

Graph: $\{x \mid x < 2\} \cup \{x \mid x \geq -2\}$

Your solution

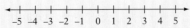

Solutions on pp. S23–S24

Copyright © Houghton Mifflin Company. All rights reserved.

9.1 Exercises

Objective A   To write a set using the roster method

1. Explain how to find the union of two sets.

2. Explain how to find the intersection of two sets.

For Exercises 3 to 8, use the roster method to write the set.

3. The integers between 15 and 22

4. The integers between -10 and -4

5. The odd integers between 8 and 18

6. The even integers between -11 and -1

7. The letters of the alphabet between a and d

8. The letters of the alphabet between p and v

For Exercises 9 to 16, find $A \cup B$.

9. $A = \{3, 4, 5\}$ $B = \{4, 5, 6\}$

10. $A = \{-3, -2, -1\}$ $B = \{-2, -1, 0\}$

11. $A = \{-10, -9, -8\}$ $B = \{8, 9, 10\}$

12. $A = \{a, b, c\}$ $B = \{x, y, z\}$

13. $A = \{a, b, d, e\}$ $B = \{c, d, e, f\}$

14. $A = \{m, n, p, q\}$ $B = \{m, n, o\}$

15. $A = \{1, 3, 7, 9\}$ $B = \{7, 9, 11, 13\}$

16. $A = \{-3, -2, -1\}$ $B = \{-1, 1, 2\}$

For Exercises 17 to 22, find $A \cap B$.

17. $A = \{3, 4, 5\}$ $B = \{4, 5, 6\}$

18. $A = \{-4, -3, -2\}$ $B = \{-6, -5, -4\}$

19. $A = \{-4, -3, -2\}$ $B = \{2, 3, 4\}$

20. $A = \{1, 2, 3, 4\}$ $B = \{1, 2, 3, 4\}$

21. $A = \{a, b, c, d, e\}$ $B = \{c, d, e, f, g\}$

22. $A = \{m, n, o, p\}$ $B = \{k, l, m, n\}$

Objective B   To write a set using set-builder notation

For Exercises 23 to 30, use set-builder notation to write the set.

23. The negative integers greater than -5

24. The positive integers less than 5

25. The integers greater than 30

26. The integers less than -70

27. The even integers greater than 5

28. The odd integers less than -2

29. The real numbers greater than 8

30. The real numbers less than 57

Copyright © Houghton Mifflin Company. All rights reserved.

Objective C **To graph an inequality on the number line**

For Exercises 31 to 40, graph.

31. $\{x \mid x > 2\}$

⟨+—+—+—+—+—+—+—+—+—+—+⟩
−5 −4 −3 −2 −1 0 1 2 3 4 5

32. $\{x \mid x \geq -1\}$

⟨+—+—+—+—+—+—+—+—+—+—+⟩
−5 −4 −3 −2 −1 0 1 2 3 4 5

33. $\{x \mid x \leq 0\}$

⟨+—+—+—+—+—+—+—+—+—+—+⟩
−5 −4 −3 −2 −1 0 1 2 3 4 5

34. $\{x \mid x < 4\}$

⟨+—+—+—+—+—+—+—+—+—+—+⟩
−5 −4 −3 −2 −1 0 1 2 3 4 5

35. $\{x \mid x > -2\} \cup \{x \mid x < -4\}$

⟨+—+—+—+—+—+—+—+—+—+—+⟩
−5 −4 −3 −2 −1 0 1 2 3 4 5

36. $\{x \mid x > 4\} \cup \{x \mid x < -2\}$

⟨+—+—+—+—+—+—+—+—+—+—+⟩
−5 −4 −3 −2 −1 0 1 2 3 4 5

37. $\{x \mid x > -2\} \cap \{x \mid x < 4\}$

⟨+—+—+—+—+—+—+—+—+—+—+⟩
−5 −4 −3 −2 −1 0 1 2 3 4 5

38. $\{x \mid x > -3\} \cap \{x \mid x < 3\}$

⟨+—+—+—+—+—+—+—+—+—+—+⟩
−5 −4 −3 −2 −1 0 1 2 3 4 5

39. $\{x \mid x \geq -2\} \cup \{x \mid x < 4\}$

⟨+—+—+—+—+—+—+—+—+—+—+⟩
−5 −4 −3 −2 −1 0 1 2 3 4 5

40. $\{x \mid x > 0\} \cup \{x \mid x \leq 4\}$

⟨+—+—+—+—+—+—+—+—+—+—+⟩
−5 −4 −3 −2 −1 0 1 2 3 4 5

APPLYING THE CONCEPTS

41. Determine whether the statement is always true, sometimes true, or never true.
  a. Given that $a > 0$ and $b < 0$, then $ab > 0$.
  b. Given that $a < 0$, then $a^2 > 0$.
  c. Given that $a > 0$ and $b < 0$, then $a^2 > b$.

42. **a.** By trying various sets, make a conjecture about whether or not finding the union of two sets is a commutative operation.
  b. By trying various sets, make a conjecture about whether or not finding the union of two sets is an associative operation.

43. **a.** By trying various sets, make a conjecture about whether or not finding the intersection of two sets is a commutative operation.
  b. By trying various sets, make a conjecture about whether or not finding the intersection of two sets is an associative operation.

Copyright © Houghton Mifflin Company. All rights reserved.

9.2 The Addition and Multiplication Properties of Inequalities

Copyright © Houghton Mifflin Company. All rights reserved.

Objective A **To solve an inequality using the Addition Property of Inequalities**

Recall that the solution set of an inequality is the set of numbers each element of which, when substituted for the variable, results in a true inequality.

The inequality at the right is true if the variable is replaced by 7, 9.3, or $\frac{15}{2}$.

$$x + 5 > 8$$

$$\left.\begin{array}{l} 7 + 5 > 8 \\ 9.3 + 5 > 8 \\ \frac{15}{2} + 5 > 8 \end{array}\right\} \text{True inequalities}$$

The inequality $x + 5 > 8$ is false if the variable is replaced by 2, 1.5, or $-\frac{1}{2}$.

$$\left.\begin{array}{l} 2 + 5 > 8 \\ 1.5 + 5 > 8 \\ -\frac{1}{2} + 5 > 8 \end{array}\right\} \text{False inequalities}$$

There are many values of the variable x that will make the inequality $x + 5 > 8$ true. The solution set of $x + 5 > 8$ is any number greater than 3.

At the right is the graph of the solution set of $x + 5 > 8$.

In solving an inequality, the goal is to rewrite the given inequality in the form *variable > constant* or *variable < constant*. The Addition Property of Inequalities is used to rewrite an inequality in this form.

> **Addition Property of Inequalities**
>
> The same term can be added to each side of an inequality without changing the solution set of the inequality.
>
> $$\text{If } a > b, \text{ then } a + c > b + c.$$
> $$\text{If } a < b, \text{ then } a + c < b + c.$$

The Addition Property of Inequalities also holds true for an inequality containing the symbol $\geq$ or $\leq$.

The Addition Property of Inequalities is used when, in order to rewrite an inequality in the form *variable > constant* or *variable < constant*, we must remove a term from one side of the inequality. Add the opposite of that term to each side of the inequality.

HOW TO Solve: $x - 4 < -3$

$$x - 4 < -3$$
$$x - 4 + 4 < -3 + 4 \qquad \bullet \text{ Add 4 to each side of the inequality.}$$
$$x < 1 \qquad \bullet \text{ Simplify.}$$

At the right is the graph of the solution set of $x - 4 < -3$.

Because subtraction is defined in terms of addition, the Addition Property of Inequalities allows the same term to be subtracted from each side of an inequality.

> **HOW TO** Solve: $5x - 6 \leq 4x - 4$
>
> $5x - 6 \leq 4x - 4$
>
> $5x - 4x - 6 \leq 4x - 4x - 4$ • **Subtract 4x** from each side of the inequality.
>
> $x - 6 \leq -4$ • **Simplify.**
>
> $x - 6 + 6 \leq -4 + 6$ • **Add 6** to each side of the inequality.
>
> $x \leq 2$ • **Simplify.**

Example 1

Solve $3 < x + 5$ and graph the solution set.

Solution

$3 < x + 5$

$3 - 5 < x + 5 - 5$ • **Subtract 5.**

$-2 < x$

$$\begin{array}{c} \xleftarrow{\qquad} \ \ (\ \ \ \ \ \ \ \ \xrightarrow{\qquad} \\ -5\ -4\ -3\ -2\ -1\ \ 0\ \ 1\ \ 2\ \ 3\ \ 4\ \ 5 \end{array}$$

You Try It 1

Solve $x + 2 < -2$ and graph the solution set.

Your solution

$$\begin{array}{c} \xleftarrow{\qquad\qquad\qquad\qquad\qquad} \\ -5\ -4\ -3\ -2\ -1\ \ 0\ \ 1\ \ 2\ \ 3\ \ 4\ \ 5 \end{array}$$

Example 2

Solve: $7x - 14 \leq 6x - 16$

Solution

$7x - 14 \leq 6x - 16$

$7x - 6x - 14 \leq 6x - 6x - 16$ • **Subtract 6x.**

$x - 14 \leq -16$

$x - 14 + 14 \leq -16 + 14$ • **Add 14.**

$x \leq -2$

You Try It 2

Solve: $5x + 3 > 4x + 5$

Your solution

Solutions on p. S24

Copyright © Houghton Mifflin Company. All rights reserved.

Objective B   **To solve an inequality using the Multiplication Property of Inequalities**

In solving an inequality, the goal is to rewrite the given inequality in the form *variable > constant* or *variable < constant*. The Multiplication Property of Inequalities is used when, in order to rewrite an inequality in this form, we must remove a coefficient from one side of the inequality.

Multiplication Property of Inequalities

Each side of an inequality can be multiplied by the same positive number without changing the solution set of the inequality.

$$\text{If } a > b \text{ and } c > 0, \text{ then } ac > bc.$$
$$\text{If } a < b \text{ and } c > 0, \text{ then } ac < bc.$$

If each side of an inequality is multiplied by the same negative number and the inequality symbol is reversed, then the solution set of the inequality is not changed.

$$\text{If } a > b \text{ and } c < 0, \text{ then } ac < bc.$$
$$\text{If } a < b \text{ and } c < 0, \text{ then } ac > bc.$$

Copyright © Houghton Mifflin Company. All rights reserved.

TAKE NOTE

Any time an inequality is multiplied or divided by a negative number, the inequality symbol must be reversed. Compare the next two examples.

$2x < -4$ Divide each
$\dfrac{2x}{2} < \dfrac{-4}{2}$ side by *posi-tive* 2.

$x < -2$ Inequality *is not* reversed.

$-2x < 4$ Divide each
$\dfrac{-2x}{-2} > \dfrac{4}{-2}$ side by *nega-tive* 2.

$x > -2$ Inequality *is* reversed.

$5 > 4$ • A true inequality

$5(2) > 4(2)$ • Multiply by *positive* 2.

$10 > 8$ • Still a true inequality

$6 < 9$ • A true inequality

$6(-3) > 9(-3)$ • Multiply by *negative* 3 and *reverse* the inequality.

$-18 > -27$ • Still a true inequality

The Multiplication Property of Inequalities also holds true for an inequality containing the symbol $\geq$ or $\leq$.

HOW TO   Solve $-\dfrac{3}{2}x \leq 6$ and graph the solution set.

$$-\frac{3}{2}x \leq 6$$

$$-\frac{2}{3}\left(-\frac{3}{2}x\right) \geq -\frac{2}{3}(6)$$

$$x \geq -4$$

• Multiply each side of the inequality by $-\dfrac{2}{3}$. Because $-\dfrac{2}{3}$ is a negative number, the inequality symbol must be reversed.

• Graph $\{x \mid x \geq -4\}$.

Because division is defined in terms of multiplication, the Multiplication Property of Inequalities allows each side of an inequality to be divided by a nonzero constant.

TAKE NOTE

As shown in the example at the right, the goal in solving an inequality can be *constant < variable* or *constant > variable*. We could have written the answer to this example as $x > -\dfrac{2}{3}$.

HOW TO   Solve: $-4 < 6x$

$$-4 < 6x$$

$$\frac{-4}{6} < \frac{6x}{6}$$

• Divide each side of the inequality by 6.

$$-\frac{2}{3} < x$$

• Simplify: $\dfrac{-4}{6} = -\dfrac{2}{3}$.

Example 3 Solve $-7x > 14$ and graph the solution set.

Solution

$-7x > 14$

$\dfrac{-7x}{-7} < \dfrac{14}{-7}$ • Divide by −7.

$x < -2$

You Try It 3 Solve $-3x > -9$ and graph the solution set.

Your solution

Example 4 Solve: $-\dfrac{5}{8}x \le \dfrac{5}{12}$

Solution

$-\dfrac{5}{8}x \le \dfrac{5}{12}$

$-\dfrac{8}{5}\left(-\dfrac{5}{8}x\right) \ge -\dfrac{8}{5}\left(\dfrac{5}{12}\right)$ • Multiply by $-\dfrac{8}{5}$.

$x \ge -\dfrac{2}{3}$

You Try It 4 Solve: $-\dfrac{3}{4}x \ge 18$

Your solution

Solutions on p. S24

Objective C **To solve application problems**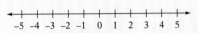

Example 5

A student must have at least 450 points out of 500 points on five tests to receive an A in a course. One student's results on the first four tests were 94, 87, 77, and 95. What scores on the last test will enable this student to receive an A in the course?

Strategy
To find the scores, write and solve an inequality using N to represent the possible scores on the last test.

Solution

Total number of points on the five tests	is greater than or equal to	450

$94 + 87 + 77 + 95 + N \ge 450$

$353 + N \ge 450$ • Simplify.

$353 - 353 + N \ge 450 - 353$ • Subtract 353.

$N \ge 97$

The student's score on the last test must be greater than or equal to 97.

You Try It 5

An appliance dealer will make a profit on the sale of a television set if the cost of the new set is less than 70% of the selling price. What selling prices will enable the dealer to make a profit on a television set that costs the dealer $314?

Your strategy

Your solution

Solution on p. S24

Copyright © Houghton Mifflin Company. All rights reserved.

9.2 Exercises

Objective A   **To solve an inequality using the Addition Property of Inequalities**

For Exercises 1 to 8, solve the inequality and graph the solution set.

1. $x + 1 < 3$

$$\begin{array}{c}\longleftrightarrow\\ -5\ -4\ -3\ -2\ -1\ \ 0\ \ 1\ \ 2\ \ 3\ \ 4\ \ 5\end{array}$$

2. $y + 2 < 2$

$$\begin{array}{c}\longleftrightarrow\\ -5\ -4\ -3\ -2\ -1\ \ 0\ \ 1\ \ 2\ \ 3\ \ 4\ \ 5\end{array}$$

3. $x - 5 > -2$

$$\begin{array}{c}\longleftrightarrow\\ -5\ -4\ -3\ -2\ -1\ \ 0\ \ 1\ \ 2\ \ 3\ \ 4\ \ 5\end{array}$$

4. $x - 3 > -2$

$$\begin{array}{c}\longleftrightarrow\\ -5\ -4\ -3\ -2\ -1\ \ 0\ \ 1\ \ 2\ \ 3\ \ 4\ \ 5\end{array}$$

5. $7 \le n + 4$

$$\begin{array}{c}\longleftrightarrow\\ -5\ -4\ -3\ -2\ -1\ \ 0\ \ 1\ \ 2\ \ 3\ \ 4\ \ 5\end{array}$$

6. $3 \le 5 + x$

$$\begin{array}{c}\longleftrightarrow\\ -5\ -4\ -3\ -2\ -1\ \ 0\ \ 1\ \ 2\ \ 3\ \ 4\ \ 5\end{array}$$

7. $x - 6 \le -10$

$$\begin{array}{c}\longleftrightarrow\\ -5\ -4\ -3\ -2\ -1\ \ 0\ \ 1\ \ 2\ \ 3\ \ 4\ \ 5\end{array}$$

8. $y - 8 \le -11$

$$\begin{array}{c}\longleftrightarrow\\ -5\ -4\ -3\ -2\ -1\ \ 0\ \ 1\ \ 2\ \ 3\ \ 4\ \ 5\end{array}$$

For Exercises 9 to 12, write an inequality that represents the set of numbers shown in the graph.

9.

$$\begin{array}{c}\longleftrightarrow\\ -5\ -4\ -3\ -2\ -1\ \ 0\ \ 1\ \ 2\ \ 3\ \ 4\ \ 5\end{array}$$

10.

$$\begin{array}{c}\longleftrightarrow\\ -5\ -4\ -3\ -2\ -1\ \ 0\ \ 1\ \ 2\ \ 3\ \ 4\ \ 5\end{array}$$

11.

$$\begin{array}{c}\longleftrightarrow\\ -5\ -4\ -3\ -2\ -1\ \ 0\ \ 1\ \ 2\ \ 3\ \ 4\ \ 5\end{array}$$

12.

$$\begin{array}{c}\longleftrightarrow\\ -5\ -4\ -3\ -2\ -1\ \ 0\ \ 1\ \ 2\ \ 3\ \ 4\ \ 5\end{array}$$

For Exercises 13 to 42, solve.

13. $y - 3 \ge -12$

14. $x + 8 \ge -14$

15. $3x - 5 < 2x + 7$

16. $5x + 4 < 4x - 10$

17. $8x - 7 \ge 7x - 2$

18. $3n - 9 \ge 2n - 8$

19. $2x + 4 < x - 7$

20. $9x + 7 < 8x - 7$

21. $4x - 8 \le 2 + 3x$

22. $5b - 9 < 3 + 4b$

23. $6x + 4 \ge 5x - 2$

24. $7x - 3 \ge 6x - 2$

25. $2x - 12 > x - 10$

26. $3x + 9 > 2x + 7$

27. $d + \dfrac{1}{2} < \dfrac{1}{3}$

Copyright © Houghton Mifflin Company. All rights reserved.

28. $x - \dfrac{3}{8} < \dfrac{5}{6}$

29. $x + \dfrac{5}{8} \geq -\dfrac{2}{3}$

30. $y + \dfrac{5}{12} \geq -\dfrac{3}{4}$

31. $x - \dfrac{3}{8} < \dfrac{1}{4}$

32. $y + \dfrac{5}{9} \leq \dfrac{5}{6}$

33. $2x - \dfrac{1}{2} < x + \dfrac{3}{4}$

34. $6x - \dfrac{1}{3} \leq 5x - \dfrac{1}{2}$

35. $3x + \dfrac{5}{8} > 2x + \dfrac{5}{6}$

36. $4b - \dfrac{7}{12} \geq 3b - \dfrac{9}{16}$

37. $3.8x < 2.8x - 3.8$

38. $1.2x < 0.2x - 7.3$

39. $x + 5.8 \leq 4.6$

40. $n - 3.82 \leq 3.95$

41. $x - 3.5 < 2.1$

42. $x - 0.23 \leq 0.47$

Objective B   **To solve an inequality using the Multiplication Property of Inequalities**

For Exercises 43 to 52, solve the inequality and graph the solution set.

43. $3x < 12$

44. $8x \leq -24$

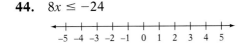

45. $15 \leq 5y$

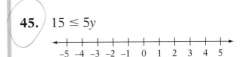

46. $-48 < 24x$

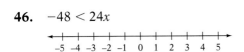

47. $16x \leq 16$

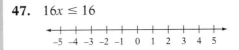

48. $3x > 0$

49. $-8x > 8$

50. $-2n \leq -8$

51. $-6b > 24$

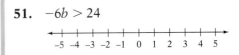

52. $-4x < 8$

Copyright © Houghton Mifflin Company. All rights reserved.

For Exercises 53 to 79, solve.

53. $-5y \geq 0$

54. $-3z < 0$

55. $7x > 2$

56. $6x \leq -1$

57. $2x \leq -5$

58. $\dfrac{5}{6}n < 15$

59. $\dfrac{3}{4}x < 12$

60. $\dfrac{2}{3}y \geq 4$

61. $10 \leq \dfrac{5}{8}x$

62. $4 \geq \dfrac{2}{3}x$

63. $-\dfrac{3}{7}x \leq 6$

64. $-\dfrac{2}{11}b \geq -6$

65. $-\dfrac{4}{7}x \geq -12$

66. $\dfrac{2}{3}n < \dfrac{1}{2}$

67. $-\dfrac{3}{5}x < 0$

68. $-\dfrac{2}{3}x \geq 0$

69. $-\dfrac{3}{8}x \geq \dfrac{9}{14}$

70. $-\dfrac{3}{5}x < -\dfrac{6}{7}$

71. $-\dfrac{4}{5}x < -\dfrac{8}{15}$

72. $-\dfrac{3}{4}y \geq -\dfrac{5}{8}$

73. $-\dfrac{8}{9}x \geq -\dfrac{16}{27}$

74. $1.5x \leq 6.30$

75. $2.3x \leq 5.29$

76. $-3.5d > 7.35$

77. $-0.24x > 0.768$

78. $4.25m > -34$

79. $-3.9x \geq -19.5$

Objective C **To solve application problems**

80. Number Sense Three-fifths of a number is greater than two-thirds. Find the smallest integer that satisfies this inequality.

81. Sports To be eligible for a basketball tournament, a basketball team must win at least 60% of its remaining games. If the team has 17 games remaining, how many games must the team win to qualify for the tournament?

82. Taxes To avoid a tax penalty, at least 90% of a self-employed person's total annual income tax liability must be paid in estimated tax payments during the year. What amount of income tax must a person with an annual income tax liability of $3500 pay in estimated tax payments?

Copyright © Houghton Mifflin Company. All rights reserved.

83. Recycling A service organization will receive a bonus of $200 for collecting more than 1850 lb of aluminum cans during its four collection drives. On the first three drives, the organization collected 505 lb, 493 lb, and 412 lb. How many pounds of cans must the organization collect on the fourth drive to receive the bonus?

84. Software Development Computer software engineers are fond of saying that software takes at least twice as long to develop as they think it will. According to that saying, how many hours will it take to develop a software product that an engineer thinks can be finished in 50 h?

85. Health A government agency recommends a minimum daily allowance of vitamin C of 60 mg. How many additional milligrams of vitamin C does a person who has already drunk a glass of orange juice with 10 mg of vitamin C need in order to satisfy the recommended daily allowance?

86. Grading To pass a course with a B grade, a student must have an average of 80 points on five tests. The student's grades on the first four tests were 75, 83, 86, and 78. What scores can the student receive on the fifth test to earn a B grade?

87. Grading A professor scores all tests with a maximum of 100 points. To earn an A grade in this course, a student must have an average of 92 on four tests. One student's grades on the first three tests were 89, 86, and 90. Can this student earn an A grade?

88. Health A health official recommends a maximum cholesterol level of 200 units. How many units must a patient with a cholesterol level of 275 units reduce her cholesterol level to satisfy the recommended maximum level?

APPLYING THE CONCEPTS

For Exercises 89 to 94, given that $a > b$ and that a and b are real numbers, determine for which real numbers c the statement is true. Use set-builder notation to write the answer.

89. $ac > bc$

90. $ac < bc$

91. $a + c > b + c$

92. $a + c < b + c$

93. $\dfrac{a}{c} > \dfrac{b}{c}$

94. $\dfrac{a}{c} < \dfrac{b}{c}$

95. In your own words, state the Addition Property of Inequalities.

96. In your own words, state the Multiplication Property of Inequalities.

Copyright © Houghton Mifflin Company. All rights reserved.

9.3 General Inequalities

Objective A   **To solve general inequalities**

Solving an inequality frequently requires application of both the Addition and the Multiplication Properties of Inequalities.

HOW TO   Solve: $4y - 3 \geq 6y + 5$

$$4y - 3 \geq 6y + 5$$

$$4y - 6y - 3 \geq 6y - 6y + 5$$ • **Subtract 6y** from each side of the inequality.

$$-2y - 3 \geq 5$$ • **Simplify.**

$$-2y - 3 + 3 \geq 5 + 3$$ • **Add 3** to each side of the inequality.

$$-2y \geq 8$$ • **Simplify.**

$$\frac{-2y}{-2} \leq \frac{8}{-2}$$ • **Divide** each side of the inequality **by −2**. Because **−2** is a negative number, the inequality symbol must be reversed.

$$y \leq -4$$

When an inequality contains parentheses, one of the steps in solving the inequality requires the use of the Distributive Property.

HOW TO   Solve: $-2(x - 7) > 3 - 4(2x - 3)$

$$-2(x - 7) > 3 - 4(2x - 3)$$

$$-2x + 14 > 3 - 8x + 12$$ • Use the **Distributive Property** to remove parentheses.

$$-2x + 14 > -8x + 15$$ • **Simplify.**

$$-2x + 8x + 14 > -8x + 8x + 15$$ • **Add 8x** to each side of the inequality.

$$6x + 14 > 15$$ • **Simplify.**

$$6x + 14 - 14 > 15 - 14$$ • **Subtract 14** from each side of the inequality.

$$6x > 1$$ • **Simplify.**

$$\frac{6x}{6} > \frac{1}{6}$$ • **Divide** each side of the inequality **by 6.**

$$x > \frac{1}{6}$$

Example 1   Solve: $7x - 3 \leq 3x + 17$

Solution

$$7x - 3 \leq 3x + 17$$

$$7x - 3x - 3 \leq 3x - 3x + 17$$ • **Subtract 3x.**

$$4x - 3 \leq 17$$

$$4x - 3 + 3 \leq 17 + 3$$ • **Add 3.**

$$4x \leq 20$$

$$\frac{4x}{4} \leq \frac{20}{4}$$ • **Divide by 4.**

$$x \leq 5$$

You Try It 1   Solve: $5 - 4x > 9 - 8x$

Your solution

Solution on p. S24

Copyright © Houghton Mifflin Company. All rights reserved.

Example 2

Solve: $3(3 - 2x) \geq -5x - 2(3 - x)$

Solution

$$3(3 - 2x) \geq -5x - 2(3 - x)$$
$$9 - 6x \geq -5x - 6 + 2x \quad \bullet \text{ Distributive Property}$$
$$9 - 6x \geq -3x - 6$$
$$9 - 6x + 3x \geq -3x + 3x - 6 \quad \bullet \text{ Add } 3x.$$
$$9 - 3x \geq -6$$
$$9 - 9 - 3x \geq -6 - 9 \quad \bullet \text{ Subtract 9.}$$
$$-3x \geq -15$$
$$\frac{-3x}{-3} \leq \frac{-15}{-3} \quad \bullet \text{ Divide by } -3.$$
$$x \leq 5$$

You Try It 2

Solve: $8 - 4(3x + 5) \leq 6(x - 8)$

Your solution

Solution on p. S24

Objective B   **To solve application problems**

Example 3

A rectangle is 10 ft wide and $(2x + 4)$ ft long. Express as an integer the maximum length of the rectangle when the area is less than 200 ft². (The area of a rectangle is equal to its length times its width.)

Strategy

To find the maximum length:

• Replace the variables in the area formula by the given values and solve for x.
• Replace the variable in the expression $2x + 4$ with the value found for x.

Solution

Length times width	is less than	200 ft²

$$10(2x + 4) < 200$$
$$20x + 40 < 200 \quad \bullet \text{ Distributive Property}$$
$$20x + 40 - 40 < 200 - 40 \quad \bullet \text{ Subtract 40.}$$
$$20x < 160$$
$$\frac{20x}{20} < \frac{160}{20} \quad \bullet \text{ Divide by 20.}$$
$$x < 8$$

The length is $(2x + 4)$ ft. Because $x < 8$, $2x + 4 < 2(8) + 4 = 20$. Therefore, the length is less than 20 ft. The maximum length is 19 ft.

You Try It 3

Company A rents cars for $8 a day and $.10 for every mile driven. Company B rents cars for $10 a day and $.08 per mile driven. You want to rent a car for 1 week. What is the maximum number of miles you can drive a Company A car if it is to cost you less than a Company B car?

Your strategy

Your solution

Solution on p. S24

Copyright © Houghton Mifflin Company. All rights reserved.

9.3 Exercises

Objective A   **To solve general inequalities**

For Exercises 1 to 20, solve.

1. $4x - 8 < 2x$

2. $7x - 4 < 3x$

3. $2x - 8 > 4x$

4. $3y + 2 > 7y$

5. $8 - 3x \leq 5x$

6. $10 - 3x \leq 7x$

7. $3x + 2 > 5x - 8$

8. $2n - 9 \geq 5n + 4$

9. $5x - 2 < 3x - 2$

10. $8x - 9 > 3x - 9$

11. $0.1(180 + x) > x$

12. $x > 0.2(50 + x)$

13. $2(2y - 5) \leq 3(5 - 2y)$

14. $2(5x - 8) \leq 7(x - 3)$

15. $5(2 - x) > 3(2x - 5)$

16. $4(3d - 1) > 3(2 - 5d)$

17. $4 - 3(3 - n) \leq 3(2 - 5n)$

18. $15 - 5(3 - 2x) \leq 4(x - 3)$

19. $2x - 3(x - 4) \geq 4 - 2(x - 7)$

20. $4 + 2(3 - 2y) \leq 4(3y - 5) - 6y$

Objective B   **To solve application problems**

21. Wages   The sales agent for a jewelry company is offered a flat monthly salary of $3200 or a salary of $1000 plus an 11% commission on the selling price of each item sold by the agent. If the agent chooses the $3200, what dollar amount does the agent expect to sell in 1 month?

22. Sports   A baseball player is offered an annual salary of $200,000 or a base salary of $100,000 plus a bonus of $1000 for each hit over 100 hits. How many hits must the baseball player make to earn more than $200,000?

Copyright © Houghton Mifflin Company. All rights reserved.

23. Comparing Services   A computer bulletin board service charges a flat fee of $10 per month or a fee of $4 per month plus $.10 for each minute the service is used. How many minutes must a person use this service to exceed $10?

24. Comparing Services   A site licensing fee for a computer program is $1500. Paying this fee allows the company to use the program at any computer terminal within the company. Alternatively, the company can choose to pay $200 for each individual computer it has. How many individual computers must a company have for the site license to be more economical for the company?

25. Health   For a product to be labeled orange juice, a state agency requires that at least 80% of the drink be real orange juice. How many ounces of artificial flavors can be added to 32 oz of real orange juice and have it still be legal to label the drink orange juice?

26. Health   Grade A hamburger cannot contain more than 20% fat. How much fat can a butcher mix with 300 lb of lean meat to meet the 20% requirement?

27. Transportation   A shuttle service taking skiers to a ski area charges $8 per person each way. Four skiers are debating whether to take the shuttle bus or rent a car for $45 plus $.25 per mile. Assuming that the skiers will share the cost of the car and that they want the least expensive method of transportation, find how far away the ski area is if they choose the shuttle service.

APPLYING THE CONCEPTS

28. Determine whether the statement is always true, sometimes true, or never true, given that a, b, and c are real numbers.
  a. If $a > b$, then $-a > -b$.
  b. If $a < b$, then $ac < bc$.
  c. If $a > b$, then $a + c > b + c$.
  d. If $a \neq 0$, $b \neq 0$, and $a > b$, then $\frac{1}{a} > \frac{1}{b}$.

For Exercises 29 and 30, use the roster method to list the set of positive integers that are solutions of the inequality.

29. $7 - 2b \leq 15 - 5b$

30. $-6(2 - d) \geq 4d - 9$

For Exercises 31 and 32, use the roster method to list the set of integers that are common to the solution sets of the two inequalities.

31. $5x - 12 \leq x + 8$
       $3x - 4 \geq 2 + x$

32. $3(x + 2) > 9x - 2$
       $4(x + 5) > 3(x + 6)$

33. Determine the solution set of $2 - 3(x + 4) < 5 - 3x$.

34. Determine the solution set of $3x + 2(x - 1) > 5(x + 1)$.

Copyright © Houghton Mifflin Company. All rights reserved.

9.4 Graphing Linear Inequalities

Objective A To graph an inequality in two variables

Point of Interest

Linear inequalities play an important role in applied mathematics. They are used in a branch of mathematics called *linear programming*, which was developed during World War II to solve problems in supplying the Air Force with the machine parts necessary to keep planes flying. Today, its applications extend to many other disciplines.

The graph of the linear equation $y = x - 2$ separates a plane into three sets:

The set of points on the line
The set of points above the line
The set of points below the line

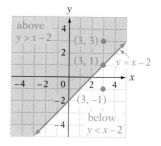

The point $(3, 1)$ is a solution of $y = x - 2$.

$$\frac{y = x - 2}{1 \mid 3 - 2}$$
$$1 = 1$$

The point $(3, 3)$ is a solution of $y > x - 2$.

$$\frac{y > x - 2}{3 \mid 3 - 2}$$
$$3 > 1$$

Any point above the line is a solution of $y > x - 2$.

Study Tip

Be sure to do all you need to do in order to be successful at graphing linear inequalities: Read through the introductory material, work through the How To examples, study the paired examples, do the You Try Its, and check your solutions against those in the back of the book. See *AIM for Success*, page xxv.

The point $(3, -1)$ is a solution of $y < x - 2$.

$$\frac{y < x - 2}{-1 \mid 3 - 2}$$
$$-1 < 1$$

Any point below the line is a solution of $y < x - 2$.

The solution set of $y = x - 2$ is all points on the line. The solution set of $y > x - 2$ is all points above the line. The solution set of $y < x - 2$ is all points below the line. The solution set of an inequality in two variables is a **half-plane.**

The following illustrates the procedure for graphing a linear inequality.

HOW TO Graph the solution set of $2x + 3y \leq 6$.

Solve the inequality for y.

$$2x + 3y \leq 6$$
$$2x - 2x + 3y \leq -2x + 6$$ • **Subtract 2x** from each side.
$$3y \leq -2x + 6$$ • Simplify.
$$\frac{3y}{3} \leq \frac{-2x + 6}{3}$$ • Divide each side by **3**.
$$y \leq -\frac{2}{3}x + 2$$ • Simplify.

Change the inequality to an equality and graph $y = -\frac{2}{3}x + 2$. If the inequality is $\geq$ or $\leq$, the line is in the solution set and is shown by a solid line. If the inequality is $>$ or $<$, the line is not a part of the solution set and is shown by a dotted line.

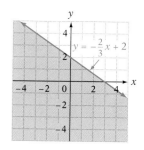

If the inequality is $>$ or $\geq$, shade the upper half-plane. If the inequality is $<$ or $\leq$, shade the lower half-plane.

Copyright © Houghton Mifflin Company. All rights reserved.

Example 1

Graph the solution set of $3x + y > -2$.

Solution

$$3x + y > -2$$
$$3x - 3x + y > -3x - 2 \quad \bullet \text{ Subtract } 3x.$$
$$y > -3x - 2$$

Graph $y = -3x - 2$ as a dotted line.
Shade the upper half-plane.

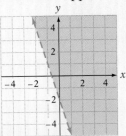

You Try It 1

Graph the solution set of $x - 3y < 2$.

Your solution

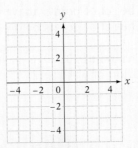

Example 2

Graph the solution set of $2x - y \geq 2$.

Solution

$$2x - y \geq 2$$
$$2x - 2x - y \geq -2x + 2 \quad \bullet \text{ Subtract } 2x.$$
$$-y \geq -2x + 2$$
$$-1(-y) \leq -1(-2x + 2) \quad \bullet \text{ Multiply by } -1.$$
$$y \leq 2x - 2$$

Graph $y = 2x - 2$ as a solid line.
Shade the lower half-plane.

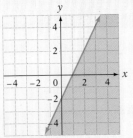

You Try It 2

Graph the solution set of $2x - 4y \leq 8$.

Your solution

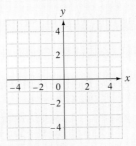

Example 3

Graph the solution set of $y > -1$.

Solution

Graph $y = -1$ as a dotted line.
Shade the upper half-plane.

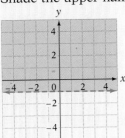

You Try It 3

Graph the solution set of $x < 3$.

Your solution

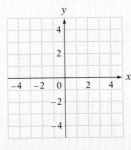

Solutions on pp. S24–S25

Copyright © Houghton Mifflin Company. All rights reserved.

9.4 Exercises

Objective A   **To graph an inequality in two variables**

For Exercises 1 to 18, graph the solution set of the inequality.

1. $x + y > 4$

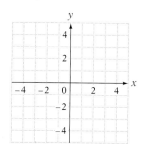

2. $x - y > -3$

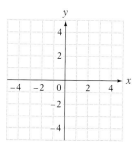

3. $2x - y < -3$

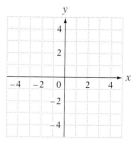

4. $3x - y < 9$

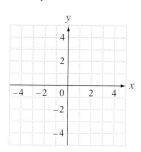

5. $2x + y \geq 4$

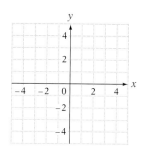

6. $3x + y \geq 6$

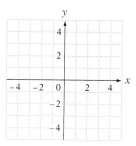

7. $y \leq -2$

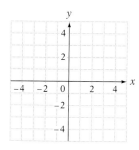

8. $y > 3$

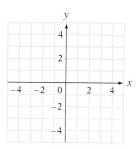

9. $3x - 2y < 8$

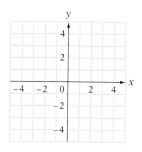

10. $5x + 4y > 4$

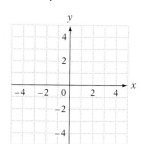

11. $-3x - 4y \geq 4$

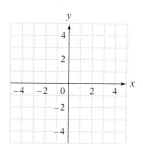

12. $-5x - 2y \geq 8$

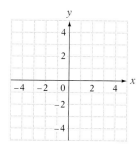

Copyright © Houghton Mifflin Company. All rights reserved.

13. $6x + 5y \le -10$

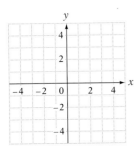

14. $2x + 2y \le -4$

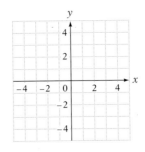

15. $-4x + 3y < -12$

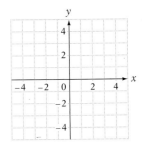

16. $-4x + 5y < 15$

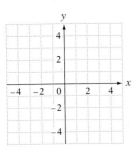

17. $-2x + 3y \le 6$

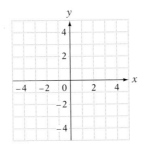

18. $3x - 4y > 12$

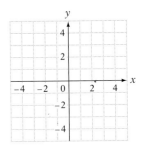

APPLYING THE CONCEPTS

For Exercises 19 to 21, graph the solution set of the inequality.

19. $\dfrac{x}{4} + \dfrac{y}{2} > 1$

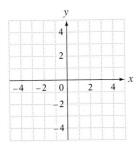

20. $2x - 3(y + 1) > y - (4 - x)$

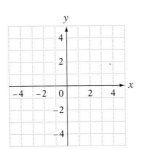

21. $4y - 2(x + 1) \ge 3(y - 1) + 3$

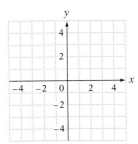

For Exercises 22 to 24, write the inequality given its graph.

22.

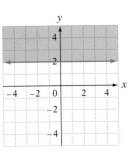

23.

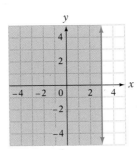

24.

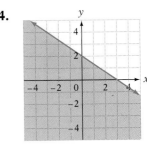

Copyright © Houghton Mifflin Company. All rights reserved.

Focus on Problem Solving

Graphing Data Graphs are very useful in displaying data. By studying a graph, we can reach various conclusions about the data.

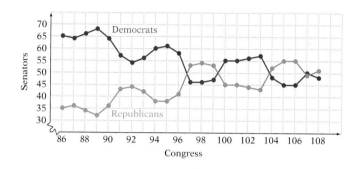

The double-line graph at the left shows the number of Democrats and the number of Republicans in the U.S. Senate for the 86th Congress (1959–1961) through the 108th Congress (2003–2005).

1. How many Democratic and how many Republican senators were in the 90th Congress?

2. In which Congress was the difference between the numbers of Democrats and Republicans the greatest?

3. In which Congress did the majority first change from Democratic to Republican?

4. Between which two Congresses did the number of Republican senators increase but the number of Democratic senators remain the same?

5. In what percent of the Congresses did the number of Democrats exceed the number of Republicans? Round to the nearest tenth.

6. In which Congresses were there a greater number of Republican senators than Democratic senators?

Year	DVD Recorders	DVD Players
2003	1	59
2004	2	62
2005	9	60
2006	18	54
2007	28	44
2008	44	34

The table at the left, based on data from Allied Business Intelligence, Inc., shows the actual and projected number (in millions) of worldwide shipments of DVD players and DVD recorders for the years 2003 through 2008.

7. Make a double-line graph of the data.

8. In which year will the number of DVD recorders shipped first exceed the number of DVD players shipped?

9. Based on this data, for the year 2009, would you expect the difference between the number of DVD recorders shipped and the number of DVD players shipped to be less than or greater than the same number in 2008?

Projects and Group Activities

Mean and Standard Deviation An automotive engineer tests the miles-per-gallon ratings of 15 cars and records the results as follows:

$$25 \quad 22 \quad 21 \quad 27 \quad 25 \quad 35 \quad 29 \quad 31 \quad 25 \quad 26 \quad 21 \quad 39 \quad 34 \quad 32 \quad 28$$

The **mean** of the data is the sum of the measurements divided by the number of measurements. The symbol for the mean is $\bar{x}$.

$$\text{Mean} = \bar{x} = \frac{\text{sum of all data values}}{\text{number of data values}}$$

Copyright © Houghton Mifflin Company. All rights reserved.

To find the mean for the data on p. 469, add the numbers and then divide by 15.

$$\bar{x} = \frac{25 + 22 + 21 + 27 + 25 + 35 + 29 + 31 + 25 + 26 + 21 + 39 + 34 + 32 + 28}{15}$$

$$= \frac{420}{15} = 28$$

The mean number of miles per gallon for the 15 cars tested was 28 mi/gal.

The mean is one of the most frequently computed averages. It is the one that is commonly used to calculate a student's performance in a class.

The scores for a history student on five tests were 78, 82, 91, 87, and 93. What was the mean score for this student?

To find the mean, add the numbers. Then divide by 5.

$$\bar{x} = \frac{78 + 82 + 91 + 87 + 93}{5}$$

$$= \frac{431}{5} = 86.2$$

The mean score for the history student was 86.2.

Consider two students, each of whom has taken five exams.

Scores for student A

84	86	83	85	87

Scores for student B

90	75	94	68	98

$$\bar{x} = \frac{84 + 86 + 83 + 85 + 87}{5} = \frac{425}{5} = 85$$

$$\bar{x} = \frac{90 + 75 + 94 + 68 + 98}{5} = \frac{425}{5} = 85$$

The mean for student A is 85.

The mean for student B is 85.

For each of these students, the mean (average) for the five exams is 85. However, student A has a more consistent record of scores than student B. One way to measure the consistency, or "clustering" near the mean, of data is to use the **standard deviation.**

To calculate the standard deviation:

Step 1. Sum the squares of the differences between each value of the data and the mean.

Step 2. Divide the result in Step 1 by the number of items in the set of data.

Step 3. Take the square root of the result in Step 2.

The calculation for student A is shown at the right.

Step 1:

x	$x - \bar{x}$	$(x - \bar{x})^2$
84	84 − 85	$(-1)^2 = 1$
86	86 − 85	$1^2 = 1$
83	83 − 85	$(-2)^2 = 4$
85	85 − 85	$0^2 = 0$
87	87 − 85	$2^2 = 4$
		Total = 10

The symbol for standard deviation is the lowercase Greek letter *sigma*, σ.

Step 2: $\frac{10}{5} = 2$

Step 3: $\sigma = \sqrt{2} \approx 1.414$

The standard deviation for student A's scores is approximately 1.414.

Copyright © Houghton Mifflin Company. All rights reserved.

Following a similar procedure for student B shows that the standard deviation for student B's scores is approximately 11.524. Because the standard deviation of student B's scores is greater than that of student A's (11.524 > 1.414), student B's scores are not as consistent as those of student A.

1. The weights in ounces of 6 newborn infants were recorded by a hospital. The weights were 96, 105, 84, 90, 102, and 99. Find the standard deviation of the weights. Round to the nearest hundredth.

2. The numbers of rooms occupied in a hotel on 6 consecutive days were 234, 321, 222, 246, 312, and 396. Find the standard deviation for the number of rooms occupied. Round to the nearest hundredth.

3. Seven coins were tossed 100 times. The numbers of heads recorded for each coin were 56, 63, 49, 50, 48, 53, and 52. Find the standard deviation of the number of heads. Round to the nearest hundredth.

4. The temperatures, in degrees Fahrenheit, for 11 consecutive days at a desert resort were 95°, 98°, 98°, 104°, 97°, 100°, 96°, 97°, 108°, 93°, and 104°. For the same days, temperatures in Antarctica were 27°, 28°, 28°, 30°, 28°, 27°, 30°, 25°, 24°, 26°, and 21°. Which location has the greater standard deviation of temperatures?

5. The scores for 5 college basketball games were 56, 68, 60, 72, and 64. The scores for 5 professional basketball games were 106, 118, 110, 122, and 114. Which scores have the greater standard deviation?

6. The weights in pounds of the 5-man front line of a college football team are 210, 245, 220, 230, and 225. Find the standard deviation of the weights. Round to the nearest hundredth.

7. One student received test scores of 85, 92, 86, and 89. A second student received scores of 90, 97, 91, and 94 (exactly 5 points more on each test). Are the means of the two students the same? If not, what is the relationship between the means of the two students? Are the standard deviations of the scores of the two students the same? If not, what is the relationship between the standard deviations of the scores of the two students?

8. Grade-point average (GPA) is a *weighted* mean. It is called a weighted mean because a grade in a 5-unit course has more influence on your GPA than a grade in a 2-unit course. GPA is calculated by multiplying the numerical equivalent of each grade by the number of units, adding those products, and then dividing by the total number of units. Calculate your GPA for the last quarter or semester.

9. If you average 40 mph for 1 h and then 50 mph for 1 h, is your average speed $\frac{40 + 50}{2}$ = 45 mph? Why or why not?

10. A company is negotiating with its employees the terms of a raise in salary. One proposal would add $500 a year to each employee's salary. The second proposal would give each employee a 4% raise. Explain how each of these proposals would affect the current mean and standard deviation of salaries for the company.

Copyright © Houghton Mifflin Company. All rights reserved.

Chapter 9 Summary

Key Words	**Examples**
The *empty set* or *null set,* written $\varnothing$, is the set that contains no elements. [9.1A, p. 447]	The set of cars that can travel faster than 1000 mph is an empty set.
The *union* of two sets, written $A \cup B$, is the set that contains the elements of A and the elements of B. [9.1A, p. 447]	Let $A = \{2, 4, 6, 8\}$ and $B = \{0, 1, 2, 3, 4\}$. Then $A \cup B = \{0, 1, 2, 3, 4, 6, 8\}$.
The *intersection* of two sets, written $A \cap B$, is the set that contains the elements that are common to both A and B. [9.1A, p. 447]	Let $A = \{2, 4, 6, 8\}$ and $B = \{0, 1, 2, 3, 4\}$. Then $A \cap B = \{2, 4\}$.
Set-builder notation uses a rule to describe the elements of a set. [9.1B, p. 448]	Using set-builder notation, the set of real numbers greater than 2 is written $\{x \mid x > 2, x \in \text{real numbers}\}$.
An *inequality* is an expression that contains the symbol $<$, $>$, $\le$, or $\ge$. The *solution set of an inequality* is a set of numbers each element of which, when substituted for the variable, results in a true inequality. The solution set of an inequality can be graphed on a number line. [9.1C, p. 449]	$3x - 1 < 5$ is an inequality. The solution set of $3x - 1 < 5$ is $\{x \mid x < 2\}$. The graph of the solution set is ... $-5\ -4\ -3\ -2\ -1\ \ 0\ \ 1\ \ 2\ \ 3\ \ 4\ \ 5$.
The solution set of a linear inequality in two variables is a *half-plane.* [9.4A, p. 465]	The solution set of $3x + 4y \ge 12$ is the half-plane shown at the right. 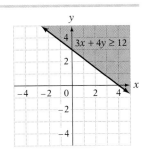

Essential Rules and Procedures	**Examples**
Addition Property of Inequalities [9.2A, p. 453] The same term can be added to each side of an inequality without changing the solution set of the inequality. If $a > b$, then $a + c > b + c$. If $a < b$, then $a + c < b + c$.	$x - 3 < -7$ $x - 3 + 3 < -7 + 3$ $x < -4$
Multiplication Property of Inequalities [9.2B, p. 455] Each side of an inequality can be multiplied by the same positive number without changing the solution set of the inequality. If $a > b$ and $c > 0$, then $ac > bc$. If $a < b$ and $c > 0$, then $ac < bc$.	$4x > -8$ $\dfrac{4x}{4} > \dfrac{-8}{4}$ $x > -2$
If each side of an inequality is multiplied by the same negative number and the inequality symbol is reversed, then the solution set of the inequality is not changed. If $a > b$ and $c < 0$, then $ac < bc$. If $a < b$ and $c < 0$, then $ac > bc$.	$-2x < 6$ $\dfrac{-2x}{-2} > \dfrac{6}{-2}$ $x > -3$

Copyright © Houghton Mifflin Company. All rights reserved.

Chapter 9 Review Exercises

1. Solve: $2x - 3 > x + 15$

2. Find $A \cap B$, given $A = \{0, 2, 4, 6, 8\}$ and $B = \{-2, -4\}$.

3. Use set-builder notation to write the set of odd integers greater than -8.

4. Find $A \cup B$, given $A = \{6, 8, 10\}$ and $B = \{2, 4, 6\}$.

5. Use the roster method to write the set of odd positive integers less than 8.

6. Solve: $12 - 4(x - 1) \leq 5(x - 4)$

7. Graph: $\{x \mid x > 3\}$

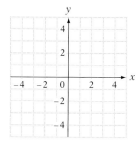

8. Solve: $3x + 4 \geq -8$

9. Graph: $3x + 2y \leq 12$

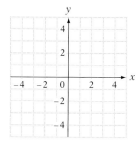

10. Graph: $5x + 2y < 6$

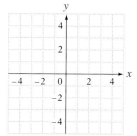

11. Use set-builder notation to write the set of real numbers greater than 3.

12. Solve and graph the solution set of $x - 3 > -1$.

13. Find $A \cap B$, given $A = \{1, 5, 9, 13\}$ and $B = \{1, 3, 5, 7, 9\}$.

14. Graph: $\{x \mid x < 2\} \cup \{x \mid x > 5\}$

Copyright © Houghton Mifflin Company. All rights reserved.

15. Graph: $\{x|x > -1\} \cap \{x|x \leq 2\}$

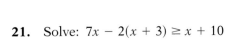

16. Solve: $-15x \leq 45$

17. Solve: $6x - 9 < 4x + 3(x + 3)$

18. Solve: $5 - 4(x + 9) > 11(12x - 9)$

19. Solve: $-\dfrac{3}{4}x > \dfrac{2}{3}$

20. Graph: $2x - 3y < 9$

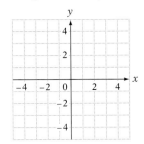

21. Solve: $7x - 2(x + 3) \geq x + 10$

22. **Floral Delivery**   Florist A charges a $3 delivery fee plus $21 per bouquet delivered. Florist B charges a $15 delivery fee plus $18 per bouquet delivered. A church wants to supply each resident of a small nursing home with a bouquet for Grandparents Day. Find the number of residents of the nursing home if using florist B is more economical than using florist A.

23. **Landscaping**   The width of a rectangular garden is 12 ft. The length of the garden is $(3x + 5)$ ft. Express as an integer the minimum length of the garden when the area is greater than 276 ft². (The area of a rectangle is equal to its length times its width.)

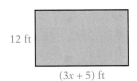

24. **Number Sense**   Six less than a number is greater than twenty-five. Find the smallest integer that will satisfy the inequality.

25. **Grading**   A student's grades on five sociology tests were 68, 82, 90, 73, and 95. What is the lowest score the student can receive on the next test and still be able to attain a minimum of 480 points?

Copyright © Houghton Mifflin Company. All rights reserved.

Chapter 9 Test

1. Graph: $\{x|x < 5\} \cap \{x|x > 0\}$

−5 −4 −3 −2 −1 0 1 2 3 4 5

2. Use set-builder notation to write the set of the positive integers less than 50.

3. Use the roster method to write the set of the even positive integers between 3 and 9.

4. Solve: $3(2x − 5) \geq 8x − 9$

5. Solve: $x + \dfrac{1}{2} > \dfrac{5}{8}$

6. Graph: $\{x\,|\,x > −2\}$

−5 −4 −3 −2 −1 0 1 2 3 4 5

7. Solve: $5 − 3x > 8$

8. Use set-builder notation to write the set of the real numbers greater than $−23$.

9. Graph the solution set of $3x + y > 4$.

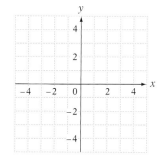

10. Graph the solution set of $4x − 5y \geq 15$.

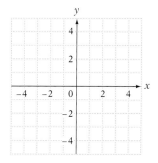

11. Find $A \cap B$, given $A = \{6, 8, 10, 12\}$ and $B = \{12, 14, 16\}$.

12. Solve and graph the solution set of $4 + x < 1$.

−5 −4 −3 −2 −1 0 1 2 3 4 5

13. Solve: $−\dfrac{3}{8}x \leq 5$

14. Solve: $6x − 3(2 − 3x) < 4(2x − 7)$

Copyright © Houghton Mifflin Company. All rights reserved.

15. Solve and graph the solution set of $\frac{2}{3}x \geq 2$.

```
←+—+—+—+—+—+—+—+—+—+—+—+→
  -5 -4 -3 -2 -1  0  1  2  3  4  5
```

16. Solve: $2x - 7 \leq 6x + 9$

17. **Safety**   To ride a certain roller coaster at an amusement park, a person must be at least 48 in. tall. How many inches must a child who is 43 in. tall grow to be eligible to ride the roller coaster?

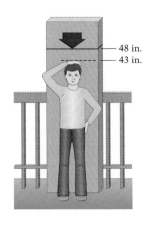

48 in.
43 in.

18. **Geometry**   A rectangle is 15 ft long and $(2x - 4)$ ft wide. Express as an integer the maximum width of the rectangle if the area is less than 180 ft². (The area of a rectangle is equal to its length times its width.)

$(2x - 4)$ ft

15 ft

19. **Machining**   A ball bearing for a rotary engine must have a circumference between 0.1220 in. and 0.1240 in. What are the allowable diameters for the bearing? Round to the nearest ten-thousandth. Recall that $C = \pi d$.

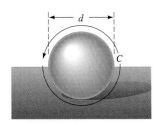

d

C

20. **Wages**   A stockbroker receives a monthly salary that is the greater of $2500 or $1000 plus 2% of the total value of all stock transactions the broker processes during the month. What dollar amounts of transactions did the broker process in a month for which the broker's salary was $2500?

Copyright © Houghton Mifflin Company. All rights reserved.

Cumulative Review Exercises

1. Simplify: $2[5a - 3(2 - 5a) - 8]$

2. Solve: $\dfrac{5}{8} - 4x = \dfrac{1}{8}$

3. Solve: $2x - 3[x - 2(x - 3)] = 2$

4. Simplify: $(-3a)(-2a^3b^2)^2$

5. Simplify: $\dfrac{27a^3b^2}{(-3ab^2)^3}$

6. Divide: $(16x^2 - 12x - 2) \div (4x - 1)$

7. Given $f(x) = x^2 - 4x - 5$, find $f(-1)$.

8. Factor: $27a^2x^2 - 3a^2$

9. Divide: $\dfrac{x^2 - 2x}{x^2 - 2x - 8} \div \dfrac{x^3 - 5x^2 + 6x}{x^2 - 7x + 12}$

10. Subtract: $\dfrac{4a}{2a - 3} - \dfrac{2a}{a + 3}$

11. Solve: $\dfrac{5y}{6} - \dfrac{5}{9} = \dfrac{y}{3} - \dfrac{5}{6}$

12. Solve $R = \dfrac{C - S}{t}$ for C.

13. Find the slope of the line that passes through the points $(2, -3)$ and $(-1, 4)$.

14. Find the equation of the line that passes through the point $(1, -3)$ and has slope $-\dfrac{3}{2}$.

15. Solve by substitution.
$$x = 3y + 1$$
$$2x + 5y = 13$$

16. Solve by the addition method.
$$9x - 2y = 17$$
$$5x + 3y = -7$$

Copyright © Houghton Mifflin Company. All rights reserved.

17. Find $A \cup B$, given $A = \{0, 1, 2\}$ and $B = \{-10, -2\}$.

18. Use set-builder notation to write the set of the real numbers less than 48.

19. Graph: $\{x | x > 1\} \cup \{x | x < -1\}$

<!-- number line -->
$$\xleftarrow{\quad|\quad|\quad|\quad|\quad|\quad|\quad|\quad|\quad|\quad|\quad|\quad}\rightarrow$$
-5 -4 -3 -2 -1 0 1 2 3 4 5

20. Graph the solution set of $\frac{3}{8}x > -\frac{3}{4}$.

<!-- number line -->
$$\xleftarrow{\quad|\quad|\quad|\quad|\quad|\quad|\quad|\quad|\quad|\quad|\quad|\quad}\rightarrow$$
-5 -4 -3 -2 -1 0 1 2 3 4 5

21. Solve: $-\frac{4}{5}x > 12$

22. Solve: $15 - 3(5x - 7) < 2(7 - 2x)$

23. **Number Sense** Three-fifths of a number is less than negative fifteen. What integers satisfy this inequality? Write the answer in set-builder notation.

24. **Rental Agencies** Company A rents cars for $6 a day and $.25 for every mile driven. Company B rents cars for $15 a day and $.10 per mile. You want to rent a car for 6 days. What is the maximum number of miles you can drive a Company A car if it is to cost you less than a Company B car?

25. **Conservation** In a lake, 100 fish are caught, tagged, and then released. Later, 150 fish are caught. Three of these 150 fish are found to have tags. Estimate the number of fish in the lake.

26. **Geometry** The measure of the first angle of a triangle is 30° more than the measure of the second angle. The measure of the third angle is 10° more than twice the measure of the second angle. Find the measure of each angle.

27. Graph: $y = 2x - 1$

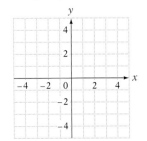

28. Graph the solution set of $6x - 3y \geq 6$.

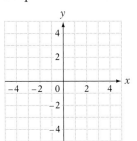

Copyright © Houghton Mifflin Company. All rights reserved.

Copyright © Houghton Mifflin Company. All rights reserved.

chapter

10 Radical Expressions

Think back to some of the carnival rides you enjoyed as a child. Was your favorite ride the Ferris wheel, or the tilt-a-whirl? Was it the scrambler, or the carousel? All of these rides involve mathematics and physics. **Exercise 29 on page 502** uses a formula to describe the speed of a child riding a merry-go-round. The formula, which uses a radical expression, is $v = \sqrt{12r}$, where v is the speed in feet per second and r is the distance in feet from the center of the merry-go-round to the rider. You will learn more about radical expressions in this chapter.

OBJECTIVES

Section 10.1

A To simplify numerical radical expressions
B To simplify variable radical expressions

Section 10.2

A To add and subtract radical expressions

Section 10.3

A To multiply radical expressions
B To divide radical expressions

Section 10.4

A To solve an equation containing a radical expression
B To solve application problems

Need help? For online student resources, such as section quizzes, visit this textbook's website at **math.college.hmco.com/students**.

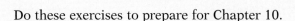

PREP TEST ● ● ●

Do these exercises to prepare for Chapter 10.

1. Evaluate: $-|-14|$

2. Simplify: $3x^2y - 4xy^2 - 5x^2y$

3. Solve: $1.5h = 21$

4. Solve: $3x - 2 = 5 - 2x$

5. Simplify: $x^3 \cdot x^3$

6. Expand: $(x + y)^2$

7. Expand: $(2x - 3)^2$

8. Multiply: $(2 - 3v)(2 + 3v)$

9. Multiply: $(a - 5)(a + 5)$

10. Simplify: $\dfrac{2x^4y^3}{18x^2y}$

GO FIGURE ● ● ●

Two pieces of lumber are used to support two walls as shown in the figure. How high above the ground do the pieces of lumber cross? (Hint: Triangles *ABC* and *EBD* are similar triangles. Therefore, the ratio of the corresponding heights equals the ratio of corresponding sides. That is, $\dfrac{y}{8 - y} = \dfrac{10}{12}$. Solve for *y* and use this value to write another equation based on different similar triangles.)

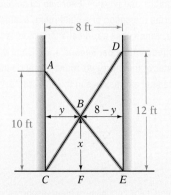

Copyright © Houghton Mifflin Company. All rights reserved.

Copyright © Houghton Mifflin Company. All rights reserved.

chapter 10

Radical Expressions

Think back to some of the carnival rides you enjoyed as a child. Was your favorite ride the Ferris wheel, or the tilt-a-whirl? Was it the scrambler, or the carousel? All of these rides involve mathematics and physics. **Exercise 29 on page 502** uses a formula to describe the speed of a child riding a merry-go-round. The formula, which uses a radical expression, is $v = \sqrt{12r}$, where v is the speed in feet per second and r is the distance in feet from the center of the merry-go-round to the rider. You will learn more about radical expressions in this chapter.

Need help? For online student resources, such as section quizzes, visit this textbook's website at **math.college.hmco.com/students**.

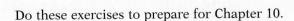

Do these exercises to prepare for Chapter 10.

1. Evaluate: $-|-14|$

2. Simplify: $3x^2y - 4xy^2 - 5x^2y$

3. Solve: $1.5h = 21$

4. Solve: $3x - 2 = 5 - 2x$

5. Simplify: $x^3 \cdot x^3$

6. Expand: $(x + y)^2$

7. Expand: $(2x - 3)^2$

8. Multiply: $(2 - 3v)(2 + 3v)$

9. Multiply: $(a - 5)(a + 5)$

10. Simplify: $\dfrac{2x^4y^3}{18x^2y}$

GO FIGURE • • •

Two pieces of lumber are used to support two walls as shown in the figure. How high above the ground do the pieces of lumber cross? (Hint: Triangles ABC and EBD are similar triangles. Therefore, the ratio of the corresponding heights equals the ratio of corresponding sides. That is, $\dfrac{y}{8 - y} = \dfrac{10}{12}$. Solve for y and use this value to write another equation based on different similar triangles.)

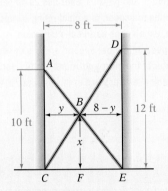

Copyright © Houghton Mifflin Company. All rights reserved.

10.1 Introduction to Radical Expressions

Objective A **To simplify numerical radical expressions**

Point of Interest

The radical symbol was first used in 1525 but was written as $\sqrt{\ }$. Some historians suggest that the radical symbol also developed into the symbols for "less than" and "greater than." Because typesetters of that time did not want to make additional symbols, the radical was rotated to the position $>$ and used as a "greater than" symbol and rotated to $<$ and used for the "less than" symbol. Other evidence, however, suggests that the "less than" and "greater than" symbols were developed independently of the radical symbol.

A **square root** of a positive number a is a number whose square is a.

A square root of 16 is 4 because $4^2 = 16$.
A square root of 16 is -4 because $(-4)^2 = 16$.

Every positive number has two square roots, one a positive and one a negative number. The symbol $\sqrt{\ }$, called a **radical sign,** is used to indicate the positive or **principal square root** of a number. For example, $\sqrt{16} = 4$ and $\sqrt{25} = 5$. The number under the radical sign is called the **radicand.**

When the negative square root of a number is to be found, a negative sign is placed in front of the radical. For example, $-\sqrt{16} = -4$ and $-\sqrt{25} = -5$.

The square of an integer is a **perfect square.** For instance, 49, 81, and 144 are perfect squares. The principal square root of a perfect-square integer is a positive integer.

$$7^2 = 49 \qquad \sqrt{49} = 7$$
$$9^2 = 81 \qquad \sqrt{81} = 9$$
$$12^2 = 144 \qquad \sqrt{144} = 12$$

If a number is not a perfect square, its square root can only be approximated. For example, 2 and 7 are not perfect squares. The square roots of these numbers are *irrational numbers.* Their decimal approximations never terminate or repeat.

$$\sqrt{2} \approx 1.4142135... \qquad \sqrt{7} \approx 2.6457513...$$

TAKE NOTE

Recall that a factor of a number divides the number evenly. For instance, 6 is a factor of 18. The perfect square 9 is also a factor of 18. It is a *perfect-square factor* of 18, whereas 6 is not a perfect-square factor of 18.

Radical expressions that contain radicands that are not perfect squares are frequently written in simplest form. A radical expression is in *simplest form* when the radicand contains no factor greater than 1 that is a perfect square. For instance, $\sqrt{50}$ is not in simplest form because 25 is a perfect-square factor of 50. The radical expression $\sqrt{15}$ is in simplest form because there are no perfect-square factors of 15 that are greater than 1.

The Product Property of Square Roots and a knowledge of perfect squares is used to simplify radicands that are not perfect squares.

> **The Product Property of Square Roots**
>
> If a and b are positive real numbers, then $\sqrt{ab} = \sqrt{a} \cdot \sqrt{b}$.

The chart below shows the square roots of some perfect squares.

Square Roots of Perfect Squares

$\sqrt{1} = 1$	$\sqrt{16} = 4$	$\sqrt{49} = 7$	$\sqrt{100} = 10$
$\sqrt{4} = 2$	$\sqrt{25} = 5$	$\sqrt{64} = 8$	$\sqrt{121} = 11$
$\sqrt{9} = 3$	$\sqrt{36} = 6$	$\sqrt{81} = 9$	$\sqrt{144} = 12$

HOW TO Simplify: $\sqrt{72}$

$$\sqrt{72} = \sqrt{36 \cdot 2}$$

- Write the radicand as the product of **a perfect square** and a factor that does not contain a perfect square.

$$= \sqrt{36}\sqrt{2}$$

- Use the Product Property of Square Roots to write the expression as a product.

$$= 6\sqrt{2}$$

- Simplify.

Copyright © Houghton Mifflin Company. All rights reserved.

Note that 72 must be written as the product of a perfect square and a *factor that does not contain a perfect square*. Therefore, it would not be correct to simplify $\sqrt{72}$ as $\sqrt{9 \cdot 8}$. Although 9 is a perfect-square factor of 72, 8 also contains a perfect-square factor $(8 = 4 \cdot 2)$. Therefore, $\sqrt{8}$ is not in simplest form. Remember to find the largest perfect-square factor of the radicand.

$$\sqrt{72} = \sqrt{9 \cdot 8}$$
$$= \sqrt{9} \cdot \sqrt{8}$$
$$= 3\sqrt{8}$$

Not in simplest form

HOW TO   Simplify: $\sqrt{147}$

$$\sqrt{147} = \sqrt{49 \cdot 3}$$ • Write the radicand as the product of **a perfect square** and a factor that does not contain a perfect square.

$$= \sqrt{49}\sqrt{3}$$ • Use the Product Property of Square Roots to write the expression as a product.

$$= 7\sqrt{3}$$ • Simplify.

HOW TO   Simplify: $\sqrt{360}$

$$\sqrt{360} = \sqrt{36 \cdot 10}$$ • Write the radicand as the product of **a perfect square** and a factor that does not contain a perfect square.

$$= \sqrt{36}\sqrt{10}$$ • Use the Product Property of Square Roots to write the expression as a product.

$$= 6\sqrt{10}$$ • Simplify.

From the last example, note that $\sqrt{360} = 6\sqrt{10}$. The two expressions are different representations of the same number. Using a calculator, we find that $\sqrt{360} \approx 18.973666$ and $6\sqrt{10} \approx 6(3.1622777) = 18.9736662$.

HOW TO   Simplify: $\sqrt{-16}$

Because the square of any real number is positive, there is no real number whose square is -16. $\sqrt{-16}$ is not a real number.

Example 1   Simplify: $3\sqrt{90}$

Solution
$3\sqrt{90} = 3\sqrt{9 \cdot 10}$ • **9 is a perfect-square factor.**
$\quad\quad = 3\sqrt{9}\sqrt{10}$ • **Product Property of Square Roots**
$\quad\quad = 3 \cdot 3\sqrt{10}$
$\quad\quad = 9\sqrt{10}$

You Try It 1   Simplify: $-5\sqrt{32}$

Your solution

Example 2   Simplify: $\sqrt{252}$

Solution
$\sqrt{252} = \sqrt{36 \cdot 7}$ • **36 is a perfect-square factor.**
$\quad\quad = \sqrt{36}\sqrt{7}$ • **Product Property of Square Roots**
$\quad\quad = 6\sqrt{7}$

You Try It 2   Simplify: $\sqrt{216}$

Your solution

Solutions on p. S25

Copyright © Houghton Mifflin Company. All rights reserved.

Copyright © Houghton Mifflin Company. All rights reserved.

Objective B **To simplify variable radical expressions**

Variable expressions that contain radicals do not always represent real numbers. For example, if $a = -4$, then

$$\sqrt{a^3} = \sqrt{(-4)^3} = \sqrt{-64}$$

and $\sqrt{-64}$ is not a real number.

Now consider the expression $\sqrt{x^2}$. Evaluate this expression for $x = -2$ and $x = 2$.

$$\sqrt{x^2}$$
$$\sqrt{(-2)^2} = \sqrt{4} = 2 = |-2|$$

$$\sqrt{x^2}$$
$$\sqrt{2^2} = \sqrt{4} = 2 = |2|$$

This suggests the following:

For any real number a, $\sqrt{a^2} = |a|$. If $a \geq 0$, then $\sqrt{a^2} = a$.

In order to avoid variable expressions that do not represent real numbers, and so that absolute value signs are not needed for certain expressions, the variables in this chapter will represent *positive* numbers unless otherwise stated.

A variable or a product of variables written in exponential form is a perfect square when each exponent is an even number.

To find the square root of a perfect square, remove the radical sign and multiply each exponent by $\frac{1}{2}$.

HOW TO Simplify: $\sqrt{a^6}$

$$\sqrt{a^6} = a^3$$ • Remove the radical sign and multiply the exponent by $\frac{1}{2}$.

A variable radical expression is in simplest form when the radicand contains no factor greater than 1 that is a perfect square.

HOW TO Simplify: $\sqrt{x^7}$

$$\sqrt{x^7} = \sqrt{x^6 \cdot x}$$ • Write x^7 as the product of **a perfect square** and x.
$$= \sqrt{x^6}\sqrt{x}$$ • Use the Product Property of Square Roots.
$$= x^3\sqrt{x}$$ • Simplify the perfect square.

HOW TO Simplify: $3x\sqrt{8x^3y^{13}}$

$$3x\sqrt{8x^3y^{13}} = 3x\sqrt{4x^2y^{12}(2xy)}$$ • Write the radicand as the product of **perfect squares** and factors that do not contain a perfect square.

$$= 3x\sqrt{4x^2y^{12}}\sqrt{2xy}$$ • Use the Product Property of Square Roots.
$$= 3x \cdot 2xy^6\sqrt{2xy}$$ • Simplify.
$$= 6x^2y^6\sqrt{2xy}$$

HOW TO Simplify: $\sqrt{25(x + 2)^2}$

$$\sqrt{25(x + 2)^2} = 5(x + 2)$$
$$= 5x + 10$$

Example 3

Simplify: $\sqrt{b^{15}}$

Solution
$\sqrt{b^{15}} = \sqrt{b^{14} \cdot b}$ • b^{14} is a perfect square.
$= \sqrt{b^{14}} \cdot \sqrt{b} = b^7\sqrt{b}$

You Try It 3

Simplify: $\sqrt{y^{19}}$

Your solution

Example 4

Simplify: $\sqrt{24x^5}$

Solution
$\sqrt{24x^5} = \sqrt{4x^4(6x)}$ • **4** and x^4 are perfect squares.
$= \sqrt{4x^4}\sqrt{6x}$
$= 2x^2\sqrt{6x}$

You Try It 4

Simplify: $\sqrt{45b^7}$

Your solution

Example 5

Simplify: $2a\sqrt{18a^3b^{10}}$

Solution
$2a\sqrt{18a^3b^{10}}$
$= 2a\sqrt{9a^2b^{10}(2a)}$ • **9**, a^2, and b^{10} are perfect squares.
$= 2a\sqrt{9a^2b^{10}}\sqrt{2a}$
$= 2a \cdot 3ab^5\sqrt{2a}$
$= 6a^2b^5\sqrt{2a}$

You Try It 5

Simplify: $3a\sqrt{28a^9b^{18}}$

Your solution

Example 6

Simplify: $\sqrt{16(x + 5)^2}$

Solution
$\sqrt{16(x + 5)^2} = 4(x + 5) = 4x + 20$

You Try It 6

Simplify: $\sqrt{25(a + 3)^2}$

Your solution

Example 7

Simplify: $\sqrt{x^2 + 10x + 25}$

Solution
$\sqrt{x^2 + 10x + 25} = \sqrt{(x + 5)^2} = x + 5$

You Try It 7

Simplify: $\sqrt{x^2 + 14x + 49}$

Your solution

Solutions on p. S25

Copyright © Houghton Mifflin Company. All rights reserved.

10.1 Exercises

Copyright © Houghton Mifflin Company. All rights reserved.

Objective A   **To simplify numerical radical expressions**

1.   Describe in your own words how to simplify a radical expression.

2.   Explain why $2\sqrt{2}$ is in simplest form and $\sqrt{8}$ is not in simplest form.

For Exercises 3 to 26, simplify.

3. $\sqrt{16}$ **4.** $\sqrt{64}$ **5.** $\sqrt{49}$ **6.** $\sqrt{144}$ **7.** $\sqrt{32}$ **8.** $\sqrt{50}$

9. $\sqrt{8}$ **10.** $\sqrt{12}$ **11.** $6\sqrt{18}$ **12.** $-3\sqrt{48}$ **13.** $5\sqrt{40}$ **14.** $2\sqrt{28}$

15. $\sqrt{15}$ **16.** $\sqrt{21}$ **17.** $\sqrt{29}$ **18.** $\sqrt{13}$ **19.** $-9\sqrt{72}$ **20.** $11\sqrt{80}$

21. $\sqrt{45}$ **22.** $\sqrt{225}$ **23.** $\sqrt{0}$ **24.** $\sqrt{210}$ **25.** $6\sqrt{128}$ **26.** $9\sqrt{288}$

For Exercises 27 to 32, find the decimal approximation rounded to the nearest thousandth.

27. $\sqrt{240}$ **28.** $\sqrt{300}$ **29.** $\sqrt{288}$ **30.** $\sqrt{600}$ **31.** $\sqrt{256}$ **32.** $\sqrt{324}$

Objective B   **To simplify variable radical expressions**

For Exercises 33 to 72, simplify.

33. $\sqrt{x^6}$ **34.** $\sqrt{x^{12}}$ **35.** $\sqrt{y^{15}}$ **36.** $\sqrt{y^{11}}$

37. $\sqrt{a^{20}}$ **38.** $\sqrt{a^{16}}$ **39.** $\sqrt{x^4y^4}$ **40.** $\sqrt{x^{12}y^8}$

41. $\sqrt{4x^4}$ **42.** $\sqrt{25y^8}$ **43.** $\sqrt{24x^2}$ **44.** $\sqrt{x^3y^{15}}$

45. $\sqrt{60x^5}$ **46.** $\sqrt{72y^7}$ **47.** $\sqrt{49a^4b^8}$ **48.** $\sqrt{144x^2y^8}$

49. $\sqrt{18x^5y^7}$ **50.** $\sqrt{32a^5b^{15}}$ **51.** $\sqrt{40x^{11}y^7}$ **52.** $\sqrt{72x^9y^3}$

53. $\sqrt{80a^9b^{10}}$ **54.** $\sqrt{96a^5b^7}$ **55.** $2\sqrt{16a^2b^3}$ **56.** $5\sqrt{25a^4b^7}$

57. $x\sqrt{x^4y^2}$ **58.** $y\sqrt{x^3y^6}$ **59.** $4\sqrt{20a^4b^7}$ **60.** $5\sqrt{12a^3b^4}$

61. $3x\sqrt{12x^2y^7}$ **62.** $4y\sqrt{18x^5y^4}$ **63.** $2x^2\sqrt{8x^2y^3}$ **64.** $3y^2\sqrt{27x^4y^3}$

65. $\sqrt{25(a+4)^2}$ **66.** $\sqrt{81(x+y)^4}$ **67.** $\sqrt{4(x+2)^4}$ **68.** $\sqrt{9(x+2)^8}$

69. $\sqrt{x^2+4x+4}$ **70.** $\sqrt{b^2+8b+16}$ **71.** $\sqrt{y^2+2y+1}$ **72.** $\sqrt{a^2+6a+9}$

APPLYING THE CONCEPTS

73. Automotive Safety Traffic accident investigators can estimate the speed S, in miles per hour, of a car from the length of its skid mark by using the formula $S = \sqrt{30fl}$, where f is the coefficient of friction (which depends on the type of road surface) and l is the length, in feet, of the skid mark. Say the coefficient of friction is 1.2 and the length of a skid mark is 60 ft.
   a. Determine the speed of the car as a radical expression in simplest form.
   b. Write the answer to part **a** as a decimal rounded to the nearest integer.

74. Travel The distance a passenger in an airplane can see to the horizon can be approximated by $d = 1.2\sqrt{h}$, where d is the distance to the horizon in miles and h is the height of the plane in feet. To the nearest tenth of a mile, what is the distance to the horizon of a passenger who is flying at an altitude of 5000 ft?

75. If a and b are positive real numbers, does $\sqrt{a+b} = \sqrt{a} + \sqrt{b}$? If not, give an example in which the expressions are not equal.

76. a. Find the two-digit perfect square that has exactly nine factors.
   b. Find two whole numbers such that their difference is 10, the smaller number is a perfect square, and the larger number is two less than a perfect square.

77. You are to grade this solution to the problem "Write $\sqrt{72}$ in simplest form." Is the solution correct? If not, what error was made? What is the correct solution?

$$\sqrt{72} = \sqrt{4}\sqrt{18}$$
$$= 2\sqrt{18}$$

78. Simplify: **a.** $\sqrt{\sqrt{16}}$ **b.** $\sqrt{\sqrt{81}}$

79. Given $f(x) = \sqrt{2x-1}$, find each of the following. Write your answer in simplest form.
   a. $f(1)$ **b.** $f(5)$ **c.** $f(14)$

Copyright © Houghton Mifflin Company. All rights reserved.

10.2 Addition and Subtraction of Radical Expressions

Objective A   **To add and subtract radical expressions**

The Distributive Property is used to simplify the sum or difference of radical expressions with like radicands.

$$5\sqrt{2} + 3\sqrt{2} = (5 + 3)\sqrt{2} = 8\sqrt{2}$$

$$6\sqrt{2x} - 4\sqrt{2x} = (6 - 4)\sqrt{2x} = 2\sqrt{2x}$$

Radical expressions that are in simplest form and have unlike radicands cannot be simplified by the Distributive Property.

$2\sqrt{3} + 4\sqrt{2}$ cannot be simplified by the Distributive Property.

HOW TO   Simplify: $4\sqrt{8} - 10\sqrt{2}$

$$
\begin{aligned}
4\sqrt{8} - 10\sqrt{2} &= 4\sqrt{4 \cdot 2} - 10\sqrt{2} \\
&= 4\sqrt{4}\sqrt{2} - 10\sqrt{2} & \bullet \text{ Use the Product Property of Square Roots.} \\
&= 4 \cdot 2\sqrt{2} - 10\sqrt{2} \\
&= 8\sqrt{2} - 10\sqrt{2} \\
&= (8 - 10)\sqrt{2} & \bullet \text{ Simplify the expression by using} \\
&= -2\sqrt{2} & \quad \text{the Distributive Property.}
\end{aligned}
$$

HOW TO   Simplify: $8\sqrt{18x} - 2\sqrt{32x}$

$$
\begin{aligned}
8\sqrt{18x} - 2\sqrt{32x} &= 8\sqrt{9 \cdot 2x} - 2\sqrt{16 \cdot 2x} \\
&= 8\sqrt{9}\sqrt{2x} - 2\sqrt{16}\sqrt{2x} & \bullet \text{ Use the Product Property} \\
& & \quad \text{of Square Roots.} \\
&= 8 \cdot 3\sqrt{2x} - 2 \cdot 4\sqrt{2x} \\
&= 24\sqrt{2x} - 8\sqrt{2x} \\
&= (24 - 8)\sqrt{2x} & \bullet \text{ Simplify the expression by using} \\
&= 16\sqrt{2x} & \quad \text{the Distributive Property.}
\end{aligned}
$$

Example 1

Simplify: $5\sqrt{2} - 3\sqrt{2} + 12\sqrt{2}$

Solution

$$
\begin{aligned}
5\sqrt{2} &- 3\sqrt{2} + 12\sqrt{2} \\
&= (5 - 3 + 12)\sqrt{2} \quad \bullet \text{ Distributive Property} \\
&= 14\sqrt{2}
\end{aligned}
$$

You Try It 1

Simplify: $9\sqrt{3} + 3\sqrt{3} - 18\sqrt{3}$

Your solution

Solution on p. S25

Copyright © Houghton Mifflin Company. All rights reserved.

Example 2

Simplify: $3\sqrt{12} - 5\sqrt{27}$

Solution

$3\sqrt{12} - 5\sqrt{27}$

$= 3\sqrt{4 \cdot 3} - 5\sqrt{9 \cdot 3}$ • Simplify $\sqrt{12}$ and $\sqrt{27}$.

$= 3\sqrt{4}\sqrt{3} - 5\sqrt{9}\sqrt{3}$

$= 3 \cdot 2\sqrt{3} - 5 \cdot 3\sqrt{3}$

$= 6\sqrt{3} - 15\sqrt{3}$

$= (6 - 15)\sqrt{3}$ • Distributive Property

$= -9\sqrt{3}$

You Try It 2

Simplify: $2\sqrt{50} - 5\sqrt{32}$

Your solution

Example 3

Simplify: $3\sqrt{12x^3} - 2x\sqrt{3x}$

Solution

$3\sqrt{12x^3} - 2x\sqrt{3x}$

$= 3\sqrt{4x^2 \cdot 3x} - 2x\sqrt{3x}$ • Simplify $\sqrt{12x^3}$.

$= 3\sqrt{4x^2}\sqrt{3x} - 2x\sqrt{3x}$

$= 3 \cdot 2x\sqrt{3x} - 2x\sqrt{3x}$

$= 6x\sqrt{3x} - 2x\sqrt{3x}$

$= (6x - 2x)\sqrt{3x}$ • Distributive Property

$= 4x\sqrt{3x}$

You Try It 3

Simplify: $y\sqrt{28y} + 7\sqrt{63y^3}$

Your solution

Example 4

Simplify: $2x\sqrt{8y} - 3\sqrt{2x^2y} + 2\sqrt{32x^2y}$

Solution

$2x\sqrt{8y} - 3\sqrt{2x^2y} + 2\sqrt{32x^2y}$

$= 2x\sqrt{4 \cdot 2y} - 3\sqrt{x^2 \cdot 2y} + 2\sqrt{16x^2 \cdot 2y}$

$= 2x\sqrt{4}\sqrt{2y} - 3\sqrt{x^2}\sqrt{2y} + 2\sqrt{16x^2}\sqrt{2y}$

$= 2x \cdot 2\sqrt{2y} - 3 \cdot x\sqrt{2y} + 2 \cdot 4x\sqrt{2y}$

$= 4x\sqrt{2y} - 3x\sqrt{2y} + 8x\sqrt{2y}$

$= 9x\sqrt{2y}$

You Try It 4

Simplify: $2\sqrt{27a^5} - 4a\sqrt{12a^3} + a^2\sqrt{75a}$

Your solution

Solutions on p. S25

Copyright © Houghton Mifflin Company. All rights reserved.

10.2 Exercises

Objective A   **To add and subtract radical expressions**

1. Which of the numbers 2, 9, 20, 25, 50, 81, and 100 are *not* perfect squares?

2. Write down a number that has a perfect-square factor that is greater than 1.

3. Write a sentence or two that you could email to a friend to explain the concept of a perfect-square factor.

4. Name the perfect-square factors of 540. What number is the largest perfect-square factor of 540?

For Exercises 5 to 60, simplify.

5. $2\sqrt{2} + \sqrt{2}$

6. $3\sqrt{5} + 8\sqrt{5}$

7. $-3\sqrt{7} + 2\sqrt{7}$

8. $4\sqrt{5} - 10\sqrt{5}$

9. $-3\sqrt{11} - 8\sqrt{11}$

10. $-3\sqrt{3} - 5\sqrt{3}$

11. $2\sqrt{x} + 8\sqrt{x}$

12. $3\sqrt{y} + 2\sqrt{y}$

13. $8\sqrt{y} - 10\sqrt{y}$

14. $-5\sqrt{2a} + 2\sqrt{2a}$

15. $-2\sqrt{3b} - 9\sqrt{3b}$

16. $-7\sqrt{5a} - 5\sqrt{5a}$

17. $3x\sqrt{2} - x\sqrt{2}$

18. $2y\sqrt{3} - 9y\sqrt{3}$

19. $2a\sqrt{3a} - 5a\sqrt{3a}$

20. $-5b\sqrt{3x} - 2b\sqrt{3x}$

21. $3\sqrt{xy} - 8\sqrt{xy}$

22. $-4\sqrt{xy} + 6\sqrt{xy}$

23. $\sqrt{45} + \sqrt{125}$

24. $\sqrt{32} - \sqrt{98}$

25. $2\sqrt{2} + 3\sqrt{8}$

26. $4\sqrt{128} - 3\sqrt{32}$

27. $5\sqrt{18} - 2\sqrt{75}$

28. $5\sqrt{75} - 2\sqrt{18}$

29. $5\sqrt{4x} - 3\sqrt{9x}$

30. $-3\sqrt{25y} + 8\sqrt{49y}$

31. $3\sqrt{3x^2} - 5\sqrt{27x^2}$

32. $-2\sqrt{8y^2} + 5\sqrt{32y^2}$

33. $2x\sqrt{xy^2} - 3y\sqrt{x^2y}$

34. $4a\sqrt{b^2a} - 3b\sqrt{a^2b}$

35. $3x\sqrt{12x} - 5\sqrt{27x^3}$

36. $2a\sqrt{50a} + 7\sqrt{32a^3}$

37. $4y\sqrt{8y^3} - 7\sqrt{18y^5}$

38. $2a\sqrt{8ab^2} - 2b\sqrt{2a^3}$

39. $b^2\sqrt{a^5b} + 3a^2\sqrt{ab^5}$

40. $y^2\sqrt{x^5y} + x\sqrt{x^3y^5}$

Copyright © Houghton Mifflin Company. All rights reserved.

41. $4\sqrt{2} - 5\sqrt{2} + 8\sqrt{2}$

42. $3\sqrt{3} + 8\sqrt{3} - 16\sqrt{3}$

43. $5\sqrt{x} - 8\sqrt{x} + 9\sqrt{x}$

44. $\sqrt{x} - 7\sqrt{x} + 6\sqrt{x}$

45. $8\sqrt{2} - 3\sqrt{y} - 8\sqrt{2}$

46. $8\sqrt{3} - 5\sqrt{2} - 5\sqrt{3}$

47. $8\sqrt{8} - 4\sqrt{32} - 9\sqrt{50}$

48. $2\sqrt{12} - 4\sqrt{27} + \sqrt{75}$

49. $-2\sqrt{3} + 5\sqrt{27} - 4\sqrt{45}$

50. $-2\sqrt{8} - 3\sqrt{27} + 3\sqrt{50}$

51. $4\sqrt{75} + 3\sqrt{48} - \sqrt{99}$

52. $2\sqrt{75} - 5\sqrt{20} + 2\sqrt{45}$

53. $\sqrt{25x} - \sqrt{9x} + \sqrt{16x}$

54. $\sqrt{4x} - \sqrt{100x} - \sqrt{49x}$

55. $3\sqrt{3x} + \sqrt{27x} - 8\sqrt{75x}$

56. $5\sqrt{5x} + 2\sqrt{45x} - 3\sqrt{80x}$

57. $2a\sqrt{75b} - a\sqrt{20b} + 4a\sqrt{45b}$

58. $2b\sqrt{75a} - 5b\sqrt{27a} + 2b\sqrt{20a}$

59. $x\sqrt{3y^2} - 2y\sqrt{12x^2} + xy\sqrt{3}$

60. $a\sqrt{27b^2} + 3b\sqrt{147a^2} - ab\sqrt{3}$

APPLYING THE CONCEPTS

61. Given $G(x) = \sqrt{x + 5} + \sqrt{5x + 3}$, write $G(3)$ in simplest form.

62. Is the equation $\sqrt{a^2 + b^2} = \sqrt{a} + \sqrt{b}$ true for all real numbers a and b?

63. ✐ Explain the steps in simplifying $4\sqrt{2a^3b} + 5\sqrt{8a^3b}$.

64. For each equation, write "ok" if the equation is correct. If the equation is incorrect, correct the right-hand side.
  a. $3\sqrt{ab} + 5\sqrt{ab} = 8\sqrt{2ab}$
  b. $7\sqrt{x^3} - 3x\sqrt{x} - x\sqrt{16x} = 0$
  c. $5 - 2\sqrt{y} = 3\sqrt{y}$

Copyright © Houghton Mifflin Company. All rights reserved.

10.3 Multiplication and Division of Radical Expressions

Objective A To multiply radical expressions

The Product Property of Square Roots is used to multiply radical expressions.

$$\sqrt{2x}\,\sqrt{3y} = \sqrt{2x \cdot 3y} = \sqrt{6xy}$$

HOW TO Simplify: $\sqrt{2x^2}\sqrt{32x^5}$

$$
\begin{aligned}
\sqrt{2x^2}\,\sqrt{32x^5} &= \sqrt{2x^2 \cdot 32x^5} && \text{• Use the Product Property of Square Roots.}\\
&= \sqrt{64x^7} && \text{• Multiply the radicands.}\\
&= \sqrt{64x^6 \cdot x} && \text{• Simplify.}\\
&= \sqrt{64x^6}\,\sqrt{x} = 8x^3\sqrt{x}
\end{aligned}
$$

HOW TO Simplify: $\sqrt{2x}(x + \sqrt{2x})$

$$
\begin{aligned}
\sqrt{2x}(x + \sqrt{2x}) &= \sqrt{2x}(x) + \sqrt{2x}\sqrt{2x} && \text{• Use the Distributive Property to}\\
&= x\sqrt{2x} + \sqrt{4x^2} && \quad \text{remove parentheses.}\\
&= x\sqrt{2x} + 2x && \text{• Simplify.}
\end{aligned}
$$

Use FOIL to multiply radical expressions with two terms.

HOW TO Simplify: $(\sqrt{2} - 3x)(\sqrt{2} + x)$

$$
\begin{aligned}
(\sqrt{2} - 3x)(\sqrt{2} + x) &= \sqrt{2 \cdot 2} + x\sqrt{2} - 3x\sqrt{2} - 3x^2 && \text{• Use the FOIL method to}\\
&= \sqrt{4} + (x - 3x)\sqrt{2} - 3x^2 && \quad \text{remove parentheses.}\\
&= 2 - 2x\sqrt{2} - 3x^2
\end{aligned}
$$

The expressions $a + b$ and $a - b$, which differ only in the sign of one term, are called **conjugates.** Recall that $(a + b)(a - b) = a^2 - b^2$.

HOW TO Simplify: $(2 + \sqrt{7})(2 - \sqrt{7})$

$$
\begin{aligned}
(2 + \sqrt{7})(2 - \sqrt{7}) &= 2^2 - (\sqrt{7})^2 && \text{• } (2 + \sqrt{7})(2 - \sqrt{7}) \text{ is the product of conjugates.}\\
&= 4 - 7 = -3
\end{aligned}
$$

TAKE NOTE

For $x > 0$,
$(\sqrt{x})^2 = x$ because
$(\sqrt{x})^2 = \sqrt{x} \cdot \sqrt{x} = \sqrt{x^2} = x.$

HOW TO Simplify: $(3 + \sqrt{y})(3 - \sqrt{y})$

$$
\begin{aligned}
(3 + \sqrt{y})(3 - \sqrt{y}) &= 3^2 - (\sqrt{y})^2 && \text{• } (3 + \sqrt{y})(3 - \sqrt{y}) \text{ is the product of conjugates.}\\
&= 9 - y
\end{aligned}
$$

Example 1 Simplify: $\sqrt{3x^4}\sqrt{2x^2y}\sqrt{6xy^2}$

Solution

$$
\begin{aligned}
&\sqrt{3x^4}\sqrt{2x^2y}\sqrt{6xy^2}\\
&= \sqrt{36x^7y^3} && \text{• Product Property of Square Roots}\\
&= \sqrt{36x^6y^2 \cdot xy} && \text{• Simplify.}\\
&= \sqrt{36x^6y^2}\sqrt{xy}\\
&= 6x^3y\sqrt{xy}
\end{aligned}
$$

You Try It 1 Simplify: $\sqrt{5a}\sqrt{15a^3b^4}\sqrt{20b^5}$

Your solution

Solution on p. S26

Copyright © Houghton Mifflin Company. All rights reserved.

Example 2

Simplify: $\sqrt{3ab}(\sqrt{3a} + \sqrt{9b})$

Solution
$\sqrt{3ab}(\sqrt{3a} + \sqrt{9b})$
$= \sqrt{9a^2b} + \sqrt{27ab^2}$ • **Distributive Property**
$= \sqrt{9a^2 \cdot b} + \sqrt{9b^2 \cdot 3a}$ • **Simplify.**
$= \sqrt{9a^2}\sqrt{b} + \sqrt{9b^2}\sqrt{3a}$
$= 3a\sqrt{b} + 3b\sqrt{3a}$

You Try It 2

Simplify: $\sqrt{5x}(\sqrt{5x} - \sqrt{25y})$

Your solution

Example 3

Simplify: $(\sqrt{a} - \sqrt{b})(\sqrt{a} + \sqrt{b})$

Solution
$(\sqrt{a} - \sqrt{b})(\sqrt{a} + \sqrt{b})$
$= (\sqrt{a})^2 - (\sqrt{b})^2$ • **Product of conjugates**
$= a - b$

You Try It 3

Simplify: $(2\sqrt{x} + 7)(2\sqrt{x} - 7)$

Your solution

Example 4

Simplify: $(2\sqrt{x} - \sqrt{y})(5\sqrt{x} - 2\sqrt{y})$

Solution
$(2\sqrt{x} - \sqrt{y})(5\sqrt{x} - 2\sqrt{y})$
$= 10(\sqrt{x})^2 - 4\sqrt{xy} - 5\sqrt{xy} + 2(\sqrt{y})^2$ • **FOIL**
$= 10x - 9\sqrt{xy} + 2y$

You Try It 4

Simplify: $(3\sqrt{x} - \sqrt{y})(5\sqrt{x} - 2\sqrt{y})$

Your solution

Solutions on p. S26

Objective B   **To divide radical expressions**

The Quotient Property of Square Roots

If a and b are positive real numbers, then $\sqrt{\dfrac{a}{b}} = \dfrac{\sqrt{a}}{\sqrt{b}}$ and $\dfrac{\sqrt{a}}{\sqrt{b}} = \sqrt{\dfrac{a}{b}}$.

This property states that the square root of a quotient is equal to the quotient of the square roots.

HOW TO   Simplify: $\sqrt{\dfrac{4x^2}{z^6}}$

$\sqrt{\dfrac{4x^2}{z^6}} = \dfrac{\sqrt{4x^2}}{\sqrt{z^6}}$ • Rewrite the radical expression as the quotient of the square roots.

$= \dfrac{2x}{z^3}$ • Simplify.

Copyright © Houghton Mifflin Company. All rights reserved.

Copyright © Houghton Mifflin Company. All rights reserved.

Point of Interest

A radical expression that occurs in Einstein's Theory of Relativity is

$$\frac{1}{\sqrt{1 - \dfrac{v^2}{c^2}}}$$

where v is the velocity of an object and c is the speed of light.

HOW TO Simplify: $\sqrt{\dfrac{24x^3y^7}{3x^7y^2}}$

$$\sqrt{\frac{24x^3y^7}{3x^7y^2}} = \sqrt{\frac{8y^5}{x^4}}$$

$$= \frac{\sqrt{8y^5}}{\sqrt{x^4}}$$

$$= \frac{\sqrt{4y^4 \cdot 2y}}{\sqrt{x^4}}$$

$$= \frac{\sqrt{4y^4}\sqrt{2y}}{\sqrt{x^4}}$$

$$= \frac{2y^2\sqrt{2y}}{x^2}$$

- Simplify the radicand.

- Rewrite the radical expression as the quotient of the square roots.

- Simplify.

The Quotient Property of Square Roots is used to divide radical expressions.

HOW TO Simplify: $\dfrac{\sqrt{4x^2y}}{\sqrt{xy}}$

$$\frac{\sqrt{4x^2y}}{\sqrt{xy}} = \sqrt{\frac{4x^2y}{xy}}$$

$$= \sqrt{4x}$$

$$= \sqrt{4}\sqrt{x}$$

$$= 2\sqrt{x}$$

- Use the Quotient Property of Square Roots.

- Simplify the radicand.

- Simplify the radical expression.

The previous examples all result in radical expressions written in simplest form.

Simplest Form of a Radical Expression

For a radical expression to be in simplest form, three conditions must be met:

1. The radicand contains no factor greater than 1 that is a perfect square.
2. There is no fraction under the radical sign.
3. There is no radical in the denominator of a fraction.

The procedure used to remove a radical from the denominator is called **rationalizing the denominator.**

HOW TO Simplify: $\dfrac{2}{\sqrt{3}}$

$$\frac{2}{\sqrt{3}} = \frac{2}{\sqrt{3}} \cdot \boxed{\frac{\sqrt{3}}{\sqrt{3}}}$$

$$= \frac{2\sqrt{3}}{(\sqrt{3})^2}$$

$$= \frac{2\sqrt{3}}{3}$$

- To rationalize the denominator, multiply the expression by $\frac{\sqrt{3}}{\sqrt{3}}$, which equals 1.

- The radicand in the denominator is a perfect square.

- Simplify.

When the denominator contains a radical expression with two terms, rationalize the denominator by multiplying the numerator and denominator by the conjugate of the denominator.

> **HOW TO** Simplify: $\dfrac{\sqrt{2y}}{\sqrt{y}+3}$

$$\dfrac{\sqrt{2y}}{\sqrt{y}+3}=\dfrac{\sqrt{2y}}{\sqrt{y}+3}\cdot\dfrac{\sqrt{y}-3}{\sqrt{y}-3}$$

$$=\dfrac{\sqrt{2y^2}-3\sqrt{2y}}{(\sqrt{y})^2-3^2}=\dfrac{y\sqrt{2}-3\sqrt{2y}}{y-9}$$

• Multiply the numerator and denominator by $\sqrt{y}-3$, the conjugate of $\sqrt{y}+3$.

Example 5 Simplify: $\dfrac{\sqrt{4x^2y^5}}{\sqrt{3x^4y}}$

Solution
$$\dfrac{\sqrt{4x^2y^5}}{\sqrt{3x^4y}}=\sqrt{\dfrac{4x^2y^5}{3x^4y}}=\sqrt{\dfrac{4y^4}{3x^2}}=\dfrac{\sqrt{4y^4}}{\sqrt{3x^2}}$$

$$=\dfrac{2y^2}{x\sqrt{3}}=\dfrac{2y^2}{x\sqrt{3}}\cdot\dfrac{\sqrt{3}}{\sqrt{3}}$$ • Rationalize the denominator.

$$=\dfrac{2y^2\sqrt{3}}{3x}$$

You Try It 5 Simplify: $\dfrac{\sqrt{15x^6y^7}}{\sqrt{3x^7y^9}}$

Your solution

Example 6 Simplify: $\dfrac{\sqrt{2}}{\sqrt{2}+\sqrt{6}}$

Solution
$$\dfrac{\sqrt{2}}{\sqrt{2}+\sqrt{6}}$$ • Multiply the numerator and denominator by the conjugate of the denominator.

$$=\dfrac{\sqrt{2}}{\sqrt{2}+\sqrt{6}}\cdot\dfrac{\sqrt{2}-\sqrt{6}}{\sqrt{2}-\sqrt{6}}$$

$$=\dfrac{(\sqrt{2})^2-\sqrt{12}}{2-6}=\dfrac{2-2\sqrt{3}}{-4}$$

$$=\dfrac{2(1-\sqrt{3})}{-4}=\dfrac{1-\sqrt{3}}{-2}=-\dfrac{1-\sqrt{3}}{2}$$

You Try It 6 Simplify: $\dfrac{\sqrt{3}}{\sqrt{3}-\sqrt{6}}$

Your solution

Example 7 Simplify: $\dfrac{3-\sqrt{5}}{2+3\sqrt{5}}$

Solution
$$\dfrac{3-\sqrt{5}}{2+3\sqrt{5}}=\dfrac{3-\sqrt{5}}{2+3\sqrt{5}}\cdot\dfrac{2-3\sqrt{5}}{2-3\sqrt{5}}$$ • Rationalize the denominator.

$$=\dfrac{6-9\sqrt{5}-2\sqrt{5}+3(\sqrt{5})^2}{4-9\cdot5}$$

$$=\dfrac{6-11\sqrt{5}+15}{4-45}$$

$$=\dfrac{21-11\sqrt{5}}{-41}=-\dfrac{21-11\sqrt{5}}{41}$$

You Try It 7 Simplify: $\dfrac{5+\sqrt{y}}{1-2\sqrt{y}}$

Your solution

Solutions on p. S26

Copyright © Houghton Mifflin Company. All rights reserved.

10.3 Exercises

Objective A   **To multiply radical expressions**

1. Explain in words and then write in symbols the Product Property of Square Roots.

2. Give an example to show that $\sqrt{a^2} \neq a$.

For Exercises 3 to 37, simplify.

3. $\sqrt{5} \cdot \sqrt{5}$

4. $\sqrt{11} \cdot \sqrt{11}$

5. $\sqrt{3} \cdot \sqrt{12}$

6. $\sqrt{2} \cdot \sqrt{8}$

7. $\sqrt{x} \cdot \sqrt{x}$

8. $\sqrt{y} \cdot \sqrt{y}$

9. $\sqrt{xy^3} \cdot \sqrt{x^5y}$

10. $\sqrt{a^3b^5} \cdot \sqrt{ab^5}$

11. $\sqrt{3a^2b^5} \cdot \sqrt{6ab^7}$

12. $\sqrt{5x^3y} \cdot \sqrt{10x^2y}$

13. $\sqrt{6a^3b^2} \cdot \sqrt{24a^5b}$

14. $\sqrt{8ab^5} \cdot \sqrt{12a^7b}$

15. $\sqrt{2}(\sqrt{2} - \sqrt{3})$

16. $3(\sqrt{12} - \sqrt{3})$

17. $\sqrt{x}(\sqrt{x} - \sqrt{y})$

18. $\sqrt{b}(\sqrt{a} - \sqrt{b})$

19. $\sqrt{5}(\sqrt{10} - \sqrt{x})$

20. $\sqrt{6}(\sqrt{y} - \sqrt{18})$

21. $\sqrt{8}(\sqrt{2} - \sqrt{5})$

22. $\sqrt{10}(\sqrt{20} - \sqrt{a})$

23. $(\sqrt{x} - 3)^2$

24. $(2\sqrt{a} - y)^2$

25. $\sqrt{3a}(\sqrt{3a} - \sqrt{3b})$

26. $\sqrt{5x}(\sqrt{10x} - \sqrt{x})$

27. $\sqrt{2ac} \cdot \sqrt{5ab} \cdot \sqrt{10cb}$

28. $\sqrt{3xy} \cdot \sqrt{6x^3y} \cdot \sqrt{2y^2}$

29. $(\sqrt{5} + 3)(2\sqrt{5} - 4)$

30. $(2 - 3\sqrt{7})(5 + 2\sqrt{7})$

31. $(4 + \sqrt{8})(3 + \sqrt{2})$

32. $(6 - \sqrt{27})(2 + \sqrt{3})$

33. $(2\sqrt{x} + 4)(3\sqrt{x} - 1)$

34. $(5 + \sqrt{y})(6 - 3\sqrt{y})$

35. $(3\sqrt{x} - 2y)(5\sqrt{x} - 4y)$

36. $(5\sqrt{x} + 2\sqrt{y})(3\sqrt{x} - \sqrt{y})$

37. $(\sqrt{x} - \sqrt{y})(\sqrt{x} + \sqrt{y})$

Copyright © Houghton Mifflin Company. All rights reserved.

Objective B **To divide radical expressions**

38. Why is $\dfrac{\sqrt{3}}{3}$ in simplest form but $\dfrac{1}{\sqrt{3}}$ not in simplest form?

39. Why can we multiply $\dfrac{2}{\sqrt{5}}$ by $\dfrac{\sqrt{5}}{\sqrt{5}}$ without changing the value of $\dfrac{2}{\sqrt{5}}$?

For Exercises 40 to 69, simplify.

40. $\dfrac{\sqrt{32}}{\sqrt{2}}$

41. $\dfrac{\sqrt{45}}{\sqrt{5}}$

42. $\dfrac{\sqrt{98}}{\sqrt{2}}$

43. $\dfrac{\sqrt{48}}{\sqrt{3}}$

44. $\dfrac{\sqrt{27a}}{\sqrt{3a}}$

45. $\dfrac{\sqrt{72x^5}}{\sqrt{2x}}$

46. $\dfrac{\sqrt{15x^3y}}{\sqrt{3xy}}$

47. $\dfrac{\sqrt{40x^5y^2}}{\sqrt{5xy}}$

48. $\dfrac{\sqrt{2a^5b^4}}{\sqrt{98ab^4}}$

49. $\dfrac{\sqrt{48x^5y^2}}{\sqrt{3x^3y}}$

50. $\dfrac{\sqrt{9xy^2}}{\sqrt{27x}}$

51. $\dfrac{\sqrt{4x^2y}}{\sqrt{3xy^3}}$

52. $\dfrac{\sqrt{16x^3y^2}}{\sqrt{8x^3y}}$

53. $\dfrac{\sqrt{2}}{\sqrt{8}+4}$

54. $\dfrac{1}{\sqrt{2}-3}$

55. $\dfrac{5}{\sqrt{7}-3}$

56. $\dfrac{3}{5+\sqrt{5}}$

57. $\dfrac{\sqrt{3}}{5-\sqrt{27}}$

58. $\dfrac{7}{\sqrt{2}-7}$

59. $\dfrac{3-\sqrt{6}}{5-2\sqrt{6}}$

60. $\dfrac{6-2\sqrt{3}}{4+3\sqrt{3}}$

61. $\dfrac{-6}{4+\sqrt{2}}$

62. $\dfrac{\sqrt{2}+2\sqrt{6}}{2\sqrt{2}-3\sqrt{6}}$

63. $\dfrac{2\sqrt{3}-\sqrt{6}}{5\sqrt{3}+2\sqrt{6}}$

64. $\dfrac{3+\sqrt{x}}{2-\sqrt{x}}$

65. $\dfrac{-\sqrt{15}}{3-\sqrt{12}}$

66. $\dfrac{\sqrt{a}-4}{2\sqrt{a}+2}$

67. $\dfrac{\sqrt{xy}}{\sqrt{x}-\sqrt{y}}$

68. $\dfrac{\sqrt{x}}{\sqrt{x}-\sqrt{y}}$

69. $\dfrac{-12}{\sqrt{6}-3}$

APPLYING THE CONCEPTS

70. In your own words, describe the process of rationalizing the denominator.

71. Show that $1+\sqrt{6}$ and $1-\sqrt{6}$ are solutions of the equation $x^2-2x-5=0$.

72. Answer true or false. If the equation is false, correct it.

  a. $(\sqrt{y})^4 = y^2$ **b.** $(2\sqrt{x})^3 = 8x\sqrt{x}$ **c.** $(\sqrt{x}+1)^2 = x+1$ **d.** $\dfrac{1}{2-\sqrt{3}} = 2+\sqrt{3}$

Copyright © Houghton Mifflin Company. All rights reserved.

10.4 Solving Equations Containing Radical Expressions

Objective A

To solve an equation containing a radical expression

An equation that contains a variable expression in a radicand is a **radical equation.**

$$\sqrt{x} = 4$$
$$\sqrt{x + 2} = \sqrt{x - 7}$$
} Radical equations

The following property of equality, which states that if two numbers are equal, the squares of the numbers are equal, is used to solve radical equations.

Property of Squaring Both Sides of an Equation

If a and b are real numbers and $a = b$, then $a^2 = b^2$.

To solve a radical equation with one radical, use the following procedure.

Solving a Radical Equation

1. Write the equation with the radical alone on one side.
2. Square both sides of the equation.
3. Solve for the variable.
4. Check the solution(s) in the original equation.

HOW TO   Solve: $\sqrt{x - 2} - 7 = 0$

$$\sqrt{x - 2} - 7 = 0$$ • Isolate the radical by adding 7
$$\sqrt{x - 2} = 7$$ to both sides of the equation.
$$(\sqrt{x - 2})^2 = 7^2$$ • Square both sides of the equation.
$$x - 2 = 49$$ • Solve the resulting equation.
$$x = 51$$

Check: $\sqrt{x - 2} - 7 = 0$

$$\begin{array}{c|c} \sqrt{51 - 2} - 7 & 0 \\ \sqrt{49} - 7 & 0 \\ 7 - 7 & 0 \\ 0 = 0 \end{array}$$ A true equation

The solution is 51.

Study Tip

When we suggest that you check a solution, substitute the solution into the original equation.

When both sides of an equation are squared, the resulting equation may have a solution that is not a solution of the original equation. Checking a proposed solution of a radical equation, as we did above, is a necessary step.

HOW TO   Solve: $\sqrt{2x - 5} + 3 = 0$

$$\sqrt{2x - 5} + 3 = 0$$ • Isolate the radical by subtracting 3
$$\sqrt{2x - 5} = -3$$ from both sides of the equation.
$$(\sqrt{2x - 5})^2 = (-3)^2$$ • Square both sides of the equation.
$$2x - 5 = 9$$ • Solve for x.
$$2x = 14$$
$$x = 7$$

Copyright © Houghton Mifflin Company. All rights reserved.

TAKE NOTE

Any time each side of an equation is squared, you must check the proposed solution of the equation.

Here is the check for the equation on the preceding page.

Check:

$$\sqrt{2x - 5} + 3 = 0$$

$$
\begin{array}{c|c}
\sqrt{2 \cdot 7 - 5} + 3 & 0 \\
\sqrt{14 - 5} + 3 & 0 \\
\sqrt{9} + 3 & 0 \\
3 + 3 & 0 \\
6 \neq 0
\end{array}
$$

7 does not check as a solution. The equation has no solution.

Example 1

Solve: $\sqrt{3x} + 2 = 5$

Solution

$$
\begin{aligned}
\sqrt{3x} + 2 &= 5 \\
\sqrt{3x} &= 3 \qquad \bullet \text{ Isolate } \sqrt{3x}. \\
(\sqrt{3x})^2 &= 3^2 \qquad \bullet \text{ Square both} \\
3x &= 9 \qquad\qquad \text{sides.} \\
x &= 3 \qquad \bullet \text{ Solve for } x.
\end{aligned}
$$

Check:

$$
\begin{array}{c|c}
\sqrt{3x} + 2 = 5 \\ \hline
\sqrt{3 \cdot 3} + 2 & 5 \\
\sqrt{9} + 2 & 5 \\
3 + 2 & 5 \\
5 = 5
\end{array}
$$

The solution is 3.

You Try It 1

Solve: $\sqrt{4x} + 3 = 7$

Your solution

Example 2

Solve: $1 = \sqrt{x} - \sqrt{x - 5}$

Solution

When an equation contains two radicals, isolate the radicals one at a time.

$$
\begin{aligned}
1 &= \sqrt{x} - \sqrt{x - 5} \\
1 + \sqrt{x - 5} &= \sqrt{x} \qquad\qquad\bullet \text{ Isolate } \sqrt{x}. \\
(1 + \sqrt{x - 5})^2 &= (\sqrt{x})^2 \qquad\bullet \text{ Square both sides.} \\
1 + 2\sqrt{x - 5} + (x - 5) &= x \qquad\qquad\bullet \text{ Square the binomial.} \\
2\sqrt{x - 5} &= 4 \qquad\qquad\bullet \text{ Simplify.} \\
\sqrt{x - 5} &= 2 \qquad\qquad\bullet \text{ Isolate } \sqrt{x - 5}. \\
(\sqrt{x - 5})^2 &= 2^2 \qquad\qquad\bullet \text{ Square both sides.} \\
x - 5 &= 4 \\
x &= 9 \qquad\qquad\bullet \text{ Solve for } x.
\end{aligned}
$$

Check:

$$
\begin{array}{c|c}
1 = \sqrt{x} - \sqrt{x - 5} \\ \hline
1 & \sqrt{9} - \sqrt{9 - 5} \\
1 & \sqrt{9} - \sqrt{4} \\
1 & 3 - 2 \\
1 = 1
\end{array}
$$

The solution is 9.

You Try It 2

Solve: $\sqrt{x} + \sqrt{x + 9} = 9$

Your solution

Solutions on p. S26

Copyright © Houghton Mifflin Company. All rights reserved.

Copyright © Houghton Mifflin Company. All rights reserved.

Objective B **To solve application problems**

A **right triangle** is a triangle that contains a 90° angle. The side opposite the 90° angle is called the **hypotenuse.** The other two sides are called **legs.**

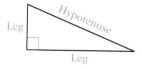

Pythagoras (c. 580 B.C.–520 B.C.)

Pythagoras, a Greek mathematician who lived around 550 B.C., is given credit for the Pythagorean Theorem. It states that the square of the hypotenuse of a right triangle is equal to the sum of the squares of the two legs. Actually, this theorem was known to the Babylonians around 1200 B.C.

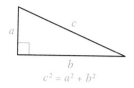

$$c^2 = a^2 + b^2$$

Point of Interest

The first known proof of this theorem occurs in a Chinese text, *Arithmetic Classic*, which was first written around 600 B.C. (but there are no existing copies) and revised over a period of 500 years. The earliest known copy of this text dates from approximately 100 B.C.

> **Pythagorean Theorem**
>
> If a and b are the lengths of the legs of a right triangle and c is the length of the hypotenuse, then $c^2 = a^2 + b^2$.

Using this theorem, we can find the hypotenuse of a right triangle when we know the two legs. Use the formula

$$\text{Hypotenuse} = \sqrt{(\text{leg})^2 + (\text{leg})^2}$$
$$c = \sqrt{a^2 + b^2}$$
$$= \sqrt{(5)^2 + (12)^2}$$
$$= \sqrt{25 + 144}$$
$$= \sqrt{169}$$
$$= 13$$

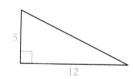

The leg of a right triangle can be found when one leg and the hypotenuse are known. Use the formula

$$\text{Leg} = \sqrt{(\text{hypotenuse})^2 - (\text{leg})^2}$$
$$a = \sqrt{c^2 - b^2}$$
$$= \sqrt{(25)^2 - (20)^2}$$
$$= \sqrt{625 - 400}$$
$$= \sqrt{225}$$
$$= 15$$

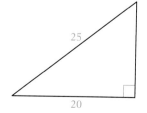

Example 3 and You Try It 3 on the following page illustrate the use of the Pythagorean Theorem. Example 4 and You Try It 4 illustrate other applications of radical equations.

Example 3

A guy wire is attached to a point 20 m above the ground on a telephone pole. The wire is anchored to the ground at a point 8 m from the base of the pole. Find the length of the guy wire. Round to the nearest tenth.

Strategy
To find the length of the guy wire, use the Pythagorean Theorem. One leg is 20 m. The other leg is 8 m. The guy wire is the hypotenuse. Solve the Pythagorean Theorem for the hypotenuse.

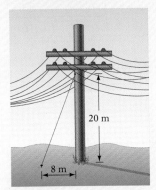

20 m

8 m

Solution
$$c = \sqrt{a^2 + b^2}$$
$$= \sqrt{(20)^2 + (8)^2} \qquad \bullet \; a = 20, b = 8$$
$$= \sqrt{400 + 64} = \sqrt{464} \approx 21.5$$

The guy wire has a length of approximately 21.5 m.

Example 4

How far would a submarine periscope have to be above the water to locate a ship 4 mi away? The equation for the distance in miles that the lookout can see is $d = \sqrt{1.5h}$, where h is the height in feet above the surface of the water. Round to the nearest hundredth.

Strategy
To find the height above the water, replace d in the equation with the given value and solve for h.

Solution
$$d = \sqrt{1.5h}$$
$$4 = \sqrt{1.5h} \qquad \bullet \; d = 4$$
$$4^2 = (\sqrt{1.5h})^2$$
$$16 = 1.5h$$
$$10.67 \approx h$$

The periscope must be approximately 10.67 ft above the water.

You Try It 3

A ladder 8 ft long is resting against a building. How high on the building will the ladder reach when the bottom of the ladder is 3 ft from the building? Round to the nearest hundredth.

Your strategy

Your solution

You Try It 4

Find the length of a pendulum that makes one swing in 2.5 s. The equation for the time for one swing is $T = 2\pi\sqrt{\dfrac{L}{32}}$, where T is the time in seconds and L is the length in feet. Use 3.14 for π. Round to the nearest hundredth.

Your strategy

Your solution

Solutions on pp. S26–S27

Copyright © Houghton Mifflin Company. All rights reserved.

10.4 Exercises

Objective A   To solve an equation containing a radical expression

For Exercises 1 to 22, solve and check.

1. $\sqrt{x} = 5$ **2.** $\sqrt{y} = 7$ **3.** $\sqrt{a} = 12$ **4.** $\sqrt{a} = 9$ **5.** $\sqrt{5x} = 5$

6. $\sqrt{4x} + 5 = 2$ **7.** $\sqrt{3x} + 9 = 4$ **8.** $\sqrt{3x - 2} = 4$ **9.** $\sqrt{5x + 6} = 1$

10. $\sqrt{2x + 1} = 7$ **11.** $\sqrt{5x + 4} = 3$ **12.** $0 = 2 - \sqrt{3 - x}$ **13.** $0 = 5 - \sqrt{10 + x}$

14. $\sqrt{5x + 2} = 0$ **15.** $\sqrt{3x - 7} = 0$ **16.** $\sqrt{3x} - 6 = -4$

17. $\sqrt{x^2 + 5} = x + 1$ **18.** $\sqrt{x^2 - 5} = 5 - x$ **19.** $\sqrt{x + 4} - \sqrt{x - 1} = 1$

20. $\sqrt{x} + \sqrt{x - 12} = 2$ **21.** $\sqrt{2x + 1} - \sqrt{2x - 4} = 1$ **22.** $\sqrt{3x + 1} - \sqrt{3x - 2} = 1$

Objective B   To solve application problems

23. **Sports** The infield of a baseball diamond is a square. The distance between successive bases is 90 ft. The pitcher's mound is on the diagonal between home plate and second base at a distance of 60.5 ft from home plate. (See the figure at the right.) Is the pitcher's mound more or less than halfway between home plate and second base?

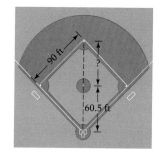

24. **Sports** The infield of a softball diamond is a square. The distance between successive bases is 60 ft. The pitcher's mound is on the diagonal between home plate and second base at a distance of 46 ft from home plate. Is the pitcher's mound more or less than halfway between home plate and second base?

25. **Periscopes** How far would a submarine periscope have to be above the water to locate a ship 5 mi away? The equation for the distance in miles that the lookout can see is $d = \sqrt{1.5h}$, where h is the height in feet above the surface of the water. Round to the nearest hundredth.

26. **Building Maintenance** A 16-foot ladder is leaning against a building. How high on the building will the ladder reach when the bottom of the ladder is 5 ft from the building? (See the figure at the right.) Round to the nearest tenth.

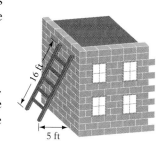

Copyright © Houghton Mifflin Company. All rights reserved.

27. Home Entertainment   The measure of a big-screen television is given by the length of a diagonal across the screen. A 36-inch television has a width of 28.8 in. Find the height of the screen to the nearest tenth of an inch.

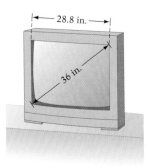

28. Home Entertainment   The measure of a television screen is given by the length of a diagonal across the screen. A 33-inch television has a width of 26.4 in. Find the height of the screen to the nearest tenth of an inch.

29. Recreation   The speed of a child riding a merry-go-round at a carnival is given by the equation $v = \sqrt{12r}$, where v is the speed in feet per second and r is the distance in feet from the center of the merry-go-round to the rider. If a child is moving at 15 ft/s, how far is the child from the center of the merry-go-round?

30. Time   Find the length of a pendulum that makes one swing in 1.5 s. The equation for the time of one swing of a pendulum is $T = 2\pi\sqrt{\dfrac{L}{32}}$, where T is the time in seconds and L is the length in feet. Use 3.14 for π. Round to the nearest hundredth.

APPLYING THE CONCEPTS

31. Geometry   In the coordinate plane, a triangle is formed by drawing lines between the points (0, 0) and (5, 0), (5, 0) and (5, 12), and (5, 12) and (0, 0). Find the perimeter of the triangle.

32. Geometry   The hypotenuse of a right triangle is $5\sqrt{2}$ cm, and one leg is $4\sqrt{2}$ cm.
a. Find the perimeter of the triangle.
b. Find the area of the triangle.

33.   If a and b are real numbers and $a^2 = b^2$, does $a = b$? Explain your answer.

34. Geometry   Can the Pythagorean Theorem be used to find the length of side c of the triangle at the right? If so, determine c. If not, explain why the theorem cannot be used.

35. Fountain Design   A circular fountain is being designed for a triangular plaza in a cultural center. The fountain is placed so that each side of the triangle touches the fountain, as shown in the diagram at the right. Find the area of the fountain. The formula for the radius of the circle is given by

$$r = \sqrt{\frac{(s - a)(s - b)(s - c)}{s}}$$

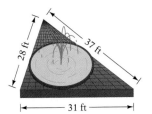

where $s = \dfrac{1}{2}(a + b + c)$ and a, b, and c are the lengths of the sides of the triangle. Round to the nearest hundredth.

36. Complete each statement using <, =, or >.
a. For an acute triangle with longest side c, $a^2 + b^2$ _____ c^2.
b. For a right triangle with longest side c, $a^2 + b^2$ _____ c^2.
c. For an obtuse triangle with longest side c, $a^2 + b^2$ _____ c^2.

Copyright © Houghton Mifflin Company. All rights reserved.

Focus on Problem Solving

Deductive Reasoning Deductive reasoning uses a rule or statement of fact to reach a conclusion. For instance, if two angles of one triangle are equal to two angles of another triangle, then the two triangles are similar. Thus any time we establish this fact about two triangles, we know that the triangles are similar. Below are two examples of deductive reasoning.

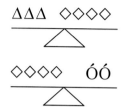

Given that ΔΔΔ = ◇◇◇◇ and ◇◇◇◇ = ÓÓ, then ΔΔΔΔΔΔ is equivalent to how many Ós?

Because 3 Δs = 4 ◇s and 4 ◇s = 2 Ós, 3 Δs = 2 Ós.

6 Δs is twice 3 Δs. We need to find twice 2 Ós, which is 4 Ós.

Therefore, ΔΔΔΔΔΔ = ÓÓÓÓ.

Lomax, Parish, Thorpe, and Wong are neighbors. Each drives a different type of vehicle: a compact car, a sedan, a sports car, or a station wagon. From the following statements, determine which type of vehicle each of the neighbors drives.

1. Although the vehicle owned by Lomax has more mileage on it than does either the sedan or the sports car, it does not have the highest mileage of all four cars. (Use X1 in the chart below to eliminate possibilities that this statement rules out.)

2. Wong and the owner of the sports car live on one side of the street, and Thorpe and the owner of the compact car live on the other side of the street. (Use X2 to eliminate possibilities that this statement rules out.)

3. Thorpe owns the vehicle with the most mileage on it. (Use X3 to eliminate possibilities that this statement rules out.)

Copyright © Houghton Mifflin Company. All rights reserved.

TAKE NOTE

To use the chart to solve this problem, write an X in a box to indicate that a possibility has been eliminated. Write a √ to show that a match has been found. When a row or column has 3 X's, a √ is written in the remaining open box in that row or column of the chart.

	Compact	Sedan	Sports Car	Wagon
Lomax	√	X1	X1	X2
Parish	X2	X2	√	X2
Thorpe	X2	X3	X2	√
Wong	X2		X2	

Lomax drives the compact car, Parish drives the sports car, Thorpe drives the station wagon, and Wong drives the sedan.

1. Given that ‡‡ = ••••• and ••••• = ΛΛ, then ‡‡‡‡‡ = how many Λs?

2. Given that □□□□□□ = ÓÓÓÓ and ÓÓÓÓ = ÎÎ, then □□□ = how many Îs?

3. Given that ¤¤¤¤ = ΩΩΩ and ΩΩΩ = ΔΔ, then ΔΔΔΔ = how many ¤s?

4. Given that ¥¥¥¥¥ = §§ and §§ = ÂÂÂ, then ÂÂÂÂÂÂ = how many ¥s?

5. Anna, Kay, Megan, and Nicole decide to travel together during spring break, but they need to find a destination where each of them will be able to participate in her favorite sport (golf, horseback riding, sailing, or tennis). From the following statements, determine the favorite sport of each student.

a. Anna and the student whose favorite sport is sailing both like to swim, whereas Nicole and the student whose favorite sport is tennis would prefer to scuba dive.

b. Megan and the student whose favorite sport is sailing are roommates. Nicole and the student whose favorite sport is golf live by themselves in singles.

6. Chang, Nick, Pablo, and Saul each take a different form of transportation (bus, car, subway, or taxi) from the office to the airport. From the following statements, determine which form of transportation each takes.

a. Chang spent more on transportation than the fellow who took the bus but less than the fellow who took the taxi.

b. Pablo, who did not travel by bus and who spent the least on transportation, arrived at the airport after Nick but before the fellow who took the subway.

c. Saul spent less on transportation than either Chang or Nick.

Projects and Group Activities

Distance to the Horizon

The formula $d = \sqrt{1.5h}$ can be used to calculate the approximate distance d (in miles) that a person can see who uses a periscope h feet above the water. That formula is derived by using the Pythagorean Theorem.

Consider the diagram (not to scale) at the right, which shows Earth as a sphere and the periscope as extending h feet above its surface. From geometry, because AB is tangent to the circle and OA is a radius, triangle AOB is a right triangle. Therefore,

$$(OA)^2 + (AB)^2 = (OB)^2$$

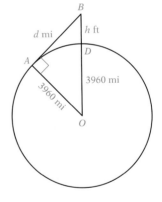

Substituting into this formula, we have

$$3960^2 + d^2 = \left(3960 + \frac{h}{5280}\right)^2$$

- Because h is in feet, $\frac{h}{5280}$ is in miles.

$$3960^2 + d^2 = 3960^2 + \frac{2 \cdot 3960}{5280}h + \left(\frac{h}{5280}\right)^2$$

$$d^2 = \frac{3}{2}h + \left(\frac{h}{5280}\right)^2$$

$$d = \sqrt{\frac{3}{2}h + \left(\frac{h}{5280}\right)^2}$$

At this point, an assumption is made that $\sqrt{\frac{3}{2}h + \left(\frac{h}{5280}\right)^2} \approx \sqrt{1.5h}$, where we have written $\frac{3}{2}$ as 1.5. Thus $d \approx \sqrt{1.5h}$ is used to approximate the distance that can be seen using a periscope h feet above the water.

Copyright © Houghton Mifflin Company. All rights reserved.

1. Write a paragraph that justifies the assumption that

$$\sqrt{\frac{3}{2}h + \left(\frac{h}{5280}\right)^2} \approx \sqrt{1.5h}$$

   (*Suggestion:* Evaluate each expression for various values of h. Because h is the height of a periscope above water, it is unlikely that $h > 25$ ft.)

2. The distance d is the distance from the top of the periscope to A. The distance along the surface of the water is given by arc AD. This distance can be approximated by the equation

$$L \approx \sqrt{1.5h} + 0.306186\left(\sqrt{\frac{h}{5280}}\right)^3$$

   Using this formula, calculate L when $h = 10$.

Chapter 10 Summary

Key Words

Examples

A *square root* of a positive number a is a number whose square is a. Every positive number has two square roots, one a positive and one a negative number. The square root of a negative number is not a real number. [10.1A, p. 481]

A square root of 49 is 7 because $7^2 = 49$.

A square root of 49 is -7 because $(-7)^2 = 49$.

$\sqrt{-9}$ is not a real number.

The symbol $\sqrt{}$ is called a *radical sign* and is used to indicate the positive or *principal square root* of a number. The negative square root of a number is indicated by placing a negative sign in front of the radical. The *radicand* is the expression under the radical sign. [10.1A, p. 481]

$\sqrt{49} = 7$

$-\sqrt{49} = -7$

In the expression $\sqrt{49xy}$, $49xy$ is the radicand.

The square of an integer is a *perfect square*. If a number is not a perfect square, its square root can only be approximated. Such square roots are *irrational numbers*. Their decimal representations never terminate or repeat. [10.1A, p. 481]

1, 4, 9, 16, 25, 36, 49, 64, ... are examples of perfect squares.

7 is not a perfect square. $\sqrt{7}$ is an irrational number.

Conjugates are expressions with two terms that differ only in the sign of one term. The expressions $a + b$ and $a - b$ are conjugates. [10.3A, p. 491]

$-5 + \sqrt{11}$ and $-5 - \sqrt{11}$ are conjugates.

$\sqrt{x} - 3$ and $\sqrt{x} + 3$ are conjugates.

A *radical equation* is an equation that contains a variable expression in a radicand. [10.4A, p. 497]

$\sqrt{2x} + 5 = 9$ is a radical equation.

$2x + \sqrt{5} = 9$ is not a radical equation.

A *right triangle* is a triangle that contains a 90° angle. The side opposite the 90° angle is the *hypotenuse*. The other two sides are called *legs*. [10.4B, p. 499]

Essential Rules and Procedures

Examples

The Product Property of Square Roots [10.1A, p. 481]
If a and b are positive real numbers, then $\sqrt{ab} = \sqrt{a} \cdot \sqrt{b}$. Use the Product Property of Square Roots and a knowledge of perfect squares to simplify radicands that are not perfect squares.

$\sqrt{28} = \sqrt{4 \cdot 7} = \sqrt{4} \cdot \sqrt{7} = 2\sqrt{7}$

$\sqrt{9x^7} = \sqrt{9x^6 \cdot x} = \sqrt{9x^6}\sqrt{x} = 3x^3\sqrt{x}$

Copyright © Houghton Mifflin Company. All rights reserved.

Adding or Subtracting Radical Expressions [10.2A, p. 487]
The Distributive Property is used to simplify the sum or difference of radical expressions with like radicands.

$$8\sqrt{2x} - 3\sqrt{2x} = (8 - 3)\sqrt{2x} = 5\sqrt{2x}$$

Multiplying Radical Expressions [10.3A, p. 491]
The Product Property of Square Roots is used to multiply radical expressions.

$$\sqrt{2y}(\sqrt{3} - \sqrt{x}) = \sqrt{6y} - \sqrt{2xy}$$

Use FOIL to multiply radical expressions with two terms.

$$(3 - \sqrt{x})(5 + \sqrt{x})$$
$$= 15 + 3\sqrt{x} - 5\sqrt{x} - (\sqrt{x})^2$$
$$= 15 - 2\sqrt{x} - x$$

The Quotient Property of Square Roots [10.3B, p. 492]
If a and b are positive real numbers, then $\sqrt{\frac{a}{b}} = \frac{\sqrt{a}}{\sqrt{b}}$ and $\frac{\sqrt{a}}{\sqrt{b}} = \sqrt{\frac{a}{b}}$.

$$\frac{\sqrt{27}}{\sqrt{3}} = \sqrt{\frac{27}{3}} = \sqrt{9} = 3$$

The Quotient Property of Square Roots is used to divide radical expressions.

$$\frac{\sqrt{3x^5y}}{\sqrt{75xy^3}} = \sqrt{\frac{3x^5y}{75xy^3}} = \sqrt{\frac{x^4}{25y^2}} = \frac{x^2}{5y}$$

Simplest Form of a Radical Expression [10.3B, p. 493]
For a radical expression to be in simplest form, three conditions must be met:

1. The radicand contains no factor greater than 1 that is a perfect square.

$\sqrt{12}$, $\sqrt{\frac{3}{4}}$, and $\frac{1}{\sqrt{3}}$ are not in simplest form.

2. There is no fraction under the radical sign.

3. There is no radical in the denominator of a fraction.

$5\sqrt{3}$ and $\frac{\sqrt{3}}{3}$ are in simplest form.

Rationalizing the Denominator [10.3B, p. 493]
The procedure used to remove a radical from the denominator is called **rationalizing the denominator.**

$$\frac{5}{\sqrt{7}} = \frac{5}{\sqrt{7}} \cdot \frac{\sqrt{7}}{\sqrt{7}} = \frac{5\sqrt{7}}{7}$$

Property of Squaring Both Sides of an Equation
[10.4A, p. 497]
If a and b are real numbers and $a = b$, then $a^2 = b^2$.

$$\sqrt{x} = 5$$
$$(\sqrt{x})^2 = 5^2$$
$$x = 25$$

Solving a Radical Equation Containing One Radical
[10.4A, p. 497]

1. Write the equation with the radical alone on one side.

2. Square both sides of the equation.

3. Solve for the variable.

4. Check the solution(s) in the original equation.

$$\sqrt{2x} - 1 = 5$$
$$\sqrt{2x} = 6 \qquad \bullet \text{ Isolate the radical.}$$
$$(\sqrt{2x})^2 = 6^2 \qquad \bullet \text{ Square both sides.}$$
$$2x = 36$$
$$x = 18 \qquad \bullet \text{ Solve for } x.$$

The solution checks.

Pythagorean Theorem [10.4B, p. 499]
If a and b are lengths of the legs of a right triangle and c is the length of the hypotenuse, then $c^2 = a^2 + b^2$.

Two legs of a right triangle measure 4 cm and 7 cm. Find the length of the hypotenuse.
$$c = \sqrt{a^2 + b^2}$$
$$c = \sqrt{4^2 + 7^2} \qquad \bullet \; a = 4, b = 7$$
$$c = \sqrt{16 + 49}$$
$$c = \sqrt{65}$$
The length of the hypotenuse is $\sqrt{65}$ cm.

Copyright © Houghton Mifflin Company. All rights reserved.

Chapter 10 Review Exercises

1. Simplify: $\sqrt{3}(\sqrt{12} - \sqrt{3})$

2. Simplify: $3\sqrt{18a^5b}$

3. Simplify: $2\sqrt{36}$

4. Simplify: $\sqrt{6a}(\sqrt{3a} + \sqrt{2a})$

5. Simplify: $\dfrac{12}{\sqrt{6}}$

6. Simplify: $2\sqrt{8} - 3\sqrt{32}$

7. Simplify: $(3 - \sqrt{7})(3 + \sqrt{7})$

8. Solve: $\sqrt{x + 3} - \sqrt{x} = 1$

9. Simplify: $\dfrac{2x}{\sqrt{3} - \sqrt{5}}$

10. Simplify: $-3\sqrt{120}$

11. Solve: $\sqrt{5x} = 10$

12. Simplify: $5\sqrt{48}$

13. Simplify: $\dfrac{\sqrt{98x^7y^9}}{\sqrt{2x^3y}}$

14. Solve: $3 - \sqrt{7x} = 5$

15. Simplify: $6a\sqrt{80b} - \sqrt{180a^2b} + 5a\sqrt{b}$

16. Simplify: $4\sqrt{250}$

17. Simplify: $2x\sqrt{60x^3y^3} + 3x^2y\sqrt{15xy}$

18. Simplify: $(4\sqrt{y} - \sqrt{5})(2\sqrt{y} + 3\sqrt{5})$

Copyright © Houghton Mifflin Company. All rights reserved.

19. Simplify: $3\sqrt{12x} + 5\sqrt{48x}$

20. Solve: $\sqrt{2x - 3} + 4 = 0$

21. Simplify: $\dfrac{8}{\sqrt{x} - 3}$

22. Simplify: $4y\sqrt{243x^{17}y^9}$

23. Simplify: $y\sqrt{24y^6}$

24. Solve: $2x + 4 = \sqrt{x^2 + 3}$

25. Simplify:
$2x^2\sqrt{18x^2y^5} + 6y\sqrt{2x^6y^3} - 9xy^2\sqrt{8x^4y}$

26. Simplify: $\dfrac{16}{\sqrt{a}}$

27. **Surveying** To find the distance across a pond, a surveyor constructs a right triangle as shown at the right. Find the distance d across the pond. Round to the nearest foot.

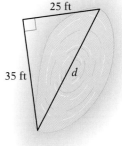

25 ft

35 ft

d

28. **Physics** The weight of an object is related to the distance the object is above the surface of Earth. An equation for this relationship is

$d = 4000\sqrt{\dfrac{W_0}{W_d}} - 4000$, where W_0 is an object's weight on the surface of Earth and W_d is the object's weight at a distance of d miles above Earth's surface. If a space explorer weighs 36 lb at a distance of 4000 mi above the surface of Earth, how much does the explorer weigh on the surface of Earth?

29. **Tsunamis** A tsunami is a great sea wave produced by underwater earthquakes or volcanic eruption. The velocity of a tsunami as it approaches land depends on the depth of the water and can be approximated by the equation $v = 3\sqrt{d}$, where d is the depth of the water in feet and v is the velocity of the tsunami in feet per second. Find the depth of the water if the velocity is 30 ft/s.

30. **Bicycle Safety** A bicycle will overturn if it rounds a corner too sharply or too fast. An equation for the maximum velocity at which a cyclist can turn a corner without tipping over is $v = 4\sqrt{r}$, where v is the velocity of the bicycle in miles per hour and r is the radius of the corner in feet. What is the radius of the sharpest corner that a cyclist can safely turn while riding at 20 mph?

Copyright © Houghton Mifflin Company. All rights reserved.

Chapter 10 Test

1. Simplify: $\sqrt{121x^8y^2}$

2. Simplify: $\sqrt{3x^2y}\sqrt{6xy^2}\sqrt{2x}$

3. Simplify: $5\sqrt{8} - 3\sqrt{50}$

4. Simplify: $\sqrt{45}$

5. Simplify: $\dfrac{\sqrt{162}}{\sqrt{2}}$

6. Solve: $\sqrt{9x} + 3 = 18$

7. Simplify: $\sqrt{32a^5b^{11}}$

8. Simplify: $\dfrac{\sqrt{98a^6b^4}}{\sqrt{2a^3b^2}}$

9. Simplify: $\dfrac{2}{\sqrt{3} - 1}$

10. Simplify: $\sqrt{8x^3y}\sqrt{10xy^4}$

Copyright © Houghton Mifflin Company. All rights reserved.

11. Solve: $\sqrt{x-5} + \sqrt{x} = 5$

12. Simplify: $3\sqrt{8y} - 2\sqrt{72x} + 5\sqrt{18y}$

13. Simplify: $\sqrt{72x^7y^2}$

14. Simplify: $(\sqrt{y} - 3)(\sqrt{y} + 5)$

15. Simplify: $2x\sqrt{3xy^3} - 2y\sqrt{12x^3y} - 3xy\sqrt{xy}$

16. Simplify: $\dfrac{2 - \sqrt{5}}{6 + \sqrt{5}}$

17. Simplify: $\sqrt{a}(\sqrt{a} - \sqrt{b})$

18. Simplify: $\sqrt{75}$

19. **Time**   Find the length of a pendulum that makes one swing in 3 s. The equation for the time of one swing of a pendulum is $T = 2\pi\sqrt{\dfrac{L}{32}}$, where T is the time in seconds and L is the length in feet. Use 3.14 for π. Round to the nearest hundredth.

20. **Camping**   A support rope for a tent is attached to the top of a pole and then secured to the ground as shown in the figure at the right. If the rope is 8 ft long and the pole is 4 ft high, how far, x, from the base of the pole should the rope be secured? Round to the nearest foot.

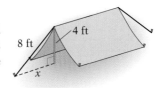

Copyright © Houghton Mifflin Company. All rights reserved.

Cumulative Review Exercises

1. Simplify:

$$\left(\frac{2}{3}\right)^2 \cdot \left(\frac{3}{4} - \frac{3}{2}\right) + \left(\frac{1}{2}\right)^2$$

2. Simplify:

$$-3[x - 2(3 - 2x) - 5x] + 2x$$

3. Solve:
$$2x - 4[3x - 2(1 - 3x)] = 2(3 - 4x)$$

4. Simplify: $(-3x^2y)(-2x^3y^4)$

5. Simplify: $\dfrac{12b^4 - 6b^2 + 2}{-6b^2}$

6. Given $f(x) = \dfrac{2x}{x - 3}$, find $f(-3)$.

7. Factor: $2a^3 - 16a^2 + 30a$

8. Multiply: $\dfrac{3x^3 - 6x^2}{4x^2 + 4x} \cdot \dfrac{3x - 9}{9x^3 - 45x^2 + 54x}$

9. Subtract: $\dfrac{x + 2}{x - 4} - \dfrac{6}{(x - 4)(x - 3)}$

10. Solve: $\dfrac{x}{2x - 5} - 2 = \dfrac{3x}{2x - 5}$

11. Find the equation of the line that contains the point $(-2, -3)$ and has slope $\frac{1}{2}$.

12. Solve by substitution:
$$4x - 3y = 1$$
$$2x + y = 3$$

13. Solve by the addition method:
$$5x + 4y = 7$$
$$3x - 2y = 13$$

14. Solve: $3(x - 7) \geq 5x - 12$

15. Simplify: $\sqrt{108}$

16. Simplify: $3\sqrt{32} - 2\sqrt{128}$

17. Simplify: $2a\sqrt{2ab^3} + b\sqrt{8a^3b} - 5ab\sqrt{ab}$

18. Simplify: $\sqrt{2a^9b}\sqrt{98ab^3}\sqrt{2a}$

Copyright © Houghton Mifflin Company. All rights reserved.

19. Simplify: $\sqrt{3}(\sqrt{6} - \sqrt{x^2})$

20. Simplify: $\dfrac{\sqrt{320}}{\sqrt{5}}$

21. Simplify: $\dfrac{3}{2 - \sqrt{5}}$

22. Solve: $\sqrt{3x - 2} - 4 = 0$

23. **Business**   The selling price of a book is $29.40. The markup rate used by the bookstore is 20%. Find the cost of the book. Use the formula $S = C + rC$, where S is the selling price, C is the cost, and r is the markup rate.

24. **Chemistry**   How many ounces of pure water must be added to 40 oz of a 12% salt solution to make a salt solution that is 5% salt?

25. **Number Sense**   The sum of two numbers is twenty-one. The product of the two numbers is one hundred four. Find the two numbers.

26. **Work**   A small water pipe takes twice as long to fill a tank as does a larger water pipe. With both pipes open, it takes 16 h to fill the tank. Find the time it would take the small pipe working alone to fill the tank.

27. Solve by graphing: $3x - 2y = 8$
$4x + 5y = 3$

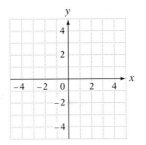

28. Graph the solution set of $3x + y \le 2$.

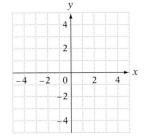

29. **Number Sense**   The square root of the sum of two consecutive integers is equal to 9. Find the smaller integer.

30. **Physics**   A stone is dropped from a building and hits the ground 5 s later. How high is the building? The equation for the distance an object falls in T seconds is $T = \sqrt{\dfrac{d}{16}}$, where d is the distance in feet.

Copyright © Houghton Mifflin Company. All rights reserved.

Copyright © Houghton Mifflin Company. All rights reserved.

chapter 11

Quadratic Equations

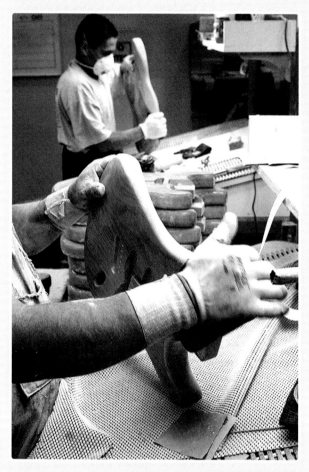

This photo shows employees at the Fender factory in Corona, California, in the process of making guitars. During this stage, the Fender Stratocaster bodies are hand-sanded. The objective of the Fender Musical Instruments Corporation, as with any business, is to earn a profit. Profit is the difference between a company's revenue (the total amount of money the company earns by selling its products or services) and its costs (the total amount of money the company spends in doing business). The **Focus on Problem Solving on page 540** uses quadratic equations to determine the maximum profit a company can earn.

OBJECTIVES

Section 11.1
A To solve a quadratic equation by factoring
B To solve a quadratic equation by taking square roots

Section 11.2
A To solve a quadratic equation by completing the square

Section 11.3
A To solve a quadratic equation by using the quadratic formula

Section 11.4
A To graph a quadratic equation of the form $y = ax^2 + bx + c$

Section 11.5
A To solve application problems

Need help? For online student resources, such as section quizzes, visit this textbook's website at **math.college.hmco.com/students**.

Do these exercises to prepare for Chapter 11.

1. Evaluate $b^2 - 4ac$ when $a = 2$, $b = -3$, and $c = -4$.

2. Solve: $5x + 4 = 3$

3. Factor: $x^2 + x - 12$

4. Factor: $4x^2 - 12x + 9$

5. Is $x^2 - 10x + 25$ a perfect square trinomial?

6. Solve: $\dfrac{5}{x - 2} = \dfrac{15}{x}$

7. Graph: $y = -2x + 3$

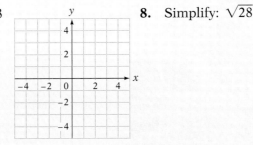

8. Simplify: $\sqrt{28}$

9. If a is *any* real number, simplify $\sqrt{a^2}$.

10. Exercising   Walking at a constant speed of 4.5 mph, Lucy and Sam walked from the beginning to the end of a hiking trail. When they reached the end, they immediately started back along the same path at a constant speed of 3 mph. If the round-trip took 2 h, what is the length of the hiking trail?

GO FIGURE • • •

Find the value of $\dfrac{1}{\sqrt{1} + \sqrt{2}} + \dfrac{1}{\sqrt{2} + \sqrt{3}} + \dfrac{1}{\sqrt{3} + \sqrt{4}} + \cdots + \dfrac{1}{\sqrt{8} + \sqrt{9}}$.

Copyright © Houghton Mifflin Company. All rights reserved.

Copyright © Houghton Mifflin Company. All rights reserved.

11.1

Solving Quadratic Equations by Factoring or by Taking Square Roots

Objective A   **To solve a quadratic equation by factoring**

An equation of the form $ax^2 + bx + c = 0$, where a, b, and c are real numbers and $a \neq 0$, is a **quadratic equation.**

$4x^2 - 3x + 1 = 0, a = 4, b = -3, c = 1$
$3x^2 - 4 = 0, a = 3, b = 0, c = -4$
$\dfrac{x^2}{2} - 2x + 4 = 0, a = \dfrac{1}{2}, b = -2, c = 4$

A quadratic equation is also called a **second-degree equation.**

A quadratic equation is in **standard form** when the polynomial is in descending order and equal to zero.

Recall that the Principle of Zero Products states that if the product of two factors is zero, then at least one of the factors must be zero.

If $a \cdot b = 0$,
then $a = 0$ or $b = 0$.

The Principle of Zero Products can be used to solve quadratic equations by factoring. Write the equation in standard form, factor the polynomial, apply the Principle of Zero Products, and solve for the variable.

HOW TO   Solve by factoring: $2x^2 - x = 1$

$$2x^2 - x = 1$$
$$2x^2 - x - 1 = 0$$ • Write the equation in standard form.
$$(2x + 1)(x - 1) = 0$$ • Factor.
$$2x + 1 = 0 \qquad x - 1 = 0$$ • Use the Principle of Zero Products to set each factor equal to zero.
$$2x = -1 \qquad\qquad x = 1$$ • Rewrite each equation in the form *variable = constant.*
$$x = -\dfrac{1}{2}$$

TAKE NOTE
You should always check your solutions by substituting the proposed solutions back into the *original* equation.

Check:

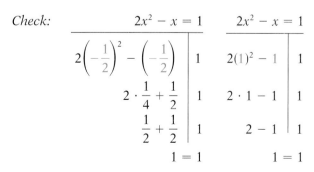

$$
\begin{array}{c|c}
2x^2 - x = 1 & 2x^2 - x = 1 \\
\hline
2\left(-\dfrac{1}{2}\right)^2 - \left(-\dfrac{1}{2}\right) \ \Big|\ 1 & 2(1)^2 - 1 \ \Big|\ 1 \\
2 \cdot \dfrac{1}{4} + \dfrac{1}{2} \ \Big|\ 1 & 2 \cdot 1 - 1 \ \Big|\ 1 \\
\dfrac{1}{2} + \dfrac{1}{2} \ \Big|\ 1 & 2 - 1 \ \Big|\ 1 \\
1 = 1 & 1 = 1
\end{array}
$$

The solutions are $-\dfrac{1}{2}$ and 1.

HOW TO   Solve by factoring: $3x^2 - 4x + 8 = (4x + 1)(x - 2)$

$3x^2 - 4x + 8 = (4x + 1)(x - 2)$

$3x^2 - 4x + 8 = 4x^2 - 7x - 2$
- Multiply the factors on the right side of the equation.

$0 = x^2 - 3x - 10$
- Write the equation in standard form.

$0 = (x - 5)(x + 2)$
- Factor.

$x - 5 = 0 \qquad x + 2 = 0$
- Use the Principle of Zero Products to set each factor equal to zero.

$x = 5 \qquad\quad x = -2$
- Rewrite each equation in the form *variable = constant*.

Check:

$3x^2 - 4x + 8 = (4x + 1)(x - 2)$	
$3(5)^2 - 4(5) + 8$	$(4[5] + 1)(5 - 2)$
$3(25) - 4(5) + 8$	$(20 + 1)(3)$
$75 - 20 + 8$	$(21)(3)$
$63 = 63$	

$3x^2 - 4x + 8 = (4x + 1)(x - 2)$	
$3(-2)^2 - 4(-2) + 8$	$(4[-2] + 1)(-2 - 2)$
$3(4) - 4(-2) + 8$	$(-8 + 1)(-4)$
$12 + 8 + 8$	$(-7)(-4)$
$28 = 28$	

The solutions are 5 and -2.

HOW TO   Solve by factoring: $x^2 - 10x + 25 = 0$

$x^2 - 10x + 25 = 0$

$(x - 5)(x - 5) = 0$
- Factor.

$x - 5 = 0 \qquad x - 5 = 0$
- Use the Principle of Zero Products.

$x = 5 \qquad\quad x = 5$
- Solve each equation for *x*.

The solution is 5.

In this last example, 5 is called a **double root** of the quadratic equation.

Example 1

Solve by factoring: $\dfrac{z^2}{2} - \dfrac{z}{4} - \dfrac{1}{4} = 0$

Solution

$\dfrac{z^2}{2} - \dfrac{z}{4} - \dfrac{1}{4} = 0$

$4\left(\dfrac{z^2}{2} - \dfrac{z}{4} - \dfrac{1}{4}\right) = 4(0)$
- Multiply each side by 4.

$2z^2 - z - 1 = 0$

$(2z + 1)(z - 1) = 0$
- Factor.

$2z + 1 = 0 \qquad z - 1 = 0$
- Principle of Zero Products

$2z = -1 \qquad\quad z = 1$

$z = -\dfrac{1}{2}$

The solutions are $-\dfrac{1}{2}$ and 1.

You Try It 1

Solve by factoring: $\dfrac{3y^2}{2} + y - \dfrac{1}{2} = 0$

Your solution

Solution on p. S27

Copyright © Houghton Mifflin Company. All rights reserved.

Objective B   **To solve a quadratic equation by taking square roots**

Consider a quadratic equation of the form $x^2 = a$. This equation can be solved by factoring.

$$x^2 = 25$$
$$x^2 - 25 = 0$$
$$(x - 5)(x + 5) = 0$$
$$x - 5 = 0 \qquad x + 5 = 0$$
$$x = 5 \qquad\qquad x = -5$$

The solutions are 5 and -5. The solutions are plus or minus the same number, which is frequently written by using $\pm$; for example, "the solutions are ± 5." An alternative method of solving this equation is suggested by the fact that ± 5 can be written as $\pm\sqrt{25}$.

Principle of Taking the Square Root of Each Side of an Equation

If $x^2 = a$, then $x = \pm\sqrt{a}$.

HOW TO   Solve by taking square roots: $x^2 = 25$

$$x^2 = 25$$
$$\sqrt{x^2} = \sqrt{25}$$
$$x = \pm\sqrt{25} = \pm 5$$

• Take the square root of each side of the equation. Then simplify.

The solutions are 5 and −5.

HOW TO   Solve by taking square roots: $3x^2 = 36$

$$3x^2 = 36$$
$$x^2 = 12$$
$$\sqrt{x^2} = \sqrt{12}$$
$$x = \pm\sqrt{12}$$
$$x = \pm 2\sqrt{3}$$

• Solve for x^2.

• Take the square root of each side.

• Simplify.

The solutions are $2\sqrt{3}$ and $-2\sqrt{3}$.

HOW TO   Solve by taking square roots: $49y^2 - 25 = 0$

$$49y^2 - 25 = 0$$
$$49y^2 = 25$$
$$y^2 = \frac{25}{49}$$
$$\sqrt{y^2} = \sqrt{\frac{25}{49}}$$
$$y = \pm\frac{5}{7}$$

• Solve for y^2.

• Take the square root of each side.

• Simplify.

The solutions are $\frac{5}{7}$ and $-\frac{5}{7}$.

Copyright © Houghton Mifflin Company. All rights reserved.

An equation that contains the square of a binomial can be solved by taking square roots.

HOW TO Solve by taking square roots: $2(x - 1)^2 - 36 = 0$

$$2(x - 1)^2 - 36 = 0$$
$$2(x - 1)^2 = 36$$ • Solve for $(x - 1)^2$.
$$(x - 1)^2 = 18$$
$$\sqrt{(x - 1)^2} = \sqrt{18}$$ • Take the square root of each side of the equation.
$$x - 1 = \pm\sqrt{18}$$
$$x - 1 = \pm 3\sqrt{2}$$ • Simplify.
$$x = 1 \pm 3\sqrt{2}$$ • Solve for x.

The solutions are $1 + 3\sqrt{2}$ and $1 - 3\sqrt{2}$.

Example 2

Solve by taking square roots:
$x^2 + 16 = 0$

Solution
$$x^2 + 16 = 0$$
$$x^2 = -16$$ • Solve for x^2.
$$\sqrt{x^2} = \sqrt{-16}$$ • Take square roots.

$\sqrt{-16}$ is not a real number.

The equation has no real number solution.

You Try It 2

Solve by taking square roots:
$x^2 + 81 = 0$

Your solution

Example 3

Solve by taking square roots:
$5(y - 4)^2 = 25$

Solution
$$5(y - 4)^2 = 25$$
$$(y - 4)^2 = 5$$ • Solve for $(y - 4)^2$.
$$\sqrt{(y - 4)^2} = \sqrt{5}$$ • Take square roots.
$$y - 4 = \pm\sqrt{5}$$ • Simplify.
$$y = 4 \pm\sqrt{5}$$ • Solve for y.

The solutions are $4 + \sqrt{5}$ and $4 - \sqrt{5}$.

You Try It 3

Solve by taking square roots:
$7(z + 2)^2 = 21$

Your solution

Solutions on p. S27

Copyright © Houghton Mifflin Company. All rights reserved.

11.1 Exercises

Objective A   To solve a quadratic equation by factoring

For Exercises 1 to 4, solve for x.

1. $(x + 3)(x - 5) = 0$

2. $x(x - 7) = 0$

3. $(2x + 5)(3x - 1) = 0$

4. $(x - 4)(2x - 7) = 0$

For Exercises 5 to 34, solve by factoring.

5. $x^2 + 2x - 15 = 0$ **6.** $t^2 + 3t - 10 = 0$ **7.** $z^2 - 4z + 3 = 0$ **8.** $s^2 - 5s + 4 = 0$

9. $p^2 + 3p + 2 = 0$ **10.** $v^2 + 6v + 5 = 0$ **11.** $x^2 - 6x + 9 = 0$ **12.** $y^2 - 8y + 16 = 0$

13. $12y^2 + 8y = 0$ **14.** $6x^2 - 9x = 0$ **15.** $r^2 - 10 = 3r$ **16.** $t^2 - 12 = 4t$

17. $3v^2 - 5v + 2 = 0$ **18.** $2p^2 - 3p - 2 = 0$ **19.** $3s^2 + 8s = 3$

20. $3x^2 + 5x = 12$ **21.** $\dfrac{3}{4}z^2 - z = -\dfrac{1}{3}$ **22.** $\dfrac{r^2}{2} = 1 - \dfrac{r}{12}$

23. $4t^2 = 4t + 3$ **24.** $5y^2 + 11y = 12$ **25.** $4v^2 - 4v + 1 = 0$

26. $9s^2 - 6s + 1 = 0$ **27.** $x^2 - 9 = 0$ **28.** $t^2 - 16 = 0$

29. $4y^2 - 1 = 0$ **30.** $9z^2 - 4 = 0$ **31.** $x + 15 = x(x - 1)$

32. $p + 18 = p(p - 2)$ **33.** $r^2 - r - 2 = (2r - 1)(r - 3)$ **34.** $s^2 + 5s - 4 = (2s + 1)(s - 4)$

Objective B   To solve a quadratic equation by taking square roots

For Exercises 35 to 61, solve by taking square roots.

35. $x^2 = 36$ **36.** $y^2 = 49$ **37.** $v^2 - 1 = 0$

38. $z^2 - 64 = 0$ **39.** $4x^2 - 49 = 0$ **40.** $9w^2 - 64 = 0$

Copyright © Houghton Mifflin Company. All rights reserved.

41. $9y^2 = 4$

42. $4z^2 = 25$

43. $16v^2 - 9 = 0$

44. $25x^2 - 64 = 0$

45. $y^2 + 81 = 0$

46. $z^2 + 49 = 0$

47. $w^2 - 24 = 0$

48. $v^2 - 48 = 0$

49. $(x - 1)^2 = 36$

50. $(y + 2)^2 = 49$

51. $2(x + 5)^2 = 8$

52. $4(z - 3)^2 = 100$

53. $9(x - 1)^2 - 16 = 0$

54. $4(y + 3)^2 - 81 = 0$

55. $49(v + 1)^2 - 25 = 0$

56. $81(y - 2)^2 - 64 = 0$

57. $(x - 4)^2 - 20 = 0$

58. $(y + 5)^2 - 50 = 0$

59. $(x + 1)^2 + 36 = 0$

60. $2\left(z - \dfrac{1}{2}\right)^2 = 12$

61. $3\left(v + \dfrac{3}{4}\right)^2 = 36$

APPLYING THE CONCEPTS

For Exercises 62 and 63, solve for x.

62. $(3x^2 - 13)^2 = 4$

63. $(6x^2 - 5)^2 = 1$

64. Investments   The value A of an initial investment of P dollars after 2 years is given by $A = P(1 + r)^2$, where r is the annual percentage rate earned by the investment. If an initial investment of $1500 grew to a value of $1782.15 in 2 years, what was the annual percentage rate?

65. Investments   An initial investment of $5000 grew to a value of $5832 in 2 years. Use the formula in Exercise 64 to find the annual percentage rate.

66. Physics   The kinetic energy of a moving body is given by $E = \dfrac{1}{2}mv^2$, where E is the kinetic energy, m is the mass, and v is the velocity in meters per second. What is the velocity of a moving body whose mass is 5 kg and whose kinetic energy is 250 newton-meters?

67. Automotive Safety   On a certain type of street surface, the equation $d = 0.0074v^2$ can be used to approximate the distance d, in feet, a car traveling v miles per hour will slide when its brakes are applied. After applying the brakes, the owner of a car involved in an accident skidded 40 ft. Did the traffic officer investigating the accident issue the car owner a ticket for speeding if the speed limit is 65 mph?

Copyright © Houghton Mifflin Company. All rights reserved.

11.2 Solving Quadratic Equations by Completing the Square

Objective A   To solve a quadratic equation by completing the square

Recall that a perfect-square trinomial is the square of a binomial.

Perfect-Square Trinomial		Square of a Binomial
$x^2 + 6x + 9$	$=$	$(x + 3)^2$
$x^2 - 10x + 25$	$=$	$(x - 5)^2$
$x^2 + 8x + 16$	$=$	$(x + 4)^2$

For each perfect-square trinomial, the square of $\frac{1}{2}$ of the coefficient of x equals the constant term.

$$x^2 + 6x + 9, \qquad \left(\frac{1}{2} \cdot 6\right)^2 = 9$$

$$x^2 - 10x + 25, \qquad \left[\frac{1}{2}(-10)\right]^2 = 25$$

$$x^2 + 8x + 16, \qquad \left(\frac{1}{2} \cdot 8\right)^2 = 16$$

Adding to a binomial the constant term that makes it a perfect-square trinomial is called **completing the square.**

HOW TO   Complete the square of $x^2 - 8x$. Write the resulting perfect-square trinomial as the square of a binomial.

$$\left[\frac{1}{2}(-8)\right]^2 = 16$$
• Find the constant term.

$$x^2 - 8x + 16$$
• Complete the square of $x^2 - 8x$ by adding the constant term.

$$x^2 - 8x + 16 = (x - 4)^2$$
• Write the resulting perfect-square trinomial as the square of a binomial.

HOW TO   Complete the square of $y^2 + 5y$. Write the resulting perfect-square trinomial as the square of a binomial.

$$\left(\frac{1}{2} \cdot 5\right)^2 = \left(\frac{5}{2}\right)^2 = \frac{25}{4}$$
• Find the constant term.

$$y^2 + 5y + \frac{25}{4}$$
• Complete the square of $y^2 + 5y$ by adding the constant term.

$$y^2 + 5y + \frac{25}{4} = \left(y + \frac{5}{2}\right)^2$$
• Write the resulting perfect-square trinomial as the square of a binomial.

Point of Interest

Early mathematicians solved quadratic equations by literally *completing the square*. For these mathematicians, all equations had geometric interpretations. They found that a quadratic equation could be solved by making certain figures into squares. See the second of the Projects and Group Activities at the end of this chapter for an idea of how this was done.

A quadratic equation that cannot be solved by factoring can be solved by completing the square. When the quadratic equation is in the form $x^2 + bx = c$, add to each side of the equation the term that completes the square on $x^2 + bx$. Factor the perfect-square trinomial, and write it as the square of a binomial. Take the square root of each side of the equation and then solve for x.

Copyright © Houghton Mifflin Company. All rights reserved.

S t u d y   T i p

This is a new skill and one that is difficult for many students. Be sure to do all you need to do in order to be successful at solving quadratic equations by completing the square: Read through the introductory material, work through the How To examples, study the paired examples, and do the You Try Its and check your solutions against the ones in the back of the book. See *AIM for Success*, page xxv.

HOW TO Solve by completing the square: $x^2 + 8x - 2 = 0$

$$x^2 + 8x - 2 = 0$$
$$x^2 + 8x = 2$$

• Add 2 to each side of the equation.

$$x^2 + 8x + \left(\frac{1}{2} \cdot 8\right)^2 = 2 + \left(\frac{1}{2} \cdot 8\right)^2$$

• Complete the square of $x^2 + 8x$. Add $\left(\frac{1}{2} \cdot 8\right)^2$ to each side of the equation.

$$x^2 + 8x + 16 = 2 + 16$$

• Simplify.

$$(x + 4)^2 = 18$$

• Factor the perfect-square trinomial.

$$\sqrt{(x + 4)^2} = \sqrt{18}$$

• Take the square root of each side of the equation.

$$x + 4 = \pm\sqrt{18}$$
$$x + 4 = \pm3\sqrt{2}$$
$$x = -4 \pm 3\sqrt{2}$$

• Solve for x.

Check:

$x^2 + 8x - 2 = 0$	
$(-4 + 3\sqrt{2})^2 + 8(-4 + 3\sqrt{2}) - 2$	0
$16 - 24\sqrt{2} + 18 - 32 + 24\sqrt{2} - 2$	0
	$0 = 0$

$x^2 + 8x - 2 = 0$	
$(-4 - 3\sqrt{2})^2 + 8(-4 - 3\sqrt{2}) - 2$	0
$16 + 24\sqrt{2} + 18 - 32 - 24\sqrt{2} - 2$	0
	$0 = 0$

The solutions are $-4 + 3\sqrt{2}$ and $-4 - 3\sqrt{2}$.

If the coefficient of the second-degree term is not 1, a step in completing the square is to multiply each side of the equation by the reciprocal of that coefficient.

HOW TO Solve by completing the square: $2x^2 - 3x + 1 = 0$

$$2x^2 - 3x + 1 = 0$$
$$2x^2 - 3x = -1$$

• Subtract 1 from each side of the equation.

$$\frac{1}{2}(2x^2 - 3x) = \frac{1}{2} \cdot (-1)$$

• For us to complete the square, the coefficient of x^2 must be 1. Multiply each side of the equation by $\frac{1}{2}$.

$$x^2 - \frac{3}{2}x = -\frac{1}{2}$$

$$x^2 - \frac{3}{2}x + \left[\frac{1}{2}\left(-\frac{3}{2}\right)\right]^2 = -\frac{1}{2} + \left[\frac{1}{2}\left(-\frac{3}{2}\right)\right]^2$$

• Complete the square. Add $\left[\frac{1}{2}\left(-\frac{3}{2}\right)\right]^2$ to each side of the equation.

$$x^2 - \frac{3}{2}x + \frac{9}{16} = -\frac{1}{2} + \frac{9}{16}$$

• Simplify.

$$\left(x - \frac{3}{4}\right)^2 = \frac{1}{16}$$

• Factor the perfect-square trinomial.

$$\sqrt{\left(x - \frac{3}{4}\right)^2} = \sqrt{\frac{1}{16}}$$

• Take the square root of each side of the equation.

$$x - \frac{3}{4} = \pm\frac{1}{4}$$

$$x = \frac{3}{4} \pm \frac{1}{4}$$

• Solve for x.

$$x = \frac{3}{4} + \frac{1}{4} = 1 \qquad x = \frac{3}{4} - \frac{1}{4} = \frac{1}{2}$$

The solutions are $\frac{1}{2}$ and 1.

Copyright © Houghton Mifflin Company. All rights reserved.

Example 1

Solve by completing the square:
$2x^2 - 4x - 1 = 0$

Solution

$2x^2 - 4x - 1 = 0$

$\quad 2x^2 - 4x = 1$ • Add 1.

$\dfrac{1}{2}(2x^2 - 4x) = \dfrac{1}{2} \cdot 1$ • Multiply by $\frac{1}{2}$.

$\quad x^2 - 2x = \dfrac{1}{2}$ • The coefficient of x^2 is 1.

Complete the square.

$x^2 - 2x + 1 = \dfrac{1}{2} + 1$ • $\left[\frac{1}{2} \cdot (-2)\right]^2 = [-1]^2 = 1$

$\quad (x - 1)^2 = \dfrac{3}{2}$ • Factor.

$\quad \sqrt{(x - 1)^2} = \sqrt{\dfrac{3}{2}}$ • Take square roots.

$\quad x - 1 = \pm\dfrac{\sqrt{6}}{2}$ • Simplify.

$\quad x = 1 \pm \dfrac{\sqrt{6}}{2}$

$x = 1 + \dfrac{\sqrt{6}}{2} \qquad x = 1 - \dfrac{\sqrt{6}}{2}$

$= \dfrac{2 + \sqrt{6}}{2} \qquad\quad = \dfrac{2 - \sqrt{6}}{2}$

Check:

$$2x^2 - 4x - 1 = 0$$

$$2\left(\dfrac{2 + \sqrt{6}}{2}\right)^2 - 4\left(\dfrac{2 + \sqrt{6}}{2}\right) - 1 \,\Big|\, 0$$

$$2\left(\dfrac{4 + 4\sqrt{6} + 6}{4}\right) - 2(2 + \sqrt{6}) - 1 \,\Big|\, 0$$

$$2 + 2\sqrt{6} + 3 - 4 - 2\sqrt{6} - 1 \,\Big|\, 0$$

$$0 = 0$$

$$2x^2 - 4x - 1 = 0$$

$$2\left(\dfrac{2 - \sqrt{6}}{2}\right)^2 - 4\left(\dfrac{2 - \sqrt{6}}{2}\right) - 1 \,\Big|\, 0$$

$$2\left(\dfrac{4 - 4\sqrt{6} + 6}{4}\right) - 2(2 - \sqrt{6}) - 1 \,\Big|\, 0$$

$$2 - 2\sqrt{6} + 3 - 4 + 2\sqrt{6} - 1 \,\Big|\, 0$$

$$0 = 0$$

The solutions are $\dfrac{2 + \sqrt{6}}{2}$ and $\dfrac{2 - \sqrt{6}}{2}$.

You Try It 1

Solve by completing the square:
$3x^2 - 6x - 2 = 0$

Your solution

Copyright © Houghton Mifflin Company. All rights reserved.

Solution on pp. S27–S28

Example 2

Solve by completing the square:
$x^2 + 4x + 5 = 0$

Solution

$x^2 + 4x + 5 = 0$

$\quad x^2 + 4x = -5$ • Subtract 5.

Complete the square.

$x^2 + 4x + 4 = -5 + 4$ • $\left(\frac{1}{2} \cdot 4\right)^2 = 2^2 = 4$

$\quad (x + 2)^2 = -1$ • Factor.

$\quad \sqrt{(x + 2)^2} = \sqrt{-1}$ • Take square roots.

$\sqrt{-1}$ is not a real number.

The quadratic equation has no real number solution.

You Try It 2

Solve by completing the square:
$x^2 + 6x + 12 = 0$

Your solution

Example 3

Solve $\frac{x^2}{4} + \frac{3x}{2} + 1 = 0$ by completing the square. Approximate the solutions to the nearest thousandth.

Solution

$\frac{x^2}{4} + \frac{3x}{2} + 1 = 0$

$4\left(\frac{x^2}{4} + \frac{3x}{2} + 1\right) = 4(0)$ • Multiply by 4.

$\quad x^2 + 6x + 4 = 0$

$\quad\quad x^2 + 6x = -4$ • Subtract 4.

Complete the square.

$x^2 + 6x + 9 = -4 + 9$ • $\left(\frac{1}{2} \cdot 6\right)^2 = 3^2 = 9$

$\quad (x + 3)^2 = 5$ • Factor.

$\quad \sqrt{(x + 3)^2} = \sqrt{5}$ • Take square roots.

$\quad\quad x + 3 = \pm\sqrt{5}$

$x + 3 = \sqrt{5} \qquad\qquad x + 3 = -\sqrt{5}$

$\quad x = -3 + \sqrt{5} \qquad\quad x = -3 - \sqrt{5}$

$\quad\quad \approx -3 + 2.236 \qquad\quad \approx -3 - 2.236$

$\quad\quad \approx -0.764 \qquad\qquad \approx -5.236$

The solutions are approximately -0.764 and -5.236.

You Try It 3

Solve $\frac{x^2}{8} + x + 1 = 0$ by completing the square. Approximate the solutions to the nearest thousandth.

Your solution

Solutions on p. S28

Copyright © Houghton Mifflin Company. All rights reserved.

11.2 Exercises

Objective A   To solve a quadratic equation by completing the square

For Exercises 1 to 4, complete the square of each binomial. Write the resulting trinomial as the square of a binomial.

1. $x^2 - 8x$

2. $x^2 + 6x$

3. $x^2 + 5x$

4. $x^2 - 3x$

For Exercises 5 to 53, solve by completing the square.

5. $x^2 + 2x - 3 = 0$

6. $y^2 + 4y - 5 = 0$

7. $z^2 - 6z - 16 = 0$

8. $w^2 + 8w - 9 = 0$

9. $x^2 = 4x - 4$

10. $z^2 = 8z - 16$

11. $v^2 - 6v + 13 = 0$

12. $x^2 + 4x + 13 = 0$

13. $y^2 + 5y + 4 = 0$

14. $v^2 - 5v - 6 = 0$

15. $w^2 + 7w = 8$

16. $y^2 + 5y = -4$

17. $v^2 + 4v + 1 = 0$

18. $y^2 - 2y - 5 = 0$

19. $x^2 + 6x = 5$

20. $w^2 - 8w = 3$

21. $\dfrac{z^2}{2} = z + \dfrac{1}{2}$

22. $\dfrac{y^2}{10} = y - 2$

23. $p^2 + 3p = 1$

24. $r^2 + 5r = 2$

25. $t^2 - 3t = -2$

26. $z^2 - 5z = -3$

27. $v^2 + v - 3 = 0$

28. $x^2 - x = 1$

29. $y^2 = 7 - 10y$

30. $v^2 = 14 + 16v$

31. $r^2 - 3r = 5$

32. $s^2 + 3s = -1$

33. $t^2 - t = 4$

34. $y^2 + y - 4 = 0$

35. $x^2 - 3x + 5 = 0$

36. $z^2 + 5z + 7 = 0$

37. $2t^2 - 3t + 1 = 0$

Copyright © Houghton Mifflin Company. All rights reserved.

38. $2x^2 - 7x + 3 = 0$ **39.** $2r^2 + 5r = 3$ **40.** $2y^2 - 3y = 9$ **41.** $2s^2 = 7s - 6$

42. $2x^2 = 3x + 20$ **43.** $2v^2 = v + 1$ **44.** $2z^2 = z + 3$ **45.** $3r^2 + 5r = 2$

46. $3t^2 - 8t = 3$ **47.** $3y^2 + 8y + 4 = 0$ **48.** $3z^2 - 10z - 8 = 0$ **49.** $4x^2 + 4x - 3 = 0$

50. $4v^2 + 4v - 15 = 0$ **51.** $6s^2 + 7s = 3$ **52.** $6z^2 = z + 2$ **53.** $6p^2 = 5p + 4$

 For Exercises 54 to 59, solve by completing the square. Approximate the solutions to the nearest thousandth.

54. $y^2 + 3y = 5$ **55.** $w^2 + 5w = 2$ **56.** $2z^2 - 3z = 7$

57. $2x^2 + 3x = 11$ **58.** $4x^2 + 6x - 1 = 0$ **59.** $4x^2 + 2x - 3 = 0$

APPLYING THE CONCEPTS

60. 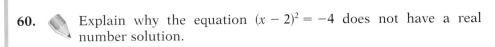 Explain why the equation $(x - 2)^2 = -4$ does not have a real number solution.

For Exercises 61 to 69, solve.

61. $\dfrac{x^2}{6} - \dfrac{x}{3} = 1$ **62.** $\sqrt{x + 2} = x - 4$ **63.** $\sqrt{3x + 4} - x = 2$

64. $\dfrac{x}{3} + \dfrac{3}{x} = \dfrac{8}{3}$ **65.** $\dfrac{x + 1}{2} + \dfrac{3}{x - 1} = 4$ **66.** $\dfrac{x - 2}{3} + \dfrac{2}{x + 2} = 4$

67. $4\sqrt{x + 1} - x = 4$ **68.** $\sqrt{2x^2 + 7} = x + 2$ **69.** $3\sqrt{x - 1} + 3 = x$

70. Sports A basketball player shoots at a basket 25 ft away. The height of the ball above the ground at time t is given by $h = -16t^2 + 32t + 6.5$. How many seconds after the ball is released does it hit the basket? *Hint:* When it hits the basket, $h = 10$ ft.

71. Sports A ball player hits a ball. The height of the ball above the ground can be approximated by the equation $h = -16t^2 + 76t + 5$. When will the ball hit the ground? *Hint:* The ball strikes the ground when $h = 0$ ft.

5 ft

Copyright © Houghton Mifflin Company. All rights reserved.

11.3 Solving Quadratic Equations by Using the Quadratic Formula

Copyright © Houghton Mifflin Company. All rights reserved.

Objective A   **To solve a quadratic equation by using the quadratic formula**

Any quadratic equation can be solved by completing the square. Applying this method to the standard form of a quadratic equation produces a formula that can be used to solve any quadratic equation.

Solve $ax^2 + bx + c = 0$ by completing the square.

$$ax^2 + bx + c = 0$$

Add the opposite of the constant term to each side of the equation.

$$ax^2 + bx + c + (-c) = 0 + (-c)$$
$$ax^2 + bx = -c$$

Multiply each side of the equation by the reciprocal of a, the coefficient of x^2.

$$\frac{1}{a}(ax^2 + bx) = \frac{1}{a}(-c)$$
$$x^2 + \frac{b}{a}x = -\frac{c}{a}$$

Complete the square by adding $\left(\frac{1}{2} \cdot \frac{b}{a}\right)^2$ to each side of the equation.

$$x^2 + \frac{b}{a}x + \left(\frac{1}{2} \cdot \frac{b}{a}\right)^2 = \left(\frac{1}{2} \cdot \frac{b}{a}\right)^2 - \frac{c}{a}$$
$$x^2 + \frac{b}{a}x + \frac{b^2}{4a^2} = \frac{b^2}{4a^2} - \frac{c}{a}$$

Simplify the right side of the equation.

$$x^2 + \frac{b}{a}x + \frac{b^2}{4a^2} = \frac{b^2}{4a^2} - \left(\frac{c}{a} \cdot \frac{4a}{4a}\right)$$
$$x^2 + \frac{b}{a}x + \frac{b^2}{4a^2} = \frac{b^2}{4a^2} - \frac{4ac}{4a^2}$$
$$x^2 + \frac{b}{a}x + \frac{b^2}{4a^2} = \frac{b^2 - 4ac}{4a^2}$$

Factor the perfect-square trinomial on the left side of the equation.

$$\left(x + \frac{b}{2a}\right)^2 = \frac{b^2 - 4ac}{4a^2}$$

Take the square root of each side of the equation.

$$\sqrt{\left(x + \frac{b}{2a}\right)^2} = \sqrt{\frac{b^2 - 4ac}{4a^2}}$$
$$x + \frac{b}{2a} = \pm\frac{\sqrt{b^2 - 4ac}}{2a}$$

Solve for x.

$$x + \frac{b}{2a} = \frac{\sqrt{b^2 - 4ac}}{2a}$$
$$x = -\frac{b}{2a} + \frac{\sqrt{b^2 - 4ac}}{2a}$$
$$= \frac{-b + \sqrt{b^2 - 4ac}}{2a}$$

$$x + \frac{b}{2a} = -\frac{\sqrt{b^2 - 4ac}}{2a}$$
$$x = -\frac{b}{2a} - \frac{\sqrt{b^2 - 4ac}}{2a}$$
$$= \frac{-b - \sqrt{b^2 - 4ac}}{2a}$$

The Quadratic Formula

The solutions of $ax^2 + bx + c = 0$, $a \neq 0$, are

$$x = \frac{-b \pm \sqrt{b^2 - 4ac}}{2a}$$

HOW TO Solve by using the quadratic formula: $2x^2 = 4x - 1$

$$2x^2 = 4x - 1$$
$$2x^2 - 4x + 1 = 0$$

$$x = \frac{-b \pm \sqrt{b^2 - 4ac}}{2a}$$

$$= \frac{-(-4) \pm \sqrt{(-4)^2 - (4 \cdot 2 \cdot 1)}}{2 \cdot 2}$$

$$= \frac{4 \pm \sqrt{16 - 8}}{4} = \frac{4 \pm \sqrt{8}}{4}$$

$$= \frac{4 \pm 2\sqrt{2}}{4} = \frac{2 \pm \sqrt{2}}{2}$$

- Write the equation in standard form.
- Subtract $4x$ from and add 1 to each side of the equation.
- The quadratic formula

- $a = 2$, $b = -4$, $c = 1$. Replace a, b, and c by their values.

- Simplify.

> TAKE NOTE
>
> $$\frac{4 \pm 2\sqrt{2}}{4} = \frac{2(2 \pm \sqrt{2})}{2 \cdot 2}$$
>
> $$= \frac{2 \pm \sqrt{2}}{2}$$

The solutions are $\dfrac{2 + \sqrt{2}}{2}$ and $\dfrac{2 - \sqrt{2}}{2}$.

Example 1

Solve by using the quadratic formula:
$2x^2 - 3x + 1 = 0$

Solution

$2x^2 - 3x + 1 = 0$ • Standard form

$$x = \frac{-(-3) \pm \sqrt{(-3)^2 - 4(2)(1)}}{2 \cdot 2}$$ • $a = 2$, $b = -3$, $c = 1$

$$= \frac{3 \pm \sqrt{9 - 8}}{4} = \frac{3 \pm \sqrt{1}}{4} = \frac{3 \pm 1}{4}$$

$$x = \frac{3 + 1}{4} \qquad\qquad x = \frac{3 - 1}{4}$$

$$= \frac{4}{4} = 1 \qquad\qquad\quad = \frac{2}{4} = \frac{1}{2}$$

The solutions are 1 and $\dfrac{1}{2}$.

You Try It 1

Solve by using the quadratic formula:
$3x^2 + 4x - 4 = 0$

Your solution

Example 2

Solve by using the quadratic formula:
$$\frac{x^2}{2} = 2x - \frac{5}{4}$$

Solution $\dfrac{x^2}{2} = 2x - \dfrac{5}{4}$

$$4\left(\frac{x^2}{2}\right) = 4\left(2x - \frac{5}{4}\right)$$ • Multiply by 4.

$$2x^2 = 8x - 5$$

$$2x^2 - 8x + 5 = 0$$ • Standard form

$$x = \frac{-(-8) \pm \sqrt{(-8)^2 - 4(2)(5)}}{2 \cdot 2}$$ • $a = 2$, $b = -8$, $c = 5$

$$= \frac{8 \pm \sqrt{64 - 40}}{4} = \frac{8 \pm \sqrt{24}}{4}$$

$$= \frac{8 \pm 2\sqrt{6}}{4} = \frac{4 \pm \sqrt{6}}{2}$$

The solutions are $\dfrac{4 + \sqrt{6}}{2}$ and $\dfrac{4 - \sqrt{6}}{2}$.

You Try It 2

Solve by using the quadratic formula:
$$\frac{x^2}{4} + \frac{x}{2} = \frac{1}{4}$$

Your solution

Solutions on p. S28

Copyright © Houghton Mifflin Company. All rights reserved.

11.3 Exercises

Objective A   **To solve a quadratic equation by using the quadratic formula**

For Exercises 1 to 38, solve by using the quadratic formula.

1. $x^2 - 4x - 5 = 0$

2. $y^2 + 3y + 2 = 0$

3. $z^2 - 2z - 15 = 0$

4. $v^2 + 5v + 4 = 0$

5. $y^2 = 2y + 3$

6. $w^2 = 3w + 18$

7. $r^2 = 5 - 4r$

8. $z^2 = 3 - 2z$

9. $2y^2 - y - 1 = 0$

10. $2t^2 - 5t + 3 = 0$

11. $w^2 + 3w + 5 = 0$

12. $x^2 - 2x + 6 = 0$

13. $p^2 - p = 0$

14. $2v^2 + v = 0$

15. $4t^2 - 9 = 0$

16. $4s^2 - 25 = 0$

17. $4y^2 + 4y = 15$

18. $6y^2 + 5y - 4 = 0$

19. $2x^2 + x + 1 = 0$

20. $3r^2 - r + 2 = 0$

21. $\dfrac{1}{2}t^2 - t = \dfrac{5}{2}$

22. $y^2 - 4y = 6$

23. $\dfrac{1}{3}t^2 + 2t - \dfrac{1}{3} = 0$

24. $z^2 + 4z + 1 = 0$

25. $w^2 = 4w + 9$

26. $y^2 = 8y + 3$

27. $9y^2 + 6y - 1 = 0$

28. $9s^2 - 6s - 2 = 0$

29. $4p^2 + 4p + 1 = 0$

30. $9z^2 + 12z + 4 = 0$

31. $\dfrac{x^2}{2} = x - \dfrac{5}{4}$

32. $r^2 = \dfrac{5}{3}r - 2$

33. $4p^2 + 16p = -11$

34. $4y^2 - 12y = -1$

35. $4x^2 = 4x + 11$

Copyright © Houghton Mifflin Company. All rights reserved.

36. $4s^2 + 12s = 3$

37. $9v^2 = -30v - 23$

38. $9t^2 = 30t + 17$

For Exercises 39 to 47, solve by using the quadratic formula. Approximate the solutions to the nearest thousandth.

39. $x^2 - 2x - 21 = 0$

40. $y^2 + 4y - 11 = 0$

41. $s^2 - 6s - 13 = 0$

42. $w^2 + 8w - 15 = 0$

43. $2p^2 - 7p - 10 = 0$

44. $3t^2 - 8t - 1 = 0$

45. $4z^2 + 8z - 1 = 0$

46. $4x^2 + 7x + 1 = 0$

47. $5v^2 - v - 5 = 0$

APPLYING THE CONCEPTS

48. Factoring, completing the square, and using the quadratic formula are three methods of solving quadratic equations. Describe each method, and cite the advantages and disadvantages of using each.

49. Explain why the equation $0x^2 + 3x + 4 = 0$ cannot be solved by the quadratic formula.

50. Solve $x^2 + ax + b = 0$ for x.

51. True or False?
 a. The equations $x = \sqrt{12 - x}$ and $x^2 = 12 - x$ have the same solutions.
 b. If $\sqrt{a} + \sqrt{b} = c$, then $a + b = c^2$.
 c. $\sqrt{9} = \pm 3$
 d. $\sqrt{x^2} = |x|$

For Exercises 52 to 57, solve.

52. $\sqrt{x + 3} = x - 3$

53. $\sqrt{x + 4} = x + 4$

54. $\sqrt{x + 1} = x - 1$

55. $\sqrt{x^2 + 2x + 1} = x - 1$

56. $\dfrac{x}{4} + \dfrac{3}{x} = \dfrac{5}{2}$

57. $\dfrac{x + 1}{5} - \dfrac{4}{x - 1} = 2$

58. Distance An L-shaped sidewalk from the parking lot to a memorial is shown in the figure at the right. The distance directly across the grass to the memorial is 650 ft. The distance to the corner is 600 ft. Find the distance from the corner to the memorial.

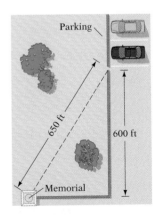

Parking

650 ft

600 ft

Memorial

59. Travel A commuter plane leaves an airport traveling due south at 400 mph. Another plane leaving at the same time travels due east at 300 mph. Find the distance between the two planes after 2 h.

Copyright © Houghton Mifflin Company. All rights reserved.

11.4 Graphing Quadratic Equations in Two Variables

Copyright © Houghton Mifflin Company. All rights reserved.

Objective A   **To graph a quadratic equation of the form $y = ax^2 + bx + c$**

TAKE NOTE

For the equation
$y = 3x^2 - x + 1$, $a = 3$,
$b = -1$, and $c = 1$.

An equation of the form $y = ax^2 + bx + c$, $a \neq 0$, is a **quadratic equation in two variables.** Examples of quadratic equations in two variables are shown at the right.

$$y = 3x^2 - x + 1$$
$$y = -x^2 - 3$$
$$y = 2x^2 - 5x$$

For these equations, y is a function of x, and we can write $f(x) = ax^2 + bx + c$. This equation represents a **quadratic function.**

HOW TO Evaluate $f(x) = 2x^2 - 3x + 4$ when $x = -2$.

$$f(x) = 2x^2 - 3x + 4$$
$$f(-2) = 2(-2)^2 - 3(-2) + 4 \qquad \bullet \text{ Replace } x \text{ by } -2.$$
$$= 2(4) + 6 + 4 = 18 \qquad \bullet \text{ Simplify.}$$

The value of the function when $x = -2$ is 18.

Point of Interest

Mirrors in some telescopes are ground into the shape of a parabola. The mirror at the Palomar Mountain Observatory is 2 ft thick at the ends and weighs 14.75 tons. The mirror has been ground to a true paraboloid (the three-dimensional version of a parabola) to within ¯0.0000015 in. A possible equation of the mirror is $y = 2640x^2$.

The graph of $y = ax^2 + bx + c$ or $f(x) = ax^2 + bx + c$ is a **parabola.** The graph is $\cup$-shaped and opens up when a is positive, and down when a is negative. The graphs of two parabolas are shown below.

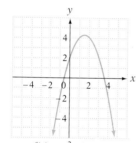

$y = 2x^2 + 3x - 2$
$a = 2$, a positive number
Parabola opens up.

$f(x) = -x^2 + 3x + 2$
$a = -1$, a negative number
Parabola opens down.

TAKE NOTE

One of the equations at the right was written as $y = 2x^2 + 3x - 2$ and the other using functional notation as $f(x) = -x^2 + 3x + 2$. Remember that y and $f(x)$ are different symbols for the same quantity.

HOW TO Graph $y = x^2 - 2x - 3$.

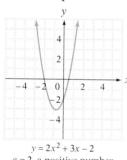

x	y
-2	5
-1	0
0	-3
1	-4
2	-3
3	0
4	5

- Find several solutions of the equation. Because the graph is not a straight line, several solutions must be found in order to determine the $\cup$-shape. Record the ordered pairs in a table.

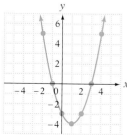

- Graph the ordered-pair solutions on a rectangular coordinate system. Draw a parabola through the points.

Integrating Technology

One of the Projects and Group Activities at the end of this chapter shows how to graph a quadratic equation by using a graphing calculator. You may want to verify the graphs you draw in this section by drawing them on a graphing calculator.

For the graph of $y = x^2 - 2x - 3$, shown here again below, note that the graph crosses the x-axis at $(-1, 0)$ and $(3, 0)$. This is also confirmed from the table for the graph (see the preceding page). From the table, note that $y = 0$ when $x = -1$ and when $x = 3$. The x-intercepts of the graph are $(-1, 0)$ and $(3, 0)$.

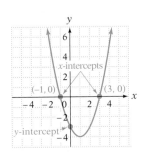

The y-intercept is the point at which the graph crosses the y-axis. At this point, $x = 0$. From the graph, we can see that the y-intercept is $(0, -3)$.

We can find the x-intercepts algebraically by letting $y = 0$ and solving for x.

$$y = x^2 - 2x - 3$$
$$0 = x^2 - 2x - 3$$
$$0 = (x + 1)(x - 3)$$
$$x + 1 = 0 \qquad x - 3 = 0$$
$$x = -1 \qquad x = 3$$

- Replace y by **0** and solve for x.
- This equation can be solved by factoring. However, it will be necessary to use the quadratic formula to solve some quadratic equations.

The x-intercepts are $(-1, 0)$ and $(3, 0)$.

Integrating Technology

The first of the Projects and Group Activities at the end of this chapter shows how to use a graphing calculator to draw the graph of a parabola and how to find the x-intercepts.

We can find the y-intercept algebraically by letting $x = 0$ and solving for y.

$$y = x^2 - 2x - 3$$
$$y = 0^2 - 2(0) - 3$$
$$= -3$$

- Replace x by **0** and simplify.

The y-intercept is $(0, -3)$.

Graph of a Quadratic Equation in Two Variables

To graph a quadratic equation in two variables, find several solutions of the equation. Graph the ordered-pair solutions on a rectangular coordinate system. Draw a parabola through the points.

To find the *x*-intercepts of the graph of a quadratic equation in two variables, let $y = 0$ and solve for x.

To find the *y*-intercept, let $x = 0$ and solve for y.

Copyright © Houghton Mifflin Company. All rights reserved.

Example 1   Graph $y = x^2 - 2x$.

Solution

x	y
-1	3
0	0
1	-1
2	0
3	3

• Find several solutions of the equation.

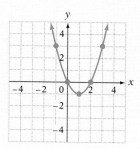

• Graph the ordered-pair solutions. Draw a parabola through the points.

You Try It 1   Graph $y = x^2 + 2$.

Your solution

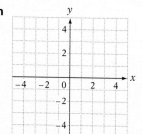

Example 2   Find the x- and y-intercepts of the graph of $y = x^2 - 2x - 5$.

Solution

To find the x-intercepts, let $y = 0$ and solve for x. This gives the equation $0 = x^2 - 2x - 5$, which is not factorable over the integers. Use the quadratic formula.

$$x = \frac{-b \pm \sqrt{b^2 - 4ac}}{2a}$$

$$= \frac{-(-2) \pm \sqrt{(-2)^2 - 4(1)(-5)}}{2(1)} \quad \bullet \; a = 1, b = -2, c = -5$$

$$= \frac{2 \pm \sqrt{24}}{2}$$

$$= \frac{2 \pm 2\sqrt{6}}{2}$$

$$= 1 \pm \sqrt{6}$$

The x-intercepts are $(1 - \sqrt{6}, 0)$ and $(1 + \sqrt{6}, 0)$.

To find the y-intercept, let $x = 0$ and solve for y.

$$y = x^2 - 2x - 5$$
$$= 0^2 - 2(0) - 5 \quad \bullet \; \text{Replace } x \text{ by } \mathbf{0}.$$
$$= -5$$

The y-intercept is $(0, -5)$.

You Try It 2   Find the x- and y-intercepts of the graph of $f(x) = x^2 - 6x + 9$.

Your solution

Solutions on pp. S28–S29

Copyright © Houghton Mifflin Company. All rights reserved.

11.4 Exercises

Objective A   To graph a quadratic equation of the form $y = ax^2 + bx + c$

For Exercises 1 to 4, determine whether the graph of the equation opens up or down.

1. $y = -\dfrac{1}{3}x^2$

2. $y = x^2 - 2x - 3$

3. $y = 2x^2 - 4$

4. $f(x) = 3 - 2x - x^2$

For Exercises 5 to 10, evaluate the function for the given value of x.

5. $f(x) = x^2 - 2x + 1; x = 3$

6. $f(x) = 2x^2 + x - 1; x = -2$

7. $f(x) = 4 - x^2; x = -3$

8. $f(x) = x^2 + 6x + 9; x = -3$

9. $f(x) = -x^2 + 5x - 6; x = -4$

10. $f(x) = -2x^2 + 2x - 1; x = -3$

For Exercises 11 to 25, graph.

11. $y = x^2$

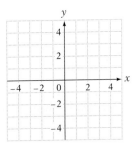

12. $y = -x^2$

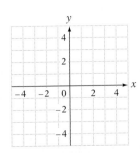

13. $y = -x^2 + 1$

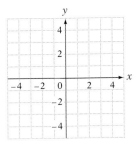

14. $y = x^2 - 1$

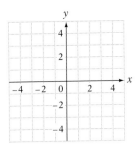

15. $y = 2x^2$

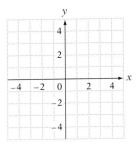

16. $y = \dfrac{1}{2}x^2$

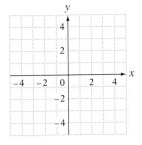

17. $y = -\dfrac{1}{2}x^2 + 1$

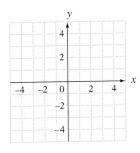

18. $y = 2x^2 - 1$

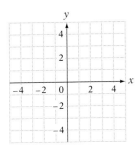

19. $y = x^2 - 4x$

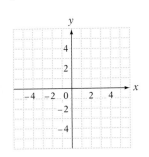

Copyright © Houghton Mifflin Company. All rights reserved.

Example 1   Graph $y = x^2 - 2x$.

Solution

x	y
-1	3
0	0
1	-1
2	0
3	3

• Find several solutions of the equation.

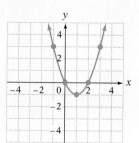

• Graph the ordered-pair solutions. Draw a parabola through the points.

You Try It 1   Graph $y = x^2 + 2$.

Your solution

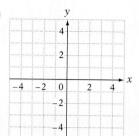

Example 2   Find the x- and y-intercepts of the graph of $y = x^2 - 2x - 5$.

Solution
To find the x-intercepts, let $y = 0$ and solve for x. This gives the equation $0 = x^2 - 2x - 5$, which is not factorable over the integers. Use the quadratic formula.

$$x = \frac{-b \pm \sqrt{b^2 - 4ac}}{2a}$$

$$= \frac{-(-2) \pm \sqrt{(-2)^2 - 4(1)(-5)}}{2(1)}$$ • $a = 1, b = -2, c = -5$

$$= \frac{2 \pm \sqrt{24}}{2}$$

$$= \frac{2 \pm 2\sqrt{6}}{2}$$

$$= 1 \pm \sqrt{6}$$

The x-intercepts are $(1 - \sqrt{6}, 0)$ and $(1 + \sqrt{6}, 0)$.

To find the y-intercept, let $x = 0$ and solve for y.

$y = x^2 - 2x - 5$
$\quad = 0^2 - 2(0) - 5$ • Replace x by 0.
$\quad = -5$

The y-intercept is $(0, -5)$.

You Try It 2   Find the x- and y-intercepts of the graph of $f(x) = x^2 - 6x + 9$.

Your solution

Copyright © Houghton Mifflin Company. All rights reserved.

Solutions on pp. S28–S29

11.4 Exercises

Objective A To graph a quadratic equation of the form $y = ax^2 + bx + c$

For Exercises 1 to 4, determine whether the graph of the equation opens up or down.

1. $y = -\dfrac{1}{3}x^2$ **2.** $y = x^2 - 2x - 3$ **3.** $y = 2x^2 - 4$ **4.** $f(x) = 3 - 2x - x^2$

For Exercises 5 to 10, evaluate the function for the given value of x.

5. $f(x) = x^2 - 2x + 1; x = 3$ **6.** $f(x) = 2x^2 + x - 1; x = -2$

7. $f(x) = 4 - x^2; x = -3$ **8.** $f(x) = x^2 + 6x + 9; x = -3$

9. $f(x) = -x^2 + 5x - 6; x = -4$ **10.** $f(x) = -2x^2 + 2x - 1; x = -3$

For Exercises 11 to 25, graph.

11. $y = x^2$

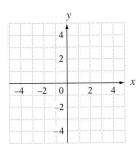

12. $y = -x^2$

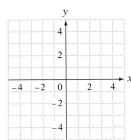

13. $y = -x^2 + 1$

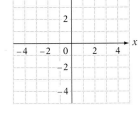

14. $y = x^2 - 1$

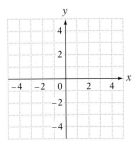

15. $y = 2x^2$

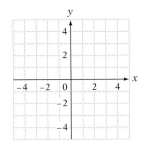

16. $y = \dfrac{1}{2}x^2$

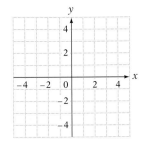

17. $y = -\dfrac{1}{2}x^2 + 1$

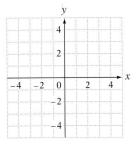

18. $y = 2x^2 - 1$

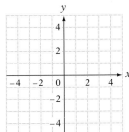

19. $y = x^2 - 4x$

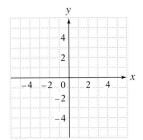

Copyright © Houghton Mifflin Company. All rights reserved.

20. $y = x^2 + 4x$

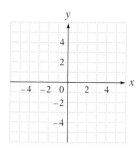

21. $y = x^2 - 2x + 3$

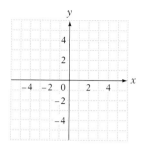

22. $y = x^2 - 4x + 2$

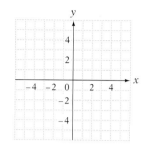

23. $y = -x^2 + 2x + 3$

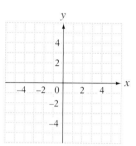

24. $y = -x^2 - 2x + 3$

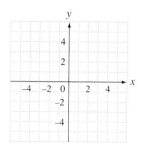

25. $y = -x^2 + 4x - 4$

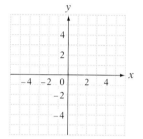

For Exercises 26 to 37, determine the x- and y-intercepts.

26. $y = x^2 - 5x + 6$

27. $y = x^2 + 5x - 6$

28. $f(x) = 9 - x^2$

29. $f(x) = x^2 + 12x + 36$

30. $y = x^2 + 2x - 6$

31. $f(x) = x^2 + 4x - 2$

32. $y = x^2 + 2x + 3$

33. $y = x^2 - x + 1$

34. $f(x) = 2x^2 - x - 3$

35. $f(x) = 2x^2 - 13x + 15$

36. $y = 4 - x - x^2$

37. $y = 2 - 3x - 3x^2$

APPLYING THE CONCEPTS

For Exercises 38 to 41, show that the equation is a quadratic equation in two variables by writing it in the form $y = ax^2 + bx + c$.

38. $y + 1 = (x - 4)^2$ **39.** $y - 2 = 3(x + 1)^2$ **40.** $y - 4 = 2(x - 3)^2$ **41.** $y + 3 = 3(x - 1)^2$

For Exercises 42 to 45, find the x-intercepts.

42. $y = x^3 - x^2 - 6x$

43. $y = x^3 - 4x^2 - 5x$

44. $y = x^3 + x^2 - 4x - 4$
    Hint: Factor by grouping.

45. $y = x^3 + 3x^2 - x - 3$
    Hint: Factor by grouping.

Copyright © Houghton Mifflin Company. All rights reserved.

11.5 Application Problems

Copyright © Houghton Mifflin Company. All rights reserved.

Objective A To solve application problems

The application problems in this section are varieties of those problems solved earlier in the text. Each of the strategies for the problems in this section will result in a quadratic equation.

HOW TO In 5 h, two campers rowed 12 mi down a stream and then rowed back to their campsite. The rate of the stream's current was 1 mph. Find the rate at which the campers rowed.

> **Strategy for Solving an Application Problem**
> 1. Determine the type of problem. For example, is it a distance-rate problem, a geometry problem, or a work problem?

The problem is a distance-rate problem.

> 2. Choose a variable to represent the unknown quantity. Write numerical or variable expressions for all the remaining quantities. These results can be recorded in a table.

The unknown rate of the campers: r

	Distance	÷	**Rate**	=	**Time**
Downstream	12	÷	$r + 1$	=	$\dfrac{12}{r + 1}$
Upstream	12	÷	$r - 1$	=	$\dfrac{12}{r - 1}$

> 3. Determine how the quantities are related.

TAKE NOTE

The time going downstream plus the time going upstream is equal to the time of the entire trip.

The total time of the trip was 5 h.

$$\frac{12}{r + 1} + \frac{12}{r - 1} = 5$$

$$(r + 1)(r - 1)\left(\frac{12}{r + 1} + \frac{12}{r - 1}\right) = (r + 1)(r - 1)5$$

$$(r - 1)12 + (r + 1)12 = (r^2 - 1)5$$

$$12r - 12 + 12r + 12 = 5r^2 - 5$$

$$24r = 5r^2 - 5$$

$$0 = 5r^2 - 24r - 5$$

$$0 = (5r + 1)(r - 5)$$

$$5r + 1 = 0 \qquad r - 5 = 0$$

$$5r = -1 \qquad r = 5$$

$$r = -\frac{1}{5}$$

TAKE NOTE

The solution $r = -\dfrac{1}{5}$ is not possible, because the rate cannot be a negative number.

The rowing rate was 5 mph.

Example 1

A painter and the painter's apprentice working together can paint a room in 2 h. Working alone, the apprentice requires 3 more hours to paint the room than the painter requires working alone. How long does it take the painter working alone to paint the room?

Strategy

- This is a work problem.
- Time for the painter to paint the room: t
  Time for the apprentice to paint the room: $t + 3$

	Rate	*Time*	*Part*
Painter	$\dfrac{1}{t}$	2	$\dfrac{2}{t}$
Apprentice	$\dfrac{1}{t + 3}$	2	$\dfrac{2}{t + 3}$

- The sum of the parts of the task completed must equal 1.

Solution

$$\frac{2}{t} + \frac{2}{t + 3} = 1$$

$$t(t + 3)\left(\frac{2}{t} + \frac{2}{t + 3}\right) = t(t + 3) \cdot 1$$

$$(t + 3)2 + t(2) = t(t + 3)$$

$$2t + 6 + 2t = t^2 + 3t$$

$$4t + 6 = t^2 + 3t$$

$$0 = t^2 - t - 6$$

$$0 = (t - 3)(t + 2)$$

$$t - 3 = 0 \qquad t + 2 = 0$$
$$t = 3 \qquad\quad t = -2$$

The solution $t = -2$ is not possible.

The time is 3 h.

You Try It 1

The length of a rectangle is 2 m more than the width. The area is 15 m². Find the width.

Your strategy

Your solution

Solution on p. S29

Copyright © Houghton Mifflin Company. All rights reserved.

11.5 Exercises

Objective A **To solve application problems**

1. **Geometry**   The height of a triangle is 2 m more than twice the length of the base. The area of the triangle is 20 m². Find the height of the triangle and the length of the base.

2. **Geometry**   The length of a rectangle is 4 ft more than twice the width. The area of the rectangle is 160 ft². Find the length and width of the rectangle.

3. **Sports**   The area of the batter's box on a major-league baseball field is 24 ft². The length of the batter's box is 2 ft more than the width. Find the length and width of the batter's box.

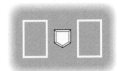

4. **Sports**   The length of the batter's box on a softball field is 1 ft less than twice the width. The area of the batter's box is 15 ft². Find the length and width of the batter's box.

5. **Sports**   The length of a swimming pool is twice the width. The area of the pool is 5000 ft². Find the length and width of the pool.

6. **Sports**   The length of a singles tennis court is 24 ft more than twice the width. The area of the tennis court is 2106 ft². Find the length and width of the court.

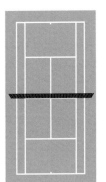

7. **Sports**   The hang time of a football that is kicked on the opening kick-off is given by $s = -16t^2 + 88t + 4$, where s is the height of the football t seconds after leaving the kicker's foot. What is the hang time of a kickoff that hits the ground without being caught? Round to the nearest tenth.

8. **Manufacturing**   A square piece of cardboard is to be formed into a box to transport pizzas. The box is formed by cutting 2-inch square corners from the cardboard and folding them up as shown in the figure at the right. If the volume of the box is 512 in³, what are the dimensions of the cardboard?

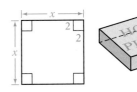

9. **Automotive Safety**   The distance, s, a car needs to come to a stop on a certain surface depends on the velocity, v, in feet per second, of the car when the brakes are applied. The equation is given by $s = 0.0344v^2 - 0.758v$. What is the maximum velocity a car can have when the brakes are applied and stop within 150 ft?

10. **Landscaping**   The perimeter of a rectangular garden is 54 ft. The area of the garden is 180 ft². Find the length and width of the garden.

11. **Food Preparation**   The radius of a large pizza is 1 in. less than twice the radius of a small pizza. The difference between the areas of the two pizzas is 33π in². Find the radius of the large pizza.

Copyright © Houghton Mifflin Company. All rights reserved.

12. **Geometry**   The hypotenuse of a right triangle is $\sqrt{13}$ cm. One leg is 1 cm shorter than twice the length of the other leg. Find the lengths of the legs of the right triangle.

13. **Computer Computations**   One computer takes 21 min longer than a second computer to calculate the value of a complex equation. Working together, these computers complete the calculation in 10 min. How long would it take each computer, working separately, to calculate the value?

14. **Plumbing**   A tank has two drains. One drain takes 16 min longer to empty the tank than does a second drain. With both drains open, the tank is emptied in 6 min. How long would it take each drain, working alone, to empty the tank?

15. **Transportation**   Using one engine of a ferryboat, it takes 6 h longer to cross a channel than it does using a second engine alone. With both engines operating, the ferryboat can make the crossing in 4 h. How long would it take each engine, working alone, to power the ferryboat across the channel?

16. **Masonry**   An apprentice mason takes 8 h longer to build a small fireplace than an experienced mason. Working together, they can build the fireplace in 3 h. How long would it take each mason, working alone, to complete the fireplace?

17. **Travel**   It took a small plane 2 h longer to fly 375 mi against the wind than to fly the same distance with the wind. The rate of the wind was 25 mph. Find the rate of the plane in calm air.

18. **Travel**   It took a motorboat 1 h longer to travel 36 mi against the current than to go 36 mi with the current. The rate of the current was 3 mph. Find the rate of the boat in calm water.

APPLYING THE CONCEPTS

19. **Food Preparation**   If a pizza with a diameter of 8 in. costs $6, what should be the cost of a pizza with a diameter of 16 in. if both pizzas cost the same amount per square inch?

20. **Geometry**   A wire 8 ft long is cut into two pieces. A circle is formed from one piece, and a square is formed from the other. The total area of both figures is given by $A = \frac{1}{16}(8 - x)^2 + \frac{x^2}{4\pi}$. What is the length of each piece of wire if the total area is 4.5 ft²?

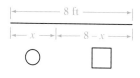

Copyright © Houghton Mifflin Company. All rights reserved.

Focus on Problem Solving

Algebraic Manipulation and Graphing Techniques

Problem solving is often easier when we have both algebraic manipulation and graphing techniques at our disposal. Solving quadratic equations and graphing quadratic equations in two variables are used here to solve problems involving profit.

A company's revenue, R, is the total amount of money the company earned by selling its products. The cost, C, is the total amount of money the company spent to manufacture and sell its products. A company's profit, P, is the difference between the revenue and the cost: $P = R - C$. A company's revenue and cost may be represented by equations.

A company manufactures and sells woodstoves. The total monthly cost, in dollars, to produce n woodstoves is $C = 30n + 2000$. Write a variable expression for the company's monthly profit if the revenue, in dollars, obtained from selling all n woodstoves is $R = 150n - 0.4n^2$.

$P = R - C$

$P = 150n - 0.4n^2 - (30n + 2000)$ • Replace R by $150n - 0.4n^2$ and C by $30n + 2000$. Then simplify.

$P = -0.4n^2 + 120n - 2000$

How many woodstoves must the company manufacture and sell in order to make a profit of $6000 a month?

$P = -0.4n^2 + 120n - 2000$

$6000 = -0.4n^2 + 120n - 2000$ • Substitute 6000 for P.

$0 = -0.4n^2 + 120n - 8000$ • Write the equation in standard form.

$0 = n^2 - 300n + 20{,}000$ • Divide each side of the equation by -0.4.

$0 = (n - 100)(n - 200)$ • Factor.

$n - 100 = 0 \qquad n - 200 = 0$ • Solve for n.

$n = 100 \qquad\quad n = 200$

The company will make a monthly profit of $6000 if either 100 or 200 woodstoves are manufactured and sold.

The graph of $P = -0.4n^2 + 120n - 2000$ is shown at the right. Note that when $P = 6000$, the values of n are 100 and 200.

Also note that the coordinates of the highest point on the graph are (150, 7000). This means that the company makes a *maximum* profit of $7000 per month when 150 woodstoves are manufactured and sold.

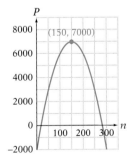

1. The total cost, in dollars, for a company to produce and sell n guitars per month is $C = 240n + 1200$. The company's revenue, in dollars, from selling all n guitars is $R = 400n - 2n^2$.

   a. How many guitars must the company produce and sell each month in order to make a monthly profit of $1200?

   b. Graph the profit equation. What is the maximum monthly profit that the company can make?

Copyright © Houghton Mifflin Company. All rights reserved.

Projects and Group Activities

Graphical Solutions of Quadratic Equations

A real number x is called a **zero of a function** if the function evaluated at x is 0. That is, if $f(x) = 0$, then x is called a zero of the function. For instance, evaluating $f(x) = x^2 + x - 6$ when $x = -3$, we have

$$f(x) = x^2 + x - 6$$
$$f(-3) = (-3)^2 + (-3) - 6 \qquad \bullet \text{ Replace } x \text{ by } -3.$$
$$f(-3) = 9 - 3 - 6 = 0$$

For this function $f(-3) = 0$, so -3 is a zero of the function.

Verify that 2 is a zero of $f(x) = x^2 + x - 6$ by showing that $f(2) = 0$.

The graph of $f(x) = x^2 + x - 6$ is shown at the left. Note that the graph crosses the x-axis at -3 and 2, the two zeros of the function. The points $(-3, 0)$ and $(2, 0)$ are x-intercepts of the graph.

Or consider the equation $0 = x^2 + x - 6$, which is $f(x) = x^2 + x - 6$ with $f(x)$ replaced by 0. Solving $0 = x^2 + x - 6$, we have

$$0 = x^2 + x - 6$$
$$0 = (x + 3)(x - 2) \qquad \bullet \text{ Solve by factoring and using the}$$
$$x + 3 = 0 \qquad x - 2 = 0 \qquad \text{Principle of Zero Products.}$$
$$x = -3 \qquad x = 2$$

Observe that the solutions of the equation are the zeros of the function. This important connection among the real zeros of a function, the x-intercepts of its graph, and the solutions of the equation is the basis for using a graphing calculator to solve an equation.

The following method of solving a quadratic equation by using a graphing calculator is based on a TI-83 or TI-83 Plus calculator. Other calculators will require a slightly different approach.

HOW TO Approximate the solutions of $x^2 + 4x = 6$ by using a graphing calculator.

Write the equation in standard form. $x^2 + 4x = 6$

1. Press [Y=] and enter $x^2 + 4x - 6$ for Y1. $x^2 + 4x - 6 = 0$

2. Press [GRAPH]. If the graph does not appear on the screen, press [ZOOM] 6.

3. Press [2nd] CALC 2. Note that the selection for 2 says **zero**. This will begin the calculation of the zeros of the function, which are the solutions of the equation.

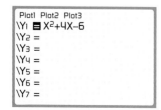

Step 1

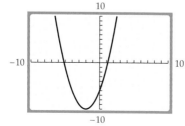

Step 2

Step 3

Copyright © Houghton Mifflin Company. All rights reserved.

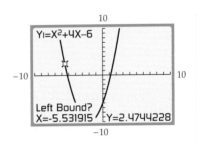

Step 4

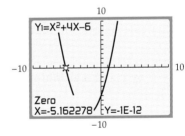

Step 5

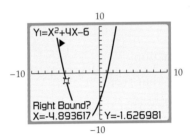

Step 6

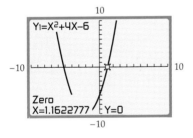

Step 7

4. At the bottom of the screen you will see **LeftBound?** This is asking you to move the blinking cursor so that it is to the *left* of the first *x*-intercept. Use the left arrow key to move the cursor to the left of the first *x*-intercept. The values of *x* and *y* that appear on your calculator may be different from the ones shown here. Just be sure that you are to the left of the *x*-intercept. When you are done, press [ENTER].

5. At the bottom of the screen you will see **RightBound?** This is asking you to move the blinking cursor so that it is to the *right* of the *x*-intercept. Use the right arrow key to move the cursor to the right of the *x*-intercept. The values of *x* and *y* that appear on your calculator may be different from the ones shown here. Just be sure that you are to the right of the *x*-intercept. When you are done, press [ENTER].

6. At the bottom of the screen you will see **Guess?** Press [ENTER].

7. The zero of the function is approximately -5.162278. Thus one solution of $x^2 + 4x = 6$ is -5.162278. Also note that the value of *y* is given as $\text{Y1} = \,^{-}1\text{E}^{-}12$. This is the way the calculator writes a number in scientific notation. We would normally write $\text{Y1} = -1 \times 10^{-12}$. This number is very close to zero.

To find the other solution, we repeat Steps 3 through 7. The screens are shown below.

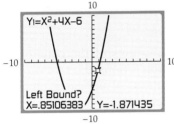

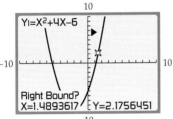

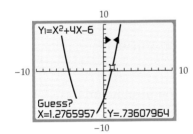

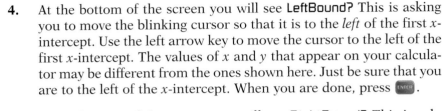

A second zero of the function is approximately 1.1622777. Thus the two solutions of $x^2 + 4x = 6$ are approximately -5.162278 and 1.1622777.

Use a graphing calculator to approximate the solutions of the following equations.

1. $x^2 + 3x - 4 = 0$

2. $x^2 - 4x - 5 = 0$

3. $x^2 + 3.4x = 4.15$

4. $2x^2 - \dfrac{5}{9}x = \dfrac{3}{8}$

5. $\pi x^2 - \sqrt{17}x - 2 = 0$

6. $\sqrt{2}x^2 + x - \sqrt{7} = 0$

Copyright © Houghton Mifflin Company. All rights reserved.

Geometric Construction of Completing the Square

Completing the square as a method for solving a quadratic equation has been known for centuries. The Persian mathematician Al-Khwarismi used this method in a textbook written around 825 A.D. The method was very geometric. That is, Al-Khwarismi literally completed a square. To understand how this method works, consider the following geometric shapes: a square whose area is x^2, a rectangle whose area is x, and another square whose area is 1.

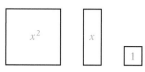

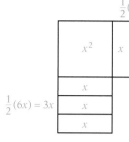

Now consider the expression $x^2 + 6x$. From our discussion in this chapter, to complete the square, we added $\left(\frac{1}{2} \cdot 6\right)^2 = 3^2 = 9$ to the expression. The geometric construction that Al-Khwarismi used is shown at the left.

Note that it is necessary to add 9 squares to the figure to "complete the square." One of the difficulties of using a geometric method such as this is that it cannot easily be extended to $x^2 - 6x$. There is no way to draw an area of $-6x$! That really did not bother Al-Khwarismi much. Negative numbers were not a significant part of mathematics until well into the 13th century.

1. Show how Al-Khwarismi would have completed the square for $x^2 + 4x$.
2. Show how Al-Khwarismi would have completed the square for $x^2 + 10x$.

Chapter 11 Summary

Key Words	**Examples**
A *quadratic equation* is an equation that can be written in the form $ax^2 + bx + c = 0$, where a, b, and c are real numbers and $a \neq 0$. [11.1A, p. 515]	$3x^2 - 5x - 3 = 0$ is a quadratic equation. For this equation, $a = 3$, $b = -5$, $c = -3$.
A quadratic equation is in *standard form* when the polynomial is in descending order and equal to zero. [11.1A, p. 515]	$2x - 4 + 5x^2 = 0$ is not in standard form. The same equation in standard form is $5x^2 + 2x - 4 = 0$.
Adding to a binomial the constant term that makes it a perfect-square trinomial is called *completing the square*. [11.2A, p. 521]	Adding to $x^2 - 8x$ the constant term 16 results in a perfect square trinomial: $x^2 - 8x + 16 = (x - 4)^2$.
An equation of the form $y = ax^2 + bx + c$, $a \neq 0$, is a *quadratic equation in two variables*. [11.4A, p. 531]	$y = 2x^2 + 3x - 4$ is a quadratic equation in two variables.

Copyright © Houghton Mifflin Company. All rights reserved.

The graph of an equation of the form $y = ax^2 + bx + c$, $a \neq 0$, is a *parabola*. The graph is U-shaped and opens up when $a > 0$ and opens down when $a < 0$. [11.4A, p. 531]

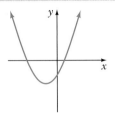

$a > 0$
Parabola opens up

$a < 0$
Parabola opens down

Essential Rules and Procedures

Examples

Solving a Quadratic Equation by Factoring [11.1A, p. 515]
Write the equation in standard form, factor the polynomial, apply the Principle of Zero Products, and solve for the variable.

$$x^2 - 3x = 10$$
$$x^2 - 3x - 10 = 0$$
$$(x + 2)(x - 5) = 0$$
$$x + 2 = 0 \qquad x - 5 = 0$$
$$x = -2 \qquad x = 5$$

Principle of Taking the Square Root of Each Side of an Equation [11.1B, p. 517]
If $x^2 = a$, then $x = \pm\sqrt{a}$.

This principle is used to solve quadratic equations by taking square roots.

$$2x^2 - 36 = 0$$
$$2x^2 = 36$$
$$x^2 = 18$$
$$\sqrt{x^2} = \sqrt{18}$$
$$x = \pm\sqrt{18} = \pm 3\sqrt{2}$$

Solving a Quadratic Equation by Completing the Square [11.2A, p. 521]
When the quadratic equation is in the form $x^2 + bx = c$, add to each side of the equation the term that completes the square on $x^2 + bx$. Factor the perfect-square trinomial, and write it as the square of a binomial. Take the square root of each side of the equation and solve for x.

$$x^2 + 6x = 5$$
$$x^2 + 6x + 9 = 5 + 9$$
$$(x + 3)^2 = 14$$
$$\sqrt{(x + 3)^2} = \sqrt{14}$$
$$x + 3 = \pm\sqrt{14}$$
$$x = -3 \pm \sqrt{14}$$

The Quadratic Formula [11.3A, p. 527]
The solutions of $ax^2 + bx + c = 0$, $a \neq 0$, are
$$x = \frac{-b \pm \sqrt{b^2 - 4ac}}{2a}.$$

$$2x^2 + 3x - 6 = 0$$
$$x = \frac{-b \pm \sqrt{b^2 - 4ac}}{2a}$$
$$= \frac{-3 \pm \sqrt{(3)^2 - 4(2)(-6)}}{2(2)}$$
$$= \frac{-3 \pm \sqrt{9 + 48}}{4} = \frac{-3 \pm \sqrt{57}}{4}$$

Graph of a Quadratic Equation in Two Variables [11.4A, p. 532]
To graph a quadratic equation in two variables, find several solutions of the equation. Graph the ordered-pair solutions on a rectangular coordinate system. Draw a parabola through the points.

To find the x-intercepts of the graph of a quadratic equation in two variables, let $y = 0$ and solve for x.

To find the y-intercept, let $x = 0$ and solve for y.

$$y = x^2 - x - 2$$

x	y
-2	4
-1	0
0	-2
1	-2
2	0
3	4

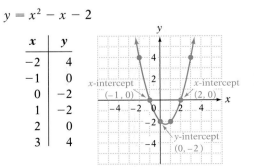

x-intercept $(-1, 0)$
x-intercept $(2, 0)$
y-intercept $(0, -2)$

Copyright © Houghton Mifflin Company. All rights reserved.

Geometric Construction of Completing the Square

Completing the square as a method for solving a quadratic equation has been known for centuries. The Persian mathematician Al-Khwarismi used this method in a textbook written around 825 A.D. The method was very geometric. That is, Al-Khwarismi literally completed a square. To understand how this method works, consider the following geometric shapes: a square whose area is x^2, a rectangle whose area is x, and another square whose area is 1.

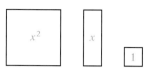

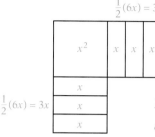

Now consider the expression $x^2 + 6x$. From our discussion in this chapter, to complete the square, we added $\left(\frac{1}{2} \cdot 6\right)^2 = 3^2 = 9$ to the expression. The geometric construction that Al-Khwarismi used is shown at the left.

Note that it is necessary to add 9 squares to the figure to "complete the square." One of the difficulties of using a geometric method such as this is that it cannot easily be extended to $x^2 - 6x$. There is no way to draw an area of $-6x$! That really did not bother Al-Khwarismi much. Negative numbers were not a significant part of mathematics until well into the 13th century.

1. Show how Al-Khwarismi would have completed the square for $x^2 + 4x$.

2. Show how Al-Khwarismi would have completed the square for $x^2 + 10x$.

9 squares were added

Chapter 11 Summary

Key Words

Examples

A *quadratic equation* is an equation that can be written in the form $ax^2 + bx + c = 0$, where a, b, and c are real numbers and $a \neq 0$. [11.1A, p. 515]

$3x^2 - 5x - 3 = 0$ is a quadratic equation. For this equation, $a = 3$, $b = -5$, $c = -3$.

A quadratic equation is in *standard form* when the polynomial is in descending order and equal to zero. [11.1A, p. 515]

$2x - 4 + 5x^2 = 0$ is not in standard form. The same equation in standard form is $5x^2 + 2x - 4 = 0$.

Adding to a binomial the constant term that makes it a perfect-square trinomial is called *completing the square*. [11.2A, p. 521]

Adding to $x^2 - 8x$ the constant term 16 results in a perfect square trinomial: $x^2 - 8x + 16 = (x - 4)^2$.

An equation of the form $y = ax^2 + bx + c$, $a \neq 0$, is a *quadratic equation in two variables*. [11.4A, p. 531]

$y = 2x^2 + 3x - 4$ is a quadratic equation in two variables.

Copyright © Houghton Mifflin Company. All rights reserved.

The graph of an equation of the form $y = ax^2 + bx + c, a \neq 0$, is a *parabola*. The graph is U-shaped and opens up when $a > 0$ and opens down when $a < 0$. [11.4A, p. 531]

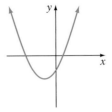

$a > 0$
Parabola opens up

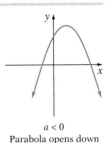

$a < 0$
Parabola opens down

Essential Rules and Procedures	**Examples**

Solving a Quadratic Equation by Factoring [11.1A, p. 515]
Write the equation in standard form, factor the polynomial, apply the Principle of Zero Products, and solve for the variable.

$$x^2 - 3x = 10$$
$$x^2 - 3x - 10 = 0$$
$$(x + 2)(x - 5) = 0$$
$$x + 2 = 0 \qquad x - 5 = 0$$
$$x = -2 \qquad x = 5$$

Principle of Taking the Square Root of Each Side of an Equation [11.1B, p. 517]
If $x^2 = a$, then $x = \pm\sqrt{a}$.

This principle is used to solve quadratic equations by taking square roots.

$$2x^2 - 36 = 0$$
$$2x^2 = 36$$
$$x^2 = 18$$
$$\sqrt{x^2} = \sqrt{18}$$
$$x = \pm\sqrt{18} = \pm 3\sqrt{2}$$

Solving a Quadratic Equation by Completing the Square [11.2A, p. 521]
When the quadratic equation is in the form $x^2 + bx = c$, add to each side of the equation the term that completes the square on $x^2 + bx$. Factor the perfect-square trinomial, and write it as the square of a binomial. Take the square root of each side of the equation and solve for x.

$$x^2 + 6x = 5$$
$$x^2 + 6x + 9 = 5 + 9$$
$$(x + 3)^2 = 14$$
$$\sqrt{(x + 3)^2} = \sqrt{14}$$
$$x + 3 = \pm\sqrt{14}$$
$$x = -3 \pm \sqrt{14}$$

The Quadratic Formula [11.3A, p. 527]
The solutions of $ax^2 + bx + c = 0, a \neq 0$, are

$$x = \frac{-b \pm \sqrt{b^2 - 4ac}}{2a}.$$

$$2x^2 + 3x - 6 = 0$$
$$x = \frac{-b \pm \sqrt{b^2 - 4ac}}{2a}$$
$$= \frac{-3 \pm \sqrt{(3)^2 - 4(2)(-6)}}{2(2)}$$
$$= \frac{-3 \pm \sqrt{9 + 48}}{4} = \frac{-3 \pm \sqrt{57}}{4}$$

Graph of a Quadratic Equation in Two Variables [11.4A, p. 532]
To graph a quadratic equation in two variables, find several solutions of the equation. Graph the ordered-pair solutions on a rectangular coordinate system. Draw a parabola through the points.

To find the x-intercepts of the graph of a quadratic equation in two variables, let $y = 0$ and solve for x.

To find the y-intercept, let $x = 0$ and solve for y.

$y = x^2 - x - 2$

x	y
-2	4
-1	0
0	-2
1	-2
2	0
3	4

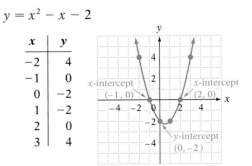

Copyright © Houghton Mifflin Company. All rights reserved.

Chapter 11 Review Exercises

1. Solve by factoring: $6x^2 + 13x - 28 = 0$

2. Solve by taking square roots:
$49x^2 = 25$

3. Solve by completing the square:
$x^2 + 2x - 24 = 0$

4. Solve by using the quadratic formula:
$x^2 + 5x - 6 = 0$

5. Solve by completing the square:
$2x^2 + 5x = 12$

6. Solve by factoring: $12x^2 + 10 = 29x$

7. Solve by taking square roots:
$(x + 2)^2 - 24 = 0$

8. Solve by using the quadratic formula:
$2x^2 + 3 = 5x$

9. Solve by factoring: $6x(x + 1) = x - 1$

10. Solve by taking square roots:
$4y^2 + 9 = 0$

11. Solve by completing the square:
$x^2 - 4x + 1 = 0$

12. Solve by using the quadratic formula:
$x^2 - 3x - 5 = 0$

13. Solve by completing the square:
$x^2 + 6x + 12 = 0$

14. Solve by factoring: $(x + 9)^2 = x + 11$

15. Solve by taking square roots:
$$\left(x - \frac{1}{2}\right)^2 = \frac{9}{4}$$

16. Solve by completing the square:
$4x^2 + 16x = 7$

Copyright © Houghton Mifflin Company. All rights reserved.

17. Solve by using the quadratic formula:
$x^2 - 4x + 8 = 0$

18. Solve by using the quadratic formula:
$2x^2 + 5x + 2 = 0$

19. Graph $y = -3x^2$.

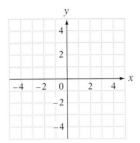

20. Graph $y = -\dfrac{1}{4}x^2$.

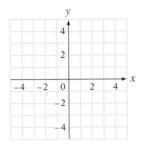

21. Graph $y = 2x^2 + 1$.

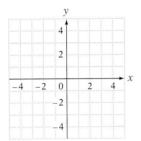

22. Graph $y = x^2 - 4x + 3$.

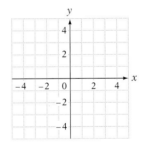

23. Graph $y = -x^2 + 4x - 5$.

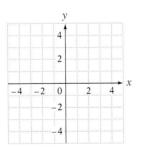

24. Find the x- and y-intercepts of the graph of
$y = x^2 - 2x - 15$.

25. **Travel**   It took a hawk half an hour longer to fly 70 mi against the wind than to fly 40 mi with the wind. The rate of the wind was 5 mph. Find the rate of the hawk in calm air.

Copyright © Houghton Mifflin Company. All rights reserved.

Chapter 11 Test

1. Solve by factoring: $x^2 - 5x - 6 = 0$

2. Solve by factoring: $3x^2 + 7x = 20$

3. Solve by taking square roots:
$2(x - 5)^2 - 50 = 0$

4. Solve by taking square roots:
$3(x + 4)^2 - 60 = 0$

5. Solve by completing the square:
$x^2 + 4x - 16 = 0$

6. Solve by completing the square:
$x^2 + 3x = 8$

7. Solve by completing the square:
$2x^2 - 6x + 1 = 0$

8. Solve by completing the square:
$2x^2 + 8x = 3$

9. Solve by using the quadratic formula:
$x^2 + 4x + 2 = 0$

10. Solve by using the quadratic formula:
$x^2 - 3x = 6$

Copyright © Houghton Mifflin Company. All rights reserved.

11. Solve by using the quadratic formula:
$2x^2 - 5x - 3 = 0$

12. Solve by using the quadratic formula:
$3x^2 - x = 1$

13. Graph $y = x^2 + 2x - 4$.

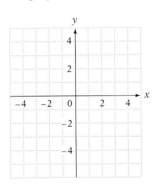

14. Find the x- and y-intercepts of the graph of $f(x) = x^2 + x - 12$.

15. **Geometry** The length of a rectangle is 2 ft less than twice the width. The area of the rectangle is 40 ft². Find the length and width of the rectangle.

16. **Travel** It took a motorboat 1 h longer to travel 60 mi against a current than it took the boat to travel 60 mi with the current. The rate of the current was 1 mph. Find the rate of the boat in calm water.

Copyright © Houghton Mifflin Company. All rights reserved.

Cumulative Review Exercises

1. Simplify: $2x - 3[2x - 4(3 - 2x) + 2] - 3$

2. Solve: $-\dfrac{3}{5}x = -\dfrac{9}{10}$

3. Solve: $2x - 3(4x - 5) = -3x - 6$

4. Simplify: $(2a^2b)^2(-3a^4b^2)$

5. Divide: $(x^2 - 8) \div (x - 2)$

6. Factor: $3x^3 + 2x^2 - 8x$

7. Divide: $\dfrac{3x^2 - 6x}{4x - 6} \div \dfrac{2x^2 + x - 6}{6x^3 - 24x}$

8. Subtract: $\dfrac{x}{2(x - 1)} - \dfrac{1}{(x - 1)(x + 1)}$

9. Simplify: $\dfrac{1 - \dfrac{7}{x} + \dfrac{12}{x^2}}{2 - \dfrac{1}{x} - \dfrac{15}{x^2}}$

10. Find the x- and y-intercepts for the graph of the line $4x - 3y = 12$.

11. Find the equation of the line that contains the point $(-3, 2)$ and has slope $-\dfrac{4}{3}$.

12. Solve by substitution:
$3x - y = 5$
$y = 2x - 3$

13. Solve by the addition method:
$3x + 2y = 2$
$5x - 2y = 14$

14. Solve: $2x - 3(2 - 3x) > 2x - 5$

15. Simplify: $(\sqrt{a} - \sqrt{2})(\sqrt{a} + \sqrt{2})$

16. Simplify: $\dfrac{\sqrt{108a^7b^3}}{\sqrt{3a^4b}}$

Copyright © Houghton Mifflin Company. All rights reserved.

17. Simplify: $\dfrac{\sqrt{3}}{5 + 2\sqrt{3}}$

18. Solve: $3 = 8 - \sqrt{5x}$

19. Solve by factoring: $6x^2 - 17x = -5$

20. Solve by taking square roots:
$2(x - 5)^2 = 36$

21. Solve by completing the square:
$3x^2 + 7x = -3$

22. Solve by using the quadratic formula:
$2x^2 - 3x - 2 = 0$

23. **Food Mixtures** Find the cost per pound of a mixture made from 20 lb of cashews that cost $3.50 per pound and 50 lb of peanuts that cost $1.75 per pound.

24. **Investments** A stock investment of 100 shares paid a dividend of $215. At this rate, how many additional shares must the investor own to earn a dividend of $752.50?

25. **Travel** A 720-mile trip from one city to another takes 3 h when a plane is flying with the wind. The return trip, against the wind, takes 4.5 h. Find the rate of the plane in still air and the rate of the wind.

26. **Grading** A student received a 70, a 91, an 85, and a 77 on four tests in a mathematics class. What scores on the last test will enable the student to receive a minimum of 400 points?

27. **Number Sense** The sum of the squares of three consecutive odd integers is 83. Find the middle odd integer.

28. **Exercise** A jogger ran 7 mi at a constant rate and then reduced the rate by 3 mph and ran an additional 8 mi at the reduced rate. The total time spent jogging the 15 mi was 3 h. Find the jogger's rate for the last 8 mi.

29. Graph the solution set of $2x - 3y > 6$.

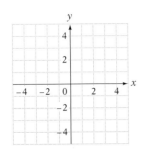

30. Graph $y = x^2 - 2x - 3$.

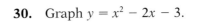
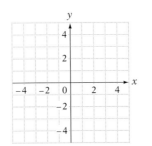

Copyright © Houghton Mifflin Company. All rights reserved.

Final Exam

1. Evaluate $-|-3|$.

2. Subtract: $-15 - (-12) - 3$

3. Simplify: $-2^4 \cdot (-2)^4$

4. Simplify: $-7 - \dfrac{12 - 15}{2 - (-1)} \cdot (-4)$

5. Evaluate $\dfrac{a^2 - 3b}{2a - 2b^2}$ when $a = 3$ and $b = -2$.

6. Simplify: $6x - (-4y) - (-3x) + 2y$

7. Simplify: $(-15z)\left(-\dfrac{2}{5}\right)$

8. Simplify: $-2[5 - 3(2x - 7) - 2x]$

9. Solve: $20 = -\dfrac{2}{5}x$

10. Solve: $4 - 2(3x + 1) = 3(2 - x) + 5$

11. Write $\dfrac{1}{8}$ as a percent.

12. Find 19% of 80.

13. Subtract: $(2x^2 - 5x + 1) - (5x^2 - 2x - 7)$

14. Simplify: $(-3xy^3)^4$

15. Multiply: $(3x^2 - x - 2)(2x + 3)$

16. Simplify: $\dfrac{(-2x^2y^3)^3}{(-4xy^4)^2}$

17. Divide: $\dfrac{12x^2y - 16x^3y^2 - 20y^2}{4xy^2}$

18. Divide: $(5x^2 - 2x - 1) \div (x + 2)$

19. Simplify: $(4x^{-2}y)^2(2xy^{-2})^{-2}$

20. Given $f(t) = \dfrac{t}{t + 1}$, find $f(3)$.

21. Factor: $x^2 - 5x - 6$

22. Factor: $6x^2 - 5x - 6$

Copyright © Houghton Mifflin Company. All rights reserved.

23. Factor: $8x^3 - 28x^2 + 12x$

24. Factor: $25x^2 - 16$

25. Factor: $2a(4 - x) - 6(x - 4)$

26. Factor: $75y - 12x^2y$

27. Solve: $2x^2 = 7x - 3$

28. Multiply: $\dfrac{2x^2 - 3x + 1}{4x^2 - 2x} \cdot \dfrac{4x^2 + 4x}{x^2 - 2x + 1}$

29. Subtract: $\dfrac{5}{x + 3} - \dfrac{3x}{2x - 5}$

30. Simplify: $x - \dfrac{1}{1 - \dfrac{1}{x}}$

31. Solve: $\dfrac{5x}{3x - 5} - 3 = \dfrac{7}{3x - 5}$

32. Solve $a = 3a - 2b$ for a.

33. Find the slope of the line that contains the points $(-1, -3)$ and $(2, -1)$.

34. Find the equation of the line that contains the point $(3, -4)$ and has slope $-\dfrac{2}{3}$.

35. Solve by substitution:
$y = 4x - 7$
$y = 2x + 5$

36. Solve by the addition method:
$4x - 3y = 11$
$2x + 5y = -1$

37. Solve: $4 - x \geq 7$

38. Solve: $2 - 2(y - 1) \leq 2y - 6$

39. Simplify: $\sqrt{49x^6}$

40. Simplify: $2\sqrt{27a} + 8\sqrt{48a}$

41. Simplify: $\dfrac{\sqrt{3}}{\sqrt{5} - 2}$

42. Solve: $\sqrt{2x - 3} + 4 = 5$

43. Solve by factoring:
$3x^2 - x = 4$

44. Solve by using the quadratic formula:
$4x^2 - 2x - 1 = 0$

Copyright © Houghton Mifflin Company. All rights reserved.

45. Number Sense   Translate and simplify "the sum of twice a number and three times the difference between the number and two."

46. Depreciation   Because of depreciation, the value of an office machine is now $2400. This is 80% of its original value. Find the original value.

47. Business   The manufacturer's cost for a laser printer is $900. The manufacturer then sells the printer for $1485. What is the markup rate?

48. Investment   An investment of $3000 is made at an annual simple interest rate of 8%. How much additional money must be invested at 11% so that the total interest earned is 10% of the total investment?

49. Food Mixtures   A grocer mixes 4 lb of peanuts that cost $2 per pound with 2 lb of walnuts that cost $5 per pound. What is the cost per pound of the resulting mixture?

50. Pharmacology   A pharmacist mixes together 20 L of a solution that is 60% acid and 30 L of a solution that is 20% acid. What is the percent concentration of the acid in the mixture?

51. Travel   At 2 P.M. a small plane had been flying 1 h when a change of wind direction doubled its average ground speed. The complete 860-kilometer trip took 2.5 h. How far did the plane travel in the first hour?

52. Geometry   The angles of a triangle are such that the measure of the second angle is 10° more than the measure of the first angle, and the measure of the third angle is 10° more than the measure of the second angle. Find the measure of each of the three angles.

Copyright © Houghton Mifflin Company. All rights reserved.

53. **Number Sense**   The sum of the squares of three consecutive integers is 50. Find the middle integer.

54. **Geometry**   The length of a rectangle is 5 m more than the width. The area of the rectangle is 50 m². Find the dimensions of the rectangle.

55. **Paint Mixtures**   A paint formula requires 2 oz of dye for every 15 oz of base paint. How many ounces of dye are required for 120 oz of base paint?

56. **Food Preparation**   It takes a chef 1 h to prepare a dinner. The chef's apprentice can prepare the dinner in 1.5 h. How long would it take the chef and the apprentice, working together, to prepare the dinner?

57. **Travel**   With the current, a motorboat travels 50 mi in 2.5 h. Against the current, it takes twice as long to travel 50 mi. Find the rate of the boat in calm water and the rate of the current.

58. **Travel**   Flying against the wind, it took a pilot $\frac{1}{2}$ h longer to travel 500 mi than it took flying with the wind. The rate of the plane in calm air is 225 mph. Find the rate of the wind.

59. Graph the line that has slope $-\frac{1}{2}$ and y-intercept $(0, -3)$.

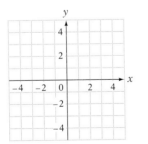

60. Graph $y = x^2 - 4x + 3$.

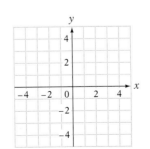

Copyright © Houghton Mifflin Company. All rights reserved.

Appendix A

Keystroke Guide for the TI-83 and TI-83 Plus

Basic Operations

Numerical calculations are performed on the **home screen.** You can always return to the home screen by pressing ⟨2nd⟩ QUIT. Pressing ⟨CLEAR⟩ erases the home screen.

To evaluate the expression $-2(3 + 5) - 8 \div 4$, use the following keystrokes.

⟨(-)⟩ 2 ⟨(⟩ 3 ⟨+⟩ 5 ⟨)⟩ ⟨−⟩ 8 ⟨÷⟩ 4 ⟨ENTER⟩

Note: There is a difference between the key to enter a negative number, ⟨(-)⟩, and the key for subtraction, ⟨−⟩. You cannot use these keys interchangeably.

The ⟨2nd⟩ key is used to access the commands in gold writing above a key. For instance, to evaluate the $\sqrt{49}$, press ⟨2nd⟩ ⟨√⟩ 49 ⟨)⟩ ⟨ENTER⟩.

The ⟨ALPHA⟩ key is used to place a letter on the screen. One reason to do this is to store a value of a variable. The following keystrokes give A the value of 5.

5 ⟨STO▸⟩ ⟨ALPHA⟩ A ⟨ENTER⟩

This value is now available in calculations. For instance, we can find the value of $3a^2$ by using the following keystrokes: 3 ⟨ALPHA⟩ A ⟨x²⟩. To display the value of the variable on the screen, press ⟨2nd⟩ RCL ⟨ALPHA⟩ A.

Note: When you use the ⟨ALPHA⟩ key, only capital letters are available on the TI-83 calculator.

TAKE NOTE

The descriptions in the margins (for example, Basic Operations and Evaluating Functions) are the same as those used in the text and are arranged alphabetically.

Evaluating Functions

There are various methods of evaluating a function but all methods require that the expression be entered as one of the ten functions Y₁ to Y₀. To evaluate $f(x) = \frac{x^2}{x - 1}$ when $x = -3$, enter the expression into, for instance, Y₁, and then press ⟨VARS⟩ ▸ 11 ⟨(⟩ ⟨(-)⟩ 3 ⟨)⟩ ⟨ENTER⟩.

Note: If you try to evaluate a function at a number that is not in the domain of the function, you will get an error message. For instance, 1 is not in the domain of $f(x) = \frac{x^2}{x - 1}$. If we try to evaluate the function at 1, the error screen at the right appears.

TAKE NOTE

Use the down arrow key to scroll past Y₇ to see Y₈, Y₉, and Y₀.

Evaluating Variable Expressions

To evaluate a variable expression, first store the values of each variable. Then enter the variable expression. For instance, to evaluate $s^2 + 2sl$ when $s = 4$ and $l = 5$, use the following keystrokes.

4 ⟨STO▸⟩ ⟨ALPHA⟩ S ⟨ENTER⟩ 5 ⟨STO▸⟩ ⟨ALPHA⟩ L ⟨ENTER⟩ ⟨ALPHA⟩ S

⟨x²⟩ ⟨+⟩ 2 ⟨ALPHA⟩ S ⟨ALPHA⟩ L ⟨ENTER⟩

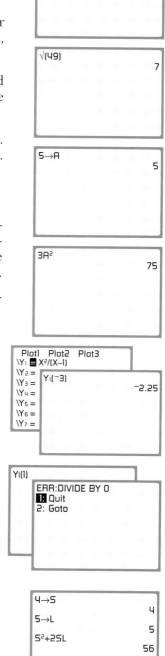

Copyright © Houghton Mifflin Company. All rights reserved.

Graph   To graph a function, use the  key to enter the expression for the function, select a suitable viewing window, and then press GRAPH. For instance, to graph $f(x) = 0.1x^3 - 2x - 1$ in the standard viewing window, use the following keystrokes.

Y= 0.1 X,T,θ,n ^ 3 − 2 X,T,θ,n − 1 ZOOM (scroll to 6) ENTER

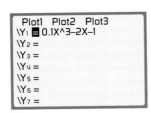 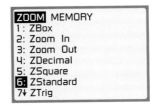 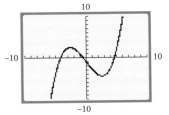

Note: For the keystrokes above, you do not have to scroll to 6. Alternatively, use ZOOM 6. This will select the standard viewing window and automatically start the graph. Use the WINDOW key to create a custom window for a graph.

Graphing Inequalities   To illustrate this feature, we will graph $y \le 2x - 1$. Enter $2x - 1$ into Y₁. Because $y \le 2x - 1$, we want to shade below the graph. Move the cursor to the left of Y₁ and press ENTER three times. Press GRAPH.

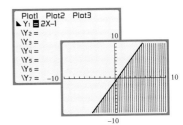

Note: To shade above the graph, move the cursor to the left of Y₁ and press ENTER two times. An inequality with the symbol $\le$ or $\ge$ should be graphed with a solid line, and an inequality with the symbol $<$ or $>$ should be graphed with a dashed line. However, the graph of a linear inequality on a graphing calculator does not distinguish between a solid line and a dashed line.

To graph the solution set of a system of inequalities, solve each inequality for y and graph each inequality. The solution set is the intersection of the two inequalities. The solution set of $\begin{array}{l} 3x + 2y > 10 \\ 4x - 3y \le 5 \end{array}$ is shown at the right.

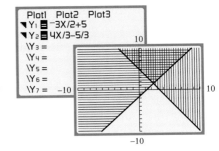

Intersect   The INTERSECT feature is used to solve a system of equations. To illustrate this feature, we will use the system of equations $\begin{array}{l} 2x - 3y = 13 \\ 3x + 4y = -6 \end{array}$.

Note: Some equations can be solved by this method. See the section "Solve an equation" below. Also, this method is used to find a number in the domain of a function for a given number in the range. See the section "Find a domain element."

Solve each of the equations in the system of equations for y. In this case, we have $y = \frac{2}{3}x - \frac{13}{3}$ and $y = -\frac{3}{4}x - \frac{3}{2}$.

Copyright © Houghton Mifflin Company. All rights reserved.

Use the Y-editor to enter $\frac{2}{3}x - \frac{13}{3}$ into Y_1 and $-\frac{3}{4}x - \frac{3}{2}$ into Y_2. Graph the two functions in the standard viewing window. (If the window does not show the point of intersection of the two graphs, adjust the window until you can see the point of intersection.)

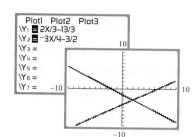

Press ⟨2nd⟩ CALC (scroll to 5, **intersect**) ⟨ENTER⟩.

Alternatively, you can just press ⟨2nd⟩ CALC 5.

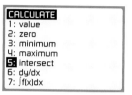

First curve? is shown at the bottom of the screen and identifies one of the two graphs on the screen. Press ⟨ENTER⟩.

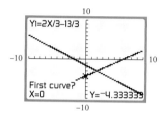

Second curve? is shown at the bottom of the screen and identifies the second of the two graphs on the screen. Press ⟨ENTER⟩.

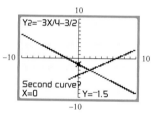

Guess? shown at the bottom of the screen asks you to use the left or right arrow key to move the cursor to the *approximate* location of the point of intersection. (If there are two or more points of intersection, it does not matter which one you choose first.) Press ⟨ENTER⟩.

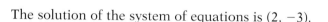
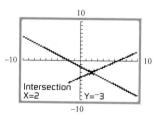

The solution of the system of equations is $(2, -3)$.

Solve an equation To illustrate the steps involved, we will solve the equation $2x + 4 = -3x - 1$. The idea is to write the equation as the system of equations $\begin{array}{l} y = 2x + 4 \\ y = -3x - 1 \end{array}$ and then use the steps for solving a system of equations.

Use the Y-editor to enter the left and right sides of the equation into Y_1 and Y_2. Graph the two functions and then follow the steps for **Intersect**.

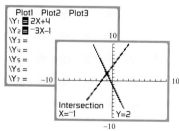

The solution is -1, the x-coordinate of the point of intersection.

Find a domain element For this example, we will find a number in the domain of $f(x) = -\frac{2}{3}x + 2$ that corresponds to 4 in the range of the function. This is like solving the system of equations $y = -\frac{2}{3}x + 2$ and $y = 4$.

Copyright © Houghton Mifflin Company. All rights reserved.

Use the Y= editor to enter the expression for the function in Y₁ and the desired output, 4, in Y₂. Graph the two functions and then follow the steps for Intersect.

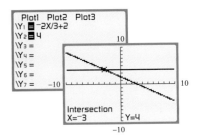

The point of intersection is $(-3, 4)$. The number -3 in the domain of f produces an output of 4 in the range of f.

Math   Pressing (MATH) gives you access to many built-in functions. The following keystrokes will convert 0.125 to a fraction: .125 (MATH) 1 (ENTER).

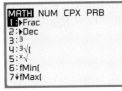

Additional built-in functions under (MATH) can be found by pressing (MATH) ▸. For instance, to evaluate $-|-25|$, press (−) (MATH) ▸ 1 (−) 25) (ENTER).

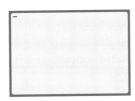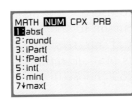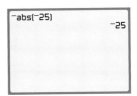

See your owner's manual for assistance with other functions under the (MATH) key.

Radical Expressions   To evaluate a square-root expression, press (2nd) √ .

For instance, to evaluate $0.15\sqrt{p^2 + 4p + 10}$ when $p = 100{,}000$, first store 100,000 in P. Then press 0.15 (2nd) √ (ALPHA) P (x²) (+) 4 (ALPHA) P (+) 10) (ENTER).

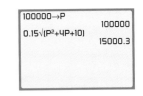

To evaluate a radical expression other than a square root, access $\sqrt[x]{\ }$ by pressing (MATH). For instance, to evaluate $\sqrt[4]{67}$, press 4 (the index of the radical) (MATH) (scroll to 5) (ENTER) 67 (ENTER).

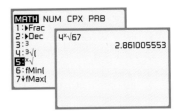

Scientific Notation   To enter a number in scientific notation, use (2nd) EE. For instance, to find $\dfrac{3.45 \times 10^{-12}}{1.5 \times 10^{25}}$, press 3.45 (2nd) EE (−) 12 (÷) 1.5 (2nd) EE 25 (ENTER). The answer is 2.3×10^{-37}.

Table   There are three steps in creating an input/output table for a function. First use the (Y=) editor to input the function. The second step is setting up the table, and the third step is displaying the table.

Copyright © Houghton Mifflin Company. All rights reserved.

To set up the table, press ⬚ ⬚ TBLSET. **TblStart** is the first value of the independent variable in the input/output table. △**Tbl** is the difference between successive values. Setting this to 1 means that, for this table, the input values are −2, −1, 0, 1, 2. . . . If △**Tbl** = 0.5, then the input values are −2, −1.5, −1, −0.5, 0, 0.5, . . .

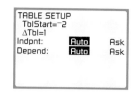

Indpnt is the independent variable. When this is set to **Auto**, values of the independent variable are automatically entered into the table. **Depend** is the dependent variable. When this is set to **Auto**, values of the dependent variable are automatically entered into the table.

To display the table, press ⬚ TABLE. An input/output table for $f(x) = x^2 - 1$ is shown at the right.

Once the table is on the screen, the up and down arrow keys can be used to display more values in the table. For the table at the right, we used the up arrow key to move to $x = -7$.

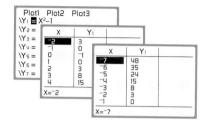

An input/output table for any given input can be created by selecting **Ask** for the independent variable. The table at the right shows an input/output table for $f(x) = \frac{4x}{x - 2}$ for selected values of x. Note the word **ERROR** when 2 was entered. This occurred because f is not defined when $x = 2$.

Note: Using the table feature in **Ask** mode is the same as evaluating a function for given values of the independent variable. For instance, from the table at the right, we have $f(4) = 8$.

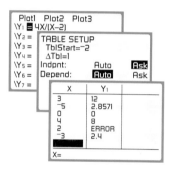

Test The TEST feature has many uses, one of which is to graph the solution set of a linear inequality in one variable. To illustrate this feature, we will graph the solution set of $x - 1 < 4$. Press **Y=** **X,T,θ,n** **−** 1 ⬚ TEST (scroll to 5) **ENTER** 4 **GRAPH**.

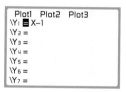
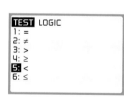
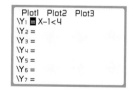
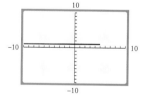

Trace Once a graph is drawn, pressing **TRACE** will place a cursor on the screen, and the coordinates of the point below the cursor are shown at the bottom of the screen. Use the left and right arrow keys to move the cursor along the graph. For the graph at the right, we have $f(4.8) = 3.4592$, where $f(x) = 0.1x^3 - 2x + 2$ is shown at the top left of the screen.

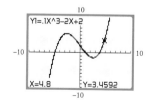

In TRACE mode, you can evaluate a function at any value of the independent variable that is within Xmin and Xmax. To do this, first graph the function. Now press **TRACE** (the value of x) **ENTER**. For the graph at the left below, we used $x = -3.5$. If a value of x is chosen outside the window, an error message is displayed.

Copyright © Houghton Mifflin Company. All rights reserved.

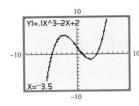
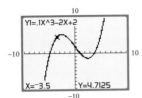
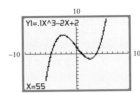

In the example above where we entered -3.5 for x, the value of the function was calculated as 4.7125. This means that $f(-3.5) = 4.7125$. The keystrokes 2nd QUIT VARS ▶ 11 MATH 1 ENTER will convert the decimal value to a fraction.

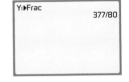

When the TRACE feature is used with two or more graphs, the up and down arrow keys are used to move between the graphs. The graphs below are for the functions $f(x) = 0.1x^3 - 2x + 2$ and $g(x) = 2x - 3$. By using the up and down arrows, we can place the cursor on either graph. The right and left arrows are used to move along the graph.

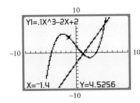
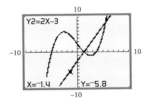

Window   The viewing window for a graph is controlled by pressing WINDOW. Xmin and Xmax are the minimum value and maximum value, respectively, of the independent variable shown on the graph. Xscl is the distance between tic marks on the x-axis. Ymin and Ymax are the minimum value and maximum value, respectively, of the dependent variable shown on the graph. Yscl is the distance between tic marks on the y-axis. Leave Xres as 1.

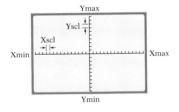

Note: In the standard viewing window, the distance between tic marks on the x-axis is different from the distance between tic marks on the y-axis. This will distort a graph. A more accurate picture of a graph can be created by using a square viewing window. See ZOOM.

Y=   The Y= editor is used to enter the expression for a function. There are ten possible functions, labeled Y_1 to Y_0, that can be active at any one time. For instance, to enter $f(x) = x^2 + 3x - 2$ as Y_1, use the following keystrokes.

Y= X,T,θ,n x^2 + 3 X,T,θ,n − 2

Note: If an expression is already entered for Y_1, place the cursor anywhere on that expression and press CLEAR.

To enter $s = \dfrac{2v - 1}{v^3 - 3}$ into Y_2, place the cursor to the right of the equals sign for Y_2. Then press (2 X,T,θ,n − 1) ÷ (X,T,θ,n ^ 3 − 3) .

Copyright © Houghton Mifflin Company. All rights reserved.

Note: When we enter an equation, the independent variable, v in the expression above, is entered using [X,T,θ,n]. The dependent variable, s in the expression above, is one of Y_1 to Y_0. Also note the use of parentheses to ensure the correct order of operations.

Observe the black rectangle that covers the equals sign for the two examples we have shown. This rectangle means that the function is "active." If we were to press [GRAPH], then the graph of both functions would appear. You can make a function inactive by using the arrow keys to move the cursor over the equals sign of that function and then pressing [ENTER]. This will remove the black rectangle. We have done that for Y_2, as shown at the right. Now if [GRAPH] is pressed, only Y_1 will be graphed.

```
Plot1  Plot2  Plot3
\Y1 ■ X²+3X−2
\Y2 = (2X−1)/(X^3−3)
\Y3 =
\Y4 =
\Y5 =
\Y6 =
\Y7 =
```

It is also possible to control the appearance of the graph by moving the cursor on the [Y=] screen to the left of any Y. With the cursor in this position, pressing [ENTER] will change the appearance of the graph. The options are shown at the right.

```
Plot1  Plot2  Plot3
\Y1 = Default graph line
\Y2 = Bold graph line
▀Y3 = Shade above graph
▄Y4 = Shade below graph
·oY5 = Draw path of graph
oY6 = Travel path of graph
':Y7 = Dashed graph line
```

Zero   The ZERO feature of a graphing calculator is used for various calculations: to find the x-intercepts of a function, to solve some equations, and to find the zero of a function.

x-intercepts   To illustrate the procedure for finding x-intercepts, we will use $f(x) = x^2 + x - 2$.

First, use the Y-editor to enter the expression for the function and then graph the function in the standard viewing window. (It may be necessary to adjust this window so that the intercepts are visible.) Once the graph is displayed, use the keystrokes below to find the x-intercepts of the graph of the function.

Press [2nd] CALC (scroll to 2 for **zero** of the function) [ENTER].

Alternatively, you can just press [2nd] CALC 2.

```
CALCULATE
1: value
2: zero
3: minimum
4: maximum
5: intersect
6: dy/dx
7: ∫f(x)dx
```

Left Bound? shown at the bottom of the screen asks you to use the left or right arrow key to move the cursor to the *left* of the desired x-intercept. Press [ENTER].

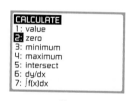

Right Bound? shown at the bottom of the screen asks you to use the left or right arrow key to move the cursor to the *right* of the desired x-intercept. Press [ENTER].

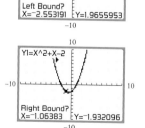

Guess? shown at the bottom of the screen asks you to use the left or right arrow key to move the cursor to the *approximate* location of the desired x-intercept. Press [ENTER].

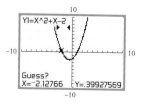

Copyright © Houghton Mifflin Company. All rights reserved.

The x-coordinate of an x-intercept is -2. Therefore, an x-intercept is $(-2, 0)$.

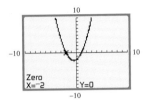

To find the other x-intercept, follow the same steps as above. The screens for this calculation are shown below.

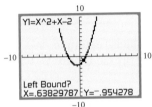
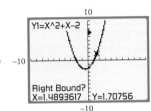
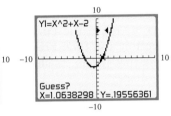
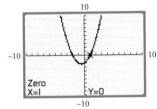

A second x-intercept is $(1, 0)$.

Solve an equation To use the ZERO feature to solve an equation, first rewrite the equation with all terms on one side. For instance, one way to solve the equation $x^3 - x + 1 = -2x + 3$ is first to rewrite it as $x^3 + x - 2 = 0$. Enter $x^3 + x - 2$ into Y_1 and then follow the steps for finding x-intercepts.

Find the real zeros of a function To find the real zeros of a function, follow the steps for finding x-intercepts.

Zoom Pressing **ZOOM** allows you to select some preset viewing windows. This key also gives you access to **ZBox**, **Zoom In**, and **Zoom Out**. These functions enable you to redraw a selected portion of a graph in a new window. Some windows used frequently in this text are shown below.

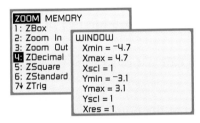
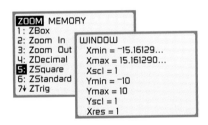
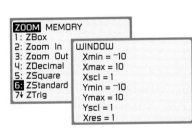
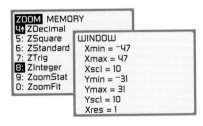

Copyright © Houghton Mifflin Company. All rights reserved.

Appendix B

Tables

Table of Symbols

$+$	add
$-$	subtract
$\cdot, \times, (a)(b)$	multiply
$\dfrac{a}{b}, \div$	divide
$(\ )$	parentheses, a grouping symbol
$[\ ]$	brackets, a grouping symbol
π	pi, a number approximately equal to $\dfrac{22}{7}$ or 3.14
$-a$	the opposite, or additive inverse, of a
$\dfrac{1}{a}$	the reciprocal, or multiplicative inverse, of a
$=$	is equal to
$\approx$	is approximately equal to
$\neq$	is not equal to

$<$	is less than
$\leq$	is less than or equal to
$>$	is greater than
$\geq$	is greater than or equal to
(a, b)	an ordered pair whose first component is a and whose second component is b
$^\circ$	degree (for angles)
$\sqrt{a}$	the principal square root of a
$\varnothing, \{\ \}$	the empty set
$|a|$	the absolute value of a
$\cup$	union of two sets
$\cap$	intersection of two sets
$\in$	is an element of (for sets)
$\notin$	is not an element of (for sets)

Table of Measurement Abbreviations

U.S. Customary System

Length		Capacity		Weight		Area	
in.	inches	oz	fluid ounces	oz	ounces	in²	square inches
ft	feet	c	cups	lb	pounds	ft²	square feet
yd	yards	qt	quarts				
mi	miles	gal	gallons				

Metric System

Length		Capacity		Weight/Mass		Area	
mm	millimeter (0.001 m)	ml	milliliter (0.001 L)	mg	milligram (0.001 g)	cm²	square centimeters
cm	centimeter (0.01 m)	cl	centiliter (0.01 L)	cg	centigram (0.01 g)	m²	square meters
dm	decimeter (0.1 m)	dl	deciliter (0.1 L)	dg	decigram (0.1 g)		
m	meter	L	liter	g	gram		
dam	decameter (10 m)	dal	decaliter (10 L)	dag	decagram (10 g)		
hm	hectometer (100 m)	hl	hectoliter (100 L)	hg	hectogram (100 g)		
km	kilometer (1000 m)	kl	kiloliter (1000 L)	kg	kilogram (1000 g)		

Time

h	hours	min	minutes	s	seconds

Copyright © Houghton Mifflin Company. All rights reserved.

Table of Equations and Formulas

Slope of a line

$$m = \frac{y_2 - y_1}{x_2 - x_1}, \; x_1 \neq x_2$$

Slope-intercept form of a straight line

$$y = mx + b$$

Point-slope formula for a line

$$y - y_1 = m(x - x_1)$$

Quadratic Formula

$$x = \frac{-b \pm \sqrt{b^2 - 4ac}}{2a}$$

$$\text{discriminant} = b^2 - 4ac$$

Perimeter and Area of a Triangle, and Sum of the Measures of the Angles

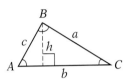

$$P = a + b + c$$

$$A = \frac{1}{2}bh$$

$$A + B + C = 180°$$

Pythagorean Theorem

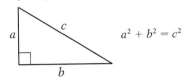

$$a^2 + b^2 = c^2$$

Perimeter and Area of a Rectangle

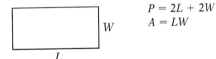

$$P = 2L + 2W$$

$$A = LW$$

Perimeter and Area of a Square

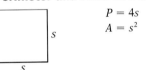

$$P = 4s$$

$$A = s^2$$

Circumference and Area of a Circle

$$C = 2\pi r \quad \text{or} \quad C = \pi d$$

$$A = \pi r^2$$

Volume of a Rectangular Solid

$$V = LWH$$

Volume of a Cube

$$V = s^3$$

Volume of a Sphere

$$V = \frac{4}{3}\pi r^3$$

Volume of a Right Circular Cylinder

$$V = \pi r^2 h$$

Copyright © Houghton Mifflin Company. All rights reserved.

Table of Properties

Properties of Real Numbers

The Associative Property of Addition
If a, b, and c are real numbers, then
$(a + b) + c = a + (b + c)$.

The Associative Property of Multiplication
If a, b, and c are real numbers, then
$(a \cdot b) \cdot c = a \cdot (b \cdot c)$.

The Commutative Property of Addition
If a and b are real numbers, then
$a + b = b + a$.

The Commutative Property of Multiplication
If a and b are real numbers, then
$a \cdot b = b \cdot a$.

The Addition Property of Zero
If a is a real number, then
$a + 0 = 0 + a = a$.

The Multiplication Property of One
If a is a real number, then
$a \cdot 1 = 1 \cdot a = a$.

The Multiplication Property of Zero
If a is a real number, then
$a \cdot 0 = 0 \cdot a = 0$.

The Inverse Property of Multiplication
If a is a real number and $a \neq 0$, then
$a \cdot \dfrac{1}{a} = \dfrac{1}{a} \cdot a = 1$.

The Inverse Property of Addition
If a is a real number, then
$a + (-a) = (-a) + a = 0$.

Distributive Property
If a, b, and c are real numbers, then
$a(b + c) = ab + ac$.

Properties of Equations

Addition Property of Equations
If $a = b$, then $a + c = b + c$.

Multiplication Property of Equations
If $a = b$ and $c \neq 0$, then $a \cdot c = b \cdot c$.

Properties of Exponents

If m and n are integers, then $x^m \cdot x^n = x^{m+n}$.
If m and n are integers, then $(x^m)^n = x^{mn}$.

If $x \neq 0$, then $x^0 = 1$.

If m and n are integers and $x \neq 0$, then $\dfrac{x^m}{x^n} = x^{m-n}$.

If m, n, and p are integers, then $(x^m \cdot y^n)^p = x^{mp}y^{np}$.
If n is a positive integer and $x \neq 0$, then

$x^{-n} = \dfrac{1}{x^n}$ and $\dfrac{1}{x^{-n}} = x^n$.

If m, n, and p are integers and $y \neq 0$, then $\left(\dfrac{x^m}{y^n}\right)^p = \dfrac{x^{mp}}{y^{np}}$.

Properties of Inequalities

Addition Property of Inequalities
If $a > b$, then $a + c > b + c$.
If $a < b$, then $a + c < b + c$.

Multiplication Property of Inequalities
If $a > b$ and $c > 0$, then $ac > bc$.
If $a < b$ and $c > 0$, then $ac < bc$.
If $a > b$ and $c < 0$, then $ac < bc$.
If $a < b$ and $c < 0$, then $ac > bc$.

Principle of Zero Products

If $a \cdot b = 0$, then $a = 0$ or $b = 0$.

Properties of Radical Expressions

If a and b are positive real numbers, then $\sqrt{ab} = \sqrt{a}\sqrt{b}$.

If a and b are positive real numbers, then $\sqrt{\dfrac{a}{b}} = \dfrac{\sqrt{a}}{\sqrt{b}}$.

Property of Squaring Both Sides of an Equation

If a and b are real numbers and $a = b$, then $a^2 = b^2$.

Copyright © Houghton Mifflin Company. All rights reserved.

Solutions to Chapter 1 "You Try It"

SECTION 1.1

You Try It 1 $A = \{1, 2, 3, 4, 5, 6\}$

You Try It 2
$-5 < -1$
$-1 = -1$
$5 > -1$

The element 5 is greater than -1.

You Try It 3
$|-5| = 5$
$-|-23| = -23$

You Try It 4
$-(-11) = 11$
$-0 = 0$
$-(8) = -8$

You Try It 5
$|-37| = 37$
$|0| = 0$
$|29| = 29$

SECTION 1.2

You Try It 1 $100 + (-43) = 57$

You Try It 2
$(-51) + 42 + 17 + (-102)$
$= -9 + 17 + (-102)$
$= 8 + (-102)$
$= -94$

You Try It 3 $-8 + 7 = -1$

You Try It 4
$19 - (-32) = 19 + 32$
$= 51$

You Try It 5
$-9 - (-12) - 17 - 4$
$= -9 + 12 + (-17) + (-4)$
$= 3 + (-17) + (-4)$
$= -14 + (-4)$
$= -18$

You Try It 6 $-11 - (-12) = -11 + 12 = 1$

You Try It 7

Strategy   To find the difference between the two average temperatures, subtract the smaller number (-130) from the larger number (-17).

Solution $-17 - (-130) = -17 + 130 = 113$

The difference is $113°$F.

SECTION 1.3

You Try It 1 $-7(32) = -224$

You Try It 2
$8(-9)10 = -72(10)$
$= -720$

You Try It 3
$(-2)3(-8)7 = -6(-8)7$
$= 48(7)$
$= 336$

You Try It 4 $-9(34) = -306$

You Try It 5 $(-135) \div (-9) = 15$

You Try It 6 $\dfrac{-72}{4} = -18$

You Try It 7
$-\dfrac{36}{-12} = -(-3)$
$= 3$

You Try It 8 $\dfrac{-72}{-8} = 9$

You Try It 9

Strategy

To find the average daily low temperature:
• Add the seven temperature readings.
• Divide the sum by 7.

Solution
$-6 + (-7) + 0 + (-5) + (-8) + (-1) + (-1) = -28$
$-28 \div 7 = -4$

The average daily low temperature was $-4°$C.

SECTION 1.4

You Try It 1 $-6^3 = -(6 \cdot 6 \cdot 6) = -216$

You Try It 2 $(-3)^4 = (-3)(-3)(-3)(-3) = 81$

You Try It 3
$(3^3)(-2)^3 = (3)(3)(3) \cdot (-2)(-2)(-2)$
$= 27(-8) = -216$

You Try It 4   The product of an odd number of negative factors is odd. Therefore, $(-1)^7 = -1$.

Copyright © Houghton Mifflin Company. All rights reserved.

You Try It 5
$$-2^2 \cdot (-1)^{12} \cdot (-3)^2 = -(2 \cdot 2) \cdot 1 \cdot (-3) \cdot (-3)$$
$$= -4 \cdot 1 \cdot 9 = -36$$

You Try It 6
$7 - 2[2 \cdot 3 - 7 \cdot 2]^2$
$= 7 - 2[6 - 14]^2$ • Perform operations inside grouping symbols.
$= 7 - 2[-8]^2$ • Simplify exponential expressions.
$= 7 - 2[64]$ • Do multiplication and division from left to right.
$= 7 - 128$
$= -121$ • Do addition and subtraction from left to right.

You Try It 7
$18 - 5[8 - 2(2 - 5)] \div 10$
$= 18 - 5[8 - 2(-3)] \div 10$ • Perform operations inside grouping symbols.
$= 18 - 5[8 + 6] \div 10$
$= 18 - 5[14] \div 10$
$= 18 - 70 \div 10$ • Do multiplication and division from left to right.
$= 18 - 7$ • Do addition and subtraction from left to right.
$= 11$

You Try It 8
$36 \div (8 - 5)^2 - (-3)^2 \cdot 2$
$= 36 \div (3)^2 - (-3)^2 \cdot 2$ • Perform operations inside grouping symbols.
$= 36 \div 9 - 9 \cdot 2$ • Simplify exponential expressions.
$= 4 - 9 \cdot 2$ • Do multiplication and division from left to right.
$= 4 - 18$ • Do addition and subtraction from left to right.
$= -14$

SECTION 1.5

You Try It 1
$24 \div 1 = 24$
$24 \div 2 = 12$
$24 \div 3 = 8$
$24 \div 4 = 6$
$24 \div 5$ Remainder is not 0.
$24 \div 6 = 4$ The factors repeat.
The factors of 24 are 1, 2, 3, 4, 6, 8, 12, and 24.

You Try It 2

315	
3	105
3	35
5	7
7	1

$315 = 3^2 \cdot 5 \cdot 7$

You Try It 3

326	
2	163
163	1

• For 163, try prime numbers up to 13 because $13^2 > 163$.

$326 = 2 \cdot 163$

You Try It 4 $20 = 2 \cdot 2 \cdot 5$ $21 = 3 \cdot 7$
LCM $= 2 \cdot 2 \cdot 3 \cdot 5 \cdot 7 = 420$

You Try It 5 $42 = 2 \cdot 3 \cdot 7$ $63 = 3 \cdot 3 \cdot 7$
GCF $= 3 \cdot 7 = 21$

SECTION 1.6

You Try It 1 $\dfrac{60}{140} = \dfrac{\overset{1}{\cancel{2}} \cdot \overset{1}{\cancel{2}} \cdot 3 \cdot \overset{1}{\cancel{5}}}{\underset{1}{\cancel{2}} \cdot \underset{1}{\cancel{2}} \cdot \underset{1}{\cancel{5}} \cdot 7} = \dfrac{3}{7}$

You Try It 2 $\dfrac{4}{9} = 4 \div 9 = 0.\overline{4}$

You Try It 3   The LCM of 9 and 12 is 36.
$$\frac{5}{9} + \left(-\frac{11}{12}\right) = \frac{20}{36} + \left(-\frac{33}{36}\right) = \frac{20 + (-33)}{36}$$
$$= \frac{-13}{36} = -\frac{13}{36}$$

You Try It 4   The LCM of 8 and 6 is 24.
$$-\frac{5}{6} + \frac{7}{8} = -\frac{20}{24} + \frac{21}{24} = \frac{-20 + 21}{24}$$
$$= \frac{1}{24}$$

You Try It 5 $-6.12 + (-12.881) = -19.001$

You Try It 6   The LCM of 8, 12, and 9 is 72.
$$\frac{7}{8} - \left(-\frac{5}{12}\right) - \frac{1}{9} = \frac{63}{72} - \left(-\frac{30}{72}\right) - \frac{8}{72}$$
$$= \frac{63 - (-30) - 8}{72}$$
$$= \frac{63 + 30 - 8}{72} = \frac{85}{72}$$

You Try It 7 $-12.03 - 19.117 = -12.03 + (-19.117)$
$$= -31.147$$

You Try It 8

Strategy

To find the difference:
• Find the sum of the numbers from the graph that correspond to oil and gas production (8.5, 7.7, 3.6, 3.4, 3.0).
• Find the sum of the numbers from the graph that correspond to oil and gas exports (7.1, 5.1, 2.4, 3.1, 2.5).
• Find the difference between the two sums.

Solution $8.5 + 7.7 + 3.6 + 3.4 + 3.0 = 26.2$
$7.1 + 5.1 + 2.4 + 3.1 + 2.5 = 20.2$
$26.2 - 20.2 = 6.0$

The difference between the amount of gas and oil produced and the amount exported is 6 million barrels per day.

Copyright © Houghton Mifflin Company. All rights reserved.

Solutions to Chapter 1 "You Try It"

SECTION 1.1

You Try It 1 $A = \{1, 2, 3, 4, 5, 6\}$

You Try It 2
$-5 < -1$
$-1 = -1$
$5 > -1$

The element 5 is greater than -1.

You Try It 3
$|-5| = 5$
$-|-23| = -23$

You Try It 4
$-(-11) = 11$
$-0 = 0$
$-(8) = -8$

You Try It 5
$|-37| = 37$
$|0| = 0$
$|29| = 29$

SECTION 1.2

You Try It 1 $100 + (-43) = 57$

You Try It 2
$(-51) + 42 + 17 + (-102)$
$= -9 + 17 + (-102)$
$= 8 + (-102)$
$= -94$

You Try It 3 $-8 + 7 = -1$

You Try It 4
$19 - (-32) = 19 + 32$
$= 51$

You Try It 5
$-9 - (-12) - 17 - 4$
$= -9 + 12 + (-17) + (-4)$
$= 3 + (-17) + (-4)$
$= -14 + (-4)$
$= -18$

You Try It 6 $-11 - (-12) = -11 + 12 = 1$

You Try It 7

Strategy To find the difference between the two average temperatures, subtract the smaller number (-130) from the larger number (-17).

Solution $-17 - (-130) = -17 + 130 = 113$

The difference is $113°F$.

SECTION 1.3

You Try It 1 $-7(32) = -224$

You Try It 2
$8(-9)10 = -72(10)$
$= -720$

You Try It 3
$(-2)3(-8)7 = -6(-8)7$
$= 48(7)$
$= 336$

You Try It 4 $-9(34) = -306$

You Try It 5 $(-135) \div (-9) = 15$

You Try It 6 $\dfrac{-72}{4} = -18$

You Try It 7
$-\dfrac{36}{-12} = -(-3)$
$= 3$

You Try It 8 $\dfrac{-72}{-8} = 9$

You Try It 9

Strategy

To find the average daily low temperature:
• Add the seven temperature readings.
• Divide the sum by 7.

Solution
$-6 + (-7) + 0 + (-5) + (-8) + (-1) + (-1) = -28$
$-28 \div 7 = -4$

The average daily low temperature was $-4°C$.

SECTION 1.4

You Try It 1 $-6^3 = -(6 \cdot 6 \cdot 6) = -216$

You Try It 2 $(-3)^4 = (-3)(-3)(-3)(-3) = 81$

You Try It 3
$(3^3)(-2)^3 = (3)(3)(3) \cdot (-2)(-2)(-2)$
$= 27(-8) = -216$

You Try It 4 The product of an odd number of negative factors is odd. Therefore, $(-1)^7 = -1$.

Copyright © Houghton Mifflin Company. All rights reserved.

You Try It 5
$$-2^2 \cdot (-1)^{12} \cdot (-3)^2 = -(2 \cdot 2) \cdot 1 \cdot (-3) \cdot (-3)$$
$$= -4 \cdot 1 \cdot 9 = -36$$

You Try It 6
$7 - 2[2 \cdot 3 - 7 \cdot 2]^2$
$= 7 - 2[6 - 14]^2$ • Perform operations inside grouping symbols.
$= 7 - 2[-8]^2$ • Simplify exponential expressions.
$= 7 - 2[64]$ • Do multiplication and division from left to right.
$= 7 - 128$
• Do addition and subtraction from left to right.
$= -121$

You Try It 7
$18 - 5[8 - 2(2 - 5)] \div 10$
$= 18 - 5[8 - 2(-3)] \div 10$ • Perform operations inside grouping symbols.
$= 18 - 5[8 + 6] \div 10$
$= 18 - 5[14] \div 10$
$= 18 - 70 \div 10$ • Do multiplication and division from left to right.
$= 18 - 7$
$= 11$ • Do addition and subtraction from left to right.

You Try It 8
$36 \div (8 - 5)^2 - (-3)^2 \cdot 2$
$= 36 \div (3)^2 - (-3)^2 \cdot 2$ • Perform operations inside grouping symbols.
$= 36 \div 9 - 9 \cdot 2$ • Simplify exponential expressions.
$= 4 - 9 \cdot 2$ • Do multiplication and division from left to right.
$= 4 - 18$
$= -14$ • Do addition and subtraction from left to right.

SECTION 1.5

You Try It 1
$24 \div 1 = 24$
$24 \div 2 = 12$
$24 \div 3 = 8$
$24 \div 4 = 6$
$24 \div 5$ Remainder is not 0.
$24 \div 6 = 4$ The factors repeat.
The factors of 24 are 1, 2, 3, 4, 6, 8, 12, and 24.

You Try It 2

$$\begin{array}{c|c} & 315 \\ \hline 3 & 105 \\ 3 & 35 \\ 5 & 7 \\ 7 & 1 \end{array}$$
$315 = 3^2 \cdot 5 \cdot 7$

You Try It 3

$$\begin{array}{c|c} & 326 \\ \hline 2 & 163 \\ 163 & 1 \end{array}$$
$326 = 2 \cdot 163$

• For 163, try prime numbers up to 13 because $13^2 > 163$.

You Try It 4 $20 = 2 \cdot 2 \cdot 5$ $21 = 3 \cdot 7$
LCM $= 2 \cdot 2 \cdot 3 \cdot 5 \cdot 7 = 420$

You Try It 5 $42 = 2 \cdot 3 \cdot 7$ $63 = 3 \cdot 3 \cdot 7$
GCF $= 3 \cdot 7 = 21$

SECTION 1.6

You Try It 1 $\dfrac{60}{140} = \dfrac{\overset{1}{\cancel{2}} \cdot \overset{1}{\cancel{2}} \cdot 3 \cdot \overset{1}{\cancel{5}}}{\underset{1}{\cancel{2}} \cdot \underset{1}{\cancel{2}} \cdot \underset{1}{\cancel{5}} \cdot 7} = \dfrac{3}{7}$

You Try It 2 $\dfrac{4}{9} = 4 \div 9 = 0.\overline{4}$

You Try It 3   The LCM of 9 and 12 is 36.
$$\frac{5}{9} + \left(-\frac{11}{12}\right) = \frac{20}{36} + \left(-\frac{33}{36}\right) = \frac{20 + (-33)}{36}$$
$$= \frac{-13}{36} = -\frac{13}{36}$$

You Try It 4   The LCM of 8 and 6 is 24.
$$-\frac{5}{6} + \frac{7}{8} = -\frac{20}{24} + \frac{21}{24} = \frac{-20 + 21}{24}$$
$$= \frac{1}{24}$$

You Try It 5 $-6.12 + (-12.881) = -19.001$

You Try It 6   The LCM of 8, 12, and 9 is 72.
$$\frac{7}{8} - \left(-\frac{5}{12}\right) - \frac{1}{9} = \frac{63}{72} - \left(-\frac{30}{72}\right) - \frac{8}{72}$$
$$= \frac{63 - (-30) - 8}{72}$$
$$= \frac{63 + 30 - 8}{72} = \frac{85}{72}$$

You Try It 7 $-12.03 - 19.117 = -12.03 + (-19.117)$
$$= -31.147$$

You Try It 8

Strategy

To find the difference:
• Find the sum of the numbers from the graph that correspond to oil and gas production (8.5, 7.7, 3.6, 3.4, 3.0).
• Find the sum of the numbers from the graph that correspond to oil and gas exports (7.1, 5.1, 2.4, 3.1, 2.5).
• Find the difference between the two sums.

Solution $8.5 + 7.7 + 3.6 + 3.4 + 3.0 = 26.2$
$7.1 + 5.1 + 2.4 + 3.1 + 2.5 = 20.2$
$26.2 - 20.2 = 6.0$

The difference between the amount of gas and oil produced and the amount exported is 6 million barrels per day.

Copyright © Houghton Mifflin Company. All rights reserved.

You Try It 9

Strategy   To find the fraction of her day spent on these activities, add the fractions $\frac{1}{6}$, $\frac{1}{8}$, and $\frac{1}{4}$.

Solution   The common denominator is 24.

$$\frac{1}{6} + \frac{1}{8} + \frac{1}{4} = \frac{4}{24} + \frac{3}{24} + \frac{6}{24}$$

- Write each fraction in terms of the common denominator.

$$= \frac{4 + 3 + 6}{24} = \frac{13}{24}$$

- Add the fractions.

She spent $\frac{13}{24}$ of her day on these activities.

SECTION 1.7

You Try It 1

$$\frac{5}{8}\left(-\frac{4}{25}\right) = -\frac{5 \cdot 4}{8 \cdot 25}$$

- The signs are different. The product is negative.

$$= -\frac{\overset{1}{\cancel{5}} \cdot \overset{1}{\cancel{2}} \cdot \overset{1}{\cancel{2}}}{\underset{1}{\cancel{2}} \cdot \underset{1}{\cancel{2}} \cdot 2 \cdot \underset{1}{\cancel{5}} \cdot 5}$$

$$= -\frac{1}{10}$$

- Write the answer in simplest form.

You Try It 2

$$-\frac{4}{5} \cdot \left(-\frac{3}{8}\right) \cdot \left(-\frac{10}{27}\right)$$

$$= -\frac{\overset{1}{\cancel{2}} \cdot \overset{1}{\cancel{2}} \cdot \overset{1}{\cancel{3}} \cdot \overset{1}{\cancel{2}} \cdot \overset{1}{\cancel{5}}}{\underset{1}{\cancel{5}} \cdot \underset{1}{\cancel{2}} \cdot \underset{1}{\cancel{2}} \cdot \underset{1}{\cancel{2}} \cdot \underset{1}{\cancel{3}} \cdot 3 \cdot 3}$$

- The product is negative.

$$= -\frac{1}{9}$$

- Write the answer in simplest form.

You Try It 3

The product is negative.
$$0.034(-2.14) = -0.07276$$

You Try It 4

The quotient is negative.

$$\frac{5}{8} \div \left(-\frac{10}{11}\right) = -\left(\frac{5}{8} \cdot \frac{11}{10}\right)$$

- Multiply by the reciprocal of the divisor.

$$= -\frac{\overset{1}{\cancel{5}} \cdot 11}{2 \cdot 2 \cdot 2 \cdot \underset{1}{\cancel{5}}}$$

$$= -\frac{11}{16}$$

- Write the answer in simplest form.

You Try It 5

The quotient is negative.

$$-\frac{1}{3} \div \frac{7}{15} = -\left(\frac{1}{3} \cdot \frac{15}{7}\right)$$

- Multiply by the reciprocal of the divisor.

$$= -\frac{1 \cdot \overset{1}{\cancel{3}} \cdot 5}{\underset{1}{\cancel{3}} \cdot 7}$$

$$= -\frac{5}{7}$$

- Write the answer in simplest form.

You Try It 6

The quotient is positive.
$$-34 \div (-9.02) \approx 3.77$$

You Try It 7

$$125\% = 125\left(\frac{1}{100}\right) = \frac{125}{100} = \frac{5}{4}$$
$$125\% = 125(0.01) = 1.25$$

You Try It 8

$$16\frac{2}{3}\% = 16\frac{2}{3}\left(\frac{1}{100}\right) = \frac{50}{3}\left(\frac{1}{100}\right) = \frac{1}{6}$$

You Try It 9

$$\frac{9}{16} = \frac{9}{16}(100\%) = \frac{900}{16}\% = 56.25\%$$
$$\text{or } 56\frac{1}{4}\%$$

You Try It 10

$$0.043 = 0.043(100\%) = 4.3\%$$

You Try It 11

Strategy   To find the number of cushions, divide 20 by $1\frac{1}{2}$.

Solution

$$20 \div 1\frac{1}{2} = 20 \div \frac{3}{2} = \frac{20}{1} \cdot \frac{2}{3}$$

$$= \frac{40}{3} = 13\frac{1}{3}$$

The number of cushions must be a whole number. Therefore, the number of cushions is 13.

SECTION 1.8

You Try It 1

To find the complement of 87°, subtract 87° from 90°.

$$90° - 87° = 3°$$

3° is the complement of 87°.

You Try It 2

To find the supplement of 87°, subtract 87° from 180°.

$$180° - 87° = 93°$$

93° is the supplement of 87°.

You Try It 3

$m\angle x$ is the sum of the measures of two angles.
$$m\angle x = 34° + 95° = 129°$$

You Try It 4

$$\text{Perimeter} = 4 \cdot \text{side}$$
$$= 4 \cdot 4.2 \text{ m} = 16.8 \text{ m}$$

You Try It 5

$$\text{Circumference} = \pi \cdot \text{diameter}$$
$$\approx 3.14 \cdot 5 \text{ in.}$$
$$= 15.7 \text{ in.}$$

Copyright © Houghton Mifflin Company. All rights reserved.

You Try It 6

Strategy To find the cost of the metal strip:
- Find the circumference of the table in inches.
- Convert inches to feet.
- Multiply the circumference by the per-foot cost of the metal strip.

Solution Circumference $= \pi \cdot$ diameter
$$\approx 3.14 \cdot 36 \text{ in.}$$
$$= 113.04$$

$$\frac{113.04}{12} = 9.42$$

Cost: $9.42(3.21) = 30.2382$

The cost is $30.24.

You Try It 7 Area $= \dfrac{1}{2} \cdot$ base $\cdot$ height

$$= \frac{1}{2} \cdot 5 \text{ ft} \cdot 3 \text{ ft} = 7.5 \text{ ft}^2$$

You Try It 8 Area $= \pi \cdot (\text{radius})^2$
$$\approx 3.14 \cdot (6 \text{ in.})^2$$
$$= 113.04 \text{ in}^2$$

You Try It 9 Area $=$ base $\cdot$ height
$$= 28 \text{ in.} \cdot 15 \text{ in.} = 420 \text{ in}^2$$

You Try It 10

Strategy To find how much more expensive the wool rug is than the nylon rug:
- Find the area of the rug.
- Multiply the area of the rug by the per-square-foot cost of the wool rug.
- Multiply the area of the rug by the per-square-foot cost of the nylon rug.
- Find the difference between the two costs.

Solution Area $=$ length $\cdot$ width
$$= 15 \text{ ft} \cdot 4 \text{ ft} = 60 \text{ ft}^2$$

Cost of wool rug $= \$1.93 \cdot 60 = \$115.80$

Cost of nylon rug $= \$1.25 \cdot 60 = \$75$

Difference in cost $= \$115.80 - \$75 = \$40.80$

Solutions to Chapter 2 "You Try It"

SECTION 2.1

You Try It 1 -4 is the constant term.

You Try It 2 $2xy + y^2$
$2(-4)(2) + (2)^2$
$$= 2(-4)(2) + 4$$
$$= (-8)(2) + 4$$
$$= (-16) + 4$$
$$= -12$$

You Try It 3 $\dfrac{a^2 + b^2}{a + b}$

$$\frac{5^2 + (-3)^2}{5 + (-3)} = \frac{25 + 9}{5 + (-3)}$$
$$= \frac{34}{2}$$
$$= 17$$

You Try It 4 $x^3 - 2(x + y) + z^2$
$(2)^3 - 2[2 + (-4)] + (-3)^2$
$$= 8 - 2(-2) + 9$$
$$= 8 + 4 + 9$$
$$= 12 + 9$$
$$= 21$$

SECTION 2.2

You Try It 1 $3a - 2b - 5a + 6b = -2a + 4b$

You Try It 2 $-3y^2 + 7 + 8y^2 - 14 = 5y^2 - 7$

You Try It 3 $-5(4y^2) = -20y^2$

You Try It 4 $-7(-2a) = 14a$

You Try It 5 $-\dfrac{3}{5}\left(-\dfrac{7}{9}a\right) = \dfrac{7}{15}a$

You Try It 6 $5(3 + 7b) = 15 + 35b$

You Try It 7 $(3a - 1)5 = 15a - 5$

You Try It 8 $-8(-2a + 7b) = 16a - 56b$

You Try It 9 $3(12x^2 - x + 8) = 36x^2 - 3x + 24$

You Try It 10 $3(-a^2 - 6a + 7) = -3a^2 - 18a + 21$

You Try It 11 $3y - 2(y - 7x) = 3y - 2y + 14x$
$$= y + 14x$$

Copyright © Houghton Mifflin Company. All rights reserved.

You Try It 12

$-2(x - 2y) - (-x + 3y) = -2x + 4y + x - 3y$
$$= -x + y$$

You Try It 13

$3y - 2[x - 4(2 - 3y)] = 3y - 2[x - 8 + 12y]$
$$= 3y - 2x + 16 - 24y$$
$$= -2x - 21y + 16$$

SECTION 2.3

You Try It 1   the <u>difference between</u> <u>twice</u> n and <u>one-third of</u> n

$$2n - \frac{1}{3}n$$

You Try It 2   the <u>quotient of</u> 7 <u>less than</u> b and 15

$$\frac{b - 7}{15}$$

Solutions to Chapter 3 "You Try It"

SECTION 3.1

You Try It 1

$$\frac{5 - 4x = 8x + 2}{5 - 4\left(\frac{1}{4}\right) \,\bigg|\, 8\left(\frac{1}{4}\right) + 2}$$
$$5 - 1 \,\bigg|\, 2 + 2$$
$$4 = 4$$

Yes, $\frac{1}{4}$ is a solution.

You Try It 2

$$\frac{10x - x^2 = 3x - 10}{10(5) - (5)^2 \,\bigg|\, 3(5) - 10}$$
$$50 - 25 \,\bigg|\, 15 - 10$$
$$25 \neq 5$$

No, 5 is not a solution.

You Try It 3

$$\frac{5}{6} = y - \frac{3}{8}$$
$$\frac{5}{6} + \frac{3}{8} = y - \frac{3}{8} + \frac{3}{8}$$
$$\frac{29}{24} = y$$

The solution is $\frac{29}{24}$.

You Try It 4

$$-\frac{2}{5}x = 6$$
$$\left(-\frac{5}{2}\right)\left(-\frac{2}{5}x\right) = \left(-\frac{5}{2}\right)(6)$$
$$x = -15$$

The solution is -15.

You Try It 3   the unknown number: n
the cube of the number: n^3
the total of ten and the cube of the number: $10 + n^3$

$$-4(10 + n^3)$$

You Try It 4   the unknown number: x
the difference between the number and sixty: $x - 60$

$$5(x - 60)$$
$$= 5x - 300$$

You Try It 5   the speed of the older model: s
the speed of the new jet plane is twice the speed of the older model: $2s$

You Try It 6   the length of the longer piece: y
the length of the shorter piece: $6 - y$

You Try It 5

$$4x - 8x = 16$$
$$-4x = 16$$
$$\frac{-4x}{-4} = \frac{16}{-4}$$
$$x = -4$$

The solution is -4.

You Try It 6

$$P \cdot B = A$$
$$\frac{1}{6}B = 18 \qquad \bullet \; 16\frac{2}{3}\% = \frac{1}{6}$$
$$6 \cdot \frac{1}{6}B = 6 \cdot 18$$
$$B = 108$$

18 is $16\frac{2}{3}\%$ of 108.

You Try It 7

Strategy   Use the percent equation. $B = 83.3$, the total revenue received by the BCS; $A = 3.1$, the amount received by the college representing the Pac-10 conference; P is the unknown percent.

Solution
$$P \cdot B = A$$
$$P(83.3) = 3.1 \qquad \bullet \; B = 83.3, \, A = 3.1$$
$$P = \frac{3.1}{83.3} \approx 0.037$$

The college representing the Pac-10 conference received approximately 3.7% of the BCS revenue.

Copyright © Houghton Mifflin Company. All rights reserved.

You Try It 8

Strategy To find how much she must deposit into the bank account:
- Find the amount of interest earned on the municipal bond by solving $I = Prt$ for I using $P = 1000$, $r = 6.4\% = 0.064$, and $t = 1$.
- Solve $I = Prt$ for P using the amount of interest earned on the municipal bond as I. $r = 8\% = 0.08$, and $t = 1$.

Solution

$I = Prt$
$= 1000(0.064)(1) = 64$

The interest earned on the municipal bond was $64.

$$I = Prt$$
$$64 = P(0.08)(1) \qquad \bullet\ I = 64,\ r = 0.08,\ t = 1.$$
$$64 = 0.08P$$
$$\frac{64}{0.08} = \frac{0.08P}{0.08}$$
$$800 = P$$

Clarissa must invest $800 in the bank account.

You Try It 9

Strategy To find the number of ounces of cereal in the bowl, solve $Q = Ar$ for A using $Q = 2$ and $r = 25\% = 0.25$.

Solution

$$Q = Ar$$
$$2 = A(0.25) \qquad \bullet\ Q = 2,\ r = 0.25.$$
$$\frac{2}{0.25} = \frac{A(0.25)}{0.25}$$
$$8 = A$$

The cereal bowl contains 8 oz of cereal.

You Try It 10

Strategy To find the distance, solve the equation $d = rt$ for d. The time is 3 h. Therefore, $t = 3$. The plane is moving against the wind, which means the headwind is slowing the actual speed of the plane. 250 mph − 25 mph = 225 mph. Thus $r = 225$.

Solution

$$d = rt$$
$$d = 225(3) \qquad \bullet\ r = 225,\ t = 3.$$
$$= 675$$

The plane travels 675 mi in 3 h.

SECTION 3.2

You Try It 1

$$5x + 7 = 10$$
$$5x + 7 - 7 = 10 - 7 \qquad \bullet\ \text{Subtract 7.}$$
$$5x = 3$$
$$\frac{5x}{5} = \frac{3}{5} \qquad \bullet\ \text{Divide by 5.}$$
$$x = \frac{3}{5}$$

The solution is $\frac{3}{5}$.

You Try It 2

$$2 = 11 + 3x$$
$$2 - 11 = 11 - 11 + 3x \qquad \bullet\ \text{Subtract 11.}$$
$$-9 = 3x$$
$$\frac{-9}{3} = \frac{3x}{3} \qquad \bullet\ \text{Divide by 3.}$$
$$-3 = x$$

The solution is -3.

You Try It 3

$$\frac{5}{8} - \frac{2x}{3} = \frac{5}{4}$$
$$\frac{5}{8} - \frac{5}{8} - \frac{2}{3}x = \frac{5}{4} - \frac{5}{8} \qquad \bullet\ \text{Recall that } \frac{2x}{3} = \frac{2}{3}x.$$
$$-\frac{2}{3}x = \frac{5}{8}$$
$$-\frac{3}{2}\left(-\frac{2}{3}x\right) = -\frac{3}{2}\left(\frac{5}{8}\right) \qquad \bullet\ \text{Multiply by } -\frac{3}{2}.$$
$$x = -\frac{15}{16}$$

The solution is $-\dfrac{15}{16}$.

You Try It 4

$$\frac{2}{3}x + 3 = \frac{7}{2}$$
$$6\left(\frac{2}{3}x + 3\right) = 6\left(\frac{7}{2}\right)$$
$$6\left(\frac{2}{3}x\right) + 6(3) = 6\left(\frac{7}{2}\right) \qquad \bullet\ \text{Distributive Property}$$
$$4x + 18 = 21$$
$$4x + 18 - 18 = 21 - 18 \qquad \bullet\ \text{Subtract 18.}$$
$$4x = 3$$
$$\frac{4x}{4} = \frac{3}{4} \qquad \bullet\ \text{Divide by 4.}$$
$$x = \frac{3}{4}$$

The solution is $\frac{3}{4}$.

Copyright © Houghton Mifflin Company. All rights reserved.

You Try It 5

$$x - 5 + 4x = 25$$
$$5x - 5 = 25$$
$$5x - 5 + 5 = 25 + 5$$
$$5x = 30$$
$$\frac{5x}{5} = \frac{30}{5}$$
$$x = 6$$

The solution is 6.

You Try It 6

Strategy   Given: $S = 986$
$r = 45\% = 0.45$
Unknown: C

Solution
$$S = C + rC$$
$$986 = C + 0.45C$$
$$986 = 1.45C$$
$$\frac{986}{1.45} = \frac{1.45C}{1.45}$$
$$680 = C$$

The cost of the outboard motor is $680.

You Try It 7

Strategy   Given: $S = 159$
$r = 25\% = 0.25$
Unknown: R

Solution
$$S = R - rR$$
$$159 = R - 0.25R$$
$$159 = 0.75R$$
$$\frac{159}{0.75} = \frac{0.75R}{0.75}$$
$$212 = R$$

The regular price of the garage door opener is $212.

You Try It 8

Strategy   Given: $P = 45$
Unknown: D

Solution
$$P = 15 + \frac{1}{2}D$$
$$45 = 15 + \frac{1}{2}D$$
$$45 - 15 = 15 - 15 + \frac{1}{2}D$$
$$30 = \frac{1}{2}D$$
$$2(30) = 2 \cdot \frac{1}{2}D$$
$$60 = D$$

The depth is 60 ft.

SECTION 3.3

You Try It 1

$$5x + 4 = 6 + 10x$$
$$5x - 10x + 4 = 6 + 10x - 10x \qquad \bullet \text{ Subtract } 10x.$$
$$-5x + 4 = 6$$
$$-5x + 4 - 4 = 6 - 4 \qquad \bullet \text{ Subtract 4.}$$
$$-5x = 2$$
$$\frac{-5x}{-5} = \frac{2}{-5} \qquad \bullet \text{ Divide by } -5.$$
$$x = -\frac{2}{5}$$

The solution is $-\frac{2}{5}$.

You Try It 2

$$5x - 10 - 3x = 6 - 4x$$
$$2x - 10 = 6 - 4x \qquad \bullet \text{ Combine like terms.}$$
$$2x + 4x - 10 = 6 - 4x + 4x \qquad \bullet \text{ Add } 4x.$$
$$6x - 10 = 6$$
$$6x - 10 + 10 = 6 + 10 \qquad \bullet \text{ Add 10.}$$
$$6x = 16$$
$$\frac{6x}{6} = \frac{16}{6} \qquad \bullet \text{ Divide by 6.}$$
$$x = \frac{8}{3}$$

The solution is $\frac{8}{3}$.

You Try It 3

$$5x - 4(3 - 2x) = 2(3x - 2) + 6$$
$$5x - 12 + 8x = 6x - 4 + 6 \qquad \bullet \text{ Distributive Property}$$
$$13x - 12 = 6x + 2$$
$$13x - 6x - 12 = 6x - 6x + 2 \qquad \bullet \text{ Subtract } 6x.$$
$$7x - 12 = 2$$
$$7x - 12 + 12 = 2 + 12 \qquad \bullet \text{ Add 12.}$$
$$7x = 14$$
$$\frac{7x}{7} = \frac{14}{7} \qquad \bullet \text{ Divide by 7.}$$
$$x = 2$$

The solution is 2.

You Try It 4

$$-2[3x - 5(2x - 3)] = 3x - 8$$
$$-2[3x - 10x + 15] = 3x - 8 \qquad \bullet \text{ Distributive}$$
$$-2[-7x + 15] = 3x - 8 \qquad \qquad \text{ Property}$$
$$14x - 30 = 3x - 8$$
$$14x - 3x - 30 = 3x - 3x - 8 \qquad \bullet \text{ Subtract } 3x.$$
$$11x - 30 = -8$$
$$11x - 30 + 30 = -8 + 30 \qquad \bullet \text{ Add 30.}$$
$$11x = 22$$
$$\frac{11x}{11} = \frac{22}{11} \qquad \bullet \text{ Divide by 11.}$$
$$x = 2$$

The solution is 2.

Copyright © Houghton Mifflin Company. All rights reserved.

You Try It 5

Strategy Given: $F_1 = 45$
$F_2 = 80$
$d = 25$
Unknown: x

Solution
$$F_1 x = F_2(d - x)$$
$$45x = 80(25 - x)$$
$$45x = 2000 - 80x$$
$$45x + 80x = 2000 - 80x + 80x$$
$$125x = 2000$$
$$\frac{125x}{125} = \frac{2000}{125}$$
$$x = 16$$

The fulcrum is 16 ft from the 45-pound force.

SECTION 3.4

You Try It 1

The smaller number: n
The larger number: $12 - n$

The total of three times the smaller number and six	amounts to	seven less than the product of four and the larger number

$$3n + 6 = 4(12 - n) - 7$$
$$3n + 6 = 48 - 4n - 7$$
$$3n + 6 = 41 - 4n$$
$$3n + 4n + 6 = 41 - 4n + 4n$$
$$7n + 6 = 41$$
$$7n + 6 - 6 = 41 - 6$$
$$7n = 35$$
$$\frac{7n}{7} = \frac{35}{7}$$
$$n = 5$$

$12 - n = 12 - 5 = 7$

The smaller number is 5.
The larger number is 7.

You Try It 2

Strategy
• First integer: n
  Second integer: $n + 1$
  Third integer: $n + 2$
• The sum of the three integers is -6.

Solution
$$n + (n + 1) + (n + 2) = -6$$
$$3n + 3 = -6$$
$$3n = -9$$
$$n = -3$$
$$n + 1 = -3 + 1 = -2$$
$$n + 2 = -3 + 2 = -1$$

The three consecutive integers are -3, -2, and -1.

You Try It 3

Strategy
To find the number of tickets that you are purchasing, write and solve an equation using x to represent the number of tickets purchased.

Solution

$3.50 plus $17.50 for each ticket	is	$161

$$3.50 + 17.50x = 161$$
$$3.50 - 3.50 + 17.50x = 161 - 3.50$$
$$17.50x = 157.50$$
$$\frac{17.50x}{17.50} = \frac{157.50}{17.50}$$
$$x = 9$$

You are purchasing 9 tickets.

You Try It 4

Strategy
To find the length, write and solve an equation using x to represent the length of the shorter piece and $22 - x$ to represent the length of the longer piece.

Solution

The length of the longer piece	is	4 in. more than twice the length of the shorter piece

$$22 - x = 2x + 4$$
$$22 - x - 2x = 2x - 2x + 4$$
$$22 - 3x = 4$$
$$22 - 22 - 3x = 4 - 22$$
$$-3x = -18$$
$$\frac{-3x}{-3} = \frac{-18}{-3}$$
$$x = 6$$

$22 - x = 22 - 6 = 16$

The length of the shorter piece is 6 in.
The length of the longer piece is 16 in.

Copyright © Houghton Mifflin Company. All rights reserved.

SECTION 3.5

You Try It 1

Strategy   The angles labeled are adjacent angles of intersecting lines and are therefore supplementary angles. To find x, write an equation and solve for x.

Solution
$$x + (3x + 20°) = 180°$$
$$4x + 20° = 180°$$
$$4x = 160°$$
$$x = 40°$$

You Try It 2

Strategy $2x = y$ because alternate exterior angles have the same measure. $y + (x + 15°) = 180°$ because adjacent angles of intersecting lines are supplementary angles. Substitute $2x$ for y and solve for x.

Solution
$$y + (x + 15°) = 180°$$
$$2x + (x + 15°) = 180°$$
$$3x + 15° = 180°$$
$$3x = 165°$$
$$x = 55°$$

You Try It 3

Strategy
- To find the measure of angle a, use the fact that $\angle a$ and $\angle y$ are vertical angles.
- To find the measure of angle b, use the fact that the sum of the measures of the interior angles of a triangle is $180°$.
- To find the measure of angle d, use the fact that the sum of an interior and an exterior angle is $180°$.

Solution $m\angle a = m\angle y = 55°$

$$m\angle a + m\angle b + 90° = 180°$$
$$55° + m\angle b + 90° = 180°$$
$$m\angle b + 145° = 180°$$
$$m\angle b = 35°$$

$$m\angle d + m\angle b = 180°$$
$$m\angle d + 35° = 180°$$
$$m\angle d = 145°$$

You Try It 4

Strategy   To find the measure of the third angle, use the fact that the sum of the measures of the interior angles of a triangle is $180°$. Write an equation using x to represent the measure of the third angle. Solve the equation for x.

Solution
$$x + 90° + 27° = 180°$$
$$x + 117° = 180°$$
$$x = 63°$$

The measure of the third angle is $63°$.

SECTION 3.6

You Try It 1

Strategy   • Pounds of $.55 fertilizer: x

	Amount	Cost	Value
$.80 fertilizer	20	.80	0.80(20)
$.55 fertilizer	x	.55	0.55x
$.75 fertilizer	20 + x	.75	0.75(20 + x)

• The sum of the values before mixing equals the value after mixing.

Solution
$$0.80(20) + 0.55x = 0.75(20 + x)$$
$$16 + 0.55x = 15 + 0.75x$$
$$16 - 0.20x = 15$$
$$-0.20x = -1$$
$$x = 5$$

5 lb of the $.55 fertilizer must be added.

You Try It 2

Strategy   • Liters of 6% solution: x

	Amount	Percent	Quantity
6% solution	x	0.06	0.06x
12% solution	5	0.12	5(0.12)
8% solution	x + 5	0.08	0.08(x + 5)

• The sum of the quantities before mixing equals the quantity after mixing.

Solution
$$0.06x + 5(0.12) = 0.08(x + 5)$$
$$0.06x + 0.60 = 0.08x + 0.40$$
$$-0.02x + 0.60 = 0.40$$
$$-0.02x = -0.20$$
$$x = 10$$

The pharmacist adds 10 L of the 6% solution to the 12% solution to get an 8% solution.

Copyright © Houghton Mifflin Company. All rights reserved.

You Try It 3

Strategy
• Rate of the first train: r
Rate of the second train: $2r$

	Rate	Time	Distance
1st train	r	3	$3r$
2nd train	$2r$	3	$3(2r)$

• The sum of the distances traveled by the two trains equals 288 mi.

Solution
$$3r + 3(2r) = 288$$
$$3r + 6r = 288$$
$$9r = 288$$
$$r = 32$$

$$2r = 2(32) = 64$$

The first train is traveling at 32 mph.
The second train is traveling at 64 mph.

You Try It 4

Strategy
• Time spent flying out: t
Time spent flying back: $5 - t$

	Rate	Time	Distance
Out	150	t	$150t$
Back	100	$5 - t$	$100(5 - t)$

• The distance out equals the distance back.

Solution
$$150t = 100(5 - t)$$
$$150t = 500 - 100t$$
$$250t = 500$$
$$t = 2 \quad \text{(The time out was 2 h.)}$$

The distance out $= 150t = 150(2)$
$$= 300 \text{ mi}$$

The parcel of land was 300 mi away.

Solutions to Chapter 4 "You Try It"

SECTION 4.1

You Try It 1
$(-4x^3 + 2x^2 - 8) + (4x^3 + 6x^2 - 7x + 5)$
$= (-4x^3 + 4x^3) + (2x^2 + 6x^2) + (-7x) + (-8 + 5)$
$= 8x^2 - 7x - 3$

You Try It 2
$$
\begin{array}{r}
6x^3 \phantom{{}+ 2x + 8} + 2x + 8 \\
-9x^3 + 2x^2 - 12x - 8 \\
\hline
-3x^3 + 2x^2 - 10x
\end{array}
$$

You Try It 3
$(-4w^3 + 8w - 8) - (3w^3 - 4w^2 - 2w - 1)$
$= (-4w^3 + 8w - 8)$
$\quad + (-3w^3 + 4w^2 + 2w + 1)$
$= -7w^3 + 4w^2 + 10w - 7$

You Try It 4
$$
\begin{array}{r}
13y^3 \phantom{{}- 6y - 7} - 6y - 7 \\
- 4y^2 + 6y + 9 \\
\hline
13y^3 - 4y^2 \phantom{{}+ 6y} + 2
\end{array}
$$

SECTION 4.2

You Try It 1
$(8m^3n)(-3n^5)$
$= [8(-3)](m^3)(n \cdot n^5)$ • Multiply coefficients. Add
$= -24m^3n^6$ exponents with same base.

You Try It 2
$(12p^4q^3)(-3p^5q^2)$
$= [12(-3)](p^4 \cdot p^5)(q^3 \cdot q^2)$ • Multiply coefficients. Add
$= -36p^9q^5$ exponents with same base.

You Try It 3
$(-3a^4bc^2)^3 = (-3)^{1 \cdot 3}a^{4 \cdot 3}b^{1 \cdot 3}c^{2 \cdot 3}$ • Rule for Simplifying
$= (-3)^3a^{12}b^3c^6$ the Power of a Product
$= -27a^{12}b^3c^6$

You Try It 4
$(-xy^4)(-2x^3y^2)^2 = (-xy^4)[(-2)^{1 \cdot 2}x^{3 \cdot 2}y^{2 \cdot 2}]$ • Rule for
$= (-xy^4)[(-2)^2x^6y^4]$ Simplifying
$= (-xy^4)(4x^6y^4)$ the Power of
$= -4x^7y^8$ a Product

SECTION 4.3

You Try It 1 $(-2y + 3)(-4y) = 8y^2 - 12y$

You Try It 2
$-a^2(3a^2 + 2a - 7) = -3a^4 - 2a^3 + 7a^2$

You Try It 3
$$
\begin{array}{r}
2y^3 + 2y^2 \phantom{{}- 3} - 3 \\
3y - 1 \\
\hline
- 2y^3 - 2y^2 \phantom{{}+ 3} + 3 = -1(2y^3 + 2y^2 - 3) \\
6y^4 + 6y^3 \phantom{{}- 9y} - 9y \phantom{{}= } = 3y(2y^3 + 2y^2 - 3) \\
\hline
6y^4 + 4y^3 - 2y^2 - 9y + 3
\end{array}
$$

Copyright © Houghton Mifflin Company. All rights reserved.

You Try It 4
$$(4y - 5)(2y - 3) = 8y^2 - 12y - 10y + 15$$
$$= 8y^2 - 22y + 15$$

You Try It 5
$$(3b + 2)(3b - 5) = 9b^2 - 15b + 6b - 10$$
$$= 9b^2 - 9b - 10$$

You Try It 6 $(2a + 5c)(2a - 5c) = 4a^2 - 25c^2$

You Try It 7 $(3x + 2y)^2 = 9x^2 + 12xy + 4y^2$

You Try It 8

Strategy   To find the area, replace the variable r in the equation $A = \pi r^2$ by $(x - 4)$ and solve for A.

Solution
$$A = \pi r^2$$
$$A = \pi(x - 4)^2$$
$$A = \pi(x^2 - 8x + 16)$$
$$A = \pi x^2 - 8\pi x + 16\pi$$

The area of the circle is $(\pi x^2 - 8\pi x + 16\pi)$ ft^2.

SECTION 4.4

You Try It 1 $(-2x^2)(x^{-3}y^{-4})^{-2}$
$$= (-2x^2)(x^6y^8) \quad \bullet \text{ Rule for Simplifying}$$
$$= -2x^8y^8 \qquad\qquad \text{the Power of a Product}$$

You Try It 2 $\dfrac{(6a^{-2}b^3)^{-1}}{(4a^3b^{-2})^{-2}}$

$$= \frac{6^{-1}a^2b^{-3}}{4^{-2}a^{-6}b^4} \quad \bullet \text{ Rule for Simplifying the}$$
$$\qquad\qquad\qquad \text{Power of a Product}$$
$$= 4^2(6^{-1}a^8b^{-7}) \quad \bullet \text{ Rule for Dividing}$$
$$\qquad\qquad\qquad \text{Exponential Expressions}$$
$$= \frac{16a^8}{6b^7} = \frac{8a^8}{3b^7}$$

You Try It 3 $\left[\dfrac{6r^3s^{-3}}{9r^3s^{-1}}\right]^{-2} = \left[\dfrac{2r^0s^{-2}}{3}\right]^{-2}$

$$= \frac{2^{-2}s^4}{3^{-2}} = \frac{9s^4}{4}$$

You Try It 4 $0.000000961 = 9.61 \times 10^{-7}$

You Try It 5 $7.329 \times 10^6 = 7,329,000$

SECTION 4.5

You Try It 1
$$\frac{24x^2y^2 - 18xy + 6y}{6xy} = \frac{24x^2y^2}{6xy} - \frac{18xy}{6xy} + \frac{6y}{6xy}$$
$$= 4xy - 3 + \frac{1}{x}$$

You Try It 2

$$\begin{array}{r} x^2 + 2x - 1 \\ 2x - 3 \overline{)\,2x^3 + x^2 - 8x - 3} \\ \underline{2x^3 - 3x^2} \\ 4x^2 - 8x \\ \underline{4x^2 - 6x} \\ -2x - 3 \\ \underline{-2x + 3} \\ -6 \end{array}$$

$$(2x^3 + x^2 - 8x - 3) \div (2x - 3)$$
$$= x^2 + 2x - 1 - \frac{6}{2x - 3}$$

You Try It 3

$$\begin{array}{r} x^2 + x - 1 \\ x - 1 \overline{)\,x^3 + 0x^2 - 2x + 1} \\ \underline{x^3 - x^2} \\ x^2 - 2x \\ \underline{x^2 - x} \\ -x + 1 \\ \underline{-x + 1} \\ 0 \end{array}$$

$$(x^3 - 2x + 1) \div (x - 1) = x^2 + x - 1$$

Solutions to Chapter 5 "You Try It"

SECTION 5.1

You Try It 1   The GCF is $7a^2$.

$$14a^2 - 21a^4b = 7a^2(2) + 7a^2(-3a^2b)$$
$$= 7a^2(2 - 3a^2b)$$

You Try It 2   The GCF is 9.

$$27b^2 + 18b + 9$$
$$= 9(3b^2) + 9(2b) + 9(1)$$
$$= 9(3b^2 + 2b + 1)$$

Copyright © Houghton Mifflin Company. All rights reserved.

You Try It 3
The GCF is $3x^2y^2$.

$6x^4y^2 - 9x^3y^2 + 12x^2y^4$
$= 3x^2y^2(2x^2) + 3x^2y^2(-3x) + 3x^2y^2(4y^2)$
$= 3x^2y^2(2x^2 - 3x + 4y^2)$

You Try It 4
$2y(5x - 2) - 3(2 - 5x)$
$= 2y(5x - 2) + 3(5x - 2)$ • **5x − 2 is the**
$= (5x - 2)(2y + 3)$ **common factor.**

You Try It 5
$a^2 - 3a + 2ab - 6b$
$= (a^2 - 3a) + (2ab - 6b)$
$= a(a - 3) + 2b(a - 3)$ • **a − 3 is the common factor.**
$= (a - 3)(a + 2b)$

You Try It 6
$2mn^2 - n + 8mn - 4$
$= (2mn^2 - n) + (8mn - 4)$ • **2mn − 1 is the**
$= n(2mn - 1) + 4(2mn - 1)$ **common factor.**
$= (2mn - 1)(n + 4)$

You Try It 7
$3xy - 9y - 12 + 4x$
$= (3xy - 9y) - (12 - 4x)$ • **−12 + 4x = −(12 − 4x)**
$= 3y(x - 3) - 4(3 - x)$ • **−(3 − x) = (x − 3)**
$= 3y(x - 3) + 4(x - 3)$ • **x − 3 is the common factor.**
$= (x - 3)(3y + 4)$

SECTION 5.2

You Try It 1
Find the positive factors of 20 whose sum is 9.

Factors	Sum
1, 20	21
2, 10	12
4, 5	9

$x^2 + 9x + 20 = (x + 4)(x + 5)$

You Try It 2
Find the factors of −18 whose sum is 7.

Factors	Sum
+1, −18	−17
−1, +18	17
+2, −9	−7
−2, +9	7
+3, −6	−3
−3, +6	3

$x^2 + 7x - 18 = (x + 9)(x - 2)$

You Try It 3
The GCF is $-2x$.

$-2x^3 + 14x^2 - 12x = -2x(x^2 - 7x + 6)$

Factor the trinomial $x^2 - 7x + 6$. Find two negative factors of 6 whose sum is −7.

Factors	Sum
−1, −6	−7
−2, −3	−5

$-2x^3 + 14x^2 - 12x = -2x(x - 6)(x - 1)$

You Try It 4
The GCF is 3.

$3x^2 - 9xy - 12y^2 = 3(x^2 - 3xy - 4y^2)$

Factor the trinomial.

Find the factors of −4 whose sum is −3.

Factors	Sum
+1, −4	−3
−1, +4	3
+2, −2	0

$3x^2 - 9xy - 12y^2 = 3(x + y)(x - 4y)$

SECTION 5.3

You Try It 1
Factor the trinomial $2x^2 - x - 3$.

Positive factors of 2: 1, 2   Factors of −3: +1, −3
                                   −1, +3

Trial Factors	Middle Term
$(x + 1)(2x - 3)$	$-3x + 2x = -x$
$(x - 3)(2x + 1)$	$x - 6x = -5x$
$(x - 1)(2x + 3)$	$3x - 2x = x$
$(x + 3)(2x - 1)$	$-x + 6x = 5x$

$2x^2 - x - 3 = (x + 1)(2x - 3)$

You Try It 2
The GCF is $-3y$.

$-45y^3 + 12y^2 + 12y = -3y(15y^2 - 4y - 4)$

Factor the trinomial $15y^2 - 4y - 4$.

Positive factors of 15: 1, 15   Factors of −4: 1, −4
                       3, 5   −1, 4
                                               2, −2

Copyright © Houghton Mifflin Company. All rights reserved.

Trial Factors	Middle Term
$(y + 1)(15y - 4)$	$-4y + 15y = 11y$
$(y - 4)(15y + 1)$	$y - 60y = -59y$
$(y - 1)(15y + 4)$	$4y - 15y = -11y$
$(y + 4)(15y - 1)$	$-y + 60y = 59y$
$(y + 2)(15y - 2)$	$-2y + 30y = 28y$
$(y - 2)(15y + 2)$	$2y - 30y = -28y$
$(3y + 1)(5y - 4)$	$-12y + 5y = -7y$
$(3y - 4)(5y + 1)$	$3y - 20y = -17y$
$(3y - 1)(5y + 4)$	$12y - 5y = 7y$
$(3y + 4)(5y - 1)$	$-3y + 20y = 17y$
$(3y + 2)(5y - 2)$	$-6y + 10y = 4y$
$(3y - 2)(5y + 2)$	$6y - 10y = -4y$

$$-45y^3 + 12y^2 + 12y = -3y(3y - 2)(5y + 2)$$

You Try It 3

Factors of -14 [$2(-7)$]	Sum
$+1, -14$	-13
$-1, +14$	13
$+2, -7$	-5
$-2, +7$	5

$$
\begin{aligned}
2a^2 + 13a - 7 &= 2a^2 - a + 14a - 7 \\
&= (2a^2 - a) + (14a - 7) \\
&= a(2a - 1) + 7(2a - 1) \\
&= (2a - 1)(a + 7)
\end{aligned}
$$

$$2a^2 + 13a - 7 = (2a - 1)(a + 7)$$

You Try It 4
The GCF is $5x$.

$$15x^3 + 40x^2 - 80x = 5x(3x^2 + 8x - 16)$$

Factors of -48 [$3(-16)$]	Sum
$+1, -48$	-47
$-1, +48$	47
$+2, -24$	-22
$-2, +24$	22
$+3, -16$	-13
$-3, +16$	13
$+4, -12$	-8
$-4, +12$	8

$$
\begin{aligned}
3x^2 + 8x - 16 &= 3x^2 - 4x + 12x - 16 \\
&= (3x^2 - 4x) + (12x - 16) \\
&= x(3x - 4) + 4(3x - 4) \\
&= (3x - 4)(x + 4)
\end{aligned}
$$

$$
\begin{aligned}
15x^3 + 40x^2 - 80x &= 5x(3x^2 + 8x - 16) \\
&= 5x(3x - 4)(x + 4)
\end{aligned}
$$

SECTION 5.4

You Try It 1
$$
\begin{aligned}
25a^2 - b^2 &= (5a)^2 - b^2 \qquad \text{• Difference of} \\
&= (5a + b)(5a - b) \qquad \text{two squares}
\end{aligned}
$$

You Try It 2
$$
\begin{aligned}
n^4 - 81 &= (n^2)^2 - 9^2 \qquad\qquad \text{• Difference of} \\
&\qquad\qquad\qquad\qquad\quad \text{two squares} \\
&= (n^2 + 9)(n^2 - 9) \qquad \text{• Difference of} \\
&= (n^2 + 9)(n + 3)(n - 3) \quad \text{two squares}
\end{aligned}
$$

You Try It 3   Because $16y^2 = (4y)^2$, $1 = 1^2$, and $8y = 2(4y)(1)$, the trinomial is a perfect-square trinomial.

$$16y^2 + 8y + 1 = (4y + 1)^2$$

You Try It 4   Because $x^2 = (x)^2$, $36 = 6^2$, and $15x \ne 2(x)(6)$, the trinomial is not a perfect-square trinomial. Try to factor the trinomial by another method.

$$x^2 + 15x + 36 = (x + 3)(x + 12)$$

You Try It 5   The GCF is $3x$.

$$
\begin{aligned}
12x^3 - 75x &= 3x(4x^2 - 25) \\
&= 3x(2x + 5)(2x - 5)
\end{aligned}
$$

You Try It 6
Factor by grouping.

$$
\begin{aligned}
a^2b - 7a^2 - b + 7 &\\
= (a^2b - 7a^2) - (b - 7)& \\
= a^2(b - 7) - (b - 7)& \quad \text{• } b - 7 \text{ is the common factor.} \\
= (b - 7)(a^2 - 1)& \quad \text{• } a^2 - 1 \text{ is the difference} \\
= (b - 7)(a + 1)(a - 1)& \quad \text{of two squares.}
\end{aligned}
$$

You Try It 7
The GCF is $4x$.

$$
\begin{aligned}
4x^3 + 28x^2 - 120x& \\
= 4x(x^2 + 7x - 30)& \quad \text{• Factor the GCF, } 4x. \\
= 4x(x + 10)(x - 3)& \quad \text{• Factor the trinomial.}
\end{aligned}
$$

SECTION 5.5

You Try It 1
$$2x(x + 7) = 0$$

$$
\begin{array}{ll}
2x = 0 \qquad & x + 7 = 0 \qquad \text{• Principle of} \\
x = 0 & x = -7 \qquad \text{Zero Products}
\end{array}
$$

The solutions are 0 and -7.

Copyright © Houghton Mifflin Company. All rights reserved.

You Try It 2

$$4x^2 - 9 = 0 \qquad \bullet \text{ Difference of two squares}$$
$$(2x - 3)(2x + 3) = 0$$

$$2x - 3 = 0 \qquad 2x + 3 = 0 \qquad \bullet \text{ Principle of Zero Products}$$
$$2x = 3 \qquad 2x = -3$$
$$x = \frac{3}{2} \qquad x = -\frac{3}{2}$$

The solutions are $\frac{3}{2}$ and $-\frac{3}{2}$.

You Try It 3

$$(x + 2)(x - 7) = 52$$
$$x^2 - 5x - 14 = 52$$
$$x^2 - 5x - 66 = 0$$
$$(x + 6)(x - 11) = 0$$

$$x + 6 = 0 \qquad x - 11 = 0 \qquad \bullet \text{ Principle of Zero Products}$$
$$x = -6 \qquad x = 11$$

The solutions are -6 and 11.

You Try It 4

Strategy First consecutive positive integer: n
Second consecutive positive integer:
$n + 1$

The sum of the squares of the two
consecutive positive integers is 61.

Solution
$$n^2 + (n + 1)^2 = 61$$
$$n^2 + n^2 + 2n + 1 = 61$$
$$2n^2 + 2n + 1 = 61$$
$$2n^2 + 2n - 60 = 0$$
$$2(n^2 + n - 30) = 0$$
$$2(n - 5)(n + 6) = 0$$

$$n - 5 = 0 \qquad n + 6 = 0 \qquad \bullet \text{ Principle of Zero Products}$$
$$n = 5 \qquad n = -6$$

Because -6 is not a positive integer,
it is not a solution.

$$n = 5$$
$$n + 1 = 5 + 1 = 6$$

The two integers are 5 and 6.

You Try It 5

Strategy Width $= x$
Length $= 2x + 4$

The area of the rectangle is 96 in².
Use the equation $A = L \cdot W$.

Solution
$$A = L \cdot W$$
$$96 = (2x + 4)x$$
$$96 = 2x^2 + 4x$$
$$0 = 2x^2 + 4x - 96$$
$$0 = 2(x^2 + 2x - 48)$$
$$0 = 2(x + 8)(x - 6)$$

$$x + 8 = 0 \qquad x - 6 = 0 \qquad \bullet \text{ Principle of}$$
$$x = -8 \qquad x = 6 \qquad \text{Zero Products}$$

Because the width cannot be a
negative number, -8 is not a
solution.

$$x = 6$$
$$2x + 4 = 2(6) + 4 = 12 + 4 = 16$$

The length is 16 in. The width is 6 in.

Solutions to Chapter 6 "You Try It"

SECTION 6.1

You Try It 1

$$\frac{6x^5y}{12x^2y^3} = \frac{\overset{1}{2} \cdot \overset{1}{3} \cdot x^5y}{\underset{1}{2} \cdot 2 \cdot \underset{1}{3} \cdot x^2y^3} = \frac{x^3}{2y^2}$$

You Try It 2

$$\frac{x^2 + 2x - 24}{16 - x^2} = \frac{\overset{-1}{(x - 4)}(x + 6)}{\underset{1}{(4 - x)}(4 + x)} \qquad \bullet \ (4 - x) = -1(x - 4)$$

$$= -\frac{x + 6}{x + 4}$$

You Try It 3

$$\frac{x^2 + 4x - 12}{x^2 - 3x + 2} = \frac{\overset{1}{(x - 2)}(x + 6)}{(x - 1)\underset{1}{(x - 2)}} = \frac{x + 6}{x - 1}$$

You Try It 4

$$\frac{12x^2 + 3x}{10x - 15} \cdot \frac{8x - 12}{9x + 18} = \frac{3x(4x + 1)}{5(2x - 3)} \cdot \frac{4(2x - 3)}{9(x + 2)} \qquad \bullet \text{ Factor.}$$

$$= \frac{\overset{1}{3}x(4x + 1) \cdot 2 \cdot 2\overset{1}{(2x - 3)}}{5\underset{1}{(2x - 3)} \cdot \underset{1}{3} \cdot 3(x + 2)}$$

$$= \frac{4x(4x + 1)}{15(x + 2)}$$

Copyright © Houghton Mifflin Company. All rights reserved.

You Try It 5

$$\frac{x^2 + 2x - 15}{9 - x^2} \cdot \frac{x^2 - 3x - 18}{x^2 - 7x + 6}$$

$$= \frac{(x - 3)(x + 5)}{(3 - x)(3 + x)} \cdot \frac{(x + 3)(x - 6)}{(x - 1)(x - 6)} \qquad \bullet \text{ Factor.}$$

$$= \frac{\overset{-1}{\cancel{(x - 3)}}(x + 5) \cdot \cancel{(x + 3)}\cancel{(x - 6)}}{\cancel{(3 - x)}\cancel{(3 + x)} \cdot (x - 1)\cancel{(x - 6)}} = -\frac{x + 5}{x - 1}$$

You Try It 6

$$\frac{a^2}{4bc^2 - 2b^2c} \div \frac{a}{6bc - 3b^2}$$

$$= \frac{a^2}{4bc^2 - 2b^2c} \cdot \frac{6bc - 3b^2}{a} \qquad \bullet \text{ Multiply by the reciprocal.}$$

$$= \frac{a^2 \cdot 3b(2c - b)}{2bc(2c - b) \cdot a} = \frac{3a}{2c}$$

You Try It 7

$$\frac{3x^2 + 26x + 16}{3x^2 - 7x - 6} \div \frac{2x^2 + 9x - 5}{x^2 + 2x - 15}$$

$$= \frac{3x^2 + 26x + 16}{3x^2 - 7x - 6} \cdot \frac{x^2 + 2x - 15}{2x^2 + 9x - 5} \qquad \bullet \text{ Multiply by the reciprocal.}$$

$$= \frac{(3x + 2)(x + 8) \cdot (x + 5)(x - 3)}{(3x + 2)(x - 3) \cdot (2x - 1)(x + 5)} = \frac{x + 8}{2x - 1}$$

SECTION 6.2

You Try It 1

$8uv^2 = 2 \cdot 2 \cdot 2 \cdot u \cdot v \cdot v$

$12uw = 2 \cdot 2 \cdot 3 \cdot u \cdot w$

LCM $= 2 \cdot 2 \cdot 2 \cdot 3 \cdot u \cdot v \cdot v \cdot w = 24uv^2w$

You Try It 2

$m^2 - 6m + 9 = (m - 3)(m - 3)$

$m^2 - 2m - 3 = (m + 1)(m - 3)$

LCM $= (m - 3)(m - 3)(m + 1)$

You Try It 3

The LCM is $36xy^2z$.

$$\frac{x - 3}{4xy^2} = \frac{x - 3}{4xy^2} \cdot \frac{9z}{9z} = \frac{9xz - 27z}{36xy^2z}$$

$$\frac{2x + 1}{9y^2z} = \frac{2x + 1}{9y^2z} \cdot \frac{4x}{4x} = \frac{8x^2 + 4x}{36xy^2z}$$

You Try It 4

The LCM is $(x + 2)(x - 5)(x + 5)$.

$$\frac{x + 4}{x^2 - 3x - 10} = \frac{x + 4}{(x + 2)(x - 5)} \cdot \frac{x + 5}{x + 5}$$

$$= \frac{x^2 + 9x + 20}{(x + 2)(x - 5)(x + 5)}$$

$$\frac{2x}{25 - x^2} = \frac{2x}{-(x^2 - 25)} = -\frac{2x}{(x - 5)(x + 5)} \cdot \frac{x + 2}{x + 2}$$

$$= -\frac{2x^2 + 4x}{(x + 2)(x - 5)(x + 5)}$$

SECTION 6.3

You Try It 1

$$\frac{2x^2}{x^2 - x - 12} - \frac{7x + 4}{x^2 - x - 12}$$

$$= \frac{2x^2 - (7x + 4)}{x^2 - x - 12} = \frac{2x^2 - 7x - 4}{x^2 - x - 12}$$

$$= \frac{(2x + 1)(x - 4)}{(x + 3)(x - 4)} = \frac{2x + 1}{x + 3}$$

You Try It 2

$$\frac{x^2 - 1}{x^2 - 8x + 12} - \frac{2x + 1}{x^2 - 8x + 12} + \frac{x}{x^2 - 8x + 12}$$

$$= \frac{(x^2 - 1) - (2x + 1) + x}{x^2 - 8x + 12} = \frac{x^2 - 1 - 2x - 1 + x}{x^2 - 8x + 12}$$

$$= \frac{x^2 - x - 2}{x^2 - 8x + 12} = \frac{(x + 1)(x - 2)}{(x - 2)(x - 6)} = \frac{x + 1}{x - 6}$$

You Try It 3

The LCM of the denominators is $24y$.

$$\frac{z}{8y} - \frac{4z}{3y} + \frac{5z}{4y}$$

$$= \frac{z}{8y} \cdot \frac{3}{3} - \frac{4z}{3y} \cdot \frac{8}{8} + \frac{5z}{4y} \cdot \frac{6}{6} \qquad \bullet \text{ Write each fraction using the LCM.}$$

$$= \frac{3z}{24y} - \frac{32z}{24y} + \frac{30z}{24y}$$

$$= \frac{3z - 32z + 30z}{24y} = \frac{z}{24y} \qquad \bullet \text{ Combine the numerators.}$$

You Try It 4

$2 - x = -(x - 2) \qquad$ Therefore, $\dfrac{3}{2 - x} = \dfrac{-3}{x - 2}$.

$$\frac{5x}{x - 2} + \frac{3}{2 - x} = \frac{5x}{x - 2} + \frac{-3}{x - 2} \qquad \bullet \text{ The LCM is } x - 2.$$

$$= \frac{5x + (-3)}{x - 2} = \frac{5x - 3}{x - 2} \qquad \bullet \text{ Combine the numerators.}$$

You Try It 5

The LCM is $(3x - 1)(x + 4)$.

$$\frac{4x}{3x - 1} + \frac{9}{x + 4} = \frac{4x}{3x - 1} \cdot \frac{x + 4}{x + 4} + \frac{9}{x + 4} \cdot \frac{3x - 1}{3x - 1}$$

$$= \frac{4x^2 + 16x}{(3x - 1)(x + 4)} + \frac{27x - 9}{(3x - 1)(x + 4)}$$

$$= \frac{(4x^2 + 16x) + (27x - 9)}{(3x - 1)(x + 4)}$$

$$= \frac{4x^2 + 16x + 27x - 9}{(3x - 1)(x + 4)}$$

$$= \frac{4x^2 + 43x - 9}{(3x - 1)(x + 4)}$$

Copyright © Houghton Mifflin Company. All rights reserved.

You Try It 6   The LCM is $x - 3$.

$$2 - \frac{1}{x - 3} = 2 \cdot \frac{x - 3}{x - 3} - \frac{1}{x - 3}$$

$$= \frac{2x - 6}{x - 3} - \frac{1}{x - 3}$$

$$= \frac{2x - 6 - 1}{x - 3}$$

$$= \frac{2x - 7}{x - 3}$$

You Try It 7

$$\frac{2}{5 - x} = \frac{-2}{x - 5}$$

The LCM is $(x + 5)(x - 5)$.

$$\frac{2x - 1}{x^2 - 25} + \frac{2}{5 - x} = \frac{2x - 1}{(x + 5)(x - 5)} + \frac{-2}{x - 5}$$

$$= \frac{2x - 1}{(x + 5)(x - 5)} + \frac{-2}{x - 5} \cdot \frac{x + 5}{x + 5}$$

$$= \frac{2x - 1}{(x + 5)(x - 5)} + \frac{-2(x + 5)}{(x + 5)(x - 5)}$$

$$= \frac{2x - 1 + (-2)(x + 5)}{(x + 5)(x - 5)}$$

$$= \frac{2x - 1 - 2x - 10}{(x + 5)(x - 5)}$$

$$= \frac{-11}{(x + 5)(x - 5)}$$

$$= -\frac{11}{(x + 5)(x - 5)}$$

You Try It 8
The LCM is $(3x + 2)(x - 1)$.

$$\frac{2x - 3}{3x^2 - x - 2} + \frac{5}{3x + 2} - \frac{1}{x - 1}$$

$$= \frac{2x - 3}{(3x + 2)(x - 1)} + \frac{5}{3x + 2} \cdot \frac{x - 1}{x - 1}$$

$$- \frac{1}{x - 1} \cdot \frac{3x + 2}{3x + 2}$$

$$= \frac{2x - 3}{(3x + 2)(x - 1)} + \frac{5x - 5}{(3x + 2)(x - 1)}$$

$$- \frac{3x + 2}{(3x + 2)(x - 1)}$$

$$= \frac{(2x - 3) + (5x - 5) - (3x + 2)}{(3x + 2)(x - 1)}$$

$$= \frac{2x - 3 + 5x - 5 - 3x - 2}{(3x + 2)(x - 1)}$$

$$= \frac{4x - 10}{(3x + 2)(x - 1)} = \frac{2(2x - 5)}{(3x + 2)(x - 1)}$$

SECTION 6.4

You Try It 1
The LCM of 3, x, 9, and x^2 is $9x^2$.

$$\frac{\dfrac{1}{3} - \dfrac{1}{x}}{\dfrac{1}{9} - \dfrac{1}{x^2}} = \frac{\dfrac{1}{3} - \dfrac{1}{x}}{\dfrac{1}{9} - \dfrac{1}{x^2}} \cdot \frac{9x^2}{9x^2} = \frac{\dfrac{1}{3} \cdot 9x^2 - \dfrac{1}{x} \cdot 9x^2}{\dfrac{1}{9} \cdot 9x^2 - \dfrac{1}{x^2} \cdot 9x^2}$$ • Multiply by the LCM.

$$= \frac{3x^2 - 9x}{x^2 - 9} = \frac{3x(\cancel{x - 3})}{(\cancel{x - 3})(x + 3)} = \frac{3x}{x + 3}$$

You Try It 2
The LCM of x and x^2 is x^2.

$$\frac{1 + \dfrac{4}{x} + \dfrac{3}{x^2}}{1 + \dfrac{10}{x} + \dfrac{21}{x^2}} = \frac{1 + \dfrac{4}{x} + \dfrac{3}{x^2}}{1 + \dfrac{10}{x} + \dfrac{21}{x^2}} \cdot \frac{x^2}{x^2}$$ • Multiply by the LCM.

$$= \frac{1 \cdot x^2 + \dfrac{4}{x} \cdot x^2 + \dfrac{3}{x^2} \cdot x^2}{1 \cdot x^2 + \dfrac{10}{x} \cdot x^2 + \dfrac{21}{x^2} \cdot x^2}$$ • Distributive Property

$$= \frac{x^2 + 4x + 3}{x^2 + 10x + 21} = \frac{(x + 1)(\cancel{x + 3})}{(\cancel{x + 3})(x + 7)}$$

$$= \frac{x + 1}{x + 7}$$

You Try It 3
The LCM is $x - 5$.

$$\frac{x + 3 - \dfrac{20}{x - 5}}{x + 8 + \dfrac{30}{x - 5}} = \frac{x + 3 - \dfrac{20}{x - 5}}{x + 8 + \dfrac{30}{x - 5}} \cdot \frac{x - 5}{x - 5}$$ • Multiply by the LCM.

$$= \frac{(x + 3)(x - 5) - \dfrac{20}{x - 5} \cdot (x - 5)}{(x + 8)(x - 5) + \dfrac{30}{x - 5} \cdot (x - 5)}$$

$$= \frac{x^2 - 2x - 15 - 20}{x^2 + 3x - 40 + 30} = \frac{x^2 - 2x - 35}{x^2 + 3x - 10}$$

$$= \frac{(\cancel{x + 5})(x - 7)}{(x - 2)(\cancel{x + 5})} = \frac{x - 7}{x - 2}$$

Copyright © Houghton Mifflin Company. All rights reserved.

SECTION 6.5

You Try It 1

$$\frac{x}{x+6} = \frac{3}{x}$$ • The LCM is $x(x+6)$.

$$\frac{x(x+6)}{1} \cdot \frac{x}{x+6} = \frac{x(x+6)}{1} \cdot \frac{3}{x}$$ • Multiply by the LCM.

$$x^2 = (x+6)3$$ • Simplify.
$$x^2 = 3x + 18$$
$$x^2 - 3x - 18 = 0$$
$$(x+3)(x-6) = 0$$ • Factor.

$$x + 3 = 0 \qquad x - 6 = 0$$ • Principle of
$$x = -3 \qquad x = 6$$ Zero Products

Both -3 and 6 check as solutions.
The solutions are -3 and 6.

You Try It 2

$$\frac{5x}{x+2} = 3 - \frac{10}{x+2}$$ • The LCM is $x+2$.

$$\frac{(x+2)}{1} \cdot \frac{5x}{x+2} = \frac{(x+2)}{1}\left(3 - \frac{10}{x+2}\right)$$ • Clear denominators.

$$\frac{x+2}{1} \cdot \frac{5x}{x+2} = \frac{x+2}{1} \cdot 3 - \frac{x+2}{1} \cdot \frac{10}{x+2}$$

$$5x = (x+2)3 - 10$$ • Solve for x.
$$5x = 3x + 6 - 10$$
$$5x = 3x - 4$$
$$2x = -4$$
$$x = -2$$

-2 does not check as a solution.
The equation has no solution.

SECTION 6.6

You Try It 1

$$\frac{2}{x+3} = \frac{6}{5x+5}$$

$$\frac{(x+3)(5x+5)}{1} \cdot \frac{2}{x+3} = \frac{(x+3)(5x+5)}{1} \cdot \frac{6}{5x+5}$$

$$\frac{(x+3)(5x+5)}{1} \cdot \frac{2}{x+3} = \frac{(x+3)(5x+5)}{1} \cdot \frac{6}{5x+5}$$

$$(5x+5)2 = (x+3)6$$ • Solve for x.
$$10x + 10 = 6x + 18$$
$$4x + 10 = 18$$
$$4x = 8$$
$$x = 2$$

The solution is 2.

You Try It 2

Strategy To find the total area that 256 ceramic tiles will cover, write and solve a proportion using x to represent the number of square feet that 256 tiles will cover.

Solution

$$\frac{9}{16} = \frac{x}{256}$$ • Write a proportion.

$$256\left(\frac{9}{16}\right) = 256\left(\frac{x}{256}\right)$$ • Clear denominators.

$$144 = x$$

An area of 144 ft^2 can be tiled using 256 ceramic tiles.

You Try It 3

Strategy To find the area of triangle AOB:
• Solve a proportion to find the length of AO (the height of triangle AOB).
• Use the formula for the area of a triangle. AB is the base and AO is the height.

Solution

$$\frac{CD}{AB} = \frac{DO}{AO}$$ • Write a proportion.

$$\frac{4}{10} = \frac{3}{AO}$$ • Substitute.

$$10 \cdot AO \cdot \frac{4}{10} = 10 \cdot AO \cdot \frac{3}{AO}$$

$$4(AO) = 30$$
$$AO = 7.5$$

$$A = \frac{1}{2}bh$$ • Area of a triangle

$$= \frac{1}{2}(10)(7.5)$$ • Substitute.

$$= 37.5$$

The area of triangle AOB is 37.5 cm^2.

SECTION 6.7

You Try It 1

$$5x - 2y = 10$$
$$5x - 5x - 2y = -5x + 10$$ • Subtract $5x$.
$$-2y = -5x + 10$$
$$\frac{-2y}{-2} = \frac{-5x + 10}{-2}$$ • Divide by -2.
$$y = \frac{5}{2}x - 5$$

Copyright © Houghton Mifflin Company. All rights reserved.

You Try It 2

$$s = \frac{A + L}{2}$$

$$2 \cdot s = 2\left(\frac{A + L}{2}\right) \qquad \bullet \text{ Multiply by 2.}$$

$$2s = A + L$$

$$2s - A = A - A + L \qquad \bullet \text{ Subtract } A.$$

$$2s - A = L$$

You Try It 3

$$S = a + (n - 1)d$$

$$S = a + nd - d$$

$$S - a = a - a + nd - d \qquad \bullet \text{ Subtract } a.$$

$$S - a = nd - d$$

$$S - a + d = nd - d + d \qquad \bullet \text{ Add } d.$$

$$S - a + d = nd$$

$$\frac{S - a + d}{d} = \frac{nd}{d} \qquad \bullet \text{ Divide by } d.$$

$$\frac{S - a + d}{d} = n$$

You Try It 4

$$S = rS + C$$

$$S - rS = rS - rS + C \qquad \bullet \text{ Subtract } rS.$$

$$S - rS = C$$

$$(1 - r)S = C \qquad \bullet \text{ Factor.}$$

$$\frac{(1 - r)S}{1 - r} = \frac{C}{1 - r} \qquad \bullet \text{ Divide by } 1 - r.$$

$$S = \frac{C}{1 - r}$$

SECTION 6.8

You Try It 1

Strategy • Time for one printer to complete the job: t

	Rate	Time	Part
1st printer	$\dfrac{1}{t}$	2	$\dfrac{2}{t}$
2nd printer	$\dfrac{1}{t}$	5	$\dfrac{5}{t}$

• The sum of the parts of the task completed must equal 1.

Solution

$$\frac{2}{t} + \frac{5}{t} = 1$$

$$t\left(\frac{2}{t} + \frac{5}{t}\right) = t \cdot 1$$

$$2 + 5 = t$$

$$7 = t$$

Working alone, one printer takes 7 h to print the payroll.

You Try It 2

Strategy • Rate sailing across the lake: r
Rate sailing back: $3r$

	Distance	Rate	Time
Across	6	r	$\dfrac{6}{r}$
Back	6	$3r$	$\dfrac{6}{3r}$

• The total time for the trip was 2 h.

Solution

$$\frac{6}{r} + \frac{6}{3r} = 2$$

$$3r\left(\frac{6}{r} + \frac{6}{3r}\right) = 3r(2) \qquad \bullet \text{ Multiply by the LCM, } 3r.$$

$$3r \cdot \frac{6}{r} + 3r \cdot \frac{6}{3r} = 6r$$

$$18 + 6 = 6r \qquad \bullet \text{ Solve for } r.$$

$$24 = 6r$$

$$4 = r$$

The rate across the lake was 4 km/h.

Solutions to Chapter 7 "You Try It"

SECTION 7.1

You Try It 1

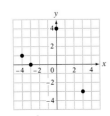

You Try It 2 $A(4, -2), B(-2, 4).$
The abscissa of D is 0.
The ordinate of C is 0.

You Try It 3

$$x - 3y = -14$$

$$\begin{array}{c|c} -2 - 3(4) & -14 \\ -2 - 12 & -14 \\ -14 = -14 \end{array}$$

Copyright © Houghton Mifflin Company. All rights reserved.

Yes, $(-2, 4)$ is a solution of
$x - 3y = -14$.

You Try It 4 $x + 2y = 4$
$2y = -x + 4$
$y = -\dfrac{1}{2}x + 2$

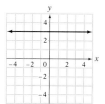

You Try It 5

$\{(145, 140), (140, 125), (150, 130), (165, 150), (140, 130), (165, 160)\}$

No, the relation is not a function. The two ordered pairs $(140, 125)$ and $(140, 130)$ have the same first coordinate but different second coordinates.

You Try It 6   Determine the ordered pairs defined by the equation. Replace x in $y = \dfrac{1}{2}x + 1$ by the given values and solve for y. $\{(-4, -1), (0, 1), (2, 2)\}$
Yes, y is a function of x.

You Try It 7 $H(x) = \dfrac{x}{x - 4}$

$H(8) = \dfrac{8}{8 - 4}$ • Replace x by 8.

$H(8) = \dfrac{8}{4} = 2$

SECTION 7.2

You Try It 1

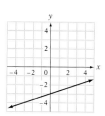

You Try It 2

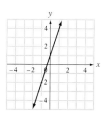

You Try It 3

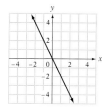

You Try It 4 $5x - 2y = 10$ • Solve for y.
$-2y = -5x + 10$
$y = \dfrac{5}{2}x - 5$

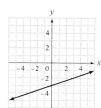

You Try It 5 $x - 3y = 9$ • Solve for y.
$-3y = -x + 9$
$y = \dfrac{1}{3}x - 3$

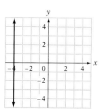

You Try It 6

You Try It 7

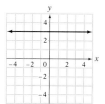

You Try It 8

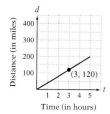

The ordered pair $(3, 120)$ means that in 3 h the car will travel 120 mi.

Copyright © Houghton Mifflin Company. All rights reserved.

SECTION 7.3

You Try It 1 x-intercept: y-intercept:

$$y = 2x - 4 \qquad (0, b)$$
$$0 = 2x - 4 \qquad b = -4$$
$$-2x = -4 \qquad (0, -4)$$
$$x = 2$$
$$(2, 0)$$

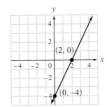

You Try It 2   Let $P_1 = (1, 4)$ and $P_2 = (-3, 8)$.

$$m = \frac{y_2 - y_1}{x_2 - x_1} = \frac{8 - 4}{-3 - 1} = \frac{4}{-4} = -1$$

The slope is -1.

You Try It 3   Let $P_1 = (-1, 2)$ and $P_2 = (4, 2)$.

$$m = \frac{y_2 - y_1}{x_2 - x_1} = \frac{2 - 2}{4 - (-1)} = \frac{0}{5} = 0$$

The slope is 0.

You Try It 4 $m = \dfrac{8650 - 6100}{1 - 4} = \dfrac{2550}{-3}$

$m = -850$

A slope of -850 means that the value of the car is decreasing at a rate of $850 per year.

You Try It 5 y-intercept $= (0, b) = (0, -1)$

$$m = -\frac{1}{4}$$

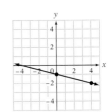

You Try It 6   Solve the equation for y.

$$x - 2y = 4$$
$$-2y = -x + 4$$
$$y = \frac{1}{2}x - 2$$

y-intercept $= (0, b) = (0, -2)$

$$m = \frac{1}{2}$$

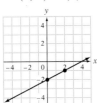

SECTION 7.4

You Try It 1   Because the slope and y-intercept are known, use the slope-intercept formula, $y = mx + b$.

$$y = mx + b$$
$$y = \frac{5}{3}x + 2 \qquad \bullet\ m = \frac{5}{3}; b = 2$$

You Try It 2 $m = \dfrac{3}{4}$ $(x_1, y_1) = (4, -2)$

$$y - y_1 = m(x - x_1)$$
$$y - (-2) = \frac{3}{4}(x - 4)$$
$$y + 2 = \frac{3}{4}x - 3$$
$$y = \frac{3}{4}x - 5$$

The equation of the line is $y = \dfrac{3}{4}x - 5$.

You Try It 3   Find the slope of the line between the two points.

$$m = \frac{y_2 - y_1}{x_2 - x_1} = \frac{1 - (-2)}{3 - (-6)} = \frac{3}{9} = \frac{1}{3}$$

Use the point-slope formula.

$$y - y_1 = m(x - x_1)$$
$$y - (-2) = \frac{1}{3}[x - (-6)] \qquad \bullet\ y_1 = -2;$$
$$\qquad\qquad\qquad\qquad\qquad\qquad x_1 = -6$$
$$y + 2 = \frac{1}{3}x + 2$$
$$y = \frac{1}{3}x$$

You Try It 4

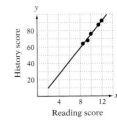

Reading score

The slope of the line means that the grade on the history test increases 8.3 points for each 1-point increase in the grade on the reading test.

Copyright © Houghton Mifflin Company. All rights reserved.

Solutions to Chapter 8 "You Try It"

SECTION 8.1

You Try It 1

$2x - 5y = 8$		$-x + 3y = -5$	
$2(-1) - 5(-2)$	8	$-(-1) + 3(-2)$	-5
$-2 + 10$	8	$1 + (-6)$	-5
$8 = 8$		$-5 = -5$	

Yes, $(-1, -2)$ is a solution of the system of equations.

You Try It 2

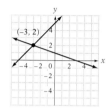

The solution is $(-3, 2)$.

You Try It 3

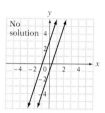

The lines are parallel. The system of equations is inconsistent and does not have a solution.

SECTION 8.2

You Try It 1

(1) $7x - y = 4$
(2) $3x + 2y = 9$

Solve Equation (1) for y.

$$7x - y = 4$$
$$-y = -7x + 4$$
$$y = 7x - 4$$

Substitute in Equation (2).

$$3x + 2y = 9$$
$$3x + 2(7x - 4) = 9 \quad \bullet \; y = 7x - 4$$
$$3x + 14x - 8 = 9$$
$$17x - 8 = 9$$
$$17x = 17$$
$$x = 1$$

Substitute in Equation (1).

$$7x - y = 4$$
$$7(1) - y = 4 \quad \bullet \; x = 1$$
$$7 - y = 4$$
$$-y = -3$$
$$y = 3$$

The solution is $(1, 3)$.

You Try It 2   (1) $3x - y = 4$
(2) $y = 3x + 2$

$$3x - y = 4$$
$$3x - (3x + 2) = 4 \quad \bullet \; y = 3x + 2$$
$$3x - 3x - 2 = 4$$
$$-2 = 4$$

This is not a true equation. The system of equations is inconsistent and therefore has no solution.

You Try It 3   (1) $y = -2x + 1$
(2) $6x + 3y = 3$

$$6x + 3y = 3$$
$$6x + 3(-2x + 1) = 3 \quad \bullet \; y = -2x + 1$$
$$6x - 6x + 3 = 3$$
$$3 = 3$$

The system of equations is dependent. The solutions are the ordered pairs that satisfy the equation $y = -2x + 1$.

You Try It 4

Strategy   • Amount invested at 6.5%: x
Amount invested at 4.5%: y

	Principal	Rate	Interest
Amount at 6.5%	x	0.065	$0.065x$
Amount at 4.5%	y	0.045	$0.045y$

• The sum of the two investments is $330,000: $x + y = 330{,}000$. The interest earned at 6.5% equals the interest earned at 4.5%: $0.065x = 0.045y$

Solution
(1) $x + y = 330{,}000$
(2) $0.065x = 0.045y$

Copyright © Houghton Mifflin Company. All rights reserved.

Solve Equation (2) for y.

$(3) \qquad y = \dfrac{13}{9}x$

Replace y by $\dfrac{13}{9}x$ in Equation (1) and solve for x.

$$x + y = 330{,}000$$
$$x + \frac{13}{9}x = 330{,}000 \qquad \bullet \; y = \frac{13}{9}x$$
$$\frac{22}{9}x = 330{,}000$$
$$x = 135{,}000$$

Replace x by $135{,}000$ in Equation (3) and solve for y.

$$y = \frac{13}{9}x$$
$$= \frac{13}{9}(135{,}000) = 195{,}000 \quad \bullet \; x = 135{,}000$$

$135{,}000 should be invested at 6.5% and $195{,}000 should be invested at 4.5%.

SECTION 8.3

You Try It 1 $(1) \qquad x - 2y = 1$
               $(2) \qquad 2x + 4y = 0$

Eliminate y.

$$2(x - 2y) = 2 \cdot 1 \qquad \bullet \; \text{Multiply by 2.}$$
$$2x + 4y = 0$$

$$2x - 4y = 2$$
$$2x + 4y = 0$$

Add the equations.

$$4x = 2$$
$$x = \frac{2}{4} = \frac{1}{2}$$

Replace x in Equation (2).

$$2\left(\frac{1}{2}\right) + 4y = 0 \qquad \bullet \; x = \frac{1}{2}$$
$$1 + 4y = 0$$
$$4y = -1$$
$$y = -\frac{1}{4}$$

The solution is $\left(\dfrac{1}{2}, -\dfrac{1}{4}\right)$.

You Try It 2 $(1) \qquad 2x - 3y = 4$
               $(2) \qquad -4x + 6y = -8$

Eliminate y.

$$2(2x - 3y) = 2 \cdot 4 \qquad \bullet \; \text{Multiply by 2.}$$
$$-4x + 6y = -8$$

$$4x - 6y = 8$$
$$-4x + 6y = -8$$

Add the equations.

$$0x + 0y = 0$$
$$0 = 0$$

The system of equations is dependent. The solutions are the ordered pairs that satisfy the equation $2x - 3y = 4$.

You Try It 3 $(1) \qquad 4x + 5y = 11$
               $(2) \qquad 3y = x + 10$

Write equation (2) in the form $Ax + By = C$.

$$3y = x + 10$$
$$-x + 3y = 10$$

Eliminate x.

$$4x + 5y = 11$$
$$4(-x + 3y) = 4 \cdot 10 \qquad \bullet \; \text{Multiply by 4.}$$

$$4x + 5y = 11$$
$$-4x + 12y = 40$$

Add the equations.

$$17y = 51$$
$$y = 3$$

Replace y in Equation (1).

$$4x + 5y = 11$$
$$4x + 5 \cdot 3 = 11 \qquad \bullet \; y = 3$$
$$4x + 15 = 11$$
$$4x = -4$$
$$x = -1$$

The solution is $(-1, 3)$.

SECTION 8.4

You Try It 1

Strategy   • Rate of the current: c
            Rate of the canoeist in calm water: r

	Rate	Time	Distance
With current	$r + c$	3	$3(r + c)$
Against current	$r - c$	5	$5(r - c)$

Copyright © Houghton Mifflin Company. All rights reserved.

• The distance traveled with the current is 15 mi.
The distance traveled against the current is 15 mi.

Solution

$$3(r + c) = 15 \qquad \frac{1}{3} \cdot 3(r + c) = \frac{1}{3} \cdot 15 \quad \bullet \text{ Multiply by } \frac{1}{3}.$$

$$5(r - c) = 15 \qquad \frac{1}{5} \cdot 5(r - c) = \frac{1}{5} \cdot 15 \quad \bullet \text{ Multiply by } \frac{1}{5}.$$

$$r + c = 5$$
$$r - c = 3$$

$$2r = 8$$
$$r = 4$$

$$r + c = 5$$
$$4 + c = 5 \qquad \bullet \ r = 4$$
$$c = 1$$

The rate of the current: 1 mph.
The rate of the canoeist in calm water: 4 mph.

You Try It 2

Strategy • Cost of an orange tree: x
Cost of a grapefruit tree: y

First purchase:

	Amount	Unit Cost	Value
Orange trees	25	x	$25x$
Grapefruit trees	20	y	$20y$

Second purchase:

	Amount	Unit Cost	Value
Orange trees	20	x	$20x$
Grapefruit trees	30	y	$30y$

• The total of the first purchase was $290.
The total of the second purchase was $330.

Solution

$$25x + 20y = 290 \qquad 4(25x + 20y) = 4 \cdot 290$$
$$\bullet \text{ Multiply by } 4.$$
$$20x + 30y = 330 \qquad -5(20x + 30y) = -5 \cdot 330$$
$$\bullet \text{ Multiply by } -5.$$

$$100x + 80y = 1160$$
$$-100x - 150y = -1650$$
$$-70y = -490$$
$$y = 7$$

$$25x + 20y = 290$$
$$25x + 20(7) = 290 \qquad \bullet \ y = 7$$
$$25x + 140 = 290$$
$$25x = 150$$
$$x = 6$$

The cost of an orange tree is $6.
The cost of a grapefruit tree is $7.

Solutions to Chapter 9 "You Try It"

SECTION 9.1

You Try It 1 $A = \{-9, -7, -5, -3, -1\}$

You Try It 2 $A = \{1, 3, 5, \ldots\}$

You Try It 3 $A \cup B = \{-2, -1, 0, 1, 2, 3, 4\}$

You Try It 4 $C \cap D = \{10, 16\}$

You Try It 5 $A \cap B = \varnothing$

You Try It 6 $\{x \mid x < 59, x \in \text{positive even integers}\}$

You Try It 7 $\{x \mid x > -3, x \in \text{real numbers}\}$

You Try It 8 The solution set is the numbers greater than -2.

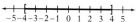

You Try It 9 The solution set is the numbers greater than -1 and the numbers less than -3.

You Try It 10 The solution set is the numbers less than or equal to 4 and greater than or equal to -4.

Copyright © Houghton Mifflin Company. All rights reserved.

You Try It 11   The solution set is the real numbers.

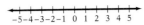

SECTION 9.2

You Try It 1

$$x + 2 < -2$$
$$x + 2 - 2 < -2 - 2 \quad \bullet \text{ Subtract 2.}$$
$$x < -4$$

You Try It 2

$$5x + 3 > 4x + 5$$
$$5x - 4x + 3 > 4x - 4x + 5 \quad \bullet \text{ Subtract } 4x.$$
$$x + 3 > 5$$
$$x + 3 - 3 > 5 - 3 \quad \bullet \text{ Subtract 3.}$$
$$x > 2$$

You Try It 3

$$-3x > -9$$
$$\frac{-3x}{-3} < \frac{-9}{-3} \quad \bullet \text{ Divide by } -3.$$
$$x < 3$$

You Try It 4

$$-\frac{3}{4}x \geq 18$$
$$-\frac{4}{3}\left(-\frac{3}{4}x\right) \leq -\frac{4}{3}(18) \quad \bullet \text{ Multiply by } -\frac{4}{3}.$$
$$x \leq -24$$

You Try It 5

Strategy   To find the selling prices, write and solve an inequality using p to represent the possible selling prices.

Solution $0.70p > 314$
$p > 448.571 \quad \bullet \text{ Divide by 0.70.}$

The dealer will make a profit with any selling price greater than or equal to $448.58.

SECTION 9.3

You Try It 1

$$5 - 4x > 9 - 8x$$
$$5 - 4x + 8x > 9 - 8x + 8x \quad \bullet \text{ Add } 8x.$$
$$5 + 4x > 9$$
$$5 - 5 + 4x > 9 - 5 \quad \bullet \text{ Subtract 5.}$$
$$4x > 4$$
$$\frac{4x}{4} > \frac{4}{4} \quad \bullet \text{ Divide by 4.}$$
$$x > 1$$

You Try It 2
$$8 - 4(3x + 5) \leq 6(x - 8)$$
$$8 - 12x - 20 \leq 6x - 48 \quad \bullet \text{ Distributive Property}$$
$$-12 - 12x \leq 6x - 48$$
$$-12 - 12x - 6x \leq 6x - 6x - 48 \quad \bullet \text{ Subtract } 6x.$$
$$-12 - 18x \leq -48$$
$$-12 + 12 - 18x \leq -48 + 12 \quad \bullet \text{ Add 12.}$$
$$-18x \leq -36$$
$$\frac{-18x}{-18} \geq \frac{-36}{-18} \quad \bullet \text{ Divide by } -18.$$
$$x \geq 2$$

You Try It 3

Strategy   To find the maximum number of miles:
- Write an expression for the cost of each car, using x to represent the number of miles driven during the week.
- Write and solve an inequality.

Solution

Cost of a Company A car	is less than	cost of a Company B car

$$8(7) + 0.10x < 10(7) + 0.08x$$
$$56 + 0.10x < 70 + 0.08x$$
$$56 + 0.10x - 0.08x < 70 + 0.08x - 0.08x \quad \bullet \text{ Subtract } 0.08x.$$
$$56 + 0.02x < 70$$
$$56 - 56 + 0.02x < 70 - 56 \quad \bullet \text{ Subtract 56.}$$
$$0.02x < 14$$
$$\frac{0.02x}{0.02} < \frac{14}{0.02} \quad \bullet \text{ Divide by 0.02.}$$
$$x < 700$$

The maximum number of miles is 699 mi.

SECTION 9.4

You Try It 1

$$x - 3y < 2$$
$$x - x - 3y < -x + 2$$
$$-3y < -x + 2 \quad \bullet \text{ Subtract } x.$$
$$\frac{-3y}{-3} > \frac{-x + 2}{-3}$$
$$y > \frac{1}{3}x - \frac{2}{3} \quad \bullet \text{ Divide by } -3.$$

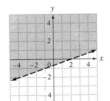

Copyright © Houghton Mifflin Company. All rights reserved.

You Try It 2

$$2x - 4y \le 8$$
$$2x - 2x - 4y \le -2x + 8 \qquad \bullet \text{ Subtract } 2x.$$
$$-4y \le -2x + 8$$
$$\frac{-4y}{-4} \ge \frac{-2x + 8}{-4} \qquad \bullet \text{ Divide by } -4.$$
$$y \ge \frac{1}{2}x - 2$$

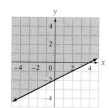

You Try It 3 $x < 3$

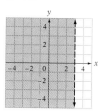

Solutions to Chapter 10 "You Try It"

SECTION 10.1

You Try It 1
$$-5\sqrt{32} = -5\sqrt{16 \cdot 2} \qquad \bullet \text{ 16 is a perfect square.}$$
$$= -5\sqrt{16}\sqrt{2}$$
$$= -5 \cdot 4\sqrt{2} = -20\sqrt{2}$$

You Try It 2
$$\sqrt{216} = \sqrt{36 \cdot 6} \qquad \bullet \text{ 36 is a perfect square.}$$
$$= \sqrt{36}\sqrt{6} = 6\sqrt{6}$$

You Try It 3
$$\sqrt{y^{19}} = \sqrt{y^{18} \cdot y} \qquad \bullet \ y^{18} \text{ is a perfect square.}$$
$$= \sqrt{y^{18}}\sqrt{y} = y^9\sqrt{y}$$

You Try It 4
$$\sqrt{45b^7} = \sqrt{9b^6 \cdot 5b} \qquad \bullet \ 9b^6 \text{ is a perfect square.}$$
$$= \sqrt{9b^6}\sqrt{5b} = 3b^3\sqrt{5b}$$

You Try It 5
$$3a\sqrt{28a^9b^{18}} = 3a\sqrt{4a^8b^{18}(7a)} \qquad \bullet \ 4a^8b^{18} \text{ is a perfect square.}$$
$$= 3a\sqrt{4a^8b^{18}}\sqrt{7a}$$
$$= 3a \cdot 2a^4b^9\sqrt{7a} = 6a^5b^9\sqrt{7a}$$

You Try It 6 $\sqrt{25(a + 3)^2} = 5(a + 3) = 5a + 15$

You Try It 7 $\sqrt{x^2 + 14x + 49} = \sqrt{(x + 7)^2} = x + 7$

SECTION 10.2

You Try It 1
$$9\sqrt{3} + 3\sqrt{3} - 18\sqrt{3} = (9 + 3 - 18)\sqrt{3} = -6\sqrt{3}$$

You Try It 2
$$2\sqrt{50} - 5\sqrt{32}$$
$$= 2\sqrt{25 \cdot 2} - 5\sqrt{16 \cdot 2} \qquad \bullet \text{ Simplify the radicands.}$$
$$= 2\sqrt{25}\sqrt{2} - 5\sqrt{16}\sqrt{2}$$
$$= 2 \cdot 5\sqrt{2} - 5 \cdot 4\sqrt{2}$$
$$= 10\sqrt{2} - 20\sqrt{2}$$
$$= (10 - 20)\sqrt{2} \qquad \bullet \text{ Distributive Property}$$
$$= -10\sqrt{2}$$

You Try It 3
$$y\sqrt{28y} + 7\sqrt{63y^3}$$
$$= y\sqrt{4 \cdot 7y} + 7\sqrt{9y^2 \cdot 7y} \qquad \bullet \text{ Simplify the radicands.}$$
$$= y\sqrt{4}\sqrt{7y} + 7\sqrt{9y^2}\sqrt{7y}$$
$$= y \cdot 2\sqrt{7y} + 7 \cdot 3y\sqrt{7y}$$
$$= 2y\sqrt{7y} + 21y\sqrt{7y}$$
$$= (2y + 21y)\sqrt{7y} \qquad \bullet \text{ Distributive Property}$$
$$= 23y\sqrt{7y}$$

You Try It 4
$$2\sqrt{27a^5} - 4a\sqrt{12a^3} + a^2\sqrt{75a}$$
$$= 2\sqrt{9a^4 \cdot 3a} - 4a\sqrt{4a^2 \cdot 3a} + a^2\sqrt{25 \cdot 3a}$$
$$= 2\sqrt{9a^4}\sqrt{3a} - 4a\sqrt{4a^2}\sqrt{3a}$$
$$\quad + a^2\sqrt{25}\sqrt{3a}$$
$$= 2 \cdot 3a^2\sqrt{3a} - 4a \cdot 2a\sqrt{3a} + a^2 \cdot 5\sqrt{3a}$$
$$= 6a^2\sqrt{3a} - 8a^2\sqrt{3a} + 5a^2\sqrt{3a} = 3a^2\sqrt{3a}$$

Copyright © Houghton Mifflin Company. All rights reserved.

SECTION 10.3

You Try It 1

$$\sqrt{5a}\sqrt{15a^3b^4}\sqrt{20b^5}$$
$$= \sqrt{1500a^4b^9} = \sqrt{100a^4b^8 \cdot 15b}$$
$$= \sqrt{100a^4b^8} \cdot \sqrt{15b}$$
$$= 10a^2b^4\sqrt{15b}$$

You Try It 2

$$\sqrt{5x}(\sqrt{5x} - \sqrt{25y})$$
$$= \sqrt{25x^2} - \sqrt{125xy} \qquad \bullet \text{ Distributive Property}$$
$$= \sqrt{25x^2} - \sqrt{25 \cdot 5xy} = \sqrt{25x^2} - \sqrt{25}\sqrt{5xy}$$
$$= 5x - 5\sqrt{5xy}$$

You Try It 3

$$(2\sqrt{x} + 7)(2\sqrt{x} - 7)$$
$$= 4(\sqrt{x})^2 - 7^2 \qquad \bullet \text{ Product of conjugates}$$
$$= 4x - 49$$

You Try It 4

$$(3\sqrt{x} - \sqrt{y})(5\sqrt{x} - 2\sqrt{y})$$
$$= 15(\sqrt{x})^2 - 6\sqrt{xy} - 5\sqrt{xy} + 2(\sqrt{y})^2 \qquad \bullet \text{ FOIL}$$
$$= 15(\sqrt{x})^2 - 11\sqrt{xy} + 2(\sqrt{y})^2$$
$$= 15x - 11\sqrt{xy} + 2y$$

You Try It 5

$$\frac{\sqrt{15x^6y^7}}{\sqrt{3x^7y^9}} = \sqrt{\frac{15x^6y^7}{3x^7y^9}} = \sqrt{\frac{5}{xy^2}} = \frac{\sqrt{5}}{\sqrt{xy^2}}$$
$$= \frac{\sqrt{5}}{y\sqrt{x}} = \frac{\sqrt{5}}{y\sqrt{x}} \cdot \frac{\sqrt{x}}{\sqrt{x}} \qquad \bullet \text{ Rationalize the denominator.}$$
$$= \frac{\sqrt{5x}}{xy}$$

You Try It 6

$$\frac{\sqrt{3}}{\sqrt{3} - \sqrt{6}} = \frac{\sqrt{3}}{\sqrt{3} - \sqrt{6}} \cdot \frac{\sqrt{3} + \sqrt{6}}{\sqrt{3} + \sqrt{6}} \qquad \bullet \text{ Rationalize the denominator.}$$
$$= \frac{3 + \sqrt{18}}{3 - 6} = \frac{3 + 3\sqrt{2}}{-3}$$
$$= \frac{3(1 + \sqrt{2})}{-3} = -1(1 + \sqrt{2})$$
$$= -1 - \sqrt{2}$$

You Try It 7

$$\frac{5 + \sqrt{y}}{1 - 2\sqrt{y}} = \frac{5 + \sqrt{y}}{1 - 2\sqrt{y}} \cdot \frac{1 + 2\sqrt{y}}{1 + 2\sqrt{y}} \qquad \bullet \text{ Rationalize the denominator.}$$
$$= \frac{5 + 10\sqrt{y} + \sqrt{y} + 2(\sqrt{y})^2}{1 - 4y}$$
$$= \frac{5 + 11\sqrt{y} + 2y}{1 - 4y}$$

SECTION 10.4

You Try It 1

$$\sqrt{4x} + 3 = 7$$
$$\sqrt{4x} = 4 \qquad \bullet \text{ Isolate } \sqrt{4x}.$$
$$(\sqrt{4x})^2 = 4^2 \qquad \bullet \text{ Square both sides.}$$
$$4x = 16$$
$$x = 4 \qquad \bullet \text{ Solve for } x.$$

Check:

$$\begin{array}{c|c} \sqrt{4x} + 3 = 7 & \\ \hline \sqrt{4 \cdot 4} + 3 & 7 \\ \sqrt{16} + 3 & 7 \\ 4 + 3 & 7 \\ 7 = 7 & \end{array}$$

The solution is 4.

You Try It 2

$$\sqrt{x} + \sqrt{x + 9} = 9$$
$$\sqrt{x} = 9 - \sqrt{x + 9} \qquad \bullet \text{ Isolate } \sqrt{x}.$$
$$(\sqrt{x})^2 = (9 - \sqrt{x + 9})^2 \qquad \bullet \text{ Square both sides.}$$
$$x = 81 - 18\sqrt{x + 9} + (x + 9)$$
$$-90 = -18\sqrt{x + 9}$$
$$5 = \sqrt{x + 9} \qquad \bullet \text{ Isolate } \sqrt{x + 9}.$$
$$5^2 = (\sqrt{x + 9})^2 \qquad \bullet \text{ Square both sides.}$$
$$25 = x + 9$$
$$16 = x \qquad \bullet \text{ Solve for } x.$$

Check:

$$\begin{array}{c|c} \sqrt{x} + \sqrt{x + 9} = 9 & \\ \hline \sqrt{16} + \sqrt{16 + 9} & 9 \\ \sqrt{16} + \sqrt{25} & 9 \\ 4 + 5 & 9 \\ 9 = 9 & \end{array}$$

The solution is 16.

You Try It 3

Strategy To find the distance, use the Pythagorean Theorem. The hypotenuse is the length of the ladder. One leg is the distance from the bottom of the ladder to the base of the building. The distance along the building from the ground to the top of the ladder is the unknown leg.

Copyright © Houghton Mifflin Company. All rights reserved.

Solution
$$a^2 = \sqrt{c^2 - b^2}$$
$$= \sqrt{(8)^2 - (3)^2} \qquad \bullet\ c = 8,\ b = 3$$
$$= \sqrt{64 - 9}$$
$$= \sqrt{55}$$
$$\approx 7.42$$

The distance is approximately 7.42 ft.

You Try It 4

Strategy To find the length of the pendulum, replace T in the equation with the given value and solve for L.

Solution
$$T = 2\pi\sqrt{\frac{L}{32}}$$
$$2.5 = 2(3.14)\sqrt{\frac{L}{32}} \qquad \bullet\ T = 2.5$$
$$2.5 = 6.28\sqrt{\frac{L}{32}}$$
$$\frac{2.5}{6.28} = \sqrt{\frac{L}{32}}$$
$$\left(\frac{2.5}{6.28}\right)^2 = \left(\sqrt{\frac{L}{32}}\right)^2$$
$$\frac{6.25}{39.4384} = \frac{L}{32}$$
$$(32)\left(\frac{6.25}{39.4384}\right) = (32)\left(\frac{L}{32}\right)$$
$$\frac{200}{39.4384} = L$$
$$5.07 \approx L$$

The length of the pendulum is approximately 5.07 ft.

Solutions to Chapter 11 "You Try It"

SECTION 11.1

You Try It 1
$$\frac{3y^2}{2} + y - \frac{1}{2} = 0$$
$$2\left(\frac{3y^2}{2} + y - \frac{1}{2}\right) = 2(0) \qquad \bullet\ \text{Multiply each side by 2.}$$
$$3y^2 + 2y - 1 = 0$$
$$(3y - 1)(y + 1) = 0 \qquad \bullet\ \text{Factor.}$$

$$
\begin{array}{ll}
3y - 1 = 0 \qquad & y + 1 = 0 \qquad \bullet\ \text{Principle of} \\
\quad 3y = 1 & \quad y = -1 \qquad \text{Zero Products} \\
\quad\ y = \dfrac{1}{3} &
\end{array}
$$

The solutions are $\dfrac{1}{3}$ and -1.

You Try It 2
$$x^2 + 81 = 0$$
$$x^2 = -81 \qquad \bullet\ \text{Solve for } x^2.$$
$$\sqrt{x^2} = \sqrt{-81} \qquad \bullet\ \text{Take square roots.}$$

$\sqrt{-81}$ is not a real number.

The equation has no real number solution.

You Try It 3
$$7(z + 2)^2 = 21$$
$$(z + 2)^2 = 3 \qquad \bullet\ \text{Solve for } (z + 2)^2.$$
$$\sqrt{(z + 2)^2} = \sqrt{3} \qquad \bullet\ \text{Take square roots.}$$
$$z + 2 = \pm\sqrt{3}$$
$$z = -2 \pm \sqrt{3} \qquad \bullet\ \text{Solve for } z.$$

The solutions are $-2 + \sqrt{3}$ and $-2 - \sqrt{3}$.

SECTION 11.2

You Try It 1
$$3x^2 - 6x - 2 = 0$$
$$3x^2 - 6x = 2 \qquad \bullet\ \text{Add 2.}$$
$$\frac{1}{3}(3x^2 - 6x) = \frac{1}{3}\cdot 2 \qquad \bullet\ \text{Multiply by } \frac{1}{3}.$$
$$x^2 - 2x = \frac{2}{3}$$

Copyright © Houghton Mifflin Company. All rights reserved.

Complete the square.

$$x^2 - 2x + 1 = \frac{2}{3} + 1$$

- $\left[\frac{1}{2}(-2)\right]^2 = [-1]^2 = 1$

$$(x - 1)^2 = \frac{5}{3}$$

- Factor.

$$\sqrt{(x - 1)^2} = \sqrt{\frac{5}{3}}$$

- Take square roots.

$$x - 1 = \pm\sqrt{\frac{5}{3}}$$

- Simplify.

$$x = 1 \pm \sqrt{\frac{5}{3}}$$

$$x = 1 \pm \frac{\sqrt{15}}{3}$$

$$x = \frac{3 \pm \sqrt{15}}{3}$$

The solutions are $\dfrac{3 + \sqrt{15}}{3}$ and $\dfrac{3 - \sqrt{15}}{3}$.

You Try It 2

$$x^2 + 6x + 12 = 0$$
$$x^2 + 6x = -12$$

- Subtract 12.

$$x^2 + 6x + 9 = -12 + 9$$

- $\left(\frac{1}{2} \cdot 6\right)^2 = 3^2 = 9$

$$(x + 3)^2 = -3$$

- Factor.

$$\sqrt{(x + 3)^2} = \sqrt{-3}$$

- Take square roots.

$\sqrt{-3}$ is not a real number.

The quadratic equation has no real number solution.

You Try It 3

$$\frac{x^2}{8} + x + 1 = 0$$

$$8\left(\frac{x^2}{8} + x + 1\right) = 8(0)$$

- Multiply by 8.

$$x^2 + 8x + 8 = 0$$
$$x^2 + 8x = -8$$

- Subtract 8.

$$x^2 + 8x + 16 = -8 + 16$$

- $\left(\frac{1}{2} \cdot 8\right)^2 = 4^2 = 16$

$$(x + 4)^2 = 8$$

- Factor.

$$\sqrt{(x + 4)^2} = \sqrt{8}$$

- Take square roots.

$$x + 4 = \pm\sqrt{8}$$
$$x + 4 = \pm 2\sqrt{2}$$
$$x = -4 \pm 2\sqrt{2}$$

$$x = -4 + 2\sqrt{2} \qquad x = -4 - 2\sqrt{2}$$
$$\approx -4 + 2(1.414) \qquad \approx -4 - 2(1.414)$$
$$\approx -4 + 2.828 \qquad \approx -4 - 2.828$$
$$\approx -1.172 \qquad \approx -6.828$$

The solutions are approximately -1.172 and -6.828.

SECTION 11.3

You Try It 1

$$3x^2 + 4x - 4 = 0$$
$$a = 3, b = 4, c = -4$$

$$x = \frac{-(4) \pm \sqrt{(4)^2 - 4(3)(-4)}}{2 \cdot 3}$$

$$= \frac{-4 \pm \sqrt{16 + 48}}{6}$$

$$= \frac{-4 \pm \sqrt{64}}{6} = \frac{-4 \pm 8}{6}$$

$$x = \frac{-4 + 8}{6} \qquad x = \frac{-4 - 8}{6}$$

$$= \frac{4}{6} = \frac{2}{3} \qquad = \frac{-12}{6} = -2$$

The solutions are $\dfrac{2}{3}$ and -2.

You Try It 2

$$\frac{x^2}{4} + \frac{x}{2} = \frac{1}{4}$$

$$4\left(\frac{x^2}{4} + \frac{x}{2}\right) = 4\left(\frac{1}{4}\right)$$

- Multiply by 4.

$$x^2 + 2x = 1$$
$$x^2 + 2x - 1 = 0$$

- Standard form

$$a = 1, b = 2, c = -1$$

$$x = \frac{-(2) \pm \sqrt{(2)^2 - 4(1)(-1)}}{2 \cdot 1}$$

$$= \frac{-2 \pm \sqrt{4 + 4}}{2} = \frac{-2 \pm \sqrt{8}}{2}$$

$$= \frac{-2 \pm 2\sqrt{2}}{2} = -1 \pm \sqrt{2}$$

The solutions are $-1 + \sqrt{2}$ and $-1 - \sqrt{2}$.

SECTION 11.4

You Try It 1

$$y = x^2 + 2$$

x	y
-2	6
-1	3
0	2
1	3
2	6

You Try It 2

To find the x-intercept, let $f(x) = 0$ and solve for x.

$$f(x) = x^2 - 6x + 9$$
$$0 = x^2 - 6x + 9$$
$$0 = (x - 3)(x - 3)$$

- Factor.

$$x - 3 = 0 \qquad x - 3 = 0$$

- Principle of Zero Products

$$x = 3 \qquad x = 3$$

Copyright © Houghton Mifflin Company. All rights reserved.

The x-intercept: $(3, 0)$.

There is only one x-intercept. The equation has a double root.

To find the y-intercept, evaluate the function at $x = 0$.

$f(x) = x^2 - 6x + 9$
$f(0) = 0^2 - 6(0) + 9 = 9$

The y-intercept: $(0, 9)$.

SECTION 11.5

You Try It 1

Strategy • This is a geometry problem.
• Width of the rectangle: W
  Length of the rectangle: $W + 2$
• Use the equation $A = L \cdot W$.

Solution

$A = L \cdot W$
$15 = (W + 2)W$ • $A = 15$, $L = W + 2$
$15 = W^2 + 2W$
$0 = W^2 + 2W - 15$
$0 = (W + 5)(W - 3)$ • Factor.

$W + 5 = 0 \qquad W - 3 = 0$ • Principle of
$\qquad W = -5 \qquad\quad W = 3$ Zero Products.

The solution -5 is not possible.
The width is 3 m.

Copyright © Houghton Mifflin Company. All rights reserved.

Answers to Chapter 1 Selected Exercises

PREP TEST

1. 127.16   **2.** 46,514   **3.** 4517   **4.** 11,396   **5.** 508   **6.** 24   **7.** 4   **8.** $3 \cdot 7$ **9.** $\frac{2}{5}$ **10.** d

SECTION 1.1

3. >   **5.** <   **7.** >   **9.** >   **11.** >   **13.** False   **15.** True   **17.** False   **19.** True   **21.** True
23. {1, 2, 3, 4, 5, 6, 7, 8}   **25.** {1, 2, 3, 4, 5, 6, 7, 8}   **27.** {−6, −5, −4, −3, −2, −1}   **29.** 5   **31.** −23, −18
33. 21, 37   **35.** −52, −46, 0   **37.** −17, 0, 4, 29   **39.** 5, 6, 7, 8, 9   **41.** −10, −9, −8, −7, −6, −5   **43.** −4
45. 9   **47.** 36   **49.** 40   **51.** −39   **53.** 74   **55.** −82   **57.** −81   **59.** >   **61.** <   **63.** <   **65.** <
67. a. 11, 7, 3, −1, −5   **b.** 11, 7, 3, 1, 5   **69.** Never true

SECTION 1.2

3. −11   **5.** −5   **7.** −83   **9.** −46   **11.** 0   **13.** −5   **15.** 9   **17.** 1   **19.** −10   **21.** −18   **23.** −41
25. −12   **27.** 0   **29.** −34   **31.** 0   **33.** −61   **35.** −65   **37.** −15   **39.** 22   **43.** 8   **45.** −7
47. −9   **49.** 9   **51.** −3   **53.** 18   **55.** −9   **57.** 11   **59.** −18   **61.** 0   **63.** 2   **65.** −138   **67.** −8
69. −12   **71.** −20   **73.** 15   **75.** −39   **77.** 16   **79.** The difference in temperature is 399°C.
81. The difference in elevation is 7046 m.   **83.** The continent with the greatest difference between the highest and lowest
elevation is Asia.   **85.** The difference in temperature is 396°C.   **87.** Yes   **89.** The difference between the heights is
20,370 m.   **91.** The difference in wind-chill factor is 6°F.   **93.** No. For example, the difference between 10 and −8 is
18, which is greater than either 10 or −8.

SECTION 1.3

3. 42   **5.** 60   **7.** −253   **9.** −114   **11.** −105   **13.** −216   **15.** −315   **17.** 336   **19.** −2772   **21.** 0
23. 350   **25.** −336   **27.** −352   **29.** −2   **31.** 8   **33.** −7   **35.** −12   **37.** −6   **39.** 11   **41.** −14
43. 15   **45.** −16   **47.** 0   **49.** −29   **51.** Undefined   **53.** −11   **55.** Undefined   **57.** −12   **59.** 24
61. 28   **63.** The average daily high temperature was −26°F.   **65.** The five-day moving average is −12, 28, 40, 28,
48, 52.   **67.** The student's score was 74.   **69.** The expression $-3x$ is greatest for $x = -6$.

SECTION 1.4

1. 36   **3.** −49   **5.** 9   **7.** 81   **9.** $-4^4 = -(4 \cdot 4 \cdot 4 \cdot 4) = 256$ **11.** $2 \cdot (-3)^2 = 2(-3)(-3) = 2 \cdot 9 = 18$
13. −27   **15.** 216   **17.** −12   **19.** 16   **21.** −864   **23.** −1008   **25.** −81   **27.** −77,760   **31.** 9   **33.** 12
35. 1   **37.** 8   **39.** −16   **41.** 12   **43.** 13   **45.** −36   **47.** 13   **49.** 4   **51.** 15   **53.** −1   **55.** 4
57. 1   **59.** 0   **61.** −172

SECTION 1.5

1. 1, 2, 4   **3.** 1, 2, 3, 4, 6, 12   **5.** 1, 2, 4, 8   **7.** 1, 13   **9.** 1, 2, 4, 7, 8, 14, 28, 56   **11.** 1, 3, 5, 9, 15, 45
13. 1, 29   **15.** 1, 2, 4, 13, 26, 52   **17.** 1, 2, 41, 82   **19.** 1, 3, 19, 57   **21.** 1, 2, 3, 4, 6, 8, 12, 16, 24, 48
23. 1, 2, 5, 10, 25, 50   **25.** 1, 7, 11, 77   **27.** 1, 2, 4, 5, 10, 20, 25, 50, 100   **29.** 1, 5, 17, 85   **31.** $2 \cdot 7$ **33.** $2^3 \cdot 3^2$
35. $2^3 \cdot 3$ **37.** $2^2 \cdot 3^2$ **39.** $2 \cdot 13$ **41.** 7^2 **43.** Prime   **45.** $2 \cdot 31$ **47.** Prime   **49.** $2 \cdot 43$ **51.** $5 \cdot 19$
53. $2 \cdot 3 \cdot 13$ **55.** $2^4 \cdot 3^2$ **57.** $5^2 \cdot 7$ **59.** $2^4 \cdot 5^2$ **61.** 24   **63.** 12   **65.** 36   **67.** 140   **69.** 36   **71.** 240
73. 720   **75.** 216   **77.** 360   **79.** 160   **81.** 24   **83.** 30   **85.** 72   **87.** 150   **89.** 108   **91.** 2   **93.** 1
95. 2   **97.** 6   **99.** 4   **101.** 10   **103.** 8   **105.** 7   **107.** 30   **109.** 1   **111.** 2   **113.** 6   **115.** 12
117. 26   **119.** 18   **121.** Answers may vary. One possibility is 3, 5; 5, 7; 11, 13.   **125.** The product of the two
numbers.

SECTION 1.6

1. $\frac{1}{3}$ **3.** $-\frac{4}{11}$ **5.** $-\frac{2}{3}$ **7.** $\frac{3}{2}$ **9.** 0   **11.** $-\frac{3}{5}$ **13.** $-\frac{7}{5}$ **15.** −15   **17.** $\frac{1}{2}$ **19.** 0.8   **21.** $0.1\overline{6}$

23. $-0.\overline{3}$ **25.** $-0.\overline{2}$ **27.** $-0.58\overline{3}$ **29.** $0.91\overline{6}$ **31.** $-0.3\overline{8}$ **33.** 0.5625   **35.** $-0.\overline{857142}$ **37.** $\frac{13}{12}$

Copyright © Houghton Mifflin Company. All rights reserved.

39. $\frac{35}{24}$ **41.** $\frac{1}{24}$ **43.** $-\frac{29}{26}$ **45.** $-\frac{19}{12}$ **47.** $\frac{25}{18}$ **49.** $-\frac{7}{48}$ **51.** $-\frac{17}{24}$ **53.** 8.022 **55.** -25.4

57. -9.964 **59.** -2.84 **61.** -6.961 **63.** 0 **65.** -108.677 **67.** $-\frac{1}{12}$ **69.** $-\frac{7}{18}$ **71.** -5.55 **73.** $-\frac{2}{27}$

75. $-\frac{1}{8}$ **77.** $-\frac{37}{24}$ **79.** $-\frac{23}{18}$ **81.** $\frac{73}{60}$ **83.** $\frac{29}{16}$ **85.** -1 **87.** $\frac{143}{72}$ **89.** -2.801 **91.** 2.088

93. 21.16 **95.** 0 **97.** 14.452 **99.** -17.606 **101.** 4.001 **103.** $-\frac{29}{24}$ **105.** $-\frac{1}{12}$ **107.** $-\frac{7}{18}$

109. From thinnest to thickest, the coins are dime, penny, quarter, nickel. **111.** The sum of the weights of the four coins is 15.438 g. **113.** The largest difference in diameter between two coins is 6.35 mm. **115.** $\frac{1}{12}$ cup of additional broth is needed. **117.** Apple Computer had negative earnings per share in 1996, 1997, 2001. **119.** Between 1996 and 1997, the decrease in earnings per share was $1.70. **121.** The five countries consume 34.9 million barrels of oil per day. **123.** The difference between the number of barrels of oil per day consumed and the number of imported is 11.5 million barrels. **125.** $\frac{13}{24}$ **127.** $-\frac{5}{24}$ **129.** $-\frac{7}{60}$ **131.** < **133.** = **135.** $\frac{1}{9}$ **137.** $\frac{5}{12}$ **139.** 2, 5 **141.** Larger

SECTION 1.7

1. $\frac{10}{21}$ **3.** $-\frac{3}{16}$ **5.** $-\frac{1}{8}$ **7.** $-\frac{4}{9}$ **9.** $\frac{9}{50}$ **11.** $\frac{9}{16}$ **13.** $\frac{5}{36}$ **15.** $\frac{3}{64}$ **17.** -7 **19.** $-\frac{11}{18}$ **21.** -1.794

23. 0.7407 **25.** -0.408 **27.** -0.22165 **29.** 27.2136 **31.** -0.097722 **33.** $-\frac{3}{5}$ **35.** 1.035 **37.** $-\frac{5}{12}$

39. $\frac{10}{9}$ **41.** $\frac{11}{14}$ **43.** $-\frac{1}{11}$ **45.** $-\frac{11}{20}$ **47.** $-\frac{2}{3}$ **49.** $\frac{14}{5}$ **51.** -2 **53.** -4.925 **55.** 1.05 **57.** -0.09

59. 0.11 **61.** 0.970 **63.** 0.007 **65.** 1.5 **67.** -0.0046 **69.** $\frac{23}{30}$ **71.** $-\frac{1}{8}$ **73.** $-\frac{11}{18}$ **75.** 0 **77.** -2

79. $-\frac{4}{27}$ **81.** -4.09 **83.** -0.1238 **85.** 0 **89.** $\frac{3}{4}$, 0.75 **91.** $\frac{16}{25}$, 0.64 **93.** $\frac{7}{4}$, 1.75 **95.** $\frac{19}{100}$, 0.19

97. $\frac{1}{20}$, 0.05 **99.** $\frac{1}{9}$ **101.** $\frac{1}{8}$ **103.** $\frac{2}{3}$ **105.** $\frac{1}{200}$ **107.** $\frac{5}{6}$ **109.** 0.073 **111.** 0.158 **113.** 0.003

115. 0.099 **117.** 1.212 **119.** 15% **121.** 5% **123.** 17.5% **125.** 115% **127.** 0.8% **129.** 54%

131. $33\frac{1}{3}\%$ **133.** $45\frac{5}{11}\%$ **135.** 87.5% **137.** $166\frac{2}{3}\%$ **139.** The center of the frame is $9\frac{1}{8}$ in. from the left side of the frame and $12\frac{1}{4}$ in. from the bottom of the frame. **141.** The cut should be made $18\frac{1}{4}$ in. from the left side of the board. **143.** The staircase is $87\frac{1}{2}$ in. high. **145.** $\frac{1}{22}$ **147.** $\frac{1}{16}$ ft

SECTION 1.8

1. 90° **3.** 28° **5.** 132° **7.** 83° **9.** 91° **11.** 132° **13.** 51° **15.** 77° **17.** 79° **19.** 292° **21.** 9.71 cm **23.** 14 ft 2 in. **25.** 52 in. **27.** 131.88 cm **29.** 3.768 m **31.** The wood framing would cost $76.96. **33.** The cost of the binding is $19.78. **35.** 32 ft² **37.** 378 cm² **39.** 50.24 in² **41.** 16.81 m² **43.** 52.5 cm² **45.** 226.865 in² **47.** 138.6 gal of water should be used. **49.** The cost to build the design is $172. **51.** The cost to plaster the room is $990.72. **53.** Perimeter: 265.6 m; area: 4056 m² **55.** Perimeter: 340 m; area: 1600 m² **57. a.** 39 in² **b.** 87 in²

CHAPTER 1 REVIEW EXERCISES*

1. -6 [1.2A] **2.** 0.28 [1.6A] **3.** -25 [1.4A] **4.** 10 [1.4B] **5.** 37° [1.8A] **6.** 0.062 [1.7C]

7. -42 [1.3A] **8.** $\frac{7}{12}$ [1.6C] **9.** 34° [1.8A] **10.** -4 [1.1A] **11.** 1, 2, 4, 7, 8, 14, 28, 56 [1.5A]

*Note: The numbers in brackets following the answers in the Chapter Review are a reference to the objective that corresponds to that problem. For example, the reference [1.2A] stands for Section 1.2, Objective A. This notation will be used for all Prep Tests, Chapter Reviews, Chapter Tests, and Cumulative Reviews throughout the text.

Copyright © Houghton Mifflin Company. All rights reserved.

12. -1.068 [1.6C] **13.** 62.5% [1.7C] **14.** $0.1\overline{3}$ [1.6A] **15.** -4 [1.2B] **16.** $-\dfrac{2}{15}$ [1.6C] **17.** 4 [1.1B]

18. 18 cm² [1.8C] **19.** -20 [1.3B] **20.** $\dfrac{159}{200}$ [1.7C] **21.** $2 \cdot 2 \cdot 2 \cdot 5 \cdot 7$ [1.5B] **22.** 31 [1.4B]

23. -13 [1.2A] **24.** $\dfrac{17}{40}$ [1.6B] **25.** $54\dfrac{2}{7}\%$ [1.7C] **26.** 28.26 m² [1.8C] **27.** -4.6224 [1.7A]

28. -5 [1.1B] **29.** 1 [1.2B] **30.** $-\dfrac{8}{15}$ [1.7B] **31.** $152°$ [1.8A] **32.** 44 in. [1.8B]

33. $-|6| < |-10|$ [1.1B] **34.** 1 [1.7B] **35.** The score for the exam was 98. [1.3C]

36. 38.2% of the PDAs were shipped by Palm. [1.7D] **37.** The difference between the boiling point and freezing point of mercury is 396°C. [1.2C] **38.** The sod cost $240.96. [1.8C]

CHAPTER 1 TEST*

1. 17 [1.3B] **2.** $83\dfrac{1}{3}\%$ [1.7C] **3.** $62°$ [1.8A] **4.** -5.3578 [1.7A] **5.** -14 [1.2B] **6.** $\dfrac{3}{8}$ [1.7C]

7. $\dfrac{1}{24}$ [1.6C] **8.** 8 [1.7B] **9.** 90 [1.3A] **10.** 84.78 in² [1.8B] **11.** -108 [1.4A] **12.** 90 cm² [1.8C]

13. $-2 > -40$ [1.1A] **14.** $-\dfrac{7}{20}$ [1.6B] **15.** -4 [1.1B] **16.** $\dfrac{9}{20}; 0.45$ [1.7C] **17.** -16 [1.2A]

18. -48 [1.3A] **19.** $2 \cdot 3 \cdot 3 \cdot 5 \cdot 11$ [1.5B] **20.** 17 [1.4B] **21.** 4 [1.2B] **22.** $-\dfrac{1}{2}$ [1.7B]

23. $47°$ [1.8A] **24.** 9 [1.4B] **25.** $0.\overline{7}$ [1.6A] **26. a.** The annual earnings would be $-\$457.6$ million. **b.** The average monthly loss would be $-\$67$ thousand. [1.7D] **27.** The cost of the new fencing is $4564. [1.8C]

Answers to Chapter 2 Selected Exercises

PREP TEST

1. 3 [1.2B] **2.** 4 [1.3B] **3.** $\dfrac{1}{12}$ [1.6B] **4.** $-\dfrac{4}{9}$ [1.7B] **5.** $\dfrac{3}{10}$ [1.7B] **6.** -16 [1.4A] **7.** $\dfrac{8}{27}$ [1.7A]

8. 48 [1.4B] **9.** 1 [1.4B] **10.** 12 [1.4B]

SECTION 2.1

1. $2x^2, 5x, \underline{-8}$ **3.** $-a^4, \underline{6}$ **5.** $7\underline{x^2y}, 6xy^2$ **7.** $1, -9$ **9.** $1, -4, -1$ **13.** 10 **15.** 32 **17.** 21 **19.** 16
21. -9 **23.** 41 **25.** -7 **27.** 13 **29.** -15 **31.** 41 **33.** 1 **35.** 5 **37.** 1 **39.** 57 **41.** 5
43. 8 **45.** -3 **47.** -2 **49.** -4 **51.** 225 **53.** 60 **55.** 4 **57.** 81 **59.** $n^x > x^n$ if $x \geq n + 1$

SECTION 2.2

3. $14x$ **5.** $5a$ **7.** $-6y$ **9.** $7 - 3b$ **11.** $5a$ **13.** $-2ab$ **15.** $5xy$ **17.** 0 **19.** $-\dfrac{5}{6}x$ **21.** $6.5x$

23. $0.45x$ **25.** $7a$ **27.** $-14x^2$ **29.** $-\dfrac{11}{24}x$ **31.** $17x - 3y$ **33.** $-2a - 6b$ **35.** $-3x - 8y$ **37.** $-4x^2 - 2x$

39. $12x$ **41.** $-21a$ **43.** $6y$ **45.** $8x$ **47.** $-6a$ **49.** $12b$ **51.** $-15x^2$ **53.** x^2 **55.** a **57.** x **59.** n

61. x **63.** y **65.** $3x$ **67.** $-2x$ **69.** $-8a^2$ **71.** $8y$ **73.** $4y$ **75.** $-2x$ **77.** $6a$ **79.** $-x - 2$

81. $8x - 6$ **83.** $-2a - 14$ **85.** $-6y + 24$ **87.** $35 - 21b$ **89.** $2 - 5y$ **91.** $15x^2 + 6x$ **93.** $2y - 18$

95. $-15x - 30$ **97.** $-6x^2 - 28$ **99.** $-6y^2 + 21$ **101.** $3x^2 - 3y^2$ **103.** $-4x + 12y$ **105.** $-6a^2 + 7b^2$

107. $4x^2 - 12x + 20$ **109.** $\dfrac{3}{2}x - \dfrac{9}{2}y + 6$ **111.** $-12a^2 - 20a + 28$ **113.** $12x^2 - 9x + 12$ **115.** $10x^2 - 20xy - 5y^2$

117. $-8b^2 + 6b - 9$ **119.** $a - 7$ **121.** $-11x + 13$ **123.** $-4y - 4$ **125.** $-2x - 16$ **127.** $14y - 45$

Copyright © Houghton Mifflin Company. All rights reserved.

129. $a + 7b$ **131.** $6x + 28$ **133.** $5x - 75$ **135.** $4x - 4$ **137.** $2x - 9$ **139. a.** False. For example, $8 \div 2 \neq 2 \div 8$ **b.** False. For example, $(12 \div 4) \div 2 \neq 12 \div (4 \div 2)$ **c.** False. For example, $(9 - 2) - 3 \neq 9 - (2 - 3)$ **d.** False. For example, $10 - 4 \neq 4 - 10$ **141.** No. 0 does not have a multiplicative inverse.

SECTION 2.3

1. $8 + y$ **3.** $t + 10$ **5.** $z + 14$ **7.** $x^2 - 20$ **9.** $\frac{3}{4}n + 12$ **11.** $8 + \frac{n}{4}$ **13.** $3(y + 7)$ **15.** $t(t + 16)$

17. $\frac{1}{2}x^2 + 15$ **19.** $5n^3 + n^2$ **21.** $r - \frac{r}{3}$ **23.** $x^2 - (x + 17)$ **25.** $9(z + 4)$ **27.** $12 - x$ **29.** $\frac{2}{3}x$ **31.** $\frac{2x}{9}$

33. $11x - 8$ **35.** $(x + 2) - 9; x - 7$ **37.** $\frac{7}{5 + x}$ **39.** $5 + \frac{1}{2}(x + 3); \frac{1}{2}x + \frac{13}{2}$ **41.** $(2x - 4) + x; 3x - 4$

43. $(x - 5)7; 7x - 35$ **45.** $\frac{2x + 5}{x}$ **47.** $x - (3x - 8); -2x + 8$ **49.** $x + 3x; 4x$ **51.** $(x + 6) + 5; x + 11$

53. $x - (x + 10); -10$ **55.** $\frac{1}{6}x + \frac{4}{9}x; \frac{11}{18}x$ **57.** $\frac{x}{3} + x; \frac{4}{3}x$ **59.** Number of emails: A; number of spam emails: $\frac{1}{2}A$

61. Length of one piece: S; length of second piece: $12 - S$ **63.** Distance traveled by the faster car: x; distance traveled by the slower car: $200 - x$ **65.** Number of bones in your body: N; number of bones in your foot: $\frac{1}{4}N$

67. Salary needed in San Francisco: S; salary needed in Daytona Beach: $\frac{1}{2}S$ **69.** $2x$

CHAPTER 2 REVIEW EXERCISES

1. $3x^2 - 24x - 21$ [2.2C]   **2.** $11x$ [2.2A]   **3.** $8a - 4b$ [2.2A]   **4.** $-5n$ [2.2B]   **5.** 79 [2.1A]
6. $10x - 35$ [2.2C]   **7.** $12y^2 + 8y - 10$ [2.2C]   **8.** $-6a$ [2.2B]   **9.** $-42x^2$ [2.2B]   **10.** $-63 - 36x$ [2.2C]
11. $-5y$ [2.2A]   **12.** -4 [2.1A]   **13.** $-6x - 1$ [2.2D]   **14.** $-40a + 40$ [2.2D]   **15.** $24y + 30$ [2.2D]
16. $9c - 5d$ [2.2A]   **17.** $20x$ [2.2B]   **18.** $7x + 46$ [2.2D]   **19.** 29 [2.1A]   **20.** $-9r + 8s$ [2.2A]
21. 50 [2.1A]   **22.** 28 [2.1A]   **23.** $-4x^2 + 6x$ [2.2A]   **24.** $-90x + 25$ [2.2D]   **25.** $-0.2x + 150$ [2.2D]
26. $-\frac{1}{12}x$ [2.2A]   **27.** $28a^2 - 8a + 12$ [2.2C]   **28.** $-4x + 20$ [2.2D]   **29.** -7 [2.1A]   **30.** $36y$ [2.2B]
31. $\frac{2}{3}(x + 10)$ [2.3A]   **32.** $4x$ [2.3A]   **33.** $x - 6$ [2.3A]   **34.** $x + 2x; 3x$ [2.3B]   **35.** $2x - \frac{1}{2}x; \frac{3}{2}x$ [2.3B]
36. $3x + 5(x - 1); 8x - 5$ [2.3B]   **37.** Number of American League cards: A; number of National League cards: $5A$ [2.3C]   **38.** Number of ten-dollar bills: T; number of five-dollar bills: $35 - T$ [2.3C]   **39.** Number of calories in an apple: a; number of calories in a candy bar: $2a + 8$ [2.3C]   **40.** Width of Parthenon: w; length of Parthenon: $1.6w$ [2.3C]   **41.** Kneeling height: h; standing height: $1.3h$ [2.3C]

CHAPTER 2 TEST

1. $5x$ [2.2A]   **2.** $-6x^2 + 21y^2$ [2.2C]   **3.** $-x + 6$ [2.2D]   **4.** $-7x + 33$ [2.2D]   **5.** $-9x - 7y$ [2.2A]
6. 22 [2.1A]   **7.** $2x$ [2.2B]   **8.** $7x + 38$ [2.2D]   **9.** $-10x^2 + 15x - 30$ [2.2C]   **10.** $-2x - 5y$ [2.2A]
11. 3 [2.1A]   **12.** $3x$ [2.2B]   **13.** y^2 [2.2A]   **14.** $-4x + 8$ [2.2C]   **15.** $-10a$ [2.2B]
16. $2x + y$ [2.2D]   **17.** $36y$ [2.2B]   **18.** $15 - 35b$ [2.2C]   **19.** $a^2 - b^2$ [2.3A]
20. $10(x - 3) = 10x - 30$ [2.3B]   **21.** $x + 2x^2$ [2.3B]   **22.** $\frac{6}{x} - 3$ [2.3B]   **23.** $b - 7b$ [2.3A]   **24.** Speed of return throw: s; speed of fastball: $2s$ [2.3C]   **25.** Shorter piece: x; longer piece: $4x - 3$ [2.3C]

CUMULATIVE REVIEW EXERCISES

1. -7 [1.2A]   **2.** 5 [1.2B]   **3.** 24 [1.3A]   **4.** -5 [1.3B]   **5.** $53°$ [1.8A]   **6.** $\frac{11}{48}$ [1.6C]   **7.** $-\frac{1}{6}$ [1.7B]

8. $\frac{1}{4}$ [1.7A]   **9.** 75% [1.7C]   **10.** -5 [1.4B]   **11.** $-\frac{27}{26}$ [1.7B]   **12.** 16 [2.1A]   **13.** $5x^2$ [2.2A]

14. $-7a - 10b$ [2.2A]   **15.** 153.86 cm^2 [1.8C]   **16.** 96 ft [1.8B]   **17.** $24 - 6x$ [2.2C]   **18.** $6y - 18$ [2.2C]

Copyright © Houghton Mifflin Company. All rights reserved.

19. $\frac{3}{8}$ [1.7C]   **20.** 0.0105 [1.7C]   **21.** $-8x^2 + 12y^2$ [2.2C]   **22.** $-9y^2 + 9y + 21$ [2.2C]

23. $-7x + 14$ [2.2D]   **24.** $5x - 43$ [2.2D]   **25.** $17x - 24$ [2.2D]   **26.** $-3x + 21y$ [2.2D]

27. $\frac{1}{2}b + b$ [2.3A]   **28.** $\frac{10}{y-2}$ [2.3A]   **29.** $8 - \frac{x}{12}$ [2.3B]   **30.** $x + (x + 2); 2x + 2$ [2.3B]

31. 3600 ft^2 [1.8C]   **32.** Speed of dial-up connection: s; speed of DSL connection: $10s$ [2.3C]

Answers to Chapter 3 Selected Exercises

PREP TEST

1. 0.09 [1.7C]   **2.** 75% [1.7C]   **3.** 63 [2.1A]   **4.** 0.65R [2.2A]   **5.** $\frac{7}{6}x$ [2.2A]   **6.** $9x - 18$ [2.2C]

7. $1.66x + 1.32$ [2.2C]   **8.** $5 - 2n$ [2.3B]   **9.** Speed of the old card: s; speed of the new card: $5s$ [2.3C]

10. $5 - x$ [2.3C]

SECTION 3.1

3. Yes   **5.** No   **7.** No   **9.** Yes   **11.** No   **13.** Yes   **15.** No   **17.** Yes   **19.** No   **23.** 2   **25.** 15

27. 6   **29.** 3   **31.** 0   **33.** -7 **35.** -7 **37.** -12 **39.** -5 **41.** 15   **43.** 9   **45.** 14   **47.** -1

49. 1   **51.** $-\frac{1}{2}$ **53.** $-\frac{3}{4}$ **55.** $\frac{1}{12}$ **57.** $-\frac{7}{12}$ **59.** 0.6529   **61.** -0.283 **63.** 9.257   **67.** -3 **69.** 0

71. -2 **73.** 9   **75.** 80   **77.** 0   **79.** -7 **81.** 12   **83.** -18 **85.** 15   **87.** -20 **89.** 0   **91.** $\frac{8}{3}$

93. $\frac{1}{3}$ **95.** $-\frac{1}{2}$ **97.** $-\frac{3}{2}$ **99.** $\frac{15}{7}$ **101.** 4   **103.** 3   **105.** 4.745   **107.** 2.06   **109.** -2.13

111. Equal to   **113.** 28   **115.** 0.72   **117.** 64   **119.** 24%   **121.** 7.2   **123.** 400   **125.** 9   **127.** 25%

129. 200%   **131.** 400   **133.** 7.7   **135.** 200   **137.** 400   **139.** 30   **141.** There are 4536 L of oxygen in the room.   **143.** The median income was $42,428.   **145.** 47.9% of the U.S. population watched Super Bowl XXXVIII.

147. You need to know the number of people three years old and older in the U.S.   **149.** Andrea must invest $1875.

151. Octavia will earn the greater amount of interest.   **153.** $1500 was invested at 8%.   **155.** There are 1.8 g of platinum in the necklace.   **157.** There are 131.25 lb of wool in the carpet.   **159.** The percent concentration of sugar is 50%.   **161.** The percent concentration of the resulting mixture is 6%.   **163.** The runner will travel 3 mi.

165. Marcella's average rate of speed is 36 mph.   **167.** It would take Palmer 2.5 h to walk the course.

169. The two joggers will meet 40 min after they start.   **171.** It will take them 0.5 h.   **173.** 40°   **175.** 10°

177. a. Answers will vary.   **b.** Answers will vary.

SECTION 3.2

1. 3   **3.** 6   **5.** -1 **7.** -3 **9.** 2   **11.** 2   **13.** 5   **15.** -3 **17.** 6   **19.** 3   **21.** 1   **23.** 6   **25.** -7

27. 0   **29.** $\frac{3}{4}$ **31.** $\frac{4}{9}$ **33.** $\frac{1}{3}$ **35.** $-\frac{1}{2}$ **37.** $-\frac{3}{4}$ **39.** $\frac{1}{3}$ **41.** $-\frac{1}{6}$ **43.** 1   **45.** 1   **47.** 0   **49.** $\frac{13}{10}$

51. $\frac{2}{5}$ **53.** $-\frac{4}{3}$ **55.** $-\frac{3}{2}$ **57.** 18   **59.** 8   **61.** -16 **63.** 25   **65.** $\frac{3}{4}$ **67.** $\frac{3}{8}$ **69.** $\frac{16}{9}$ **71.** $\frac{1}{18}$

73. $\frac{15}{2}$ **75.** $-\frac{18}{5}$ **77.** 2   **79.** 3   **81.** $x = 7$ **83.** $y = 3$ **85.** 19   **87.** -1 **89.** -11 **91.** The markup rate is 60%.   **93.** The cost of the basketball is $59.   **95.** The markup rate is 44.4%.   **97.** The cost of the CD is $8.50.   **99.** The discount rate is 23.2%.   **101.** The regular price of the tool set was $300.   **103.** The markdown rate is 38%.   **105.** The regular price of the telescope is $275.   **107.** The initial velocity is 8 ft/s.   **109.** The depreciated value will be $38,000 after 2 years.   **111.** The approximate length is 31.8 in.   **113.** The distance the car will skid is 168 ft.   **115.** The estimated population is 51,000 people.   **117.** $a = 7$ **119.** The regular price is $317.65.

121. 6 m   **123.** 385

Copyright © Houghton Mifflin Company. All rights reserved.

SECTION 3.3

1. 2   **3.** 3   **5.** −1   **7.** 2   **9.** −2   **11.** −3   **13.** 0   **15.** −1   **17.** −3   **19.** −1   **21.** 4   **23.** $\frac{2}{3}$

25. $\frac{5}{6}$ **27.** $\frac{3}{4}$ **29.** −17   **31.** 41   **33.** 8   **35.** 1   **37.** 4   **39.** −1   **41.** −1   **43.** 24   **45.** 495

47. $\frac{1}{2}$ **49.** $-\frac{1}{3}$ **51.** $\frac{10}{3}$ **53.** $-\frac{1}{4}$ **55.** 0   **57.** −1   **59.** A force of 25 lb must be applied to the other end.

61. The fulcrum is 6 ft from the 180-pound person.   **63.** The fulcrum is 10 ft from the 128-pound acrobat.

65. The minimum force to move the rock is 34.6 lb.   **67.** The break-even point is 260 barbecues.   **69.** The break-even point is 520 desk lamps.   **71.** The break-even point is 3000 softball bats.   **73.** No solution   **75.** 0

SECTION 3.4

1. $x - 15 = 7$; 22   **3.** $7x = -21$; −3   **5.** $9 - x = 7$; 2   **7.** $5 - 2x = 1$; 2   **9.** $2x + 5 = 15$; 5   **11.** $4x - 6 = 22$; 7   **13.** $3(4x - 7) = 15$; 3   **15.** $3x = 2(20 - x)$; 8, 12   **17.** $2x - (14 - x) = 1$; 5, 9   **19.** 15, 17, 19   **21.** −1, 1, 3   **23.** 4, 6   **25.** 5, 7   **27.** The processor speed of the newer personal computer is 4.2$\overline{6}$ GHz.   **29.** The lengths of the sides are 6 ft, 6 ft, and 11 ft.   **31.** The union member worked 168 h during March.   **33.** 37 h of labor was required to paint the house.   **35.** There are 1024 vertical pixels.   **37.** The length is 13 m; the width is 8 m.   **39.** The shorter piece is 3 ft; the longer piece is 9 ft.   **41.** The larger scholarship is $5000.

SECTION 3.5

1. 116°   **3.** 20°   **5.** 20°   **7.** 20°   **9.** 106°   **11.** 11°   **13.** $m\angle a$ is 38°; $m\angle b$ is 142°   **15.** $m\angle a$ is 47°; $m\angle b$ is 133°   **17.** 20°   **19.** 47°   **21.** 141°   **23.** $m\angle x$ is 155°; $m\angle y$ is 70°   **25.** $m\angle a$ is 45°; $m\angle b$ is 135°   **27.** 90° − x **29.** 60°   **31.** 35°   **33.** 102°   **35.** 120°   **37.** 360°

SECTION 3.6

1. The amount of $1 herbs is 20 oz.   **3.** The mixture will cost $1.84 per pound.   **5.** The amount of caramel is 3 lb.   **7.** The amount of olive oil is 2 c; the amount of vinegar is 8 c.   **9.** The cost of the mixture is $3.00 per ounce.   **11.** To make the mixture, 16 oz of the alloy are needed.   **13.** The amount of almonds is 37 lb; the amount of walnuts is 63 lb.   **15.** There were 228 adult tickets sold.   **17.** The cost per pound of the sugar-coated cereal is $.70.   **19.** The percent concentration of gold in the mixture is 24%.   **21.** The amount of the 15% acid is 20 gal.   **23.** The amount of the 25% wool yarn is 30 lb.   **25.** The amount of 9% nitrogen plant food is 6.25 gal.   **27.** The percent concentration of sugar in the mixture is 19%.   **29.** 20 lb of 40% java bean coffee must be used.   **31.** The amount of the 7% solution is 100 ml; the amount of the 4% solution is 200 ml.   **33.** 150 oz of pure chocolate must be added.   **35.** The percent concentration of the resulting alloy is 50%.   **37.** The first plane is traveling at a rate of 105 mph; the second plane is traveling at a rate of 130 mph.   **39.** The planes will be 3000 km apart at 11 A.M.   **41.** In 2 h the cabin cruiser will be alongside the motorboat.   **43.** The corporate offices are 120 mi from the airport.   **45.** The rate of the car is 68 mph.   **47.** The distance between the airports was 300 mi.   **49.** The planes will pass each other in 2.5 h after the plane leaves Seattle.   **51.** The cyclists will meet after 1.5 h.   **53.** The bus overtakes the car in 180 mi.   **55.** 75 g of pure water must be added.   **57.** 3.75 gal of 20% antifreeze must be drained.   **59.** The bicyclist's average speed is $13\frac{1}{3}$ mph.

CHAPTER 3 REVIEW EXERCISES

1. 21 [3.1B]   **2.** 10 [3.3B]   **3.** 7 [3.2A]   **4.** No [3.1A]   **5.** 20 [3.1C]   **6.** −2 [3.3B]   **7.** 250% [3.1D]   **8.** 4 [3.3A]   **9.** −1 [3.3B]   **10.** 4 [3.3A]   **11.** $671.25 [3.2B]   **12.** 35° [3.5A]   **13.** 26° [3.5A]   **14.** The force is 24 lb. [3.3C]   **15.** The average speed on the winding road was 32 mph. [3.6C]   **16.** The discount rate is $33\frac{1}{3}$%. [3.2B]   **17.** $m\angle x = 22°$, $m\angle y = 158°$ [3.5B]   **18.** The amount of cranberry juice is 7 qt; the amount of apple juice is 3 qt. [3.6A]   **19.** The three integers are −1, 0, 1. [3.4A]

Copyright © Houghton Mifflin Company. All rights reserved.

20. The angles measure 75°, 60°, and 45°. [3.5B] **21.** $5n - 4 = 16; 4$ [3.4A] **22.** The height of the Eiffel Tower is 1063 ft. [3.4B] **23.** 25° [3.5A] **24.** 60° [3.5A] **25.** The jet overtakes the propeller-driven plane 600 mi from the starting point. [3.6C] **26.** The numbers are 8 and 13. [3.4A] **27.** The mixture is 14% butterfat. [3.6B]

CHAPTER 3 TEST

1. -5 [3.3A] **2.** -5 [3.1B] **3.** -3 [3.2A] **4.** 2 [3.3B] **5.** No [3.1A] **6.** 5 [3.2A] **7.** 0.04 [3.1D]
8. $-\dfrac{1}{3}$ [3.3B] **9.** 2 [3.3A] **10.** -12 [3.1C] **11.** The amount of rye is 10 lb; the amount of wheat is 5 lb. [3.6A]
12. 19° [3.5A] **13.** The discount rate is 20%. [3.2B] **14.** 200 calculators were produced. [3.2B]
15. The measure of one of the equal angles is 70°. [3.5B] **16.** The numbers are 10, 12, and 14. [3.4A] **17.** 1.25 gal of water must be added. [3.6B] **18.** $m\angle a = 138°, m\angle b = 42°$ [3.5A] **19.** $3x - 15 = 27; 14$ [3.4A] **20.** The rate of the snowmobile was 6 mph. [3.6C] **21.** The company makes 110 25-inch TVs each day. [3.4B]
22. The smaller number is 8; the larger number is 10. [3.4A] **23.** The distance between the airports is 360 mi. [3.6C]
24. $m\angle x = 138°; m\angle y = 130°$ [3.5B] **25.** The final temperature is 60°C. [3.3C]

CUMULATIVE REVIEW EXERCISES

1. 6 [1.2B] **2.** -48 [1.3A] **3.** $-\dfrac{19}{48}$ [1.6C] **4.** -2 [1.7B] **5.** 54 [1.7A] **6.** 24 [1.4B]
7. 6 [2.1A] **8.** $-17x$ [2.2A] **9.** $-5a - 2b$ [2.2A] **10.** $2x$ [2.2B] **11.** $36y$ [2.2B]
12. $2x^2 + 6x - 4$ [2.2C] **13.** $-4x + 14$ [2.2D] **14.** $6x - 34$ [2.2D] **15.** Yes [3.1A] **16.** No [3.1A]
17. 19.2 [3.1D] **18.** -25 [3.1C] **19.** -3 [3.2A] **20.** 3 [3.2A] **21.** 13 [3.3B] **22.** 2 [3.3B]
23. -3 [3.3A] **24.** $\dfrac{1}{2}$ [3.3A] **25.** The final temperature is 60°C. [3.3C] **26.** $12 - 5x = -18; 6$ [3.4A]
27. The area of the garage is 600 ft². [3.4B] **28.** 20 lb of oat flour are needed for the mixture. [3.6A] **29.** 25 g of pure gold must be added. [3.6B] **30.** The length is 12 ft; the width is 10 ft. [1.8B] **31.** 131° [3.5A]
32. The measure of one of the equal angles is 60°. [3.5B] **33.** The length of the track is 120 m. [3.6C]

Answers to Chapter 4 Selected Exercises

PREP TEST

1. 1 [1.2B] **2.** -18 [1.3A] **3.** $\dfrac{2}{3}$ [1.6A] **4.** 48 [2.1A] **5.** 0 [1.6A] **6.** No [2.2A]
7. $5x^2 - 9x - 6$ [2.2A] **8.** 0 [2.2A] **9.** $-6x + 24$ [2.2C] **10.** $-7xy + 10y$ [2.2D]

SECTION 4.1

1. Yes **3.** No **5.** Yes **7.** Yes **9.** Binomial **11.** Trinomial **13.** None of these **15.** Binomial
17. $-2x^2 + 3x$ **19.** $y^2 - 8$ **21.** $5x^2 + 7x + 20$ **23.** $x^3 + 2x^2 - 6x - 6$ **25.** $2a^3 - 3a^2 - 11a + 2$ **27.** $5x^2 + 8x$
29. $7x^2 + xy - 4y^2$ **31.** $3a^2 - 3a + 17$ **33.** $5x^3 + 10x^2 - x - 4$ **35.** $3r^3 + 2r^2 - 11r + 7$ **37.** $4x$
39. $3y^2 - 4y - 2$ **41.** $-7x - 7$ **43.** $4x^3 + 3x^2 + 3x + 1$ **45.** $y^3 + 5y^2 - 2y - 4$ **47.** $-y^2 - 13xy$
49. $2x^2 - 3x - 1$ **51.** $-2x^3 + x^2 + 2$ **53.** $3a^3 - 2$ **55.** $4y^3 + 2y^2 + 2y - 4$ **57.** $x^2 + 9x - 11$

SECTION 4.2

3. $30x^3$ **5.** $-42c^6$ **7.** $9a^7$ **9.** x^3y^4 **11.** $-10x^9y$ **13.** $12x^7y^8$ **15.** $-6x^3y^5$ **17.** x^4y^5z **19.** $a^3b^5c^4$
21. $-30a^5b^8$ **23.** $6a^5b$ **25.** $40y^{10}z^6$ **27.** $x^3y^3z^2$ **29.** $-24a^3b^3c^3$ **31.** $8x^7yz^6$ **33.** $30x^6y^8$ **35.** $-36a^3b^2c^3$
37. x^{15} **39.** x^{14} **41.** x^8 **43.** y^{12} **45.** $-8x^6$ **47.** x^4y^6 **49.** $9x^4y^2$ **51.** $-243x^{15}y^{10}$ **53.** $-8x^7$
55. $24x^8y^7$ **57.** a^4b^6 **59.** $64x^{12}y^3$ **61.** $-18x^3y^4$ **63.** $-8a^7b^5$ **65.** $-54a^9b^3$ **67.** $12x^2$ **69.** $2x^6y^2 + 9x^4y^2$

Copyright © Houghton Mifflin Company. All rights reserved.

71. 0   **73.** $17x^4y^8$ **75.** True   **77.** False. $(x^2)^5 = x^{2 \cdot 5} = x^{10}$ **79.** No. $2^{(3^2)}$ is larger.

SECTION 4.3

1. 3, 3, $12x - 15$ **3.** $x^2 - 2x$ **5.** $-x^2 - 7x$ **7.** $3a^3 - 6a^2$ **9.** $-5x^4 + 5x^3$ **11.** $-3x^5 + 7x^3$ **13.** $12x^3 - 6x^2$
15. $6x^2 - 12x$ **17.** $3x^2 + 4x$ **19.** $-x^3y + xy^3$ **21.** $2x^4 - 3x^2 + 2x$ **23.** $2a^3 + 3a^2 + 2a$ **25.** $3x^6 - 3x^4 - 2x^2$
27. $-6y^4 - 12y^3 + 14y^2$ **29.** $-2a^3 - 6a^2 + 8a$ **31.** $6y^4 - 3y^3 + 6y^2$ **33.** $x^3y - 3x^2y^2 + xy^3$ **35.** $x^3 + 4x^2 + 5x + 2$
37. $a^3 - 6a^2 + 13a - 12$ **39.** $-2b^3 + 7b^2 + 19b - 20$ **41.** $-6x^3 + 31x^2 - 41x + 10$ **43.** $x^3 - 3x^2 + 5x - 15$
45. $x^4 - 4x^3 - 3x^2 + 14x - 8$ **47.** $15y^3 - 16y^2 - 70y + 16$ **49.** $5a^4 - 20a^3 - 5a^2 + 22a - 8$
51. $y^4 + 4y^3 + y^2 - 5y + 2$ **53.** $x^2 + 4x + 3$ **55.** $a^2 + a - 12$ **57.** $y^2 - 5y - 24$ **59.** $y^2 - 10y + 21$
61. $2x^2 + 15x + 7$ **63.** $3x^2 + 11x - 4$ **65.** $4x^2 - 31x + 21$ **67.** $3y^2 - 2y - 16$ **69.** $9x^2 + 54x + 77$
71. $21a^2 - 83a + 80$ **73.** $6a^2 - 25ab + 14b^2$ **75.** $2a^2 - 11ab - 63b^2$ **77.** $100a^2 - 100ab + 21b^2$
79. $15x^2 + 56xy + 48y^2$ **81.** $14x^2 - 97xy - 60y^2$ **83.** $56x^2 - 61xy + 15y^2$ **85.** $y^2 - 25$ **87.** $4x^2 - 9$
89. $9x^2 - 49$ **91.** $16 - 9y^2$ **93.** $x^2 + 2x + 1$ **95.** $9a^2 - 30a + 25$ **97.** $4a^2 + 4ab + b^2$ **99.** $x^2 - 4xy + 4y^2$
101. $25x^2 + 20xy + 4y^2$ **103.** The area of the rectangle is $(10x^2 - 35x)$ ft^2.   **105.** The area is $(4x^2 + 4x + 1)$ km^2.
107. The area of the triangle is $(4x^2 + 10x)$ m^2.   **109.** The area is $(60w + 3000)$ yd^2.   **111.** $x^4 + 2x^3 - 5x^2 - 6x + 9$
113. $12x^2 - x - 20$ **115.** $x^3 - 7x^2 - 7$

SECTION 4.4

1. 7, 5, 2   **3.** $\dfrac{1}{25}$ **5.** 64   **7.** $\dfrac{1}{27}$ **9.** 2   **11.** $\dfrac{1}{x^2}$ **13.** a^6 **15.** $\dfrac{4}{x^7}$ **17.** $\dfrac{2}{3z^2}$ **19.** $5b^8$ **21.** $\dfrac{x^2}{3}$ **23.** 1

25. -1 **27.** y^4 **29.** a^3 **31.** p^4 **33.** $2x^3$ **35.** $2k$ **37.** m^5n^2 **39.** $\dfrac{3r^2}{2}$ **41.** $-\dfrac{2a}{3}$ **43.** $\dfrac{1}{y^5}$ **45.** $\dfrac{1}{a^6}$

47. $\dfrac{1}{3x^3}$ **49.** $\dfrac{2}{3x^5}$ **51.** $\dfrac{y^4}{x^2}$ **53.** $\dfrac{2}{5m^3n^8}$ **55.** $\dfrac{1}{p^3q}$ **57.** $\dfrac{1}{2y^3}$ **59.** $\dfrac{7xz}{8y^3}$ **61.** $\dfrac{p^2}{2m^3}$ **63.** $-\dfrac{8x^3}{y^6}$ **65.** $\dfrac{9}{x^2y^4}$

67. $\dfrac{2}{x^4}$ **69.** $-\dfrac{5}{a^8}$ **71.** $-\dfrac{a^5}{8b^4}$ **73.** $\dfrac{10y^3}{x^4}$ **75.** $\dfrac{1}{2x^3}$ **77.** $\dfrac{3}{x^3}$ **79.** $\dfrac{1}{2x^2y^6}$ **81.** $\dfrac{1}{x^6y}$ **83.** $\dfrac{a^4}{y^{10}}$ **85.** $-\dfrac{1}{6x^3}$

87. $-\dfrac{a^2b}{6c^2}$ **89.** $-\dfrac{7b^6}{a^2}$ **91.** $\dfrac{s^8t^4}{4r^{12}}$ **93.** $\dfrac{125p^3}{27m^{15}n^6}$ **97.** 3.24×10^{-9} **99.** 3×10^{-18} **101.** 3.2×10^{16}

103. 1.22×10^{-19} **105.** 5.47×10^8 **107.** 0.000167 **109.** $68{,}000{,}000$ **111.** 0.0000305 **113.** 0.00000000102

115. 6.023×10^{23} **117.** 3.7×10^{-6} **119.** 1×10^{-9} **121.** 1.6×10^{-19} **123.** $\dfrac{1}{4}, \dfrac{1}{2}$, 1, 2, 4   **125.** 4, 2, 1, $\dfrac{1}{2}, \dfrac{1}{4}$

127. False. $(2a)^{-3} = \dfrac{1}{8a^3}$ **129.** False. $(2 + 3)^{-1} = (5)^{-1} = \dfrac{1}{5}$

SECTION 4.5

1. $2a - 5$ **3.** $6y + 4$ **5.** $x - 2$ **7.** $-x + 2$ **9.** $x^2 + 3x - 5$ **11.** $x^4 - 3x^2 - 1$ **13.** $xy + 2$ **15.** $-3y^3 + 5$
17. $3x - 2 + \dfrac{1}{x}$ **19.** $-3x + 7 - \dfrac{6}{x}$ **21.** $4a - 5 + 6b$ **23.** $9x + 6 - 3y$ **25.** $(x + 2), (x - 3)$ **27.** $b - 7$
29. $y - 5$ **31.** $2y - 7$ **33.** $2y + 6 + \dfrac{25}{y - 3}$ **35.** $x - 2 + \dfrac{8}{x + 2}$ **37.** $3y - 5 + \dfrac{20}{2y + 4}$ **39.** $6x - 12 + \dfrac{19}{x + 2}$
41. $b - 5 - \dfrac{24}{b - 3}$ **43.** $3x + 17 + \dfrac{64}{x - 4}$ **45.** $5y + 3 + \dfrac{1}{2y + 3}$ **47.** $4a + 1$ **49.** $2a + 9 + \dfrac{33}{3a - 1}$
51. $x^2 - 5x + 2$ **53.** $x^2 + 5$ **55.** $3ab$

CHAPTER 4 REVIEW EXERCISES

1. $8b^2 - 2b - 15$ [4.3C]   **2.** $21y^2 + 4y - 1$ [4.1A]   **3.** $x^4y^8z^4$ [4.2A]   **4.** $\dfrac{2x^3}{3}$ [4.4A]

5. $-8x^3 - 14x^2 + 18x$ [4.3A]   **6.** $-\dfrac{1}{2a}$ [4.4A]   **7.** $16u^{12}v^{16}$ [4.2B]   **8.** 64   [4.2B]   **9.** $2x^2 + 3x - 8$ [4.1B]

Copyright © Houghton Mifflin Company. All rights reserved.

10. $\dfrac{b^6}{a^4}$ [4.4A]   **11.** $-108x^{18}$ [4.2B]   **12.** $25y^2 - 70y + 49$ [4.3D]   **13.** $100a^{15}b^{13}$ [4.2B]

14. $4b^4 + 12b^2 - 1$ [4.5A]   **15.** $-\dfrac{1}{16}$ [4.4A]   **16.** $13y^3 - 12y^2 - 5y - 1$ [4.1B]   **17.** $-x + 2 + \dfrac{1}{x + 3}$ [4.5B]

18. $2ax - 4ay - bx + 2by$ [4.3C]   **19.** $6y^3 + 17y^2 - 2y - 21$ [4.3B]   **20.** $b^2 + 5b + 2 + \dfrac{7}{b - 7}$ [4.5B]

21. $8a^3b^3 - 4a^2b^4 + 6ab^5$ [4.3A]   **22.** $4a^2 - 25b^2$ [4.3D]   **23.** $12b^5 - 4b^4 - 6b^3 - 8b^2 + 5$ [4.3B]
24. $2x^3 + 9x^2 - 3x - 12$ [4.1A]   **25.** $-4y + 8$ [4.5A]   **26.** $a^2 - 49$ [4.3D]   **27.** 3.756×10^{10} [4.4B]

28. 14,600,000 [4.4B]   **29.** $-54a^{13}b^5c^7$ [4.2A]   **30.** $2y - 9$ [4.5B]   **31.** $\dfrac{x^4y^6}{9}$ [4.4A]

32. $10a^2 + 31a - 63$ [4.3C]   **33.** 1.27×10^{-7} [4.4B]   **34.** 0.0000000000032 [4.4B]
35. The area is $(2w^2 - w)$ ft². [4.3E]   **36.** The area is $(9x^2 - 12x + 4)$ in². [4.3E]

CHAPTER 4 TEST

1. $4x^3 - 6x^2$ [4.3A]   **2.** $4x - 1 + \dfrac{3}{x^2}$ [4.5A]   **3.** $-\dfrac{4}{x^6}$ [4.4A]   **4.** $-6x^3y^6$ [4.2A]   **5.** $x - 1 + \dfrac{2}{x + 1}$ [4.5B]

6. $x^3 - 7x^2 + 17x - 15$ [4.3B]   **7.** $-8a^6b^3$ [4.2B]   **8.** $\dfrac{9y^{10}}{x^{10}}$ [4.4A]   **9.** $a^2 + 3ab - 10b^2$ [4.3C]

10. $4x^4 - 2x^2 + 5$ [4.5A]   **11.** $x + 7$ [4.5B]   **12.** $6y^4 - 9y^3 + 18y^2$ [4.3A]   **13.** $-4x^4 + 8x^3 - 3x^2 - 14x + 21$
[4.3B]   **14.** $16y^2 - 9$ [4.3D]   **15.** a^4b^7 [4.2A]   **16.** $8ab^4$ [4.4A]   **17.** $4a - 7$ [4.5A]
18. $-5a^3 + 3a^2 - 4a + 3$ [4.1B]   **19.** $4x^2 - 20x + 25$ [4.3D]   **20.** $2x + 3 + \dfrac{2}{2x - 3}$ [4.5B]
21. $-2x^3$ [4.4A]   **22.** $10x^2 - 43xy + 28y^2$ [4.3C]   **23.** $3x^3 + 6x^2 - 8x + 3$ [4.1A]   **24.** 3.02×10^{-9} [4.4B]
25. The area of the circle is $(\pi x^2 - 10\pi x + 25\pi)$ m². [4.3E]

CUMULATIVE REVIEW EXERCISES

1. $\dfrac{5}{144}$ [1.6C]   **2.** $\dfrac{5}{3}$ [1.7A]   **3.** $\dfrac{25}{11}$ [1.7B]   **4.** $-\dfrac{22}{9}$ [2.1A]   **5.** $5x - 3xy$ [2.2A]   **6.** $-9x$ [2.2B]

7. $-18x + 12$ [2.2D]   **8.** -16 [3.1C]   **9.** -16 [3.3A]   **10.** 15 [3.3B]   **11.** 22% [3.1D]
12. $4b^3 - 4b^2 - 8b - 4$ [4.1A]   **13.** $3y^3 + 2y^2 - 10y$ [4.1B]   **14.** a^9b^{15} [4.2B]   **15.** $-8x^3y^6$ [4.2A]

16. $6y^4 + 8y^3 - 16y^2$ [4.3A]   **17.** $10a^3 - 39a^2 + 20a - 21$ [4.3B]   **18.** $15b^2 - 31b + 14$ [4.3C]   **19.** $\dfrac{1}{2b^2}$ [4.4A]

20. $a - 7$ [4.5B]   **21.** 0.0000609 [4.4B]   **22.** $8x - 2x = 18$; 3 [3.4B]   **23.** The percent concentration of orange
juice in the mixture is 28%. [3.6B]   **24.** The car overtakes the cyclist 25 mi from the starting point. [3.6C]
25. The length is 15 m and the width is 6 m. [3.1D]

Answers to Chapter 5 Selected Exercises

PREP TEST

1. $2 \cdot 3 \cdot 5$ [1.5B]   **2.** $-12y + 15$ [2.2C]   **3.** $-a + b$ [2.2C]   **4.** $-3a + 3b$ [2.2D]   **5.** 0 [3.1C]

6. $-\dfrac{1}{2}$ [3.2A]   **7.** $x^2 - 2x - 24$ [4.3C]   **8.** $6x^2 - 11x - 10$ [4.3C]   **9.** x^3 [4.4A]   **10.** $3x^3y$ [4.4A]

SECTION 5.1

3. $5(a + 1)$ **5.** $8(2 - a^2)$ **7.** $4(2x + 3)$ **9.** $6(5a - 1)$ **11.** $x(7x - 3)$ **13.** $a^2(3 + 5a^3)$ **15.** $y(14y + 11)$

17. $2x(x^3 - 2)$ **19.** $2x^2(5x^2 - 6)$ **21.** $4a^5(2a^3 - 1)$ **23.** $xy(xy - 1)$ **25.** $3xy(xy^3 - 2)$ **27.** $xy(x - y^2)$

29. $5y(y^2 - 4y + 1)$ **31.** $3y^2(y^2 - 3y - 2)$ **33.** $3y(y^2 - 3y + 8)$ **35.** $a^2(6a^3 - 3a - 2)$ **37.** $ab(2a - 5ab + 7b)$

39. $2b(2b^4 + 3b^2 - 6)$ **41.** $x^2(8y^2 - 4y + 1)$ **43.** $(a + z)(y + 7)$ **45.** $(a - b)(3r + s)$ **47.** $(m - 7)(t - 7)$

49. $(4a - b)(2y + 1)$ **51.** $(x + 2)(x + 2y)$ **53.** $(p - 2)(p - 3r)$ **55.** $(a + 6)(b - 4)$ **57.** $(2z - 1)(z + y)$

Copyright © Houghton Mifflin Company. All rights reserved.

59. $(2v - 3y)(4v + 7)$ **61.** $(2x - 5)(x - 3y)$ **63.** $(y - 2)(3y - a)$ **65.** $(3x - y)(y + 1)$ **67.** $(3s + t)(t - 2)$

69. a. 28 **b.** 496 **71. a.** $r^2(\pi - 2)$ **b.** $2r^2(4 - \pi)$ **c.** $r^2(4 - \pi)$

SECTION 5.2

1. The same **3.** $(x + 1)(x + 2)$ **5.** $(x + 1)(x - 2)$ **7.** $(a + 4)(a - 3)$ **9.** $(a - 1)(a - 2)$ **11.** $(a + 2)(a - 1)$

13. $(b - 3)(b - 3)$ **15.** $(b + 8)(b - 1)$ **17.** $(y + 11)(y - 5)$ **19.** $(y - 2)(y - 3)$ **21.** $(z - 5)(z - 9)$

23. $(z + 8)(z - 20)$ **25.** $(p + 3)(p + 9)$ **27.** $(x + 10)(x + 10)$ **29.** $(b + 4)(b + 5)$ **31.** $(x + 3)(x - 14)$

33. $(b + 4)(b - 5)$ **35.** $(y + 3)(y - 17)$ **37.** $(p + 3)(p - 7)$ **39.** Nonfactorable over the integers

41. $(x - 5)(x - 15)$ **43.** $(p + 3)(p + 21)$ **45.** $(x + 2)(x + 19)$ **47.** $(x + 9)(x - 4)$ **49.** $(a + 4)(a - 11)$

51. $(a - 3)(a - 18)$ **53.** $(z + 21)(z - 7)$ **55.** $(c + 12)(c - 15)$ **57.** $(p + 9)(p + 15)$ **59.** $(c + 2)(c + 9)$

61. $(x + 15)(x - 5)$ **63.** $(x + 25)(x - 4)$ **65.** $(b - 4)(b - 18)$ **67.** $(a + 45)(a - 3)$ **69.** $(b - 7)(b - 18)$

71. $(z + 12)(z + 12)$ **73.** $(x - 4)(x - 25)$ **75.** $(x + 16)(x - 7)$ **77.** $3(x + 2)(x + 3)$ **79.** $-(x - 2)(x + 6)$

81. $a(b + 8)(b - 1)$ **83.** $x(y + 3)(y + 5)$ **85.** $-2a(a + 1)(a + 2)$ **87.** $4y(y + 6)(y - 3)$ **89.** $2x(x^2 - x + 2)$

91. $6(z + 5)(z - 3)$ **93.** $3a(a + 3)(a - 6)$ **95.** $(x + 7y)(x - 3y)$ **97.** $(a - 5b)(a - 10b)$ **99.** $(s + 8t)(s - 6t)$

101. Nonfactorable over the integers **103.** $z^2(z + 10)(z - 8)$ **105.** $b^2(b + 2)(b - 5)$ **107.** $3y^2(y + 3)(y + 15)$

109. $-x^2(x + 1)(x - 12)$ **111.** $3y(x + 3)(x - 5)$ **113.** $-3x(x - 3)(x - 9)$ **115.** $(x - 3y)(x - 5y)$

117. $(a - 6b)(a - 7b)$ **119.** $(y + z)(y + 7z)$ **121.** $3y(x + 21)(x - 1)$ **123.** $3x(x + 4)(x - 3)$

125. $4z(z + 11)(z - 3)$ **127.** $4x(x + 3)(x - 1)$ **129.** $5(p + 12)(p - 7)$ **131.** $p^2(p + 12)(p - 3)$

133. $(t - 5s)(t - 7s)$ **135.** $(a + 3b)(a - 11b)$ **137.** $y(x + 6)(x - 9)$ **139.** $-36, 36, -12, 12$

141. $22, -22, 10, -10$ **143.** 6, 10, 12 **145.** 6, 10, 12 **147.** 4, 6

SECTION 5.3

1. $(x + 1)(2x + 1)$ **3.** $(y + 3)(2y + 1)$ **5.** $(a - 1)(2a - 1)$ **7.** $(b - 5)(2b - 1)$ **9.** $(x + 1)(2x - 1)$

11. $(x - 3)(2x + 1)$ **13.** $(t + 2)(2t - 5)$ **15.** $(p - 5)(3p - 1)$ **17.** $(3y - 1)(4y - 1)$ **19.** Nonfactorable over

the integers **21.** $(2t - 1)(3t - 4)$ **23.** $(x + 4)(8x + 1)$ **25.** Nonfactorable over the integers

27. $(3y + 1)(4y + 5)$ **29.** $(a + 7)(7a - 2)$ **31.** $(b - 4)(3b - 4)$ **33.** $(z - 14)(2z + 1)$ **35.** $(p + 8)(3p - 2)$

37. $2(x + 1)(2x + 1)$ **39.** $5(y - 1)(3y - 7)$ **41.** $x(x - 5)(2x - 1)$ **43.** $b(a - 4)(3a - 4)$ **45.** Nonfactorable

over the integers **47.** $-3x(x + 4)(x - 3)$ **49.** $4(4y - 1)(5y - 1)$ **51.** $z(2z + 3)(4z + 1)$ **53.** $y(2x - 5)(3x + 2)$

55. $5(t + 2)(2t - 5)$ **57.** $p(p - 5)(3p - 1)$ **59.** $2(z + 4)(13z - 3)$ **61.** $2y(y - 4)(5y - 2)$

63. $yz(z + 2)(4z - 3)$ **65.** $3a(2a + 3)(7a - 3)$ **67.** $y(3x - 5y)(3x - 5y)$ **69.** $xy(3x - 4y)(3x - 4y)$

71. $(2x - 3)(3x - 4)$ **73.** $(b + 7)(5b - 2)$ **75.** $(3a + 8)(2a - 3)$ **77.** $(z + 2)(4z + 3)$ **79.** $(2p + 5)(11p - 2)$

81. $(y + 1)(8y + 9)$ **83.** $(6t - 5)(3t + 1)$ **85.** $(b + 12)(6b - 1)$ **87.** $(3x + 2)(3x + 2)$ **89.** $(2b - 3)(3b - 2)$

91. $(3b + 5)(11b - 7)$ **93.** $(3y - 4)(6y - 5)$ **95.** $(3a + 7)(5a - 3)$ **97.** $(2y - 5)(4y - 3)$ **99.** $(2z + 3)(4z - 5)$

101. Nonfactorable over the integers **103.** $(2z - 5)(5z - 2)$ **105.** $(6z + 5)(6z + 7)$ **107.** $(x + y)(3x - 2y)$

109. $(a + 2b)(3a - b)$ **111.** $(y - 2z)(4y - 3z)$ **113.** $-(z - 7)(z + 4)$ **115.** $-(x - 1)(x + 8)$

117. $3(x + 5)(3x - 4)$ **119.** $4(2x - 3)(3x - 2)$ **121.** $a^2(5a + 2)(7a - 1)$ **123.** $5(b - 7)(3b - 2)$

125. $(x - 7y)(3x - 5y)$ **127.** $3(8y - 1)(9y + 1)$ **129.** $-(x - 1)(x + 21)$ **133.** $x(x - 1)$ **135.** $(2y + 1)(y + 3)$

137. $(4y - 3)(y - 3)$ **139.** $-5, 5, -1, 1$ **141.** $-5, 5, -1, 1$ **143.** $-9, 9, -3, 3$

SECTION 5.4

1. a. Answers will vary. For instance, $x^2 - 25$. **b.** Answers will vary. For instance, $x^2 + 6x + 9$. **3.** $(x + 2)(x - 2)$

5. $(a + 9)(a - 9)$ **7.** $(y + 1)^2$ **9.** $(a - 1)^2$ **11.** $(2x + 1)(2x - 1)$ **13.** $(x^3 + 3)(x^3 - 3)$ **15.** Nonfactorable over

the integers **17.** $(x + y)^2$ **19.** $(2a + 1)^2$ **21.** $(3x + 1)(3x - 1)$ **23.** $(1 + 8x)(1 - 8x)$ **25.** Nonfactorable over

the integers **27.** $(3a + 1)^2$ **29.** $(b^2 + 4a)(b^2 - 4a)$ **31.** $(2a - 5)^2$ **33.** $(3a - 7)^2$ **35.** $(5z + y)(5z - y)$

37. $(ab + 5)(ab - 5)$ **39.** $(5x + 1)(5x - 1)$ **41.** $(2a - 3b)^2$ **43.** $(2y - 9z)^2$ **45.** $\left(\frac{1}{x} + 2\right)\left(\frac{1}{x} - 2\right)$

47. $(3ab - 1)^2$ **49.** $2(2y + 1)(2y - 1)$ **51.** $3a(a + 1)^2$ **53.** $(m^2 + 16)(m + 4)(m - 4)$ **55.** $(x + 1)(9x + 4)$

57. $4y^2(2y + 3)^2$ **59.** $(y^4 + 9)(y^2 + 3)(y^2 - 3)$ **61.** $(5 - 2p)^2$ **63.** $(4x - 3 + y)(4x - 3 - y)$

65. $(x - 2 + y)(x - 2 - y)$ **67.** $5(x + 1)(x - 1)$ **69.** $x(x + 2)^2$ **71.** $x^2(x + 7)(x - 5)$ **73.** $5(b + 3)(b + 12)$

75. Nonfactorable over the integers **77.** $2y(x + 11)(x - 3)$ **79.** $x(x^2 - 6x - 5)$ **81.** $3(y^2 - 12)$

Copyright © Houghton Mifflin Company. All rights reserved.

83. $(2a + 1)(10a + 1)$ **85.** $y^2(x + 1)(x - 8)$ **87.** $5(a + b)(2a - 3b)$ **89.** $-2(x + 5)(x - 5)$ **91.** $b^2(a - 5)^2$ **93.** $ab(3a - b)(4a + b)$ **95.** $3a(2a - 1)^2$ **97.** $3(81 + a^2)$ **99.** $2a(2a - 5)(3a - 4)$ **101.** $a(2a + 5)^2$ **103.** $3b(3a - 1)^2$ **105.** $-6(x - 2)(x + 4)$ **107.** $x^2(x + y)(x - y)$ **109.** $2a(3a + 2)^2$ **111.** $-b(3a - 2)(2a + 1)$ **113.** $2x^2(x - 8)(2x - 3)$ **115.** $x^2(x + 5)(x - 5)$ **117.** $(a^2 + 4)(a + 2)(a - 2)$ **119.** $-3y^2(2y + 5)(4y - 3)$ **121.** $2(x - 3)(2a - b)$ **123.** $(a - b)(y + 1)(y - 1)$ **125.** $(a + b)(a - b)(x - y)$ **127.** $12, -12$ **129.** $16, -16$ **131.** $10, -10$

SECTION 5.5

3. $-3, -2$ **5.** $7, 3$ **7.** $0, 5$ **9.** $0, 9$ **11.** $0, -\dfrac{3}{2}$ **13.** $0, \dfrac{2}{3}$ **15.** $-2, 5$ **17.** $-9, 9$ **19.** $-\dfrac{7}{2}, \dfrac{7}{2}$ **21.** $-\dfrac{1}{3}, \dfrac{1}{3}$ **23.** $-2, -4$ **25.** $-7, 2$ **27.** $-\dfrac{1}{2}, 5$ **29.** $-\dfrac{1}{3}, -\dfrac{1}{2}$ **31.** $0, 3$ **33.** $0, 7$ **35.** $-1, -4$ **37.** $2, 3$ **39.** $\dfrac{1}{2}, -4$ **41.** $\dfrac{1}{3}, 4$ **43.** $3, 9$ **45.** $-2, 9$ **47.** $-1, -2$ **49.** $-9, 5$ **51.** $-7, 4$ **53.** $-2, -3$ **55.** $-8, 9$ **57.** $1, 4$ **59.** $-5, 2$ **61.** The number is 6. **63.** The numbers are 2 and 4. **65.** The numbers are 4 and 5. **67.** The numbers are 3 and 7. **69.** There will be 12 consecutive numbers. **71.** There are 6 teams in the league. **73.** The object will hit the ground 3 s later. **75.** The golf ball will return to the ground 3.75 s later. **77.** The length is 15 in. The width is 5 in. **79.** The height of the triangle is 14 m. **81.** The dimensions of the type area are 4 in. by 7 in. **83.** The radius of the original circular lawn was approximately 3.81 ft. **85.** $3, 48$ **87.** $-\dfrac{3}{2}, -5$ **89.** $0, 9$

CHAPTER 5 REVIEW EXERCISES

1. $(b - 10)(b - 3)$ [5.2A] **2.** $(x - 3)(4x + 5)$ [5.1B] **3.** Nonfactorable over the integers [5.3A] **4.** $5x(x^2 + 2x + 7)$ [5.1A] **5.** $7y^3(2y^6 - 7y^3 + 1)$ [5.1A] **6.** $(y + 9)(y - 4)$ [5.2A] **7.** $(2x - 7)(3x - 4)$ [5.3A] **8.** $3ab(4a + b)$ [5.1A] **9.** $(a^3 + 10)(a^3 - 10)$ [5.4A] **10.** $n^2(n - 3)(n + 1)$ [5.2B] **11.** $(6y - 1)(2y + 3)$ [5.3A] **12.** $2b(2b - 7)(3b - 4)$ [5.4B] **13.** $(3y^2 + 5z)(3y^2 - 5z)$ [5.4A] **14.** $(c + 6)(c + 2)$ [5.2A] **15.** $(6a - 5)(3a + 2)$ [5.3B] **16.** $\dfrac{1}{4}, -7$ [5.5A] **17.** $4x(x - 6)(x + 1)$ [5.2B] **18.** $3(a - 7)(a + 2)$ [5.2B] **19.** $(a - 12)(2a + 5)$ [5.3B] **20.** $7, -3$ [5.5A] **21.** $(3a - 5b)(7x + 2y)$ [5.1B] **22.** $(ab + 1)(ab - 1)$ [5.4A] **23.** $(2x + 5)(5x + 2y)$ [5.1B] **24.** $5(x - 3)(x + 2)$ [5.2B] **25.** $3(x + 6)^2$ [5.4B] **26.** $(x - 5)(3x - 2)$ [5.3B] **27.** The length is 100 yd. The width is 60 yd. [5.5B] **28.** The distance is 20 ft. [5.5B] **29.** The width of the frame is 1.5 in. [5.5B] **30.** The side of the original garden plot was 20 ft. [5.5B]

CHAPTER 5 TEST

1. $(b + 6)(a - 3)$ [5.1B] **2.** $2y^2(y - 8)(y + 1)$ [5.2B] **3.** $4(x + 4)(2x - 3)$ [5.3B] **4.** $(2x + 1)(3x + 8)$ [5.3A] **5.** $(a - 16)(a - 3)$ [5.2A] **6.** $2x(3x^2 - 4x + 5)$ [5.1A] **7.** $(x + 5)(x - 3)$ [5.2A] **8.** $-\dfrac{1}{2}, \dfrac{1}{2}$ [5.5A] **9.** $5(x^2 - 9x - 3)$ [5.1A] **10.** $(p + 6)^2$ [5.4A] **11.** $3, 5$ [5.5A] **12.** $3(x + 2y)^2$ [5.4B] **13.** $(b + 4)(b - 4)$ [5.4A] **14.** $3y^2(2x + 1)(x + 1)$ [5.3B] **15.** $(p + 3)(p + 2)$ [5.2A] **16.** $(x - 2)(a + b)$ [5.1B] **17.** $(p + 1)(x - 1)$ [5.1B] **18.** $3(a + 5)(a - 5)$ [5.4B] **19.** Nonfactorable over the integers [5.3A] **20.** $(x - 12)(x + 3)$ [5.2A] **21.** $(2a - 3b)^2$ [5.4A] **22.** $(2x + 7y)(2x - 7y)$ [5.4A] **23.** $\dfrac{3}{2}, -7$ [5.5A] **24.** The two numbers are 7 and 3. [5.5B] **25.** The length is 15 cm. The width is 6 cm. [5.5B]

CUMULATIVE REVIEW EXERCISES

1. 7 [1.2B] **2.** 4 [1.4B] **3.** -7 [2.1A] **4.** $15x^2$ [2.2B] **5.** 12 [2.2D] **6.** $\dfrac{2}{3}$ [3.1C] **7.** $\dfrac{7}{4}$ [3.3A] **8.** 3 [3.3B] **9.** 45 [3.1D] **10.** $9a^6b^4$ [4.2B] **11.** $x^3 - 3x^2 - 6x + 8$ [4.3B] **12.** $4x + 8 + \dfrac{21}{2x - 3}$ [4.5B] **13.** $\dfrac{y^6}{x^8}$ [4.4A] **14.** $(a - b)(3 - x)$ [5.1B] **15.** $5xy^2(3 - 4y^2)$ [5.1A] **16.** $(x - 7y)(x + 2y)$ [5.2A] **17.** $(p - 10)(p + 1)$ [5.2A] **18.** $3a(3a + 2)(2a + 5)$ [5.4B] **19.** $(6a + 7b)(6a - 7b)$ [5.4A]

Copyright © Houghton Mifflin Company. All rights reserved.

20. $(2x + 7y)^2$ [5.4A]   **21.** $(3x + 7)(3x - 2)$ [5.3A]   **22.** $2(3x - 4y)^2$ [5.4B]   **23.** $(x - 3)(3y - 2)$ [5.1B]

24. $\dfrac{2}{3}$, -7 [5.5A]   **25.** The shorter piece is 4 ft long. The longer piece is 6 ft long. [3.4B]   **26.** The discount rate is 40%. [3.2B]   **27.** $m\angle a = 72°$; $m\angle b = 108°$ [3.5A]   **28.** The distance to the resort is 168 mi. [3.6C]   **29.** The integers are 10, 12, and 14. [3.4A]   **30.** The length of the base of the triangle is 12 in. [5.5B]

Answers to Chapter 6 Selected Exercises

PREP TEST

1. 36 [1.5C]   **2.** $\dfrac{3x}{y^3}$ [4.4A]   **3.** $-\dfrac{5}{36}$ [1.6C]   **4.** $-\dfrac{10}{11}$ [1.7B]   **5.** No [1.7B]   **6.** $\dfrac{19}{8}$ [3.2A]

7. 130° [3.5B]   **8.** $(x - 6)(x + 2)$ [5.2A]   **9.** $(2x - 3)(x + 1)$ [5.3A]   **10.** 9:40 A.M. [3.6C]

SECTION 6.1

3. $\dfrac{3}{4x}$ **5.** $\dfrac{1}{x + 3}$ **7.** -1 **9.** $\dfrac{2}{3y}$ **11.** $-\dfrac{3}{4x}$ **13.** $\dfrac{a}{b}$ **15.** $-\dfrac{2}{x}$ **17.** $\dfrac{y - 2}{y - 3}$ **19.** $\dfrac{x + 5}{x + 4}$ **21.** $\dfrac{x + 4}{x - 3}$

23. $-\dfrac{x + 2}{x + 5}$ **25.** $\dfrac{2(x + 2)}{x + 3}$ **27.** $\dfrac{2x - 1}{2x + 3}$ **29.** $-\dfrac{x + 7}{x + 6}$ **31.** $\dfrac{35ab^2}{24x^2y}$ **33.** $\dfrac{4x^3y^3}{3a^2}$ **35.** $\dfrac{3}{4}$ **37.** ab^2

39. $\dfrac{x^2(x - 1)}{y(x + 3)}$ **41.** $\dfrac{y(x - 1)}{x^2(x + 10)}$ **43.** $-ab^2$ **45.** $\dfrac{x + 5}{x + 4}$ **47.** 1   **49.** $-\dfrac{n - 10}{n - 7}$ **51.** $\dfrac{x(x + 2)}{2(x - 1)}$ **53.** $-\dfrac{x + 2}{x - 6}$

55. $\dfrac{x + 5}{x - 12}$ **59.** $\dfrac{7a^3y^2}{40bx}$ **61.** $\dfrac{4}{3}$ **63.** $\dfrac{3a}{2}$ **65.** $\dfrac{x^2(x + 4)}{y^2(x + 2)}$ **67.** $\dfrac{x(x - 2)}{y(x - 6)}$ **69.** $-\dfrac{3by}{ax}$ **71.** $\dfrac{(x + 6)(x - 3)}{(x + 7)(x - 6)}$

73. 1   **75.** $-\dfrac{x + 8}{x - 4}$ **77.** $\dfrac{2n + 1}{2n - 3}$ **81.** $\dfrac{x}{x + 8}$ **83.** $\dfrac{n - 2}{n + 3}$

SECTION 6.2

1. $24x^3y^2$ **3.** $30x^4y^2$ **5.** $8x^2(x + 2)$ **7.** $6x^2y(x + 4)$ **9.** $36x(x + 2)^2$ **11.** $6(x + 1)^2$

13. $(x - 1)(x + 2)(x + 3)$ **15.** $(2x + 3)^2(x - 5)$ **17.** $(x - 1)(x - 2)$ **19.** $(x + 2)(x - 3)(x + 4)$

21. $(x + 1)(x + 4)(x - 7)$ **23.** $(x + 4)(x - 6)(x + 6)$ **25.** $(x + 3)(x - 10)(x - 8)$ **27.** $(x - 3)(3x - 2)(x + 2)$

29. $(x + 2)(x - 3)$ **31.** $(x + 1)(x - 5)$ **33.** $(x - 1)(x - 2)(x - 3)(x - 6)$ **35.** $\dfrac{5}{ab^2}$, $\dfrac{6b}{ab^2}$ **37.** $\dfrac{15y^2}{18x^2y}$, $\dfrac{14x}{18x^2y}$

39. $\dfrac{ay + 5a}{y^2(y + 5)}$, $\dfrac{6y}{y^2(y + 5)}$ **41.** $\dfrac{a^2y + 7a^2}{y(y + 7)^2}$, $\dfrac{ay}{y(y + 7)^2}$ **43.** $\dfrac{b}{y(y - 4)}$, $-\dfrac{b^2y}{y(y - 4)}$ **45.** $\dfrac{-3y - 21}{(y - 7)^2}$, $\dfrac{2}{(y - 7)^2}$ **47.** $\dfrac{2y^2}{y^2(y - 3)}$,

$\dfrac{3}{y^2(y - 3)}$ **49.** $\dfrac{x^3 + 4x^2}{(2x - 1)(x + 4)}$, $\dfrac{2x^2 + x - 1}{(2x - 1)(x + 4)}$ **51.** $\dfrac{3x^2 + 15x}{(x + 5)(x - 5)}$, $\dfrac{4}{(x + 5)(x - 5)}$ **53.** $\dfrac{x^2 - 1}{(x - 3)(x + 5)(x + 1)}$,

$\dfrac{x^2 - 3x}{(x - 3)(x + 5)(x + 1)}$ **55.** $\dfrac{800}{10^5}$, $\dfrac{9}{10^5}$ **57.** $\dfrac{x^3 - x}{x^2 - 1}$, $\dfrac{x}{x^2 - 1}$ **59.** $\dfrac{3c^2 - 3cd}{3(6c + d)(c + d)(c - d)}$, $\dfrac{6cd + d^2}{3(6c + d)(c + d)(c - d)}$

SECTION 6.3

1. $\dfrac{11}{y^2}$ **3.** $-\dfrac{7}{x + 4}$ **5.** $\dfrac{8x}{2x + 3}$ **7.** $\dfrac{5x + 7}{x - 3}$ **9.** $\dfrac{2x - 5}{x + 9}$ **11.** $\dfrac{-3x - 4}{2x + 7}$ **13.** $\dfrac{1}{x + 5}$ **15.** $\dfrac{1}{x - 6}$ **17.** $\dfrac{3}{2y - 1}$

19. $\dfrac{1}{x - 5}$ **23.** $\dfrac{4y + 5x}{xy}$ **25.** $\dfrac{19}{2x}$ **27.** $\dfrac{5}{12x}$ **29.** $\dfrac{19x - 12}{6x^2}$ **31.** $\dfrac{52y - 35x}{20xy}$ **33.** $\dfrac{13x + 2}{15x}$ **35.** $\dfrac{7}{24}$

37. $\dfrac{x + 90}{45x}$ **39.** $\dfrac{x^2 + 2x + 2}{2x^2}$ **41.** $\dfrac{2x^2 + 3x - 10}{4x^2}$ **43.** $\dfrac{-x^2 - 4x + 4}{x + 4}$ **45.** $\dfrac{4x + 7}{x + 1}$ **47.** $\dfrac{4x^2 + 9x + 9}{24x^2}$

49. $\dfrac{3x - 1 - 2xy - 3y}{xy^2}$ **51.** $\dfrac{20x^2 + 28x - 12xy + 9y}{24x^2y^2}$ **53.** $\dfrac{9x^2 - 3x - 2xy - 10y}{18xy^2}$ **55.** $\dfrac{7x - 23}{(x - 3)(x - 4)}$

Copyright © Houghton Mifflin Company. All rights reserved.

57. $\dfrac{-y - 33}{(y + 6)(y - 3)}$ **59.** $\dfrac{3x^2 + 20x - 8}{(x - 4)(x + 6)}$ **61.** $\dfrac{3(4x^2 + 5x - 5)}{(x + 5)(2x + 3)}$ **63.** $\dfrac{-4x + 5}{x - 6}$ **65.** $\dfrac{2(y + 2)}{(y + 4)(y - 4)}$ **67.** $-\dfrac{4x}{(x + 1)^2}$

69. $\dfrac{2x - 1}{(1 + x)(1 - x)}$ **71.** $\dfrac{14}{(x - 5)^2}$ **73.** $\dfrac{-2(x + 7)}{(x + 6)(x - 7)}$ **75.** $\dfrac{x - 4}{x - 6}$ **77.** $\dfrac{2x + 1}{x - 1}$ **79.** $\dfrac{-3(x^2 + 8x + 25)}{(x - 3)(x + 7)}$

81. a. $\dfrac{20{,}400}{x}$ dollars **b.** $\dfrac{102{,}000}{x(x + 5)}$ dollars **c.** \$136

SECTION 6.4

1. $\dfrac{x}{x - 3}$ **3.** $\dfrac{2}{3}$ **5.** $\dfrac{y + 3}{y - 4}$ **7.** $\dfrac{2(2x + 13)}{5x + 36}$ **9.** $\dfrac{x + 2}{x + 3}$ **11.** $\dfrac{x - 6}{x + 5}$ **13.** $\dfrac{-x + 2}{x + 1}$ **15.** $x - 1$ **17.** $\dfrac{1}{2x - 1}$

19. $\dfrac{x - 3}{x + 5}$ **21.** $\dfrac{x - 7}{x - 8}$ **23.** $\dfrac{2y - 1}{2y + 1}$ **25.** $\dfrac{x - 2}{2x - 5}$ **27.** $\dfrac{-x - 1}{4x - 3}$ **29.** $\dfrac{x + 1}{2(5x - 2)}$ **31.** $\dfrac{5}{3}$ **33.** $-\dfrac{1}{x - 1}$

35. $\dfrac{y + 4}{2(y - 2)}$ **37.** $\dfrac{x + 1}{x - 1}$ **39.** $\dfrac{y^2 + x^2}{xy}$

SECTION 6.5

3. 3 **5.** 1 **7.** 9 **9.** 1 **11.** $\dfrac{1}{4}$ **13.** 1 **15.** -3 **17.** $\dfrac{1}{2}$ **19.** 8 **21.** 5 **23.** -1 **25.** 5

27. No solution **29.** 4, 2 **31.** $-\dfrac{3}{2}$, 4 **33.** 3 **35.** 4 **37.** 0 **39.** $-\dfrac{2}{5}$ **41.** 0, $-\dfrac{2}{3}$

SECTION 6.6

3. 9 **5.** 12 **7.** 7 **9.** 6 **11.** 1 **13.** -6 **15.** 4 **17.** $-\dfrac{2}{3}$ **19.** 20,000 voters voted in favor of the amendment. **21.** The distance between the two cities is 175 mi. **23.** The sales tax will be \$97.50 higher. **25.** The person is 67.5 in. tall. **27.** There are approximately 75 elk in the preserve. **29.** The length of side AC is 6.7 cm. **31.** The height is 2.9 m. **33.** The perimeter is 22.5 ft. **35.** The area is 48 m². **37.** The length of BC is 6.25 cm. **39.** The length of DA is 6 in. **41.** The length of OP is 13 cm. **43.** The distance across the river is 35 m. **45.** The first person won \$1.25 million. **47.** The player made 210 foul shots.

SECTION 6.7

1. $y = -3x + 10$ **3.** $y = 4x - 3$ **5.** $y = -\dfrac{3}{2}x + 3$ **7.** $y = \dfrac{2}{5}x - 2$ **9.** $y = -\dfrac{2}{7}x + 2$ **11.** $y = -\dfrac{1}{3}x + 2$

13. $y = 3x + 8$ **15.** $y = -\dfrac{2}{3}x - 3$ **17.** $x = -6y + 10$ **19.** $x = \dfrac{1}{2}y + 3$ **21.** $x = -\dfrac{3}{4}y + 3$ **23.** $x = 4y + 3$

25. $t = \dfrac{d}{r}$ **27.** $T = \dfrac{PV}{nR}$ **29.** $l = \dfrac{P - 2w}{2}$ **31.** $b_1 = \dfrac{2A - hb_2}{h}$ **33.** $h = \dfrac{3V}{A}$ **35.** $S = C - Rt$ **37.** $P = \dfrac{A}{1 + rt}$

39. $w = \dfrac{A}{S + 1}$ **41. a.** $S = \dfrac{F + BV}{B}$ **b.** The required selling price is \$180. **c.** The required selling price is \$75.

SECTION 6.8

3. It will take 2 h to fill the fountain with both sprinklers working. **5.** With both skiploaders working together, it would take 3 h to remove the earth. **7.** With both computers working, it would take 30 h to solve the problem. **9.** It would take 30 min to cool the room with both air conditioners working. **11.** It would take the second pipeline 90 min to fill the tank. **13.** It would take the apprentice 15 h to construct the wall. **15.** It will take the second technician 3 h to complete the wiring. **17.** It would have taken one of the welders 40 h to complete the welds. **19.** It would have taken one machine 28 h to fill the boxes. **21.** The jogger ran 16 mi in 2 h. **23.** The rate of travel in the congested area was 20 mph. **25.** The rate of the jogger was 8 mph. The rate of the cyclist was 20 mph. **27.** The rate of the jet is 360 mph. **29.** Camille's walking rate is 4 mph. **31.** The rate of the car is 48 mph. **33.** The rate of the wind is 20 mph. **35.** The rate of the gulf current is 6 mph. **37.** The rate of the trucker for the first 330 mi was 55 mph. **39.** The bus usually travels 60 mph.

Copyright © Houghton Mifflin Company. All rights reserved.

CHAPTER 6 REVIEW EXERCISES

1. $\dfrac{b^3y}{10ax}$ [6.1C]   **2.** $\dfrac{7x + 22}{60x}$ [6.3B]   **3.** $\dfrac{2xy}{5}$ [6.1B]   **4.** $\dfrac{2xy}{3(x + y)}$ [6.1C]   **5.** $\dfrac{x - 2}{3x - 10}$ [6.4A]

6. $-\dfrac{x + 6}{x + 3}$ [6.1A]   **7.** $\dfrac{2x^4}{3y^7}$ [6.1A]   **8.** 62 [6.6A]   **9.** $\dfrac{(3y - 2)^2}{(y - 1)(y - 2)}$ [6.1C]   **10.** $x = \dfrac{5}{3a - 1}$ [6.7A]

11. 8 [6.5A]   **12.** $\dfrac{x^2 + 3y}{xy}$ [6.3B]   **13.** $y = -\dfrac{5}{4}x + 5$ [6.7A]   **14.** $\dfrac{by^3}{6ax^2}$ [6.1B]   **15.** $\dfrac{x}{x - 7}$ [6.4A]

16. $\dfrac{3x^2 - x}{(6x - 1)(2x + 3)(3x - 1)}, \dfrac{24x^3 - 4x^2}{(6x - 1)(2x + 3)(3x - 1)}$ [6.2B]   **17.** $a = \dfrac{T - 2bc}{2b + 2c}$ [6.7A]   **18.** 2 [6.5A]

19. $\dfrac{2x + 1}{3x - 2}$ [6.4A]   **20.** $\dfrac{x^2 + 5}{(x - 5)(x - 2)}$ [6.3B]   **21.** $c = \dfrac{100m}{i}$ [6.7A]   **22.** No solution [6.5A]   **23.** $\dfrac{1}{x^2}$ [6.1C]

24. $\dfrac{2y - 3}{5y - 7}$ [6.3B]   **25.** $\dfrac{1}{x + 3}$ [6.3A]   **26.** $(5x - 3)(2x - 1)(4x - 1)$ [6.2A]   **27.** $y = -\dfrac{4}{9}x + 2$ [6.7A]

28. $\dfrac{2x + 1}{x + 2}$ [6.1B]   **29.** 5 [6.5A]   **30.** $\dfrac{3x - 1}{x - 5}$ [6.3B]   **31.** 10 [6.6A]   **32.** 12 [6.6A]   **33.** The length of QO

is 15 cm. [6.6C]   **34.** The area is $\dfrac{256}{3}$ in². [6.6C]   **35.** It would take 6 h to fill the pool. [6.8A]   **36.** The rate of

the car is 45 mph. [6.8B]   **37.** The rate of the wind is 20 mph. [6.8B]   **38.** The ERA is 1.35. [6.6B]

CHAPTER 6 TEST

1. $\dfrac{x^2 - 4x + 5}{(x + 3)(x - 2)}$ [6.3B]   **2.** -1 [6.6A]   **3.** $\dfrac{(2x - 1)(x - 5)}{(x + 3)(2x + 5)}$ [6.1B]   **4.** $\dfrac{2x^3}{3y^3}$ [6.1A]   **5.** $t = \dfrac{d - s}{r}$ [6.7A]

6. 2 [6.5A]   **7.** $-\dfrac{x + 5}{x + 1}$ [6.1A]   **8.** $3(2x - 1)(x + 1)$ [6.2A]   **9.** $\dfrac{5}{(2x - 1)(3x + 1)}$ [6.3B]   **10.** $\dfrac{x + 5}{x + 4}$ [6.1C]

11. $\dfrac{x - 3}{x - 2}$ [6.4A]   **12.** $\dfrac{3x + 6}{x(x - 2)(x + 2)}, \dfrac{x^2}{x(x - 2)(x + 2)}$ [6.2B]   **13.** $\dfrac{2}{x + 5}$ [6.3A]   **14.** $y = \dfrac{3}{8}x - 2$ [6.7A]

15. No solution [6.5A]   **16.** $\dfrac{x + 1}{x^3(x - 2)}$ [6.1B]   **17.** The length of CE is 12.8 ft. [6.6C]   **18.** An additional 2 lb of

salt are needed. [6.6B]   **19.** It would take 4 h to fill the pool. [6.8A]   **20.** The rate of the wind is 20 mph. [6.8B]

21. 54 sprinklers are needed for a 3600 square-foot lawn. [6.6B]

CUMULATIVE REVIEW EXERCISES

1. $\dfrac{31}{30}$ [1.7A]   **2.** 21 [2.1A]   **3.** $5x - 2y$ [2.2A]   **4.** $-8x + 26$ [2.2D]   **5.** $-\dfrac{9}{2}$ [3.2A]   **6.** -12 [3.3B]

7. 10 [3.1D]   **8.** a^3b^7 [4.2A]   **9.** $a^2 + ab - 12b^2$ [4.3C]   **10.** $3b^3 - b + 2$ [4.5A]   **11.** $x^2 + 2x + 4$ [4.5B]

12. $(3x - 1)(4x + 1)$ [5.3A]   **13.** $(y - 6)(y - 1)$ [5.2A]   **14.** $a(a + 5)(2a - 3)$ [5.3A]

15. $4(b + 5)(b - 5)$ [5.4B]   **16.** $-3, \dfrac{5}{2}$ [5.5A]   **17.** $\dfrac{2x^3}{3y^5}$ [6.1A]   **18.** $-\dfrac{x - 2}{x + 5}$ [6.1A]   **19.** 1 [6.1C]

20. $\dfrac{3}{(2x - 1)(x + 1)}$ [6.3B]   **21.** $\dfrac{x + 3}{x + 5}$ [6.4A]   **22.** 4 [6.5A]   **23.** 3 [6.6A]   **24.** $t = \dfrac{f - v}{a}$ [6.7A]

25. $5x - 13 = -8; x = 1$ [3.4A]   **26.** The silver alloy is 70% silver. [3.6B]   **27.** The base is 10 in. The height is

6 in. [5.5B]   **28.** The cost of a $5000 policy is $80. [6.6B]   **29.** It would take both pipes 6 min to fill the

tank. [6.8A]   **30.** The rate of the current is 2 mph. [6.8B]

Answers to Chapter 7 Selected Exercises

PREP TEST

1. 3 [1.4B]   **2.** -1 [2.1A]   **3.** $-3x + 12$ [2.2C]   **4.** -2 [3.2A]   **5.** $x = 5$ [3.2A]   **6.** $y = -2$ [3.2A]

7. $-4x + 5$ [4.5A]   **8.** 4 [6.6A]   **9.** $y = \dfrac{3}{5}x - 3$ [6.7A]   **10.** $y = -\dfrac{1}{2}x - 5$ [6.7A]

Copyright © Houghton Mifflin Company. All rights reserved.

SECTION 7.1

1.

3.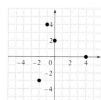

5.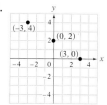

7. $A(2, 3), B(4, 0), C(-4, 1), D(-2, -2)$

9. $A(-2, 5), B(3, 4), C(0, 0), D(-3, -2)$ **11. a.** $2, -4$ **b.** $1, -3$ **15.** Yes   **17.** No   **19.** No   **21.** No

23.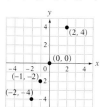

25.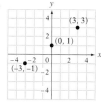

27.

29. $\{(24, 600), (32, 750), (22, 430), (15, 300), (4.4, 68), (17, 370), (15, 310), (4.4, 55)\}$; No

31. $\{(390, 0.115), (591, 0.073), (517, 0.077), (576, 0.068), (605, 0.064)\}$; Yes   **33.** Yes   **35.** No   **37.** Yes   **39.** 8

41. 9   **43.** 2   **45.** -1 **47.** 22   **49.** $-\dfrac{3}{2}$ **51.** -7

SECTION 7.2

1.

3.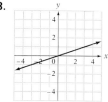

5.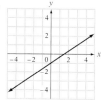

7.

9.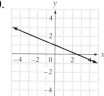

11.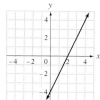

13.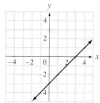

15.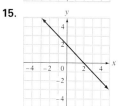

17.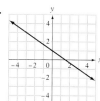

19.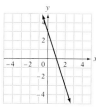

21.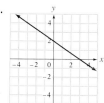

23.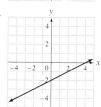

25.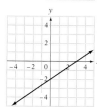

27.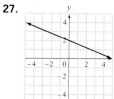

29.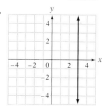

31.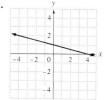

33.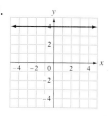

35.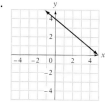

Copyright © Houghton Mifflin Company. All rights reserved.

37. After flying for 3 min, the helicopter is 3.5 mi away from the victims.

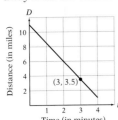

39. A dog 6 years is equivalent in age to a human 40 years old.

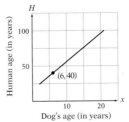

43. Increases; 3; 3   **45. a.** The cost is $.99.   **b.** The cost is $1.74.

SECTION 7.3

1. $(3, 0), (0, -3)$ **3.** $(2, 0), (0, -6)$ **5.** $(10, 0), (0, -2)$ **7.** $(-4, 0), (0, 12)$ **9.** $(0, 0), (0, 0)$ **11.** $(6, 0), (0, 3)$

13.
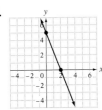

15.
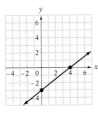

17.
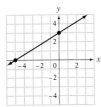

21. -2 **23.** $\dfrac{1}{3}$ **25.** $-\dfrac{5}{2}$

27. Undefined   **29.** 0   **31.** $-\dfrac{1}{3}$ **33.** Neither   **35.** Neither   **37.** Parallel   **39.** Neither   **41.** $m = 33$. The worldwide sales of camera-phones are increasing by 33 million units per year.   **43.** $m = -180$. The value of the car is decreasing $180 for each additional 1000 miles the car is driven.   **45.** $m = \dfrac{2}{3}$; $(0, -2)$ **47.** $m = -\dfrac{2}{5}$; $(0, 2)$

49. $m = \dfrac{1}{4}$; $(0, 0)$ **51.**
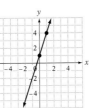

53.
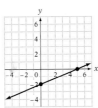

55.
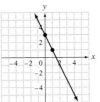

57.
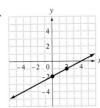

59.
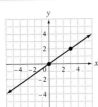

61.
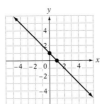

63.
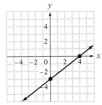

65.
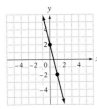

67. Yes

SECTION 7.4

3. $y = 2x + 2$ **5.** $y = -3x - 1$ **7.** $y = \dfrac{1}{3}x$ **9.** $y = \dfrac{3}{4}x - 5$ **11.** $y = -\dfrac{3}{5}x$ **13.** $y = \dfrac{1}{4}x + \dfrac{5}{2}$

15. $y = 2x - 3$ **17.** $y = -2x - 3$ **19.** $y = \dfrac{2}{3}x$ **21.** $y = \dfrac{1}{2}x + 2$ **23.** $y = -\dfrac{3}{4}x - 2$ **25.** $y = \dfrac{3}{4}x + \dfrac{5}{2}$

Copyright © Houghton Mifflin Company. All rights reserved.

27. The tennis player is using 1.55 g of carbohydrates per minute.

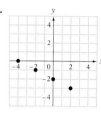

29. The percent of music purchased in stores is decreasing 3% per year.

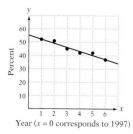

Year ($x = 0$ corresponds to 1997)

31. No   **33.** Yes   **35.** $-\dfrac{3}{2}$ **37.** -5 **39.** $y = -\dfrac{2}{3}x + \dfrac{5}{3}$

CHAPTER 7 REVIEW EXERCISES

1. a.

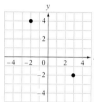

2.

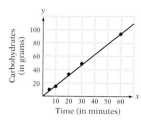

[7.1B]

3. $y = -\dfrac{8}{3}x + \dfrac{1}{3}$ [7.4B]   **4.** $y = -\dfrac{5}{2}x + 16$ [7.4A]

b. -2

c. -4 [7.1A]

5.

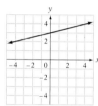

[7.2A]

6.

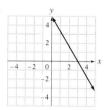

[7.2B]

7. Neither   [7.3B]   **8.** -1 [7.1D]

9. $y = -\dfrac{2}{3}x + \dfrac{11}{3}$ [7.4B]   **10.** Yes   [7.1C]   **11.** $\dfrac{7}{11}$ [7.3B]   **12.** $(8, 0), (0, -12)$ [7.3A]   **13.** 0   [7.3B]

14.

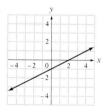

[7.3C]

15.

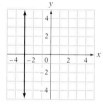

[7.2B]

16.

[7.3C]

17.

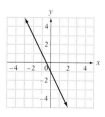

[7.2A]

18.

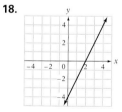

[7.3C]

19.

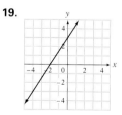

[7.2B]

20. $\{(55, 95), (57, 101), (53, 94), (57, 98), (60, 100), (61, 105), (58, 97), (54, 95)\}$; No   [7.1C]

21. The cost of 50 min of access time for one month is $97.50.

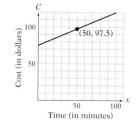

[7.2C]

22. The average annual telephone bill for a family is increasing by $34 per year.

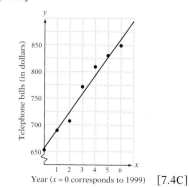

Year ($x = 0$ corresponds to 1999)   [7.4C]

Copyright © Houghton Mifflin Company. All rights reserved.

CHAPTER 7 TEST

1. $(3, -3)$ [7.1B]   **2.**

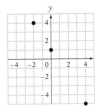

[7.1B]

3. Yes [7.1C]   **4.** 6 [7.1D]   **5.** 3 [7.1D]

6. $\{(3.5, 25), (4.0, 30), (5.2, 45), (5.0, 38), (4.0, 42), (6.3, 12), (5.4, 34)\}$; No [7.1C]   **7.**

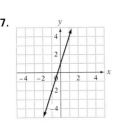

[7.2A]

8.

[7.2A]

9.

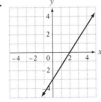

[7.2B]

10.

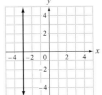

[7.2B]

11.

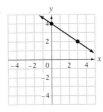

[7.3C]

12.

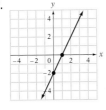

[7.3C]

13. After 1 s, the speed of the ball is 96 ft/s.   [7.2C]

14. $m = 0.46$. The average hourly wage is increasing by $.46 per year.   [7.3B]

15.

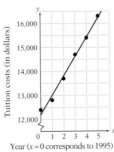

The average annual tuition for a private 4-year college is increasing $809 per year.   [7.4C]

16. $(2, 0), (0, -3)$ [7.3A]   **17.** $(-2, 0), (0, 1)$ [7.3A]   **18.** 2 [7.3B]   **19.** Parallel [7.3B]

20. Undefined [7.3B]   **21.** $-\dfrac{2}{3}$ [7.3B]   **22.** $y = 3x - 1$ [7.4A]   **23.** $y = \dfrac{2}{3}x + 3$ [7.4A]

24. $y = -\dfrac{5}{8}x - \dfrac{7}{8}$ [7.4B]   **25.** $y = -\dfrac{2}{7}x - \dfrac{4}{7}$ [7.4B]

CUMULATIVE REVIEW EXERCISES

1. -12 [1.4B]   **2.** $-\dfrac{5}{8}$ [2.1A]   **3.** $f(-2) = -\dfrac{2}{3}$ [7.1D]   **4.** $\dfrac{3}{2}$ [3.2A]   **5.** $\dfrac{19}{18}$ [3.3B]   **6.** $\dfrac{1}{15}$ [1.7C]

7. $-32x^8y^7$ [4.2B]   **8.** $-3x^2$ [4.4A]   **9.** $x + 3$ [4.5B]   **10.** $5(x + 2)(x + 1)$ [5.2B]   **11.** $(a + 2)(x + y)$ [5.1A]

12. 4 and -2 [5.5A]   **13.** $\dfrac{x^3(x + 3)}{y(x + 2)}$ [6.1B]   **14.** $\dfrac{3}{x + 8}$ [6.3A]   **15.** 2 [6.5A]   **16.** $y = \dfrac{4}{5}x - 3$ [6.7A]

17. $(-2, -5)$ [7.1B]   **18.** Zero [7.3B]   **19.** $y = \dfrac{1}{2}x - 2$ [7.4A]   **20.** $y = -3x + 2$ [7.4A]

Copyright © Houghton Mifflin Company. All rights reserved.

21. $y = 2x + 2$ [7.4A]   **22.** $y = \frac{2}{3}x - 3$ [7.4A]   **23.** \$62.30 [3.2B]   **24.** 46°, 43°, 91° [3.5B]   **25.** The value

of the home is \$110,000. [6.6B]   **26.** It would take $3\frac{3}{4}$ h for both, working together, to wire the garage. [6.8A]

27. 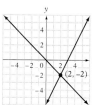   **28.**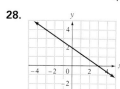

[7.2A]   [7.3C]

Answers to Chapter 8 Selected Exercises

PREP TEST

1. $y = \frac{3}{4}x - 6$ [6.7A]   **2.** \$1000 [3.3B]   **3.** $33y$ [2.2D]   **4.** $10x - 10$ [2.2D]   **5.** Yes [7.1B]

6. $(4, 0), (0, -3)$ [7.3A]   **7.** Yes [7.3B]   **8.** 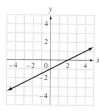   [7.3C]

9. 5 ml of 55% acetic acid are needed. [3.6B]   **10.** It will take 1.5 h after the second hiker starts for the hikers to be
side-by-side. [3.6C]

SECTION 8.1

1. I: c; II: a; III: b   **3.** $(2, -1)$ **5.** The ordered-pair solutions of $y = -\frac{3}{2}x + 1$ **7.** No solution   **9.** $(-2, 4)$

11. Yes   **13.** No   **15.** Yes   **17.** Yes

19. 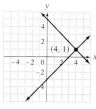   **21.** 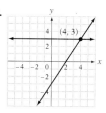   **23.** 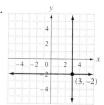   **25.**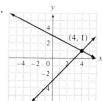

27. 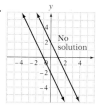   **29.** 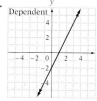   **31.** 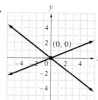   The ordered-pair solutions of $y = 2x - 2$

33. 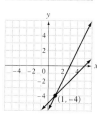   **35.** 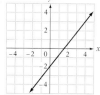   **37.**   **39. a.** Sometimes true   **b.** Always true
     c. Never true   **d.** Always true
     43. Answers will vary.

Copyright © Houghton Mifflin Company. All rights reserved.

SECTION 8.2

3. $(2, 1)$ **5.** $(4, 1)$ **7.** $(-1, 1)$ **9.** No solution   **11.** No solution   **13.** $\left(-\dfrac{3}{4}, -\dfrac{3}{4}\right)$ **15.** $(1, 1)$ **17.** $(2, 0)$

19. $(1, -2)$ **21.** $(0, 0)$ **23.** Dependent. The solutions satisfy the equation $2x - y = 2$.   **25.** $(-4, -2)$ **27.** $(10, 31)$

29. $(3, -10)$ **31.** $(-22, -5)$ **33.** The amounts invested should be $1900 at 5% and $1600 at 7.5%.

35. The amounts invested were $2400 at 9% and $3600 at 6%.   **37.** The amounts invested should be $4400 at 8% and

$1600 at 11%.   **39.** The amount invested at 6.5% was $21,000.   **41.** The amounts invested were $12,000 at 8% and

$8000 at 7%.   **43.** The amount invested in the trust deed was $3750.   **45.** The gas dryer becomes more economical

after 185 loads of clothes.   **47.** 1   **49.** The assertion is not correct. The system of equations is independent. The

solution is $(0, 2)$.   **51.** The research consultant's investment is $45,000.   **53.** Simple interest: $400; Compounded

monthly: $415.00; Compounded daily: $416.39

SECTION 8.3

1. $(5, -1)$ **3.** $(1, 3)$ **5.** $(1, 1)$ **7.** $(3, -2)$ **9.** Dependent. The solutions satisfy the equation $2x - y = 1$.

11. $(3, 1)$ **13.** Dependent. The solutions satisfy the equation $2x - 3y = 1$.   **15.** $\left(\dfrac{2}{3}, \dfrac{1}{2}\right)$ **17.** $(2, 0)$ **19.** $(0, 0)$

21. $(5, -2)$ **23.** $\left(\dfrac{32}{19}, -\dfrac{9}{19}\right)$ **25.** $\left(\dfrac{7}{4}, -\dfrac{5}{16}\right)$ **27.** $(1, -1)$ **29.** No solution   **31.** $(3, 1)$ **33.** $(-1, 2)$

35. $(1, 1)$ **39.** $A = 3; B = -1$ **41. a.** 1   **b.** $\dfrac{3}{2}$ **c.** 4

SECTION 8.4

1. The rate of the whale in calm water was 35 mph. The rate of the current was 5 mph.   **3.** The rate rowing in calm

water was 14 km/h. The rate of the current was 6 km/h.   **5.** The rate of the Learjet was 525 mph. The rate of the wind

was 35 mph.   **7.** The rate of the helicopter in calm air was 225 mph. The rate of the wind was 45 mph.   **9.** The rate

of the canoeist in calm water was 6 mph. The rate of the current was 1 mph.   **11.** The cost per pound of the wheat flour

was $.65. The cost per pound of the rye flour was $.70.   **13.** Reagent I is 25% hydrochloric acid. Reagent II is 35%

hydrochloric acid.   **15.** 12 gal of 87-octane gasoline and 6 gal of 93-octane gasoline must be used.   **17.** 1 nickel and

2 dimes or 3 nickels and 1 dime are in the bank.   **19.** 12.5 acres of good land and 87.5 acres of bad land were bought.

CHAPTER 8 REVIEW EXERCISES

1. Yes [8.1A]   **2.** No [8.1A]   **3.**

[8.1A]

4.

The solutions are the ordered-pair solutions of $y = 2x - 4$.

[8.1A]

5.

[8.1A]   **6.** $(-1, 1)$ [8.2A]   **7.** $(1, 6)$ [8.2A]   **8.** $(-3, 1)$ [8.3A]   **9.** $\left(-\dfrac{5}{6}, \dfrac{1}{2}\right)$ [8.3A]

10. No solution [8.2A]   **11.** $(1, 6)$ [8.2A]   **12.** $(1, -5)$ [8.3A]   **13.** No solution [8.3A]   **14.** Dependent.

The solutions satisfy the equation $y = -\dfrac{4}{3}x + 4$. [8.2A]   **15.** $(-1, -3)$ [8.2A]   **16.** Dependent. The solutions satisfy

the equation $3x + y = -2$. [8.3A]   **17.** $\left(\dfrac{2}{3}, -\dfrac{1}{6}\right)$ [8.3A]   **18.** The rate of the sculling team in calm water was

9 mph. The rate of the current was 3 mph. [8.4A]   **19.** 1300 $6 shares were purchased, and 200 $25 shares were

Copyright © Houghton Mifflin Company. All rights reserved.

purchased. [8.4B]   **20.** The rate of the flight crew in calm air was 125 km/h. The rate of the wind was 15 km/h. [8.4A]
21. The rate of the plane in calm air was 105 mph. The rate of the wind was 15 mph. [8.4A]   **22.** The number of ads
requiring \$.25 postage was 130. The number of ads requiring \$.45 postage was 60. [8.4B]   **23.** The amounts invested
are \$7000 at 7% and \$5000 at 8.5%. [8.2B]   **24.** There were originally 350 bushels of lentils and 200 bushels of corn in
the silo. [8.4B]   **25.** The amounts invested were \$165,000 at 5.4% and \$135,000 at 6.6%. [8.2B]

CHAPTER 8 TEST

1. Yes [8.1A]   **2.** Yes [8.1A]   **3.**

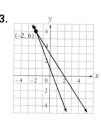

[8.1A]

4. (3, 1) [8.2A]   **5.** (1, −1) [8.2A]

6. (2, −1) [8.2A]   **7.** $\left(\dfrac{22}{7}, -\dfrac{5}{7}\right)$ [8.2A]   **8.** No solution [8.2A]   **9.** (2, 1) [8.3A]   **10.** $\left(\dfrac{1}{2}, -1\right)$ [8.3A]
11. Dependent. The solutions satisfy the equation $x + 2y = 8$. [8.3A]   **12.** (2, −1) [8.3A]   **13.** (1, −2) [8.3A]
14. The rate of the plane in calm air is 100 mph. The rate of the wind is 20 mph. [8.4A]   **15.** The price of a reserved-
seat ticket was \$10. The price of a general-admission ticket was \$6. [8.4B]   **16.** The amounts invested were \$15,200 at
6.4% and \$12,800 at 7.6%. [8.2B]

CUMULATIVE REVIEW EXERCISES

1. $\dfrac{3}{2}$ [2.1A]   **2.** $-\dfrac{3}{2}$ [3.1C]   **3.** 7 [7.1D]   **4.** $-6a^3 + 13a^2 - 9a + 2$ [4.3B]   **5.** $-2x^5y^2$ [4.4A]

6. $2b - 1 + \dfrac{1}{2b - 3}$ [4.5B]   **7.** $-\dfrac{4y}{x^3}$ [4.4A]   **8.** $4y^2(xy - 4)(xy + 4)$ [5.4B]   **9.** 4, −1 [5.5A]

10. $x - 2$ [6.1C]   **11.** $\dfrac{x^2 + 2}{(x + 2)(x - 1)}$ [6.3B]   **12.** $\dfrac{x - 3}{x + 1}$ [6.4A]   **13.** $-\dfrac{1}{5}$ [6.5A]   **14.** $r = \dfrac{A - P}{Pt}$ [6.7A]

15. x-intercept: (6, 0); y-intercept: (0, −4) [7.3A]   **16.** $-\dfrac{7}{5}$ [7.3B]   **17.** $y = -\dfrac{3}{2}x$ [7.4A]

18. Yes [8.1A]   **19.** (−6, 1) [8.2A]   **20.** (4, −3) [8.3A]   **21.** The amounts invested should be \$3750 at 9.6% and
\$5000 at 7.2%. [8.2B]   **22.** The rate of the passenger train is 56 mph. The rate of the freight train is 48 mph. [3.6C]
23. The side of the original square is 8 in. [5.5B]   **24.** The rate of the wind is 30 mph. [8.4A]

25.

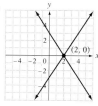

[7.2B]

26.

[8.1A]

27. The rate of the motorboat in calm water is 14 mph. [8.4A]   **28.** The percent concentration of sugar in the mixture
is 35.3%. [3.6A]

Answers to Chapter 9 Selected Exercises

PREP TEST

1. < [1.1A]   **2.** $-7x + 15$ [2.2D]   **3.** The same number can be added to each side of an equation without changing
the solution of the equation. [3.1B]   **4.** Each side of an equation can be multiplied by the same nonzero number without changing the solution of the equation. [3.1C]   **5.** There are 0.45 lb of fat in 3 lb of this grade of hamburger. [3.1D]

Copyright © Houghton Mifflin Company. All rights reserved.

6. $-\dfrac{1}{2}$ [3.2A]   **7.** $-\dfrac{8}{3}$ [3.2A]   **8.** 2 [3.3B]   **9.**

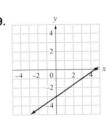

[7.2A]   **10.**

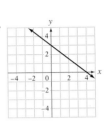

[7.2B]

SECTION 9.1

3. $A = \{16, 17, 18, 19, 20, 21\}$ **5.** $A = \{9, 11, 13, 15, 17\}$ **7.** $A = \{b, c\}$ **9.** $A \cup B = \{3, 4, 5, 6\}$
11. $A \cup B = \{-10, -9, -8, 8, 9, 10\}$ **13.** $A \cup B = \{a, b, c, d, e, f\}$ **15.** $A \cup B = \{1, 3, 7, 9, 11, 13\}$
17. $A \cap B = \{4, 5\}$ **19.** $A \cap B = \varnothing$ **21.** $A \cap B = \{c, d, e\}$ **23.** $\{x | x > -5, x \in \text{negative integers}\}$
25. $\{x | x > 30, x \in \text{integers}\}$ **27.** $\{x | x > 5, x \in \text{even integers}\}$ **29.** $\{x | x > 8, x \in \text{real numbers}\}$
31. ++++++(++++ −5 −4 −3 −2 −1 0 1 2 3 4 5   **33.** +++++++]+++ −5 −4 −3 −2 −1 0 1 2 3 4 5   **35.** +)++(+++++ −5 −4 −3 −2 −1 0 1 2 3 4 5
37. +++(++++)++ −5 −4 −3 −2 −1 0 1 2 3 4 5   **39.** +++++++++++ −5 −4 −3 −2 −1 0 1 2 3 4 5   **41. a.** Never true   **b.** Always true
c. Always true   **43. a.** Yes   **b.** Yes

SECTION 9.2

1. $x < 2$ +++++++)+++ −5 −4 −3 −2 −1 0 1 2 3 4 5   **3.** $x > 3$ ++++++++(++ −5 −4 −3 −2 −1 0 1 2 3 4 5
5. $n \geq 3$ ++++++++[+++ −5 −4 −3 −2 −1 0 1 2 3 4 5   **7.** $x \leq -4$ ++]++++++++ −5 −4 −3 −2 −1 0 1 2 3 4 5   **9.** $x < 1$ **11.** $x \leq -3$
13. $y \geq -9$ **15.** $x < 12$ **17.** $x \geq 5$ **19.** $x < -11$ **21.** $x \leq 10$ **23.** $x \geq -6$ **25.** $x > 2$ **27.** $d < -\dfrac{1}{6}$
29. $x \geq -\dfrac{31}{24}$ **31.** $x < \dfrac{5}{8}$ **33.** $x < \dfrac{5}{4}$ **35.** $x > \dfrac{5}{24}$ **37.** $x < -3.8$ **39.** $x \leq -1.2$ **41.** $x < 5.6$
43. ++++++++)+ −5 −4 −3 −2 −1 0 1 2 3 4 5 $x < 4$ **45.** +++++++[+++ −5 −4 −3 −2 −1 0 1 2 3 4 5 $y \geq 3$
47. ++++++]++++ −5 −4 −3 −2 −1 0 1 2 3 4 5 $x \leq 1$ **49.** +++)+++++++ −5 −4 −3 −2 −1 0 1 2 3 4 5 $x < -1$
51. +)+++++++++ −5 −4 −3 −2 −1 0 1 2 3 4 5 $b < -4$ **53.** $y \leq 0$ **55.** $x > \dfrac{2}{7}$ **57.** $x \leq -\dfrac{5}{2}$ **59.** $x < 16$ **61.** $x \geq 16$
63. $x \geq -14$ **65.** $x \leq 21$ **67.** $x > 0$ **69.** $x \leq -\dfrac{12}{7}$ **71.** $x > \dfrac{2}{3}$ **73.** $x \leq \dfrac{2}{3}$ **75.** $x \leq 2.3$ **77.** $x < -3.2$
79. $x \leq 5$ **81.** The team must win 11 or more games to be eligible for the tournament.   **83.** The service organization
must collect more than 440 lb on the fourth drive to collect the bonus.   **85.** The person needs 50 or more milligrams of
additional vitamin C to satisfy the recommended daily allowance.   **87.** No, the student cannot earn an A grade.
89. $\{c | c > 0\}$ **91.** $\{c | c > \in \text{real numbers}\}$ **93.** $\{c | c > 0\}$

SECTION 9.3

1. $x < 4$ **3.** $x < -4$ **5.** $x \geq 1$ **7.** $x < 5$ **9.** $x < 0$ **11.** $x < 20$ **13.** $y \leq \dfrac{5}{2}$ **15.** $x < \dfrac{25}{11}$ **17.** $n \leq \dfrac{11}{18}$
19. $x \geq 6$ **21.** In one month the agent expects to make sales totaling $20,000 or less.   **23.** A person must use more
than 60 min to exceed $10.   **25.** The amount of artificial flavors that can be added is less than or equal to 8 oz.
27. The distance to the ski resort must be greater than 38 mi.   **29.** $\{1, 2\}$ **31.** $\{3, 4, 5\}$ **33.** $\{x | x \in \text{real numbers}\}$

SECTION 9.4

1.

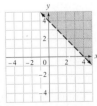

3.

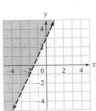

5.

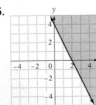

7.

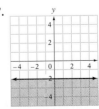

Copyright © Houghton Mifflin Company. All rights reserved.

9. 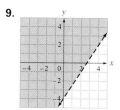   **11.** 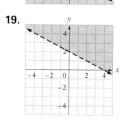   **13.** 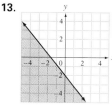   **15.**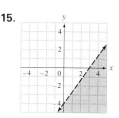

17. 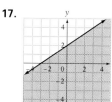   **19.** 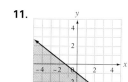   **21.** 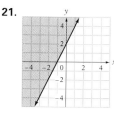   **23.** $x \leq 3$

CHAPTER 9 REVIEW

1. $x > 18$ [9.2A]   **2.** $A \cap B = \varnothing$ [9.1A]   **3.** $\{x \mid x > -8, x \in \text{odd integers}\}$ [9.1B]

4. $A \cup B = \{2, 4, 6, 8, 10\}$ [9.1A]   **5.** $A = \{1, 3, 5, 7\}$ [9.1A]   **6.** $x \geq 4$ [9.3A]

7. 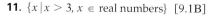 [9.1C]   **8.** $x \geq -4$ [9.3A]   **9.** 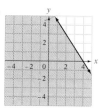   [9.4A]

10. 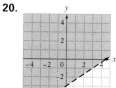   [9.4A]   **11.** $\{x \mid x > 3, x \in \text{real numbers}\}$ [9.1B]   **12.** $x > 2$  [9.2A]

13. $A \cap B = \{1, 5, 9\}$ [9.1A]   **14.** 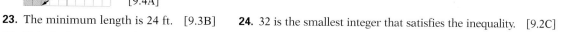 [9.1C]   **15.** 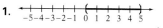 [9.1C]

16. $x \geq -3$ [9.2B]   **17.** $x > -18$ [9.3A]   **18.** $x < \dfrac{1}{2}$ [9.3A]   **19.** $x < -\dfrac{8}{9}$ [9.2B]

20. 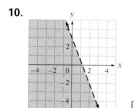   [9.4A]   **21.** $x \geq 4$ [9.3A]   **22.** For florist B to be more economical, there must be 5 or more residents in the nursing home. [9.3B]

23. The minimum length is 24 ft. [9.3B]   **24.** 32 is the smallest integer that satisfies the inequality. [9.2C]

25. 72 is the lowest score that the student can receive and still attain a minimum of 480 points. [9.2C]

CHAPTER 9 TEST

1.   [9.1C]   **2.** $\{x \mid x < 50, x \in \text{positive integers}\}$ [9.1B]   **3.** $A = \{4, 6, 8\}$ [9.1A]

4. $x \leq -3$ [9.3A]   **5.** $x > \dfrac{1}{8}$ [9.2A]   **6.**   [9.1C]   **7.** $x < -1$ [9.3A]

8. $\{x \mid x > -23, x \in \text{real numbers}\}$ [9.1B]   **9.** 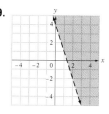   [9.4A]   **10.** 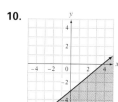   [9.4A]

Copyright © Houghton Mifflin Company. All rights reserved.

11. $A \cap B = \{12\}$ [9.1A] **12.** $x < -3$![number line from -5 to 5] [9.2A] **13.** $x \geq -\dfrac{40}{3}$ [9.2B]

14. $x < -\dfrac{22}{7}$ [9.3A] **15.** $x \geq 3$![number line from -5 to 5] [9.2B] **16.** $x \geq -4$ [9.3A] **17.** The child must grow 5 in. or more. [9.2C] **18.** The width must be less than or equal to 11 ft. [9.3B] **19.** The diameter must be between 0.0389 in. and 0.0395 in. [9.2C] **20.** The total value of the stock processed by the broker was less than or equal to $75,000. [9.3B]

CUMULATIVE REVIEW EXERCISES

1. $40a - 28$ [2.2D] **2.** $\dfrac{1}{8}$ [3.2A] **3.** 4 [3.3B] **4.** $-12a^7b^4$ [4.2B] **5.** $-\dfrac{1}{b^4}$ [4.4A]

6. $4x - 2 - \dfrac{4}{4x - 1}$ [4.5B] **7.** 0 [7.1D] **8.** $3a^2(3x + 1)(3x - 1)$ [5.4B] **9.** $\dfrac{1}{x + 2}$ [6.1C]

10. $\dfrac{18a}{(2a - 3)(a + 3)}$ [6.3B] **11.** $-\dfrac{5}{9}$ [6.5A] **12.** $C = S + Rt$ [6.7A] **13.** $-\dfrac{7}{3}$ [7.3B]

14. $y = -\dfrac{3}{2}x - \dfrac{3}{2}$ [7.4A] **15.** $(4, 1)$ [8.2A] **16.** $(1, -4)$ [8.3A] **17.** $A \cup B = \{-10, -2, 0, 1, 2\}$ [9.1A]

18. $\{x \mid x < 48, x \in \text{real numbers}\}$ [9.1B] **19.** ![number line from -5 to 5] [9.1C]

20. ![number line from -5 to 5] [9.2B] **21.** $x < -15$ [9.2B] **22.** $x > 2$ [9.3A]

23. $\{x \mid x \leq -26, x \in \text{integers}\}$ [9.2C] **24.** The maximum number of miles is 359 mi. [9.3B] **25.** There are an estimated 5000 fish in the lake. [6.6B] **26.** The angle measures are 65°, 35°, and 80°. [3.5B]

27.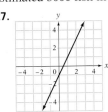

[7.2A]

28.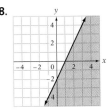

[9.4A]

Answers to Chapter 10 Selected Exercises

PREP TEST

1. -14 [1.1B] **2.** $-2x^2y - 4xy^2$ [2.2A] **3.** 14 [3.1C] **4.** $\dfrac{7}{5}$ [3.3A] **5.** x^6 [4.2A]

6. $x^2 + 2xy + y^2$ [4.3D] **7.** $4x^2 - 12x + 9$ [4.3D] **8.** $4 - 9v^2$ [4.3D] **9.** $a^2 - 25$ [4.3D] **10.** $\dfrac{x^2y^2}{9}$ [4.4A]

SECTION 10.1

3. 4 **5.** 7 **7.** $4\sqrt{2}$ **9.** $2\sqrt{2}$ **11.** $18\sqrt{2}$ **13.** $10\sqrt{10}$ **15.** $\sqrt{15}$ **17.** $\sqrt{29}$ **19.** $-54\sqrt{2}$ **21.** $3\sqrt{5}$
23. 0 **25.** $48\sqrt{2}$ **27.** 15.492 **29.** 16.971 **31.** 16 **33.** x^3 **35.** $y^7\sqrt{y}$ **37.** a^{10} **39.** x^2y^2 **41.** $2x^2$
43. $2x\sqrt{6}$ **45.** $2x^2\sqrt{15x}$ **47.** $7a^2b^4$ **49.** $3x^2y^3\sqrt{2xy}$ **51.** $2x^5y^3\sqrt{10xy}$ **53.** $4a^4b^5\sqrt{5a}$ **55.** $8ab\sqrt{b}$
57. x^3y **59.** $8a^2b^3\sqrt{5b}$ **61.** $6x^2y^3\sqrt{3y}$ **63.** $4x^3y\sqrt{2y}$ **65.** $5a + 20$ **67.** $2x^2 + 8x + 8$ **69.** $x + 2$
71. $y + 1$ **73. a.** The speed of the car was $12\sqrt{15}$ mph. **b.** 46 mph **75.** No. For example, let $a = 16$ and $b = 9$.
$\sqrt{a + b} = \sqrt{16 + 9} = \sqrt{25} = 5$. $\sqrt{a} + \sqrt{b} = \sqrt{16} + \sqrt{9} = 4 + 3 = 7$. **77.** No. 18 contains a perfect-square factor. $6\sqrt{2}$
79. a. 1 **b.** 3 **c.** $3\sqrt{3}$

SECTION 10.2

1. 2, 20, and 50 **5.** $3\sqrt{2}$ **7.** $-\sqrt{7}$ **9.** $-11\sqrt{11}$ **11.** $10\sqrt{x}$ **13.** $-2\sqrt{y}$ **15.** $-11\sqrt{3b}$ **17.** $2x\sqrt{2}$
19. $-3a\sqrt{3a}$ **21.** $-5\sqrt{xy}$ **23.** $8\sqrt{5}$ **25.** $8\sqrt{2}$ **27.** $15\sqrt{2} - 10\sqrt{3}$ **29.** $\sqrt{x}$ **31.** $-12x\sqrt{3}$

Copyright © Houghton Mifflin Company. All rights reserved.

33. $2xy\sqrt{x} - 3xy\sqrt{y}$ **35.** $-9x\sqrt{3x}$ **37.** $-13y^2\sqrt{2y}$ **39.** $4a^2b^2\sqrt{ab}$ **41.** $7\sqrt{2}$ **43.** $6\sqrt{x}$ **45.** $-3\sqrt{y}$
47. $-45\sqrt{2}$ **49.** $13\sqrt{3} - 12\sqrt{5}$ **51.** $32\sqrt{3} - 3\sqrt{11}$ **53.** $6\sqrt{x}$ **55.** $-34\sqrt{3x}$ **57.** $10a\sqrt{3b} + 10a\sqrt{5b}$
59. $-2xy\sqrt{3}$ **61.** $5\sqrt{2}$

SECTION 10.3

3. 5 **5.** 6 **7.** x **9.** x^3y^2 **11.** $3ab^6\sqrt{2a}$ **13.** $12a^4b\sqrt{b}$ **15.** $2 - \sqrt{6}$ **17.** $x - \sqrt{xy}$ **19.** $5\sqrt{2} - \sqrt{5x}$
21. $4 - 2\sqrt{10}$ **23.** $x - 6\sqrt{x} + 9$ **25.** $3a - 3\sqrt{ab}$ **27.** $10abc$ **29.** $-2 + 2\sqrt{5}$ **31.** $16 + 10\sqrt{2}$
33. $6x + 10\sqrt{x} - 4$ **35.** $15x - 22y\sqrt{x} + 8y^2$ **37.** $x - y$ **41.** 3 **43.** 4 **45.** $6x^2$ **47.** $2x^2\sqrt{2y}$
49. $4x\sqrt{y}$ **51.** $\dfrac{2\sqrt{3x}}{3y}$ **53.** $\dfrac{-1 + \sqrt{2}}{2}$ **55.** $-\dfrac{5\sqrt{7} + 15}{2}$ **57.** $-\dfrac{5\sqrt{3} + 9}{2}$ **59.** $3 + \sqrt{6}$ **61.** $\dfrac{-12 + 3\sqrt{2}}{7}$
63. $\dfrac{14 - 9\sqrt{2}}{17}$ **65.** $\sqrt{15} + 2\sqrt{5}$ **67.** $\dfrac{x\sqrt{y} + y\sqrt{x}}{x - y}$ **69.** $4\sqrt{6} + 12$

SECTION 10.4

1. 25 **3.** 144 **5.** 5 **7.** No solution **9.** -1 **11.** 1 **13.** 15 **15.** $\dfrac{7}{3}$ **17.** 2 **19.** 5 **21.** 4

23. The pitcher's mound is less than halfway between home plate and second base. **25.** The periscope must be 16.67 ft above the water. **27.** The height of the screen is 21.6 in. **29.** The distance of the child from the center is 18.75 ft. **31.** The perimeter of the triangle is 30 units. **35.** The area of the fountain is approximately 244.78 ft².

CHAPTER 10 REVIEW EXERCISES

1. 3 [10.3A] **2.** $9a^2\sqrt{2ab}$ [10.1B] **3.** 12 [10.1A] **4.** $3a\sqrt{2} + 2a\sqrt{3}$ [10.3A] **5.** $2\sqrt{6}$ [10.3B]
6. $-8\sqrt{2}$ [10.2A] **7.** 2 [10.3A] **8.** 1 [10.4A] **9.** $-x\sqrt{3} - x\sqrt{5}$ [10.3B] **10.** $-6\sqrt{30}$ [10.1A]
11. 20 [10.4A] **12.** $20\sqrt{3}$ [10.1A] **13.** $7x^2y^4$ [10.3B] **14.** No solution [10.4A]
15. $18a\sqrt{5b} + 5a\sqrt{b}$ [10.2A] **16.** $20\sqrt{10}$ [10.1A] **17.** $7x^2y\sqrt{15xy}$ [10.2A] **18.** $8y + 10\sqrt{5y} - 15$ [10.3A]
19. $26\sqrt{3x}$ [10.2A] **20.** No solution [10.4A] **21.** $\dfrac{8\sqrt{x} + 24}{x - 9}$ [10.3B] **22.** $36x^8y^5\sqrt{3xy}$ [10.1B]
23. $2y^4\sqrt{6}$ [10.1B] **24.** -1 [10.4A] **25.** $-6x^3y^2\sqrt{2y}$ [10.2A] **26.** $\dfrac{16\sqrt{a}}{a}$ [10.3B] **27.** The distance across the pond is approximately 43 ft. [10.4B] **28.** The explorer weighs 144 lb on the surface of Earth. [10.4B]
29. The depth of the water is 100 ft. [10.4B] **30.** The radius of the corner is 25 ft. [10.4B]

CHAPTER 10 TEST

1. $11x^4y$ [10.1B] **2.** $6x^2y\sqrt{y}$ [10.3A] **3.** $-5\sqrt{2}$ [10.2A] **4.** $3\sqrt{5}$ [10.1A] **5.** 9 [10.3B]
6. 25 [10.4A] **7.** $4a^2b^5\sqrt{2ab}$ [10.1B] **8.** $7ab\sqrt{a}$ [10.3B] **9.** $\sqrt{3} + 1$ [10.3B] **10.** $4x^2y^2\sqrt{5y}$ [10.3A]
11. 9 [10.4A] **12.** $21\sqrt{2y} - 12\sqrt{2x}$ [10.2A] **13.** $6x^3y\sqrt{2x}$ [10.1B] **14.** $y + 2\sqrt{y} - 15$ [10.3A]
15. $-2xy\sqrt{3xy} - 3xy\sqrt{xy}$ [10.2A] **16.** $\dfrac{17 - 8\sqrt{5}}{31}$ [10.3B] **17.** $a - \sqrt{ab}$ [10.3A] **18.** $5\sqrt{3}$ [10.1A]
19. The length of the pendulum is 7.30 ft. [10.4B] **20.** The rope should be secured about 7 ft from the base of the pole. [10.4B]

CUMULATIVE REVIEW EXERCISES

1. $-\dfrac{1}{12}$ [1.7A] **2.** $2x + 18$ [2.2D] **3.** $\dfrac{1}{13}$ [3.3B] **4.** $6x^5y^5$ [4.2A] **5.** $-2b^2 + 1 - \dfrac{1}{3b^2}$ [4.5A]

6. 1 [7.1D] **7.** $2a(a - 5)(a - 3)$ [5.2B] **8.** $\dfrac{1}{4(x + 1)}$ [6.1B] **9.** $\dfrac{x + 3}{x - 3}$ [6.3B] **10.** $\dfrac{5}{3}$ [6.5A]

11. $y = \dfrac{1}{2}x - 2$ [7.4A] **12.** $(1, 1)$ [8.2A] **13.** $(3, -2)$ [8.3A] **14.** $x \le -\dfrac{9}{2}$ [9.3A] **15.** $6\sqrt{3}$ [10.1A]
16. $-4\sqrt{2}$ [10.2A] **17.** $4ab\sqrt{2ab} - 5ab\sqrt{ab}$ [10.2A] **18.** $14a^5b^2\sqrt{2a}$ [10.3A] **19.** $3\sqrt{2} - x\sqrt{3}$ [10.3A]
20. 8 [10.3B] **21.** $-6 - 3\sqrt{5}$ [10.3B] **22.** 6 [10.4A] **23.** The cost of the book is $24.50. [3.2B]
24. 56 oz of water must be added. [3.6B] **25.** The numbers are 8 and 13. [5.5B] **26.** It would take the small pipe,

Copyright © Houghton Mifflin Company. All rights reserved.

working alone, 48 h.   [6.8A]   **27.**

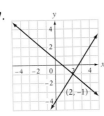

[8.1A]

28.

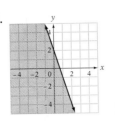

[9.4A]

29. The smaller integer is 40.   [10.4B]   **30.** The height of the building is 400 ft.   [10.4B]

Answers to Chapter 11 Selected Exercises

PREP TEST

1. 41   [2.1A]   **2.** $-\dfrac{1}{5}$ [3.2A]   **3.** $(x + 4)(x - 3)$ [5.2A]   **4.** $(2x - 3)^2$ [5.4A]   **5.** Yes   [5.4A]   **6.** 3   [6.5A]

7.

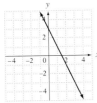

[7.2A]

8. $2\sqrt{7}$ [10.1A]   **9.** $|a|$ [10.1B]   **10.** 3.6 mi   [3.6C]

SECTION 11.1

1. $-3, 5$ **3.** $-\dfrac{5}{2}, \dfrac{1}{3}$ **5.** $-5, 3$ **7.** $1, 3$ **9.** $-2, -1$ **11.** 3   **13.** $-\dfrac{2}{3}, 0$ **15.** $-2, 5$ **17.** $\dfrac{2}{3}, 1$

19. $-3, \dfrac{1}{3}$ **21.** $\dfrac{2}{3}$ **23.** $-\dfrac{1}{2}, \dfrac{3}{2}$ **25.** $\dfrac{1}{2}$ **27.** $-3, 3$ **29.** $-\dfrac{1}{2}, \dfrac{1}{2}$ **31.** $-3, 5$ **33.** $1, 5$ **35.** $-6, 6$

37. $-1, 1$ **39.** $-\dfrac{7}{2}, \dfrac{7}{2}$ **41.** $-\dfrac{2}{3}, \dfrac{2}{3}$ **43.** $-\dfrac{3}{4}, \dfrac{3}{4}$ **45.** No real number solution   **47.** $-2\sqrt{6}, 2\sqrt{6}$ **49.** $-5, 7$

51. $-7, -3$ **53.** $-\dfrac{1}{3}, \dfrac{7}{3}$ **55.** $-\dfrac{12}{7}, -\dfrac{2}{7}$ **57.** $4 - 2\sqrt{5}, 4 + 2\sqrt{5}$ **59.** No real number solution

61. $-\dfrac{3}{4} - 2\sqrt{3}, -\dfrac{3}{4} + 2\sqrt{3}$ **63.** $-1, -\dfrac{\sqrt{6}}{3}, \dfrac{\sqrt{6}}{3}, 1$ **65.** The annual percentage rate is 8%.   **67.** Yes, because

$v \approx 73.5$ mph.

SECTION 11.2

1. $x^2 - 8x + 16, (x - 4)^2$ **3.** $x^2 + 5x + \dfrac{25}{4}, \left(x + \dfrac{5}{2}\right)^2$ **5.** $-3, 1$ **7.** $-2, 8$ **9.** 2   **11.** No real number

solution   **13.** $-4, -1$ **15.** $-8, 1$ **17.** $-2 - \sqrt{3}, -2 + \sqrt{3}$ **19.** $-3 - \sqrt{14}, -3 + \sqrt{14}$ **21.** $1 - \sqrt{2}, 1 + \sqrt{2}$

23. $\dfrac{-3 - \sqrt{13}}{2}, \dfrac{-3 + \sqrt{13}}{2}$ **25.** $1, 2$ **27.** $\dfrac{-1 - \sqrt{13}}{2}, \dfrac{-1 + \sqrt{13}}{2}$ **29.** $-5 - 4\sqrt{2}, -5 + 4\sqrt{2}$

31. $\dfrac{3 - \sqrt{29}}{2}, \dfrac{3 + \sqrt{29}}{2}$ **33.** $\dfrac{1 - \sqrt{17}}{2}, \dfrac{1 + \sqrt{17}}{2}$ **35.** No real number solution   **37.** $\dfrac{1}{2}, 1$ **39.** $-3, \dfrac{1}{2}$

41. $\dfrac{3}{2}, 2$ **43.** $-\dfrac{1}{2}, 1$ **45.** $-2, \dfrac{1}{3}$ **47.** $-2, -\dfrac{2}{3}$ **49.** $-\dfrac{3}{2}, \dfrac{1}{2}$ **51.** $-\dfrac{3}{2}, \dfrac{1}{3}$ **53.** $-\dfrac{1}{2}, \dfrac{4}{3}$ **55.** $-5.372, 0.372$

57. $-3.212, 1.712$ **59.** $-1.151, 0.651$ **61.** $1 - \sqrt{7}, 1 + \sqrt{7}$ **63.** $-1, 0$ **65.** $4 - \sqrt{3}, 4 + \sqrt{3}$ **67.** $0, 8$

69. $\dfrac{15 + 3\sqrt{17}}{2}$ **71.** The ball hits the ground approximately 4.81 s after it is hit.

SECTION 11.3

1. $-1, 5$ **3.** $-3, 5$ **5.** $-1, 3$ **7.** $-5, 1$ **9.** $-\dfrac{1}{2}, 1$ **11.** No real number solution   **13.** $0, 1$ **15.** $-\dfrac{3}{2}, \dfrac{3}{2}$

Copyright © Houghton Mifflin Company. All rights reserved.

17. $-\dfrac{5}{2}, \dfrac{3}{2}$ **19.** No real number solution   **21.** $1 - \sqrt{6}, 1 + \sqrt{6}$ **23.** $-3 - \sqrt{10}, -3 + \sqrt{10}$

25. $2 - \sqrt{13}, 2 + \sqrt{13}$ **27.** $\dfrac{-1 - \sqrt{2}}{3}, \dfrac{-1 + \sqrt{2}}{3}$ **29.** $-\dfrac{1}{2}$ **31.** No real number solution   **33.** $\dfrac{-4 - \sqrt{5}}{2}, \dfrac{-4 + \sqrt{5}}{2}$

35. $\dfrac{1 - 2\sqrt{3}}{2}, \dfrac{1 + 2\sqrt{3}}{2}$ **37.** $\dfrac{-5 - \sqrt{2}}{3}, \dfrac{-5 + \sqrt{2}}{3}$ **39.** $-3.690, 5.690$ **41.** $-1.690, 7.690$ **43.** $-1.089, 4.589$

45. $-2.118, 0.118$ **47.** $-0.905, 1.105$ **51. a.** False   **b.** False   **c.** False   **d.** True   **53.** $-4, -3$

55. No solution   **57.** $-1, 11$ **59.** The planes are 1000 mi apart after 2 h.

SECTION 11.4

1. Down   **3.** Up   **5.** 4   **7.** -5 **9.** -42 **11.**   **13.**

15. 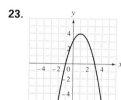   **17.** 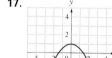   **19.** 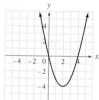   **21.**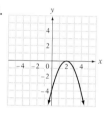

23.   **25.**   **27.** $(-6, 0), (1, 0); (0, -6)$ **29.** $(-6, 0); (0, 36)$

31. $(-2 - \sqrt{6}, 0), (-2 + \sqrt{6}, 0); (0, -2)$ **33.** No x-intercepts; $(0, 1)$ **35.** $\left(\dfrac{3}{2}, 0\right), (5, 0); (0, 15)$

37. $\left(\dfrac{-3 - \sqrt{33}}{6}, 0\right), \left(\dfrac{-3 + \sqrt{33}}{6}, 0\right); (0, 2)$ **39.** $y = 3x^2 + 6x + 5$ **41.** $y = 3x^2 - 6x$ **43.** $(-1, 0), (0, 0), (5, 0)$

45. $(-3, 0), (-1, 0), (1, 0)$

SECTION 11.5

1. The length is 4 m. The height is 10 m.   **3.** The width is 4 ft. The length is 6 ft.   **5.** The width is 50 ft. The length is 100 ft.   **7.** The hang time of the football is approximately 5.5 s.   **9.** The maximum velocity is 78 ft/s.   **11.** The large pizza has a radius of 7 in.   **13.** The first computer can solve the equation in 35 min. The second computer can solve the equation in 14 min.   **15.** Using the first engine it would take 12 h. Using the second engine it would take 6 h.   **17.** The rate of the plane in calm air is 100 mph.   **19.** The 16-inch pizza should cost $24.

CHAPTER 11 REVIEW EXERCISES

1. $-\dfrac{7}{2}, \dfrac{4}{3}$ [11.1A]   **2.** $-\dfrac{5}{7}, \dfrac{5}{7}$ [11.1B]   **3.** $-6, 4$ [11.2A]   **4.** $-6, 1$ [11.3A]   **5.** $-4, \dfrac{3}{2}$ [11.2A]

6. $\dfrac{5}{12}, 2$ [11.1A]   **7.** $-2 - 2\sqrt{6}, -2 + 2\sqrt{6}$ [11.1B]   **8.** $1, \dfrac{3}{2}$ [11.3A]   **9.** $-\dfrac{1}{2}, -\dfrac{1}{3}$ [11.1A]

Copyright © Houghton Mifflin Company. All rights reserved.

10. No real number solution [11.1B]   **11.** $2 - \sqrt{3}, 2 + \sqrt{3}$ [11.2A]   **12.** $\dfrac{3 - \sqrt{29}}{2}, \dfrac{3 + \sqrt{29}}{2}$ [11.3A]

13. No real number solution [11.2A]   **14.** $-10, -7$ [11.1A]   **15.** $-1, 2$ [11.1B]

16. $\dfrac{-4 - \sqrt{23}}{2}, \dfrac{-4 + \sqrt{23}}{2}$ [11.2A]   **17.** No real number solution [11.3A]   **18.** $-2, -\dfrac{1}{2}$ [11.3A]

19.
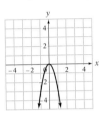
[11.4A]

20.
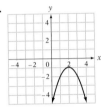
[11.4A]

21.
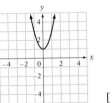
[11.4A]

22.
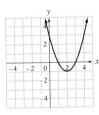
[11.4A]

23.
[11.4A]

24. $(-3, 0), (5, 0); (0, -15)$ [11.4A]   **25.** The rate of the hawk in calm air is 75 mph. [11.5A]

CHAPTER 11 TEST

1. $-1, 6$ [11.1A]   **2.** $-4, \dfrac{5}{3}$ [11.1A]   **3.** $0, 10$ [11.1B]   **4.** $-4 - 2\sqrt{5}, -4 + 2\sqrt{5}$ [11.1B]

5. $-2 - 2\sqrt{5}, -2 + 2\sqrt{5}$ [11.2A]   **6.** $\dfrac{-3 - \sqrt{41}}{2}, \dfrac{-3 + \sqrt{41}}{2}$ [11.2A]   **7.** $\dfrac{3 - \sqrt{7}}{2}, \dfrac{3 + \sqrt{7}}{2}$ [11.2A]

8. $\dfrac{-4 - \sqrt{22}}{2}, \dfrac{-4 + \sqrt{22}}{2}$ [11.2A]   **9.** $-2 - \sqrt{2}, -2 + \sqrt{2}$ [11.3A]   **10.** $\dfrac{3 - \sqrt{33}}{2}, \dfrac{3 + \sqrt{33}}{2}$ [11.3A]

11. $-\dfrac{1}{2}, 3$ [11.3A]   **12.** $\dfrac{1 - \sqrt{13}}{6}, \dfrac{1 + \sqrt{13}}{6}$ [11.3A]   **13.**
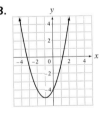
[11.4A]

14. $(-4, 0), (3, 0); (0, -12)$ [11.4A]   **15.** The width is 5 ft. The length is 8 ft. [11.5A]

16. The rate of the boat in calm water is 11 mph. [11.5A]

CUMULATIVE REVIEW EXERCISES

1. $-28x + 27$ [2.2D]   **2.** $\dfrac{3}{2}$ [3.1C]   **3.** 3 [3.3B]   **4.** $-12a^8 b^4$ [4.2B]   **5.** $x + 2 - \dfrac{4}{x - 2}$ [4.5B]

6. $x(3x - 4)(x + 2)$ [5.3A/5.3B]   **7.** $\dfrac{9x^2(x - 2)^2}{(2x - 3)^2}$ [6.1C]   **8.** $\dfrac{x + 2}{2(x + 1)}$ [6.3B]   **9.** $\dfrac{x - 4}{2x + 5}$ [6.4A]

10. $(3, 0); (0, -4)$ [7.3A]   **11.** $y = -\dfrac{4}{3}x - 2$ [7.4A]   **12.** $(2, 1)$ [8.2A]   **13.** $(2, -2)$ [8.3A]   **14.** $x > \dfrac{1}{9}$ [9.3A]

15. $a - 2$ [10.3A]   **16.** $6ab\sqrt{a}$ [10.3B]   **17.** $\dfrac{-6 + 5\sqrt{3}}{13}$ [10.3B]   **18.** 5 [10.4A]   **19.** $\dfrac{1}{3}, \dfrac{5}{2}$ [11.1A]

20. $5 - 3\sqrt{2}, 5 + 3\sqrt{2}$ [11.1B]   **21.** $\dfrac{-7 - \sqrt{13}}{6}, \dfrac{-7 + \sqrt{13}}{6}$ [11.2A]   **22.** $-\dfrac{1}{2}, 2$ [11.3A]   **23.** The cost of the

mixture is $2.25 per pound. [3.6A]   **24.** 250 additional shares are required. [6.6B]   **25.** The rate of the plane in still

air is 200 mph. The rate of the wind is 40 mph. [8.4A]   **26.** The score on the last test must be 77 or better. [9.2C]

Copyright © Houghton Mifflin Company. All rights reserved.

17. $-\dfrac{5}{2}, \dfrac{3}{2}$ **19.** No real number solution   **21.** $1 - \sqrt{6}, 1 + \sqrt{6}$ **23.** $-3 - \sqrt{10}, -3 + \sqrt{10}$

25. $2 - \sqrt{13}, 2 + \sqrt{13}$ **27.** $\dfrac{-1 - \sqrt{2}}{3}, \dfrac{-1 + \sqrt{2}}{3}$ **29.** $-\dfrac{1}{2}$ **31.** No real number solution   **33.** $\dfrac{-4 - \sqrt{5}}{2}, \dfrac{-4 + \sqrt{5}}{2}$

35. $\dfrac{1 - 2\sqrt{3}}{2}, \dfrac{1 + 2\sqrt{3}}{2}$ **37.** $\dfrac{-5 - \sqrt{2}}{3}, \dfrac{-5 + \sqrt{2}}{3}$ **39.** $-3.690, 5.690$ **41.** $-1.690, 7.690$ **43.** $-1.089, 4.589$

45. $-2.118, 0.118$ **47.** $-0.905, 1.105$ **51. a.** False   **b.** False   **c.** False   **d.** True   **53.** $-4, -3$

55. No solution   **57.** $-1, 11$ **59.** The planes are 1000 mi apart after 2 h.

SECTION 11.4

1. Down   **3.** Up   **5.** 4   **7.** -5 **9.** -42 **11.** 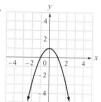   **13.**

15. 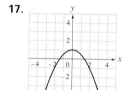   **17.** 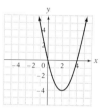   **19.**   **21.**

23. 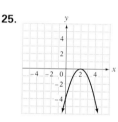   **25.**   **27.** $(-6, 0), (1, 0); (0, -6)$ **29.** $(-6, 0); (0, 36)$

31. $(-2 - \sqrt{6}, 0), (-2 + \sqrt{6}, 0); (0, -2)$ **33.** No x-intercepts; $(0, 1)$ **35.** $\left(\dfrac{3}{2}, 0\right), (5, 0); (0, 15)$

37. $\left(\dfrac{-3 - \sqrt{33}}{6}, 0\right), \left(\dfrac{-3 + \sqrt{33}}{6}, 0\right); (0, 2)$ **39.** $y = 3x^2 + 6x + 5$ **41.** $y = 3x^2 - 6x$ **43.** $(-1, 0), (0, 0), (5, 0)$

45. $(-3, 0), (-1, 0), (1, 0)$

SECTION 11.5

1. The length is 4 m. The height is 10 m.   **3.** The width is 4 ft. The length is 6 ft.   **5.** The width is 50 ft. The length is 100 ft.   **7.** The hang time of the football is approximately 5.5 s.   **9.** The maximum velocity is 78 ft/s.   **11.** The large pizza has a radius of 7 in.   **13.** The first computer can solve the equation in 35 min. The second computer can solve the equation in 14 min.   **15.** Using the first engine it would take 12 h. Using the second engine it would take 6 h.   **17.** The rate of the plane in calm air is 100 mph.   **19.** The 16-inch pizza should cost $24.

CHAPTER 11 REVIEW EXERCISES

1. $-\dfrac{7}{2}, \dfrac{4}{3}$ [11.1A]   **2.** $-\dfrac{5}{7}, \dfrac{5}{7}$ [11.1B]   **3.** $-6, 4$ [11.2A]   **4.** $-6, 1$ [11.3A]   **5.** $-4, \dfrac{3}{2}$ [11.2A]

6. $\dfrac{5}{12}, 2$ [11.1A]   **7.** $-2 - 2\sqrt{6}, -2 + 2\sqrt{6}$ [11.1B]   **8.** $1, \dfrac{3}{2}$ [11.3A]   **9.** $-\dfrac{1}{2}, -\dfrac{1}{3}$ [11.1A]

Copyright © Houghton Mifflin Company. All rights reserved.

10. No real number solution   [11.1B]   **11.** $2 - \sqrt{3}, 2 + \sqrt{3}$ [11.2A]   **12.** $\dfrac{3 - \sqrt{29}}{2}, \dfrac{3 + \sqrt{29}}{2}$ [11.3A]

13. No real number solution   [11.2A]   **14.** $-10, -7$ [11.1A]   **15.** $-1, 2$ [11.1B]

16. $\dfrac{-4 - \sqrt{23}}{2}, \dfrac{-4 + \sqrt{23}}{2}$ [11.2A]   **17.** No real number solution   [11.3A]   **18.** $-2, -\dfrac{1}{2}$ [11.3A]

19. 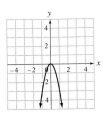   [11.4A]

20. 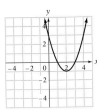   [11.4A]

21. 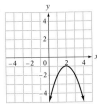   [11.4A]

22. [11.4A]

23. [11.4A]

24. $(-3, 0), (5, 0); (0, -15)$ [11.4A]   **25.** The rate of the hawk in calm air is 75 mph.   [11.5A]

CHAPTER 11 TEST

1. $-1, 6$ [11.1A]   **2.** $-4, \dfrac{5}{3}$ [11.1A]   **3.** $0, 10$ [11.1B]   **4.** $-4 - 2\sqrt{5}, -4 + 2\sqrt{5}$ [11.1B]

5. $-2 - 2\sqrt{5}, -2 + 2\sqrt{5}$ [11.2A]   **6.** $\dfrac{-3 - \sqrt{41}}{2}, \dfrac{-3 + \sqrt{41}}{2}$ [11.2A]   **7.** $\dfrac{3 - \sqrt{7}}{2}, \dfrac{3 + \sqrt{7}}{2}$ [11.2A]

8. $\dfrac{-4 - \sqrt{22}}{2}, \dfrac{-4 + \sqrt{22}}{2}$ [11.2A]   **9.** $-2 - \sqrt{2}, -2 + \sqrt{2}$ [11.3A]   **10.** $\dfrac{3 - \sqrt{33}}{2}, \dfrac{3 + \sqrt{33}}{2}$ [11.3A]

11. $-\dfrac{1}{2}, 3$ [11.3A]   **12.** $\dfrac{1 - \sqrt{13}}{6}, \dfrac{1 + \sqrt{13}}{6}$ [11.3A]   **13.**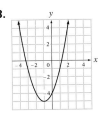

[11.4A]

14. $(-4, 0), (3, 0); (0, -12)$ [11.4A]   **15.** The width is 5 ft. The length is 8 ft.   [11.5A]

16. The rate of the boat in calm water is 11 mph.   [11.5A]

CUMULATIVE REVIEW EXERCISES

1. $-28x + 27$ [2.2D]   **2.** $\dfrac{3}{2}$ [3.1C]   **3.** 3 [3.3B]   **4.** $-12a^8b^4$ [4.2B]   **5.** $x + 2 - \dfrac{4}{x - 2}$ [4.5B]

6. $x(3x - 4)(x + 2)$ [5.3A/5.3B]   **7.** $\dfrac{9x^2(x - 2)^2}{(2x - 3)^2}$ [6.1C]   **8.** $\dfrac{x + 2}{2(x + 1)}$ [6.3B]   **9.** $\dfrac{x - 4}{2x + 5}$ [6.4A]

10. $(3, 0); (0, -4)$ [7.3A]   **11.** $y = -\dfrac{4}{3}x - 2$ [7.4A]   **12.** $(2, 1)$ [8.2A]   **13.** $(2, -2)$ [8.3A]   **14.** $x > \dfrac{1}{9}$ [9.3A]

15. $a - 2$ [10.3A]   **16.** $6ab\sqrt{a}$ [10.3B]   **17.** $\dfrac{-6 + 5\sqrt{3}}{13}$ [10.3B]   **18.** 5 [10.4A]   **19.** $\dfrac{1}{3}, \dfrac{5}{2}$ [11.1A]

20. $5 - 3\sqrt{2}, 5 + 3\sqrt{2}$ [11.1B]   **21.** $\dfrac{-7 - \sqrt{13}}{6}, \dfrac{-7 + \sqrt{13}}{6}$ [11.2A]   **22.** $-\dfrac{1}{2}, 2$ [11.3A]   **23.** The cost of the mixture is $2.25 per pound.   [3.6A]   **24.** 250 additional shares are required.   [6.6B]   **25.** The rate of the plane in still air is 200 mph. The rate of the wind is 40 mph.   [8.4A]   **26.** The score on the last test must be 77 or better.   [9.2C]

Copyright © Houghton Mifflin Company. All rights reserved.

27. The middle integer can be -5 or 5. [11.5A]   **28.** The rate for the last 8 mi is 4 mph. [11.5A]

29. 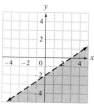 [9.4A]   **30.** [11.4A]

FINAL EXAM

1. -3 [1.1B]   **2.** -6 [1.2B]   **3.** -256 [1.4A]   **4.** -11 [1.4B]   **5.** $-\dfrac{15}{2}$ [2.1A]   **6.** $9x + 6y$ [2.2A]

7. $6z$ [2.2B]   **8.** $16x - 52$ [2.2D]   **9.** -50 [3.1C]   **10.** -3 [3.3B]   **11.** 12.5% [1.4B]   **12.** 15.2 [3.1D]

13. $-3x^2 - 3x + 8$ [4.1B]   **14.** $81x^4y^{12}$ [4.2B]   **15.** $6x^3 + 7x^2 - 7x - 6$ [4.3B]   **16.** $-\dfrac{x^4y}{2}$ [4.4A]

17. $\dfrac{3x}{y} - 4x^2 - \dfrac{5}{x}$ [4.5A]   **18.** $5x - 12 + \dfrac{23}{x + 2}$ [4.5B]   **19.** $\dfrac{4y^6}{x^6}$ [4.4A]   **20.** $\dfrac{3}{4}$ [7.1D]

21. $(x - 6)(x + 1)$ [5.2A]   **22.** $(3x + 2)(2x - 3)$ [5.3A/5.3B]   **23.** $4x(2x - 1)(x - 3)$ [5.4B]

24. $(5x + 4)(5x - 4)$ [5.4A]   **25.** $2(a + 3)(4 - x)$ [5.1B]   **26.** $3y(5 + 2x)(5 - 2x)$ [5.4B]   **27.** $\dfrac{1}{2}, 3$ [5.5A]

28. $\dfrac{2(x + 1)}{x - 1}$ [6.1B]   **29.** $\dfrac{-3x^2 + x - 25}{(x + 3)(2x - 5)}$ [6.3B]   **30.** $\dfrac{x^2 - 2x}{x - 1}$ [6.4A]   **31.** 2 [6.5A]   **32.** $a = b$ [6.7A]

33. $\dfrac{2}{3}$ [7.3B]   **34.** $y = -\dfrac{2}{3}x - 2$ [7.4A]   **35.** $(6, 17)$ [8.2A]   **36.** $(2, -1)$ [8.3A]   **37.** $x \le -3$ [9.2B]

38. $y \ge \dfrac{5}{2}$ [9.3A]   **39.** $7x^3$ [10.1B]   **40.** $38\sqrt{3a}$ [10.2A]   **41.** $\sqrt{15} + 2\sqrt{3}$ [10.3B]   **42.** 2 [10.4A]

43. $-1, \dfrac{4}{3}$ [11.1A]   **44.** $\dfrac{1 - \sqrt{5}}{4}, \dfrac{1 + \sqrt{5}}{4}$ [11.3A]   **45.** $2x + 3(x - 2), 5x - 6$ [2.3B]   **46.** The original value

is $3000. [3.1D]   **47.** The markup rate is 65%. [3.2B]   **48.** $6000 must be invested at 11%. [8.2B]   **49.** The cost

for the mixture is $3 per pound. [3.6A]   **50.** The percent concentration of acid in the mixture is 36%. [3.6B]

51. The distance traveled in the first hour was 215 km. [3.6C]   **52.** The angles measure 50°, 60°, and 70°. [3.5B]

53. The middle integer can be -4 or 4. [11.5A]   **54.** The width is 5 m. The length is 10 m. [5.5B]   **55.** 16 oz of dye

are required. [6.6B]   **56.** Working together, it would take them 36 min or 0.6 h. [6.8A]   **57.** The rate of the boat in

calm water is 15 mph. The rate of the current is 5 mph. [8.4A]   **58.** The rate of the wind is 25 mph. [11.5A]

59. [7.3C]   **60.** 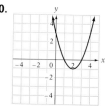 [11.4A]

Copyright © Houghton Mifflin Company. All rights reserved.

Glossary

abscissa The first number in an ordered pair. It measures a horizontal distance and is also called the first coordinate. [7.1]

absolute value of a number The distance of the number from zero on the number line. [1.1]

acute angle An angle whose measure is between $0°$ and $90°$. [3.5]

addend In addition, a number being added. [1.2]

addition The process of finding the total of two numbers. [1.2]

addition method An algebraic method of finding an exact solution of a system of linear equations. [8.3]

additive inverses Numbers that are the same distance from zero on the number line, but on opposite sides; also called opposites. [1.1/2.2]

adjacent angles Two angles that share a common side. [3.5]

alternate exterior angles Two angles that are on opposite sides of the transversal and outside the parallel lines. [3.5]

alternate interior angles Two angles that are on opposite sides of the transversal and between the parallel lines. [3.5]

analytic geometry Geometry in which a coordinate system is used to study the relationships between variables. [7.1]

angle Figure formed when two rays start from the same point. [1.8]

arithmetic mean of values Average determined by calculating the sum of the values and then dividing that result by the number of values. [1.3]

axes The two number lines that form a rectangular coordinate system; also called coordinate axes. [7.1]

base In exponential notation, the factor that is multiplied the number of times shown by the exponent. [1.4]

basic percent equation Percent times base equals amount. [3.1]

binomial A polynomial of two terms. [4.1]

binomial factor A factor that has two terms. [5.1]

center of a circle The point from which all points on the circle are equidistant. [1.8]

circle Plane figure in which all points are the same distance from its center. [1.8]

circumference The perimeter of a circle. [1.8]

clearing denominators Removing denominators from an equation that contains fractions by multiplying each side of the equation by the LCM of the denominators. [3.2/6.5]

coefficient The number part of a variable term. [2.1]

combining like terms Using the Distributive Property to add the coefficients of like variable terms; adding like terms of a variable expression. [2.2]

complementary angles Two angles whose sum is $90°$. [1.8]

completing the square Adding to a binomial the constant term that makes it a perfect-square trinomial. [11.2]

complex fraction A fraction whose numerator or denominator contains one or more fractions. [6.4]

composite number A natural number greater than 1 that is not a prime number. [1.5]

conjugates Binomial expressions that differ only in the sign of a term. The expressions $a + b$ and $a - b$ are conjugates. [10.3]

consecutive even integers Even integers that follow one another in order. [3.4]

consecutive integers Integers that follow one another in order. [3.4]

consecutive odd integers Odd integers that follow one another in order. [3.4]

constant term A term that includes no variable part; also called a constant. [2.1]

coordinate axes The two number lines that form a rectangular coordinate system; also simply called axes. [7.1]

coordinates of a point The numbers in an ordered pair that is associated with a point. [7.1]

corresponding angles Two angles that are on the same side of the transversal and are both acute angles or are both obtuse angles. [3.5]

cost The price that a business pays for a product. [3.2]

decimal notation Notation in which a number consists of a whole-number part, a decimal point, and a decimal part. [1.6]

degree A unit used to measure angles. [1.8]

degree of a polynomial in one variable The largest exponent that appears on the variable. [4.1]

dependent system A system of equations that has an infinite number of solutions. [8.1]

dependent variable In a function, the variable whose value depends on the value of another variable known as the independent variable. [7.1]

descending order The terms of a polynomial in one variable arranged so that the exponents on the variable decrease from left to right. The polynomial $9x^5 - 2x^4 + 7x^3 + x^2 - 8x + 1$ is in descending order. [4.1]

diameter Line segment across a circle that passes through the circle's center. [1.8]

difference of two squares A polynomial of the form $a^2 - b^2$. [5.4]

discount The amount by which a retailer reduces the regular price of a product for a promotional sale. [3.2]

discount rate The percent of the regular price that the discount represents. [3.2]

domain The set of first coordinates of the ordered pairs in a relation. [7.1]

double root The two equal roots of a quadratic equation, which occurs when the discriminant $b^2 - 4ac$ equals zero. [11.1]

Copyright © Houghton Mifflin Company. All rights reserved.

element of a set One of the objects in a set. [1.1/9.1]

empty set The set that contains no elements; also called the null set. [9.1]

equation A statement of the equality of two mathematical expressions. [3.1]

equilateral triangle A triangle in which all three sides are of equal length. [1.8]

equivalent equations Equations that have the same solution. [3.1]

evaluating a function Replacing x in $f(x)$ with some value and then simplifying the numerical expression that results. [7.1]

evaluating a variable expression Replacing each variable by its value and then simplifying the resulting numerical expression. [2.1]

even integer An integer that is divisible by 2. [3.4]

exponent In exponential notation, the elevated number that indicates how many times the base occurs in the multiplication. [1.4]

exponential form The expression 2^5 is in exponential form. Compare *factored form*. [1.4]

exterior angle of a triangle Angle adjacent to an interior angle of a triangle. [3.5]

factor by grouping Process of grouping and factoring terms in a polynomial in such a way that a common binomial factor is found. [5.1]

factor completely Refers to writing a polynomial as a product of factors that are nonfactorable over the integers. [5.2]

factor of a number In multiplication, a number being multiplied. [1.3]

factor a polynomial To write the polynomial as a product of other polynomials. [5.1]

factor a trinomial of the form $x^2 + bx + c$ To express the trinomial as the product of two binomials. [5.2]

factored form The expression $2 \cdot 2 \cdot 2 \cdot 2 \cdot 2$ is in factored form. Compare *exponential form*. [1.4]

first coordinate The first number in an ordered pair. It measures a horizontal distance and is also called the abscissa. [7.1]

first-degree equation in two variables An equation of the form $y = mx + b$, where m is the coefficient and b is a constant; also called a linear equation in two variables or a linear function. [7.2]

FOIL A method of finding the product of two binomials; the letters stand for First, Outer, Inner, and Last. [4.3]

formula A literal equation that states rules about measurements. [6.7]

function A relation in which no two ordered pairs that have the same first coordinate have different second coordinates. [7.1]

functional notation A function designated by $f(x)$, which is the value of the function at x. [7.1]

graph a point in the plane To place a dot at the location given by the ordered pair; also called plotting a point in the plane. [7.1]

graph of a relation The graph of the ordered pairs that belong to the relation. [7.1]

graph of an equation in two variables A graph of the ordered-pair solutions of the equation. [7.2]

graph of an integer A heavy dot directly above that number on the number line. [1.1]

graph of an ordered pair The dot drawn at the coordinates of the point in the plane. [7.1]

greater than A number a is greater than another number b, written $a > b$, if a is to the right of b on the number line. [1.1]

greater than or equal to The symbol $\geq$ means "is greater than or equal to." [1.1]

greatest common factor (GCF) The greatest common factor of two or more integers is the greatest integer that is a factor of all the integers. [1.5]

greatest common factor (GCF) of two or more monomials The product of the GCF of the coefficients and the common variable factors. [5.1]

half-plane The solution set of an inequality in two variables. [9.4]

hypotenuse In a right triangle, the side opposite the 90° angle. [10.4]

inconsistent system A system of equations that has no solution. [8.1]

independent system A system of equations that has one solution. [8.1]

independent variable In a function, the variable that varies independently and whose value determines the value of the dependent variable. [7.1]

inequality An expression that contains the symbol $>$, $<$, $\geq$ (is greater than or equal to), or $\leq$ (is less than or equal to). [9.1]

integers The numbers $\ldots$, -3, -2, -1, 0, 1, 2, 3, $\ldots$. [1.1]

interior angle of a triangle Angle within the region enclosed by a triangle. [3.5]

intersecting lines Lines that cross at a point in a plane. [1.8]

intersection of sets A and B The set that contains the elements that are common to both A and B. [9.1]

irrational number The decimal representation of an irrational number never repeats or terminates and can only be approximated. [1.6/10.1]

isosceles triangle A triangle that has two equal angles and two equal sides. [1.8]

least common denominator The smallest number that is a multiple of each denominator in question. [1.6]

least common multiple (LCM) The LCM of two or more numbers is the smallest number that contains the prime factorization of each number. [1.5]

least common multiple (LCM) of two or more polynomials The polynomial of least degree that contains all the factors of each polynomial. [6.2]

legs The sides opposite the hypotenuse in a right triangle. [10.4]

Copyright © Houghton Mifflin Company. All rights reserved.

less than A number a is less than another number b, written $a < b$, if a is to the left of b on the number line. [1.1]

less than or equal to The symbol $\leq$ means "is less than or equal to". [1.1]

like terms Terms of a variable expression that have the same variable part. [2.2]

line Having no width, it extends indefinitely in two directions in a plane. [1.8]

line of best fit A line drawn to approximate data that are graphed as points in a coordinate system. [7.4]

line segment Part of a line that has two endpoints. [1.8]

linear equation in two variables An equation of the form $y = mx + b$, where m and b are constants; also called a linear function or a first-degree equation in two variables. [7.2]

linear function An equation of the form $y = mx + b$, where m and b are constants; also called a linear equation in two variables or a first-degree equation in two variables. [7.2]

linear model A first-degree equation that is used to describe a relationship between quantities. [7.4]

literal equation An equation that contains more than one variable. [6.7]

markdown The amount by which a retailer reduces the regular price of a product for a promotional sale. [3.2]

markup The difference between selling price and cost. [3.2]

markup rate The percent of retailer's cost that the markup represents. [3.2]

monomial A number, a variable, or a product of numbers and variables; a polynomial of one term. [4.1]

moving average The arithmetic mean of the changes in the value of a stock for a given number of days. [1.3]

multiplicative inverse The reciprocal of a number. [2.2]

natural numbers The numbers 1, 2, 3, …. [1.1]

negative integers The numbers …, -4, -3, -2, -1. [1.1]

negative slope A property of a line that slants downward to the right. [7.3]

nonfactorable over the integers A polynomial that does not factor using only integers. [5.2]

null set The set that contains no elements; also called the empty set. [9.1]

numerical coefficient The number part of a variable term. When the numerical coefficient is 1 or -1, the 1 is usually not written. [2.1]

obtuse angle An angle whose measure is between 90° and 180°. [3.5]

odd integer An integer that is not divisible by 2. [3.4]

opposite of a polynomial The polynomial created when the sign of each term of the original polynomial is changed. [4.1]

opposites Numbers that are the same distance from zero on the number line, but on opposite sides; also called additive inverses. [1.1]

ordered pair Pair of numbers of the form that can be used to identify a point in the plane determined by the axes of a rectangular coordinate system. [7.1]

Order of Operations Agreement A set of rules that tell us in what order to perform the operations that occur in a numerical expression. [1.4]

ordinate The second number in an ordered pair. It measures a vertical distance and is also called the second coordinate. [7.1]

origin The point of intersection of the two coordinate axes that form a rectangular coordinate system. [7.1]

parabola The graph of a quadratic equation in two variables. [11.4]

parallel lines Lines that never meet; the distance between them is always the same. Parallel lines have the same slope. [1.8/7.3]

parallelogram Four-sided plane figure with opposite sides parallel. [1.8]

percent Parts of 100. [1.7]

perfect square The square of an integer. [10.1]

perfect-square trinomial A trinomial that is a product of a binomial and itself. [5.4]

perimeter The distance around a plane geometric figure. [1.8]

perpendicular lines Intersecting lines that form right angles. [1.8]

plane Flat surface that extends indefinitely. [1.8/7.1]

plane figure Figure that lies entirely in a plane. [1.8]

plot a point in the plane To place a dot at the location given by the ordered pair; to graph a point in the plane. [7.1]

point-slope formula If (x_1, y_1) is a point on a line with slope m, then $y - y_1 = m(x - x_1)$. [7.4]

polynomial A variable expression in which the terms are monomials. [4.1]

positive integers The integers, 1, 2, 3, 4, …. [1.1]

positive slope A property of a line that slants upward to the right. [7.3]

prime factorization Expressing a number as a product of its prime factors. [1.5]

prime number Number whose only factors are 1 and the number. [1.5]

prime polynomial A polynomial that is nonfactorable over the integers. [5.2]

principal square root The positive square root of a number. [10.1]

product In multiplication, the result of multiplying two numbers. [1.3]

proportion An equation that states the equality of two ratios or rates. [6.6]

Pythagorean Theorem The square of the hypotenuse of a right triangle is equal to the sum of the squares of the two legs. [10.4]

Copyright © Houghton Mifflin Company. All rights reserved.

quadrant One of the four regions into which the two axes of a rectangular coordinate system divide the plane. [7.1]

quadratic equation An equation of the form $ax^2 + bx + c = 0$, where a, b, and c are constants and a is not equal to zero; also called a second-degree equation. [5.5/11.1]

quadratic equation in two variables An equation of the form $y = ax^2 + bx + c$, where a is not equal to zero. [11.4]

quadratic function A quadratic function is given by $f(x) = ax^2 + bx + c$, where a is not equal to zero. [11.4]

radical equation An equation that contains a variable expression in a radicand. [10.4]

radical sign The symbol $\sqrt{}$, which is used to indicate the positive, or principal, square root of a number. [10.1]

radicand In a radical expression, the expression under the radical sign. [10.1]

radius Line segment from the center of a circle to a point on the circle. [1.8]

range The set of second coordinates of the ordered pairs in a relation. [7.1]

rate The quotient of two quantities that have different units. [6.6]

rate of work That part of a task that is completed in one unit of time. [6.8]

ratio The quotient of two quantities that have the same unit. [6.6]

rational expression A fraction in which the numerator or denominator is a polynomial. [6.1]

rational number A number that can be written in the form a/b, where a and b are integers and b is not equal to zero. [1.6]

rationalizing the denominator The procedure used to remove a radical from the denominator of a fraction. [10.3]

ray Line that starts at a point and extends indefinitely in one direction. [1.8]

real numbers The rational numbers and the irrational numbers. [1.6]

reciprocal of a fraction Fraction that results when the numerator and denominator of a fraction are interchanged. [1.7]

reciprocal of a rational expression A rational expression in which the numerator and denominator have been interchanged. [6.1]

rectangle Parallelogram that has four right angles. [1.8]

rectangular coordinate system System formed by two number lines, one horizontal and one vertical, that intersect at the zero point of each line. [7.1]

relation Any set of ordered pairs. [7.1]

repeating decimal Decimal that is formed when dividing the numerator of its fractional counterpart by the denominator results in a decimal part wherein a block of digits repeats infinitely. [1.6]

right angle An angle whose measure is 90°. [1.8]

right triangle A triangle that contains a 90° angle. [10.4]

roster method Method of writing a set by enclosing a list of the elements in braces. [1.1/9.1]

scatter diagram A graph of collected data as points in a coordinate system. [7.4]

scientific notation Notation in which each number is expressed as the product of two factors, one a number between 1 and 10 and the other a power of ten. [4.4]

second coordinate The second number in an ordered pair. It measures a vertical distance and is also called the ordinate. [7.1]

second-degree equation An equation of the form $ax^2 + bx + c = 0$, where a, b, and c are constants and a is not equal to zero; also called a quadratic equation. [11.1]

selling price The price for which a business sells a product to a customer. [3.2]

set A collection of objects. [1.1/9.1]

set-builder notation A method of designating a set that makes use of a variable and a certain property that only elements of that set possess. [9.1]

similar objects Similar objects have the same shape but not necessarily the same size. [6.6]

simplest form of a fraction A fraction in which the numerator and denominator have no common factors other than 1. [1.6]

simplest form of a rational expression A rational expression is in simplest form when the numerator and denominator have no common factors. [6.1]

slope The measure of the slant of a line. The symbol for slope is m. [7.3]

slope-intercept form The slope-intercept form of an equation of a straight line is $y = mx + b$. [7.3]

solid An object that exists in space. [1.8]

solution of a system of equations in two variables An ordered pair that is a solution of each equation of the system. [8.1]

solution of an equation A number that, when substituted for the variable, results in a true equation. [3.1]

solution of an equation in two variables An ordered pair whose coordinates make the equation a true statement. [7.1]

solution set of an inequality A set of numbers, each element of which, when substituted for the variable, results in a true inequality. [9.1]

solving an equation Finding a solution of the equation. [3.1]

square Rectangle with four equal sides. [1.8]

square root A square root of a positive number x is a number a for which $a^2 = x$. [10.1]

standard form A quadratic equation is in standard form when the polynomial is in descending order and equal to zero. $ax^2 + bx + c = 0$ is in standard form. [5.5/11.1]

standard form of a linear equation in two variables An equation of the form $Ax + By = C$, where A and B are coefficients and C is a constant. [7.2]

straight angle An angle whose measure is 180°. [1.8]

Copyright © Houghton Mifflin Company. All rights reserved.

substitution method An algebraic method of finding an exact solution of a system of equations. [8.2]

sum In addition, the total of two or more numbers. [1.2]

supplementary angles Two angles whose sum is 180°. [1.8]

system of equations Equations that are considered together. [8.1]

terminating decimal Decimal that is formed when dividing the numerator of its fractional counterpart by the denominator results in a remainder of zero. [1.6]

terms of a variable expression The addends of the expression. [2.1]

transversal A line intersecting two other lines at two different points. [3.5]

triangle A three-sided closed figure. [1.8]

trinomial A polynomial of three terms. [4.1]

undefined slope A property of a vertical line. [7.3]

uniform motion The motion of a moving object whose speed and direction do not change. [3.1/6.8]

union of sets A and B The set that contains all the elements of A and all the elements of B. [9.1]

value of a function at x The result of evaluating a variable expression, represented by the symbol $f(x)$. [7.1]

value of a variable The number assigned to the variable. [2.1]

variable A letter of the alphabet used to stand for a number that is unknown or that can change. [1.1]

variable expression An expression that contains one or more variables. [2.1]

variable part In a variable term, the variable or variables and their exponents. [2.1]

variable term A term composed of a numerical coefficient and a variable part. [2.1]

vertex Point at which the rays that form an angle meet. [1.8]

vertical angles Two angles that are on opposite sides of the intersection of two lines. [3.5]

x-coordinate The abscissa in an xy-coordinate system. [7.1]

x-intercept The point at which a graph crosses the x-axis. [7.3]

xy-coordinate system A rectangular coordinate system in which the horizontal axis is labeled x and the vertical axis is labeled y. [7.1]

y-coordinate The ordinate in an xy-coordinate system. [7.1]

y-intercept The point at which a graph crosses the y-axis. [7.3]

zero slope A property of a horizontal line. [7.3]

Copyright © Houghton Mifflin Company. All rights reserved.

Index

Copyright © Houghton Mifflin Company. All rights reserved.

Copyright © Houghton Mifflin Company. All rights reserved.

Copyright © Houghton Mifflin Company. All rights reserved.

Copyright © Houghton Mifflin Company. All rights reserved.

Copyright © Houghton Mifflin Company. All rights reserved.

Copyright © Houghton Mifflin Company. All rights reserved.

Copyright © Houghton Mifflin Company. All rights reserved.

TI-30X IIS

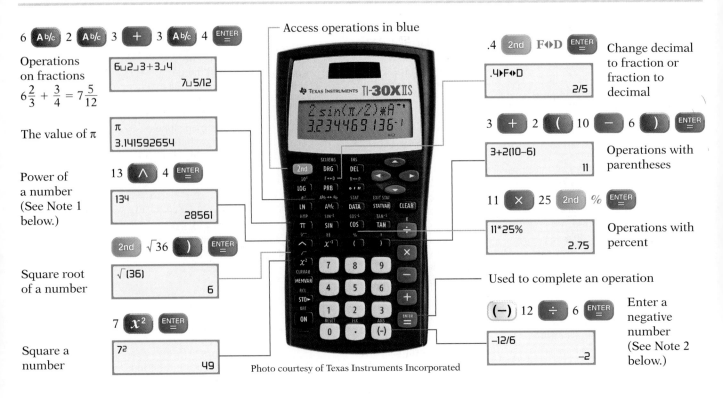

Access operations in blue

6 [Ab/c] 2 [Ab/c] 3 [+] 3 [Ab/c] 4 [ENTER]

Operations on fractions

$6\frac{2}{3} + \frac{3}{4} = 7\frac{5}{12}$

```
6⌐2⌐3+3⌐4
            7⌐5/12
```

The value of π

```
π
3.141592654
```

Power of a number (See Note 1 below.)

13 [∧] 4 [ENTER]

```
13⁴
            28561
```

[2nd] √ 36 [)] [ENTER]

Square root of a number

```
√(36)
            6
```

7 [x²] [ENTER]

Square a number

```
7²
            49
```

Photo courtesy of Texas Instruments Incorporated

.4 [2nd] [F◀▶D] [ENTER]

Change decimal to fraction or fraction to decimal

```
.4▶F◀▶D
            2/5
```

3 [+] 2 [(] 10 [−] 6 [)] [ENTER]

```
3+2(10−6)
            11
```

Operations with parentheses

11 [×] 25 [2nd] [%] [ENTER]

```
11*25%
            2.75
```

Operations with percent

Used to complete an operation

[(−)] 12 [÷] 6 [ENTER]

Enter a negative number (See Note 2 below.)

```
−12/6
            −2
```

fx-300MS

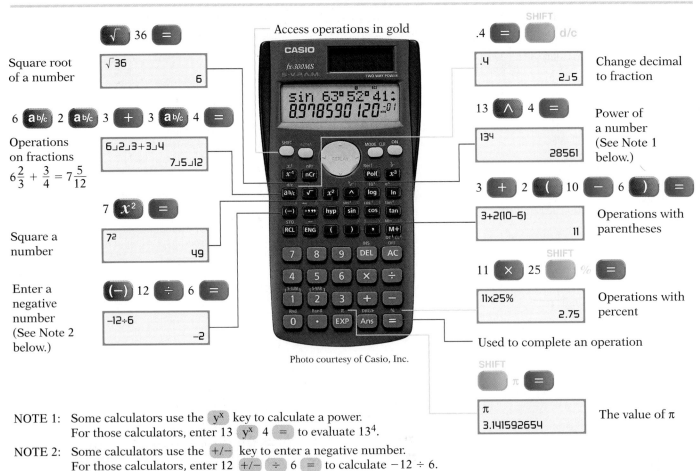

[√] 36 [=]

Square root of a number

```
√36
            6
```

6 [ab/c] 2 [ab/c] 3 [+] 3 [ab/c] 4 [=]

Operations on fractions

$6\frac{2}{3} + \frac{3}{4} = 7\frac{5}{12}$

```
6⌐2⌐3+3⌐4
            7⌐5⌐12
```

7 [x²] [=]

Square a number

```
7²
            49
```

[(−)] 12 [÷] 6 [=]

Enter a negative number (See Note 2 below.)

```
−12÷6
            −2
```

Photo courtesy of Casio, Inc.

Access operations in gold

.4 [=] [d/c]

Change decimal to fraction

```
.4
            2⌐5
```

13 [∧] 4 [=]

Power of a number (See Note 1 below.)

```
13⁴
            28561
```

3 [+] 2 [(] 10 [−] 6 [)] [=]

Operations with parentheses

```
3+2(10−6)
            11
```

11 [×] 25 [%] [=]

Operations with percent

```
11x25%
            2.75
```

Used to complete an operation

[π] [=]

The value of π

```
π
3.141592654
```

NOTE 1: Some calculators use the y^x key to calculate a power. For those calculators, enter 13 y^x 4 [=] to evaluate 13^4.

NOTE 2: Some calculators use the +/− key to enter a negative number. For those calculators, enter 12 +/− ÷ 6 [=] to calculate $-12 \div 6$.

Math Study Skills Workbook

Second Edition

Your Guide to Reducing Test Anxiety
and Improving Study Strategies

Paul D. Nolting, Ph.D.
Learning Specialist

Contents

Preface

Wouldn't it be nice if all we had to do was listen to a lecture on math and read the textbook in order to learn it? That would be paradise. However, most math courses take place on Earth, and students have to do much more than just take notes and read a textbook. They need a system of study skills that help them understand and master mathematics.

Many students in all levels of math courses have difficulty learning math because it is one of the most difficult subjects in college. First, many students who have struggled with math before going to college continue to struggle when they take their first college math courses, even developmental math courses. Second, some students who have done well in the past begin to struggle when they take upper level math like college algebra or calculus. They made A's and B's in previous math courses, but all of a sudden they are failing upper level math courses. These students probably lived off their intelligence until they took a math course that challenged them. Since they never had to study for math, they did not know where to begin to study. Students in all levels of math courses benefit from designing a system of study skills.

Does this sound far fetched? Not really. I have watched calculus students read this workbook and improve their grades in order to be more competitive candidates for engineering programs. For example, I asked two calculus II students why they were taking my math study skills course. I thought they needed a one credit hour course to allow them to graduate, but I was wrong. Both students made a C in calculus I and said that they needed to make an A or B in calculus II because they were engineering majors. In most cases, students who make C's in calculus are not admitted to engineering schools. Both students made an A in my course and a B in their calculus II course and went on to engineering school. Success for them!

Another success story belongs to a student who failed a beginning algebra course three times. Yes, three times. She came to my office ready to quit college all together. I convinced her to focus on designing a system of study skills before taking the course one more time. She did. With her new system of study skills, she passed with a B and stayed in college. Success for her!

Earning better math grades doesn't have to be the only benefit from spending time developing a system for studying math. According to many students, they were able to take the math study skills and adapt them to their other courses. In fact, some students claim it is easier than adapting general study skills to learning math. They reported that their other course grades also went up. What a good experience!

So, what kind of assistance do most students want? Students want tips and procedures they can use to help them improve their math grades. The math study suggestions, however, have to be based on research and be statistically proven to improve student learning and grades. **Math Study Skills Workbook** is based on **Winning at Math: Your Guide to Learning Mathematics Through Successful Study Skills** (2002), which is the only math study skills text that can boast statistical evidence demonstrating an improvement in students' ability to learn math and make better grades. Learning and using these study skills techniques will improve your grades.

Math Study Skills Workbook is designed to supplement your math course or study skills course or to be used as part of a math anxiety workshop. The workbook is designed for independent work or to be supplemented by lectures. To gain the most benefit, the workbook needs to be completed by midterm.

This workbook is designed to enhance learning by teaching math learning skills in small "chunks." After each section, you are required to recall the most important information by writing it down. The writing exercises are especially designed for you to personalize learning techniques, such as the ten steps to better test taking. Each chapter review is designed to reinforce your learning and to give you the opportunity to immediately select and use the best learning strategy from that chapter. Lecture Notes pages have been added, where possible, to allow you to include your notes in this workbook.

As you adapt and practice each study skill, place them into a larger system for studying math. Decide which study skills will help with each of the following tasks you must do when learning math:

1. Taking and reworking lecture notes
2. Learning vocabulary
3. Completing homework in a way that helps you learn and master the math

4. Mastering and memorizing the concepts
5. Preparing for tests
6. Taking tests
7. Managing any anxiety whether math or text anxiety

Remember, to reap the most benefit, you need to complete this workbook and be using your system of study skills by midterm. Practice and master! Then, maybe we can use your success story in our next edition of the workbook!

1

What You Need to Know to Study Math

Mathematics courses are not like other courses. Because they are different, they require different study procedures. Passing most of your other courses requires only that you read, understand, and recall the subject material. To pass math, however, an extra step is required: You must use the information you have learned to correctly solve math problems.

Learning general study skills can help you pass most of your courses, but special math study skills are needed to help you learn more and get better grades in math. In this chapter, you will find out

- why learning math is different from learning other subjects,
- what the differences are between high school and college math, and
- why your first math test is very important.

Why learning math is different from learning other subjects

In a math course, you must be able to do four things:

1. *Understand* the material
2. *Process* the material
3. *Apply* what you have learned to correctly solve a problem, and
4. *Remember* what you have learned to learn new material.

Of these four tasks, applying what you have learned to correctly solve a problem is the hardest.

Examples: Political science courses require that you learn about politics and public service. But your instructor will not make you run for

continued

governor to pass the course. Psychology courses require you to understand the concepts of different psychology theories. But you will not have to help a patient overcome depression to pass the course. In math, however, you must be able to correctly solve problems to pass the course.

Sequential Learning Pattern

Another reason learning math is different than learning other subjects is that it follows a *sequential learning pattern*, which simply means that the material learned on one day is used the next day and the next day, and so forth. This building-block approach to learning math is the reason it is difficult to catch up when you fall behind. *All* building blocks must be included to be successful in learning math.

You can compare learning math to building a house. Like a house, which must be built foundation first, walls second, and roof last, math must be learned in a specific order. Just as you cannot build a house roof first, you cannot learn to solve complex problems without first learning to solve simple ones.

Example: In a history class, if you study Chapters 1 and 2, do not understand Chapter 3, and end up studying and having a test on Chapter 4, you *could* pass. Understanding Chapter 4 in history is not totally based

continued

on comprehending Chapters 1, 2, and 3. To succeed in math, however, each previous chapter has to be completely understood before you can continue to the next chapter.

Sequential learning affects your ability to study for math tests, as well. If you study Chapter 1 and understand it, study Chapter 2 and understand it, and study Chapter 3 and *do not* understand it, then when you study for a test on Chapter 4, you are not going to understand it either, and you probably will not do well on the test.

REMEMBER

To learn the new math material for the test on Chapter 5, you must first go back and learn the material in Chapter 4. This means you will have to go back and learn Chapter 4 while learning Chapter 5. (The best of us can fall behind under these circumstances.) However, if you do not understand the material in Chapter 4, you will not understand the material in Chapter 5 either, and you will fail the test on Chapter 5. This is why the sequential learning of math concepts if so important.

The sequential learning pattern is also affected by

- your previous math course grade,
- your math placement test scores, and
- the time elapsed since your last math course.

Sequential learning is influenced by how much math knowledge you have at the beginning of your course. Students who forgot or never acquired the necessary skills from their previous math course will have difficulty with their current math course. If you do not remember what you learned in your last math course, you will have to relearn the math concepts from the previous course as well as the new material for the current course. In most other courses, such as the humanities, previous course knowledge is not required. However, in math you must remember what the last course taught you so that you are prepared for the current course. Measuring previous course knowledge will be explained in Chapter 2, "How to Discover Your Math-Learning Strengths and Weaknesses."

Math placement scores also affect sequential learning. If you barely scored high enough to be placed into a math course, then you will have math-learning gaps. Learning problems will occur when new math material is based on one of your learning gaps. The age of the placement test score also affects sequential learning. Placement test scores are designed to measure your *current* math knowledge and are to be used immediately.

Sequential learning is interrupted if math courses are taken irregularly. Math courses are designed to be taken one after another. By taking math courses each semester, without semester breaks in between courses, you are less likely to forget the concepts required for the next course. Research has shown that the more time between math courses, the more likely that a student will fail the math course.

Now that you understand the "building block" nature of math, think about your math history.

- What are your previous math grades?
- How well did you do on the math placement test at your college?

- How long has it been since you took a math course?
- When you look at your math history, are there semesters when you did not take math?

These questions are important because if there was too much time in between your different math courses, you may have forgotten important math concepts that you need in your current class. To use the building block analogy, the blocks may not be as strong any more.

Now that you understand how learning math is a building experience, what should you do? *First,* don't get anxious. Stay calm. *Second,* if your college has a diagnostic math inventory in the tutoring center or math lab, take it to see what math concepts you have forgotten. Then ask your instructor where you can go to relearn these math concepts. *Third,* take the time to follow through. Many students just give up too easily, or they think they will catch up a little at a time. Don't think that and don't give up. The energy put into your class at the beginning of the semester will be more productive than energy put into class at the end of the semester when you try to learn everything during the last week before the final exam. *Fourth,* study and really learn the math; don't practice mimicking it. *Finally,* when it is time to register for the next semester, register immediately so you will be able to get into the math class you need. Why do all this? Because math is sequential!

Math as a Foreign Language

Another helpful technique for studying math is to consider it a foreign language. Looking at math as a foreign language can improve your study procedures. If you do not practice a foreign language, what happens? You forget it. If you do not practice math, what happens? You are likely to forget it, too.

Students who excel in a foreign language must practice it *at least* every other day. The same study skills apply to math, because it is considered a foreign language.

Like a foreign language, math has unfamiliar vocabulary words or terms that must be put in sentences called *equations*. Understanding and solving a math equation is similar to speaking and understanding a sentence in a foreign language.

> **Example:** Math sentences use symbols (which stand for spoken words), such as
>
> = (for which you *say*, "equal"),
> - (for which you *say*, "less"), and
> *a* (for which you *say*, "unknown").

Learning *how* to speak math as a language is the key to math success. Currently, most universities consider computer and statistics (a form of math) courses as foreign languages. Some universities have now gone so far as to actually classify math as a foreign language.

Math as a Skill Subject

Math is a *skill subject*, which means you have to practice actively the skills involved to master it. Learning math is similar to learning to play a sport, learning to play a musical instrument, or learning auto mechanics skills. You can listen and watch your coach or instructor all day, but unless you *practice* those skills yourself, you will not learn.

Examples: In basketball, the way to improve your free throw is to *see and understand* the correct shooting form and then *practice* the shots yourself. Practicing the shots improves your free-throwing percentage. However, if you simply listen to your coach describe the correct form and see him demonstrate it, but you do not practice the correct form yourself, you will not increase your shooting percentage.

Suppose you want to play the piano, and you hired the best available piano instructor. You sit next to your instructor on the piano bench and watch the instructor demonstrate beginning piano-playing techniques. You see and understand how to place your hand on the keys and play. But what does it take to learn to play the piano? You have to place your hands on the keys and *practice*.

Math works the same way. You can go to class, listen to your instructor, watch the instructor demonstrate skills, and understand everything that is said (and feel that you are quite capable of solving the problems). However, if you leave the class *and do not practice* — by working and successfully solving the problems — you will not learn math.

Many of your other courses can be learned by methods other than practicing. In social studies, for example, information can be learned by listening to your instructor, taking good notes, and participating in class discussions. Many students mistakenly believe that math can be learned the same way.

REMEMBER

Math is different. If you want to learn math, you must practice. Practice not only means doing your homework but also means spending the time it takes to understand the reasons for each step in each problem.

Math as a Speed Subject

Math is a *speed subject,* which means, in most cases, it is taught faster than your other subjects. Math instructors have a certain amount of material that must be covered each semester. They have to finish certain chapters because the next math course is based on the information taught in their courses. In many cases a common math department final exam is given to make sure you know the material for the next course. Instructors are under pressure to make sure you are ready for the final exam because it demonstrates how prepared you are for the next level in math. This is different from, let's say, a sociology course where if an instructor doesn't teach the last chapter it will not cause students too many problems in the next sociology or social science course. So don't complain to the math instructor about the speed of the course. Instead, improve your study skills so you can keep up!

Another way math is a speed subject is that most of the tests are timed, and many students think that they will run out of time. THIS CAUSES PANIC AND FEAR! This is different than most of your other courses where you generally have enough time to complete your tests or in your other courses that have multiple choice tests where you can start bubbling the responses on the scantron sheet if you start running out of time. Students must not only understand how to do the math problems but also must learn the math well enough to complete the problems with enough speed to finish the test.

What makes me curious is, if students feel like they don't have enough time to complete the math test, why are most of them gone before the test time is over? Sure, students who have learned the math thoroughly may complete the test early. That makes sense. Some students leave, however, because they either don't know the material, want to leave the anxious environment, or carelessly work through the test.

So, since speed is an issue in learning math, what should you do? *First,* to use an analogy, start a daily workout program to stay in shape. Review, review, and review as you learn new material. *Second,* practice doing problems within a time constraint. Give yourself practice tests.

A Bad Math "Attitude"

Students' attitudes about learning math are different from their attitudes about learning their other subjects. Many students who have had bad experiences with math do not like math and have a bad attitude about learning it. In fact, some students actually *hate* math, even though these same students have positive experiences with their other subjects and look forward to going to those classes.

Society, as a whole, reinforces students' negative attitudes about math. It has become socially acceptable not to do well in math. This negative attitude has become evident even in popular comic strips, such as *Peanuts*. The underlying message is that math should be feared and hated and that it is all right not to learn math.

This "popular" attitude toward math may reinforce your belief that it is all right to fail math. Such a belief is constantly being reinforced by others.

The bad math attitude is not a major problem, however. Many students who hate math pass it anyway, just as many students who hate history still pass it. The major problem concerning the bad math attitude is how you use this attitude. If a bad math attitude leads to poor class attendance, poor concentration, and poor study skills, then you have a bad math attitude *problem*.

REMEMBER

Passing math is your goal, regardless of your attitude.

Section Review

1. How does a sequential learning pattern affect math learning?

_____

_____

_____

_____

_____

Give two examples:

Example 1: _____

_____

_____

_____

_____

Example 2: _____

_____

_____

_____

_____

_____

2. List two examples of how learning math is similar to learning a foreign language.

Example 1: _____

_____

_____

_____

Example 2: _____

_____

_____

_____

3. How is math similar to a skills subject?

_____

_____

_____

_____

_____

_____

4. List three ways a bad math attitude could affect your mathematics learning.

First Way: _____

_____

_____

Second Way: _____

_____

Third Way: _____

_____

_____

5. Why is math study skills important at all levels of math (see Preface)?

_____

_____

_____

_____

6. Math study skills can help you in other subjects but general study skills usually cannot improve math learning. Why is this statement important? (see Preface)

_____

_____

_____

_____

_____

_____

_____

_____

The differences between high school and college math

Math, as a college-level course, is almost two to three times as difficult as high school-level math courses. There are many reasons for the increased difficulty: course class time allowance, the amount of material covered in a course, the length of a course, and the college grading system.

The first important difference between high school and college math courses is the length of time devoted to instruction each week. Most college math instruction, for the fall and spring semesters, has been cut to three hours per week; high school math instruction is provided five hours per week. Additionally, college courses cover twice the material in the same time frame as do high school courses. What is learned in one year of high school math is learned in one semester (four months) of college math.

Simply put, in college math courses you are receiving less instructional time per week and covering twice the ground per course as you were in high school math courses. The responsibility for learning in college is the student's. As a result, most of your learning (and *practicing*) will have to occur outside of the college classroom.

College Summer Semester Versus Fall or Spring Semester and the Difference Between Night and Day

College math courses taught during summer semesters are more difficult than those taught during fall or spring. Further, math taught during night courses is more difficult than math taught during day courses.

Students attending a six-week summer math session must learn the information — and master the skills — two and a half times as fast as students attending regular, full-semester math sessions. Though you receive the same amount of instructional classroom time, there is less time to understand and *practice the skills* between class sessions.

Summer classes are usually two hours per day, four days per week (nighttime summer classes are four hours per night, two nights per week).

Example: If you do not understand the lecture on Monday, then you have only Monday night to learn the material before progressing to more difficult material on Tuesday. During a night course, you have to learn and understand the material before the break; after the break, you will move on to the more difficult material — *that night.*

Because math is a sequential learning experience, where every building block must be understood before proceeding to the next block, you can quickly fall behind, and you may never catch up. In fact, some students become lost during the first half of a math lecture and never understand the rest of the lecture (this can happen during just one session of night class). This is called "kamikaze" math because most students do not survive summer courses.

If you *must* take a summer math course, take a 10- or 12-week *daytime* session so that you will have more time to process the material between classes.

Course Grading System

The grading system for math is different in college than in high school.

Example: While in high school, if you make a D or borderline D/F, the teacher more than likely will give you

continued

a D, and you may continue to the next course. However, in some college math courses, a D is not considered a passing grade, or if a D is made, the course will not count toward graduation.

College instructors are more likely to give the grade of N (no grade), W (withdrawal from class), or F for barely knowing the material. This is because the instructor knows that you will be unable to pass the next course if you barely know the current one.

Most colleges require students to pass two college-level algebra courses to graduate. In most high schools, you may graduate by passing one to three math courses. In some college-degree programs, you may even have to take four math courses and make a C in all of them to graduate.

The grading systems for math courses are very precise compared to the grading systems for humanities courses.

Example: In a math course, if you have a 79 percent average and you need 80 percent to get a B, you will get a C in the course. But if you make a 79 percent in English class, you may be able to talk to your instructor and do extra credit work to earn a B.

Because math is an exact science and not as subjective as English, do not expect your math instructor to let you do extra work to earn a better grade. In college, there usually is not a grade given for "daily work," as is often offered in high school.

In fact, *your test scores may be the only grades that will count toward your final grade.* Therefore, you should not assume that you will be able to "make up" for a bad test score.

The Ordering of College Math Courses

College math courses should be taken, *in order,* from the fall semester to the spring semester. If at all possible, avoid taking math courses from the spring to fall semesters. There is less time between the fall and spring semester for you to forget the information. During the summer break, you are more likely to forget important concepts required for the next course and therefore experience greater difficulty.

Section Review

1. Compare the amount of college class time to high school class time.

   _____

   _____

   _____

   _____

   _____

2. How do college summer math courses differ from courses offered in the spring or fall?

_____

_____

_____

_____

_____

_____

3. How can the order of taking math courses affect your learning?

_____

_____

_____

_____

_____

_____

_____

Why your first math test is very important

Making a high grade on the first major math test is more important than making a high grade on the first major test in other subjects. The first major math test is the easiest and, most often, the one the student is least prepared for.

Students often feel that the first major math test is mainly a review and that they can make a B or C without much study. These students are overlooking an excellent opportunity to make an A on the easiest major math test of the semester. (Do not forget that this test counts the same as the more difficult remaining math tests.)

At the end of the semester, these students sometimes do not pass the math course or do not make an A because of their first major test grade. In other words, the first math test score was not high enough to "pull up" a low test score on one of the remaining major tests.

Studying hard for the first major math test and obtaining an A offers you several advantages:

- A high score on the first test can compensate for a low score on a more difficult fourth or fifth math test. All major tests have equal value in the final grade calculations.

- A high score on the first test can provide assurance that you have learned the basic math skills required to pass the course. This means you will not have to spend time relearning the misunderstood material covered on the first major test while learning new material for the next test.

- A high score on the first test can motivate you to do well. Improved motivation can cause you to increase your math study time, which will allow you to master the material.

- A high score on the first test can improve your confidence for higher test scores. With more confidence, you are more likely to work harder on the difficult math homework assignments, which will increase your chances of doing well in the course.

What happens if after all your studying, you make a low score on your first math test? You can still use this test experience to help you improve your next grade or to help determine if you are in the right math course. Your first math test, no matter what you make on it, can be used as a diagnostic test. Your teacher can review your test with you to see which type of math problems you got right and which ones you need to learn how to solve. It may be that you missed only a few concepts that caused the low score, and you can learn how to do these problems by getting help from the teacher, a learning resource center, or the math lab. However, you need to learn how to do these problems immediately so that you don't fall behind in the course. After meeting with your teacher, ask for the best way you can learn the concepts that are the bases of the missed problems and how to prepare for the next test. Even students who made a B on the first math test can benefit by seeing the teacher.

If you made below a 50 on your first math test, I suggest that it might be a good idea to drop back to a lower level math course. Even though it might be beyond the first week of drop and add, many colleges/ universities will let you drop back to a lower math class after the first major math test. Students who drop back get a good foundation in mathematics that helps them in their next math courses. On the other hand, I have seen students who insisted on staying in the math course and repeated it several times before passing. Some of the students stopped taking the math course and dropped out of college. Dropping back to a lower level math course and passing it is the smartest move. These students went on to become more successful in their math courses.

REMEMBER

It does not matter where you start as long as you graduate. Discuss this option with your teacher.

Section Review

1. Why is your first math test supposed to be the easiest?

_____

_____

_____

_____

_____

_____

2. Give three reasons why your first math test is so important.

First Reason: _____

_____

_____

_____

_____

Second Reason: _____

_____

_____

_____

_____

_____

continued

Third Reason: _____

_____

_____

_____

_____

_____

3. What do you do if you fail your first test?

Option One: _____

_____

_____

_____

_____

_____

Option Two: _____

_____

_____

_____

_____

_____

Option Three: _____

_____

_____

_____

_____

_____

Chapter 1 Review

1. In a math course, you must be able to do three things. You must _____ the material, _____ the material, and _____ what you have learned to solve the problem.

2. Math requires _____ _____, which means one concept builds on the next concept.

3. Placement test scores are designed to measure your _____ math knowledge and are to be used _____.

4. Like a foreign language, math has unfamiliar vocabulary words or terms that must be put in sentences called _____.

5. Learning how to speak math as a _____ _____ is one key to math success.

6. Keeping a _____ attitude about math will help you study more efficiently.

7. College math courses are _____ to _____ times as difficult as high school math courses.

8. Math course are even more difficult than other courses because a grade of _____ or better is usually required to take the next course.

9. The math grading is exact; in many cases, you cannot do _____ _____ to improve your grade.

10. If you fail your first math test, you need to make an appointment with your _____ to review your _____. Then you can decide if you should _____ _____ to a lower math course.

What is the most important information you learned from this chapter?

_____

_____

_____

_____

_____

_____

_____

_____

How can you immediately use it?

_____

_____

_____

_____

_____

_____

_____

_____

Lecture Notes

How to Discover Your
Math-Learning Strengths and Weaknesses

Math-learning strengths and weaknesses affect students' grades. You need to understand your strengths and weaknesses to improve your math-learning skills.

Just as a mechanic does a diagnostic test on a car before repairing it, you need to do your own testing to learn what you need to improve upon. You do not want the mechanic to charge you for something that does not need repairing, nor do you want to work on learning areas that do not need improvement. You want to identify learning areas you need to improve or better understand.

In this chapter, you will learn

- how what you know about math affects your grades,
- how quality of math instruction affects your grades,
- how affective student characteristics affect your math grades,
- how to determine your learning style,
- how to develop a math-learning profile of your strengths and weaknesses, and
- how to improve your math knowledge.

Areas of math strengths and weaknesses include math knowledge, level of test anxiety, study skills, study attitudes, motivation, and test-taking skills. Before we start identifying your math strengths

and weaknesses, you need to understand what contributes to math academic success.

Dr. Benjamin Bloom, a famous researcher in the field of educational learning, discovered that your IQ (intelligence) and your cognitive entry skills account for 50 percent of your course grade.

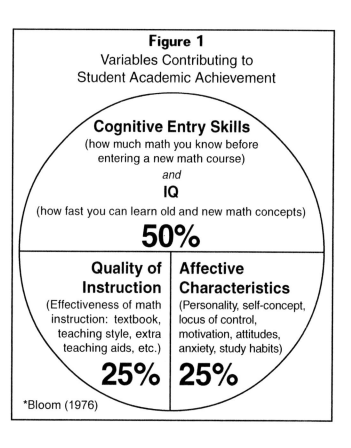

Figure 1
Variables Contributing to
Student Academic Achievement

Cognitive Entry Skills
(how much math you know before entering a new math course)
and
IQ
(how fast you can learn old and new math concepts)
50%

Quality of Instruction	**Affective Characteristics**
(Effectiveness of math instruction: textbook, teaching style, extra teaching aids, etc.)	(Personality, self-concept, locus of control, motivation, attitudes, anxiety, study habits)
25%	**25%**

*Bloom (1976)

Quality of instruction represents 25 percent of your course grade, while affective student characteristics reflect the remaining 25 percent of your grade.

- *Intelligence* may be considered, for our purpose, how fast you can learn or relearn math concepts.
- *Cognitive entry skills* refer to how much math you knew before entering your current math course. For example, previous math knowledge can be based on previous math grades, the length of time since the last math course, the type of math courses completed (algebra I, algebra II, intermediate algebra), and placement test scores.
- *Quality of instruction* is concerned with the effectiveness of math instructors when presenting material to students in the classroom and math lab. This effectiveness depends on the course textbook, curriculum, teaching style, extra teaching aids (videos, audiocassettes), and other assistance. Quality of instruction may also include math lab resources, quality of tutoring based on tutor training and Supplemental Instruction.
- *Affective student characteristics* are characteristics you possess that affect your course grades — excluding how much math you knew before entering your math course. Some of these affective characteristics are anxiety, study skills, study attitudes, self-concept, motivation, and test-taking skills.

How what you know about math affects your grades

Poor math knowledge can cause low grades. A student placed in a math course that requires a more extensive math background than the student possesses will probably fail that course. Without the correct math background, you will fall behind and never catch up.

Placement Tests and Previous Course Grades

The math you need to know to enroll in a particular math course can be measured by a placement test (ACT, SAT) or by the grade received in the prerequisite math course. Some students are incorrectly placed in math courses by placement tests.

If, by the second class meeting, everything looks like Greek and you do not understand what is being explained, move to a lower-level course. In the lower-level math course you will have a better chance to understand the material and to pass the course. A false evaluation of math ability and knowledge can only lead to frustration, anxiety, and failure.

Requests by Students for Higher Placement

Some students try to encourage their instructors to move them to a higher-level math course because they believe they have received an inaccurate placement score. Many students try to avoid noncredit math courses, while other students do not want to repeat a course that they have previously failed. These moves can also lead to failure.

Some older students imagine that their math skills are just as good as when they completed their last math course, which was five to ten years ago. If they have not been practicing their math skills, they are just fooling themselves. Still other students believe they do not need the math skills obtained in a prerequisite math course to pass the next course. This is also incorrect thinking. Research indicates that students who were placed correctly in their prerequisite math course, and who subsequently failed it, will not pass the next math course.

What My Research Shows

I have conducted research on thousands of students who have either convinced their instructors to place them in higher-level math courses or have placed themselves in higher-level math courses. The results? These students failed their math courses many times before realizing they did not possess the prerequisite math knowledge needed to pass the course. Students who, without good reason, talk their instructors into moving them up a course level are setting themselves up to fail.

To be successful in a math course, you must have the appropriate math knowledge. If you think you may have difficulty passing a higher-level math course, you probably do not have an adequate math background. Even if you do pass the math course with a D or C, research indicates that you will most likely fail the next higher math course.

It is better to be conservative and pass a lower-level math course with an A or B instead of making a C or D in a higher-level math course and failing the next course at a higher level.

This is evident when many students repeat a higher-level math course up to five times before repeating the lower-level math course that was barely passed. After repeating the lower-level math course with an A or B, these students passed their higher-level math course.

How quality of math instruction affects your grades

Quality of instruction accounts for 25 percent of your grade. Quality of instruction includes such things as classroom atmosphere, instructor's teaching style, lab instruction, and textbook content and format. All of these "quality" factors can affect your ability to learn in the math classroom.

Interestingly enough, probably the most important "quality" variable is the compatibility of an instructor's teaching style with your learning style. You need to discover your learning style and compare it to the instructional style. Noncompatibility can be best solved by finding an instructor who better matches your learning style. However, if you cannot find an instructor to match your learning style, improving your math study skills and using the math lab or learning resource center (LRC) can compensate for most of the mismatch.

Use of the math lab or learning resource center (LRC) can dramatically improve the quality of instruction. With today's new technologies, students are able to select their best learning aids. These learning aids could be videotapes, CD-ROMs, computer programs, math study skills computer programs, math texts, and websites.

The quality of tutors is also a major factor in the math lab or learning resource center. A low student-to-tutor ratio and trained tutors are essential for good tutorial instruction. Otherwise, the result is just a math study hall with a few helpers.

The math textbook should be up to date with good examples and a solutions manual. This increases the amount of learning from the textbook compared to older, poorly designed texts.

The curriculum design affects the sequence of math courses, which could cause learning problems.

Some math courses have gaps between them that cause learning problems for all students.

How affective student characteristics affect your math grades

Affective student characteristics account for about 25 percent of your grade. These affective characteristics include math study skills, test anxiety, motivation, locus of control, learning style, and other variables that affect your personal ability to learn math.

Most students do not have this 25 percent of the grade in their favor. In fact, most students have never been taught *any* study skills, let alone how to study math, specifically. Students also do not know their best learning style, which means they may study ineffectively by using their least effective learning style. However, it is not your fault that you were not taught math study skills or made aware of your learning style.

By working to improve your affective characteristics, you will reap the benefits by learning math more effectively and receiving higher grades. Thousands of students have improved their math affective characteristics and thereby their grades by using this workbook.

Variable 2: _____

_____

_____

_____

_____

_____

Variable 3: _____

_____

_____

_____

_____

_____

Section Review

1. List and explain the three variables that contribute to your academic success.

Variable 1: _____

2. How does your placement test or previous math course grade affect your math learning?

_____

_____

_____

_____

_____

_____

3. How does the quality of math instructors affect your grade?

_____

_____

_____

_____

_____

_____

4. How do your affective characteristics affect your grade?

_____

_____

_____

_____

_____

_____

How to determine your learning style

Research has shown that matching a student's best learning style with the instructional style improves learning. Research has also shown that students who understand their learning style can improve their learning

effectiveness. A learning disadvantage will occur for students who do not know or do not understand their learning style. Students should talk to their instructor or counselor about taking one or more learning style inventories.

Taking Stock of Your Learning Style

There are different types of learning styles assessments. One type of learning style assessment focuses on learning modalities, whereas other assessments focus on cognitive or environmental learning styles. We will focus on learning modalities and cognitive learning styles.

Learning modalities
(using your senses)

If it is available to you, take a learning style inventory that measures _learning modalities_. Learning modalities focus on the best way your brain receives information — that is, learning _visually_ (seeing), _auditorially_ (hearing), or _kinesthetically_ (touching).

If possible, take an inventory that offers math-learning modalities, such as _Learning Styles Inventory_ (Brown and Cooper, 1978). Other learning style inventories can measure learning modalities, but they focus on English- or reading-learning modes. Sometimes students' math-learning modes may be different from their language-learning modes. Your instructor can go to www.academicsuccess.com to learn more about using the Learning Styles Inventory.

If you cannot locate a modality-type inventory, then you may want to do a self-assessment on your math-learning modality. Do you learn math best by seeing it, by hearing it, or by having hands-on learning experiences? Hands-on learning means you need to touch or feel things to best learn about them. Group learning is when two to four students get together to discuss math problems and talk about the solutions. Individual learning is studying by yourself, using your book, CD or web resources. Do you learn best in groups or by yourself?

This is not the most accurate way to determine your learning modality; however, you may use this informal assessment until you have an opportunity to take a more formal modality learning style inventory.

Cognitive learning styles
(processing what you sense)

The second type of learning-styles inventory focuses on *cognitive learning styles.* Cognitive learning styles describe how you process information, once you have heard, seen or felt the information. Based on McCarthy (1981), you can process information four different ways: Innovative, Analytic, Common sense, or Dynamic. You can find information on the McCarthy inventory by going to www.AboutLearning.com.

If you do not have the opportunity to take a cognitive learning-style inventory, you may want to do a self-assessment of your cognitive learning style. Which of the following best describes you? Remember that you may not have all the characteristics mentioned for each learning style. Look for the style that best matches you.

Innovative learners solve problems or make decisions by personally relating to them and using their feelings. They are interested in people and like cultural events. They want to work towards peace and harmony. They work towards self-improvement. Some of their careers are in counseling, social work, personnel, humanities, social sciences, English and most fields dealing with people. They function through life by using their values.

They learn best by listening and discussing their ideas with others, while searching for the meaning. Their best learning activities are motivational stories, simulations, group discussion, journal writing, integrative lectures, group projects, and discussion lectures. They usually like subjective tests. They are students with good ideas, and they find unique solutions to problems. They ask the question, "Why do I need to learn this?" (McCarthy, 1981). In the math course they may ask for the underlying value of learning the course material as it relates to them. Their favorite question is, "Why?" Is this your cognitive learning style?

Analytic learners look for facts and ask experts for advice when solving problems. They want to know what the experts think and the research to back up the facts. In some cases they are more interested in ideas and concepts than people. They want to collect information before making logical decisions. If they are having difficulty with a decision, they will re-examine the facts or go to experts. They are very uncomfortable with subjective decisions. They generally major in math and the hard sciences; they enjoy careers involving research (McCarthey, 1981).

They learn by thinking about the concepts and solving problems by using logic. They learn best in the traditional lecture classroom by instructors giving them facts. Their best learning activities are lecture with visual aids, textbook reading assignments, problem solving and demonstration by the instructor (expert), independent research and gathering data. They usually like objective tests. They want to know, "What exactly is the content or skill?" (McCarthey, 1981). They learn best by taking exact and well organized notes and consider class discussion an interruption in learning. They usually do not like group learning and prefer studying individually. They usually major in areas like math, physics, chemistry and accounting. Their favorite question is "What?" Is this your learning style?

Common-sense learners solve problems by knowing how things work. They want to know how to use the material today and how it can be used in real life. They want to use facts to build concepts and then test the practicality of the concepts. In their discussions they get right to the point. They are skills oriented and like to tinker with things. They use their sensory experiences to help make decisions (McCarthy, 1981).

They learn by testing concepts in a practical way and make decisions by applying theories. They want to know practical ways to use theories and concepts. They need hands on experiences as part of their learning such as labs and manipulatives. Some additional learning activities are homework problems, field trips, computer-aided instruction, and lecture with demonstration, individual reports and computer simulations. They usually like objective tests (Harb, Terry, Hurt, & Williamson, 1981). They major in applied sciences, health fields, computer science, engineering and technologies. Their favorite question is "How?" Is this your learning style?

Dynamic learners solve problems by looking at hidden possibilities and processing information concretely. They like self-discovery. They like new things and can adapt to change. They can be very flexible. They can also reach correct conclusions and answers without knowing how they did it. They need to know what else can be done with this information. They seek to answer this question, "If I do this, what possibilities will it create?" (McCarthy, 1981).

They learn best by independent self-instruction and trial-and-error practice. They are flexible and will take risks. They function in life by acting out and testing experiences. Some of their best learning activities are Socratic lecture, role playing, open-ended laboratories, student lectures, outside projects, simulations and group problem solving (Harb, Terry, Hurt, & Williamson, 1981). They major in business, marketing and sales occupations. Their favorite question is, "What can this become?" Is this your learning style?

By now you may have identified your cognitive learning style. If not, take time to examine how you are learning and then revisit these descriptions. Also you may have recognized that innovative learners are the opposite of common sense learners and analytic learners are the opposite of dynamic learners. This does not mean that you cannot learn from someone teaching in the opposite learning style, but it will take more concentration and better study skills.

REMEMBER

It is important to recognize that students who do not have learning styles that fit easily with learning math can still learn math. Good study skills and awareness on the part of the math instructor and student can compensate a mismatch of modality/ learning styles. Working together to find strategies to help improve the teaching and learning is most productive for everyone.

Section Review

1. My learning modality is _____

   because I learn math best by _____

   _____

   _____

   _____

2. My cognitive learning style is _____

   because I process math information by _____

   _____

   _____

   _____

Assessing your math strengths and weaknesses

Students have learning strengths and weaknesses that affect math learning and their course grades. They must identify and understand these strengths and weaknesses to know which strengths support them and which weaknesses to improve upon. After reading this section, you will be asked to complete the Math-Learning Profile Sheet (Figure 2).

> **Examples**: Having good math study skills is a *positive* math-learning characteristic, whereas having high test anxiety is a *negative* math-learning characteristic.

Measuring Your Math Study Skills

Math study skills is the main part of the affective characteristics that represents about 25% of your math grade. You can determine your math study skills expertise by taking the free Math Study Skills Evaluation at www.academicsuccess.com, or you can do a self determination about your math study skills. If you take the online Math Study Skills Evaluation, the results will indicate your score. If you score below 70, you need to really focus on developing a set of study skills. A score between 70 and 80 means you have average study skills and a score above 90 means that you have good study skills. A self determination can be made by asking yourself if you have ever been taught math study skills. If you

have, then you probably have either good or excellent math study skills. If you have not, then you most likely need to develop good math study skills.

Scoring low on the study skills inventory is not the end of the world. In fact, it might indicate that your problems with learning math may be due to your study habits instead of math intelligence.

A low score on the math study skills evaluation means that most of your learning problems are because you have not been taught math study skills. This workbook can teach you math study skills. If you had previous poor math grades, it was probably due to lack of math study skills, and this is not your fault.

You can improve your math study skills and become a successful math student. If you have good math study skills, you can still improve them to make A's in math. If you have excellent math study skills, you can still learn some math study skills that can help you in future courses.

Measuring Your Math Anxiety Level

It is also important to assess any math anxiety you might have, so that you can start managing it immediately. If you took the online Math Study Skills Evaluation, your answer should indicate if you "seldom", "often" or "almost always" become anxious and forget important concepts during a test.

If you have medium or high math test anxiety, it can be reduced. This problem can be fixed if you follow the suggestions in the next chapter. This also means that some of your problems learning math (if you have any) are due to anxiety instead of your intellectual ability to learn math.

If you have high text anxiety, then you will probably need to use the *How to Reduce Test Anxiety* audio cassette tape. You can order this tape by going to www.academicsuccess.com.   It is

Figure 2
Math-Learning Profile Sheet

After completing this profile, use your answers to select which areas you would like to improve first. If you have a serious problem with math or test anxiety, get help in this area immediately. Circle the X that best describes your belief.

	Strongly Disagree	Disagree	Agree	Strongly Agree
1. I have good math study skills.	X	X	X	X
2. Math tests don't make me anxious.	X	X	X	X
3. I have a strong basic knowledge of math to pass this class.	X	X	X	X
4. I have a positive attitude about learning math.	X	X	X	X
5. It has been less than a year since I took my last math class.	X	X	X	X
6. I have taken a math study skills course before this class.	X	X	X	X
7. I have good reading skills to be successful in this class.	X	X	X	X
8. I know if I improve my study habits, I can improve my math grades.	X	X	X	X

This informal survey is to help you decide what kind of shape you are in to do well in your course.

Which areas are strong and help you do well in a math class?

Which areas keep you from doing the best you can in a math class? How can you change them?

important that you find support and guidance in managing this anxiety immediately!

You can reduce and learn to manage test anxiety! It just takes work!

Measuring Previous Math Knowledge

Now it's time to measure your previous math knowledge. Your previous math knowledge is an excellent predictor of future math success. In fact, the best predictor of math success is your previous math grade (if taken within one year of your current course).

The measurement of your math knowledge is based on your previous math grade or your math placement test. If your college does not require a placement test, contact the chairperson of the math department to help you with your course placement. Selecting a math course yourself is like playing Russian roulette with five bullets in the chamber. Without help, you will probably commit mathematical suicide.

Visit your counselor or instructor to discuss your math placement test score. Ask your counselor or instructor how your placement score relates to the scores for the math course below and above your course.

Find out if your score barely placed you into your course, if your score was in the middle, or if your score almost put you into the next course. If your placement score barely put you into your current math course, you will have difficulty learning the math required in that course.

A barely passing placement score means that you are missing some of the math knowledge you require to be successful, and you will really struggle to pass the course.

A middle placement score indicates that you probably have most of the math knowledge required for the course; however, you still may have difficulty passing the course. A high placement score means you have most of the math knowledge needed to pass

and may have a good chance of passing the course. (Still, if you have poor math study skills and high test anxiety, you may not pass the course.)

Students who have low placement test scores need to improve their math knowledge as soon as possible. Students with middle placement scores should also consider improving their math knowledge.

You need to improve your math skills within the first few weeks of the semester. If you wait too long, you will become hopelessly lost. Suggestions for improving your math knowledge are included in a later section in this chapter.

If you are taking your second math course, use your previous math course grade to determine your math knowledge. Your previous course grade is the best predictor of success in your next math course. (However, this is only true if your last math course was taken within the last year.)

Students who made an A or B in their previous math course have a good chance of passing their next math course. However, students who made a C in their previous math course have a poor chance of passing their next math course. Students who made below a C or withdrew from the previous math course have virtually no chance of passing their next math course.

What level of math knowledge preparation do you possess right now for this course? If you have poor or good previous math knowledge, you need to build up those math concepts and skills as soon as possible. Go to your instructor and discuss how to build up your skills. Some suggestions are to review the chapters of the previous math text or ask your instructor about some websites that help students review math. Students with poor math knowledge have about three to four weeks to improve their math before it can become a major problem. Don't wait. Start now to review your math skills in order to improve your mathematics learning and grades. Additional information on improving your math knowledge is in the next section.

Determining Your Math Attitude

Math attitude can play a major part in being successful in your course. Some students who have a poor attitude toward math may not attend class as often as they should or may procrastinate in doing their homework. Other students avoid math until they *have* to take it. What is your math attitude — good, neutral, or negative? Remember, students with negative math attitudes can pass math. It depends on what you do with your attitude that counts. Don't let your negative attitude cause problems in your math course. Also, if you have a good math attitude, don't forget to study. A good math attitude can only get you so far!

The Amount of Time Since Your Last Math Course

Another indicator of math success can be the length of time since your last math course. The longer the time since your last math course, the less likely you will be to pass the current course. The exception to this rule is if you were practicing math while you were not taking math courses. Math is very easily forgotten if not practiced.

How long has it been since your last math course? If it has been two or more years since your last math class, your math may be similar to someone who has a low placement score or made a C or lower in the last math class. You can improve your math knowledge by following the same suggestions made for those students and applying the information in the next section (How to improve your math knowledge).

Previous Study Skills Courses or Training

General study skills training that does not focus on math may help some students in time management, reading techniques, learning styles, and overcoming procrastination. Even though math requires specific study techniques, you can learn many other helpful general study skills and personal habits that will help you in math classes. If you have had no study skills training, then this workbook will help you in your math and other courses. Students with no past study skills training should also take advantage of any other opportunities to learn about college success skills in addition to learning from this workbook. Contact your learning centers, instructors or counselors about study skills workshops and when they will be offered.

Previous/Current Required Reading Courses

College reading skills are also needed for success in math courses. This is especially true when you are trying to solve story or word problems.

At community colleges and some universities, there are three levels of reading courses. Usually, the first two levels are noncredit reading courses, and the third level is for college credit.

If you are enrolled in the lowest reading level, you will have difficulty reading and understanding the math text. Students who are enrolled in the second level may also have difficulty reading the text. Students who are not required to take a reading class or who are reading at grade level may still have some difficulty reading the math text.

REMEMBER

Math texts are not written like English or history texts. Even students with college-level reading skills may experience difficulty understanding their math text.

Students not at college level reading can pass math courses; however, if you need help in reading, get that help as soon as possible. Or, many learning or study skills centers help students with reading skills for specific courses. Work with a learning specialist to design a system for reading your math textbook. If you do not know your reading level, ask to take a reading test. If in doubt about your reading level, take a reading class.

Your Locus of Control

"Locus of control" is a concept that describes how much control you feel you have over both your life and your course grades. Some students feel they have a lot of control over their lives and learning, while other students feel they have very little control over their lives or grades. For example, when a student with external locus of control does poorly on a test, the blame is put on the instructor, when in reality the student did not study well enough. On the other hand, the student with internal locus of control admits that studying more and paying more attention in class would have helped. Students can improve their internal locus of control and be successful in mathematics at the same time.

Your Best Learning Modality

Learning modalities are best described by your preference in how you receive lecture information, either visually, auditory or kinesthetically (hands-on). As discussed, some of these modalities match your math instructor and some do not. Most math instructors teach through a visual mode, with the distant second modality being auditory.

Even if your learning style is different from your math instructor's, there is no excuse for not learning math. The awareness of your learning modality and cognitive style can help you become a better learner in and outside of the classroom. This information will be discussed in Chapter 6, "How to Remember What You Have Learned."

Your Cognitive Learning Style

Cognitive learning styles are best described as how information is processed once it is received by your brain. Most math instructors process information *analytically*, and they therefore teach math the same way. Using *common sense* is the second most popular way math instructors cognitively process information.

If you do not process information analytically or through common sense, then you may be at a learning disadvantage. However, improved math study skills can compensate for a mismatch in learning modes/cognitive styles. Chapter 6 suggests different study skills that can help you learn math through your preferred learning modality and cognitive style.

Evaluating Your Findings

Now that you understand some of the general characteristics that help students be successful, complete the informal Math-Learning Profile Sheet (Figure 2).

When you look at your profile, think of it as similar to going to a fitness center for the first time. A professional trainer asks questions, runs you through equipment, analyzing your strengths and weaknesses, and then designs a program for you. You are doing the same thing here.

Section Review

1. List your strengths for learning math.

   _____

   _____

   _____

   _____

   _____

   _____

   _____

2. List three areas you will start improving right now.

   Area One: _____

   Area Two: _____

   Area Three: _____

3. How will you improve these areas?

   Area One: _____

   _____

   _____

   _____

Area Two: _____

   _____

   _____

   _____

   _____

Area Three: _____

   _____

   _____

   _____

   _____

How to improve your math knowledge

Instructors always operate on the premise that you finished your previous math course just last week; they do not wait for you to catch up on current material. It does not matter if your previous math course was a month ago or five years ago. Instructors expect you to know the previous course material — period.

Review Your Previous Math Course Material and Tests

There are several ways to improve your math knowledge. Review your previous math course material before attending your present math course. Look closely at your final exam to determine your weak areas. Work on your weak areas as soon as possible so they can become building blocks (instead of stumbling blocks) for your current course.

If it has been some time since your last math course, visit the math lab or learning resource center to locate review material. Ask the instructor if there are any computer programs that will assess your math skills to determine your strengths and weaknesses for your course. Review math videotapes on the math course below your level. Also review any computer software designed for the previous math course.

Another way to enhance learning is to review the previous math course text by taking all of the chapter review tests. If you score above 80 percent on one chapter review test, move on to the next chapter. A score below 80 percent means you need to work on that chapter before moving on to the next chapter. Get a tutor to help you with those chapters if you have trouble. Make sure you review all the chapters required in the previous course as soon as possible. If you wait more than two weeks to conclude this exercise, it may be too late to catch up (while learning new material at the same time).

Employ a Tutor

One last way to improve your cognitive entry skills is to employ a private tutor. If you have a history of not doing well in math courses, you may need to start tutorial sessions *the same week class begins*. This will give the tutor a better chance of helping you regain those old math skills.

You still need to work hard to relearn old math skills while continuing to learn the new material. If you wait four to five weeks to employ a tutor, it will probably be too late to catch up and do well or even pass the course.

REMEMBER

Tutorial sessions work best when the sessions begin during the first two weeks of a math course.

Schedule Math Courses "Back to Back"

Another way to maintain appropriate math knowledge is to take your math courses "back to back." It is better to take math courses every semester — even if you do not like math — so that you can maintain sequential (linear) learning.

I have known students who have made B's or C's in a math class and who then waited six months to a year to take the next math course. Inevitably, many failed. These students did not complete any preparatory math work before the math course and were lost after the second chapter. This is similar to having one semester of Spanish, not speaking it for a year, then visiting Spain and not understanding what is being said.

The only exception to taking math courses "back to back" is taking a six-week "kamikaze" math course (an ultracondensed version of a regular course), which should be avoided.

If you are one of the unfortunate many who are currently failing a math course, you need to ask yourself, "Am I currently learning any math or just becoming more confused?" If you are learning some math, stay in the course. If you are getting more confused, withdraw from the course. Improve your math knowledge prior to enrolling in a math course during the next semester.

Example: You have withdrawn from a math course after midterm due to low grades. Instead of waiting until next semester, attend a math lab or seek a tutor and learn Chapters 1, 2, and 3 *to perfection*. Also use this time to improve your math study skills. You will enter the same math course next semester with excellent math knowledge and study skills. In fact, you can make an A on the first test and complete the course with a high grade. Does this sound farfetched? It may, but I know hundreds of students who have used this learning procedure and passed their math course instead of failing it again and again.

Finding Your Best Instructor

Finding an instructor who best matches your learning style can be a difficult task. Your learning style is important; your learning style is how you best acquire information.

Example: Auditory learners do better when hearing the information over and over again instead of carefully reading the information. If an auditory learner is taught by a visual-style instructor who prefers that students read materials on their own and who prefers working problems instead of

continued

describing them, the mismatch could cause the student to do worse than if the student were taught by an auditory instructor.

Most students are placed in their first math course by an academic advisor. Usually, academic advisors know who are the most popular and least popular math instructors. However, advisors can be reluctant to discuss teacher popularity. And unfortunately, students may want the counselor to devise a course schedule based on the student's time limits instead of teacher selection.

To learn who are the best math instructors, ask the academic advisor which math instructor's classes fill up first. This does not place the academic advisor in the position of making a value judgment; neither does it guarantee the best instructor. But it will increase the odds in your favor.

Another way to acquire a good math instructor is to ask your friends about their current and previous math instructors. However, if another student says that an instructor is excellent, make sure your learning style matches your friend's learning style. Ask you friend, "Exactly what makes the instructor so good?" Then compare the answer with how you learn best. If you have a different learning style than your friend, look for another instructor, or ask another friend whose learning style more closely matches your own.

To obtain the most from an instructor, discover your best learning style and match it to the instructor's teaching style. Most learning centers or student personnel offices will have counselors who can measure and explain your learning style. Interview or observe the instructor while the instructor is teaching. This process is time consuming, but it is well worth the effort!

Once you have found your best instructor, *do not* change. Remain with the same instructor for every math class whenever possible.

REMEMBER

The first step in becoming a better math student is knowing your learning strengths and weaknesses. Now you can focus on what you need to improve.

Whether your learning styles are a natural fit for learning math or not, you still own the responsibility to be smart in selecting your study system for math class. You can be successful!

Section Review

1. Give four examples of how you can review math course materials.

   Example 1: _____

   _____

   _____

   _____

   Example 2: _____

   _____

   _____

   _____

   Example 3: _____

   _____

Example 4: _____

_____

_____

_____

_____

2. What is the advantage of taking math courses back to back?

   _____

   _____

   _____

   _____

3. Give two examples of how to find your best math instructor.

   Example 1: _____

   _____

   _____

   _____

   Example 2: _____

   _____

   _____

   _____

Chapter 2 Review

1. _____ _____ _____ refers to how much math you knew before entering your current math course.

2. _____ _____ _____ are the learning skills you possess that affect your grades.

3. Students who talk their instructors into moving up a course level usually _____ that course.

4. The _____ of tutors is a major part of the math lab or learning resource center.

5. Affective student characteristics account for about _____ percent of your grade.

6. _____ learners look for facts and ask experts for their advice.

7. Hands-on learners are called _____ learners.

8. Assessing your math-learning _____ and _____ will help you improve your studying and learning.

9. Students having previous difficulty in mathematics should start tutorial sessions the _____ week class begins.

10. Once you find your best math instructor _____ _____ change instructors for your next math course.

What is the most important information you learned from this chapter?

_____

_____

_____

_____

_____

_____

_____

_____

_____

_____

_____

How can you immediately use it?

_____

_____

_____

_____

_____

_____

_____

_____

_____

_____

_____

Lecture Notes

3

How to Reduce Math Test Anxiety

In this chapter you will first learn about math anxiety and then explore test anxiety. They are different issues that can affect each other. It is best to learn about them as separate entities first.

Math anxiety is a common problem for many high school, college and university students. It is especially difficult for students in developmental courses who normally have more math anxiety than other students. However, there are students who are in higher level math courses that also struggle with this problem. It is very common for students to have anxiety only about math and not in their other subjects.

Most students think that math anxiety only affects them when taking a test, but it also affects other areas. It can affect the way you do your homework, learn in the classroom or through distance learning courses, and the way you choose a career. Students who have math anxiety may procrastinate in doing their homework or put off sitting down and completing an online lesson. This can lead to math failure. Students also select a major based on the amount of math that is required, which could lead to lower paying or dissatisfying careers. However, most students with math anxiety meet it face to face during the test, experiencing test anxiety as well.

Mild test anxiety can be a motivational factor in that it can make students properly prepare for a test. However, high test anxiety can cause major problems in both learning and test taking, as students avoid studying for the test when anxiety begins to afflict their thought processes. Reducing test anxiety is the key for many students to become successful in math. Such students need to learn the causes of test anxiety and how to reduce the test anxiety that affects their learning and grades.

Several techniques have proven helpful in reducing both math anxiety and math test anxiety. However, reducing them does not guarantee good math grades. It must be coupled with effective study skills and a desire to do well in math.

In this chapter, you will learn about math anxiety:

* definitions of math anxiety,
* types of math anxiety,
* causes of math anxiety, and
* effects of math anxiety on learning.

And you will learn about test anxiety:

* definitions of test anxiety

- why math tests create anxiety
- causes of test anxiety
- different types of test anxiety
- how to reduce test anxiety

Understanding math anxiety

Definition of Math Anxiety

Math anxiety is a relatively new concept in education. During the 1970s, certain educators began using the terms "mathophobia" and "mathemaphobia" as a possible cause for children's unwillingness to learn math. Math anxiety is an extreme emotional and/or physical reaction to a very negative attitude toward math. There is a strong relationship between low math confidence and high math test anxiety (Fox, 1977).

Math anxiety is the feeling of tension and anxiety that interferes with the manipulation of numbers and the solving of math problems during tests. (Richardson and Sulnn, 1972). Math anxiety is a state of panic, helplessness, paralysis, and mental disorganization that occurs in some students when they are required to solve math problems. This discomfort varies in intensity and is the outcome of numerous previous experiences students have had in their past learning situations (Tobias, 1986).

Today, math anxiety is accepted as one of the major problems students have in completing their math courses. It is real, but it can be overcome.

Types of Math Anxiety

Math anxiety can be divided into three separate anxieties: Math Test Anxiety, Numerical Anxiety and Abstraction Anxiety. Math Test Anxiety involves anticipation, completion and feedback of math tests. Numerical Anxiety refers to everyday situations requiring working with numbers and performing arithmetic calculations. Numerical anxiety can also include students who are trying to figure out the amount for a tip, thinking about mathematics, doing math homework or listening/seeing math instruction. Abstraction Anxiety deals with working with variables and mathematical concepts used to solve equations. Students can have all three math anxieties or only one anxiety. Most often, the students I have worked with have Math Test Anxiety and Abstraction Anxiety. These students don't have any anxiety working with numbers, but once they start learning algebra, they develop both conditions. This may have happened in high school or college.

The Causes of Math Anxiety

Since math anxiety is a learned condition, its causes are unique with each student, but they are all rooted in individuals' past experiences. Bad experiences in elementary school are one of the most common sources for students' math anxiety: coming in last in math races at the blackboard, watching a classmate next to them finish a problem twice as fast as they do, teachers saying, "That's okay. You just aren't good in math. You are better in English," or classmates and teachers calling them stupid. These words and experiences remain with people; they can still hear the words and eventually begin telling themselves the same thing. When these students walk into the classroom or open a math book, or take a test, these "mental tapes" play in their minds. When asked, many students indicated that they were made fun of when trying to solve math problems at the chalkboard. When they could not solve the problem, the teacher and/or students would call them "stupid."

Teacher and peer embarrassment and humiliation become the conditioning experience that causes some students' math anxiety. Over the years, this math anxiety is reinforced and even increases in magnitude. In fact, many math anxious students — now 30, 40 and 50 years old — *still* have extreme fear about working math problems on the board. One 56 year old indicated that he had a great deal of fear that the instructor would call him to the board. Even if he knew how to do the problem, displaying that knowledge to his peers was very difficult. Some students have said that they absolutely refused to go to the board.

Being embarrassed by family members can also cause math anxiety. According to students who have been interviewed on the matter, their parents tried to help them with math and this sometimes led to serious trauma. These students claim that the tutoring from their guardians, mainly their fathers, often resulted in scolding when they were not able to complete the problems. One student reported that his father hit him every time he got a multiplication problem wrong. Brothers and sisters can also tease one another about being dumb in math. This is particularly true of boys telling girls that they cannot do math. When people hear these statements enough times, they may start to believe them and associate these bad feelings with the word math. So, for students many years later just hearing the word "math" triggers a response of anxiety, consciously or unconsciously recalling the bad feelings, and becoming uneasy.

A good example of this is a student who I worked with who had completed her BS degree fifteen years ago at a college that did not require much math. She was returning to college to be an elementary school teacher, which required her to take math and a math placement test. As soon as I mentioned that she had to take math, she said, "I can't do math and I will have to wait a few days to get psychologically ready to take the math placement test." She indicated her old feeling of not being able

to do math rushed through her and she almost had an anxiety attack. This is an extreme case but a true example of math anxiety. In most cases math anxiety is not this bad, but it is disruptive enough to cause learning and testing problems.

If you have math anxiety, try to remember the first time you had anxiety about learning math. This doesn't include anxiety when taking a math test, which will be discussed later in this chapter. Was it in elementary, middle or high school? Can you recall a specific incident? Was it being called on in class or getting a poor grade on a homework assignment? Was it a parent saying he/she cannot do math and neither can you? If you can't remember a specific incident, then when was the last time you told yourself that you couldn't learn math? Today? Yesterday? A year ago?

Some students with math anxiety also have had positive experiences with learning math. Now, try to remember the first positive experience with math. Was it a teacher, parent, friend, or other person who praised you? Was it getting back a good grade on a homework assignment or test? Was it making a good grade in a math course? Think back to your last positive experience with math one more time. How did it make you feel? Those who don't have math anxiety still need to understand it, so they can help classmates that might have it and need support. Also, it is possible that they will develop it in the future.

One way to overcome math anxiety is to try to find out when and why it first occurred and how it is still affecting you. Right now, reread the questions in the previous two paragraphs and on separate paper, brainstorm your answers to the questions. You do not have to write complete sentences. Just record the information. If you think of other memories as you brainstorm, write those down too.

If some of the memories are tough ones, just make a short note of them right now. Don't dwell on them because they are in the past. This can be frustrating for some students, but it is worth working through it. You will realize that math anxiety is usually a result of events, not because of being "dumb" or other personal flaws.

If you have had a very positive experience in the past when you were studying math, brainstorm answers to the appropriate questions. You may want to keep in mind friends you know who struggle with math and think about the other questions.

This brainstorming will help you write your math autobiography assignment at the end of this chapter.

How Math Anxiety Affects Learning

Math Anxiety can cause learning problems in several ways. It can affect how you do your homework and your participation in the classroom and study groups. Let's first start by looking at how math anxiety could affect your homework. Students with high math anxiety may have difficulty starting or completing their homework. Doing homework reminds some students of their learning problems in math. More specifically, it reminds them of their previous math failures, which causes further anxiety. This anxiety can lead to total avoidance of homework or "approach-avoidance" behavior.

Total homework avoidance is called procrastination. The very thought of doing their homework causes these students anxiety, which causes them to put off tackling their homework. This makes them feel better for a short amount of time — *until test day.*

Math anxiety can also affect your classroom participation and learning. Usually students with math anxiety are afraid to speak out in class and ask questions. They remember that in the past they were made fun of when giving the wrong answer. They are also afraid of asking a question that others, including the teacher, will consider dumb. So they sit in class fearful of being asked a question, looking like they understand the lecture so they will not be

called on. They also take a lot of notes even though they don't understand them, in order to give the illusion of comprehension. If you are one of these students, these are hard habits to break. However, these habits may cause you to be unsuccessful in your math class. Here are some suggestions to break these habits:

1. Make an appointment to talk to your math instructor. Math instructors want you to talk to them. When I do my consulting around the country one of the major complaints I get from math instructors is that the students don't come and see them. Make an appointment to see your math instructor before the first major test to discuss you math history and to ask for learning suggestions. In most cases it is easier to talk to the instructor before you get your first grade.

2. Before class, ask the instructor to work one homework problem. You might want to write the problem on the board before the instructor arrives. This is less stressful because you are not asking the question in front of the whole class. In fact, one of my good friends, Dr. Mike Hamm, suggests that his students put the problems they do not know on the board before class. Other students go to the board and solve the problems. Dr. Hamm then solves the ones the students cannot do.

3. Prepare one question from your homework and ask it within the first 15 minutes of class. Instructors are more likely to answer questions in the first part of class when they have time instead of the end of class when time is running out.

4. Ask a question that you already know the answer. That way if the instructor starts to ask you a question about the problem, you will know the answer. This is good practice for asking those questions to which you don't know the answers.

5. Use email to send questions to your instructor. This way you can still get the answer with very little anxiety.

By using these suggestions you can reduce your math anxiety and learn more mathematics. A question unanswered could be a missed test question.

REMEMBER

The instructor's job is to answer your questions and you are paying for the course.

Math anxious students sometimes avoid doing additional math outside of the classroom. They avoid study groups and Supplemental Instruction. It is like asking a person with aquafobia (fear of water) to take vacation at the beach. However, a person with aquafobia can go to the beach and enjoy himself or herself and not get wet. In other words, students can still attend study groups and Supplemental Instruction and just listen. When they are ready to get their feet wet, they can ask a few questions. Don't let these great opportunities go by.

Math anxiety can affect how you learn mathematics. It can be overcome with your effort. You don't have to live in the past with your math fears. Today is a new day and you can change how math affects you. The next step is to understand how test anxiety can affect your demonstration of math knowledge.

Section Review

1. List and explain the three types of math anxiety.

    First Type: _____

    _____

    _____

    _____

    _____

    _____

    Second Type: _____

    _____

    _____

    _____

    _____

    Third Type: _____

    _____

    _____

    _____

    _____

    _____

2. List and explain two causes of math anxiety.

    First Cause: _____

    _____

    _____

    _____

    _____

    _____

    Second Cause: _____

    _____

    _____

continued

_____

_____

_____

3. List three ways math anxiety can cause learning problems.

First Way: _____

_____

_____

_____

_____

Second Way: _____

_____

_____

_____

_____

_____

Third Way: _____

_____

_____

_____

_____

4. List and explain three ways students can manage or avoid the affects of math anxiety.

First Way: _____

_____

_____

_____

_____

_____

Second Way: _____

_____

_____

_____

_____

_____

_____

Third Way: _____

_____

_____

_____

_____

_____

_____

How to recognize test anxiety

Test anxiety has existed for as long as tests have been used to evaluate student performance. Because it is so common and because it has survived the test of time, test anxiety has been carefully studied over the last 50 years. Pioneering studies indicate that test anxiety generally leads to low test scores.

At the University of South Florida (Tampa), Dr. Charles Spielberger investigated the relationship between test anxiety and intellectual ability. The study results suggested that anxiety coupled with high ability can improve academic performance; but anxiety coupled with low or average ability can interfere with academic performance. That is:

Anxiety + high ability = improvement
Anxiety + low or average ability = no improvement

Example: Students with average ability and low test anxiety had better performance and higher grades than did students with average ability and high test anxiety. However, there are students who make good grades, take calculus, and still have test anxiety.

Test anxiety is a *learned* response; a person is not born with it. An environmental situation brings about test anxiety. The good news is that because it is a learned response, it can be *unlearned*. Test anxiety is a special kind of general stress. General stress is considered "strained exertion," which can lead to physical and psychological problems.

Defining Test Anxiety

There are several definitions of test anxiety. One definition states, "Test anxiety is a conditioned emotional habit to either a single terrifying experience, recurring experience of high anxiety, or a continuous condition of anxiety" (Wolpe, 1958).

Another definition of test anxiety relates to the educational system. The educational system develops evaluations that measure one's mental performance, and this creates test anxiety. This definition suggests that test anxiety is the *anticipation* of some realistic or nonrealistic situational threat (Cattell, 1966). The "test" can be a research paper, an oral report, work at the chalkboard, a multiple-choice exam, a written essay, or a math test.

Math test anxiety is a relatively new concept in education. *Ms.* magazine published "Math Anxiety: Why Is a Smart Girl Like You Counting on Your Fingers?" (Tobias, 1976) and coined the phrase *math anxiety*. During the 1970s, other educators began referring to *mathophobia* and *mathemaphobia* as a possible cause for children's unwillingness to learn math. Additional studies on the graduate level discovered that math anxiety was common among adults as well as children.

One of my students once described math test anxiety as "being in a burning house with no way out." No matter how you define it, math test anxiety is real, and it affects millions of students.

Why Math Tests Create Anxiety

It has been shown that math anxiety exists among many students who usually do not suffer from other tensions. Counselors at a major university reported that one-third of the students who enrolled in behavior therapy programs offered through counseling centers had problems with math anxiety (Sulnn, 1988).

Educators know that math anxiety is common among college students and is more prevalent in women than in men. They also know that math anxiety frequently occurs in students with a poor high school math background. These students were found to have the greatest amount of anxiety.

Approximately half of the students in college prep math courses (designed for students with inadequate high school math background or low placement scores) could be considered to have math anxiety. However, math anxiety also occurs in students in high-level math courses, such as college algebra and calculus.

Educators investigating the relationship between anxiety and math have indicated that anxiety contributes to poor grades in math. They also found that simply *reducing* math test anxiety does not guarantee higher math grades. Students often have other problems that affect their math grades, such as poor study skills, poor test-taking skills, or poor class attendance.

Section Review

1. What is your personal definition of test anxiety?

   _____

   _____

   _____

   _____

   _____

2. What type of student does math anxiety affect most?

   _____

   _____

   _____

   _____

   _____

The causes of test anxiety

The causes of test anxiety can be different for each student, but they can be explained by seven basic concepts. See Figure 3 on the next page.

The causes of math test anxiety can be different for each student. It could possibly have first occurred in middle or high school. However, for many students it first occurs in college when passing tests is the only way to pass the course. Homework and extra credit in most college courses don't count toward your grade. Now students must have a passing average and in some cases must pass the departmental final exam. Additional pressure also exists because not passing algebra means you won't graduate and you might not get the job you want. As you can see, there are more reasons to have math test anxiety in college than in high school.

If you have math test anxiety, I want you to try to remember the first time it surfaced. Was it in middle school, high school or college? Can you recall a specific incident? Was it your first algebra test? Was it your first math test after being out of school for a long time? Was it after you decided to get serious about college? Was it a college algebra course that was required for your major? Was it an instructor who told you that if you don't pass the next math test you would fail the course? Was it when you needed to pass the next math test to pass the course so you could keep your financial aid? Was it your children asking you why you failed your last math test? If you cannot remember a specific incident when you had test anxiety, do you expect to have any major anxiety on the next test?

Students with math test anxiety also have had positive experiences with taking math tests. Now try to remember your first positive experience with taking a math test. Was it in middle school, high school or in college? Was it after studying many hours for the test? Now think back to your last positive experience with a math test. How did it make you feel?

Since we have already explored your experiences with taking math tests, let's look at some of the direct causes of your math test anxiety. If you do have test anxiety, what is the main cause? If you don't know, then review the seven Causes of Test Anxiety in Figure 3. Does one of these reasons fit you? If you don't have test anxiety, what would be a reason that could cause test anxiety?

Like you did before, brainstorm answers to the questions listed in the previous three paragraphs. You will also use this brainstorming to help you write your autobiography.

REMEMBER

For some students, just writing about their previous math history helps them.

Section Review

1. List and describe five causes of test anxiety.

First Cause: _____

_____

_____

_____

_____

_____

Figure 3
Causes of Test Anxiety

- Test anxiety can be a learned behavior resulting from the expectations of parents, teachers, or other significant people in the student's life.

- Test anxiety can be caused by the association between grades and a student's personal worth.

- Test anxiety develops from fear of alienating parents, family, or friends due to poor grades.

- Test anxiety can stem from a feeling of lack of control and an inability to change one's life situation.

- Test anxiety can be caused by a student's being embarrassed by the teacher or other students when trying to do math problems.

- Test anxiety can be caused by timed tests and the fear of not finishing the test, even if one can do all the problems.

- Test anxiety can be caused by being put in math courses above one's level of competence.

What are the cause(s) of *your* test anxiety?

Second Cause: _____

_____

_____

_____

_____

_____

Third Cause: _____

_____

_____

_____

_____

_____

Fourth Cause: _____

_____

_____

_____

_____

_____

Fifth Cause: _____

_____

_____

_____

_____

_____

2. Put into your own words your cause(s) of test anxiety. If you don't have test anxiety, list what you think is the major cause of test anxiety.

_____

_____

_____

_____

_____

_____

_____

_____

_____

_____

_____

The different types of test anxiety

The two basic types of test anxiety are emotional (educators call this *somatic*) and worry (educators call this *cognitive*). Students with high test anxiety have *both* emotional and worry anxiety.

Signs of emotional anxiety are upset stomach, nausea, sweaty palms, pain in the neck, stiff shoulders, high blood pressure, rapid, shallow breathing, rapid heartbeat, or general feelings of nervousness. As anxiety increases, these feelings intensify. Some students even run to the bathroom to throw up or have diarrhea.

Even though these *feelings* are caused by anxiety, the physical response is real. These feelings and physical inconveniences can affect your concentration and your testing speed, and they can cause you to completely "draw a blank."

Worry anxiety causes the student to think about failing the test. These negative thoughts can happen either before or during the test. This negative "self-talk" causes students to focus on their anxiety instead of recalling math concepts.

The effects of test anxiety range from a "mental block" on a test to avoiding homework. One of the most common side effects of test anxiety is getting the test and immediately forgetting information that you know. Some students describe this event as having a "mental block," "going blank," or finding that the test looks like a foreign language.

After five or ten minutes into the test, some of these students can refocus on the test and start working the problems. They have, however, lost valuable time. For other students, anxiety persists throughout the test and they cannot recall the needed math information. It is only after they walk out the door that they can remember how to work the problems.

Sometimes math anxiety does not cause students to "go blank" but slows down their mental processing speed. This means it takes longer to recall formulas and concepts and to work problems. The result is frustration and loss of time, leading to *more* anxiety. Since, in most cases, math tests are *speed* tests (those in which you have a certain amount of time to complete the test), you may not have enough time to work all the problems or to check the answers if you have mentally slowed down. The result is a lower test score, because even though you know the material, you do not complete all of the questions before test time runs out.

Not using all of the time allotted for the test is another problem caused by test anxiety. Students know that they should use all of the test time to check their answers. In fact, math is one of the few subjects in which you can check test problems to find out if your work is correct. However, most students do not use all of the test time, and this results in lower test scores. Why does this happen?

Students with high test anxiety do not want to stay in the classroom. This is especially true of students whose test anxiety increases as the test progresses. The test anxiety gets so bad that they would rather leave early and receive a lower grade than stay in that "burning house."

Figure 4
The 12 Myths About Test Anxiety

Students are born with test anxiety.

Test anxiety is a mental illness.

Test anxiety cannot be reduced.

Any level of test anxiety is bad.

All students who are not prepared have test anxiety.

Students with test anxiety cannot learn math.

Students who are well prepared will not have test anxiety.

Very intelligent students and students taking high-level courses, such as calculus, do not have test anxiety.

Attending class and doing homework should reduce all test anxiety.

Being told to relax during a test will make a student relaxed.

Doing nothing about test anxiety will make it go away.

Reducing test anxiety will guarantee better grades.

Students have another reason for leaving the test early: the fear of what the instructor and other students will think about them for being the last one to hand in the test. These students refuse to be the last ones to finish the test because they think that the instructor or other students will think they are dumb. This is middle-school thinking, but the feelings are still real — no matter what the age of the student. These students do not realize that some students who turn in their tests first fail, whereas many students who turn in their tests last make A's and B's.

Another effect of test anxiety relates to completing homework assignments. Students who have high test anxiety may have difficulty starting or completing their math homework. Doing homework reminds some students of their learning problems in math. More specifically, it reminds them of their previous math failures, which causes further anxiety. This anxiety can lead to total homework avoidance or "approach-avoidance" behavior.

Total homework avoidance is called procrastination. The very thought of doing their homework causes these students anxiety, which causes them to put off tackling their homework. This makes them feel better for a short amount of time — *until test day.*

Example: Some students begin their homework and work some problems successfully. They then get stuck on an problem that causes them anxiety, so they take a break. During their break the anxiety disappears until they start doing their homework again. Doing their homework causes more anxiety, which leads to another break. The breaks become more frequent.

continued

Finally, the student ends up taking one long break and not doing the homework. Quitting, to them, means *no more anxiety* until the next homework assignment.

The effects of math test anxiety can be different for each student. Students can have several of the mentioned characteristics that can interfere with math learning and test taking. However, there are certain myths about math that each student needs to know. Review Figure 4, The 12 Myths About Test Anxiety, to see which ones you believe. If you have test anxiety, which of the mentioned characteristics are true for you?

Section Review

1. List and describe the two basic types of test anxiety.

   First Type: _____

   _____

   _____

   _____

   Second Type: _____

   _____

   _____

   _____

2. List two reasons students leave the test room early instead of checking their answers.

First Reason: _____

_____

_____

Second Reason: _____

_____

_____

3. List 6 of the 12 myths about test anxiety that you most believed.

First Myth: _____

_____

_____

Second Myth: _____

_____

_____

Third Myth: _____

_____

_____

Fourth Myth: _____

_____

_____

Fifth Myth: _____

_____

_____

Sixth Myth: _____

_____

_____

How to reduce test anxiety

To reduce math test anxiety, you need to understand both the relaxation response and how negative self-talk undermines your abilities.

Relaxation Techniques

The relaxation response is any technique or procedure that helps you to become relaxed. It will take the place of an anxiety response. Someone simply telling you to relax or even telling yourself to relax, however, does little to reduce your test anxiety. There are both short-term and long-term relaxation response techniques that help control emotional (somatic) math test anxiety. These techniques will also help reduce worry (cognitive) anxiety. Effective *short-term* techniques include the tensing and differential relaxation method, the palming method, and deep breathing.

Short-Term Relaxation Techniques

The Tensing and Differential Relaxation Method

The tensing and differential relaxation method helps you relax by tensing and relaxing your muscles all at once. Follow these procedures while you are sitting at your desk before taking a test:

1. Put your feet flat on the floor.
2. With your hands, grab underneath the chair.
3. Push down with your feet and pull up on your chair at the same time for about five seconds.
4. Relax for five to ten seconds.
5. Repeat the procedure two or three times.
6. Relax all your muscles except the ones that are actually used to take the test.

The Palming Method

The palming method is a visualization procedure used to reduce test anxiety. While you are at your desk before or during a test, follow these procedures:

1. Close and cover your eyes using the center of the palms of your hands.
2. Prevent your hands from touching your eyes by resting the lower parts of your palms on your cheekbones and placing your fingers on your forehead. Your eyeballs must not be touched, rubbed, or handled in any way.
3. Think of some real or imaginary relaxing scene. Mentally visualize this scene. Picture the scene as if you were actually there, looking through your own eyes.
4. Visualize this relaxing scene for one to two minutes.
5. Open your eyes and wait about two minutes and repeat the visualization scene again. This time also imagine any sounds or smells that can enhance your scene. For example, if you are imagining a beach scene, hear waves on the beach and smell the salty air. This technique can also be completed without having your hands over your eyes by just closing your eyes.

Practice visualizing this scene several days before taking a test and the effectiveness of this relaxation procedure will improve.

Deep Breathing

Deep breathing is another short-term relaxation technique that can help you relax. Proper breathing is a way to reduce stress and decrease test anxiety. When breathing properly, enough oxygen gets into your bloodstream to nourish your body and mind. A lack of oxygen in your blood contributes to an anxiety state that makes it more difficult to react to stress. Proper deep breathing can help you control your test anxiety.

Deep breathing can replace the rapid, shallow breathing that sometimes accompanies test anxiety, or it can prevent test anxiety. Here are the steps to deep breathing:

1. Sit straight up in your chair in a good posture position.
2. Slowly inhale through your nose.
3. As you inhale, first fill the lower section of your lungs and work your way up to the upper part of your lungs.
4. Hold your breath for a few seconds.
5. Exhale slowly through your mouth.
6. Wait a few seconds and repeat the cycle.
7. Keep doing this exercise for four or five minutes. This should involve going through about ten breathing cycles. Remember to take two normal breaths between each cycle. If you start to feel light-headed during this exercise, stop for thirty – forty-five seconds and then start again.
8. Throughout the entire exercise, make sure you keep breathing smoothly and in a regular rhythm without gulping air or suddenly exhaling.
9. As an extra way to improve your relaxation, say "relax" or "be calm" to yourself as you exhale. This can start a conditioned response that can trigger relaxation when you repeat the words during anxious situations. As you keep practicing, this conditioned response will strengthen. Practice is the key to success.

You need to practice this breathing exercise for several weeks before using the technique during tests. If you don't practice this technique, then it will not work. After taking your first test keep doing the exercise several times a week to strengthen the relaxation response.

Side one of the audiocassette *How to Reduce Test Anxiety* (Nolting, 1986) further explains test anxiety and discusses these and other short-term relaxation response techniques. Short-term relaxation techniques can be learned quickly but are not as successful as the long-term relaxation technique. Short-term techniques are intended to be used while learning the long-term technique.

Long-Term Relaxation Techniques

The cue-controlled relaxation response technique is the best long-term relaxation technique. It is presented on side two of the audiocassette *How To Reduce Test Anxiety* (Nolting, 1986). Cue-controlled relaxation means you can induce your own relaxation based on repeating certain cue words to yourself. In essence, you are taught to relax and then silently repeat cue words, such as "I am relaxed."

After enough practice, you can relax during math tests. The cue-controlled relaxation technique has worked with thousands of students. For a better understanding of test anxiety and how to reduce it, listen to *How to Reduce Test Anxiety* (Nolting, 1986).

Negative Self-Talk

According to cognitive psychologists, self-talk is what we say to ourselves as a response to an event or situation. These dialogues determine our feelings about that event or situation. Sometimes we say it

so quickly and automatically that we don't even hear ourselves. We then think it is the situation that causes the feeling, but in reality it is our interactions or thoughts about the experience that are controlling our emotions. This sequence of events can be represented by this timeline:

External Events
(math test)
↓
Interpretation of Events and Self-Talk
(how you feel about the test
and what you are telling yourself)
↓
Feelings and Emotions
(happy, glad, angry, mad, upset)

In most cases this means that you are responsible for how and what you feel. You can have positive self-talk or negative self-talk about taking a math test. Yes, some students have positive self-talk about math tests and see it as a challenge and something to accomplish that makes them feel good, while others see it as an upsetting event that leads to anger and anxiety. In other words, you are what you tell yourself.

Negative self-talk is a form of worry (cognitive) anxiety. This type of worrying can interfere with your test preparation and can keep you from concentrating on the test. Worrying can motivate you to study, but too much worrying may prevent you from studying at all.

Negative self-talk is defined as the negative statements you tell yourself before and during tests. Negative self-talk causes students to lose confidence and to give up on tests. Further, it can give you an inappropriate excuse for failing math and cause you to give up on learning math.

Examples of Negative Self-Talk

"No matter what I do, I will not pass this course."

"I failed this course last semester, and I will fail it again."

"I am no good at math, so why should I try?"

"I cannot do it; I cannot do the problems, and I am going to fail this test."

"I have forgotten how to do the problems, and I am going to fail."

"I am going to fail this test and never graduate."

"If I can't pass this test, I am too dumb to learn math and will flunk out."

Students who have test anxiety are prone to negative self talk. Test anxiety can be generated or heightened by repeatedly making statements to yourself that usually begin with "what if." For example, "What if I fail the test?" or "What if I fail this class again?" These "what if" statements generate more anxiety that can cause students to feel sick. These statements tell them to be anxious. Some other aspects of self-talk are:

- Self–talk can be in telegraphic form with short words or images.
- Self-talk can be illogical but at the time the person believes it.
- Negative self-talk can lead to avoidance like not taking the test or skipping classes.
- Negative self-talk can cause more anxiety.
- Negative self-talk can lead to depression and a feeling of helplessness.
- Negative self-talk is a bad habit that can be changed.

There are different types of negative self-talk. If you have negative self-talk, then review the types below and see which one fits you best. You may use a combination of them:

1. *The Critic* is the person inside us who is always trying to put us down. It constantly judges behaviors and finds fault even if it is not there. It jumps on any mistake and exaggerates it to cause more anxiety. The Critic puts us down for mistakes on the test and blames us for not controlling the anxiety. The Critic reminds us of previous comments from real people who have criticized us. It compares us to other students who are doing better in the class. It loves to say, "That was a stupid mistake!" or "You are a total disappointment. You can't pass this math class like everyone else can!" The Critic's goal is to promote low self-esteem.

2. *The Worrier* is the person inside us who looks at the worst-case scenario. It wants to scare us with the ideas of disasters and complete failure. When it sees the first sign of anxiety, it "blows it out of proportion" to the extent that we will not remember anything and totally fail the test. The Worrier creates more anxiety than normal. The Worrier anticipates the worst, underestimates our ability, and sees us not only failing the test but "failing life." The Worrier loves to ask "What if?" For example, "What if I fail the math test and don't graduate?" or "What if I can't control my anxiety and throw up in the math class?" The goal of the Worrier is to cause more anxiety so we will quit.

3. *The Victim* is the person inside us who wants us to feel helpless or hopeless. It wants us to believe that no matter what we do, we will not be successful in math. The Victim does not blame other events (poor schooling) or people (bad teachers) for math failures. It blames us. It dooms us and puts us into a learned helpless mode,

meaning that if we try to learn math, we will fail or if we don't try to learn math we will fail. So why try? The Victim likes to say, "I can't learn math." The goal of the Victim is to cause depression and make us not try.

4. *The Perfectionist* is similar to the Critic, but is the opposite of the Victim. It wants us to do the best we can and will guide us into doing better. It tells us that we are not studying enough for the math test and that a B is not good enough and that we must make an A. In fact, sometimes an A is not good enough unless it is a 100%. So, if we make a B on the test, the Perfectionist says, "A or B is just like making an F." The Perfectionist is the hard-driving part of us that wants the best but cannot stand mistakes or poor grades. It can drive us to mental and physical exhaustion to make that perfect grade. It is not interested in self-worth, just perfect grades. Students driven by the Perfectionist often drop a math course because they only have a B average. The Perfectionist loves to repeat, "I should have . . ." or "I must . . ." The goal of the Perfectionist is to cause chronic anxiety that leads to burnout.

Review these types of personalities to see which one may fit you best. We may have a little of each one in us, but what is the dominant one for you in math? Now we can look at how to stop these negative thoughts.

Managing Negative Self-Talk

Students need to change their negative self-talk to positive self-talk without making unrealistic statements.

Positive self-statements can improve your studying and test preparation. During tests, positive self-talk can build confidence and decrease your test anxiety. These positive statements (see examples) can help reduce your test anxiety and improve your grades. Some more examples of positive self-statements are on the cassette *How to Reduce Test Anxiety* (Nolting, 1986). Before the test, make up some positive statements to tell yourself.

There are several ways to counter and control negative self-talk. Students can replace the negative self-talk with positive statements and questions that make them think in a realistic way about the situation. Another way is to develop thought stopping techniques to reduce or eliminate the negative thoughts. Try each way or a combination to see what works best for you.

Countering self-talk involves writing down and practicing positive statements that can replace negative statements. Students can develop their own positive statements. Some rules for writing positive statements are:

1. Use the first person present tense, for example, "I can control my anxiety and relax myself."
2. Don't use negatives in the statement. For example, don't say, "I am not going to have anxiety." Instead say, "I will be calm during the test."
3. Have a positive and realistic belief in the statement. For example say, "I am going to pass this test," instead of saying, "I am going to make an A on this test" when you have not made an A on any of the tests.

The statements used to counter the negative thoughts can be based on the type of negative self-talk. The Critic who puts down by saying, "Your test anxiety is going to cause you to fail the test" can be countered with, "I have test anxiety, but I am learning to control it." The Worrier who says, "What if I fail the test" can be countered with "So what? I will do better on the next test." The Victim who thinks things are hopeless and says, "I will never be able to pass math" can be countered with, "I have studied differently for this test, and I can pass the math course." The Perfectionist who says, "I need to make an A on the test or I will drop out of school" can be countered with, "I don't need to make an A to please anyone. All I need is to pass the math course to get the career I want." These are some of the examples of counter statements that can control anxiety.

Examples of Positive Self-Talk

"I failed the course last semester, but I can now use my math study skills to pass this course."

"I went blank on the last test, but I now know how to reduce my test anxiety."

"I know my poor math skills are due to poor study skills, not my own ability, and since I am working on my study skills, my math skills will improve."

"I know that with hard work, I will pass math."

"I prepared for this test and will do the best I can. I will reduce my test anxiety and use the best test-taking procedures. I expect some problems will be difficult, but I will not get discouraged."

"I am solving problems and feel good about myself. I am not going to worry about that difficult problem; I am going to work the problems that I can do. I am going to use all the test time and check for careless errors. Even if I do not get the grade I want on this test, it is not the end of the world."

Thought-Stopping Techniques

Some students have difficulty stopping their negative self-talk. These students cannot just tell themselves to eliminate those thoughts. These students need to use a thought-stopping technique to overcome their worry and become relaxed.

Thought stopping involves focusing on the unwanted thoughts and, after a few seconds, suddenly stopping those thoughts by emptying your mind. Using the command "Stop!" or a loud noise like clapping your hands can effectively interrupt the negative self-talk. In a homework situation, you may be able to use a loud noise to stop your thoughts, but don't use it during a test.

To stop your thoughts in the classroom or during a test, silently shout to yourself "Stop!" or "Stop thinking about that." After your silent shout, either relax yourself or repeat one of your positive self-talk statements. You may have to shout to yourself several times during a test or while doing homework to control negative self-talk. After every shout, use a different relaxation technique/scene or positive self-talk statement.

Thought stopping works because it interrupts the worry response before it can cause high negative emotions. During that interruption, you can replace the negative self-talk with positive statements or relaxation. However, students with high worry anxiety should practice this technique three days to one week before taking a test. Contact your counselor if you have additional questions about the thought-stopping techniques.

Writing the Math Autobiography

An autobiography relates how you remember and feel about your past experiences. In addition, many autobiographies explore how these past feelings and experiences shape current life. While some people write autobiographies for others to learn about their lives, many people write private autobiographies in order to understand what is going on in their lives.

Use the Appendix as a guide to write your Math Autobiography. Your final product will be a typed (or handwritten if your instructor approves) paper that summarizes your experiences learning math, how you feel about these experiences, and how they shape your current perspective on learning math.

Section Review

1. Describe your best short-term relaxation technique.

   _____

   _____

   _____

   _____

   _____

   _____

2. You can use the palming method by closing your eyes and visualizing a scene without putting your hands to your face. Describe a very relaxing scene that you could visualize. Make sure to include some sounds and visual images in your scene.

   _____

   _____

   _____

_____

_____

_____

3. Practice your relaxation scene for three to five minutes for the next five days. List the times and dates you practiced your scene.

   Date: _____ Time: _____
   Date: _____ Time: _____
   Date: _____ Time: _____
   Date: _____ Time: _____
   Date: _____ Time: _____

4. How does negative self-talk cause you to have text anxiety?

   _____

   _____

   _____

   _____

   _____

   _____

5. Make up three positive self-talk statements that are not listed in this text.

   Statement 1: _____

   _____

   Statement 2: _____

   _____

Statement 3: _____
_____
_____

6. Describe the thought-stopping technique in your own words.

_____

_____

_____

_____

_____

_____

7. What words will you use as your silent shout?

_____

_____

_____

_____

8. What will you do after your silent shout?

_____

_____

_____

9. How does the thought-stopping technique work?

_____

Chapter 3 Review

1. Reducing math test anxiety _____ _____ guarantee good math grades.

2. Test anxiety is a _____ response; a person is not _____ with it.

3. Math test anxiety involves _____, _____, and _____ of math tests.

4. One cause of test anxiety is that a student goes to the board to work a problem but is called _____ when he or she cannot work the problem.

5. The two basic types of test anxiety are _____ and _____.

6. The effects of test anxiety range from a "_____ _____" on a test to _____ homework.

7. The _____ and _____ _____ method helps you relax by tensing and relaxing your muscles all at once.

8. The palming method is a _____ procedure used to reduce text anxiety.

9. Negative self-talk is a form of _____ anxiety.

10. Thought stopping works because it interrupts the _____ response before it can cause _____ emotions.

What is the most important information you learned from this chapter?

_____

_____

_____

_____

_____

_____

_____

_____

_____

_____

_____

_____

_____

_____

How can you immediately use it?

_____

_____

_____

_____

_____

_____

_____

_____

_____

_____

_____

_____

_____

Lecture Notes

4

How to Improve Your Listening and Note-Taking Skills

Listening and note-taking skills in a math class are very important, since most students either do not read the math text or have difficulty understanding it. In most of your other classes, if you do not understand the lecture, you can read the text and get almost all the information. In math class, however, the instructor can usually explain the textbook better than the students can read and understand it.

Students who do not have good listening skills or note-taking skills will be at a disadvantage in learning math. Most math understanding takes place in the classroom. Students must learn how to take advantage of learning math in the classroom by becoming effective listeners, calculator users, and notetakers.

In this chapter you will learn

- how to become an effective listener,
- how to become a good notetaker,
- when to take notes,
- the seven steps to math note-taking, and
- how to rework your notes.

How to become an effective listener

Becoming an effective listener is the foundation for good note-taking. You can become an effective listener using a set of skills that you can learn and practice. To become an effective listener, you must prepare yourself both physically and mentally.

Sitting in the Golden Triangle

The physical preparation for becoming an effective listener involves *where you sit* in the classroom. Sit in the best area to obtain high grades, that is, in "the golden triangle of success." The golden triangle of success begins with seats in the front row facing the instructor's desk (see Figure 5 on the next page).

Students seated in this area (especially in the front row) directly face the teacher and are most likely to pay attention to the lecture. This is a great seating location for visual learners. There is also less tendency for them to be distracted by activities outside the classroom or by students making noise within the classroom.

The middle seat in the back row is another point in the golden triangle for students to sit, especially those who are auditory (hearing) learners. You can hear the instructor better because the instructor's voice is projected to that point. This means that there is less chance of misunderstanding the instructor, and you can hear well enough to ask appropriate questions.

By sitting in the golden triangle of success, you can force yourself to pay more attention during class and be less distracted by other students. This is very important for math students because math instructors usually go over a point once and continue on to the next point. If you miss that point in the lesson, then you could be lost for the remainder of the class.

This may seem obvious, but I know too many students who do the following if they have early morning classes. Roll out of bed, put on the closest clothes to them, brush their hair, maybe their teeth, and race to class. Don't do this. You won't wake up until the end of your class! Train yourself to be somewhere on campus to review your notes at least fifteen minutes before class starts and do some of the other strategies suggested in the next section. So, another physical way to be ready is to be awake and nourished!

Warming Up for Math Class

The mental preparation for note-taking involves "warming up" before class begins and becoming an active listener. Just as an athlete must warm up before a game begins, you must warm up before taking notes. Warm up by

- reviewing the previous day's notes,
- reviewing the reading material,
- reviewing the homework,
- preparing questions, and
- working one or two unassigned homework problems.

This mental warm up before the lecture allows you to refresh your memory and prepare pertinent questions, making it easier to learn the new lecture material.

Figure 5
The Golden Triangle of Success

Instructor

Visual Learners

Auditory Learners

Back of the Classroom

How to Become an Active Listener

Becoming an active listener is the second part of the mental preparation for note-taking.

Do not anticipate what the instructor is going to say or immediately judge the instructor's information before the point is made. This will distract you from learning the information.

> **Examples:** Watch the speaker, listen for main ideas, and nod your head or say to yourself, "I understand," when agreeing with the instructor.

Expend energy looking for interesting topics in the lecture. When the instructor discusses information that you need to know, immediately repeat it to yourself to begin the learning process.

You can practice this exercise by viewing math videotapes and repeating important information. This is an especially good learning technique for auditory learners.

Ask appropriate questions when you have them. Sometimes if you are so confused you do not know what question to ask, still let the instructor know you are confused. Also, mark your notes where you got lost so that if the instructor doesn't have time to answer your question, you can visit during office hours to get an answer.

REMEMBER

> Class time is an intense study period that should not be wasted.

Listening and Learning

Some students think that listening to the instructor and taking notes is a waste of valuable time. Students too often sit in class and use only a fraction of their learning ability. Class time should be considered a valuable study period where you can listen, take notes, and learn at the same time. One way to do this is by memorizing important facts when the instructor is talking about material you already know. Another technique is to repeat back to yourself the important concepts right after the instructor says them in class. Using class time to learn math is an efficient learning system.

Section Review

1. Explain how sitting in the golden triangle of success makes you a better listener.

   _____

   _____

   _____

   _____

   _____

2. List five ways you can warm up before math class begins.

   First Way: _____

   _____

   _____

   _____

Second Way: _____

_____

_____

_____

Third Way: _____

_____

_____

_____

Fourth Way: _____

_____

_____

_____

Fifth Way: _____

_____

_____

_____

3. Select your best warm-up process and try it before your next math class. What was your warm-up process, and how did it help you?

_____

_____

_____

_____

4. How can you listen and learn at the same time?

_____

_____

_____

_____

_____

_____

_____

_____

_____

_____

How to become a good notetaker

Becoming a good notetaker requires two basic strategies. One strategy is to be specific in detail. In other words, *copy* the problems down, step by step. The second strategy is to *understand* the general principles, general concepts, and general ideas.

Copying from the Board

While taking math notes, you need to copy each and every step of the problem even though you may already know every step of the problem. While in the classroom, you might understand each step, but a week later you might not remember how to do the problem unless all the steps were written down. In addition, as you write down each step, you are memorizing it. Make sure to copy every step for each problem written on the board.

There will be times when you will get lost while listening to the lecture. Nevertheless, you should keep taking notes, even though you do not understand the problem. This will provide you with a reference point for further study. Put a question mark (?) by those steps that you do not understand;

then, after class, review the steps you did not understand with the instructor, your tutor, or another student.

If you seem to think many other students are lost too, ask questions or let the instructor know.

Taking Notes

The goal of note-taking is to take the least amount of notes and get the greatest amount of information on your paper. This could be the opposite of what most instructors have told you. Some instructors tell you to take down everything. This is not necessarily a good note-taking system, since it is very difficult to take precise, specific notes while trying to understand the instructor.

What you need to develop is a note-taking system in which you write the least amount possible and get the most information down while still understanding what the instructor is saying.

Develop an Abbreviation List

To reduce the amount of written notes, an abbreviation system is needed. An abbreviation system is your own system of reducing long words to shorter versions that you still can understand. By writing less, you can listen more and have a better understanding of the material.

Example: When the instructor starts explaining the commutative property, you need to write "commutative property" out the first time. After that, use "COM." You should develop abbreviations for all the most commonly used words in math.

Figure 6
Abbreviations

E.G.	(for example)
CF.	(compare, remember in context)
N.B.	(note well, this is important)
\	(therefore)
∴	(because)
É	(implies, it follows from this)
>	(greater than)
<	(less than)
=	(equals, is the same)
ı	(does not equal, is not the same)
( )	(parentheses in the margin, around a sentence or group of sentences, indicates an important idea)
?	(used to indicate you do not understand the material)
O	(a circle around a word may indicate that you are not familiar with it; look it up)
E	(marks important materials likely to be used in an exam)
1, 2, 3,	(to indicate a series of facts)
D	(shows disagreement with statement or passage)
REF	(reference)
et al.	(and others)
bk	(book)
p	(page)
etc.	(and so forth)
V	(see)
VS	(see above)
SC	(namely)
SQ	(the following)
Com.	(commutative)
Dis.	(distributive)
APA	(associative property of addition)
AI	(additive inverse)
IPM	(identity property of multiplication)

Figure 6 on the previous page provides a list of abbreviations. Add your own abbreviations to this list. By using abbreviations as much as possible, you can obtain the same meaning from your notes and have more time to listen to the instructor.

When to take notes

To become a better notetaker, you must know when to take notes and when not to take notes. The instructor will give cues that indicate what material is important. These cues include

- presenting usual facts or ideas,
- writing on the board,
- summarizing,
- pausing,
- repeating statements,
- enumerating, such as "1, 2, 3" or "A, B, C,"
- working several examples of the same type of problem on the chalkboard,
- saying, "This is a tricky problem." Most students will miss it (for example, 5/0 is "undefined" instead of "zero").
- saying, "This is the most difficult step in the problem,"
- indicating that certain types of problems will be on the test, such as coin or age problems, and
- explaining bold-face words.

You must learn the cues your instructor gives to indicate important material. If you are in doubt about the importance of the class material, do not hesitate to ask the instructor about its importance.

While taking notes, you may become confused about math material. At that point, take as many notes as possible, and do not give up on note-taking. One thing you shouldn't do—bother the student next to you for an explanation. This can be annoying, and both of you will miss important information while you are talking.

As you take notes on confusing problem steps, leave extra space; then go back and fill in information that clarifies your misunderstanding of the steps in question. Ask your tutor or instructor for help with the uncompleted problem steps, and write down the reasons for each step in the space provided.

Another procedure to save time while taking notes is to stop writing complete sentences. Write your main thoughts in phrases. Phrases are easier to jot down and easier to memorize.

Section Review

1. Note-taking requires two basic strategies:

   _____ and

   _____.

2. List five abbreviations (and their meanings) that you use in your math class (one abbreviation and meaning per line).

   _____

   _____

   _____

   _____

   _____

3. List five cues that your instructor gives that indicate what material is important (one cue per line).

_____

_____

_____

_____

_____

The seven steps to math note-taking

No matter what format you use to take notes, a good set of notes contains the following:

- examples of problems
- explanations of how to complete the problems
- appropriate math vocabulary and rules that are used for each problem

In addition, it is wise to make them as organized as possible. It may get on your nerves for a while, but when you see that organized notes save time when studying, you will be more willing to take the time to organize them.

Many successful math students, whether in developmental math courses or calculus, use a form of the three column note-taking system. It organizes the information, making it easy to review. While creating the three column system, students are also learning the math, so it is not busy work. It does take a few weeks to feel comfortable with it, but it is worth sticking with it. Many students have adopted this system and improved their studying time.

The seven steps lead you through several important stages in the learning process. Steps one through three help you *record the information* accurately, completely, and in an organized fashion. Steps four through seven help you *rehearse and understand the information*. These steps also help you learn to *recall the information*, helping you prepare for the test.

To prepare for taking notes this way, get a good chunk of paper ready. Here are some choices:

- Use regular paper, portrait or landscape style, and draw two lines to create three columns. They should be proportioned as follows:

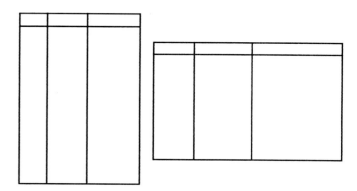

- Some students purchase graph paper because it helps them keep numbers lined up correctly.
- Other students design the page on the computer and print out copies.
- Choose two columns next to each other to be the "Examples" and the "Explanations" columns. Label them as such. These are the two columns you will use all the time during class.
- Label the other column "Key Words and Rules." You might fill this column in during class but will probably complete it outside of class.

If you do this ahead of time, you do not have to frantically draw lines while the instructor starts talking. If you forget or run out of paper, it might be quicker to fold the paper into three columns.

Figure 7

Three Column Note-taking Method

Key Words/Rules	Example	Explanation
Solve linear equation	$4(x+4) + 3(x-4) = 2(x-1)$	Have to get x on one side of the = and numbers on other side of the =
Distributive Property	$4x + 16 + 3x - 12 = 2x - 2$	Multiply numbers to the left of the ( ) with each variable and number in the ( ).
Community Property	$4x + 3x + 16 - 12 = 2x - 2$	Regroup numbers and variables.
Combine like terms	$7x + 4 = 2x - 2$	
Additive inverse	$7x - 2x + 4 = 2x - 2x - 2$ $5x + 4 = -2$	Subtract $2x$ from both sides in order to get variables all on the right of the =
Additive inverse	$5x + 4 - 4 = -2 - 4$ $5x = -6$	Subtract 4 from both sides to get numbers all on the right of the =
Mulitiplicative inverse	$\dfrac{5x}{5} = \dfrac{-6}{5}$	Divide both sides by 5 to get x by itself on the left side of the =
Simplify	$x = -1\ 1/5$	Put as a whole number and fraction.
	Insert new problem	

Draw a line dividing the end of one problem from the beginning of the next. Or, do one problem per page to give plenty of room for notes. Graph paper can help students keep numbers and words lined up. Turn the paper landscape.

During class you will use two columns most of the time—the "Example" and the "Explanations" columns. Sometimes, when instructors really want their students to use this method, they will organize their board work to reflect what the notes should look like. Most of the time, however, this does not happen.

The "Examples" column is where you will place the examples the instructor uses during class.

The "Explanations" column is where you put the instructor's verbal explanations of the examples. If the instructor tends to explain by just doing the problem without talking about it, ask questions that will force the instructor to explain the problem verbally.

When the instructor simply explains a vocabulary word or rule, write the vocabulary word or a phrase naming the rule in the 'Key Words and Rules" column. Write the explanation in the "Explanation" column. If the instructor provides an example, obviously it goes in the "Example" column. If no example is provided, ask for one. Then if your wish is not granted, look one up in the textbook after class.

Here are the "Seven Steps to Taking Math Notes." Remember, these are the basic steps. You can adapt them to work for you.

Step 1   *Record the instructor's example problems in the "Example" column.*

Step 2   *Record the reasons for each step in the "Explanations" column.* After completing a problem, draw a horizontal line, separating it from the next problem.

If you get lost during the lecture, place a question mark in this column and move on.

Step 3   *If possible, record the key words and rules in the other column.* Here is an opportunity for you to ask what rules are being used if the instructor is assuming the class knows them and you do not. Otherwise, you will need to refer to the

textbook, work with a tutor, or use online resources after class to complete this column.

It is important to understand the vocabulary and rules –not just know how to do them!

Step 4   *In order to keep everything in your heads, not letting anything slip out, use the notes to review.* Cover up the "Examples" and "Explanations" columns and quiz yourself on the vocabulary and rules. Or, fold back the "Key Words and Rules" column, look at the examples and explanations and choose the correct key words and rules used to complete the examples.

Step 5   *Place a check mark by the key words and concepts that you did not know.*

Step 6   *Review the information that you checked until you understand and remember it.*

Step 7   *Develop a glossary of the math terms and rules that are difficult for you to remember.*

For a while you might spend more time organizing this system outside of class. Some students have to still depend on their current process, concentrating on getting the information down. Then they transfer the information into the three column method. However, it is best to learn how to do most of the process during math class.

Remember the learning modalities and learning styles? This process can use all of them!

- *Kinesthetic:* Writing out the notes
- *Auditory:* Saying information out loud
- *Visual:* Looking at the notes and thinking; color code each column with highlighters, making sure you do not smear the pencil writing
- *Group:* Working on the notes with one or two other people to make sure they are complete and accurate

• *Individual:* Reading them alone while waiting to meet someone.

Current research is discovering that using language (vocabulary, discussion) to learn math helps students to understand the concepts more deeply. Then completing many problems helps master the skills and increase the speed which is important for preparing for tests. These notes will help you understand the math so that it will stick with you. You will remember more information. Homework will help you master the skills and increase the speed at which you can complete problems.

A Math Glossary

The math glossary is created to define a math vocabulary in your own words. Since math is considered a foreign language, understanding the math vocabulary becomes the key to comprehending math. Creating a glossary for each chapter of your textbook will help you understand math.

Your glossary should include all words printed in bold face in the text and any words you do not understand. If you cannot explain the math vocabulary in your own words, ask your instructor or tutor for help. You may want to use the last pages in your notebook to develop a math glossary for each chapter in your textbook. Some students use note cards. Review your math glossary before each test.

Section Review

1. List and describe the seven steps to note-taking.

   First Step: _____

_____

_____

_____

Second Step: _____

_____

_____

_____

Third Step: _____

_____

_____

_____

Fourth Step: _____

_____

_____

_____

Fifth Step: _____

_____

_____

_____

Sixth Step: _____

_____

_____

_____

Seventh Step: _____

_____

_____

_____

2. Make several copies of the modified three-column note-taking system, and take notes using this system in your next math class. Or develop the modified three-column note-taking system in your own notebook.

3. List two benefits of a math glossary.

Benefit 1: _____

_____

_____

_____

Benefit 2: _____

_____

_____

_____

4. From your current math chapter, list and define five words that you can put in your math glossary.

Word 1: _____

_____

_____

Word 2: _____

_____

Word 3: _____

_____

_____

Word 4: _____

_____

_____

Word 5: _____

_____

_____

How to rework your notes

The note-taking system does not stop when you leave the classroom. As soon as possible after class, rework your notes. You can rework the notes between classes or as soon as you get home. By reworking your notes as soon as possible, you can decrease the amount of forgetting. This is an excellent procedure to transfer math information from short-term memory to long-term memory.

REMEMBER

Most forgetting occurs right after learning the material. You need to rework the notes as soon as possible. Waiting means that you probably will not understand what was written.

Here are important steps in reworking your notes:

Step 1   *Rewrite the material you cannot read or will not be able to understand a few weeks later.* If you do not rework your notes, you will be frustrated when studying for a test if you come across notes you cannot read. Another benefit of rewriting the notes is that you immediately learn the new material. Waiting means it will take more time to learn the material.

Step 2   *Fill in the gaps.* Most of the time, when you are listening to the lecture, you cannot write down everything. It is almost impossible to write down everything, even if you know shorthand. Locate the portions of your notes that are incomplete. Fill in the concepts that were left out. In the future, skip two or three lines in your notebook page for anticipated lecture gaps.

Step 3   *Add additional key words and rules in the keywords and rules column.* These key words or ideas were the ones not recorded during the lecture.

> **Example:** You did not know that you should add the *opposite* of 18 to solve a particular problem, and you incorrectly added 18. Put additional important key words and ideas (such as "opposite" and "negative of") in the notes; these are the words that will improve your understanding of math.

Step 4   *Make a problem log on those problems that the teacher worked in class.* The problem log is a separate section of your notebook that contains a listing of the problems (without explanations — just problems) that your teacher worked in class. If your teacher chose those problems to work in class, you can bet that they are considered important. The problems in this log can be used as a practice test for the next exam. Your regular class notes will not only contain the solutions but also all the steps involved in arriving at those solutions and can be used as a reference when you take your practice test.

Write the problem on the left half of the page. Then save some room on the right side for writing out what the test instructions might be for the problem. Sometimes we forget what the test instructions are, and we miss points on the test by following the directions wrong. Skip several lines to complete the problem when you use the problem log as a practice test.

Step 5   *Make a calculator handbook and put in it your keystroke sequences.* The calculator handbook can be a spiral-bound set of note cards or a separate section of your notebook that holds only calculator-related information. Your handbook should also include an explanation of when that particular set of keystrokes is to be used.

Step 6   *Reflection and synthesis.* Once you have finished going over your notes, review the major points in your mind. Combine your new notes with your previous knowledge to have a better understanding of what you have learned today.

If you are a group learner and even if you are not, it helps to review and reflect with other students.

Use a Tape Recorder

If you have a math class during which you cannot get all the information down while listening to the lecture, ask your instructor about using a tape recorder. To ensure success, the tape recorder must have a tape counter and must be voice activated.

The tape counter displays a number indicating the amount of tape to which you have listened. When you find you are in an area of confusing information, write the beginning and ending tape counter number in the left margin of your notes. When reviewing your notes, the tape count number will be a reference point for obtaining information to work the problem. You can also reduce the time it takes to listen to the tape by using the pause button to stop the recording of unnecessary material.

Ask Questions

To obtain the most from a lecture, you must ask questions in class. By asking questions, you improve your understanding of the material and decrease your homework time. By *not* asking questions, you create for yourself unnecessary confusion during the remainder of the class period. Also, it is much easier to ask questions in class about potential homework problems than it is to spend hours trying to figure out the problems on your own at a later time.

If you are shy about asking questions in class, write down the questions and read them to your instructor. If the instructor seems confused about the questions, tell him or her that you will discuss the problem after class. To encourage yourself to ask questions, remember

- you have paid for the instructor's help,
- five other students probably have the same question,
- the instructor needs feedback on his or her teaching to help the class learn the material, and
- there is no such thing as a "stupid" question.

Record Each Problem Step

The final suggestion in note-taking is to record each step of every problem written or verbally explained. By recording each problem step, you begin *overlearning* how to work the problems. This will increase your problem-solving speed during future tests. If you get stuck on the homework, you will also have complete examples to review.

The major reason for recording every step of a problem is to understand how to do the problems while the instructor is explaining them instead of trying to remember unwritten steps. Although it may seem time consuming, it pays off during homework and test time.

Section Review

1. List and describe the six steps to reworking your notes.

First Step: _____

_____

_____

_____

Second Step: _____

_____

_____

_____

_____

Third Step: _____

_____

_____

_____

_____

Fourth Step: _____

_____

_____

_____

_____

Fifth Step: _____

_____

_____

_____

_____

Sixth Step: _____

_____

_____

_____

_____

2. How can using a tape recorder in class improve
    your learning?

    _____

    _____

    _____

3. List three reasons for asking questions in class.

    First Reason: _____

    _____

    _____

    _____

    Second Reason: _____

    _____

    _____

    _____

    Third Reason: _____

    _____

    _____

    _____

4. What is the major reason for recording each
    problem step?

    _____

    _____

    _____

    _____

    _____

    _____

    _____

Chapter 4 Review

1. Sitting in the golden triangle of success can help you pay more _____ during class.

2. Class time is an intense _____ _____ that should not be wasted.

3. Auditory learners should review the math videotape and _____ back the important information.

4. The goal of note-taking is to take the _____ amount of notes and get the _____ amount of information.

5. Using abbreviations can help you write _____ and listen _____.

6. Your _____ should include any words you don't know and their explanations.

7. Rework your notes to transfer information from _____ - _____ memory to _____ - _____ memory.

8. A _____ _____ is a separate section in your notebook that contains problems that your teacher worked in class.

9. Asking _____ in class will improve your understanding and decrease your homework time.

10. _____ how to work problems will increase your problem-solving speed and memory.

What is the most important information you learned from this chapter?

_____

_____

_____

_____

_____

_____

_____

_____

_____

How can you immediately use it?

_____

_____

_____

_____

_____

_____

_____

Lecture Notes

How to Improve Your Reading, Homework, and Study Techniques

Reading a math textbook is more difficult than reading other textbooks. Math textbooks are written differently than your English or social science textbooks. Math textbooks contain condensed material and therefore take longer to read.

Mathematicians can reduce a page of writing to one paragraph using math formulas and symbols. To make sure you understood that same information, an English instructor would take that original page of writing and expand it into two pages. Mathematicians pride themselves on how little they can write and still cover the concept. This is one reason it may take you two to three times as long to read your math text as it would any other text.

REMEMBER

Reading your math text will take longer than reading your other texts.

Math students are expected to know how to do their homework; however, most math students do not have a homework system. Most students begin their homework by going directly to the problems and trying to work them. When they get stuck, they usually quit. This is not a good homework system.

A good homework system will improve your homework success and math learning at the same time.

In this chapter, you will learn

- how to read a math textbook,
- how to do your homework,
- how to solve word problems,
- how to work with a study buddy.

How to read a math textbook

The way you read a math textbook is different from the traditional way students are taught to read textbooks in high school or college. Students are taught to read quickly or skim the material. If you do not understand a word, you are supposed to keep on reading. Instructors of other courses want students to continue to read so they can pick up the unknown words and their meanings from context.

This reading technique may work with your other classes, but using it in your math course will be totally confusing. By skipping some major concept words or bold-face words, you will not understand the math textbook or be able to do the homework. Reading a math textbook takes more time and concentration than reading your other textbooks.

If you have a reading problem, it would be wise to take a developmental reading course before taking math. This is especially true with reform math courses, where reading and writing are emphasized.

Reform math classes deal more with word problems than traditional math courses do. If you cannot take the developmental reading course before taking math, then take it during the same semester as the math course.

Ten Steps to Understanding Reading Materials

There are several appropriate steps in reading a math textbook:

Step 1   *Skim the assigned reading material.* Skim the material to get the general idea about the major topics. Read the chapter introduction and each section summary. You do not want to learn the material at this time; you simply want to get an overview of the assignment. Then think about similar math topics that you already know.

> **Example:** Skimming will allow you to see if problems presented in one chapter section are further explained in later chapter sections.

Step 2   *As you skim the chapter, circle (using*

pencil) the new words that you do not understand. If you do not understand these new words after reading the assignment, then ask the instructor for help. Skimming the reading assignments should take only five to ten minutes.

Step 3   *Put all your concentration into reading.* While reading the textbook, highlight the material that is important to you. However, do not highlight more than 50 percent of a page because the material is not being narrowed down enough for future study. Especially highlight the material that is also discussed in the lecture. Material discussed in both the textbook and lecture usually appears on the test. The purpose of highlighting is to emphasize the important material for future study. Do not skip reading assignments.

> REMEMBER
>
> *Reading a math textbook is very difficult. It might take you half an hour to read and understand just one page.*

Step 4   *When you get to the examples, go through each step.* If the example skips any steps, make sure you write down each one of those skipped steps in the textbook for better understanding. Later on, when you go back and review, the steps are already filled in. You will understand how each step was completed. Also, by filling in the extra steps, you are starting to overlearn the material for better recall on future tests.

Step 5   *Mark the concepts and words that you do not know.* Maybe you marked them the first time while skimming. If you understand them now, erase the marks. If you do not understand the words or concepts, then reread the page or look them up in the glossary. Try not to read any further until you understand all the words and concepts.

Step 6 *If you do not clearly understand some words or concepts, add these words to the note-taking glossary in the back of your notebook.* Your glossary will contain the bold-face words that you do not understand. If you have difficulty understanding the bold-face words, ask the instructor for a better explanation. You should know all the words and concepts in your notebook's glossary before taking the test.

Step 7 *If you do not understand the material, follow these eight points, one after the other, until you do understand the material:*

Point 1 — Go back to the previous page and reread the information to maintain a train of thought.

Point 2 — Read ahead to the next page to discover if any additional information better explains the misunderstood material.

Point 3 — Locate and review any diagrams, examples, or rules that explain the misunderstood material.

Point 4 — Read the misunderstood paragraph(s) several times aloud to better understand their meaning.

Point 5 — Refer to your math notes for a better explanation of the misunderstood material.

Point 6 — Refer to another math textbook, computer software program, or videotape that expands the explanation of the misunderstood material.

Point 7 — Define exactly what you do not understand and call your study buddy for help.

Point 8 — Contact your math tutor or math instructor for help in understanding the material.

Step 8 *Recall important material.* Try to recall some of the important concepts and rules. If you have difficulty recalling this information, look at the highlighted areas of the text, bold print words or formulas. If you are a visual learner, write down these words several times on a separate sheet of paper. If you are an auditory learner, repeat these word several times out loud.

If you are a kinesthetic (hands on) learner, write out the formulas twice on two separate cards. Cut one of them up into puzzle pieces. Mix up the parts of the puzzle. Turn over the other formula card so you cannot see it. Put the puzzle back together. Do this several times.

For bold print words, write the bold print word on one card and its explanation twice on two other separate cards. Make another puzzle. Mix up the pieces of the explanation and put them back together.

Then add these words and formulas to your note-taking glossary. Recalling important material right after you read it improves you ability to remember that information.

Step 9   *Reflect on what you have read.* Combine what you already know with the new information that you just read. Think about how this new information enhances your math knowledge. Prepare questions for your instructor on the confusing information. Ask those questions at the next class meeting.

Step 10   *Write anticipated test questions.* Research has noted that students have about 80 percent accuracy in predicting test questions. Think about what is the most important concept you just read and what problems the instructor could give you that would test the knowledge of that concept. Make up three to five problems and add them to your problem log (this was part of your note-taking system). Indicate in the log that these are your questions, not the instructors.

By using this reading technique, you have narrowed down the important material to be learned. You have skimmed the textbook to get an overview of the assignment. You have carefully read the material and highlighted the important parts. You then added to your note-taking glossary unknown words or concepts.

REMEMBER

The highlighted material should be reviewed before doing the homework problems, and the glossary has to be learned 100 percent before taking the test.

How Reading Ahead Can Help

Reading ahead is another way to improve learning. If you read ahead, do not expect to understand everything. Read ahead two or three sections and put question marks (in pencil) by the material you do not understand.

When the instructor starts discussing that material, have your questions prepared and take good notes. Also, if the lecture is about to end, ask the instructor to explain the confusing material in the textbook. Reading ahead will take more time and effort, but it will better prepare you for the lectures.

How to Establish Study Period Goals

Before beginning your homework, establish goals for the study period. Do not just do the homework problems.

Ask yourself this question: "What am I going to do tonight to become more successful in math?"

By setting up short-term homework goals and reaching them, you will feel more confident about math. This also improves your self-esteem and helps you become a more internally motivated student. Set up homework tasks that you can complete. Be realistic.

Study period goals are set up either on a time-line basis or an item-line basis. Studying on a time-line basis is studying math for a certain amount of time.

Example: You may want to study math for an hour, then switch to another subject. You will study on a time-line basis.

Studying on an item-line basis means you will study your math until you have completed a certain number of homework problems.

Example: You might set a goal to study math until you have completed all the odd-numbered problems in the chapter review. The odd-numbered problems are the most important problems to work. These, in most texts, are answered in the answer section in the back of the book. Such problems provide the opportunity to recheck your work if you do not get the answer correct. Once you have completed these problems, do the even-numbered problems.

No matter what homework system you use, remember this important rule: Always finish a homework session by understanding a concept or doing a homework problem correctly.

Do not end a homework session with a problem you cannot complete. You will lose confidence, since all you will think about is the last problem you could not solve instead of the 50 problems you correctly solved. If you did quit on a problem you could not solve, return and rework problems you have done correctly.

REMEMBER

Do not end your study period with a problem you could not complete.

Section Review

1. List and describe the ten steps to reading a math textbook.

First Step: _____

_____

_____

_____

Second Step: _____

_____

_____

_____

Third Step: _____

_____

_____

_____

Fourth Step: _____

_____

_____

_____

Fifth Step: _____

_____

_____

_____

Sixth Step: _____

_____

_____

_____

Seventh Step: _____

_____

_____

_____

Eighth Step: _____

_____

_____

_____

Ninth Step: _____

_____

_____

_____

Tenth Step: _____

_____

_____

_____

2. After trying each of the ten steps to reading your math text, make up, list, and try your own condensed version.

First Step: _____

_____

_____

Second Step: _____

_____

_____

Third Step: _____

_____

Fourth Step: _____

_____

_____

Fifth Step: _____

_____

_____

Sixth Step: _____

_____

_____

Seventh Step: _____

_____

_____

_____

Eighth Step: _____

_____

_____

_____

Ninth Step: _____

_____

_____

_____

Tenth Step: _____

_____

_____

_____

3. List two reasons why reading ahead can improve your learning.

First Reason: _____

_____

_____

_____

Second Reason: _____

_____

_____

_____

4. Describe the two types of study period goals. Which one do you use?

First Goal: _____

_____

_____

Second Goal: _____

_____

_____

How to do your homework

Doing your homework can be frustrating or rewarding. Most students jump right into their homework, become frustrated, and stop studying. These students usually go directly to the math problems and start working them without any preparation. When they get stuck on one problem, they flip to the back of the text for the answer. Then they either try to work the problem backward to understand the problem steps or they just copy down the answer.

Other students go to the solution guide and just copy the steps. After getting stuck several times, these students will inevitably quit doing their homework assignment. Their homework becomes a frustrating experience, and they may even quit doing their math homework altogether.

10 Steps to Doing Your Homework

To improve your homework success and learning, refer to these ten steps:

Step 1 *Review the textbook material that relates to the homework.* A proper review will increase the chances of successfully completing your homework. If you get stuck on a problem, you will have a better chance of remembering the location of similar problems. If you do not review prior to doing your homework, you could get stuck and not know where to find help in the textbook.

REMEMBER

To be successful in learning the material and in completing homework assignments, you must first review your textbook.

Step 2 *Review your lecture notes that relate to the homework.* If you could not understand the explanation in the textbook on how to complete the homework assignment, then review your notes.

REMEMBER

Reviewing your notes will give you a better idea about how to complete your homework assignment.

Step 3   *Do your homework as neatly as possible.* Doing your homework neatly has several benefits. When approaching your instructor about problems with your homework, he or she will be able to understand your previous attempts to solve the problem. The instructor will easily locate the mistakes and show you how to correct the steps without having to decipher your handwriting. Another benefit is that when you review for midterm or final exams, you can quickly relearn the homework material without having to decipher your own writing.

REMEMBER

Neatly prepared homework can help you now and in the future.

Step 4   *When doing your homework, write down every step of the problem.* Even if you can do the step in your head, write it down anyway. This will increase the amount of homework time, but you are overlearning how to solve problems, which improves your memory. Doing every step is an easy way to memorize and understand the material. Another

advantage is that when you rework the problems you did wrong, it is easy to review each step to find the mistake.

REMEMBER

In the long run, doing every step of the homework will save you time and frustration.

Step 5   *Understand the reasons for each problem step and check your answers.* Do not get into the bad habit of memorizing how to do problems without knowing the reasons for each step. Many students are smart enough to memorize procedures required to complete a set of homework problems. However, when similar homework problems are presented on a test, the student cannot solve the problems. To avoid this dilemma, keep reminding yourself about the rules, laws, or properties used to solve problems.

Example: *Problem:* $2(a + 5) = 0$. What property allows you to change the equation to $2a + 10 = 0$? *Answer:* The distributive property.

Once you know the correct reason for going from one step to another in solving a math problem, you can solve any problem requiring that property. Students who simply memorize how to do problems instead of understanding the reasons for correctly working the steps will eventually fail their math course.

How to Check Your Answers

Checking your homework answers should be part of your homework duties. Checking your answers can improve your learning and help you prepare for tests.

Check the answers of the problems for which solutions are not given. This may be the even-numbered or odd-numbered problems or the problems not answered in the solutions manual.

First, check your answer by estimating the correct answer.

> **Example:** If you are multiplying 2.234 by 5.102, remember that 2 times 5 is 10. The answer should be a little over 10.

You can also check your answers by substituting the answer back into the equation or doing the opposite function required to answer the question. The more answers you check, the faster you will become. This is very important because increasing your answer-checking speed can help you catch more careless errors on future tests.

Step 6   *If you do not understand how to do a problem, refer to the following points:*

Point 1 — Review the textbook material that relates to the problem.

Point 2 — Review the lecture notes that relate to the problem.

Point 3 — Review any similar problems, diagrams, examples, or rules that explain the misunderstood material.

Point 4 — Refer to another math textbook, solutions guide, math computer program software, or videotape to obtain a better understanding of the material.

Point 5 — Call your study buddy.

Point 6 — Skip the problem and contact your tutor or math instructor as soon as possible for help.

Step 7   *Always finish your homework by successfully completing problems.* Even if you get stuck, go back and successfully complete previous problems before quitting. You want to end your homework assignment with feelings of success.

Step 8   *After finishing your homework assignment, recall to yourself or write down the most important learned concepts.* Recalling this information will increase your ability to learn these new concepts. Additional information about step 8 will be presented later in this chapter.

Step 9   *Make up note cards containing hard-to-remember problems or concepts.* Note cards are an excellent way to review material for a test. More information on the use of note cards as learning tools is presented later in this chapter.

Step 10   *Do not fall behind.* As mentioned in Chapter 1, math is a sequential learning process. If you get behind, it is difficult to catch up because each topic builds on the next. It would be like going to Spanish class without learning the last set of vocabulary words. The teacher would be talking to you using the new vocabulary, but you would not understand what was being said.

Do Not Fall Behind

To keep up with your homework, it is necessary to complete the homework every school day and even on weekends. Doing your homework one-half hour each day for two days in a row is better than one hour every other day.

If you have to get behind in one of your courses, *make sure it is not math.* Fall behind in a course that does not have a sequential learning process, such as psychology or history. After using the ten steps to doing your homework, you may be able to combine two steps into one. Find your best combination of homework steps and use them.

REMEMBER

Getting behind in math homework is the fastest way to fail the course.

Section Review

1. List the ten steps to doing your math homework.

First Step: _____

_____

_____

Second Step: _____

_____

_____

Third Step: _____

_____

_____

Fourth Step: _____

_____

_____

Fifth Step: _____

_____

_____

Sixth Step: _____

_____

_____

Seventh Step: _____

_____

_____

_____

Eighth Step: _____

_____

_____

_____

Ninth Step: _____

_____

_____

_____

Tenth Step: _____

_____

_____

2. After trying the ten steps to doing your math homework, make up, list, and try your own condensed version.

First Step: _____

_____

_____

Second Step: _____

_____

_____

Third Step: _____

_____

_____

Fourth Step: _____

_____

_____

Fifth Step: _____

_____

_____

Sixth Step: _____

_____

_____

Seventh Step: _____

_____

_____

Eighth Step: _____

_____

_____

Ninth Step: _____

_____

_____

Tenth Step: _____

_____

_____

3. How will reviewing your notes or math text help you complete your homework more successfully?

_____

_____

_____

_____

_____

_____

_____

_____

_____

4. List two ways you can check your homework answers.

First Way: _____

_____

_____

_____

Second Way: _____

_____

_____

_____

5. What happens to students who fall behind in their homework?

_____

_____

_____

_____

_____

_____

How to solve word problems

The most difficult homework assignment for most math students is working story/word problems. Solving word problems requires excellent reading comprehension and translating skills.

Students often have difficulty substituting algebraic symbols and equations for English terms. But once an equation is written, it is usually easily solved. To help you solve word problems, follow these ten steps:

Step 1   *Read the problem three times.* Read the problem quickly the first time as a scanning procedure. As you are reading the problem the second time, answer these three questions:

1. *What is the problem asking me?* (This is usually at the end of the problem.)
2. *What is the problem telling me that is useful?* (Cross out unneeded information.)
3. *What is the problem implying?* (This is usually something you have been told to remember.) Read the problem a third time

to check that you fully understand its meaning.

Step 2   *Draw a simple picture of the problem to make it more real to you* (e.g., a circle with an arrow can represent travel in any form — by train, by boat, by plane, by car, or by foot).

Step 3   *Make a table of information and leave a blank space for the information you are not told.*

Step 4   *Use as few unknowns in your table as possible.* If you can represent all the unknown information in terms of a single letter, do so! When using more than one unknown, use a letter that reminds you of that unknown. Then write down what your unknowns represent. This eliminates the problem of assigning the right answer to the wrong unknown. Remember that you have to create as many separate equations as you have unknowns.

Step 5   *Translate the English terms into an algebraic equation using the list of terms in Figures 8 and 9 on the next page.* Remember, the English terms are sometimes stated in a different order than the algebraic terms.

Step 6   *Immediately retranslate the equation, as you now have it, back into English.* The translation will not sound like a normal English phrase, but the meaning should be the same as the original problem. If the meaning is not the same, the equation is incorrect and needs to be rewritten. Rewrite the equation until it means the same as the English phrase.

Step 7   *Review the equation to see if it is similar to equations from your homework and if it makes sense.* Some formulas dealing with specific word problems may need to

Figure 8

Translating English Terms
into Algebraic Symbols

Sum	+
Add	+
In addition	+
More than	+
Increased	+
In excess	+
Greater	+
Decreased by	-
Less than	-
Subtract	-
Difference	-
Diminished	-
Reduce	-
Remainder	-
Times as much	x
Percent of	x
Product	x
Interest on	x
Per	/
Divide	/
Quotient	/
Quantity	( )
Is	=
Was	=
Equal	=
Will be	=
Results	=
Greater than	>
Greater than or equal to	$\geq$
Less than	<
Less than or equal to	$\leq$

Figure 9

Translating English Words
into Algebraic Expressions

English Words	Algebraic Expressions
Ten more than x	$x + 10$
A number added to 5	$5 + x$
A number increased by 13	$x + 13$
5 less than 10	$10 - 5$
A number decreased by 7	$x - 7$
Difference between x and 3	$x - 3$
Difference between 3 and x	$3 - x$
Twice a number	$2x$
Ten percent of x	$0.10x$
Ten times x	$10x$
Quotient of x and 3	$x/3$
Quotient of 3 and x	$3/x$
Five is three more than a number	$5 = x + 3$
The product of 2 and a number is 10	$2x = 10$
One-half a number is 10	$x/2 = 10$
Five times the sum of x and 2	$5(x + 2)$
Seven is greater than x	$7 > x$
Five times the difference of a number and 4	$5(x - 4)$
Ten subtracted from 10 times a number is that number plus 5	$10x - 10 = x + 5$
The sum of $5x$ and 10 is equal to the product of x and 15	$5x + 10 = 15x$
The sum of two consecutive integers	$(x) + (x + 1)$
The sum of two consecutive even integers	$(x) + (x + 2)$
The sum of two consecutive odd integers	$(x) + (x + 2)$

be rewritten. Distance problems, for example, may need to be written solving for each of the other variables in the formula. Distance = rate x time; therefore, time = distance/rate, and rate = distance/time. Usually, a distance problem will identify the specific variable to be solved.

Step 8   *Solve the equation using the rules of algebra.*

REMEMBER

> *Whatever is done to one side of the equation must be done to the other side of the equation. The unknown must end up on one side of the equation, by itself. If you have more than one unknown, then use the substitution or elimination method to solve the equations.*

Step 9   *Look at your answer to see if it makes sense.*

> **Example:** If tax was added to an item, it should cost more; if a discount was applied to an item, it should cost less. Is there more than one answer? Does your answer match the original question? Does your answer have the correct units?

Step 10   *Put your answer back into the original equation to see if it is correct.* If one side of the equation equals the other side of the equation, then you have the correct answer. If you do not have the correct answer, go back to step 5.

The most difficult part of solving word problems is translating part of a sentence into algebraic symbols and then into algebraic expressions. Review Figure 7 (Translating English Terms into Algebraic Symbols) and Figure 8 (Translating English Words into Algebraic Expressions), both on the previous page.

How to Recall What You Have Learned

After completing your homework problems, a good visual learning technique is to make note cards. Note cards are 3" by 5" index cards on which you place information that is difficult to learn or material you think will be on the test.

On the front of the note cards write a math problem or information that you need to know. Color-code the important information in red or blue. On the back of the note card, write how to work the problem or explain important information.

> **Example:** If you are having difficulty remembering the rules for multiplying positive and negative numbers, you would write some examples on the front of the note card with the answers on the back.

Make note cards on important information you might forget. Every time you have five spare minutes, pull out your note cards and review them. You can glance at the front of the card, repeat to yourself the answer, and check yourself with the back of the card. If you are correct and know the information on a card, do not put it back in the deck. Mix up the cards you do not know and pick another card to test yourself on the information. Keep doing this until there are no cards left that you do not know.

If you are an auditory learner, then use the tape recorder like the note cards. Record the important information as you would on the front of the note card. Then leave a blank space on the recording before recording the answer. Play the tape back. When you hear the silence, put the tape on pause. Then say the answer out loud to yourself. Take the tape player off pause and see if you were correct.

Review What You Have Learned

After finishing your homework, close the textbook and try to remember what you have learned. Ask yourself these questions: "What major concepts did I learn tonight?" and "What test questions might the instructor ask on this material?"

Recall for about three to four minutes the major points of the assignment, especially the areas you had difficulty understanding. Write down questions for the instructor or tutor. Since most forgetting occurs right after learning the material, this short review will help you retain the new material.

Section Review

1. List the ten steps to solving word problems.

First Step: _____

_____

_____

Second Step: _____

_____

_____

Third Step: _____

_____

_____

Fourth Step: _____

_____

_____

Fifth Step: _____

_____

_____

Sixth Step: _____

_____

_____

Seventh Step: _____

_____

_____

Eighth Step: _____

_____

_____

Ninth Step: _____

_____

_____

Tenth Step: _____

_____

_____

2. What is your best way to recall what you have learned?

_____

_____

_____

_____

_____

3. How does reviewing right after doing your homework help you remember?

_____

_____

_____

_____

_____

_____

How to work with a study buddy

You need to have a study buddy when you miss class or when you do your homework. A study buddy is a friend or classmate who is taking the same course. You can find a study buddy by talking to your classmates or making friends in the math lab.

Try to find a study buddy who knows more about math than you do. Tell the class instructor that you are trying to find a study buddy and ask which students make the best grades. Meet with your study buddy several times a week to work on

problems and to discuss math. If you miss class, get the notes from your study buddy so you will not get behind.

Call your study buddy when you get stuck on your homework. You can solve math problems over the phone. Do not sit for half an hour or an hour trying to work one problem; that will destroy your confidence, waste valuable time, and possibly alienate your study buddy. Think how much you could have learned by trying the problem for 15 minutes and then calling your study buddy for help. Spend, at the maximum, 15 minutes on one problem before going on to the next problem or calling your study buddy.

REMEMBER

A study buddy can improve your learning while helping you complete the homework assignment. Just do not overuse your study buddy or expect that person to do your homework for you.

Section Review

1. How can you select your study buddy?

_____

_____

_____

_____

_____

_____

2. How can a study buddy help you learn math?

_____

_____

_____

_____

_____

_____

3. Who is your study buddy?

_____

The benefits of study breaks

Psychologists have discovered that learning decreases if you do not take study breaks. Therefore, take a break after studying math for 45 minutes to one hour.

If you have studied for only 15 or 20 minutes and feel you are not retaining the information or your mind is wandering, take a break. If you continue to force yourself to study, you will not learn the material. After taking a break, return to studying.

If you still cannot study after taking a break, review your purpose for studying and your educational goals. Think about what is required to graduate. It will probably come down to the fact that you will have to pass math. Think about how studying math today will help you pass the next test; this will increase your chances of passing the course and of graduating.

Write on an index card three positive statements about yourself and three positive statements about studying. Look at this index card every time you have a study problem. Use every opportunity available to reinforce your study habits.

Section Review

1. How can study breaks improve your learning?

_____

_____

_____

_____

_____

_____

2. What do you do when you don't want to study? How do you get started?

_____

_____

_____

_____

_____

_____

Chapter 5 Review

1. Reading one page of a math text might take you _____ _____ _____.

2. After reading your textbooks,_____ important material can improve your ability to remember that information.

3. Write anticipated tests question in your _____ _____ and indicate that they are your questions.

4. Checking your homework answers can improve your _____ and help prepare you for a _____.

5. Matching:

(for students in pre-algebra class)

_____ Product	A. +
_____ Increased	B. -
_____ Sum	C. x
_____ Per	D. /
_____ Product	E. =
_____ Is	
_____ Reduced	
_____ Remainder	

(for students taking elementary algebra or higher-level math course)

_____ Difference between x and 3	A. 10x
_____ Quotient of x and 3	B. x/3
_____ Five times the sum of x and 2	C. 5(x -4)
_____ A number added to 5	D. 5(x + 2)
	E. 5 = x + 3
	F. 5 + x

_____ Five less than 10   G. 10 -5

_____ Ten times x H. x -3

_____ Five is three more than a number

_____ Five times the difference of a number and four

6. Make sure you finish your homework by working problems you _____ do.

7. Using _____ _____ is a visual way to recall what you learned in math.

8. Using a _____ _____ is an auditory way to recall what you learned in math.

9. Most _____ occurs right after learning; a short review will help you _____ the new material.

10. A _____ _____ can improve your learning while helping you complete the homework assignment.

What is the most important information you learned from this chapter?

_____

_____

_____

_____

_____

_____

How can you immediately use it?

_____

_____

_____

_____

_____

6

How to Remember What You Have Learned

To understand the learning process, you must understand how your memory works. You learn by conditioning and thinking. But memorization is different from learning. For memorization, your brain must perform several tasks, including receiving the information, storing the information, and recalling the information.

By understanding how your memory works, you will be better able to learn at which point your memory is failing you. Most students usually experience memory trouble between the time their brain receives the information and the time the information is stored.

There are many techniques for learning information that can help you receive and store information without losing it in the process. Some of these techniques may be more successful than others, depending on how you learn best.

In this chapter, you will discover

- how you learn,
- how short-term memory affects what you remember,
- how working memory affects what you remember,
- how long-term memory/reasoning affects what you remember,
- how to use learning styles to improve memory,
- how to use memory techniques,

- how to develop practice tests, and
- how to use number sense.

How you learn

Educators tell us that learning is the process of "achieving competency." More simply put, it is how you become good at something. The three ways of learning are by conditioning, by thinking, and by a combination of conditioning and thinking.

Learning by Conditioning and Thinking

Conditioning is learning things with a maximum of physical and emotional reaction and a minimum of thinking.

Examples: Repeating the word *pi* to yourself and practicing where the

continued

symbol is found on a calculator are two forms of conditioned learning. You are learning using your voice and your eye-hand coordination (physical activities), and you are doing very little thinking.

Thinking is defined as learning with a maximum of thought and a minimum of emotional and physical reaction.

Example: Learning about *pi* by thinking is different from learning about it by conditioning. To learn "pi" by thinking, you would have to do the calculations necessary to result in the numeric value that the word *pi* represents. You are learning *using your mind* (thought activities), and you are using very little emotional or physical energy to learn *pi* in this way.

The most successful way to learn is to combine thinking and conditioning. The best learning combination is to learn by thinking first and conditioning second.

Learning by thinking means that you learn by

- *observing*,
- *processing*, and
- *understanding* the information.

How Your Memory Works

Memory is different from learning; it requires reception, storage, and retrieval of information. You

receive information through your five senses (*sensory input*): what you see, feel, hear, smell, and taste.

Examples: In math class, you will use your sense of *vision* both to watch the instructor demonstrate problems on the chalkboard and to read printed materials. You will use your sense of *hearing* to listen to the instructor and other students discuss the problems. Your sense of *touch* will be used to operate your calculator and to appreciate geometric shapes. In chemistry and other classes, however, you may additionally use your senses of *smell* and *taste* to identify substances.

The sensory register briefly holds an exact image or sound of each sensory experience until it can be processed. If the information is not processed immediately, it is forgotten. The sensory register helps us go from one situation to the next without cluttering up our minds with trivial information.

Processing the information involves placing it in either short-term memory, working memory, or long-term memory.

How short-term memory affects what you remember

Information that passes through the sensory register is stored in short-term memory. Remembering something for a short time is not hard to do for most students. By conscious effort, you can remember math laws, facts, and formulas received by the sensory register (your five senses) and put them in short-term memory. You can recognize them and register them in your mind as something to remember for a short time.

> **Example:** When you are studying math, you can tell yourself the distributive property is illustrated by $a(b + c) = ab + ac$. By deliberately telling yourself to remember that fact (by using conditioning — repeating or writing it again and again), you can remember it, at least for a while, because you have put it in short-term memory.

Psychologists have found that short-term memory cannot hold an unlimited amount of information. You may be able to use short-term memory to remember one phone number or a few formulas but not five phone numbers or ten formulas.

Items placed into short-term memory usually fade fast, as the name suggests.

> **Examples:** Looking up a telephone number in the directory, remembering it long enough to dial, then forgetting it immediately; learning the name of a person at a large party or in a class but forgetting it completely within a few seconds; cramming for a test and forgetting most of it before taking the test.

Short-term memory involves the ability to recall information immediately after it is given (without any interruptions). It is useful in helping you to concentrate on a few concepts at a time, but it is not the best way to learn. This is especially true of students who have attention problems or problems with short-term memory.

Students with short-term memory problems forget a math step as soon as the instructor explains it. To remember more facts or ideas and keep them in memory a longer period of time — especially at test time — use of a better system than short-term memory is required.

Changes in the way math is taught include the idea that conditioning (the most often used teaching form) leads only to short-term memory. Conditioning is often used for short-term recall of concept skills. The information learned in this way can only be stored for longer periods of time by thinking. Thinking takes place when applying the skills to working the problems.

How working memory affects what you remember

Working memory (or long-term retrieval) is that process in the brain that works on problems for a longer period of time than short-term memory. Working memory, then, offers an increase in the amount of *time* information is held in memory. (An increase in the *volume* of information that can be held requires long-term memory.)

Working memory is like the amount of RAM in a computer. Working memory uses the information (such as multiplication tables) recalled from long-term memory, along with new information to learn new concepts. It is the ability to think about and use many pieces of information at the same time. For instance, when you solve a linear equation you must use all the math that you learned in elementary school like addition, subtraction, and multiplication. You then have to add this to the new rules for linear equations.

Working memory can be compared to a mental workspace or an internal chalkboard. Just like a chalkboard, working memory has limited space, which can cause a "bottleneck" in learning. It involves the ability to recall information after learning has been consistently interrupted over a period of several minutes. Students with working memory problems may listen to a math lecture and understand each step as it is explained. When the instructor goes back to a previous step discussed several minutes prior, however, the student has difficulty explaining or remembering the reasons for the steps. These students have difficulty remembering series of steps long enough to understand the concept.

Working memory can go both ways in the memory process. First, it can go into long-term memory and abstract reasoning. Second, working memory can bring information out from long-term memory and abstract memory to use in learning new concepts. When learning a mathematical concept, working memory goes into abstract memory. When learning just facts such as the multiplication tables or definition of words, information goes directly into long-term memory. When solving math problems, in most cases, information is brought in from abstract reasoning and long-term memory into working memory. Students use their working memory to do their homework and when answering test questions. The amount of space in working memory is critical to answering test questions just like the amount of RAM is critical to running computer programs.

Example: In solving 26 x 32, you would put the intermediate products

continued

52 (*from 2 x 26*) and 780 (*from 30 x 26* — remember, 3 is in the 10s place, so make it 30) into working memory and add them together. The more automatic the multiplication, the less working memory you use. If you cannot remember your multiplication, you use up working memory trying to solve the multiplication problem. This leaves you with less working memory to solve the resulting addition problem.

How long-term memory/reasoning affects what you remember

Long-term memory is a storehouse of material that is retained for long periods of time. Working memory places information in long-term memory and long-term memory is recalled into working memory to solve problems. It is *not* a matter of trying harder and harder to remember more and more unrelated facts or ideas; it is a matter of organizing your short-term memories and working memories into meaningful information.

In most cases long-term memory is immeasurable. Students have so much room in their long-term memory that no one has measured its total capacity. Long-term memory also relates more to language skills rather than abstract skills. Students with good long-term memory and poor abstract skills can sometimes do well in every subject except math and the physical sciences. These students can use their long-term memory language skills by learning the vocabulary. By understanding the language of mathematics they can put into words how to solve math problems and recall these words during the test instead of depending mainly on their abstract memory. This information must be reviewed many times for it to get into long-term memory. This concept will be explored more in later chapters.

Reasoning or abstract memory is thinking about memories, comprehending their meanings and understanding their concepts. Abstract reasoning involves learning how the rules and laws apply to solving math problems. Students who understand the principals of a certain mathematical concept can use these principles to solve any similar problem. That is why the focus in mathematics is learning how to solve the problem based on concepts instead of just memorizing how to do a problem. In most cases you cannot memorize enough problems and hope they appear on the test. Without understanding the concept, the information cannot be transferred into abstract reasoning. Many students with excellent abstract reasoning major in math related careers.

The main problem most students face is converting information from working memory to long-term and/or abstract reasoning – and understanding it. To place information into long-term memory students must understand math vocabulary and practice problems. To place information into abstract reasoning students must understand the concept and remember it. In most cases student must use part of the long-term memory and abstract reasoning to solve math problems. It will depend on the skills in each area as to which one they use the best to solve math problems.

REMEMBER

Securing math information into long-term memory is not accomplished by just doing the homework — you must also understand it.

The Role of "Memory Output" in Testing

Memory output is what educators call a "retrieving process." It is necessary for verbal or written examinations. The retrieving process is used when answering questions in class, doing math homework or doing math tests. It is the method by which you recall information stored into long-term memory and by using abstract reasoning are able to place it into words or on paper. This retrieval process can come directly from long-term memory. For example, "What are whole numbers?" This is a fact question that comes from long-term memory. The retrieval process can also come from abstract reasoning. For example, "Write down the distributive property and substitute numbers to prove that it works." However, most math problems are solved through working memory by using information from both long-term memory and abstract reasoning. For example, solve: $3y - 10 = 9y + 21$. For this problem you are pulling in your number facts from long-term memory and the rules for solving equations from abstract reasoning.

Memory output can be blocked by three things:

1. insufficient processing of information into long-term memory,
2. test anxiety, or
3. poor test-taking skills.

If you did not completely place all of the information you learned into long-term memory, you may not be able to give complete answers on tests.

Test anxiety can decrease your ability to recall important information or it can cause you to block out information totally. Ways to decrease test anxiety were discussed in Chapter 3. Students who work on their test-taking skills can improve their memory output. Techniques to improve both your memory output and your test-taking skills will be explained in Chapter 7.

Understanding the Stages of Memory

Understanding the stages of memory will help you answer this common question about learning math: "How can I understand the procedures to solve a math problem one day and forget how to solve a similar problem two days later?"

There are three good answers to this question. First, after initially learning how to solve the problem, you did not *rehearse* the solving process enough for it to enter your long-term memory. Second, you did get the information into long-term memory, but the information was not *reviewed* frequently enough and was therefore forgotten. Third, you memorized how to work the problem but did not *understand* the concept.

REMEMBER

Locating where your memory breaks down and compensating for those weaknesses will improve your math learning.

Section Review

1. Label the boxes in Figure 9 on the facing page with stages of memory.

2. List, define and give examples for each stage of memory by explaining and giving examples of the following terms:

Stage One: _____

_____

_____

_____

Example: _____

_____

_____

_____

Stage Two: _____

_____

_____

_____

Example: _____

_____

_____

_____

Stage Three: _____

_____

_____

_____

continued on page 102

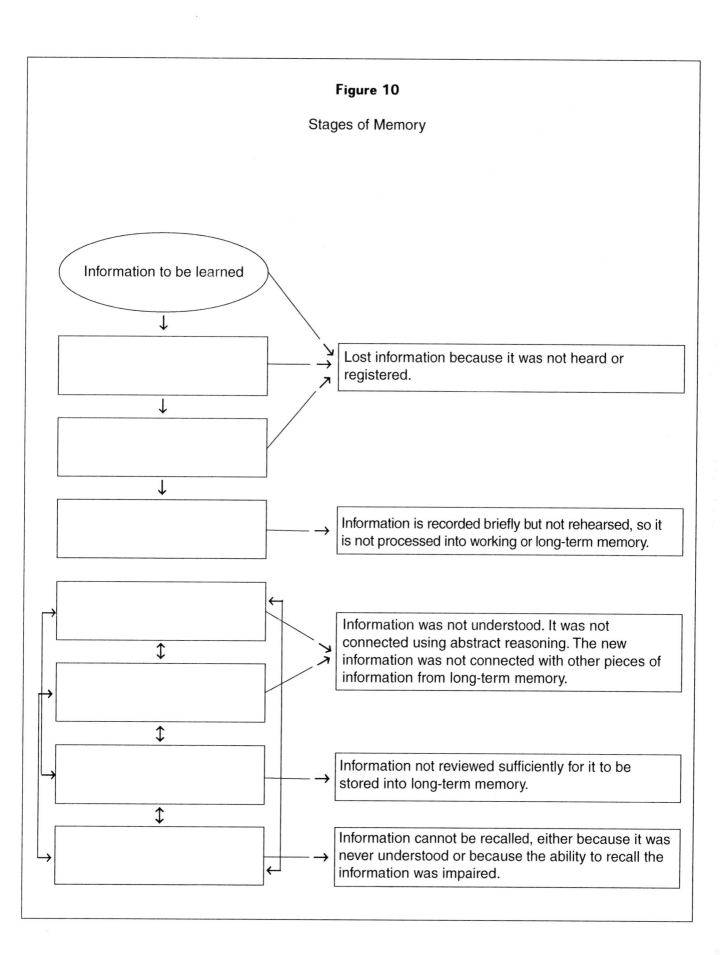

Figure 10

Stages of Memory

Information to be learned

Lost information because it was not heard or registered.

Information is recorded briefly but not rehearsed, so it is not processed into working or long-term memory.

Information was not understood. It was not connected using abstract reasoning. The new information was not connected with other pieces of information from long-term memory.

Information not reviewed sufficiently for it to be stored into long-term memory.

Information cannot be recalled, either because it was never understood or because the ability to recall the information was impaired.

Example: _____

_____

_____

_____

Stage Four: _____

_____

_____

_____

Example: _____

_____

_____

_____

Stage Five: _____

_____

_____

_____

Example: _____

_____

_____

_____

Stage Six: _____

_____

_____

_____

Example: _____

_____

_____

_____

Stage Seven: _____

_____

_____

_____

Example: _____

_____

_____

_____

3. What are three conditions that can block your memory output?

    First Condition: _____

    _____

    _____

    Second Condition: _____

    _____

    _____

    Third Condition: _____

    _____

    _____

4. Review the stages of memory and list the stages at which *your* memory breaks down. For each stage of breakdown, say how you can prevent it.

    Stage: _____

    Prevention: _____

    _____

    _____

    _____

Stage: _____

Prevention: _____

_____

_____

_____

Stage: _____

Prevention: _____

_____

_____

_____

5. What is the stage of memory you do well? Why?

Stage: _____

I do well because _____

_____

_____

_____

How to use learning styles to improve memory

There are many different techniques, which can help you store information in your long-term memory and reasoning. Using your learning sense or learning style and decreasing distraction while studying are very efficient ways to learn. Using your best *learning sense* (what educators call your "predominate learning modality") can improve how well you learn and enhance the transfer of knowledge into long-term memory/reasoning. Your learning senses are vision, hearing, touching, etc. Ask yourself if you learn best by watching (vision), listening (hearing), or touching (feeling).

Another helpful tool is the Learning Styles Inventory. It is the only inventory that has separate math learning style information. You can go to www.AcademicSuccess.com to learn more about the Learning Styles Inventory. Based on your preferred learning style practice those learning suggestions first.

REMEMBER

Learning styles are neither good nor bad and are based on genetics and environment. Knowing your best learning styles and using them effectively can dramatically improve your math learning and grades.

Visual (watching) Learner

Knowing that you are a visual math learner can help you select the memory technique that will work best for you. Repeatedly reading and writing down math materials being studied is the best way for a visual learner to study.

Based on the Learning Styles Inventory, students who learn math best by seeing it written are Visual Numerical Learners. If you are a visual numerical learner you will learn best by following the 10 suggestions in Figure 10, on the next page. Try as many of these suggestions as possible and then select and practice those that are most helpful.

A visual way to decrease distractions is by using the "my mind is full" concept. Imagine that your mind is completely filled with thoughts of

Figure 11
Visual Numerical Learners

These students learn math best by seeing it written. If you are a visual numerical learner, you may learn best by following these suggestions:

1. Studying a variety of written materials, such as additional handouts and math texts.

2. Playing games with and being involved in activities with visible printed number problems.

3. Using visually orientated computer programs, CD-ROMs, and math websites.

4. Reworking your notes using suggestions from this workbook.

5. Visualizing numbers and formulas, in detail.

6. Checking out videocassette tapes from the math lab or LRC.

7. Making 3x5 note (flash) cards, in color.

8. Using different colors of ink to emphasize different parts of the math formula.

9. Asking your tutor to *show you* how to do the problems instead of *telling you* how to do the problems.

10. Writing down each problem step the tutor tells you to do. Highlight the important steps or concepts, which cause you difficulty.

learning math, and other distracting thoughts cannot enter. Your mind has one-way input and output, which only responds to thinking about math when you are doing homework or studying.

Auditory (hearing) Learner

If you are an *auditory learner* (one who learns best by hearing the information) then learning formulas is best accomplished by repeating them back to yourself, or recording them on a tape recorder and listening to them. Reading out loud is one of the best auditory ways to get important information into long-term memory. Stating facts and ideas out loud improves your ability to think and remember. If you cannot recite out loud, recite the material to yourself, emphasizing the key words.

Based on the Learning Styles Inventory students who learn math best by hearing it are Auditory Numerical Learners. If you are an auditory numerical learner you may learn best by following the 10 suggestions in Figure 11 on the next page. Try as many of these suggestions as possible and then select and practice those that are most helpful.

An auditory way to improve your concentration is by becoming aware of your distractions and telling yourself to concentrate. If you are in a location where talking out loud will cause a disturbance, mouth the words "start concentrating" as you say them in your mind. Your concentration periods should increase.

Tactile/Concrete (touching) Learner

A *tactile/concrete learner* needs to feel and touch the material to learn it. Tactile concrete learners, who are also called *kinesthetic* learners, tend to learn best when they can concretely manipulate the information to be learned. Unfortunately, most math instructors do not use this learning sense. As a result, students who depend heavily upon feeling and touching for learning will usually have the most difficulty developing effective

Figure 12
Auditory Numerical Learners

1. Say numbers to yourself or moving your lips as you read problems.

2. Tape record your class and play it back while reading your notes.

3. Read aloud any written explanations.

4. Explain to your tutor how to work the math problem.

5. Make sure all important facts are spoken aloud with auditory repetition.

6. Remember important facts by auditory repetition.

7. Read math problems aloud and try solutions verbally and sub-verbally as you talk yourself through the problem.

8. Record directions to difficult math problems on audiotape and refer to them when solving a specific type of problem.

9. Have your tutor explain how to work the problems instead of just showing you how to solve them.

10. Record math laws and rules in your own words, by chapters, and listen to them every other day (auditory highlighting).

math learning techniques. This learning style creates a problem with math learning because math is more abstract than concrete. Also, most math instructors are visual abstract learners and have difficulty teaching math tactilely. Ask for the math instructors and tutors who give the most practical examples and who may even "act out" the math problems.

As mentioned before a tactile concrete learner will probably learn most efficiently by hands-on learning. For example, if you want to learn the FOIL method, you would take your fingers and trace the "face" to remember the steps. See Figure 13 (The FOIL Method). Also, learning is most effective when physical involvement with manipulation is combined with sight and sound. For example, as you trace the face you also say the words out loud.

Based on the Learning Styles Inventory, Tactile/Concrete Learners best learn math by manipulating the information that is to be taught. If you are a tactile/concrete learner, you may learn best by following the 10 suggestions in Figure 12. Try as many of these suggestions as possible and then select and practice the best suggestions that help. If you do not have these manipulatives or don't know how to use them, ask the math lab supervisor or instructor if they have any manipulative materials or models. If the math lab does not have any manipulative materials, you may have to ask for help to develop your own.

Tactile/concrete learners can also use graphing calculators to improve their learning. By entering the keystrokes it is easier to remember how to solve the problems. This practice is also an excellent way to remember how to solve the problem when using a calculator while taking a test.

Another way tactile/concrete learners can learn is to trace the graph with their fingers when it appears on the calculator. They should say out loud and trace every equation to "feel" how the graph changes when using different equations. For example, if you add 2 to one side of the equation, move your finger to where the graph changes and say out loud how much it moved.

A tactile/concrete way to improve your study concentration is by counting the number of distractions for each study session. Place a sheet of paper by your book when doing homework. When you catch yourself not concentrating put the letter

Figure 13
Tactile Concrete Learners

1. Cut up a paper plate to represent a fraction of a whole.

2. Fold up a piece of paper several times and cut along the fold marks to represent a fraction of a whole.

3. In order to understand math concepts, ask to be shown how to use qusinar rods or algebra tiles as manipulatives.

4. Try to use your hands and body to "act out" a solution. For example, you may "become" the car in a rate-and-distance word problem.

5. Obtain diagrams, objects or manipulatives and incorporate activities such as drawing and writing into your study time. You may also enhance your learning by doing some type of physical activity such as walking.

6. Try to get involved with at least one other student, tutor or instructor that uses manipulatives to help you learn math.

7. Ask to use the Hands-on Equations Learning System using manipulatives to learn basic algebra. You can go to their Web site (www.Borenson.com) to learn more about this system and other systems to help you learn math.

8. Go to one of the "learning stores" usually in your local mall to see if they have manipulatives.

9. Go to a K-12 learning resource store to see if they have manipulatives, such as magnetic boards with letters and numbers.

10. Talk to the coordinator of students with disabilities to see if they use manipulatives when tutoring their students.

Figure 14
The FOIL Method

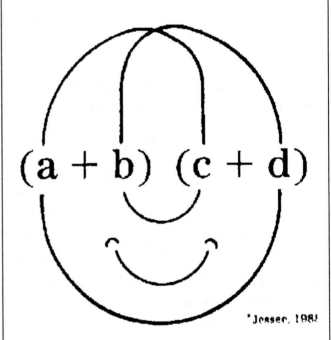

F (a) (c)
O (a) (d)
I (b) (c)
L (b) (d)

$(a + b) (c + d)$

Jesser, 1985

FOIL is used to remember the procedure to multiply two binomials. The letters in FOIL stand for First, Outside, Inside and Last. To use FOIL, multiply the following

- the First terms ((a) (c)),
- the Outside terms ((a) (d)),
- the Inside terms ((b) (c)),
- the Last terms ((b) (d)).

To learn FOIL, trace your finger along the FOIL route.

"C" on the sheet of paper. This will remind you to concentrate and get back to work. After each study period, count up the number of "C's" and watch the number decrease.

Social Individual Learner

If you are a social individual learner, learning math may best be done individually. You may learn best by yourself, working with computer programs and being individually tutored. In some cases, social individuals may have to meet in groups to develop practice tests but leave socializing to a minimum. If you are a social individual learner and visual learner, using the computer may be one of the best learning tools available. If you are a social individual learner, based on the Learning Styles Inventory, you may learn best by following the eight suggestions in Figure 14. Try as many of these suggestions as possible and then select those that are most helpful.

A problem that a social individual learner may encounter is working too long on a problem for which they could have received help. Social individual learners must understand that getting help is okay, especially if it saves study time and makes them more study efficient.

Social Group Learners

If you are a social group learner (one who best learns in a group) then learning math may best be done in study groups and in math classes that have collaborative learning (group learning). Social group learners may learn best by discussing information. They can usually develop their own study groups and discuss how to solve problems over the phone. If you are a social group learner and an auditory learner then you definitely learn best by talking to people. If you are a social group learner,

---

Figure 15
Social Individual Learners

1. Study math, English or other subjects alone.

2. Utilize videocassette tapes or auditory tapes to learn by yourself.

3. Prepare individual questions for your tutor or instructor.

4. Obtain individual help from the math lab or hire your own tutor.

5. Set up a study schedule and study area so other people will not bother you.

6. Study in the library or in some other private, quiet place.

7. Use group study times only as a way to ask questions, obtain information and take pre-tests on your subject material.

8. Use math websites to help you learn.

---

based on the Learning Styles Inventory, you may learn best by following the eight suggestions in Figure 15, above. Try as many of these suggestions as possible and then select and practice those that are most helpful.

A learning problem that a social group learner may have is talking too much about other subjects when in a study group. This is called being off task. You may want to have a student serve as a discussion monitor to let the students know when they need to get back on task. Also, social group learners need to know that they still must study math individually to be successful. During this individual study session prepare questions for the group.

Figure 16
Social Group Learners

1. Study math, English or your other subjects in a study group.

2. Sign up for math course sections which use cooperative learning (learning in small groups).

3. Review your notes with a small group.

4. Obtain help in the math lab or other labs where you can work in group situations.

5. Watch math videocassette tapes with a group and discuss the subject matter.

6. Listen to audiocassette tapes on the lecture and discuss them with the group.

7. Obtain several "study buddies" so you can discuss with them the steps to solving math problems.

8. Form a study group. Each member brings 10 test questions with explanations on the back. The group completes all the test questions.

Multiple Senses

If you have difficulty learning material from one sense (learning style), you might want to try learning material through two or three senses. Involving two or more senses in learning improves your learning and remembering. Review the figures in this section on using your learning styles and whenever possible combine the learning styles.

If your primary sense is visual and your secondary sense is auditory, you may want to write down equations while saying them out loud to yourself. Writing and reciting the material at the same time combines visual, auditory, and some tactile/concrete styles of learning.

Studying with a pen or highlighter is a visual as well as a tactile/concrete way to improve your concentration. Placing the pen or highlighter in your hand and using it will force you to concentrate more on what you are reading. After you write and recite the material back to yourself, do it five or ten more times to over learn it.

Section Review

1. List five ways visual learners can improve their memory:

   First Way: _____

   _____

   Second Way: _____

   _____

   Third Way: _____

   _____

   Fourth Way: _____

   _____

   Fifth Way: _____

   _____

2. List five ways auditory learners can improve their memory:

   First Way: _____

   _____

Second Way: _____

_____

Third Way: _____

_____

Fourth Way: _____

_____

Fifth Way: _____

_____

3. List five ways tactile/concrete learners can improve their memory:

First Way: _____

_____

Second Way: _____

_____

Third Way: _____

_____

Fourth Way: _____

_____

Fifth Way: _____

_____

4. List three ways social individual learners can improve their memory:

First Way: _____

_____

Second Way: _____

_____

Third Way: _____

_____

5. List three ways social group learners can improve their memory:

First Way: _____

_____

Second Way: _____

_____

Third Way: _____

_____

6. What is your best modality learning style for math? Visual, auditory or tactile/concrete?

_____

7. What is your second best modality learning style for math? _____

8. Are you more of an individual or group learner? _____

Why? _____

_____

9. How can you use multiple senses to improve your memory?

_____

_____

_____

10. What is your best combination of learning styles to learn math? _____

_____

Give an example of how you can combine your learning styles to learn a math concept (if you cannot think of a concept ask your instructor for one).

_____

_____

_____

_____

_____

_____

How to use memory techniques

There are many different techniques that can help store information in your long-term memory: a positive attitude about studying, selective learning, organization, visual imagery, mnemonic devices, and acronyms.

A Good Study/Math Attitude

A positive attitude about studying will help you concentrate and improve your retention of information. This means you should have at least a neutral math attitude (you neither like nor dislike it) and reserve the right to actually learn to like math. If you still don't like math, just pretend that you do while studying it. View studying as an opportunity to learn rather than an unpleasant task. Tell yourself that you can learn the material, and it will help you pass the course and graduate.

Be a Selective Learner

Being selective in your math learning will improve your memory. Prioritize the materials you are studying; decide which facts you need to know and which ones you can ignore. Narrow down information into laws and principles that can be generalized. Learn the laws and principles 100 percent.

Example: If you have been given a list of math principles and laws to learn for a test, put each one on an index card. As you go through them, create two piles: a "I already know this" pile and a "I don't know this" pile. Then study *only* the "I don't know this" pile. Study the "I don't know this" pile until it is completely memorized and understood.

Become an Organizer

Organizing your math material into idea/fact clusters or groups will help you learn and memorize it. Grouping similar material in a problem log or calculator log is an example of categorizing information. Do not learn isolated facts; always try to connect them to other similar material.

Use Visual Imagery

Using mental pictures or diagrams to help you learn the material is especially helpful for the visual learners and those who are right-hemisphere dominant (who tend to learn best by visual and spatial methods). Mental pictures and actual diagrams involve 100 percent of your brain power. Picture the steps to solve difficult math problems in your mind.

Example: Use the FOIL method (see Figure X, on the facing page) to visually learn how to multiply binomials. Memorize the face until you can sketch it from memory. If you need to use it during a test, you can then sketch the face onto your scratch paper and refer to it.

Examples: When learning the *commutative property*, remember that the word *commutative* sounds like the word *community*. A community is made up of different types of people who could be labeled as an *a* group and a *b* group. However, in a community of *a* people and *b* people, it does not matter if we count the *a* people first or the *b* people first; we still have the same total number of people in the community. Thus, $a + b = b + a$.

When learning the *distributive law of multiplication over addition*, such as $a(b + c)$, remember that "distributive" sounds like "distributor," which is associated with giving out a product. The distributor *a* is giving its products to *b* and *c*.

Make Associations

Association learning can help you remember better. Find a link between new facts and some well-established old facts and study them together. The recalling of old facts will help you remember the new facts and strengthen a mental connection between the two. Make up your own associations to remember math properties and laws.

REMEMBER

The more ridiculous the association, the more likely you are to remember it.

Use Mnemonic Devices

The use of mnemonic devices is another way to help you remember. Mnemonic devices are easily remembered words, phrases, or rhymes associated with difficult-to-remember principles or facts.

Example: Many students become confused with the *order of operations*. These students mix up the order of the steps in solving a problem, such as dividing before adding the numbers

continued

in the parentheses. A mnemonic device to remember the order of operations is "Please Excuse My Dear Aunt Sally." The first letter in each of the words represents the math function to be completed from the first to the last. Thus, the order of operations is Parentheses (*Please*), Exponents (*Excuse*), Multiplication (*My*), Division (*Dear*), Addition (*Aunt*), and Subtraction (*Sally*). Remember to multiply and/or divide whatever comes first, from left to right. Also, add or subtract whatever comes first, from left to right.

expression) and 7 (in the second expression). The Inside quantities are 3 (in the first expression) and x (in the second expression). The Last quantities are 3 (in the first expression) and 7 (in the second expression). This results in

F $(2x)(x)$ + **O** $(2x)(7)$ + **I** $(3)(x)$ + **L** $(3)(7)$.

Do the multiplication to get $2x^2 + 14x + 3x + 21$, which adds up to $2x^2 + 17x + 21$. See Figure 13 (The FOIL Method).

Use Acronyms

Acronyms are another memory device to help you learn math. Acronyms are word forms created from the first letters of a series of words.

Example: FOIL is one of the most common math acronyms. FOIL is used to remember the procedure for multiplying two binomials. Each letter in the word *FOIL* represents a math operation. **FOIL** stands for First, Outside, Inside, and Last, as it applies to multiplying two binomials such as $(2x + 3)(x + 7)$. The First quantities are $2x$ (in the first expression) and x (in the second expression). The Outside quantities are $2x$ (in the first

continued

Section Review

1. How can a good math attitude help you learn?

   _____

   _____

   _____

   _____

2. Give an example of being a selective learner in your math class.

   _____

   _____

   _____

   _____

3. From your current math lessons, make up and explain one association remembering device that is not in this workbook.

_____

_____

_____

_____

4. For your next major math test, make up and explain one mnemonic device that is not in this workbook.

_____

_____

_____

_____

5. For your next major math test, make up and explain one acronym that is not in this workbook.

_____

_____

_____

_____

How to develop practice tests

Developing a practice test is one of the best ways to evaluate your memory and math skills before taking a real test. You want to find out what you do not know *before* the real test instead of *during* the test. Practice tests should be as real as possible and should include time constraints.

You can create a practice test by reworking all the problems since your last test that you have recorded in your problem log. Another practice test can be developed using every other problem in the textbook chapter tests. Further, you can use the solutions manual to generate other problems with which to test yourself. You can also use old exams from the previous semester. Check to see if the math lab/LRC or library has tests on file from previous semesters, or ask your instructor for other tests. For some students, a better way to prepare for a test is the group method.

Example: Hold a group study session several days before the test. Have each student prepare a test with ten questions. On the back of the test, have the answers listed, worked out step by step. Have each member of the study group exchange his or her test with another member of the group. Once all the tests have been completed, have the author of each test discuss with the group the procedures used to solve those problems.

If group work improves your learning, you may want to hold a group study session at least once a week. Make sure the individual or group test is completed at least three days before the real test.

Completing practice math tests will help you increase your testing skills. It will also reveal your test problem weaknesses in enough time for you to learn how to solve the problem before the real test. If you have difficulty with any of the problems during class or after taking the practice test, be sure to see your tutor or instructor.

After taking the practice test(s), you should know what parts you do not understand (and need to study) and what is likely to be on the test. Put this valuable information on one sheet of paper. This information needs to be understood and memorized. It may include formulas, rules, or steps to solving a problem.

Use the learning strategies discussed in this chapter to remember this information. A good example of how this information should look is what students might call a mental "cheat sheet." Obviously, you cannot use the written form of this sheet during the real test.

If you cannot take a practice test, put down on your mental cheat sheet the valuable information you will need for the test. Work to understand and memorize your mental cheat sheet. Chapter 7, "How to Improve Your Math Test-Taking Skills," will discuss how to use the information on the mental cheat sheet — *without cheating.*

Section Review

1. List three different ways you can make up a practice test.

First Way: _____

_____

_____

Second Way: _____

_____

_____

Third Way: _____

_____

_____

2. List ten different problems you would put on your practice test for the next exam.

First Problem: _____

_____

Second Problem: _____

_____

Third Problem: _____

_____

Fourth Problem: _____

_____

Fifth Problem: _____

_____

Sixth Problem: _____

_____

Seventh Problem: _____

_____

Eighth Problem: _____

_____

Ninth Problem: _____

_____

Tenth Problem: _____

_____

3. Give these ten problems to your study buddy and get ten problems from him or her. Compare your answers with those of your study buddy. How many did you get right? _____ How can you correct the wrong answers?

_____

_____

_____

_____

_____

_____

How to use number sense

When taking your practice tests, you should use "number sense," or estimations, to make sure your answer is reasonable. Number sense is like common sense but it applies to math. Number sense is the ability to see if your answer makes sense without using algorithms. (Algorithms are the sequential math steps used to solve problems.) These following examples demonstrate solving a math problem (from a national math test given to high school students) using algorithms and number sense.

Example: Solve 3.04 x 5.3. Students use algorithms to solve this problem by multiplying the number 3.04 by 5.3, in sequence. Seventy-two percent of the students answered the problem

continued

correctly using algorithms.

Example: Estimate the product 3.04 x 5.3, and choose from the following answers.

A)   1.6   C) 160
B)   16   D) 1600

Only 15 percent of the students chose B, which is the correct answer. Twenty-eight percent of the students chose A. Using *estimating* to solve the answer, a whopping 85 percent of the students got the problem wrong.

These students were incorrectly using their "mental chalkboard" instead of using number sense. In using number sense to answer, you would multiply the numbers to the left of the decimal in each number to get an estimate of the answer. To estimate the answer, then, you would multiply 3 (the number to the left of the decimal in 3.04) by 5 (the number to the left of the decimal in 5.3) and expect the answer to be a little larger than 15.

It appears that the students' procedural processing (the use of algorithms) was good, but when asked to solve a non-routine problem using estimating (which is easier than using algorithms), the results were disappointing.

Another example of using number sense, or estimating, is in "rounding off."

Taking the time to estimate the answer to a math problem is a good way to check your answer. Another way to check your answer is to see if it is reasonable. Many students forget this important step and get the answer wrong. This is especially true of word or story problems.

Example: Solve 48 + 48 by rounding off. Rounding off means mentally changing the number (up or down) to make it more manageable to you, without using algorithms. By rounding off, 48 becomes 50 (easier to work with). 50 + 50 = 100. If the choices for answers were 104, 100, 98 and 96, you would then subtract 4 from the 100 (since each number was rounded up by 2) and you would get 96.

Also, when dealing with an equation, make sure to put the answer back into the equation to see if one side of the equation equals the other. If the two sides are not equal, you have the wrong answer. If you have extra time left over after you have completed a test, you should check answers using this method.

Examples: When solving a rate-and-distance problem, use your common sense to realize that one car cannot go 500 miles per hour to catch the other car. However, the car can go 50 miles per hour.

The same common sense rule applies to age-word problems, where the age of a person cannot be 150 years, but it can be 15 years.

Further, in solving equations, x is *usually* a number that is less than 20. When you solve a problem for x and get 50, then, this isn't reasonable, and you should recheck your calculations.

continued

Example: In solving the equation $x + 3 = 9$, you calculated that $x = 5$. To check your answer, substitute x with 5 and see if the problem works out correctly. 5 + 3 does not equal 9, so you know you have made a mistake and need to recalculate the problem. The correct answer, by the way, is $x = 6$.

REMEMBER

Number sense is a way to get more math problems correct by estimating your answer to determine if it is reasonable.

Section Review

1. Give an explanation of number sense.

_____

_____

_____

_____

_____

_____

_____

_____

2. How does number sense compare to common sense?

_____

_____

_____

_____

3. Give an example of a recent math problem you missed due to poor number sense.

_____

_____

_____

_____

4. List two ways you can improve your number sense.

First Way: _____

_____

_____

Second Way: _____

_____

_____

Name: _____ Date: _____

Chapter 6 Review

1. The way you receive information is through your five senses, which are _____, _____, _____, _____, and _____.

2. The main problem most students have is converting learned material into _____ _____ and _____ it.

3. Repeatedly reading and writing math material is one of the best ways for a _____ learner to study.

4. Reading _____ _____ is one of the best auditory ways to learn material.

5. A _____/_____ learner needs to feel and touch material to learn it.

6. Being a selective learner means _____ the material to study and learning the laws and principles _____ percent.

7. Mnemonic devices are easy-to-remember _____, _____, or _____ associated with difficult-to-remember principles or facts.

8. _____ are word forms created from the first letters of a series of words.

9. The reason to develop a practice test is to find out what you do not know _____ the test instead of _____ the test.

10. Social learners benefit from forming _____ _____.

What is the most important information you learned from this chapter?

_____

_____

_____

_____

_____

_____

_____

How can you immediately use it?

_____

_____

_____

_____

_____

_____

_____

7

How to Improve Your Math Test-Taking Skills

Taking a math test is different from taking tests in other subjects. Math tests require not only that you to recall the information, but that you apply the information. Multiple-choice tests, for example, usually test you on recall, and if you do not know the answer, you can guess.

Math tests build on each other, whereas history tests often do not test you on previous material. Most math tests are speed tests, where the faster your are, the better grade you can receive; most social science tests are designed for everyone to finish.

Math test preparation and test-taking skills are different from preparation and skills needed for other tests. You need to have a test-taking plan and a test analysis plan to demonstrate your total knowledge on math tests. Students with these plans make better grades compared to students without them. Math instructors want to measure your math knowledge, not your poor test-taking skills.

In this chapter, you will learn

- why attending class and doing your homework may not be enough to pass,
- the general pretest rules,
- the ten steps to better test taking,
- the six types of test-taking errors, and
- how to prepare for the final exam.

Why attending class and doing your homework may not be enough to pass

Most students and some instructors believe that attending class and doing all the homework ensures an A or B on tests. This is far from true. Doing all the homework and getting the correct answers is very different in many ways from taking tests:

1. While doing homework, there is little anxiety. A test situation is just the opposite.

2. You are not under a time constraint while doing your homework; you may have to complete a test in 55 minutes or less.

3. If you get stuck on a homework problem, your textbook and notes are there to assist you. This is not true for most tests.

4. Once you learn how to do several problems in a homework assignment, the rest are similar. In a test, the problems may be in random order.

5. In doing homework, you have the answers to at least half the problems in the back of the text and answers to all the problems in the solutions guide. This is not true for tests.

6. While doing homework, you have time to figure out how to correctly use your calculator. During the test, you can waste valuable time figuring out how to use your calculator.

7. When doing homework, you can call your study buddy or ask the tutor for help, something you cannot do on the test.

8. While doing your homework, you can go to websites for online tutoring to help you solve problems you do not understand.

Do not develop a false sense of security by believing you can make an A or B by doing just your homework. Tests measure more than just your math knowledge.

The general pretest rules

General rules are important when taking any type of test:

1. *Get a good night's sleep before taking a test.* This is true for the ACT, the SAT, and your math tests. If you imagine you are going to cram all night and perform well on your test with three to four hours of sleep, you are wrong. It would be better to get seven or eight hours of sleep and be fresh enough to use your memory to recall information needed to answer the questions.

2. *Start studying for the test at least three days ahead of time.* Make sure you take a practice test to find out, before the test, what you do not know. Review your problem log and work the problems. Review the concept errors you made on the last test. (How to identify and correct your concept errors will be discussed later in this chapter.) Meet with the instructor or tutor for help on those questions you cannot solve.

3. *Review only already-learned material the night before a test.*

4. *Make sure you know all the information on your mental cheat sheet.* Review your notebook and glossary to make sure you understand the concepts. Work a few problems and recall the information on your mental cheat sheet right before you go to bed. Go directly to bed; do not watch television, listen to the radio, or party. While you are asleep, your mind will work on and remember the last thing you did before going to sleep.

5. *Get up in the morning at your usual time and review your notes and problem log.* Do not do any new problems, but make sure your calculator is working.

Section Review

1. List three reasons why only attending class and doing your homework may not be enough to pass your math course.

First Reason: _____

_____

_____

Second Reason: _____

_____

_____

Third Reason: _____

_____

_____

2. List and explain three general pretest rules that best apply to you.

First Rule: _____

_____

_____

Second Rule: _____

_____

_____

Third Rule: _____

_____

_____

The ten steps to better test taking

You need to have a game plan to take a math test. This plan is different from taking history, English, humanities and some science tests. The game plan is to get the most points in the least amount of time. Many students lose test points because they use the wrong test-taking strategies for math. By following these ten steps you can demonstrate more knowledge on the test and get more problems right.

Step 1 *Use a memory data dump.* When you get your test, turn it over and write down the information that you might forget. Remember, this is your mental cheat sheet that you should have already made while preparing for the test. Your mental cheat sheet has now turned into a mental list, and writing down this information is not cheating. Do not put your name on the test, do not skim it, just turn it over and write down those facts, figures, and formulas from your mental cheat sheet or other information you might not remember during the test. This is called your *first memory data dump*. The data dump provides memory cues for test questions.

Example: It might take you a while to remember how to do a coin problem. However, if you had immediately turned your test over and written down different ways of solving coin problems, it would be easier to solve the coin problem.

Step 2 *Preview the test.* Put your name on the test and start previewing. Previewing the test requires you to look through the entire test to find different types of problems and their point values. Put a mark by the questions that you can do without thinking. These are the questions that you will solve first.

Step 3 *Do a second memory data dump.* The secondary data dump is for writing down material that was jarred from your memory while previewing the test. Write this information on the back of the test.

Step 4   *Develop a test progress schedule.* When you begin setting up a test schedule, determine the point value for each question. Some test questions might be worth more points than others.

In some tests, word problems are worth 5 points and other questions might be worth 2 or 3 points. You must decide the best way to get the most points in the least amount of time. This might mean working the questions worth 2 or 3 points first and leaving the more difficult word problems for last.

Decide how many problems should be completed halfway though the test. You should have more than half the problems completed by that time.

Step 5   *Answer the easiest problems first.* Solve, in order, the problems you marked while previewing the test. Then review the answers to see if they make sense. Start working through the test as fast as you can while being accurate. Answers should be reasonable.

> **Example:** The answer to a problem asking you to find the area of a rectangle cannot be negative, and the answer to a land-rate-distance problem cannot be 1000 miles per hour.

Clearly write down each step to get partial credit. Even if you end up missing the problem, you might get some credit. In most math tests, the easier problems are near the beginning of the first page; you need to answer them efficiently and quickly. This will give you both more time for the harder problems and time to review.

Step 6   *Skip difficult problems.* If you find a problem that you do not know how to work, read it twice and automatically skip it. Reading it twice will help you understand the problem and put it into your working memory. While you are solving other problems, your mind is still working on that problem. Difficult problems could be the type of problem you have never seen before or a problem in which you get stuck on the second or third step. In either case, skip the problem and go on to the next one.

Step 7   *Review the skipped problems.* When working the skipped problems, think how you have solved other, similar problems as a cue to solving the skipped ones. Also try to remember how the instructor solved that type of problem on the board.

While reviewing skipped problems, or at any other time, you may have the "Aha!" response. The "Aha!" response is your remembering how to do a skipped problem. Do not wait to finish your current problem. Go to the problem on which you had the "Aha!" and finish that problem. If you wait to finish your current problem, your "Aha!" response could turn into an "Oh, no!" response.

Step 8   *Guess at the remaining problems.* Do as much work as you can on each problem, even if it is just writing down the first step. If you cannot write down the first step, rewrite the problem. Sometimes rewriting the problem can jar your memory enough to do the first step or the entire problem. This is true of students, particularly of tactile/concrete learners where writing can trigger the memory process of solving the problem. Remember, you did not learn how to solve math problems with your hands tied behind your back. If you leave the problem blank, you will get a zero. Do not waste too much time on guessing or trying to work the problems you cannot do.

Step 9 *Review the test.* Look for careless errors or other errors you may have made. Students usually lose 2 to 5 test points on errors that could have been caught in review. Do not talk yourself out of an answer just because it may not look right. This often happens when an answer does not come out even. It is possible for the answer to be a fraction or decimal.

REMEMBER

Answers in math do not have "dress codes." Research reveals that the odds of changing a right answer to a wrong answer are greater than the odds of changing a wrong answer to a right one.

Step 10 *Use all the allowed test time.* Review each problem by substituting the answer back into the equation or doing the opposite function required to answer the question. If you cannot check the problem by the two ways mentioned, rework the problem on a separate sheet of paper and compare the answers. Do not leave the test room unless you have reviewed each problem two times or until the bell rings.

Even though we encourage students to work until the end of the test period most students leave the classroom before the end of the period. These students state that even though they know they should use all the test time, they cannot stay in the room until the end of the test time. These students also know that their grades would probably improve if they kept checking their answers or kept working the problems they are having difficulty with. After talking to hundreds of these students, I discovered two different themes for leaving the classroom early. First, test anxiety gets so overwhelming that they cannot stay in the room. The relief from the test anxiety (leaving the room) is worth more than getting a better grade. If you are one of these students, you must learn how to decrease your test anxiety by following the suggestions in Chapter 4 or by using the *How to Reduce Test Anxiety* cassette tape. Don't let the anxiety control your grades!

The other reason for leaving the test early is that they do not want to be the last or one of the last few students to turn in their tests. They still believe that students who turn their tests in last are "dumb and stupid." These students also believe that students who turn their tests in first make "A's" and "B's" and those students who turn their tests in last make "D's" and "F's." If you are one of these students, you don't need to care about what other students think about you (it's usually wrong anyway). YOU need to fight the urge to leave early and use all the test time. Remember, passing mathematics is the best way to get a high paying job and support yourself or your family. DO IT NOW!

REMEMBER

There is no prize for handing your test in first, and students who turn their papers in last do make A's.

Stapling your scratch paper to the math test when handing it in has several advantages:

- If you miscopied the answer from the scratch paper, you will probably get credit for the answer.
- If you get the answer incorrect due to a careless error, your work on the scratch paper could give you a few points.
- If you do get the problem wrong, it will be easier to locate the errors when the instructor reviews the test. This will prevent you from making the same mistakes on the next math test.

REMEMBER

Handing in your scratch paper may get you extra points or improve your next test score.

Section Review

1. List and explain the ten steps to test taking.

   First Step: _____

   _____

   Second Step: _____

   _____

   _____

Third Step: _____

_____

Fourth Step: _____

_____

Fifth Step: _____

_____

Sixth Step: _____

_____

Seventh Step: _____

_____

Eighth Step: _____

_____

Ninth Step: _____

_____

Tenth Step: _____

_____

2. After trying the ten steps to test taking, develop your own personalized test-taking steps.

   First Step: _____

   _____

   Second Step: _____

   _____

   Third Step: _____

   _____

   Fourth Step: _____

   _____

Fifth Step: _____

_____

Sixth Step: _____

_____

Seventh Step: _____

_____

Eighth Step: _____

_____

Ninth Step: _____

_____

Tenth Step: _____

_____

Additional steps

_____

_____

_____

_____

_____

The six types of test-taking errors

To improve future test scores, you must conduct a test analysis of previous tests. In analyzing your tests, you should look for the following kinds of errors:

1. misread-directions errors
2. careless errors
3. concept errors
4. application errors
5. test-taking errors
6. study errors

Students who conduct math test analyses will improve their total test scores.

Misread-directions errors occur when you skip directions or misunderstand directions, but do the problem anyway.

Examples: Suppose you have this type of problem to solve:

$$(x + 1)(x + 1)$$

Some students will try to solve for x, but the problem only calls for multiplication. You would solve for x only if you have an equation such as $(x + 1)(x + 1) = 0$.

Another common mistake is not reading the directions before doing several word problems or statistical problems. All too often, when a test is returned, you find only three out of the five problems had to be completed. Even if you did get all five of them correct, it cost you valuable time that could have been used obtaining additional test points.

To avoid misread-direction errors, carefully read and interpret all the directions. Look for anything that is unusual, or if the directions have two parts. If you do not understand the directions, ask the instructor for clarification. If you feel uneasy about asking the instructor for interpretation of the question, remember the instructor in most cases does not want to test you on the interpretation of the question but how you answer it. Also, you don't want to make the mistake of assuming that the instructor will not interpret the question. Let the instructor make the decision to interpret the question, not you.

Careless errors are mistakes that you can catch automatically upon reviewing the test. Both good and poor math students make careless errors. Such errors can cost a student a higher letter grade on a test.

> **Examples:** *Dropping the sign*:
> $-3(2x) = 6x$, instead of $-6x$
> which is the correct answer.
>   *Not simplifying your answer:*
> Leaving $(3x - 12)/3$ as your answer instead of simplifying it to $x - 4$.
>   *Adding fractions*: $1/2 + 1/3 = 2/5$, instead of $5/6$, which is the correct answer.
>   *Word problems*: $x = 15$ instead of "The student had 15 tickets."

However, many students want all their errors to be careless errors. This means that the students *did* know the math, but simply made silly mistakes. In such cases, I ask the student to solve the problem immediately, while I watch.

If the student can solve the problem or point out his or her mistake in a few seconds, it is a careless error. If the student cannot solve the problem immediately, it is not a careless error and is probably a concept error.

When working with students who make careless errors, I ask them two questions: First,

"How many points did you lose due to careless errors?" Then I follow with, "How much time was left in the class period when you handed in your test?" Students who lose test points to careless errors are giving away points if they hand in their test papers before the test period ends.

To reduce careless errors, you must realize the types of careless errors made and recognize them when reviewing your test. If you cannot solve the missed problem immediately, it is not a careless error. If your major error is not simplifying the answer, review each answer as if it were a new problem and try to reduce it.

Concept errors are mistakes made when you do not understand the properties or principles required to work the problem. Concept errors, if not corrected, will follow you from test to test, causing a loss of test points.

> **Examples:** Some common concept errors are not knowing:
> $(-)(-)x = x$, *not* $-x$
> $-1(2) > x(-1) = 2 < x$, *not* $2 > x$
> $5/0$ is undefined, *not* 0
> $(a + x) / x$ is *not* reduced to a
> the order of operations

Concept errors must be corrected to improve your next math test score. Students who have numerous concept test errors will fail the next test and the course if concepts are not understood. Just going back to rework the concept error problems is not good enough. You must go back to your textbook or notes and learn why you missed those types of problems, not just the one problem itself.

The best way to learn how to work those types of problems is to set up a concept problem error page in the back of your notebook. Label the first page "Test One Concept Errors." Write down all your concept errors and how to solve the problems. Then, work five more problems that use the same concept. Now, in your own words, write the concepts that you are using to solve these problems.

If you cannot write the concept in your own words, you do not understand it. Get assistance from your instructor if you need help finding similar problems using the same concept or cannot understand the concept. Do this for every test.

Application errors occur when you know the concept but cannot apply it to the problem. Application errors usually are found in word problems, deducing formulas (such as the quadratic equation), and graphing. Even some better students become frustrated with application errors; they understand the material but cannot apply it to the problem.

To reduce application errors, you must predict the type of application problems that will be on the test. You must then think through and practice solving those types of problems using the concepts.

> **Example:** If you must derive the quadratic formula, you should practice doing that backward and forward while telling yourself the concept used to move from one step to the next.

Application errors are common with word problems. After completing the word problem, reread the question to make sure you have applied the answer to the intended question. Application errors can be avoided with appropriate practice and insight.

Test-taking errors apply to the specific way you take tests. Some students consistently make the same types of test-taking errors. Through recognition, these bad test-taking habits can be replaced by good test-taking habits. The result will be higher test scores. The list that follows includes the test-taking errors that can cause you to lose many points on an exam:

1. *Missing more questions in the first third, second third, or last third of a test* is considered a test-taking error.

   Missing more questions in the first third of a test can be due to carelessness when doing easy problems or due to test anxiety.

   Missing questions in the last part of the test can be due to the fact that the last problems are more difficult than the earlier questions or due to increasing your test speed to finish the test.

   If you consistently miss more questions in a certain part of the test, use your remaining test time to review that section of the test first. This means you may review the last part of your test first.

2. *Not completing a problem to its last step* is another test-taking error. If you have this bad habit, review the last step of the test problem first, before doing an in-depth test review.

3. *Changing test answers from correct ones to incorrect ones* is a problem for some students. Find out if you are a good or bad answer-changer by comparing the number of answers changed to correct and to incorrect answers. If you are a bad answer-changer, write on your test, "Don't change answers." Change answers only if you can prove to yourself or the instructor that the changed answer is correct.

4. *Getting stuck on one problem and spending too much time on it* is another test-taking error. You need to set a time limit on each

problem before moving to the next problem. Working too long on a problem without success will increase your test anxiety and waste valuable time that could be used in solving other problems or in reviewing your test.

5. *Rushing through the easiest part of the test and making careless errors* is a common test-taking error for the better student. If you have the bad habit of getting more points taken off for the easy problems than for the hard problems, first review the easy problems, and later review the hard problems.

6. *Miscopying an answer from your scratch work to the test* is an uncommon test-taking error, but it does cost some students points. To avoid these kinds of errors, systematically compare your last problem step on scratch paper with the answer written on the test. In addition, always hand in your scratch work with your test.

7. *Leaving answers blank* will get you zero points. If you look at a problem and cannot figure out how to solve it, do not leave it blank. Write down some information about the problem, rewrite the problem, or try to do at least the first step.

REMEMBER

Writing down the first step of a problem is the key to solving the problem and obtaining partial credit.

8. *Solving only the first step of a two-step problem* causes problems for some students. These students get so excited when answering the first step of the problem that they forget about the second part. This is especially true on two-step word problems. To correct this test-taking error, write "two" in the margin of the problem. That will remind you that there are two steps or two answers to this problem.

9. *Not understanding all the functions of your calculator* can cause major testing problems. Some students only barely learn how to use the critical calculator functions. They then forget or have to relearn how to use their calculator, which costs test points and test time. Do not wait to learn how to use your calculator on the test. Overlearn the use of your calculator *before* the test.

10. *Leaving the test early without checking all your answers is a costly habit.* Do not worry about the first person who finishes the test and leaves. Many students start to get nervous when students start to leave after finishing the test. This can lead to test anxiety, mental blocks, and loss of recall.

    According to research, the first students finishing the test do not always get the best grades. It sometimes is the exact opposite. Ignore the exiting students, and always use the full time allowed.

Make sure you follow the ten steps to better test taking. Review your test taking procedures for discrepancies in following the ten steps to better test taking. Deviating from these proven ten steps will cost you points.

Study errors, the last type of mistake to look for in test analysis, occur when you study the wrong type of material or do not spend enough time on pertinent material. Review your test to find out if you missed problems because you did not practice that type of problem or because you did practice it but forgot how to do it during the test. Study errors

Figure 17

Math Test for Prealgebra

The answers are listed in bold. The correct answer on missed questions is shaded. Identify the type of error based on the "6 Types of Test-taking Errors." The student's test score is 70.

1. Write in words: 32.685

   Thirty-two and six hundred eighty-five thousandths

2. Write as a fraction and simplify: 0.078

   $\dfrac{\textbf{78}}{\textbf{1000}}$ $\dfrac{\textbf{39}}{\textbf{500}}$

3. Round to the nearest hundredth: 64.8653

   64.865   **64.87**

4. Combine like terms:

   $6.78x - 3.21 + 7.23x - 6.19$

   $= \textbf{6.78}x + \textbf{7.23}x + \textbf{(-3.21)} + \textbf{(-6.19)}$
   $= \textbf{14.01}x - \textbf{9.4}$

5. Divide and round to the nearest hundredth:
   68.1357   2.1

   32.445 → 32.45

6. Write as a decimal: $\dfrac{5}{16}$

   0.3125

7. Insert < or > to make a true statement:
   $\dfrac{3}{8}$ < $\dfrac{6}{13}$

8. Solve: $\dfrac{3}{x} = \dfrac{9}{12}$

   $9x = 3(12)$

   $\dfrac{9x}{9} = \dfrac{36}{9}$

   $x = 5$ **x = 4**

9. What number is 35% of 60?

   2100   **21.00**

10. 20.8 is 40% of what number?   **52**

11. 567 is what percent of 756?

    $\dfrac{756}{567} = \dfrac{9}{100}$

    = 133.3%   **75%**

12. Multiply:

    $(-6.03)\,(-2.31) = 13.9$ **13.9293**

continued

Answer Key for Exercise

1. Correct
2. *Misread directions* — Forgot to simplify by reducing the fraction.
3. *Concept error* — Did not know that hundredths is two places to the right of the decimal.
4. Correct
5. Correct
6. Correct
7. Correct
8. *Careless error* — Divided incorrectly in the last step.
9. *Test-taking error* — Not following step 5 in test-taking steps: Reviewing answers to see if they make sense. The number that equals 35% can't be larger than 60.
10. *Test-taking error* — Did not follow step 7 in test taking steps: Don't leave an answer blank
11. *Application error* — Solved the equation correctly but did not apply the correct equation.
12. *Concept error* — Did not know that when you multiply with one number in the hundredths, the answer must include the hundredths column.

will take some time to track down. But correcting study errors will help you on future tests.

Most students, after analyzing one or several tests, will recognize at least one major, common test-taking error. Understanding the effects of this test-taking error should change your study techniques or test-taking strategy.

Example: If there are seven minutes left in the test, should you review for careless errors or try to answer those two problems you could not totally solve? This is a trick question. The real question is, Do you miss more points due to careless errors or concept errors, or are the missed points about even? The answer to this question should determine how you will spend the last minutes of the test. If you miss more points due to careless errors or miss about the same number of points due to careless/concept errors, review for careless errors.

Careless errors are easier to correct than concept errors. However, if you make very few or no careless errors, you should be working on those last two problems to get the greatest number of test points. Knowing your test-taking errors can add more points to your test by changing your test-taking procedure.

Section Review

1. List the six types of test taking errors.

   Error One: _____

   Error Two: _____

   Error Three: _____

   Error Four: _____

   Error Five: _____

   Error Six: _____

2. Which of these errors have you made in the past?

   _____

   _____

   _____

   _____

   _____

   _____

3. How can you avoid these types of errors on the next test?

   _____

   _____

   _____

   _____

   _____

4. Review the Math Test for Prealgebra if you are in a prealgebra course. Review for type of errors, the points lost for each error, and example of each error.

   Error Type: _____

   Points Lost for Error: _____

   Example of Error: _____

   _____

   Error Type: _____

   Points Lost for Error: _____

   Example of Error: _____

   _____

   Error Type: _____

   Points Lost for Error: _____

   Example of Error: _____

   _____

   Error Type: _____

   Points Lost for Error: _____

   Example of Error: _____

   _____

   Error Type: _____

   Points Lost for Error: _____

   Example of Error: _____

   _____

   Error Type: _____

   Points Lost for Error: _____

   Example of Error: _____

   _____

Error Type: _____

Points Lost for Error: _____

Example of Error: _____

_____

5. Review your last test to list the type of errors, the points lost for each error, and examples of the errors. You may have more errors than allowed for in the list below. Use a separate sheet of paper if you need to list more. (Remember: Careless errors are problems you can solve immediately.)

Error Type: _____

Points Lost for Error: _____

Example of Error: _____

_____

Error Type: _____

Points Lost for Error: _____

Example of Error: _____

_____

Error Type: _____

Points Lost for Error: _____

Example of Error: _____

_____

Error Type: _____

Points Lost for Error: _____

Example of Error: _____

_____

6. List any additional concept error problems that were not included in question 2.

Points Lost for Error: _____

Example of Error: _____

_____

Points Lost for Error: _____

Example of Error: _____

_____

Using the pages in the back of your math notebook, rework the concept error problems and do five math problems just like them. For each concept error, write why you could not solve the problem on the test and what new information you learned to solve the problem. Do this for every major test. Give an example of one of your concept error(s) and the reasons you can now solve it.

Concept Error Problem(s):

_____

_____

_____

_____

_____

Reasons I Can Now Solve It:

_____

_____

_____

_____

_____

_____

_____

How to prepare for the final exam

The first day of class is when you start preparing for the final exam. Look at the syllabus or ask the instructor if the final exam is cumulative. A cumulative exam covers everything from the first chapter to the last chapter. Most math final exams are cumulative.

The second question you should ask is if the final exam is a departmental exam or if it is made up by your instructor. In most cases, departmental exams are more difficult and need a little different preparation. If you have a departmental final, you need to ask for last year's test and ask other students what their instructors say will be on the test.

The third question is, How much will the final exam count? Does it carry the same weight as a regular test or, as in some cases, will it count a third of your grade? If the latter is true, the final exam will usually make a letter grade difference on your final grade. The final exam could also determine if you pass or fail the course. Knowing this information before the final exam will help you prepare.

Preparing for the final exam is similar to preparing for each chapter test. You must create a pretest to discover what you have forgotten. You can use questions from the textbook chapter tests or questions from your study group.

Review the concept errors that you recorded in the back of your notebook labeled "Test One," "Test Two," and so on. Review your problem log for questions you consistently miss. Even review material that you knew for the first and second test but which was not used on any other tests. Students forget how to work some of these problems.

If you do not have a concept error page for each chapter and did not keep a problem log you need to develop a pre-test before taking the final exam. If you are individually preparing for the final, then copy each chapter test and do every fourth problem to see what errors you may make. However, it is better to have a study group where each of the four members brings in ten problems with the answers worked out on a separate page. Then each group member can take the 30-question test to find out what they need to study. You can refer to the answers to help you solve the problems that you miss. Remember you want to find out what you don't know before the test not during the test. Developing and using these methods before each test will improve your grades.

Make sure to use the 10 steps to better test taking and the information gained from your test analysis. Use all the time on the final exam because you could improve your final grade by a full letter if you make an "A" on the final.

Section Review

1. What are the three qustions you need to ask about your final exam?

   First Question: _____

   _____

   _____

   Second Question: _____

   _____

   _____

   Third Question: _____

   _____

   _____

Chapter 7 Review

1. You should start studying for a test at least _____ days ahead of time.

2. You should be reviewing only already-_____ material the night before the test.

3. When previewing the test, put a _____ by the questions you can solve without thinking, and do those problems _____.

4. When taking a test, you must decide the best way to get the _____ points in the _____ amount of time.

5. Halfway through the test, you should have _____ than half of the problems completed.

6. If you cannot write down the first step of a problem, you should _____ the problem to help you remember how to solve it.

7. Do not leave the test room until you have reviewed each problem _____ times or until the bell rings.

8. In reviewing the test, if you cannot solve the problem immediately (within a few seconds), it is not a _____ error and is probably a _____ error.

9. Students who have numerous _____ errors will fail the next test and the course if the _____ are not understood.

10. The _____ day of class is when you start preparing for the final exam.

What is the most important information you learned from this chapter?

_____

_____

_____

_____

_____

_____

_____

_____

How can you immediately use it?

_____

_____

_____

_____

_____

_____

_____

Appendix: Math Autobiography

This is the purpose for your math autobiography—to reflect on your past experiences when learning math, how they made you feel then, and how these experiences and feelings shape how you feel about what you are doing in your current math classes. It will close with solutions you believe will help you improve your experiences in math classes.

The following writing process will simplify this task while improving the final paper.

1. Use the questions in the sections on types of math anxiety and causes of math anxiety in Chapter 3 to brainstorm ideas for your paper. Record these answers on paper.

2. By each experience you record, write how the experience made you feel.

3. To further your brainstorming, complete the following sentences:

   My favorite math instructor was

   _____,

   because he/she

   _____

_____
_____,

I always felt

_____
_____
_____.

One math instructor who bothered me was

_____,

because he/she

_____
_____
_____,

I always felt

_____
_____
_____

continued

When I hear the word math, I think of

_____

_____

_____

_____.

I did well in math until

_____

_____

_____

_____.

The first time I experienced math anxiety was when

_____

_____

_____

_____.

The first time I was anxious before or during a math test was _____

_____

_____

_____

_____.

The most recent time I was anxious about math class was when

_____

_____

_____

_____.

The most recent time I was anxious before or during a test was when

_____

_____

_____

_____

_____

_____.

4. Since an autobiography is in chronological order, identify the important information in your brainstorming and place into the correct time category in the tables on the following pages. These tables will be the outline for your paper.

5. You will write the introduction last, after you have thought through the meaning of your experiences. Then, when you type or write the final paper, you will place the introduction paragraph first.

Elementary School
Main Paragraph One

Experiences	Feelings

As a result of these experiences, now when I am either in math class, stydying for math, or taking a math test, I . . .

Middle School
Main Paragraph Two

Experiences	Feelings

As a result of these experiences, now when I am either in math class, studying for math, or taking a math test, I . . .

High School
Main Paragraph Three

Experiences	Feelings

As a result of these experiences, now when I am either in math class, studying for math, or taking a math test, I . . .

College and/or Work Experience
Main Paragraph Four (If this applies)

Experiences	Feelings

As a result of these experiences, now when I am either in math class, studying for math, or taking a math test, I . . .

Conclusion Paragraph

When I look back on my experiences learning or using math, I understand why I have developed attitudes like

_____

_____

_____.

To develop a more positive perspective toward math, I need to improve certain behaviors.

Behaviors	*Solutions for Change*
1.	
2.	
3.	
4.	
As a result, I will become . . .	

Introduction Paragraph

(When you write your final paper, this paragraph will be first. This paragraph should introduce the main lesson you have learned through this activity. Look at what you have written for your conclusion paragraph for some ideas.)

6. Now, using the tables as your outline, either handwrite (if allowed by the instructor) or type the final paper to be turned in to your instructor.

Bibliography

Bloom, B. 1976. *Human Characteristics and School Learning.* New York: McGraw-Hill Book Company.

Brown, J. F., and R. M. Cooper. 1978. *Learning Styles Inventory.* Freeport, NY: Educational Activities, Inc.

Cattell, R. B. 1 966. "The Screen Test for the Number of Factors." *Multivariate Behavioral Research*, no. 1, pp. 245-276.

Fox, L. H. 1977. *Women and Mathematics: Research Perspective for Change.* Washington, DC: National Institute of Education.

Kolb, D. 1985. *Learning-Styles Inventory.* Boston, MA: McBer and Company.

McCarthy, B. 1981. *The 4-Mat System: Teaching to Learning Styles with Right/Left Mode Techniques.* Barrington, IL: Excel.

Nolting, P. D. 1987. *How to Reduce Test Anxiety.* An audiocassette tape. Bradenton, FL: Academic Success Press. (www.academicsuccess.com)

Richardson. F. C., and R. M. Suinn. 1973. "A comparison of Traditional Systematic Desensitization, Accelerated Mass Desensitization, and Mathematics Anxiety." *Behavior Therapy*, no. 4, pp. 212-218.

Suinn, R. M. 1988. *Mathematics Anxiety Rating Scale (MARS).* Fort Collins, CO: RMBSI, Inc.

Tobias, S. 1978. "Who's Afraid of Math and Why?" *Atlantic Monthly*, September, pp. 63-65.

Wolpe, J. 1958. *Psychotherapy by Reciprocal Inhibition.* Stanford, CA: Stanford University Press.

About the Author

Over the past 15 years, Learning Specialist Dr. Paul Nolting has helped thousands of students improve their math learning and obtain better grades. Dr. Nolting is a national expert in assessing math learning problems — from study skills to learning disabilities — and developing effective learning strategies and testing accommodations.

Dr. Nolting is also a nationally recognized consultant and trainer of math study skills and of learning and testing accommodations for students with learning disabilities. He has conducted national training grant workshops on math learning for the Association on Higher Education and Disabilities and for a two-year University of Wyoming 1991 Training Grant of Basic Skills.

Dr. Nolting has conducted numerous national conference workshops on math learning for the National Developmental Education Association, the National Council of Educational Opportunity Association, and the American Mathematical Association of Two-Year Colleges. He is a consultant for the American College Test (ACT) and the Texas Higher Education Coordinating Board. He is also a consultant for Houghton Mifflin faculty development programs, conducting workshops on math study skills for math faculty. His text *Winning at Math: Your Guide to Learning Mathematics Through Successful Study Skills* is used throughout the United States, Canada, and the world as the definitive text for math study skills.

Dr. Nolting has consulted with numerous universities and colleges. Some of the universities with which Dr. Nolting has consulted are the University of Massachusetts, the University of Colorado-Boulder, Texas Tech. University, Black Hills State University, Tennessee Tech. University, and the University of Connecticut.

Some of the colleges with which he has consulted are San Antonio College, St. Louis Community College, J. Sargeant Reynolds College, Montgomery College, Broward Community College, Miami-Dade Community College, Northeast State Technical Community College, Landmark College, Denver Community College, Valencia Community College, and Pensacola Junior College. Dr. Nolting has consulted with over 75 colleges, universities, and high schools over the last 15 years.

Dr. Nolting holds a Ph.D. in education from the University of South Florida. His Ph.D. dissertation was "The Effects of Counseling and Study Skills Training on Mathematics Academic Achievement." He is an adjunct instructor for the University of South

Florida, teaching assessment and appraisal courses at the graduate level.

His book *Winning at Math: Your Guide to Learning Mathematics Through Successful Study Skills* was selected Book of the Year by the National Association of Independent Publishers. "The strength of the book is the way the writer leads a reluctant student through a course from choosing a teacher to preparing for the final examination," says *Mathematics Teacher*, a publication of the National Council of Teachers of Mathematics. *Winning at Math* and Dr. Nolting's other titles are available at http://www.academicsuccess.com.

His two audiocassettes, *How to Reduce Test Anxiety* and *How to Ace Tests,* were also winners of awards in the National Association of Independent Publishers competition. "Dr. Nolting," says *Publisher's Report*, "is an innovative and outstanding educator and learning specialist."

A key speaker at numerous regional and national education conferences and conventions, Dr. Nolting has been widely acclaimed for his ability to communicate with faculty and students on the subject of improving math learning.

Books and Materials by Academic Success Press and Dr. Paul Nolting

How to Ace Tests. Academic Success Press, 1987.
How to Reduce Test Anxiety. Academic Success Press, 1987. (audiocassette)

How to Develop Your Own Math Study Skills Workshop and Course. Academic Success Press, 1991.

Improving Mathematics Studying and Test-Taking Skills. Boston, MA: Houghton Mifflin Company, 1994. (videotape, ISBN 0-669-33291-7)
Math and the Learning Disabled Student: A Practical Guide for Accommodations. Academic Success Press, 1991.
Math and Students with Learning Disabilities: A Practical Guide to Course Substitutions. Academic Success Press, 1993.
Strategy Cards for Higher Grades. Academic Success Press, 1989. Twenty 3-inch by 5-inch cards. (card deck)
Winning at Math — Computer Evaluation Software. Academic Success Press.
Winning at Math: Your Guide to Learning Mathematics Through Successful Study Skills. Academic Success Press, 1997.

Ordering Information

For ordering information on *Winning at Math* and other books and learning materials listed above, call toll free: 1-800-247-6553 or visit www.academicsuccess.com. Credit card, checks, and purchase orders are accepted. Shipping charges are $6.00 for the first item and $2.00 for any additional items. Academic Success Press can be contacted at 6023 26th Street West, PMB 132, Bradenton, FL 34207. For a free catalog, call (941) 359-2819.

To order Houghton Mifflin videos, contact customer service toll free at 1-800-225-1464.